The Penguin Dictionary of

Physical Geography

John B. Whittow

PENGUIN BOOKS

PENGUIN BOOKS

Published by the Penguin Group
Penguin Books Ltd, 27 Wrights Lane, London W8 5TZ, England
Penguin Putnam Inc., 375 Hudson Street, New York, New York 10014, USA
Penguin Books Australia Ltd, Ringwood, Victoria, Australia
Penguin Books Canada Ltd, 10 Alcorn Avenue, Toronto, Ontario, Canada M4V 3B2
Penguin Books (NZ) Ltd, Private Bag 102902, NSMC, Auckland, New Zealand

Penguin Books Ltd, Registered Offices: Harmondsworth, Middlesex, England

First published 1984
Second edition 2000
10 9 8 7 6 5 4 3 2 1

Set in 7/9 pt ITC Stone
Typeset by Rowland Phototypesetting Ltd, Bury St Edmunds, Suffolk
Printed in England by Clays Ltd, St Ives plc

To my wife,
without whose constant encouragement and considerable assistance
this volume would never have been completed

Preface

It is thirty-four years since *A Dictionary of Geography* (W. G. Moore) first appeared, under the Penguin imprint, and it has long remained a basic O-level reference book. Meanwhile, over the last twenty years the subject of geography has undergone such substantial changes in methodology and direction as to warrant the term revolution – some would say several revolutions. Such changes have been reflected in the appearance of new geography dictionaries within the last decade that represent attempts to keep pace with the growing number of geographical terms. All have demonstrated that today '. . . Geography is concerned with the real world in which students live and is therefore capable of raising questions and developing skills which are seen to be relevant to their present and future lives' (Schools Council, *Curriculum Development Project: Geography 16–19 Years*, University of London, 1977). With this in mind the Schools Council report recommended that of the numerous paradigms now developed in geographical thought the most suitable and relevant for a geography curriculum for the age group sixteen to nineteen was the Man–Environment paradigm.

While recognizing and accepting the importance of this integrated approach to the subject the present author is also aware of the dichotomy that has gradually developed in academic geography between the physical and human aspects of the discipline. Such a division reflects not only the depth of specialism which has grown within each part of the subject but also the considerable amount of detail which each now encompasses. Hence, instead of producing a single, all-embracing dictionary of modern geographical usage, it has been decided to write two dictionaries, one of physical geography, the other of human geography.

Each of these companion volumes has been planned along similar lines, so that their formats are broadly comparable. Each adopts a systematic approach to the subject, while each retains the useful Penguin method of cross-referencing (namely *see; see also*). Many new terms have been introduced in order to cover the fundamental methodological and conceptual changes which have taken place within the subject during the last twenty years. While care has been taken to base the new dictionaries on the modern quantitative nomenclature, it has also been recognized that since the 1970s there has been a reaction against over-quantification, reflected for example by the humanistic approach to geography.

Thus it is hoped that the scope of the dictionaries will have been expanded to reflect not only the systems approach, but also the humanistic, idealistic and Marxist approaches to the subject. In addition, the range of terms included is intended to extend their use beyond A-level studies to those students proceeding to universities, polytechnics and other institutes of higher education.

In the *Penguin Dictionary of Physical Geography* an attempt has been made to include all the terms which are currently used by physical geographers working with the age group referred to above, in the hope that in a subject which is developing so rapidly no serious omissions have occurred. Metrication and SI units have been adopted throughout, but in many instances imperial measurements are also used. Certain of the terms that are included are now regarded as obsolete by some academic geographers but they are retained because they have lingered on in particular textbooks. An attempt has been made to explain the differences between British and American usage wherever they occur, and a small number of foreign words has been included when they are in common international use. Words with only a local connotation have been cut to a minimum, whence many terms relating to regional phenomena and place-names may have been omitted.

J.B.W.
1984

Since the *Penguin Dictionary of Physical Geography* was first published in 1984 there have been several important advances in the subject, especially in the fields of Biogeography, Cartography, GIS and Remote Sensing. Thus, an update was needed and it became necessary to delete certain obsolete entries. Some 500 new entries have been added, many existing entries amended and 18 new diagrams and tables included, in the belief that the modern geography student will require a more thorough understanding of the scientific terms which underpin the subject as we enter the new millennium.

J.B.W.
1999

Acknowledgements

I am grateful to a number of people who have been of assistance during the preparation of this volume. They include: Dr P. Curran, Mr D. Foot, Dr B. Halstead, Dr B. J. Knapp, Dr A. Mannion, Mr A. Millington, Dr C. W. Mitchell, Mr M. Parry, Mr R. B. Parry, Dr R. D. Thompson and Dr J. R. G. Townshend. I should like to thank, in particular, Mrs M. Birch, who typed the manuscript, Mrs Kathleen King and Judith Fox, who drew the diagrams, Mrs C. Holland and Mrs F. McCrostie, who assisted with some of the corrections, and Peter Phillips, who prepared the typescript for the printer.

For kind permission to include some of the diagram material on which figures in this volume are based I gratefully acknowledge the following:

Figs. 14, 225 and 247, F. J. Monkhouse and R. J. Small, *A Dictionary of the Natural Environment*, Edward Arnold Ltd; Fig. 29, P. Furley and W. Newey, *Geography of the Biosphere*, Butterworths (Fig. 3.7); Figs. 30, 106, 192 and 276, R. H. Bryant, *Physical Geography Made Simple*, W. H. Allen & Co. (by kind permission of the author); Figs. 33, 51, 52, 57, 61, 96, 114, 123, 129, 173, 206, 210, 263 and 264, D. G. A. Whitten and J. V. R. Brooks, *The Penguin Dictionary of Geology*, Penguin Books Ltd; Fig. 36, M. Begon *et al.*, *Ecology: Individuals, Population and Communities*, Blackwell Scientific; Figs. 50, 68, 215 and 250, R. G. Barry and R. J. Chorley, *Atmosphere, Weather and Climate*, Methuen & Co.; Figs. 54, 124, 204 and 262, M. M. Sweeting, *Karst Landforms*, Macmillan Ltd; Fig. 80, C. Embleton and C. A. M. King, *Glacial and Periglacial Geomorphology*, Edward Arnold Ltd; Fig. 86, T. Robinson, *Setting Foot on the Shores of Connemara and other writings*, Lilliput Press; Figs. 89 and 182, R. U. Cooke and J. C. Doornkamp, *Geomorphology in Environmental Management: An Introduction*, Oxford University Press; Figs. 92, 143, 219 and 265, R. J. Price, *Glacial and Fluvioglacial Landforms*, Longman Group Ltd; Figs. 99 and 271, B. Booth and F. Fitch, *Earthshock*, J. M. Dent & Sons Ltd; Fig. 110, A. H. Lachenbruch, Special Paper 60, *Geol. Soc. Am.*, 1962; Fig. 117, B. J. Mason, 'Future developments in meteorology: an outlook to the year 2000', *Q. J. Roy. Met. Soc.*, 96, 1970; Fig. 126, Butler, 'Periodic phenomena in landscapes as a basis for soil studies', *Soil Publ.* (CSIRO, Aust.), 14, 1959; Figs. 135, 245, 252, 266 and 268, A. N. Strahler, *Physical Geography*, 4th edn, John Wiley & Sons Inc. (by kind permission of the author); Fig. 150, B. W. Atkinson, *Meso-scale Atmospheric Circulation*, Academic Press; Fig. 154, A. Goudie, *Encyclopedic Dictionary of Physical Geography*, 2nd edn, Blackwells; Fig. 170 (based on a map in), C. E. Carson and K. M. Hussey, 'The oriented lakes of Arctic Alaska', *J. Geol.*, 70 (Fig. 2), 1962; Fig. 180, A. Young, *Slopes*, Longman Group Ltd; Fig. 187, J. R. Mackay, 'The world of underground ice', *Ass. Am. Geogr. Ann.*, 62 (Fig. 22), 1972; Figs. 199, 226, 228, 238 and 260, I. Statham, *Earth Surface Sediment Transport*, Oxford University Press; Fig. 211 (in part), L. B. Leopold, M. G. Wolman and J. P. Miller, *Fluvial Processes in Geomorphology*, W. H. Freeman & Co. (copyright © 1964); Fig. 213, M. J. Selby, 'A rock mass strength classification for geomorphic purposes: with tests from Antarctica and New Zealand', *Zeitschrift für Geomorphologie*, 24, 1980; Fig. 229, I. Statham, *Earth Surface Sediment Transport*, Clarendon Press; Fig. 236, E. A. Fitzpatrick, *An Introduction to Soil Science*, Oliver & Boyd; Fig. 237, M. Meybeck, *Atmospheric Inputs and River Transport of Dissolved Substances*, IAHS Publications, 14.1; Fig. 253, T. Czudek and J. Demek, 'Thermokarst in Siberia and its influence on the development of lowland relief', *Quat. Res.*, 1 (Fig. 9), 1970; Fig. 259, J. M. Gray, 'The Main rock platform of the Firth of Lorn, western Scotland', *Trans. Inst. Br. Geogr.*, No. 61, March 1974, The Institute of British Geographers; Fig. 269, R. W. Herschy (ed.), *Hydrometry, Principles and Practice*, John Wiley & Sons Inc.; Fig. 270, A. M. MacEachen and M. Moumonier, *Cartography and Geographic Information Systems*, 19, 4 (Figs. 1 and 2), 1992; Figs. 273 and 277, R. U. Cooke and A. Warren, *Geomorphology in Deserts*, B. T. Batsford Ltd; Fig. 275, D. Tolmazin, *Elements of Dynamic Oceanography*, Allen & Unwyn.

Note

References in the text to figures and tables are printed in italic numbers within square brackets.

A

aa lava A type of LAVA characterized by its jagged, angular and sharply pointed blocks. Its black, clinker-like appearance results from the explosively escaping bursts of gas through the rapidly congealing surface as the molten lava beneath tugs at the hardening crust, causing it to break into irregular masses. The word *aa* is Hawaiian. *See also* LAVA, PAHOEHOE.

abîme A vertical shaft in KARST terrain. The deep shaft through the limestone is of such magnitude that it is often referred to as 'bottomless', but in reality it opens into a network of subterranean passages at depth. *See also* GOUFFRE, POT-HOLE, SWALLOW-HOLE.

abiotic The non-living factors which influence an ecosystem, e.g. the geology, are termed the abiotic factors.

ablation **1** The wastage or removal of surface snow or ice by melting and evaporation, achieved by: (a) transfer of SENSIBLE HEAT and LATENT HEAT (aided by VAPOUR PRESSURE and WIND VELOCITY) and by INSOLATION; (b) CONDUCTION from morainic debris or surrounding rock walls; (c) meltwater streams; (d) rainfall; (e) SUBLIMATION; (f) wind removal. The *ablation factor* is the rate at which the ice or snow surface wastes. *See also* ABLATION TILL. **2** The removal of rock debris by wind action (DEFLATION).

ablation till TILL.

Abney-level A simple, lightweight surveying instrument for measuring slope-angle (angle of inclination), where no great accuracy is required. It comprises a handheld sighting-tube linked to a spirit-level, the bubble of which is reflected in the eyepiece. When the observer sights an object, adjustment is made to the tilting spirit-level and once the bubble image coincides with the object sight the slope-angle can be read from an attached scale.

abrasion The mechanical wearing effect on rocks caused by CORROSION. The abrading agent can take a variety of forms, e.g. sand, pebbles, boulders, but all must be moving across the rock surface in order to scratch, grind or polish. These frictional agents may be transported in various ways: by wind, running water, ocean waves and currents or by glacier ice. The act of abrasion will, therefore, vary in both its speed and its magnitude according to the type of agents involved. Sand-blasting by wind creates VENTIFACTS. *See also* ABRASION PLATFORM, ATTRITION.

abrasion platform A virtually smooth marine platform cut by ocean waves at a coastline. If it is currently being fashioned it will be exposed only at low tide, but there is a possibility that the *wave-cut platform* will be hidden sporadically by a mantle of beach shingle, which is the abrading agent. If the platform is permanently exposed above high-water mark it is probably a RAISED-BEACH PLATFORM.

abscissa The horizontal axis (*x* axis) in a graph. *See also* ORDINATE.

absolute age Denotes the dating of rocks in the strictest sense, in actual rather than relative terms, usually by the number of years. *See also* GEOCHRONOLOGY.

absolute co-ordinates A GIS term referring to a pair of CO-ORDINATES that are measured directly from the origin of the

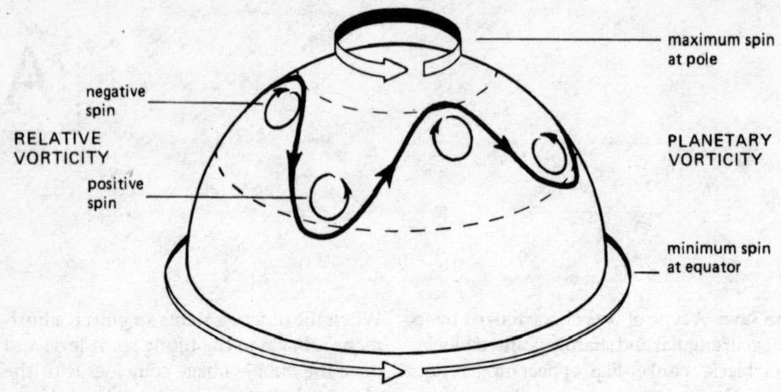

Figure 1 *Absolute vorticity*

co-ordinate system in which it lies and not to any other point in the system. *See also* RELATIVE CO-ORDINATES.

absolute drought A period of at least fifteen consecutive days on each of which less than 0.25 mm (0.01 in) of rainfall is recorded – British definition. In the USA a DRY SPELL refers to fourteen consecutive days without measurable rainfall.

absolute humidity (vapour concentration) The amount of water vapour per unit volume of air, expressed in g cm^{-3}. Sometimes wrongly applied to the pressure of water vapour in the atmosphere (VAPOUR PRESSURE). Air of a given temperature and pressure is capable of holding a specific amount of water vapour, above which point, the DEW-POINT, it becomes saturated. Cold air has a lower absolute humidity than warm air (e.g. at –1° C, absolute humidity = 2.2 g cm^{-3} compared with 9.15 g cm^{-3} at 21° C). *See also* CONDENSATION, RELATIVE HUMIDITY.

absolute instability The state of an air parcel if the existing ENVIRONMENTAL LAPSE-RATE is greater than the DRY ADIABATIC LAPSE-RATE, thereby producing INSTABILITY. In contrast to CONDITIONAL INSTABILITY, there is no dependence on moisture content during a state of absolute instability. *See also* ABSOLUTE STABILITY.

absolute permeability A measure of the possible flow under fixed conditions of a standard liquid through connected pore spaces when there is no reaction between the liquid and the solids. *See also* PERMEABILITY.

absolute stability The state of an air parcel if the existing ENVIRONMENTAL LAPSE-RATE is less than the SATURATED ADIABATIC LAPSE-RATE, thereby producing STABILITY. *See also* ABSOLUTE INSTABILITY.

absolute temperature The Kelvin temperature scale based on absolute zero (0° K), the point at which thermal molecular motion ceases (*see also* KELVIN SCALE). Used in meteorology to express upper-air temperatures. 0° K corresponds to –273.15° C.

absolute vorticity The sum of RELATIVE VORTICITY and PLANETARY VORTICITY, which remains constant through wave motion in the atmosphere. *See also* ROSSBY WAVE. [*1*].

absorbed water Water held mechanically in a soil and having physical properties similar to ordinary water at the same pressure and temperature.

absorption 1 The process by which a substance retains radiant energy (heat- and light-waves) instead of reflecting, refracting or transmitting it. Dark and matt colours absorb higher proportions than lighter shiny colours, such as ice or snow. Oxygen and nitrogen, the main atmosphere gases,

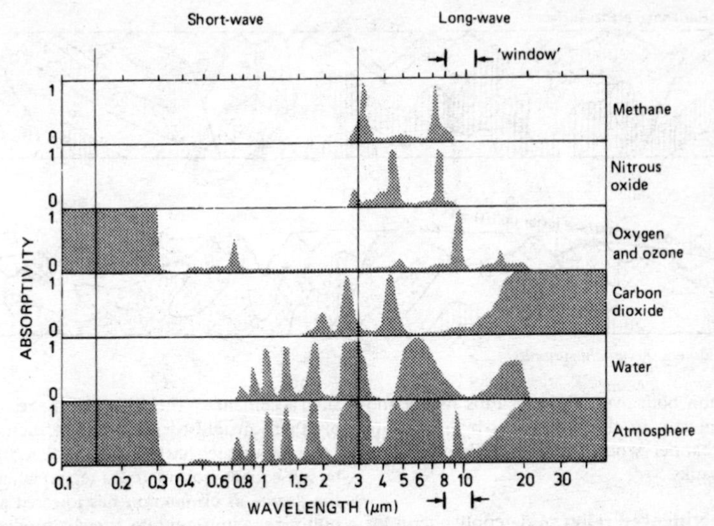

Figure 2 *Absorption of radiant energy*

do not absorb very much radiation (except in the far ultraviolet), unlike the minor gases, ozone, nitrous oxide, water vapour and carbon dioxide, which have a significant effect on atmospheric absorption. *See also* ATMOSPHERIC WINDOW, BLACK-BULB THERMOMETER. [*2*] **2** The process by which the energy of electromagnetic radiation is taken up by a molecule and transformed into a different form of energy. *See also* ALBEDO.

abstraction 1 A term used in geomorphology to denote *river capture* (CAPTURE, RIVER) or *piracy of streams*. Although not in common usage, the action involves the widening and/or lowering of a stream or river valley more rapidly than those of its neighbours, thereby capturing or abstracting the adjoining drainage. **2** A term used to describe the process by which ice-sheets grow by 'locking up' water within the HYDROLOGICAL CYCLE. Water abstracted in this way is temporarily prevented from returning to the oceans, thus leading to a glacio-eustatic (GLACIO-EUSTASY) fall of sea-level.

abundance The total number of individuals of a particular SPECIES in an area. Difficulties of counting large numbers are generally overcome by SAMPLING, often by means of a QUADRAT in a vegetation survey.

abyss From the Latin *abyssus*/Greek *abussos* (literally ('bottomless', ABÎME). It refers to an extremely deep chasm or ocean trough (DEEP, OCEAN).

abyssal plain ABYSSAL ZONE.

abyssal zone The bottom zone of the ocean, below 1,000 fathoms (1,800 m) and extending down to an abyssal plain at depths of 4,000 m, on which abyssal deposits occur (*see also* OOZE, RED CLAY). At these depths the sea temperatures do not exceed 4° C. [*162*]

accelerated erosion The speeding up of erosive processes, either directly or indirectly, by the intervention of man. Its most obvious effects can be seen in the alteration of a river's regime by careless land use or by the introduction of man's artefacts into the catchment area. Deforestation or overgrazing, for example, can increase surface runoff and lead to SOIL EROSION. Urbanization, with its attendant layers of concrete and tarmac, also increases surface runoff by inhibiting PERCOLATION, and since the soil is

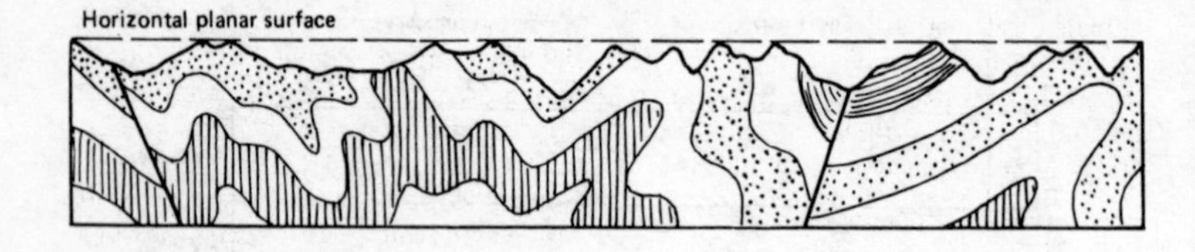

Figure 3 *Accordant summits*

now built over rainwater runs swiftly and directly to the rivers thereby increasing channel scour and the undercutting of banks.

accidented relief A descriptive term for an irregular or highly dissected terrain. It may have been derived from a French phrase, *relief accidenté*, meaning broken ground.

acclimatization The process of adjustment by an animal, normally with significant physiological effects, to a marked change in climate, related to a large change of altitude or temperature extremes.

accordant drainage A term applied to a drainage pattern when it exhibits a direct relationship with the underlying geology and structure. If the main trunk stream flows in the same direction as the DIP of the rock it is adjusted to the structure and is said to be *accordant*. If, however, the trunk stream bears little or no relationship to the underlying structure it is termed DISCORDANT.

accordant summits Those summits of hill or mountain groups which rise to a uniform surface at approximately the same elevation above sea-level. The uniform surface need not be horizontal (it may be gently curved, tilted or even undulating), but the important criterion is that the summits of a mountain massif rise and fall together, regardless of the underlying geology and structure. Such accordance is thought to be the result of either **1** an uplifted planar surface, such as a PENEPLAIN, which has been greatly dissected by subsequent erosion, or **2** a region where the degree of DENUDATION has lowered all hilltops at a uniform rate, thereby producing a *summit plane* of equal elevation. The summit accordance exhibited by the European Alps has been termed the *gipfelflur* by A. Penck, a German geomorphologist. In this example any isolated summit which rises above the general uniformity of the summit plane was termed a *gipfel*, a RESIDUAL remaining from an earlier summit plane now almost totally destroyed. In Britain, the high residual of Plynlimon (Pumlumon Fawr) rises above the accordant summit plane, the high plateau of Wales. [*3*]

accretion A process of accumulation. **1** In meteorology, the incremental growth of an ice particle each time it collides with a supercooled water droplet (BERGERON–FINDEISEN THEORY OF PRECIPITATION). **2** In geomorphology, the accumulation of mineral material in a particular locality or environment, e.g. lacustrine muds in a lake basin. **3** In astronomy and cosmogeny, a theory which suggests that the planets grew by the condensation of small cosmic dust particles into larger bodies.

accumulated temperature The sum of DEGREE-DAYS above a critical threshold value (e.g. 5.5° C/42° F) is a critical threshold value for vegetation growth in a European climate). Calculations may be made of the number of hours per day and of days per month (DEGREE-DAYS) when the mean temperature exceeds a critical threshold value and the data summed or accumulated.

accumulation The processes of snow and ice nourishment, including: PRECIPITATION (from an atmospheric DEPRESSION and enhanced by uplift of moist air, OROGRAPHIC CLOUD); RIME ice (from SUBLIMATION associated with supercooled droplets in clouds); HOAR-FROST (sublimation associated with RADIATION cooling at the surface); freezing rain; avalanche material and drifting snow. *See also* GLACIER, MASS BALANCE.

accumulation slope A type of SLOPE in which material is being gained, leading to net increase in ground. The form of the slope is controlled by the rate of accumulation. *See also* ACCRETION, DENUDATION SLOPE, FOOT SLOPE, TRANSPORTATION SLOPE.

accumulation zone **1** In geomorphology, the term refers to that part of a slope which, over a period of time, undergoes a net gain of material, leading to a progressive raising of the ground surface in that zone. **2** In glaciology, the term refers to that part of a glacier or ice-sheet which exhibits a net gain of snowfall. The accumulation zone therefore lies contiguous with but above the zone of ABLATION. This is the zone in which snow is transformed into NÉVÉ and then ice by the process of FIRNIFICATION. The main difference between subpolar and polar glaciers is that in the former some melting takes place in the accumulation zone in the summer (even though the ice remains below the PRESSURE MELTING-POINT at depth). No such melting takes place in the accumulation zone of true polar glaciers.

Acheulian A stratigraphic stage name, based on an early PALAEOLITHIC culture, for part of the European Lower Pleistocene.

acidity profile, acidity record The concentration of acid in CORE SAMPLING of glaciers and ice sheets. This is used to estimate ACID PRECIPITATION and volcanic dust fallout.

acid lava LAVA rich in SILICA which has a high melting-point (*c.* 850° C). Thus it cools very quickly on exposure to the air and flows slowly in a stiff or viscous (VISCOSITY) stream only short distances from the volcanic vent. Because of this lack of mobility acid lava tends to build steep-sided volcanic cones in contrast to the low-angled slopes of a SHIELD VOLCANO, in which a BASIC LAVA is the formative agent. Acid lava forms only about 10% of all lavas. *See also* DACITE, OBSIDIAN, RHYOLITE.

acidophile A plant which tolerates and flourishes on ACID SOILS.

acid precipitation A colloquial term used to describe precipitation in a polluted environment, where the rain and snow become contaminated by either sulphur oxides – caustic gases emitted mostly by coal combustion – or by a combination of sulphur dioxide and nitrogen oxide (SMOG) emitted from motor vehicle exhaust. Acid rain ($pH < 5.6$) is capable of burning holes in leafy vegetation, and causing considerable damage to the built environment. The effects of acid rain on the health of man and animals is currently disputed. *See also* POLLUTION, DRY DEPOSITION, OCCULT DEPOSITION.

acid rocks Igneous rocks containing more than 10% free quartz. Acidity of rocks was once classified by the amount of SILICA (SiO_2) present (acid rocks = >66% silica; intermediate rocks = 53–66% silica; basic rocks = 45–52% silica; ultrabasic rocks = <45% silica) but this arbitrary division is now obsolete. *See also* BASIC ROCKS, GRANITE.

acid soil A base-deficient soil with a PH below 7, possibly due to the parent material or to LEACHING of the soluble bases (especially sodium and calcium). *See also* ALKALINE SOIL, PODZOL.

aclinic line MAGNETIC EQUATOR.

acre An English unit of areal measurement. 1 acre = 4,840 sq yd = 0.4047 hectares (ha), 2.4711 acres = 1 ha, 247.11 acres = 1 km^2. [*15*]

acre-foot A hydrological term, used particularly in irrigation engineering in the USA. It represents the amount of water that would be required to cover an ACRE of land to a depth of 1 ft (i.e. 1,219 m^3 or 43,560 cu ft).

actinometer An instrument which measures solar radiation. The corresponding term for a recording instrument is *actinograph*.

actinomycetes Soil micro-organisms thriving in well-aerated soils with a PH between 6.0 and 7.5. They are more prolific than fungi and bacteria and help to break down resistant organic matter in the soil. It is these organisms that give soil its characteristic odour.

active acidity The activity of hydrogen IONS in the aqueous content of a soil. It is measured as a PH value.

active fault A FAULT along which movement occurs fairly regularly, thereby causing an earth tremor or an EARTHQUAKE.

active ice Ice in a glacier or an ice-sheet which remains mechanically mobile, in contrast to stagnant or dead ice.

active layer The top layer of soil in a PERMAFROST zone, subjected to seasonal freezing and thawing and which, during the melt season, becomes very mobile. *See also* MOLLISOL.

active remote-sensing system A REMOTE-SENSING system having its own source of ELECTROMAGNETIC RADIATION as is the case for RADAR sensors. This system measures the electromagnetic radiation, produced by the sensor, that has been reflected from the ground surface. *See also* PASSIVE REMOTE-SENSING SYSTEM.

activity ratio An empirical relationship used to measure the ability of a soil to take up moisture. It is defined as the PLASTICITY INDEX divided by percentage weight less than 2 μm in size. There are three classes: active, normal and inactive, with most British soils occurring in the normal or inactive classes. Active soils usually have a high CATION EXCHANGE CAPACITY. *See also* AGGREGATION RATIO.

actualism UNIFORMITARIANISM.

actual isotherm An ISOTHERM for which the plotted values relate to actual mean temperatures, not those reduced to sea-level by altitudinal correction.

adaptation An adjustment of organisms to their environments, in which a new or better-functioning system is developed in order that the organisms are more perfectly fitted for existence in those environments.

adaptive radiation The evolutionary diversification of a group of organisms responding to the ecological pressures of different HABITATS.

additivity The condition when variables under statistical investigation have no mutual effect or interaction upon each other and therefore do not have an effect in combination different from the sum of their separate effects.

adhesion A process in which dry sand blown on to a damp surface is held by the surface tension of water rising between the grains by CAPILLARITY.

adhesion ripple A small (30–40 cm long) irregular sand ridge with the STOSS (windward) side steeper than the lee side and transverse to wind direction. It forms when dry sand is moved across a smooth moist surface. *See also* RIPPLE MARKS and [*210*].

adiabatic Refers to the changes of temperature which occur in a mass of gas (air) when it is compressed (heated) or expanded (cooled) without the aid of any external sources of heating or cooling. The rate at which this occurs is termed the *lapse-rate* (DRY AND SATURATED ADIABATIC LAPSE-RATES). Generally, when a parcel of air rises into zones of lower atmospheric pressure, it expands at the expense of its internal energy and becomes warmer. For a saturated rising parcel, the fall of temperature is checked by the release of latent heat of condensation. *See also* CHINOOK, DIABATIC, FÖHN, THERMODYNAMIC EQUATION.

adjusted stream A stream which flows parallel to the STRIKE of the underlying rocks.

adobe **1** An unburnt sun-dried brick. **2** A term for clayey and silty deposits in Mexico and SW USA (especially in the BASIN-AND-

RANGE TERRAIN) used for making sun-dried bricks.

adolescence CYCLE OF EROSION.

adret A southward-facing mountain or hill slope (N hemisphere) favourable to settlement because of the high amount of sunshine received, and associated heat concentration. *See also* ASPECT, EXPOSURE, UBAC.

adsorbed water Water held in a soil by physico-chemical forces and having physical properties substantially different from ABSORBED WATER or chemically combined water at the same pressure and temperature. *See also* ADSORPTION.

adsorption The linking of a particle of a particular substance to another by adhesion, or penetration. This is a physical not a chemical linkage. In pedology, the adsorption process describes the manner in which mineral particles in the soil become surrounded by CATIONS in colloidal (COLLOID) form. *See also* ABSORBED WATER.

adsorption complex The group of substances in soil capable of adsorbing other materials (ADSORPTION). Both the inorganic and organic COLLOIDS make up the greater part of the adsorption complex for they have a considerably greater capacity for adsorption than the non-colloidal materials such as silt or sand.

advancing coast PROGRADATION.

advection The movement of air, water and other fluids in a horizontal as opposed to a vertical plane. Contrast CONVECTION. The transfer of heat from low to high latitudes is the most obvious example of advection, in both the atmospheric and oceanographic contexts (e.g. Gulf Stream/N Atlantic Drift).

advection fog A fog formed when a relatively warm, moist and stable mass of air moves laterally over a cooler surface, thereby reducing the temperature of the lower layers of the air mass (causing an INVERSION OF TEMPERATURE) until it reaches the DEW-POINT, when CONDENSATION takes place. This may occur when warm maritime air moves over a cold land mass or when warm continental air moves offshore across a cold ocean current; or when warm maritime air crosses into the realms of a cold ocean current (e.g. Gulf Stream air meeting air above the cold Labrador current near the Grand Banks of Newfoundland). It sometimes develops over a frozen, snow-covered land mass in conjunction with RADIATION FOG. *See also* ARCTIC SEA SMOKE, STEAM FOG.

adventitious A term describing the growth of plant roots and buds in unusual parts of the plant, e.g. roots growing from branches.

adventive cone A small volcanic cone or crater which appears on the flanks of a major volcano. Sometimes referred to as a 'parasitic' cone. *See also* VOLCANO.

aegir (aegre) EAGRE.

aeolian (US **eolian)** A term pertaining to the wind; hence wind-borne, wind-blown or wind-deposited materials are often referred to as aeolian. *Aeolian erosion* is a process of CORRASION, similar to sand-blasting, which leads to the wearing away of the rock surface (VENTIFACT, YARDANG, ZEUGEN) prior to the aerial transport of the abraded material (DEFLATION) and its deposition in the form of DUNES, LOESS, etc. The aeolian processes are most common in arid environments (hot and cold deserts) and on exposed shorelines. *See also* SALTATION.

aeolianite A type of rock formed when dune sand becomes cemented, usually by calcium carbonate. Under semi-arid conditions it is suggested that a minimum of 8 per cent calcium carbonate is required for cementation to occur. *See also* CALCRETE.

aeon (eon) A term which was once used to describe a very long but indefinite period of time but is now being replaced by a precise usage for a period of 10^9 years. Thus the age of the Earth is about 4.7 aeons. *See also* CHRONOSTRATIGRAPHY, CRYPTOZOIC, PHANEROZOIC.

aeration of soil The process by which atmospheric air replaces air in the soil. In a well-aerated soil the two air types are

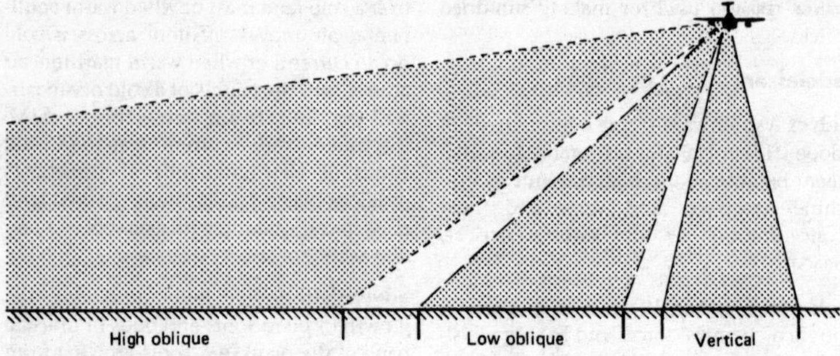

Figure 4 *Aerial photograph*

similar, but poorly aerated soil usually contains a lower oxygen content and a higher percentage of carbon dioxide than the atmosphere. The rate of aeration depends largely on the PERMEABILITY of the soil.

aerial camera A camera designed to take an AERIAL PHOTOGRAPH. In REMOTE SENSING both black and white/near infrared and FALSE COLOUR/near infrared film are commonly used because the spectral sensitivity extends beyond visible wavelengths to the region near infrared ELECTROMAGNETIC RADIATION. *See also* INFRARED PHOTOGRAPHY.

aerial photograph A vertical or oblique photograph of the surface of the Earth taken from the air. The scale of the photograph depends on the height of the aircraft or space vehicle and on the focal length of the camera. Photographs are taken in runs or strips, termed *sorties*, of overlapping prints and these are later assembled into *mosaics* (US = *print lay-down*). The majority of aerial photographs are verticals, taken when the optical axis of the camera is approximately perpendicular to the Earth's surface. A distinction is made between high-oblique and low-oblique photographs, both taken when the optical axis of the camera is directed between the horizontal and the vertical. [*4*] The high-oblique photograph includes the apparent horizon, the low-oblique photograph does not show the apparent horizon. Aerial photographs are used in resource mapping, land-use planning, archaeology, etc. *See also* PHOTOGRAMMETRY.

aerobic Pertaining to those organisms which exist only where there is free oxygen. The term is applied particularly to well-drained soils. *See also* ANAEROBIC.

aerobiology The study of the behaviour of AEROSOLS, both living and non-living organisms. It includes ACID PRECIPITATION and POLLEN ANALYSIS.

aerodynamic roughness, smoothness A physical boundary is *aerodynamically rough* when fluid flow is turbulent (TURBULENCE) down to the boundary, compared with LAMINAR FLOW over an *aerodynamically smooth* surface.

aerography A term introduced in 1973 by E. H. Rapoport to describe studies of the geographical range of taxa (TAXON), e.g. subspecies, species, genera, families, etc.

aerology The scientific study of all the ATMOSPHERE above the surface layers. *See also* AERONOMY, CLIMATOLOGY.

aeronomy The scientific study of the upper atmosphere (above 50 km) where dissociation and ionization of gas molecules occur.

aerosols Suspended minute particles (solid or liquid) of dust, sea salt and, in urban environments, carbon, lead and aluminium compounds produced by man's combustion of fuels. Because of their presence in the atmosphere as a dust veil, aerosols lower the amount of solar radiation reaching the Earth. Thus their overall thermal effect is probably one of cooling, thereby counteracting in part the GREENHOUSE EFFECT caused by rises in the amount of carbon dioxide and water vapour. *See also* DUST, HYGROSCOPIC NUCLEI, POLLUTION.

aerosphere The entire gaseous envelope surrounding planet Earth (ATMOSPHERE).

aestivation The state of inactivity adopted by desert animals during the hottest and driest seasons in order to survive.

afforestation The planting of trees on land, formerly under a different land use, to create a FOREST. Such measures have important ecological, hydrological, pedological and microclimatic effects on the area, ultimately influencing the geomorphological processes of that area.

afterglow A faint radiance occasionally visible in the western sky particularly in mountainous regions after sunset, probably caused by the scattering effect of atmospheric dust-particles on sunlight. *See also* ALPINE GLOW.

aftershock A relatively minor vibration of the Earth's crust following the main series of EARTHQUAKE shock waves. Generally, it originates at or near the FOCUS of the main earthquake and represents the subsequent readjustment of rocks that have been overstressed during the main rupture. According to the magnitude of the main earthquake the aftershocks may continue for hours, days or months, often causing the collapse of buildings or structures damaged by the initial earthquake. *See also* SEISMIC WAVES, FORESHOCK.

Aftonian A post-NEBRASKAN interglacial in the USA.

agate A semi-precious stone, consisting of a type of silica (chalcedony), which forms concentric bands of variegated colours when seen in section. It occurs in cavities in volcanic and some other rocks.

age An interval of geological time in the Chronomeric Standard scale of chronostratigraphic classification (CHRONOSTRATIGRAPHY). *See also* CHRON, CHRONOZONE, EPOCH, STAGE.

agglomerate A cemented mixture of angular, fragmented material of volcanic origin. Strictly, it should be formed by rocks ejected from a volcano (PYROCLASTIC MATERIAL) and the fragments should be larger than 2 cm in diameter, although the cementing material is much finer grained. *See also* BRECCIA.

aggradation The building-up of the land surface by the deposition of FLUVIAL or marine deposits. In the case of river-borne deposits, the material accumulates because of the inability of the river to continue to transport its LOAD, owing to a loss of velocity or volume of flow, increase of sediment or rise in base-level, usually the sea. Aggradation can also be caused by the building of a dam across a river which artificially creates a new base-level. Owing to the post-glacial rise of world sea-level (EUSTASY), following the melting of the Pleistocene ice-sheets, most rivers have aggraded their courses by depositing material in their lower reaches, thereby burying their former channels. In semi-arid areas the growth of fan-like plains of rock waste is due to periodic shifting of streams and their ultimate disappearance because of evaporation. Such features have been termed plains of aggradation. The term aggradation is also used to describe the accumulation of material in marine beach formation. *See also* DEGRADATION.

aggradational ice Ice which forms in horizontal layers underground at the base of the ACTIVE LAYER in a PERMAFROST environment. It develops as the upper surface of the permafrost layer rises.

aggregate **1** In pedology, a single mass or cluster of soil particles which adhere together in such a way that they behave mechanically as a unit (CRUMB STRUCTURE OF SOIL). **2** In civil engineering, the inert material which forms a substantial part of concrete or road metal. It can vary in size from broken stone or gravel to sand.

aggregation ratio A term used in soil science to describe the ratio of the percentage weight of clay minerals (given by mineralogical analysis) to the percentage weight of clay particles (determined by sedimentation methods). It relates to the activity potential of a soil. *See also* ACTIVITY RATIO, AGGREGATE.

aggressivity The ability of water to dissolve calcium carbonate during the process of limestone SOLUTION. Aggressivity ceases when the carbonic acid (formed by the water dissolving CO_2 from the air) is completely diluted when its CO_2 is used up.

Aghnadarragh interstadial A warm phase of the Irish Pleistocene sequence, of the early MIDLANDIAN. It postdates the FERMANAGH STADIAL.

agonic line A line drawn on a map along which the MAGNETIC DECLINATION is zero, since it joins the N MAGNETIC POLE of the Earth to the S magnetic pole. *See also* ISOGONIC LINE.

agro-climatology The study of those aspects of climate which are relevant to agricultural problems, e.g. earth temperature and ACCUMULATED TEMPERATURE data.

agro-forestry A land-use system in which trees are deliberately left (e.g. as shade), planted or encouraged (e.g. for fuel or food) on land where animals are grazed or crops grown.

agronomy The branch of agriculture that deals with the theory and practice of crop production and the scientific management of soils.

A-horizon The top zone in a SOIL PROFILE from which LEACHING removes soluble salts and COLLOIDS but where an admixture of mineral and organic matter is retained. The A_2 horizon is now termed the E-horizon (eluvial horizon). *See also* ELUVIATION, SOIL HORIZON [*see also 235*].

aiguille A French term which has been widely adopted to describe narrow, needle-shaped rocks. They are particularly well developed in the Chamonix region of the European Alps.

air-dry **1** The state of dryness of a soil at equilibrium with the moisture content of the surrounding atmosphere. The moisture content depends on the RELATIVE HUMIDITY and temperature of the atmosphere. **2** The process of soil drying in order to reach a moisture equilibrium with the surrounding atmosphere.

air frost Air at STEVENSON SCREEN level (1.2 m) with a temperature at or below 0° C (32° F). *See also* GROUND FROST.

airglow A weak and variable radiation of light from the upper atmosphere, producing the faint light of the night sky, even when moonless.

air mass An almost homogeneous mass of air of great lateral extent (generally hundreds or thousands of km), with marked horizontal temperature and humidity uniformity acquired from prolonged contact in its place of origin or source region. It is separated from an adjacent air mass by well-defined FRONTS which move as the air mass travels long distances, thereby modifying its original characteristics to a greater or lesser degree. According to its source region an air mass will be classified as: Arctic (A), Antarctic (AA), Polar (P), Tropical (T) or Equatorial (E), based on its temperature characteristics. With reference to its humidity it will be classified as: maritime (m), continental (c) or monsoonal (M). A combination of these characteristics gives: Tropical maritime (mT), Polar continental (cP) and Polar continental monsoon (cPM), etc. An indication of the air mass *stability* is given by the suffix S (stable) or U (unstable) where necessary. Whether an air mass is being warmed by equatorward movement or cooled by poleward movement is indicated by the suffixes W (warm) or k (German: *kalt*) respectively. Thus mTW(M)U refers to an unstable, monsoonal air mass of Tropical maritime characteristics being warmed as it moves equatorwards. (NB In the UK and USA air masses (for example) are generally referred to as mT rather than as Tm.) [*5*]

air pollution POLLUTION, SMOG.

air porosity A term used by soil scientists when referring to the portion of the bulk volume of a soil filled with air at a given

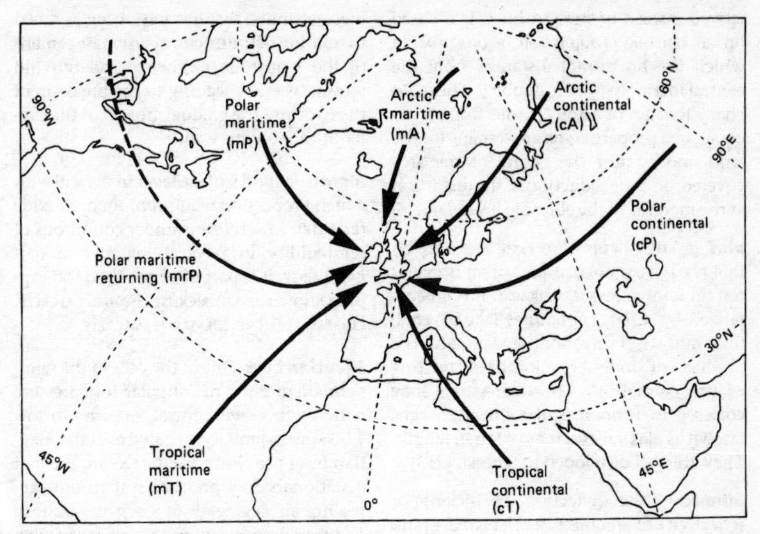

Figure 5 *Air masses affecting the British Isles*

time or under a given condition (i.e. a specified SOIL WATER potential).

air-quality standard The prescribed level of a pollutant in the outside air that should not be exceeded during a specified time in a given area. *See also* POLLUTION.

air stream An air flow which has a distinctive source or place of origin but which is not necessarily homogeneous. Thus an air stream is distinguished by its direction of approach rather than any inherent temperature or moisture characteristics. *See also* AIR MASS.

ait (eyot) A small island in a river (e.g. Chiswick Eyot on the R. Thames in London). The abbreviated form *ey* survives in many British place-names, denoting a former isle.

Aitoff's equal-area projection A MAP PROJECTION resembling a MOLLWEIDE PROJECTION but in which PARALLELS of LATITUDE (except for the equator) and MERIDIANS (except for the 0° meridian) are depicted as

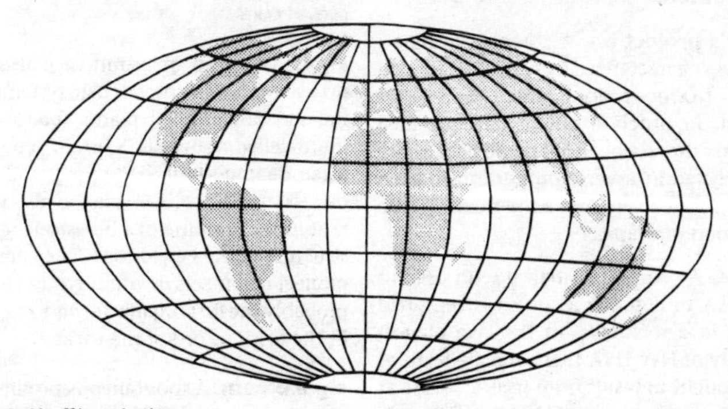

Figure 6 *Aitoff's projection*

curved instead of straight lines. It is based on a ZENITHAL EQUIDISTANT PROJECTION in which the horizontal distances from the central meridian (0°) are doubled. The main characteristics of Aitoff's projection are: its equal-area properties; its good shape for the land masses near the centre of the area covered; and the reduction of the distortion at the margins of the ellipsoid boundary. [*6*]

alas A large depression (km in diameter) in a PERMAFROST area, as part of an irregular terrain known as THERMOKARST. It is characterized by a flat, sometimes lake-covered, floor and steep surrounding walls, and is a product of long-term localized melting of the permafrost. The depressions may coalesce to form irregular linear troughs known as alas valleys, tens of km in length. They are well developed in Siberia. [*253*]

albedo The reflection coefficient or reflectivity of an object. It refers to the ratio between the total solar electromagnetic RADIATION (short wavelengths, 0.15–3.0 μm) falling upon a surface and the amount reflected, expressed as a decimal or percentage. The average *albedo* of the Earth is 0.34 (34%), but varies according to the colour and texture of the surface: fresh snow has an albedo of 0.85 (85%); dark soil 0.03 (3%); grass about 0.25 (25%); forest 0.05–0.10 (5–10%); concrete 0.17–0.27 (17–27%). The albedo of water varies from 0.7 (70%) with a low sun on a rough sea to 0.05 (5%) with a high sun over a calm sea. *See also* INSOLATION.

albedometer SOLARIMETER.

Alber's projection A CONICAL PROJECTION of a map characterized by two standard PARALLELS of LATITUDE along which the scale is correct. In order to obtain its equal-area property the MERIDIAN and parallel scales are constructed in inverse proportion to each other. At a continental scale there is little distortion of shape.

alcove A geomorphological term used in the USA to describe a curved, steep-sided cavity in a rock cliff. In the lava plateau country of NW USA these arcuate features are thought to result from SPRING SAPPING at the head of a stream. In the Colorado Plateau similar features have been ascribed to solution of the calcium carbonate cement of the sandstone following underground SEEPAGE, thereby leading to the break-up of the rock and the gradual growth of the cavity in the canyon wall.

alcrete A hard surface layer in the soil with a high proportion of aluminium hydroxide (BAUXITE), which forms under conditions of semi-aridity in tropical regions through CAPILLARITY. It is part of a rock-hard soil capping known as a DURICRUST. *See also* CALCRETE, CUIRASS, FERRICRETE, LATERITE, SILCRETE.

Aleutian Low One of the cells in the semi-permanent zone of subpolar low pressure with an average central pressure below 1,000 mb in January. Located over the Aleutian Is. of the North Pacific Ocean. It is not a stationary low pressure system but represents an aggregate of deep depressions occasionally interrupted by RIDGES OF HIGH PRESSURE and ANTICYCLONES. It is more marked in winter than in summer. *See also* ICELANDIC LOW.

alfisol One of the soil orders in the SEVENTH APPROXIMATION of soil classification. It occurs under deciduous woodland or grassland in the world's humid areas where there is some LEACHING and ELUVIATION with some colour change. Alfisol is a modern term which is equivalent to the following older terms: degraded PLANOSOL and SOL LESSIVÉ. It is generally a productive soil for common agricultural crops. In Britain the equivalents are the BROWN EARTHS.

algae A group of primitive plants with no leaves, flowers or vascular system. They include single-celled plants (DIATOM) and multi-celled seaweeds. Certain blue-green algae are important in the process of NITROGEN FIXATION in soils. Algae are also important in the building of a limestone reef-like structure termed a BIOHERM. They are often pioneer colonizers of virgin ground and are probably the first organic forms to be found in the fossil record of the rocks.

algal bloom A spontaneous proliferation of ALGAE in water bodies as a result of alter-

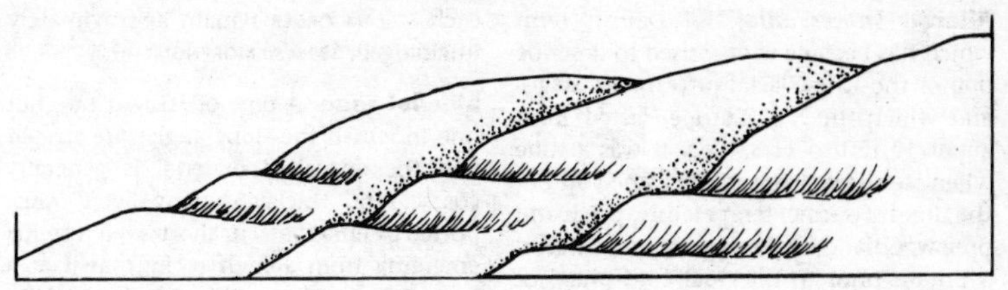

Figure 7 *Aligned sequence*

ations in water chemistry and/or temperature. *See also* EUTROPHIC.

algorithm A series of instructions or procedural steps for the solution of a specific problem.

alidade A simple sight-rule used in SURVEYING. It is utilized in conjunction with a PLANE-TABLE to determine the location of distant topographical features by angular measurements. Sophisticated models incorporate a telescope.

aliens A biogeographic term referring to the deliberate or accidental introduction by humans of organisms into regions outside the range of their natural distribution. Examples range from the introduction of the European rabbit into Australia to the accidental escape of the Oxford ragwort from the Oxford Botanical Garden along Britain's railway network.

aligned sequence A series of glacial MELTWATER CHANNELS which have a marked alignment when traced across neighbouring spurs or interfluves. [*7*]

alimentation, glacial A term relating to the gross accumulation of atmospheric snowfall, avalanche snow and ice, together with any meltwater that has refrozen in the ACCUMULATION ZONE of a glacier. The glacier will advance, retreat or remain stationary according to whether the alimentation is, respectively, greater than, less than, or in balance with the wastage by ABLATION. *See also* FIRN, NÉVÉ.

alkali In chemistry, the soluble hydroxide of metal which will react with an acid to form a salt solution, e.g. calcium, sodium and potassium.

alkali flat A dried-out lake floor where intense evaporation has left behind a level plain of alkaline sediments, e.g. in the Jordan valley of Israel. *See also* SALT FLAT.

alkaline soils Soils which exhibit a pH of 7.0 or more. They are especially common in arid areas where EVAPORATION is the dominant process and where the sporadic rainfall has been insufficient to leach (LEACHING) or wash away the soluble salts in the soil. In more humid areas the alkaline minerals (e.g. calcium carbonate, calcium sulphate and sodium chloride) are washed down from the A-HORIZON of the soil into the B-HORIZON where they accumulate. *See also* ALKALI SOILS, RENDZINAS.

alkali soils **1** Soils having a high degree of *alkalinity* (pH > 8.5) and/or having a high exchangeable sodium content (15% or more of the exchange capacity). **2** Soils which contain enough alkali (sodium) to interfere with the growth of most plants. *See also* ALKALINE SOILS, pH.

alkalization The process by which the exchangeable sodium content of a soil is increased.

allelopathy The production of chemicals by plants to inhibit the growth of competing flora or grazing fauna, e.g. heather can inhibit spruce growth on moorland; tannin in oak leaves makes them less palatable to certain caterpillars.

Allen's rule A rule which states that warm-blooded animals have a tendency for shorter legs, tails and ears in colder climates than in warmer climates. *See also* BERGMANN'S RULE.

Allerød Interstadial A Danish term which has become widely used to describe one of the Late-Glacial INTERSTADIAL stages and which, in NW Europe, lasted from about 12,150 to 11,350 BP. It was a time when ice-sheets temporarily waned and in the slightly warmer temperatures birch and pine woodland made a brief appearance in S Britain prior to the final cold phase of the Pleistocene (YOUNGER DRYAS). This stage is also referred to as Zone II in the LATE-GLACIAL pollen-zone chronology. *See also* BØLLING INTERSTADIAL, OLDER DRYAS, WINDERMERE INTERSTADIAL, BLYTT-SERNANDER MODEL.

allochthonous A term applied to the material forming rocks which have been transported to the site of deposition (CLASTIC SEDIMENTS). *See also* ALLOGENIC.

allogenic **1** A geological and geomorphological term to describe a phenomenon whose genesis lies elsewhere. It is particularly used to describe those parts of sedimentary rocks which were eroded from elsewhere prior to being transported, re-deposited and compacted into new rocks; e.g. the liver-coloured quartzite pebbles of the BUNTER pebble beds of central England are thought to have been derived originally from SW England. The term is also applied to (a) rivers which derive their water supply in a different climatic zone and are able to cross desert areas (e.g. the Nile) or (b) streams which rise on a different rock type prior to entering a limestone terrain. These latter cases are termed *allogenic streams*. **2** A biological term relating to a SUCCESSION of plants. It refers to the replacement of one ASSOCIATION by another within a SERE largely because of changes in the substratum independent of the plants themselves. *See also* AUTHIGENIC, AUTOGENIC.

allometric growth The growth of any organic system in dynamic equilibrium such that the ratios remain constant between each part of the system and each other part and also between each part of the system and the whole. E.g. as the number of stream segments in a network increases, the proportions of the segments falling into each STREAM ORDER remain approximately unchanged. *See also* MORPHOMETRY.

alluvial cone A type of ALLUVIAL FAN, but one in which the slope angles are steeper and the deposited material is generally coarser and thicker, having been transported by ephemeral or short-lived torrents emerging from a high rocky massif at a mountain front or valley side. In the USA the term is restricted to features occurring in arid or semi-arid areas but in European usage it is applied specifically to describe a fan-shaped deposit formed by Alpine torrents, where the term *cone of dejection* is used.

alluvial fan A fan- or cone-shaped mass of material, usually of sand and gravel (i.e. finer material than in an *alluvial cone*), deposited by a stream where it emerges from the constriction of a narrow valley at a mountain front and debouches on to a plain or into a wide trunk valley. The apex of the fan points upstream and this marks the thickest part of the mass (i.e. at the point of origin). The fan deposits become thinner as they are traced outwards and downwards and as the stream breaks up into a number of DISTRIBUTARIES. Because the stream velocity is checked by the change of gradient at the mountain front and because the stream splits into distributaries, more of its energy is used in overcoming friction, leaving less energy for sediment transport. Thus deposition occurs in the form of a fan, analogous to DELTA formation. Over a period of time adjacent fans may coalesce and extend some distance from the mountain front (BAHADA). *See also* ALLUVIAL CONE, TERMINAL FAN [*8; see also 189*]

alluvial fill A term defined by Lyell (1830) relating to sedimentary material deposited in a river CHANNEL. *See also* CHANNEL-FILL DEPOSIT, ALLUVIUM.

alluvial soils One of the subgroups of the AZONAL SOILS, comprising those immature soils which form on recent alluvial deposits. *See also* FLUVISOL.

alluvial terrace In general, an alluvial terrace is regarded as being synonymous

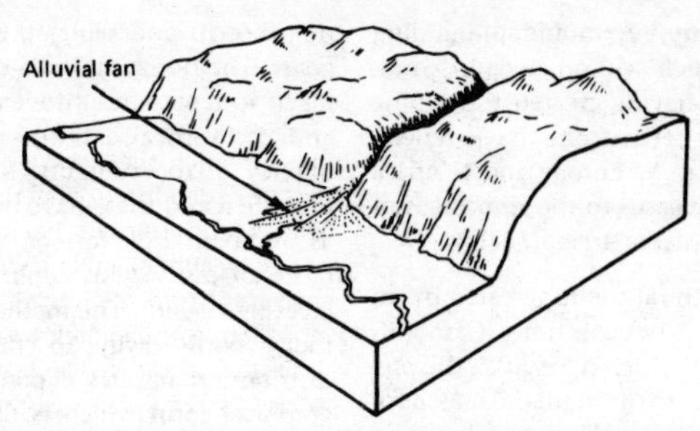

Figure 8 *Alluvial fan*

with a RIVER TERRACE, but this interpretation is only possible if one regards the term ALLUVIUM as including the coarse sands and gravels as well as the finer-grained deposits. In topographic form, however, the alluvial terrace resembles the river terrace and was formed in the same way – by the downcutting action of a river following REJUVENATION, leaving at higher levels portions of its former floodplain, which may be paired on either side of the valley as terraces. [*211*]

alluviation The process of infilling by alluvium, deposited by rivers anywhere along their courses. *See also* AGGRADATION.

alluvium The sedimentary deposits resulting from the action of rivers, thus including those laid down in river channels and floodplains, but not including lake or marine sediments. As one traces the material downstream it is seen to exhibit a marked decrease in PARTICLE SIZE, probably due to SORTING, although ATTRITION may also play a part by progressively wearing away the particles as they are carried downstream. There is some dispute over the exact definition of the term *alluvium*: in a historical sense it includes all unconsolidated fragmental material from the coarsest gravels and sands down to the finest clay and silt-sized particles; in a more restricted view (including that of the Geological Survey of Great Britain) only the finer SILTS are regarded as alluvium (particle size from 0.006 to 0.02 mm). Many important ore minerals (gold, platinum, tin) are found locally concentrated in alluvial deposits. *See also* COLLUVIUM, ELUVIAL DEPOSITS.

almwind A local wind of the FÖHN variety which blows S across the Tatra Mts and descends as a warmer wind on to the plains of S Poland in spring, thereby causing avalanches.

alp A topographical term to describe the high-altitude bench or shoulder standing above the glaciated trough in a mountainous terrain. The relatively gentle slope, though snow-covered in winter, provides valuable summer pasture for grazing animals.

alphanumeric A GIS term to describe any letters, numbers and punctuation marks that are treated as characters. It is not the same as NUMERIC, which refers to numbers that have a value and can be used in mathematical calculations.

Alpine An adjectival term describing any high mountains, which has also been used more specifically to describe: **1** a mountainous terrain characterized by a considerable degree of glacial erosion; **2** a type of mountain plant, related to a cold but not dry mountain environment; **3** a type of mountain climate; **4** a glacier in a mountainous terrain (mountain glacier); **5** a mountain soil occurring above the tree-line. *See also* ALPINE OROGENY.

Alpine glow A series of phenomena observed over mountains around sunrise and sunset. At sunset, it begins when the sun is 2° above the horizon and when snow-covered peaks assume yellow-pink-purple tints. *See also* AFTERGLOW.

Alpine orogeny A mountain-building movement which culminated in mid-Tertiary times, having created the Alpine ranges which stretch from the Pyrenees, Atlas and Alps in W Europe and N Africa through the Caucasus to the Himalayas of Asia (*Alpides*). *See also* NAPPE. [*104*]

Altaides Structural trends, created by an OROGENY in Late Palaeozoic times (Carboniferous–Permian), which have given the predominant 'grain' to the highland massifs of central Europe and northern Asia. Running in great arcuate lines, generally from W to E, these worn-down mountain ranges comprise the ARMORICAN chains in central Europe and many of the ranges of S Russia and N China, where the type-site – the Altai Mts – may be found. At present the stumps of these ranges are largely buried by newer rocks and protrude mainly as HORSTS.

alteration A geological term to describe the change in the mineralogical composition of a rock, typically brought about by the action of hydrothermal solutions during METAMORPHISM.

alternative energy Sources of ENERGY that are not reliant on FOSSIL FUELS. Such alternatives include SOLAR ENERGY, GEOTHERMAL ENERGY and energy from hydroelectricity, wind, wave and tidal power, but not nuclear power.

altimeter An instrument which indicates height above sea-level, based on the average decrease in atmospheric pressure with increasing height. In general this averages 34 mb (1 in of mercury) for every 300 m but variations occur owing to differences of air temperature and latitude. Used mainly in aircraft but also by ground surveyors (ANEROID BAROMETER). An *altigraph* is a self-recording altimeter.

altimetric frequency analysis A technique for analysing the relief of an area by map analysis, particularly in attempts to recognize and correlate EROSION SURFACES at high levels. The method involves noting all the summit spot-heights shown on the map (some authorities also include all enclosed contours where summit spot-heights are not shown) and plotting them on a bar chart. The horizontal axis on the graph is taken to represent altitude above sea-level and the vertical axis the percentage frequency of the summits. Maxima can be identified and these have been interpreted as remnants of former summit plains (ACCORDANT SUMMITS) or uplifted erosion surfaces (PENEPLAIN). The method is little used today, partly owing to criticism that the distribution implies a precision of topographical form which is not apparent in the actual relief and partly owing to the suggestion that it invited spurious correlations over wide areas, thereby ignoring local tectonic movements. *See also* AREA–HEIGHT DIAGRAM, CLINOGRAPHIC CURVE, HYPSOGRAPHIC CURVE.

altiplanation A process of levelling by WEATHERING and MASS-MOVEMENT in a PERIGLACIAL environment, during which benches or terraces are cut in solid rock. The planation is achieved by a combination of the FREEZE–THAW PROCESS, SOLIFLUCTION and CONGELITURBATION, in which rock material is broken up by frost action before being transported downslope. Thus, the altiplanation terrace is partly rock-cut but usually includes aggradational features of rock debris (GOLETZ TERRACES). The term is being replaced by CRYOPLANATION.

altiplano An elevated plateau in the Andes, particularly in Bolivia. It is generally surrounded by mountains which generate streams (and formerly glaciers) contributing to the inland drainage (ENDORHEIC) system of the altiplano. L. Titicaca and L. Poopó occupy centres of these high plateaux, for example.

altithermal A US term for the post-glacial period from approximately 6,000 BC to 1,000 BC (8,000–3,000 BP) when mid-summer mean temperatures in western N America were up to 4° F warmer than at present. This led to widespread drought and aggradation of the trunk river valleys owing to headward erosion in the tributaries.

altitude ELEVATION.

altocumulus A type of cloud (symbol: Ac), white and/or grey in colour, occurring in a patch, sheet or layer and consisting of bands, rolls or waves of globular masses, often separated by blue sky, although they may be close enough to merge, thereby obliterating the blue sky. In temperate latitudes, its limits are 2,000–7,000 m (6,500–23,000 ft), i.e. at middle altitudes. Is often an indicator of fair weather. *See also* CLOUD. [*43*]

altostratus A type of cloud (symbol: As), greyish in colour, occurring in sheets or layers, and of fibrous or uniform appearance. It has areas thin enough to reveal a hazy indication of a 'watery'-looking sun as viewed through ground glass. Its base can occur between 2,000 and 7,000 m (6,500–23,000 ft), i.e. at middle altitudes. It forms from a thickening of CIRROSTRATUS, is associated with a WARM FRONT, and is often a precursor of rainy weather. *See also* CLOUD. [*43*]

alveolate relief A terrain of dome-shaped hills (CUPOLA), particularly common in granite areas of the humid tropics. The domes are thought to be the protruding tops of buried INSELBERGS which would be uncovered by the stripping of regolith should the drainage pattern become rejuvenated.

alveole A small-scale (centimetric) hollow forming part of a network on rock surfaces due to CHEMICAL WEATHERING. *See also* HONEYCOMB WEATHERING.

ambient temperature The temperature of any part of the atmosphere which immediately surrounds an entity such as a CLOUD.

amendment of the soil An alteration of soil properties by the artificial addition of fertilizers.

amino acid An organic compound containing nitrogen which links together with other amino acids to form protein.

amino-acid racemization An important means of dating of organic remains, especially that of QUATERNARY materials. The technique is based on the discovery that protein preserved in the skeletal remains of animals is affected by a number of chemical reactions, many of which are time dependent. But unless the temperature conditions of the earlier time are known there can be serious errors of dating. *See also* RADIOCARBON DATING.

ammonium fixation The ABSORPTION or ADSORPTION of ammonium ions by the soil in such a way that they become relatively insoluble in water and largely unexchangeable by the usual processes of CATION exchange (BASE EXCHANGE).

amphibian A member of the vertebrate class Amphibia, consisting of animals with soft scale-less skin and whose larvae have gills adapted to living in aquatic environments, e.g. frogs, toads, salamanders and newts. *See also* REPTILE.

amphibole One of the ferromagnesian silicate mineral group, e.g. hornblende.

amphidromic system A tidal system in which the high water rotates round a central point (the nodal or amphidromic point), the rotation being caused by the CORIOLIS FORCE of the Earth's rotation. The rotationary movement is anticlockwise in the N hemisphere and clockwise in the S hemisphere. It is one of the consequences of the modifying influence that the Earth's rotation has on a standing oscillation created by tide-producing forces. At the amphidromic point itself the tidal range is almost zero but the tidal ranges increase as they are traced outwards from this point along CO-TIDAL LINES. The times of low and high water progress as they are traced around the amphidromic point. [*9*]

amplitude **1** The amount of elevation of the crest of a wave or ripple above the base of the adjacent trough. **2** In ecology, the term refers to the range of tolerance of a species.

anabatic wind An upslope wind formed when air on hill-sides is heated by INSOLATION conduction to a greater extent than air at the same horizontal level but vertically above the valley floor. This causes convectional rising of the heated air, which is replaced by cooler air from the valley floor. *See also* KATABATIC WIND. [*125*]

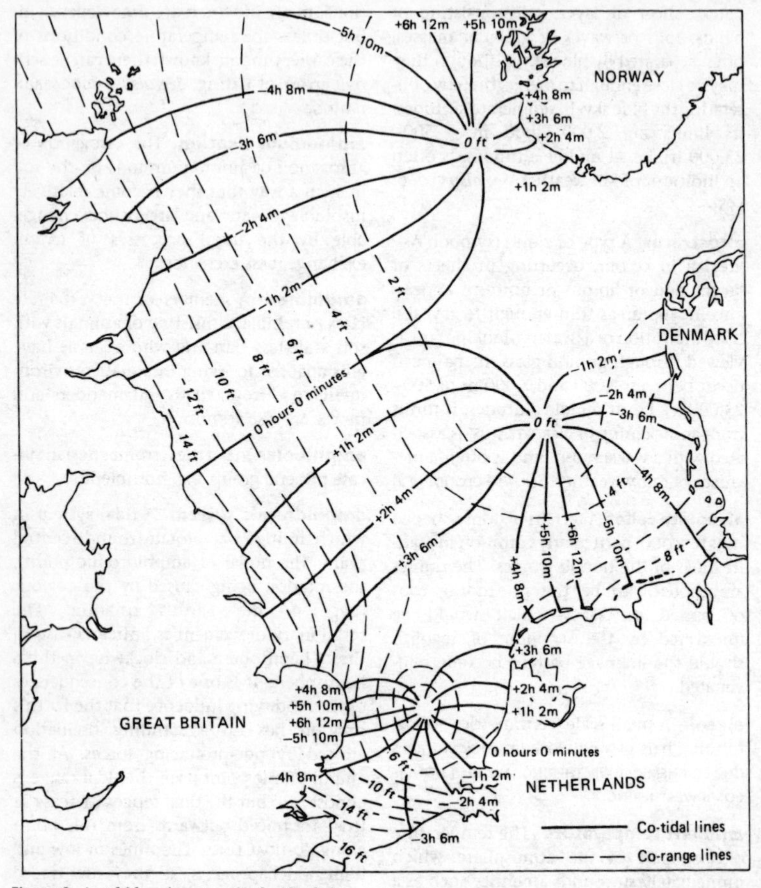

Figure 9 *Amphidromic system in the North Sea*

anabranch One of the branching streams in a BRAIDED river system. *See also* DISTRIBUTARY.

anaclinal A rarely used term to describe any topographical feature which is opposed to the general geological DIP. Thus, anaclinal streams run counter to the dip of the rocks. *See also* CATACLINAL, DIACLINAL, OBSEQUENT STREAM.

anaerobic Pertaining to those organisms which exist without free oxygen. The term is applied particularly to waterlogged soils. *See also* AEROBIC.

anafront A FRONT along which a warmer AIR MASS rises over a layer of cold air. *See also* FRONTOGENESIS, KATAFRONT.

anaglyph A method of obtaining a three-dimensional image of topography by viewing two adjoining AERIAL PHOTOGRAPHS that have been printed in red and green, by means of special lenses of which one is tinted red and the other green.

analemma EQUATION OF TIME.

analogue 1 In general terms, the similarity between two apparently unlike objects or processes. **2** In the context of REMOTE SENSING and mapping, it refers to information in graphical or pictorial form.

analogue weather forecasting A means of deducing a series of future weather situations, which may occur in repetitive patterns, by analysing the atmospheric pressure patterns of previous years. The assumption of analogue weather forecasting is that, if two pressure systems are identical at the same time of year, the weather sequence of the latter will closely follow that which succeeded the former system. *See also* NUMERICAL WEATHER FORECASTING.

analysis of variance VARIANCE, ANALYSIS OF.

anamolistic cycle The tidal cycle related to the varying distance between the Moon and the Earth. It is normally taken as 27.5 days. *See also* PERIGEE.

anaseism The vertical component of a SEISMIC WAVE moving up from an earthquake FOCUS.

anastomosing A geomorphological term relating to stream patterns in which, owing to excessive deposition in the main stream, the channels bifurcate, branch and rejoin irregularly to create a net-like formation. More recently it has been adopted to describe underground cave networks. *See also* BRAIDED STREAM.

anchor ice A type of ICE which forms on the bed of a body of moving water which itself remains unfrozen. *See also* FRAZIL.

andesite A fine-grained volcanic rock composed of andesine, similar in mineralogy to a DIORITE. Named from the Andes.

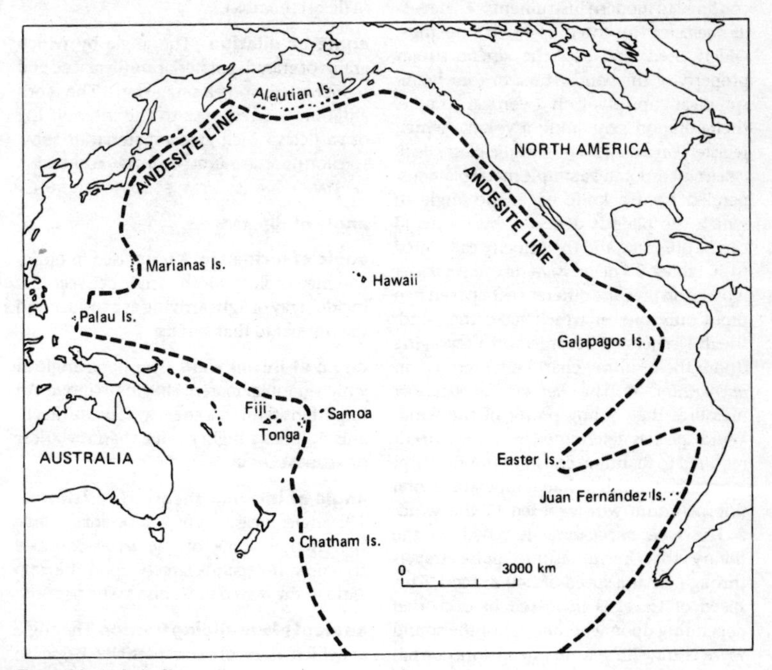

Figure 10 *Andesite line*

andesite line A boundary line drawn on a map of the Pacific Ocean delimiting the BASIC ROCKS of the true ocean basin of the central Pacific from those of greater acidity which lie on the continental side of the line. Thus, no igneous INTERMEDIATE ROCKS (ANDESITE, DACITE, RHYOLITE) are to be found *in situ* inside the andesite line. [*10*]

andosol A soil developed on newly weathered base-rich volcanic material. It is characterized by a dark, organic A-HORIZON and a barely altered B-HORIZON. As time elapses, however, increased LEACHING will change the andosol into a GRUMUSOL in humid tropical areas and a PODZOL in the humid temperate zone. *See also* INCEPTISOL.

anemometer Any instrument that measures wind-speed, which is recorded on an *anemograph*. A wind-sock at an airfield is a very simple variety (this also indicates wind direction) but the term is generally confined to modern instruments: **1** The *cup-anemometer* (invented 1846) is the most widely used, and uses the KINETIC ENERGY property of the wind. Three or four hemispherical cups pivot on a vertical spindle, their rotation generating a voltage which registers on a dial. **2** The *pressure-plate anemometer* incorporates a simple metal plate suspended from a knife edge; the angle to which the plate is deflected by the wind can be observed and the velocity calculated from tables. **3** The *pressure-tube anemometer* is based on pressure differences between two pipes only one of which faces the wind; the differences are recorded on a revolving drum, the resulting chart being termed an *anemogram*. **4** The *hot-wire anemometer* measures the cooling power of the wind. Wind-speed is determined from the current required to maintain constant the electrical resistance (and therefore temperature) of a fine platinum wire exposed to the wind. **5** The *sonic anemometer* is based on the theory that an ultrasonic pulse travels through air at a speed of 340 m sec^{-1}. This speed of travel is increased or decreased depending upon whether or not the sound wave is travelling into a head wind or tail wind.

aneroid barometer An instrument for measuring atmospheric pressure (invented in 1843). It consists of a shallow, thin corrugated metal box, evacuated of air, the flexible sides of which contract and expand respectively with rising and falling atmospheric pressure. Movements due to pressure changes are conveyed and magnified through a train of levers to a chain which actuates a pointer on a dial, or a pen on chart paper (BAROGRAPH). *See also* BAROMETER.

angiosperms A floral class, which includes the flowering plants which first appeared during the TRIAS. There are two main divisions: the Monocotyledons and the Dicotyledons. By the end of the CRETACEOUS they had replaced the GYMNOSPERMS as the dominant land plants.

angle of declination The angle between TRUE NORTH (geographical north) and the direction of the magnetic meridian (MAGNETIC DECLINATION).

angle of dilation The angle by which grains of granular material are displaced and reorientated along a SHEAR PLANE. The reorientation is a response to the interlocking of particles which give a material its SHEAR STRENGTH. *See also* ANGLE OF INTERNAL SHEARING RESISTANCE, ANGLE OF PLANE SLIDING FRICTION.

angle of dip DIP.

angle of incidence A term used in optics and meteorology for the angle between the incident ray of light arriving at a surface and the normal to that surface.

angle of initial yield The slope angle at which granular material begins to move. An important term in a snow AVALANCHE mechanism. It has a higher value than the ANGLE OF RESIDUAL SHEAR.

angle of internal shearing resistance The angle, measured by a SHEARBOX, to give the friction angle of the MOHR-COULOMB EQUATION. It depends largely upon the VOID RATIO of the material. *See also* STATIC FRICTION.

angle of plane-sliding friction The angle at which non-cohesive particles begin to slide down a surface. Also known as the

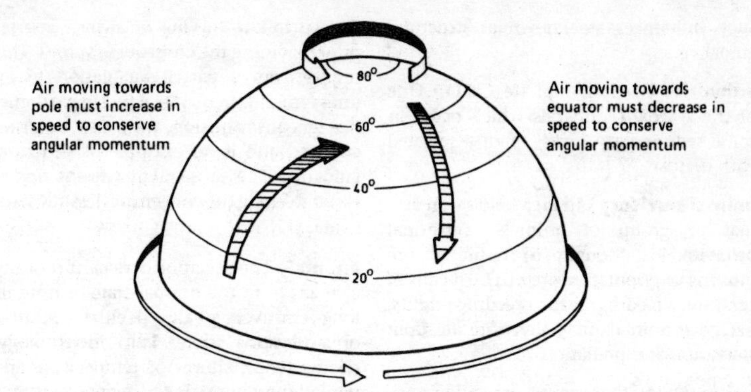

Figure 11 *Angular momentum*

static angle; the angle at which the particles stop moving is termed the dynamic angle.

angle of reflection The angle between a ray of light reflected from a surface and the normal to that surface.

angle of refraction The angle between a ray of light that has been refracted, e.g. through a body of water, and the normal from the surface at which the ray is refracted.

angle of repose (rest) The maximum slope gradient at which a mass of unconsolidated material (SCREE, TALUS) will remain stable. The angle will vary according to the character of the material. If the slope angle becomes steeper than the angle of repose it will become unstable, leading to a LANDSLIDE or EARTH-FLOW until the slope angle returns to a state of stability. The coarser the material the higher the angle of rest. *See also* COLLUVIAL SLOPE, CONSTANT SLOPE, DYNAMIC FRICTION, STATIC FRICTION, THRESHOLD ANGLE.

angle of residual shear The angle at which granular material comes to a halt after slope movement. It is comparable to the ANGLE OF INTERNAL SHEARING RESISTANCE and is less than the ANGLE OF INITIAL YIELD.

Anglian The glaciation or glacial stage of the PLEISTOSCENE which immediately preceded the great interglacial (HOXNIAN) in Britain; the antepenultimate cold stage, i.e. preceding the WOLSTONIAN which in turn preceded the DEVENSIAN. It is thought to be equivalent in age to the ELSTER GLACIATION of N Europe and the MINDEL glaciation of the Alps. [*197*]

angular momentum The product of the linear velocity of a point on the surface of the Earth (as it moves from W to E owing to the Earth's rotation) and the perpendicular distance of that point from the Earth's axis of rotation, a line drawn through the N and S poles. Its value is highest at the equator (where the perpendicular distance is greatest) and decreases systematically with latitude until it becomes zero at the poles. [*11*]

angular unconformity UNCONFORMITY.

angular velocity The rate at which a rotating body turns, expressed in revolutions per unit of time, degrees or radians (RADIAN). The angular velocity of the Earth is 15° longitude per hour or 7.29×10^{-5} radians per second. It is used in meteorology to indicate the horizontal rate of air rotating around a DEPRESSION.

angulate drainage pattern A type of TRELLIS DRAINAGE in which the tributary streams, instead of entering the trunk stream at (or nearly at) right angles, join it at very acute or obtuse angles either because of the underlying fault pattern or because of the dominant joint orientations of the rocks in a region of homogeneous strata.

Such influences are known as structural guidance.

anhydrite Calcium sulphate ($CaSO_4$). One of the EVAPORITE minerals which occur in some sedimentary rocks, often associated with GYPSUM.

animal territory An area held by an animal or group of animals. Territorial behaviour is thought to result in the limiting of population growth by means of exclusive feeding and breeding rights. Excluded animals may therefore die from starvation and predation.

anion An ION that moves, or would move, towards an anode. It is virtually synonymous with the term NEGATIVE ION.

anistrophy **1** The condition of a MINERAL or geological stratum having different physical or optical properties in different directions. **2** RADIATION that is scattered preferentially (usually in the forward direction), as distinct from that scattered equally in all directions (isotrophy). *See also* ISOTROPIC.

anistropy The presence of preferred orientation in the constituents of a clay rock or a TILL. *See also* FABRIC, TILL.

annual series A term used in FLOOD forecasting which is calculated by the selection of the series of maximum instantaneous river discharges from each year of the period under study.

annular drainage pattern That part of a drainage pattern in which the SUBSEQUENT STREAMS follow curving or arcuate courses prior to joining the CONSEQUENT STREAM. This results from a partial adaptation to an underground circular structure (e.g. a dome-like igneous intrusion, BATHOLITH; or a RING COMPLEX) and is an example of structural guidance. The subsequent streams find it easier to erode the concentric, less-resistant strata. [*12*]

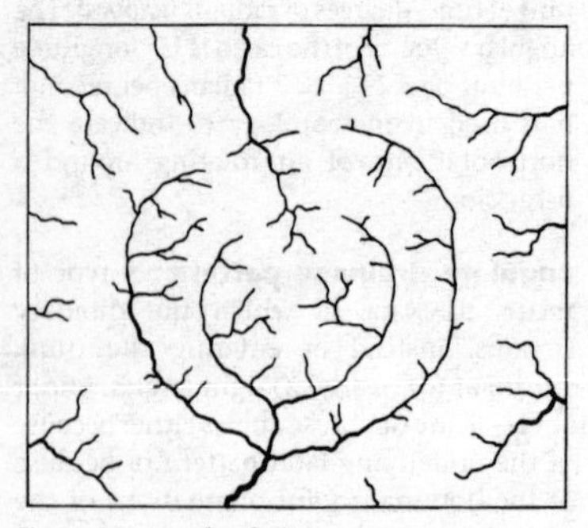

Figure 12 *Annular drainage pattern*

anomaly The deviation or departure of any element from the normal state or from its long-term average value, in either a positive or a negative sense. **1** In meteorology: related to departures of temperature and rainfall from normal. **2** In geomagnetism: a gravity anomaly is the difference between the observed gravity reading and that computed for an idealized globe. **3** In oceanography: a salinity anomaly is the difference between the observed salinity at a given point and the mean salinity of all oceans. *See also* BOUGUER ANOMALY, MAGNETIC ANOMALY.

Antarctic An adjectival term relating to the S polar region, as opposed to the N polar region (ARCTIC), e.g. the Antarctic Ocean and the Antarctic ice-cap. It is also used as a noun to describe the whole of the Earth's surface which lies within the Antarctic Circle, a parallel of latitude at 66°32'S. The land mass of Antarctica is almost entirely covered by the world's largest ice-sheet, 5 million sq miles (13 million km^2) in area and with a maximum thickness of approx. 2,300 m (7,500 ft).

antecedent drainage, antecedence A drainage system that has been able to maintain its direction of flow despite the tectonic uplift of land across its course. Thus, the rate of downcutting of the river has been approximately equal to the rate of crustal uplift. Those rivers which display antecedence generally cross the subsequently uplifted mountains or plateaux by means of exceptionally deep gorges, e.g. the gorges of the R. Colorado in SW USA were the first antecedent drainage features to be recognized. The headwaters of the Brahmaputra, Ganges and Indus cross the Himalayan ranges in spectacular antecedent gorges. *See also* INCONSEQUENT DRAINAGE.

antecedent moisture The SOIL WATER condition of an area before a further fall of rain. *See also* ANTECEDENT PRECIPITATION INDEX, WATER BALANCE.

antecedent precipitation index An index relating to the moisture conditions in a catchment area. It is useful in calculating the amount of EFFECTIVE PRECIPITATION that will form direct surface RUNOFF. It is expressed as:

API t = *k*. AP1 t − 1

where AP1 t is the index t days after the starting point. The value *k* has a seasonal variation between 0.85 and 0.98 depending on the exponential loss of moisture.

anthracite A type of COAL, characterized by a very high carbon content (85–98%) and a low degree of volatile matter (2–8%). It is hard, shiny coal and burns with no smoke or flame, while generating considerable heat.

anthropogeomorphology The study of landforms and processes generated by man, e.g. artificial lakes, quarries, and a wide variety of earthworks.

anthropomorphic soils Intrazonal soils formed as a direct result of man's activities, e.g. farming practices that have created soils of different pedological characteristics.

Antian An early stage of the PLEISTOCENE in Britain. The deposits are known only in East Anglia and comprise shelly sands of marine origin in which molluscan and foraminiferal evidence indicates a temperate environment. [*197*]

anticentre The point on the Earth's surface at the antipodal position (ANTIPODES) of the EPICENTRE of an earthquake.

anticline An arched fold or upfold in the strata of the Earth's crust. The two sides or limbs of the fold DIP in opposite directions away from a crestline or central AXIS. If the dip of each limb is equal it is a symmetric anticline; if the angle of dip of one limb is greater than the other the anticline will be asymmetric (ASYMMETRIC FOLD) or overfolded. Because denudation generally attacks the arches of folded structures faster than it does the downfolds (SYNCLINE) it is common for these to become the location of valleys which run along the anticlinal axis, thereby exposing older rocks in the core of the anticline. *See also* ANTICLINORIUM.

anticlinorium A series of smaller anticlines and synclines which form part of a larger arched structure in the Earth's crust; thus a complex upfold with numerous subordinate folds, in contrast to a complex downfold (SYNCLINORIUM). *See also* ANTICLINE.

anticyclogenesis The process by which anticyclonic activity is developed and strengthened, especially the semi-permanent thermal/glacial highs and dynamic sub-tropical highs. *See also* CYCLOGENESIS.

anticyclone A system of atmospheric pressure in which the ISOBARS on a SYNOPTIC CHART indicate a relatively high pressure in the centre and decreasingly low pressures outwards to the periphery of the system. The isobars are generally widely spaced, indicating light winds which may be absent near the centre. Air movement is clockwise in the N hemisphere and anticlockwise in the S hemisphere (BUYS BALLOT'S LAW). The term 'high' is frequently used as a synonym in modern parlance. The associated weather is settled and stable, generally warm, sunny

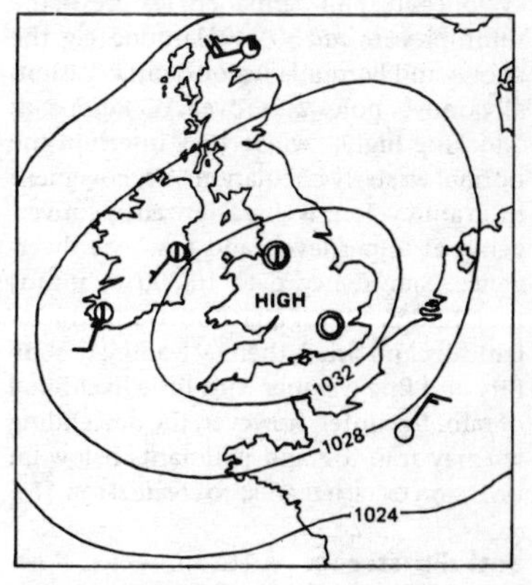

Figure 13 *Anticyclone*

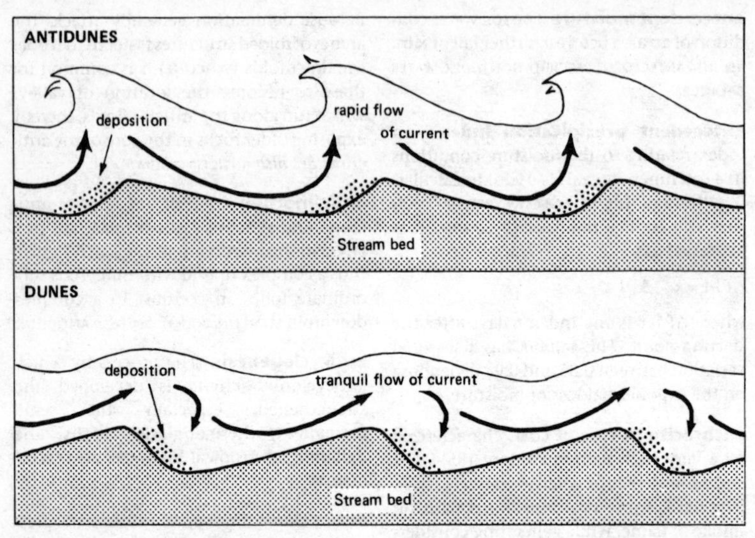

Figure 14 *Formation of antidunes*

and dry in summer and cold, frosty and clear (or foggy FOG) in winter. Cold or continental anticyclones form beneath a low TROPOPAUSE and warm stratosphere and are therefore shallow features (<3,000 m) produced by strong radiational surface cooling or in the cold air behind a DEPRESSION. Warm anticyclones are associated with higher temperatures throughout the deep TROPOSPHERE, with a cold STRATOSPHERE above. Warm anticyclones are semi-permanent features of the subtropics around 30°–40° latitude (e.g. the Azores and Bermuda regions) but occasionally move polewards over W Europe as 'blocking highs', where they interrupt the normal westerly circulation. Air movement in an anticyclone is characterized by convergence at upper levels and low-level divergence. Subsidence over 10,000 m means that the descending air is both warmed adiabatically and dried, thereby leading to stability and fine weather with little likelihood of rain. In winter, however, the descending air may trap fog and pollutants below an INVERSION OF TEMPERATURE to create SMOG. [*13*]

anti-dip stream A stream which flows in the opposite direction to the DIP of the surface rocks. But see OBSEQUENT STREAM, a term which some geomorphologists believe is synonymous with anti-dip stream. *See also* ANACLINAL.

antidune A type of small-scale CROSS-BEDDING feature formed from a sand deposit on a river bed. It develops as a sand wave from a so-called 'normal' dune as the current velocity increases in a highly loaded river. During the antidune phase ridge-like structures are created by the fast-flowing stream and these are characterized by erosion on their gentler downstream slopes and deposition on their steeper upstream slopes. Consequently, as the individual sand grains move down-current so the *antidune* moves up-current. It is, therefore, a transitory feature, forming, being eroded and re-forming. *See also* DUNE (SUBAQUEOUS), RIPPLE MARKS. [*14*]

antiform A general term denoting an upfold of strata in the Earth's crust, but one in which the precise stratigraphic relationships of the rocks are not known. Thus, the rocks exposed at the core of the upfold need not be, as in an ANTICLINE, the oldest. If an area of complex overfolding (NAPPE) is worn down by denudation it is possible for the stratigraphical succession to be inverted, even though the fold exposed at the surface is arched (convex upwards). *See also* SYNFORM.

antimeridian Any MERIDIAN which is 180° of longitude from another meridian.

antipodal bulge A term referring to the tidal effects at a point at the ANTIPODES to the so-called *tidal bulge* that occurs where lunar attraction is greatest on the side of the Earth facing the Moon. On the side furthest from the Moon, where lunar attraction is weakest, the crust is, in effect, 'pulled away', thereby producing in the ocean waters an apparent antipodal bulge.

antipodes A place on the Earth's surface which is directly opposite to another when measured through the diameter of the Earth. This means that they are both 180° of longitude apart and one of them as many degrees from the S pole as the other is from the N pole. The nearest land antipodal to the British Isles is Antipodes Island, which lies SE of New Zealand at 49°42'S and 178°50'E. It is antipodal to the Channel Is.

anti-trades Anti-trade winds (i.e. WESTERLIES) blow at high levels (above 2,000 m/ 6,500 ft) in the upper atmosphere in the opposite direction to the surface TRADE WINDS (which have an easterly component), in low latitudes only. Older textbooks referred to them as high-altitude return currents transporting to higher latitudes rising air at the INTERTROPICAL CONVERGENCE ZONE, but modern usage restricts the term to describe part of the troposphere WESTERLIES. NB JET STREAM has now replaced the term 'counter-trades'.

anvil cloud A wedge-shaped CLOUD produced when the ice crystal top of a large CUMULONIMBUS cloud produced by CONVECTION is flattened as it reaches the base of the STRATOSPHERE. The ice crystals are extended from the main cloud columns by wind changes with height, and a projecting point or anvil-shaped wedge develops.

aphelion The furthest point of a heavenly body from the Sun during its orbit. In the case of the Earth this is 152 million km (94.5 million miles) on 4 July, when the Sun–Earth distance is 1.5 per cent greater than the annual mean distance. *See also* PERIHELION.

aphotic zone A zone below 200 m depth in very deep lakes and oceans into which light does not penetrate and where PHOTOSYNTHESIS is therefore impossible. *See also* DIPHOTIC ZONE, PHOTIC ZONE.

apogean tides A tidal effect when the Moon is at its APOGEE and when its lunar attraction is decreased. The resulting low tides are higher and the high tides are lower, with an accompanying reduction in the TIDAL RANGE.

apogee **1** The meridional altitude of the Sun at midday on the longest day of the year. **2** The point in the orbit of a planet when it is furthest away from the Earth, with particular reference to the Moon. **3** The furthest point of a SATELLITE, used in REMOTE SENSING, from the point about which it is orbiting. **4** The highest point reached by a rocket, used in remote sensing, when fired upwards from the Earth's surface. *See also* PERIGEE.

apparent dip The angle between the horizontal and the tilt of the STRATUM or BEDDING PLANE measured vertically in any direction except that which is at right angles (90°) to the STRIKE. The angular value of the apparent dip will always be less than that of the TRUE DIP. [*62*]

apparent time TIME.

Appleton layer A zone in the upper atmosphere (THERMOSPHERE) (300 km/190 miles above the Earth's surface) which reflects short-wave radio-waves back to Earth. Also known as the F2 layer. *See also* HEAVISIDE–KENNELLY LAYER.

applied geomorphology Studies which involve the application of geomorphological methods of survey and analysis towards the solution of physical problems occurring within the human environment. These include such problems as flood control, irrigation, stabilization of slopes, land reclamation and coastal erosion mitigation. *See also* NATURAL HAZARDS.

applied meteorology Studies which use both archival and current atmospheric data to assist in the solution of physical problems within the human environment. This is particularly true in the field of NATURAL HAZARDS. These studies have been greatly assisted by the advent of REMOTE SENSING and computers.

apposed glacier A glacier formed when two separate glaciers coalesce.

apron A very low-angle outwash spread in front of an ALLUVIAL FAN.

apse line A theoretical line joining the points of APHELION and PERIHELION, the times of which become progressively later (*c.* 1.25 seconds per annum), because the line moves very slowly around the ORBIT OF THE EARTH in the same direction as its orbital movement.

aquaculture An alternative term for fish farming.

aquiclude A hydrological expression denoting a rock layer of low PERMEABILITY.

aquifer A rock layer which will absorb water and allow it to pass freely through. The term is also applied to any water-saturated stratum of earth or gravel that has sufficient POROSITY and PERMEABILITY to yield ample supplies of GROUNDWATER in the form of wells or springs. If it is underlain by a layer of impermeable rock it will prevent water passing downwards, thereby directing it laterally. If the stratum is bounded on its upper surface by an impermeable rock layer it is termed a *confined aquifer* (ARTESIAN BASIN). *See also* AQUICLUDE, AQUIFUGE.

aquifuge A US term (rarely used in Britain) denoting a rock layer or rock type which is completely impermeable (PERMEABILITY), i.e. it will not hold water nor will it allow water to pass through; e.g. obsidian, quartzite. Sometimes the term AQUICLUDE is used as a synonym for aquifuge although the terms have somewhat different meanings. *See also* AQUIFER, ARTESIAN BASIN.

arboreal A term pertaining to trees.

arc **1** In geometry, any part of a curve. **2** In topology, any line – which need not be part of a curve – joining two nodes. **3** In geology, a curvilinear line of islands (ISLAND ARC).

arch A natural opening through a mass of rock or boulder clay. It is most commonly seen on the sea coast where waves have cut through a promontory, e.g. Durdle Door, Dorset, the Green Bridge of Wales, the Bow Fiddle, Banff, Scotland; or where sea-caves have coalesced and partly collapsed. When the keystone of the marine arch collapses the feature will become a STACK. It is also formed in limestone areas when sections of the roof of an underground stream collapse, e.g. Marble Arch, Co. Fermanagh, N Ireland, and at Gordale Scar in the Pennines (Yorkshire). Spectacular natural arches reflecting various types of weathering and river erosion can be seen in Utah, USA, including the remarkable Rainbow Bridge.

Archaean (US **Archean**) A term which has been used in several different contexts to denote a span of geological time. **1** Originally, it was thought to refer simply to all the igneous and metamorphic rocks in the PRE-CAMBRIAN older than the PROTEROZOIC. But it is now thought unlikely that this time division is consistent in all parts of the world. **2** Some geologists regard it as equivalent only to the time division represented by the ARCHAEOZOIC, i.e. the second of the three eras of Pre-Cambrian time. **3** In N America, geologists recognize only two eras of Pre-Cambrian time, the Archaean being the earlier of the two. **4** Its usage as a synonym for the whole of the Pre-Cambrian is currently commonplace but is slowly being abandoned.

archaeological dating The dating of articles or events in history or ancient history by reference to the works of man (ARTEFACTS). *See also* GEOCHRONOLOGY, POTASSIUM–ARGON DATING, RADIOCARBON DATING.

Archaeozoic There are two different usages of this term denoting a span of geological time: **1** Specifically, the second of three eras of PRE-CAMBRIAN time. The EOZOIC precedes it, the PROTEROZOIC succeeds it. **2** In general, the entire span of Pre-Cambrian time. *See also* ARCHAEAN.

archipelago A closely grouped cluster of islands, e.g. the Tuamotu archipelago of the S Pacific, the Cyclades of the Aegean Sea.

arc of the meridian A measurement made along a meridian in order to determine the exact size and shape of the Earth.

Arctic An adjectival term relating to the N polar region, as opposed to the S polar region (ANTARCTIC), e.g. the Arctic Ocean. It is also used adjectivally to describe animals and plants characteristic of these northern regions and also in relation to weather and climatic types (ARCTIC AIR MASS, ARCTIC FRONT, ARCTIC SEA SMOKE). The whole of the Earth's surface which lies within the Arctic Circle, a parallel of latitude at 66°32'N, is frequently referred to as the Arctic (i.e. using Arctic as a noun).

Arctic air mass An extremely cold AIR MASS originating in the snow- and ice-covered Arctic, and travelling almost directly S. *See also* POLAR AIR MASS. [*5*].

Arctic front A frontal zone (FRONTOGENESIS) along which colder ARCTIC AIR MASSES meet moderately cool polar maritime or polar continental air in latitudes 50°N to 60°N. Since temperature contrasts between Arctic and POLAR AIR MASSES are not very marked the Arctic front is not very active. It often extends from S Greenland to the N of Norway in winter or spring. [*18*]

Arctic sea smoke (Arctic smoke) A type of ADVECTION FOG formed in high latitudes when cold air passes over a warmer water surface. The VAPOUR PRESSURE at the water surface exceeds the SATURATION vapour pressure at the particular air temperature, so that evaporation from the water surface proceeds at a greater rate than can be accommodated by the air at saturation-point. The excess water vapour condenses and is carried continuously upwards to evaporate in the drier air above so that the water surface appears to 'smoke'. *See also* STEAM FOG.

arcuate delta A DELTA in which the outermost margin exhibits an arc-life form, convex towards the sea; e.g. the Nile delta. [*58*]

area The extent of a surface, measured in square units, e.g. 640 acres = 1 sq mile. To convert units per square mile to units per km^2 one has to multiply by a factor of 2.58999; to convert units per km^2 to units per square mile one has to multiply by a factor of 0.3861. [*15*]

area–height diagram A graph showing the relationships between area and altitude. The vertical axis indicates elevation above a given datum (usually sea-level) while the horizontal axis denotes either the area between each pair of selected contours, or the percentage of the total area occupied by the area between each pair of selected contours. *See also* ALTIMETRIC FREQUENCY ANALYSIS, CLINOGRAPHIC CURVE, HYPSOGRAPHIC CURVE.

areal data Data which refer to the physical extent of a surface, measured in square units. [*15*]

Figure 15 *Area*

Unit	*sq metre*	*sq inch*	*sq foot*	*sq yard*	*acre*	*sq mile*
1 sq metre =	1	1,550	10.76	1.196	2.471×10^{-4}	3.861×10^{-7}
1 sq inch =	6.452×10^{-4}	1	6.944×10^{-3}	7.716×10^{-7}	1.594×10^{-7}	2.491×10^{-10}
1 sq foot =	0.0929	144	1	0.111	2.296×10^{-5}	3.587×10^{-8}
1 sq yard =	0.8361	1,296	9	1	2.066×10^{-4}	3.228×10^{-7}
1 acre =	4.047×10^{3}	6.273×10^{6}	4.355×10^{4}	4,840	1	1.563×10^{-3}
1 sq mile =	259.0×10^{4}	4.015×10^{9}	2.788×10^{7}	3.098×10^{6}	640	1

1 are = 100 sq metres = 0.01 hectare.
1 circular mil = 5.067×10^{-10} sq metre = 7.854×10^{-7} sq in.

arena A morphological term to describe a shallow, generally circular, basin surrounded or almost enclosed by a rim of higher land. It is used specifically in Uganda where intrusive granites have been denuded and lowered at a faster rate than their surrounding rims of more resistant rocks, often folded quartzites in the METAMORPHIC AUREOLE.

arenaceous Denoting a rock containing sand and having a sandy texture. The arenaceous rocks are one of the groups of detrital SEDIMENTARY ROCKS and are characterized by particles ranging in size from 2 mm to 0.062 mm. The term SANDSTONE is often regarded as being synonymous with arenaceous rocks, while the term PSAMMITIC, although exactly equivalent to arenaceous, is currently reserved to describe metamorphosed arenaceous rocks. The size and shape of the particles differ, enabling a classification to be made: >0.5 mm diameter = coarse and very coarse; 0.125–0.5 mm = medium; <0.125 mm = fine and very fine; the angular and subangular particles make up the GRITSTONE variety of the arenaceous rocks, while the rounded particles are the sandstones. Arenaceous rocks are also differentiated according to the MINERALOGY of the grains; e.g. a high percentage of feldspar will produce a coarse-grained feldspathic sandstone or grit known as *arkose*, a high percentage of GLAUCONITE will produce a GREENSAND. The mineralogy of the cement which binds the grains also differs, so that arenaceous rocks can also be classified, for example, into calcareous sandstones with a cement of a dolomitic type, ferruginous sandstones cemented by LIMONITE, and siliceous sandstones, such as ORTHOQUARTZITE. *See also* GREYWACKE.

arenite Any ARENACEOUS rock.

arenization A weathering process in solid rocks which produces a deep sandy REGOLITH. It is most active in tropical areas where high temperatures and plentiful groundwater lead to rotting by CORROSION.

areography The study of the form and size of the areal (geographical) ranges of floral and faunal taxa. *See also* BIOGEOGRAPHY, ZOOGEOGRAPHY.

arête A French term which has been widely accepted to describe a narrow, rocky and often jagged ridge which divides the steep walls of two adjacent CIRQUES. It may be on the actual crestline of a mountain or merely on a subsidiary ridge. In the USA it is termed a *combe-ridge* and in German a *Grat*.

aretic drainage A drainage pattern which is confined to an inland basin, with no outlet to the sea, e.g. the Tarim Basin drainage of central Asia, Cooper's Creek in central Australia. Also termed internal drainage. *See also* ENDORHEIC DRAINAGE.

argillaceous Denoting a rock composed largely of CLAY. The argillaceous rocks are one of the groups of detrital SEDIMENTARY ROCKS characterized by an abundance of CLAY MINERALS. The term PELITIC, although exactly equivalent to argillaceous, is now currently reserved to describe metamorphosed argillaceous rocks (pelites). The rocks are classified according to particle size: <0.002 mm is termed the CLAY grade: 0.002–0.006 mm is termed the *silt* grade. *See also* MARL, MUDSTONE, SHALE, SILTSTONE.

argillic brown earths One of the subdivisions of the BROWN EARTHS, which are themselves part of the SOIL CLASSIFICATION OF ENGLAND AND WALES. They are characteristically brown or reddish in colour and have a loamy horizon overlying a marked subsurface horizon of clay accumulation. *See also* ILLUVIATION.

arheic (areic) A term derived from the Greek verb (*rhein*) to flow, which means literally, 'without flow'. It is used by geomorphologists to describe those arid areas where no stream valleys have developed. *See also* ENDORHEIC DRAINAGE.

arid, aridity Dry, parched, deficient of moisture. A climate having insufficient precipitation to support vegetation is defined as arid, while a more statistical definition denotes an arid area as having less than 250 mm (10 in) of annual rainfall. A further definition regards the arid zone as the area where EVAPORATION potential exceeds actual precipitation. Several formulae, mainly empirical, have been devised as an Index of Aridity, foremost among them being: **1** E. de Martonne: $I = P/(T + 10)$, where I = index, P = precipitation in mm, T = mean annual temperature in °C; **2** R. Lang: $I = P/T$ (the Lang Rain Factor); **3** W. Köppen: $R = t$ (Rain mainly in winter), $R = t + 7$ (Rain evenly distributed through the year), $R = t + 14$ (Rain mainly in summer), where t = the mean annual temperature in °C; if the annual rainfall (R) in cm is less than $2t$ and greater than t, the climate is regarded as arid; **4** C. W. Thornthwaite: $I = 100\ d/n$ where d is water deficiency and n is water need. *See also* MOISTURE INDEX, PRECIPITATION EFFICIENCY.

arid cycle of erosion One of the three climatically related cycles of erosion recognized by the US geomorphologist, W. M. Davis, in the early years of the 20th cent. (NORMAL CYCLE OF EROSION, GLACIAL CYCLE OF EROSION). It is an idealized cycle which applies only to desert areas where BLOCK-FAULTING has produced many enclosed basins – a situation not present in all desert areas. Thus, it is ideally suited to the deserts of the SW USA but because it cannot be applied universally the concept now has few adherents. Davis regarded it as a climatic modification of the humid cycle (normal cycle) and the chief differences between the two may be summarized as: **1** contrasts in RUNOFF; **2** maximum relief achieved in the youthful stage of the *arid* cycle, not in maturity; **3** decrease not increase in relief as cycle progresses; **4** a prevalence of CONSEQUENT drainage into enclosed basins; **5** considerable AGGRADATION of the basins in the youthful stage when streams are actively dissecting the mountains; **6** predominance of local base-levels of erosion with few streams showing exorheic tendencies. The ultimate or 'old-age' stage was largely conjectural, with a type of 'desert peneplain' being achieved (PENEPLAIN) not by rivers but by gradual lowering of the area, largely by wind removal (DEFLATION). *See also* BAHADA, INSELBERG, PEDIMENT, PEDIPLAIN, PLAYA.

aridisol One of the ten orders defined in the SEVENTH APPROXIMATION soil classification. It includes infertile alkaline and saline soils of desert areas and is characterized by a thick accumulation of basic mineral salts at or near the surface owing to CAPILLARITY (e.g. calcium and sodium). This capillary action may cause ground-water to concentrate sodium to such a degree that the soil becomes toxic. Except for a few tolerant species the vegetation is extremely sparse on aridisols, which renders them liable to soil erosion. *See also* CALICHE, SIEROZEM, SOLONCHAK, SOLONETZ.

arkose ARENACEOUS.

Armorican orogeny The mountain-building episode of Carbo-Permian times. It is the W European equivalent of the VARISCAN OROGENY of central Europe, named by E. Suess in 1888 after the old name for Brittany (Armorica). It is part of the HERCYNIAN OROGENY of Eur-Asia which created mountain chains from Ireland to China (ALTAIDES). Since it folded and faulted the Carboniferous COAL MEASURES, a great deal is known about Armorican structures. The suffix -oid (Armoricanoid) indicates that direction only is implied, whereas the termination -ian or -an should strictly indicate that the fold belonged to that particular orogeny. Compare usage in Britain of the orogenies Caledonian (Caledonoid), Charnian (Charnoid), Malvernian (Malvernoid). *See also* ALPINE OROGENY. [*104*]

armouring A term used in FLUVIAL studies referring to heterogeneous river bed material in which coarse grains are concentrated sufficiently at the bed surface to stabilize the bed, thereby inhibiting the TRANSPORTATION of underlying finer material.

array A regular arrangement of information, especially of numbers, as in a MATRIX.

arroyo A Spanish term commonly used in Latin America to describe a stream-bed which is periodically dry, in contrast to a *rio*, which refers to a permanently flowing stream. *See also* CREEK, DRAW, GULCH, NULLAH, WADI, WASH.

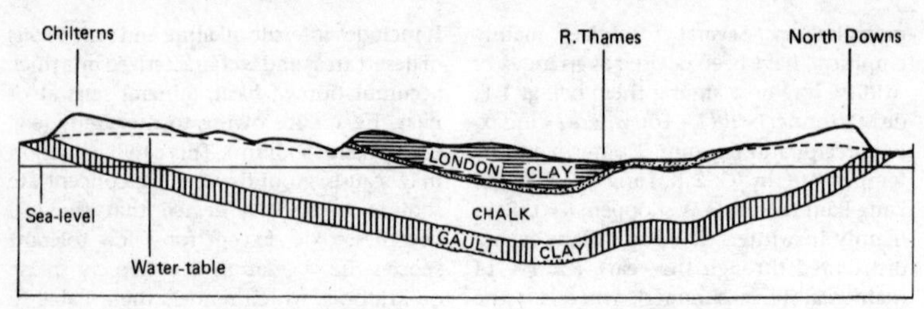

Figure 16 *Artesian basin: the London Basin*

artefact Any article produced by man, but a term used specifically in archaeology in relation to Stone Age tools.

artesian basin A structural basin of SEDIMENTARY ROCKS in the crust which produces a constant supply of water that rises to the ground surface, by means of an ARTESIAN WELL, from the subterranean AQUIFER. The aquifer holds water under a pressure head and is sandwiched between two layers of impermeable strata (PERMEABILITY). One of the world's largest artesian basins is in E-central Australia where the aquifer is charged by rainfall occurring on the Eastern Highlands. The name is derived from Artois in France, but one of the best-known artesian basins is the London Basin, where rain falling on the North Downs and the Chilterns charges the aquifer of the CHALK (and some Lower Eocene sandstones) which is enclosed by the impermeable Gault Clay (below) and the London Clay (above). [*16*]

artesian well A well achieved by boring down into an AQUIFER enclosed by impermeable strata (PERMEABILITY) within an ARTESIAN BASIN. The water in the aquifer is under a head of pressure (HYDROSTATIC PRESSURE) and once penetrated by the boring should rise to the surface spontaneously if the well-head is at a lower level than the WATER-TABLE. When wells were first sunk in the London Basin water flowed easily to the surface, thereby satisfying the city's needs, but subsequent withdrawals (currently some 1,200 million litres/day) have not only lowered the water-table, but also created a fall in hydrostatic pressure, leading to the installations of pumps to assist the supply. Pumps are always required in the sub-artesian wells, occurring in basin-like structures in which water is contained but where hydrostatic pressure is insufficient to raise it all the way to the surface. Only about half London's supply is now derived from this source.

artificial intelligence A GIS term referring to a means of formalizing advanced cognitive knowledge into a computer-manageable format to be used in problem solving.

asbestos A term for the fibrous varieties of several distinct mineral species, all of which are silicates. They include such minerals as chrysotile (fibrous SERPENTINE), and are important because, like other fibres, they can be woven and have the added advantage of being fireproof.

aseismic plates Any of the Earth's crustal plates (PLATE TECTONICS) within which there are relatively few earthquakes.

ash The unconsolidated fine-grained material ejected from the crater of a volcano during volcanic eruptions. It is formed from tiny lava fragments. *See also* PYROCLASTIC MATERIAL.

ash cone A volcanic cone made entirely, or almost entirely, of ASH. In general, ash cones are small in stature, although there are a few of major proportions, e.g. Volcán de Fuego, Guatemala, which reaches an elevation of 11,000 ft (3,360 m). *See also* CINDER CONE.

ash fall A rain of airborne volcanic ash falling from an eruption cloud. [*271*]

ash flow NUÉE ARDENTE.

ashlar A block of building stone with straight edges. *See also* FREESTONE.

aspect The direction in which a slope faces, especially in the context of EXPOSURE or different degrees of INSOLATION. *See also* ADRET, UBAC.

asphalt A HYDROCARBON of natural occurrence and with a very VISCOUS character. Most asphalts are virtually solid, although a number are sufficiently fluid to be able to be poured without their temperature being raised. They are often formed as a residual deposit when the lighter fraction of an oil pool has evaporated. The most famous examples occur in Trinidad (the Pitch Lake), La Brea in Los Angeles and at the Athabasca Tar Sands in Canada.

assemblage zone A biostratigraphic (BIOSTRATIGRAPHY) unit defined and identified by a group of associated fossils rather than by a single index fossil.

association **1** In general, a grouping or combination. **2** In biological usage the term is given the more specific meaning of an assemblage of plants living in close interdependence in a similar ECOLOGICAL NICHE. They exhibit similar habitat and growth requirements, although there are usually one or more DOMINANT species. A precise botanical definition regards it as a climax community that is the largest sub-division of a CLIMAX, BIOME or FORMATION. *See also* SOIL ASSOCIATION.

association-analysis A method which sub-divides a classification of vegetation into a statistical hierarchy: the normal association-analysis subdivides a plant population so that all association disappears. The population is sampled and first divided into two broad groups related to one species which exhibits the highest degree of association – those in which this species occurs and those without it. The procedure is then repeated for each group, i.e. of discovering which species now has the highest degree of association which, when determined, forms the basis for further subdivision.

asteroids A belt of some 1,500 heavenly bodies which revolve in the solar system between Jupiter and Mars, formed possibly from the disintegration of a former planet. None is larger in diameter than 800 km (500 miles).

asthenosphere A geological term for the zone of the Earth's MANTLE which lies immediately below the LITHOSPHERE. Since it is thought to exhibit plastic properties it has been suggested that it is capable of prolonged deformation because of its poor rigidity and low rock strength. Geophysicists suggest that it may mark the zone in which major horizontal convection flow occurs, leading to a movement of the plates of the outer lithosphere (PLATE TECTONICS). It occurs at depths between 100 and 240 km below the surface and is thought to be composed of partially molten PERIDOTITE, in which the velocity of seismic waves is considerably reduced. *See also* CRUST OF THE EARTH.

astrobleme A very ancient crater-like feature on the surface of the Earth, thought to be the result of a collision with an extra-terrestrial body. *See also* METEORITE.

asymmetrical fold A fold (either an ANTICLINE or a SYNCLINE) in which the axial plane is not vertical and which results in one of the limbs dipping more steeply than the other. *See also* FOLD. [*84*]

asymmetrical valley A valley in which the slope on one of the sides is greater than that on the other. An asymmetrical valley can be formed by: **1** structural control due to differential undercutting of valley sides by streams working along a gently dipping stratum of rock (UNICLINAL SHIFTING); or **2** differential weathering of slopes under PERIGLACIAL conditions. This occurs when valley alignments lead to differing degrees of exposure to direct sunshine and shadow. In Britain it has been suggested that the asymmetry of certain DRY VALLEYS in the CHALK of southern England may be explained by former periglacial processes which operated during the Pleistocene, in which those slopes with a southerly ASPECT experienced almost continuous weathering by FREEZE–THAW processes. These so-called 'active'

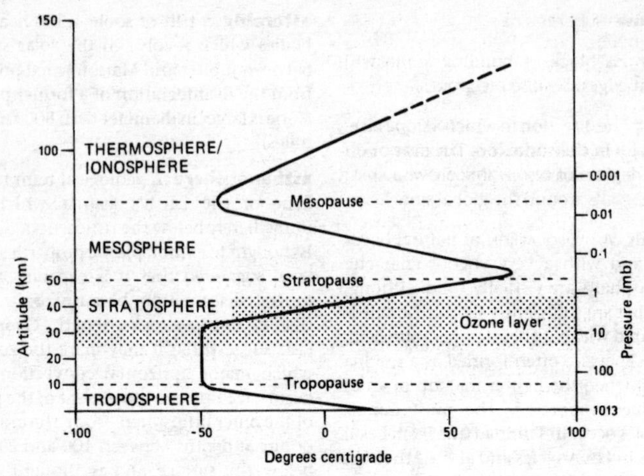

Figure 17 *Atmosphere*

slopes became gradually reduced in angle, but at a greater rate than the 'inactive' slopes, thereby leading to asymmetry. Studies have shown, however, that the patterns of asymmetry vary from place to place so that a set of hard and fast rules is difficult to apply universally.

at-a-station The term applied by hydrologists to all observations made at one point on a river by gauging its DISCHARGE. Such observations will provide information of changes of flow over a time period, and should be distinguished from observations made at one time but at many places along the river (these are termed *downstream* observations). *See also* FLOOD ROUTING.

Atlantic Period A term used to describe the warm and moist climatic phase which occurred in NW Europe between 7,450 BP and 4,450 BP, in post-glacial (FLANDRIAN) times. Sometimes referred to as the Climatic Optimum or the Megathermal Period. It has been calculated that mean annual temperatures were some 2° to 3° C above those of today. In Britain there was widespread peat formation, a growth of mixed forest up to elevations of 750 m (2,500 ft), and the spread of MESOLITHIC communities of Sauveterrian and Maglemosian culture. The marked rise in precipitation which generated peat growth (termed the Lower Turbarian phase in Scotland) coincided with a small rise of sea-level which led to the flooding of the Straits of Dover, as world ice-sheets waned and water returned to the oceans. This was the last occasion on which Britain was joined to the continent of Europe and henceforth the climate became more MARITIME. *See also* BLYTT–SERNANDER MODEL, HOLOCENE, MEDIOCRATIC, NEOLITHIC. [*81*]

Atlantic Polar Front POLAR FRONT.

Atlantic-type coast DISCORDANT COAST.

atmometer EVAPORIMETER.

atmosphere A relatively thin layer of odourless, tasteless and colourless gases (air) and dust surrounding the Earth (AEROSPHERE). It is subdivided into concentric shells according to characteristics of temperature lapse rates: the lowest is the TROPOSPHERE, separated by the TROPOPAUSE from the STRATOSPHERE, above which is the MESOSPHERE and the outermost layer, the THERMOSPHERE. Its dry gases comprise 78.09% nitrogen, 20.95% oxygen, 0.93% argon, 0.03% carbon dioxide and tiny proportions of neon, krypton, helium, methane, xenon, hydrogen and ozone, together with water vapour, which varies between 0 and 4%. Its movement, in the form of wind-systems and pressure cells, is referred to as the ATMOSPHERIC CIRCULATION. *See also* HETEROSPHERE, HOMOSPHERE. [*17*]

atmospheric circulation The movement of air in the form of PLANETARY WINDS and pressure-cells around the global surface and at higher levels in the atmosphere. The circulation is 'powered' by the global imbalance between incoming and outgoing RADIATION in which low latitudes receive more than high latitudes. Thus heat is transferred polewards by AIR MASSES and OCEAN CURRENTS. *See also* ABSOLUTE VORTICITY, ANGULAR MOMENTUM, FRICTION, HEAT BALANCE. [*18*]

atmospheric energy The ability of the Earth's atmosphere to continue its essential mechanism of transporting heat, in which it is dependent on several kinds of energy: **1** *solar radiant* energy transferred by radiation (INSOLATION) from the sun; **2** *latent energy* released by the condensation of water vapour; **3** *geopotential energy* due to gravity and height above the surface; **4** *kinetic energy*, which is continually being converted from heat energy by movement of air molecules. The equator receives 2.5 times as much annual solar radiant energy as the poles and because of this global imbalance great lateral transfers of heat energy take place, with the atmosphere acting as a gigantic heat engine. The continuous heat transfer provides the energy to maintain the ATMOSPHERIC CIRCULATION and the PLANETARY WINDS. The conversion of heat energy into kinetic energy must involve rising and descending air, although these vertical movements are less noticeable than the horizontal air movements whereby, in general, wind-speeds are approximately 100 times greater than average vertical air movements. *See also* ENERGY.

atmospheric pollution POLLUTION.

atmospheric pressure The pressure exerted by the weight of the atmosphere on the Earth's surface. The average pressure at sea-level is 1,013.25 millibars (1,000 mb = 1 bar = 10^5 pascals). The highest atmospheric pressure recorded in the world = 1,079 mb; the lowest recorded in the world = 877 mb. *See also* ANEROID BAROMETER, BAROGRAPH, BAROMETER.

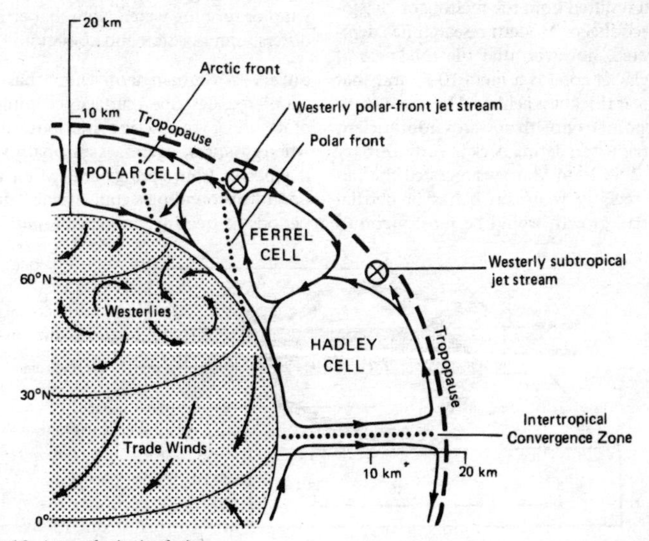

Figure 18 *Atmospheric circulation*

atmospheric window Those parts of the long-wave, infrared radiation spectrum, with wavelengths of 8.5–14 microns, in which radiation is not absorbed by water vapour and CO_2. Thus, when there is little cloud cover, terrestrial radiation of this wavelength is lost into space, allowing a cooling of the Earth's surface. REMOTE-SENSING systems view the Earth's surface through these windows. *See also* ABSORPTION, GREENHOUSE EFFECT, INFRARED THERMOGRAPHY, RADIATION, TERRESTRIAL RADIATION. [*2*]

atoll A ring-shaped island REEF that encircles (sometimes completely surrounding) a central lagoon of sea water in which detrital material collects. It is composed largely of CORAL but in some oceanic atolls certain calcareous algae may form the bulk of the reef. It is a particularly common feature of the Pacific Ocean. Several hypotheses have been proposed to account for its formation: **1** According to Charles Darwin's subsidence theory the growth of coral around a gradually submerging island kept pace with the rate of submergence (either by land subsidence or by sea-level rise). **2** R. A. Daly believed that the submergence was entirely post-glacial, owing to the rise of sea-level which resulted from the melting of Pleistocene ice-sheets. Modern research has demonstrated, however, that the thickness of post-glacial coral is a mere 10 m and that the great thickness achieved by some atolls must point to growth upwards from ancient reefs (BIOHERM) dating back to early Tertiary time. **3** Sir John Murray suggested that, as coral reefs grew up on banks of detrital material, growth would be most vigorous on the outer perimeter, thereby leading to outward growth. It is known that the surrounding rim lies in shallow water while the inner slope shelves steeply towards the deeper water of the lagoon. [*19*] *See also* BARRIER REEF, FARO.

attached-dune A sand-DUNE which occurs in deserts as a result of accumulation around a rock or other obstacle.

attenuation **1** In general, the removal of wave peaks. **2** A term which describes the loss of electromagnetic energy (solar radiation or radio waves) as it passes through the atmosphere, owing to ABSORPTION and SCATTERING by atmospheric particles and molecules.

Atterberg limits LIQUID LIMIT, PLASTIC LIMIT.

Atterberg scale A grade scale for the classification of sediments, the subdivision following the logarithmic rule. *See also* LIQUID LIMIT, PLASTIC LIMIT.

attributes The properties of an element.

attrition The mechanism by which the particle size of any material is reduced by friction during transport – by ocean waves, wind or running water. Note that attrition differs from ABRASION and CORRASION.

aufeis A German term which has been adopted to describe a surface accumulation of ice under PERIGLACIAL conditions, such as when groundwater freezes seasonally where it emerges from springs and when rivers flood surrounding countryside during periods of freezing. In general, *aufeis* is a

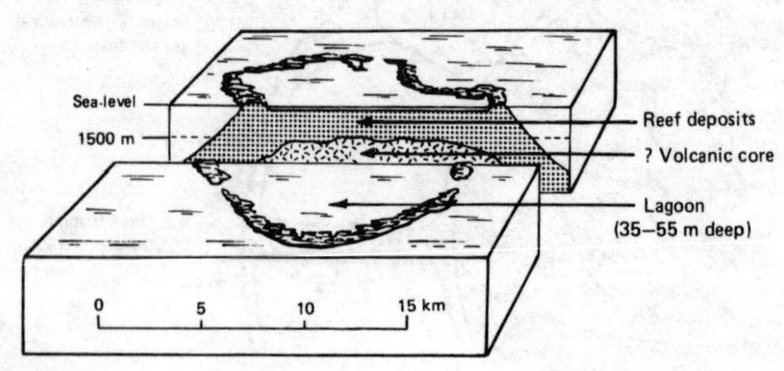

Figure 19 *Atoll*

phenomenon of winter, disappearing during the spring thaw.

auge A hot, dry wind, which blows from the South of France to the Bay of Biscay.

auger, soil A tool for boring into the soil and obtaining a small sample for analysis. There are several varieties of which the most common is the screw type.

aureole METAMORPHIC AUREOLE.

aurora Popularly called the Northern (*Aurora borealis*) or Southern (*Aurora australis*) Lights. The luminous effect of electromagnetic phenomena in the THERMOSPHERE. It is thought to result from magnetic storms and solar discharges during periods of high sun-spot activity, when electron particles may be projected into the Earth's magnetic field and there accelerated to high energy sufficient to cause ionization of gases. Manifested as streamers, veils, sheets and rays of white, red and green at heights of about 100 km in the night sky. Especially marked in polar latitudes.

authigenic A geological term to describe a phenomenon which has developed *in situ*. It is particularly used to describe sedimentary material which is produced after the formation of the rock in which it occurs, e.g. the growth of calcite on shell fragments and the secondary overgrowth of quartz on sand grains. Contrast ALLOGENIC.

autochthonous, autochthon Denoting any feature which is non-transported, i.e. is *in situ*. An autochthonous sediment is one in which the main constituents have been formed *in situ* (e.g. COAL, EVAPORITE). In studies of Alpine structures the term *autochthon* has been introduced to describe rock formations that have not been displaced by major thrusting, although they have been folded and faulted. Thus, an autochthonous NAPPE is a recumbent fold (FOLD) which can be traced back to its root zone, in contrast to a PARAUTOCHTHONOUS nappe in which the structure is now a short distance from its root zone, having been transported by major tangential thrusting. It is incorrect to use the term ALLOCHTHONOUS as the antonym of autochthonous in a structural sense, but it may be used to describe rocks which have been formed from materials which have been transported from elsewhere.

autoconsequence A theory accepted by some geomorphologists that during a fall in base-level (e.g. sea-level) rivers may be superimposed on to underlying solid rocks from their own deposits of ALLUVIUM.

autocorrelation CORRELATION, STATISTICAL.

autoecology, autecology The study of the ECOLOGY of single species and individual organisms. The term has recently been expanded from its original usage, which referred more simply to the relationship between individual species and their environment, in particular to their HABITATS. *See also* SYNECOLOGY.

autogenic Pertaining to a plant SUCCESSION that has developed in an environment which has been created by the vegetation itself; e.g. the growth of a heather moor on a peat BOG.

automated cartography The process of drawing maps with the aid of such computer-driven devices as PLOTTERS. It does not imply any information processing or interpretation. *See also* CARTOGRAPHY.

automated digitizing Conversion of a map to digital form (DIGITIZING) using a method which involves little or no operator intervention, often by use of a SCANNER.

automatic picture transmission (APT) A term used in REMOTE-SENSING studies referring to the continuous transmission of picture information from an orbiting satellite to any suitably equipped receiving station within the line of sight. It is of great importance in weather forecasting.

autometamorphism The type of metamorphic change that takes place during the cooling of an igneous rock (PNEUMATOLYSIS). *See also* CONTACT (THERMAL) METAMORPHISM, DYNAMIC METAMORPHISM, METAMORPHISM, REGIONAL METAMORPHISM.

autotrophic The ability to utilize inorganic carbon (e.g. CO_2) as the main source of carbon and to obtain energy for life processes from the oxidation of inorganic elements (*chemotrophic*) or from radiant energy (*phototrophic*). An autotrophic organism is independent of external sources of organic substances and can manufacture its own food – it is therefore 'self-nourishing'. *See also* HETEROTROPHIC.

autumn Strictly, the transitional period between the autumnal EQUINOX and the winter SOLSTICE (21 September to 22 December in the N hemisphere; 21 March to 21 June in the S hemisphere). In colloquial use it refers to the two months of September and October in the UK, while in the USA (termed 'the fall') it includes September, October and November.

available nutrients The elements which occur in the soil solution and that can readily be taken up by plants.

available relief The vertical distance between the highest points of a dissected land surface and the valley floors of the local rivers. Sometimes referred to as the *amplitude* or relief. The information is used in MORPHOMETRY.

available water That portion of soil water that can readily be absorbed by plant roots. *See also* FIELD CAPACITY, MOISTURE TENSION.

avalanche A rapid gravitational movement of snow and ice *en masse* down steep slopes, generally in a mountainous terrain. Avalanches are classified according to a number of variables: **1** type of breakaway (loose snow or slab avalanche); **2** position of sliding surface (surface or full-depth avalanche); **3** humidity of snow (wet or dry); **4** form of track (unconfined or channelled); **5** form of movement (airborne powder or ground flow). A wet-snow avalanche is the most powerful of all and is the one which generates the worst disasters. It usually occurs in spring when melting snow produces large quantities of water which are initially absorbed by the snow cover. An airborne-powder-snow avalanche is also hazardous because of the powerful shock waves which precede it (air blasts equivalent to 0.5 tonnes/m^2) – strong enough to cause buildings to 'explode' because of the sudden change of air pressure.

avalanche wind The blast of air which precedes an avalanche; it can be very destructive, causing buildings to 'explode' before the falling masses of snow and/or ice reach them.

aven A French term which has been universally adopted to describe a deep shaft-like hole in limestone terrain, leading down into extensive cave systems (KARST). In Britain the term is confined to a vertical joint in a cave roof which narrows upwards and need not necessarily give access from the surface. *See also* POT-HOLE, SWALLOW-HOLE.

avulsion The abandonment of a river CHANNEL and the establishment of a new channel at a lower elevation on its floodplain as a result of floodplain AGGRADATION.

axis **1** The Earth's axis is the diameter around which the planet rotates, tilted at an angle of 66½° to the plane of the ECLIPTIC or 23½° from a line drawn perpendicular to that plane. **2** One of two lines at right angles on a graph – the *axes* of the graph. **3** One of the lines which determine the dimensions of a geometrical figure. **4** The central line of a structural FOLD (either the crest or the trough) from which the strata dip (downwards in an ANTICLINE, upwards in a SYNCLINE) in opposing directions. [*84*]

azimuth **1** In British surveying, the term relates to the horizontal angle measured in a clockwise direction from TRUE NORTH (true azimuth) or from MAGNETIC NORTH (magnetic azimuth) to another point; e.g. the azimuth of a point due east = 90°, and of a point due south is 180°. It should be noted, however, that the US Coast and Geodetic Survey measure the azimuth clockwise from a zero, which in their case is taken as due south *not* due north: thus a US azimuth of 90° = due west and an azimuth of 180° = true north. **2** In astronomy, the term relates to the angle between the plane of the MERIDIAN of the observer and the vertical plane passing through a heavenly body.

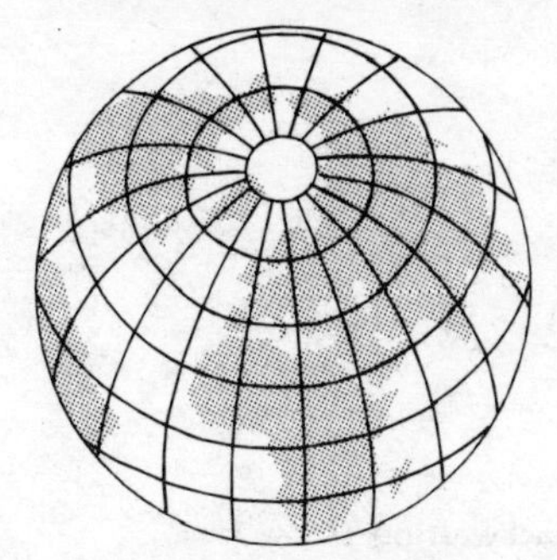

Figure 20 *Azimuthal equal-area projection*

azimuthal equal-area projection A map projection which differs from the AZIMUTHAL PROJECTION and the AZIMUTHAL EQUIDISTANT PROJECTION because its spacing of the PARALLELS of LATITUDE decreases with increasing distance from the centre of the projection, thereby producing the equal-area property. [*20*]

azimuthal equidistant projection A MAP PROJECTION that is the same as the AZIMUTHAL PROJECTION in that the straight lines radiating from the centre of the projection all exhibit their true bearings. But it has the additional property that the distances along these lines are true to scale. This projection does not possess the property of equal-area. Its PARALLELS of LATITUDE are concentric circles, depicted at their true scale distances apart. It is a useful projection for airline routes since all destinations are in their correct directions and at their true distances apart.

azimuthal projection Also termed a *zenithal* projection. A type of MAP PROJECTION constructed as if a plane was to be placed tangential to the surface of the Earth and the portions of the surface covered were to be projected on to the plane. Thus, all points have their true compass bearings.

azoic A term meaning without life. **1** The period of earth history before organic life. **2** Those parts of the oceans where organisms cannot exist.

azonal soil A soil which lacks a B-HORIZON because it is too immature for the soil-forming processes to have had time to create one. Thus, the A-HORIZON lies immediately above the C-HORIZON of weathered parent material. This is the ENTISOL of the SEVENTH APPROXIMATION soil classification. It is commonplace on volcanic soils, newly deposited glacial drift, windblown sand, cliff-foot screes, marine mud-flats and river alluvium freshly laid down. Azonal soils are subdivided into LITHOSOL, REGOSOL, and ALLUVIAL SOILS. *See also* ENTISOL.

Azores anticyclone The subtropical (warm) anticyclone located almost permanently over the eastern side of the North Atlantic Ocean. It is more extensive during the N hemisphere summer when its central pressure is about 1,027 mb at 35° N, with the axis of a ridge extending northwards to W Europe. In winter, the central pressure is about 1,024 mb at about 30° N, with the ridge axis lying across southern Spain.

B

backing (of wind) A term used to describe a change of wind direction when it moves progressively in an anticlockwise direction, e.g. from N to NW to SW. *See also* VEERING.

backset bed A US term denoting a sand deposit which accumulates to the windward of a dune, i.e. on the gentler slope.

backshore That section of a beach which extends back from the level of normal high spring tides, as far as the cliff-line or sand-dune if one exists. The backshore is only affected by waves during severe storms or exceptionally high tides.

back slope The gentler slope of a CUESTA, in contrast to the steeper ESCARPMENT. Not to be confused with DIP SLOPE, a term with which it is often used synonymously but wrongly, for the two may not be the same. [*See also 222*]

back-swamp A marshy area occurring outside the river channel's LEVEE on a FLOODPLAIN. Particularly used in connection with the R. Mississippi.

backwall The steep, precipitous rock wall at the back of a glacially eroded hollow (CIRQUE).

backwash The movement of marine water down a beach, under gravitational influence, after the breaking of a wave (SWASH). Waves with a steep profile in relation to their wavelength break vertically on to a beach, thereby causing a strong backwash and a net seaward movement of beach material. *See also* LONGSHORE DRIFT. [*140*]

backwasting BACK-WEARING.

backwater That part of a river in which current velocity is low and the water virtually unmoving or stagnant. It is commonly found in BRAIDED rivers. *See also* OXBOW.

back-wearing, backwasting A process of lateral recession of slopes (PARALLEL RETREAT OF SLOPES) in which there is no loss of steepness. The concept is in contrast to that of DOWN-WEARING (downwasting), a process suggested by W. M. Davis. *See also* BASAL SAPPING, SPRING SAPPING.

bacteria Tiny, single-celled organisms that are fundamental to soil productivity, especially those soils in which oxygen is lacking. Two broad groups are recognized: **1** the less prolific *autotrophic* bacteria, responsible for nitrification and sulphur oxidation; and **2** *heterotrophic* bacteria, which obtain their energy from organic matter in the soil, and which are much more abundant. *See also* NITROGEN FIXATION.

badlands A term originally used to describe part of South Dakota, USA, which was a terrain difficult to traverse. It is now used universally to describe any landscape characterized by deep dissection, ravines, gullies and sharp-edged ridges which have been created by fluvial erosion on rocks of relatively low resistance occurring in a semi-arid environment. Badland topography can be created artificially by over-grazing and SOIL EROSION.

bahada (bajada) A term derived from Spanish used to describe the gentle, sloping surface leading down from a mountain front to an inland basin, in an arid or semi-arid region. It is composed of unconsolidated materials, such as sand, gravel and angular scree, which together mantle the underlying rock-cut PEDIMENT [*180*]. The material, which may occur in the form of an ALLUVIAL CONE or coalescing ALLUVIAL FAN, is derived from surface RUNOFF or the intermittent stream torrents which follow infrequent but heavy rainstorms. [*189*]

Bai-u season A season of rainfall maximum (June–July) in Japan during the period of the SE monsoon. It is the first of the two rainfall maxima which characterize the annual rainfall regime in Japan. *See also* SHURIN SEASON.

bajada (Spanish) BAHADA.

balance EQUILIBRIUM.

ball clay (pipe clay) A fine-grained CLAY comprising up to 70 per cent KAOLINITE, together with smaller amounts of ILLITE, CHLORITE, MONTMORILLONITE and QUARTZ. An important raw material in the ceramics industry, and in Britain it is mined extensively in south Devon.

bamboo A genus of grass which grows to considerable heights (up to 35 m) and attains an individual stalk diameter of up to 20 cm. It flourishes in moist, warm conditions throughout the tropics and in more sheltered places in the humid temperate zone.

banded structure A term applied to a rock which exhibits 'striping', in which the individual stripes have varying physical properties and/or chemical compositions. An extremely thin band is termed a *lamination* (LAMINA).

bank **1** A colloquial term for a slope or hillside. **2** A shoal-like feature covered intermittently by shallow sea water, composed of muddy, sandy or shelly (but not rocky) deposits, e.g. Grand Banks of Newfoundland. **3** The margin of a river channel, beyond which lies the FLOODPLAIN, which in turn occupies a river VALLEY. Where fluvial erosion is active the river bank may be vertical if it is being attacked by the current. This is especially true on the outside of a curve where undercutting will cause slumping of the alluvial material into the channel where it is carried away – such collapse is termed *bank-caving. See also* LATERAL EROSION.

bank-caving BANK.

bank erosion LATERAL EROSION.

banket An Afrikaans term which relates to the gold-bearing pebble conglomerates of the African continent, especially the Transvaal, in which the individual reefs are found.

bank-full discharge stage The state of flow in a river when, because of the high volume of water, no river banks are exposed. This stage is reached just prior to the flooding stage when the brim-full river outflows its banks. It is thought that the river exhibits a constant average velocity along its entire length during a bank-full stage. *See also* CHANNEL CAPACITY.

bank storage A term referring to GROUNDWATER retained in permeable deposits marginal to river channels and whose seepage into the channel at low water helps to maintain the flow. It is especially important in semi-arid climates or in ALLOGENIC river systems.

banner cloud A stationary cloud stream which is attached to and extends away from the summit of a mountain on its leeward side. It results from condensation as a moist air-current is forced upwards on the windward side of a sharp peak, e.g. Matterhorn and Everest. *See also* OROGRAPHIC CLOUD.

bar **1** a deposit of sand or mud in a river channel (RIFFLE). **2** An elongated deposit of sand, shingle or mud, occurring in the sea, more or less parallel to the shoreline and sometimes linked to it. It can be intermittently covered by tides (*tidal bar*) or permanently submerged (*submerged bar*), especially across the mouth of a river or harbour (*harbour bar*). If the bar is permanently exposed above water level it is strictly a BARRIER-BEACH, although a bar which links two coastal headlands is referred to as a *bay-bar. See also* SPIT, TOMBOLO. **3** A unit of atmospheric pressure (MILLIBAR).

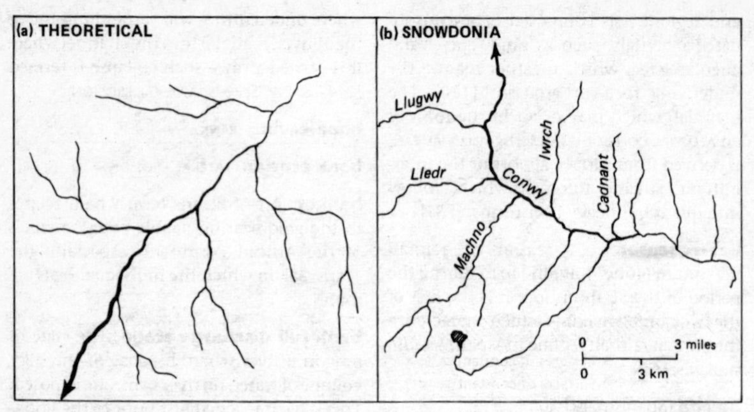

Figure 21 *Barbed drainage*

bar-and-swale A term used to describe that part of a river FLOODPLAIN in which the micro-relief of the alluvial surface is characterized by bars and troughs (swales) formed in earlier depositional phases in an area of MEANDER growth. [*147*]

barbed drainage A pattern of drainage in which the confluence of the tributary streams with the main river is characterized by a discordant junction, as if the tributaries intend to flow upstream and not downstream. The pattern is thought to result from CAPTURE of the main river which completely reverses its direction of flow, while the tributaries continue to point in the direction of former flow. [*21*]

barchan (barkhan) A crescent-shaped sand-dune in which the convex gentler windward side extends laterally to the two distal 'horns' or 'wings' which curve downwind on either side of the steeper concave slip-face of the leeward side. A barchan can only form in a desert region in which winds blow almost constantly from one direction. The wind carries the blown sand to a maximum height of 30 m up the gentler windward slope from which it slumps forward from the crest down the unstable slip-

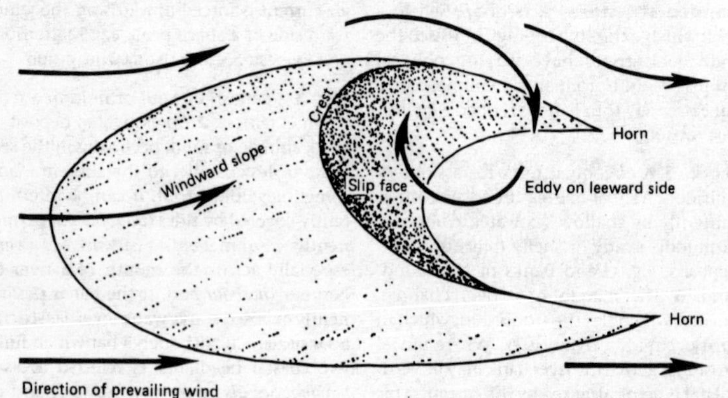

Figure 22 *Barchan*

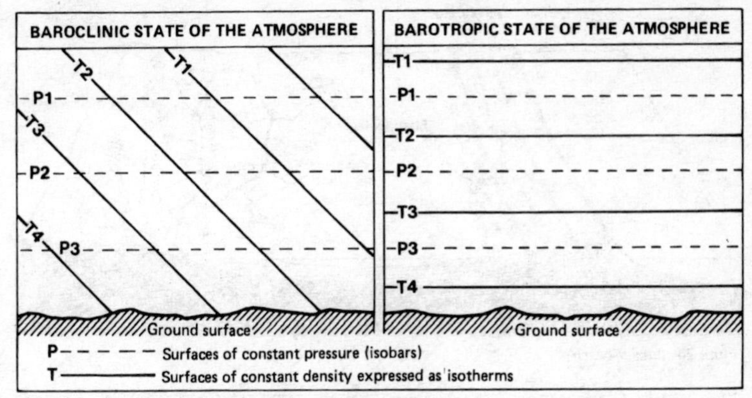

Figure 23 *Baroclinic and barotropic states of the atmosphere*

face, where eddy motions help to scour the slope and maintain its concave shape. The wings develop because the rate of advance of the barchan is more rapid at the lower extremities than at the centre, i.e. the rate of advance of the dune is inversely proportional to its height. Although barchans may be found in isolation they usually occur in groups or belts in a constantly shifting 'sand-sea'. [*22*]

baroclinic The state of a gas or fluid whereby surfaces of constant density intersect surfaces of constant pressure, implying major atmospheric INSTABILITY, probably at the frontal zone. Strong baroclinicity implies the presence of large horizontal temperature gradients and thus strong THERMAL WINDS. [*23*]

barograph A self-recording ANEROID BAROMETER.

barometer An instrument for measuring ATMOSPHERIC PRESSURE, invented in 1643 by E. Torricelli, who discovered that the weight of the atmosphere would balance a column of mercury supported in a tube. *See also* ALTIMETER, ANEROID BAROMETER, BAROGRAPH, FORTIN BAROMETER, KEW BAROMETER.

barometric gradient GEOSTROPHIC FLOW.

barometric pressure ATMOSPHERIC PRESSURE.

barometric tendency The degree or amount of increasing or decreasing atmospheric pressure over a particular time span – generally three hours. *See also* ISALLOBAR.

barotropic The state of a gas or fluid whereby surfaces of constant density lie parallel to or coincide with surfaces of constant pressure, i.e. zero baroclinicity. Although theoretically possible the barotropic state rarely, if ever, occurs in the atmosphere. BAROCLINIC. [*23*]

barranca A Spanish term, used particularly in the USA and Latin America, to describe a deep ravine cut by stream action on the slope of a volcano.

barren lands An early description of the TUNDRA regions of N Canada, characterized by PERMAFROST, sparse vegetation, long winters and inhospitable climate. Now rarely used.

barrier-bar BARRIER-BEACH.

barrier-beach, barrier island An elongated sand or shingle bank which lies parallel to the coastline and is not submerged by the tide (BAR). If it is high enough to permit dune growth it is termed a barrier island. Both are separated from the mainland by a lagoon or sound in which coastal marshes or MANGROVE swamps may occur. Examples

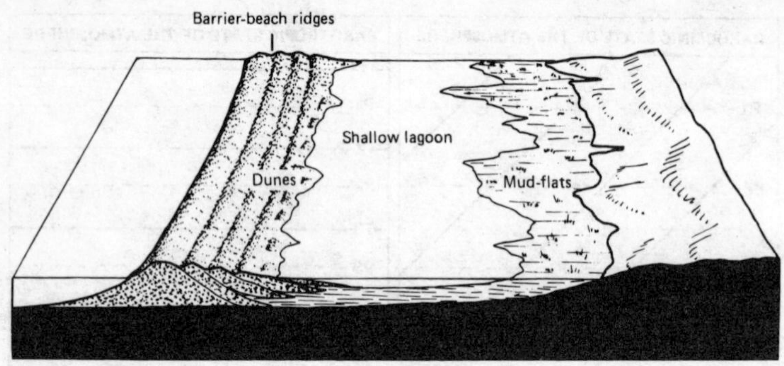

Figure 24 *Barrier-beach*

of barrier-beaches include those on the eastern coast of the USA from Florida to New Jersey (including Cape Hatteras and Cape Kennedy); barrier islands include the Lido of Venice and the Frisian Is. off Denmark, N Germany and the Netherlands. They may be formed either by the breaching of a SPIT or by the landward progression of a bar built by CONSTRUCTIVE WAVES. [*24*]

barrier-lake Any lake created by the impounding of a river by a natural dam. It may be created by: **1** a volcanic lava flow; **2** a landslide; **3** an ice-dam; **4** a morainic dam; **5** deltaic deposits; **6** a vegetation dam; **7** a calcium carbonate deposit in a karstic region; **8** an avalanche. *See also* PRO-GLACIAL LAKE.

barrier reef An elongated accumulation of CORAL lying at low-tide level parallel to the coast but separated from it by a wide and deep lagoon or strait. The coral is thought to have formed initially on a flat surface: then as the sea-level rose in post-glacial times, thereby submerging the irregular wave-cut platform, the coral growth kept pace with the rising ocean level, so creating the great thickness witnessed today in such places as the Great Barrier Reef off the E coast of Queensland, Australia. This stretches for more than 1,900 km (1,200 miles) and varies in width from about 30 km (18 miles) to 150 km (95 miles). *See also* ATOLL, FARO.

barysphere In strictest usage the term refers only to all of the Earth's interior which lies beneath the LITHOSPHERE. It therefore includes the CORE, the MANTLE and the ASTHENOSPHERE. Some writers have used the term loosely to describe only the core or only the mantle.

barytes (barite) A high-density, whitish mineral which represents an ore of barium ($BaSO_4$). It is found either in limestone cavities or in veins, associated with GALENA, FLUORITE or CALCITE, and is the product of HYDROTHERMAL ACTIVITY. Its economic uses include pharmaceuticals, paint, etc.

basal complex A general term for the rocks which make up the ancient 'shield' areas of PRE-CAMBRIAN age.

basal conglomerate A CONGLOMERATE occurring at the lowest part of a stratigraphical unit, usually resting above older rocks in an uncomformable relationship (UNCONFORMITY).

basal ice The layer of debris-laden ice at the base of a GLACIER, created at and interacting with the underlying rock surface. It is generally relatively thin (some tens of metres).

basal melting The process during which the bottom surface (hence 'bottom melting') of a glacier exceeds the melting-point, causing meltwater to be generated and incorporated debris (TILL) to be deposited. It is associated with stagnant and decaying ice.

basal platform BASAL SURFACE OF WEATHERING.

basal sapping A term used by geomorphologists to describe all the processes of erosion and transport that act separately or in unison to remove debris from the lower parts of a slope. It is the major factor in the concept of PARALLEL RETREAT OF SLOPES, in which debris is removed from the foot of the slope, thereby maintaining slope steepness and preventing any decline in slope angle. By maintaining a steep slope or FREE FACE, basal sapping often causes MASS-MOVEMENT in scarpland areas due to slippage in and subsequent removal of underlying clay strata by SPRING SAPPING, leading to undermining of higher and more competent strata in the scarp face. This mechanism of slope retreat is operative in all types of climate, from the LATERITE-capped tablelands of central Africa to the scarplands of southern England, but is thought to be more active in tropical environments where CHEMICAL WEATHERING is more rapid (in addition to MECHANICAL WEATHERING). Basal sapping is therefore thought to be an important part of INSELBERG formation. Although some geomorphologists restrict the term to spring sapping, SEEPAGE and WASH processes, others would extend it to include LATERAL EROSION by a river at the foot of a bluff, the disintegration of rocks on the backwall of a cirque by FREEZE–THAW action, and the wave action at the foot of a sea cliff. *See also* BACK-WEARING.

basal slip, basal sliding The sliding movement of a glacier over its rock-floor, resulting from the gradient of the slope and the weight of the ice. It was formerly thought that EXTRUSION FLOW was responsible for basal slip in valley glaciers, but the term extrusion flow is now reserved only for larger ice-sheets. Basal slip in glaciers normally accounts for about half of their surface velocity – the remainder is by internal deformation. *See also* STICK-SLIP.

basal surface of weathering A line of variable depth beneath the surface which marks the lower limit of active WEATHERING; the transition which marks the change from sound rock (beneath) to weathered rock (above). In parts of tropical Africa the surface lies at great depth (30–50 m) but may be very uneven or even exposed at the surface where tectonic movement or climatic change have caused rejuvenated rivers to strip away the weathered material (REGOLITH). The surface is thought to be synonymous with the term *basal platform*, a term devised by D. L. Linton to describe the partly exposed unweathered granitic platform of Dartmoor from which the TORS rise. As the chemical weathering action proceeds downwards from the surface it has sometimes been termed the *weathering front*, the depth of which will vary according to the character of the rock structure, its chemical composition and the disposition of the WATER-TABLE. It is thought that weathering can progress below the water-table and that the regolith can support a PERCHED WATER-TABLE. *See also* INSELBERG, SHIELD INSELBERG. [*25*]

basalt A very common fine-grained, usually dark-coloured basic igneous rock which

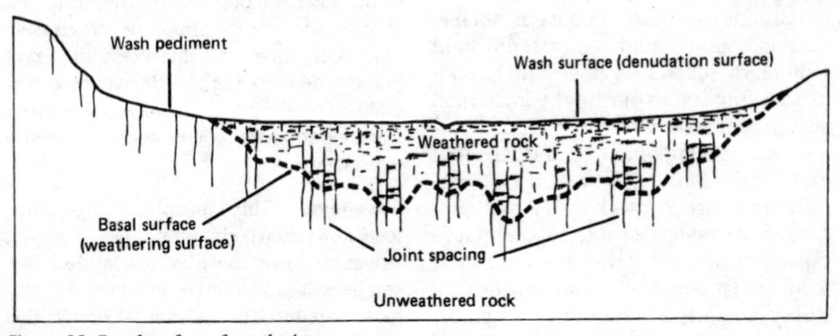

Figure 25 *Basal surface of weathering*

is made up essentially of PLAGIOCLASE and PYROXENE (usually *augite*) minerals, with or without OLIVINE. Basalt is characterized by a low silica content (45–50%) but it has a high content of ferromagnesian minerals. It is the fine-grained (volcanic) equivalent of the coarse-grained (plutonic) rock GABBRO, and the medium-grained (hypabyssal) rock DOLERITE. An *amygdaloidal basalt* is one with cavities (amygdales) in which silicates termed zeolites have crystallized during HYDROTHERMAL ACTIVITY. Basalts are commonly found as lava flows, extruded from either fissures or central vent volcanoes. Because of their fluidity they are capable of flowing great distances and often form thick lava piles as one flow builds upon another, occasionally reaching hundreds of metres in thickness (e.g. Iceland, Deccan of India, Columbia River region of NW USA, Antrim Plateau of N Ireland, Skye, Mull of the Scottish Hebrides, Faeroe Is. of the N Atlantic). On cooling it forms characteristic hexagonal columns owing to the development of joints at right angles to the cooling surface (COLUMNAR STRUCTURE), well illustrated at the Giant's Causeway (N Ireland) and the island of Staffa (Scotland). More than 90% of basic igneous rocks are basaltic, as are more than 90% of all volcanic rocks.

basal till Till carried at or deposited from the under surface of a glacier. TILL.

basal wreck CALDERA.

base **1** A substance containing the OH radical which dissociates to form OH^- ions when dissolved in water. In soil science it is the alkaline constituent in the soil, notably calcium, magnesium, potassium and sodium. These bases, together with the acidity-inducing cations (notably hydrogen and aluminium) make up the soil cations (CATION) and are important in the chemical exchange of ions termed cation exchange (BASE EXCHANGE). **2** A rock with a high proportion of base-metal oxides. **3** A substance capable of combining with silica in a rock. **4** An initial length of measurement in a survey (BASELINE). **5** In statistics, the number of digits used in a positional method of counting (e.g. base ten = zero plus nine digits).

base exchange (cation exchange) The chemical replacement of cations (CATION) within the soil. The cation, held on the surface of either a clay or a humus particle (COLLOID), can be exchanged for another cation from the surrounding soil moisture. For example, one of the soil bases (BASE) such as calcium, held by its positive charge to the negatively charged surface of the colloid, may be replaced by a hydrogen ion from the surrounding fluid. The positively charged bases of calcium, magnesium, potassium and sodium are held loosely in the soil in an exchangeable position ready to be called upon when required as nutrients by plants. Some bases, such as potassium and sodium (the metallic cations), are given up more readily than others, to be replaced by hydrogen ions. Such an exchange tends to make the soil become progressively more acid unless the bases are replenished, either artificially by fertilizer or naturally, by the decomposition of plants and animals. The base-exchange capacity of a soil is expressed in milli-equivalents (m.e.) per 100 grams of clay, and this capacity varies according to the type of colloid: the humus colloid has the highest capacity (>200 m.e. per 100 g), while the heavily weathered clay KAOLIN has the lowest capacity (< 15 m.e. per 100 g).

base-flow recession curve A hydrologic graph depicting a condition of stream discharge when its supporting streams continually drain the local AQUIFER. Such a drainage condition is referred to as a state of nil water PERCOLATION. The resulting discharge, therefore, is one of progressive reduction (shown by the recession curve) but at a decelerating rate. It may be calculated by producing a composite curve from the RECESSION LIMB OF A HYDROGRAPH. *See also* HYDROGRAPH.

base-level The controlling theoretical level down to which, but not below which, a river can lower its valley, or a land-surface can be reduced by fluvial erosion. The ultimate base-level is sea-level, although this may be replaced by a local base-level such as a hard layer of rock or a lake. Together with geology and climate, base-level forms are one of the independent variables in the CYCLE OF EROSION, so that a change in base-level due to tectonic movements or a change in height of sea-level (EUSTASY) will change the character of the geomorphological system. A fall in sea-level or rise of the land surface (positive movement of base-level) will initiate a new cycle of erosion, with renewed downcutting by the river systems; a rise of sea-level or a subsidence of the land surface (negative movement of base-level) will lead to increased deposition and burial of the former river channels. Since streams require a gradient in order to flow it has been suggested that base-level cannot be a horizontal surface within the land masses.

baseline **1** A line on the ground, measured with considerable accuracy since it is the primary stage of a TRIANGULATION survey. Angular measurements taken from each end of the baseline are built up into a series of triangles from which a topographical map can be constructed. So careful is the measurement of the initial baseline that an accuracy of 1 in 300,000 is claimed. The current baselines used in the UK triangulation survey are on the Berkshire Downs in S England and near Lossiemouth on the Moray Firth coast in Scotland. **2** An E–W surveyed line in US land surveys. It passes through the initial starting-point and follows the parallel of latitude. Settlements, etc. are numbered north and south from the baseline. **3** In an aeromagnetic survey the baseline is the reference line of magnetic intensities obtained by flying at least twice in opposite directions and at the same altitude.

base map **1** A set of topographical data displayed in a map form. **2** A BINARY digital map in a GIS which defines the area within which analysis of other spatial data is undertaken.

basement Any mass of ancient igneous and metamorphic rocks occurring beneath an unconformable 'cover' of stratified, unmetamorphosed sedimentary rocks. Most commonly the term is used to described a basement complex of Pre-Cambrian age but should not be restricted simply by age.

base net The initial figure in a TRIANGULATION survey, obtained from the triangle formed by sighting a third point from the two ends of a BASELINE.

base saturation The extent to which the ADSORPTION complex of a soil is saturated with exchangeable CATIONS other than hydrogen and aluminium. It is expressed as a percentage of the total CATION EXCHANGE CAPACITY. *See also* BASE EXCHANGE.

basic grassland A name given to the grasslands of limestone terrain, where the dominant species are the fescues (e.g. *Festuca ovina* and *Festuca rubra*).

basic lava A flow of molten igneous material from a FISSURE ERUPTION or a central vent of a VOLCANO, in which the silica content is low and the ferromagnesian elements are high (BASALT). Since it is very fluid, owing to its low melting-point, it flows great distances to form lava plains or low-angle volcanic cones (SHIELD VOLCANO). Basic lava constitutes some 90% of all lavas. Contrast ACID LAVA.

basic rocks Quartz-free igneous rocks containing FELDSPAR which is more calcic than sodic. They were originally defined by the amount of SILICA (SiO_2) present (i.e. a basic rock has 45–55% silica) and as the antithesis of ACID ROCKS, but this definition is now obsolete. Of the ferromagnesian minerals pyroxene and olivine are most common but hornblende and biotite may occur in small amounts. It is incorrect to use basic as synonymous with alkaline. Basic rocks grade into ULTRABASIC ROCKS as the amount of feldspar decreases and into INTERMEDIATE ROCKS as the sodium content of the feldspar increases. Because of the confusion which has arisen over the use of the terms 'basic' and 'acid' to describe rock character, it has been suggested by US geologists that the term 'basic' rock should be replaced by 'mafic' (to describe its base content), 'subsilicic' (because of its low silica) or 'melanocratic' (in respect of its dark-coloured

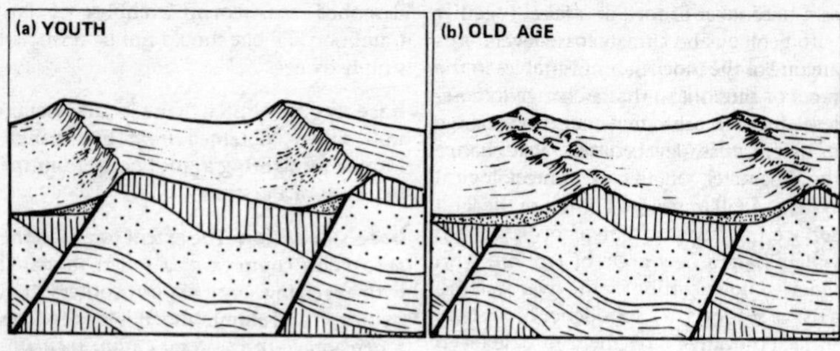

Figure 26 *Basin-and-range terrain*

minerals). *See also* ACID ROCKS, BASALT, DOLERITE, GABBRO.

basic spatial unit (BSU) A fundamental areal unit of homogeneous properties in a GIS.

basin Several uses of the term include both structural and topographical connotations: **1** a structural downfold (SYNCLINE) in the Earth's crust in which the younger rocks occupy the centre and older rocks may be exposed by erosion on the flanks; **2** a small subsidence depression in the land surface due to solution of underlying deposits such as salt or gypsum, artificially or naturally created; **3** a very large depression occupied by sea water, i.e. an ocean basin; **4** a large sediment-filled depression, enclosed by higher land, with or without an outlet; **5** the catchment area of a river-system. NB. The structural downfold of the London Basin is *not* synonymous with the river basin of the Thames.

basin-and-range terrain A landscape characterized by a number of faulted block mountains interspersed with basins. Typically, the FAULT BLOCKS are tilted to form asymmetrical ranges with steeper (the FAULT-SCARP face) and gentler slopes. The states of Nevada and Utah, USA, have a basin-and-range topography. [*26*]

basin discharge The total flow of water through a river cross-section at a given point. It depends on the HYDROLOGICAL-BALANCE BUDGET and the physical characteristics of the catchment area of the RIVER BASIN above that point, e.g. area, slope, rock-type and soil.

basin order A term used by geomorphologists to describe the hierarchy of stream drainage networks in a region. Thus, a first-order basin contains all of the drainage area of a first-order stream (STREAM ORDER); a second-order basin contains all of the drainage area of a second-order stream. [*245*]

basisol A type of tropical soil, characterized by its blackness, its low organic content and its accumulation of calcium carbonate in the form of a CONCRETION horizon. It forms from the weathering of BASALT under humid tropical conditions and is common in the TRAP landscape of the Indian Deccan.

basket-of-eggs terrain A colloquial term for a DRUMLIN landscape.

basophilous The response exhibited by organisms adapted to life in an alkaline medium, e.g. many legumes such as alfalfa.

batholith (bathylith) A large body of intrusive igneous rock (usually GRANITE) in which there is no observable bottom to the structure. Its surface manifestation may be in the form of a single PLUTON but often in the form of several distinct plutons linked at depth, e.g. the granite domes of SW England (Dartmoor, Bodmin Moor, St Austell, etc.) are surface exposures of the same batholith which has been partly 'unroofed' by erosion. The batholiths are formed either by gradual igneous replacement of the country-rock at great depth (GRANITIZATION) or by STOPING, during a mountain-building period (OROGENY). The junction of the igneous body and the surrounding country-rock is sometimes a sharp one but more frequently is marked by a transitional zone of altered rock (METAMORPHIC AUREOLE) in which varying degrees of mineralization usually take place. [*151*]

Bath stone An oolitic honey-coloured FREESTONE of Jurassic age, extensively used in buildings in the English Cotswolds. *See also* OOLITIC LIMESTONE.

bathyal zone The zone of the ocean on the CONTINENTAL SLOPE between approximately 100 and 1,000 fathoms (180–1,800 m), below the level of light penetration. It lies between the shallow NERITIC ZONE and the deeper ABYSSAL ZONE. Despite the absence of light there is a flourishing and varied animal life (BENTHOS). Blue, red, green and coral muds are found in this zone. [*162*]

bathymetric Pertaining to the depth of a body of water and its measurement (bathymetry). Maps showing depths of water above the floor utilize lines of equal depth (*isobaths*) in a similar way to those of contours on a land surface. *See also* BATHY-OROGRAPHICAL.

bathy-orographical A term applied to maps which depict both depths of water (generally, sea water) and heights of adjoining land. They are usually depicted by the technique known as LAYER TINTING, with water in shades of blue and white, and with land in shades of green, yellow and brown.

bathythermograph An automatic thermometer which can be used for temperature measurements at great depths in the oceans.

battue ice A term for large ice-floes which hinder navigation on the St Lawrence estuary during winter.

batture bed A US term for an elevated river-bed, i.e. when a river is confined by its natural LEVEES above its FLOODPLAIN level.

bauxite A clay containing the mineral ore aluminium hydroxide ($Al_2O_3.2H_2O$) from which aluminium is obtained. It is formed from the tropical weathering of feldspathic rocks by the breakdown of FELDSPAR into clay minerals and the removal of silica by leaching, leaving residual bauxite. *See also* LATERITE.

Baventian An early stage of the British PLEISTOCENE named from its type-site at Easton Bavents, Suffolk. Its pollen evidence suggests cold conditions. [*197*]

bay An open, curving indentation made by the sea or a lake into a coastline; sometimes also used in a local context to describe the extension of lowland into an upland area. *See also* BIGHT, COVE, EMBAYMENT.

bay-bar An elongated bank of sand, shingle or mud which links two headlands across the mouth of a BAY. If the bar is breached at one end that part of the bar remaining attached to the coast is termed a SPIT. It is formed in one of three ways: **1** when an offshore or *tidal bar* is driven onshore; **2** by the convergence of two spits from either side of a bay; **3** by a single spit growing in a constant direction by LONGSHORE DRIFT. *See also* BAR, NEHRUNG.

baydjarakh (baydzharakh) A conical hillock characteristic of a THERMOKARST environment in Arctic latitudes. It is formed by the melting of ICE WEDGES. [*253*]

bay-head beach An accumulation of sand and/or shingle at the back of a sea cove or bay.

bay-head delta A fan-shaped area of deposition formed when a stream abandons much of its load on reaching sea-level (BASE-LEVEL) at the head of a bay.

bay-ice An accumulation of floating sea-ice formed in a bay during an exceptionally cold period.

bayou An area of sluggish, marshy water or a cut-off meander (OXBOW LAKE) which forms a backwater beside the main river channel. It is characteristic of the lower Mississippi, USA, especially in its delta.

beach The littoral environment at the junction of the land and the sea is sometimes in the form of an accumulation of unconsolidated materials (sand, shingle),

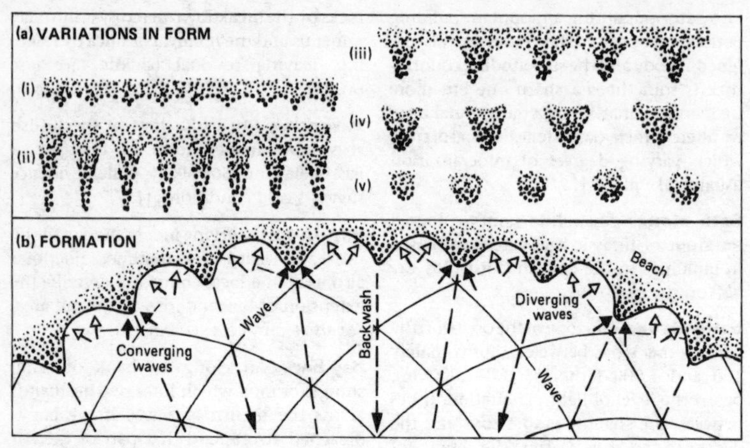

Figure 27 *Beach cusps*

lying between the lowest level of spring tides and the highest level reached by storm waves (STORM BEACH). A beach is located on a wave-cut platform of solid rock (ABRASION PLATFORM) and is generally of a low gradient with a gently concave profile. The character of the beach deposits and the rate of accretion or removal will depend on the process of LONGSHORE DRIFT and the composition of the surrounding coastal rocks. *See also* DISSIPATIVE BEACH, RAISED BEACH, REFLECTIVE BEACH.

beach cusp A crescent-shaped accumulation of sand or shingle surrounding a semicircular depression on a BEACH. The coarser material is found on the seaward-facing 'horns' or promontories between the depressions. Beach cusps always occur in combination, more commonly in shingle or cobbles than in sand, and create a characteristic scalloped shape, varying in size from a few centimetres to several metres. They are thought to be formed by a powerful SWASH and BACKWASH action of marine waves, especially when waves are breaking directly on to the beach. The size and spacing of the cusps are believed to relate to the periodicity and length of the swash. [*27*]

beach-ridge A linear accumulation of shingle on a beach, parallel with high-water mark and rising to an upstanding crest. It marks the limit of SWASH at each high tide, so that a beach-ridge will form at the uppermost limit of a spring high tide and may not be reached again until the next period of high spring tides. Thus, below the highest ridge a descending series of ridges will be produced during the periodic fall from spring high tides to the neap high tides, but these will be destroyed during the following tidal cycle. Some large beach-ridges are only periodically reached, by exceptional storm waves, but these are examples of either a STORM BEACH or a RAISED BEACH.

A beach-ridge is sometimes referred to as a *berm*.

beach rock A littoral deposit of consolidated sand which can develop in the intertidal zone of certain beaches. Although its mode of origin is uncertain it has been suggested that under warm water conditions the sand grains are cemented by a calcareous mineral, usually CALCITE or aragonite, aided by the presence of ALGAE and BACTERIA. Once it has formed, beach rock is extremely resistant to erosion and its reef-like form acts as a barrier to movement of sediment across the beach.

beaded drainage A term given to a type of drainage pattern developed in a PERMAFROST environment owing to differential surface melting. It is characterized by interconnected pools and short linking streams, having the overall appearance of a necklace of beads. It has been suggested that the pools may form by melting at the junction of ICE-WEDGE polygons and that the streams may follow aligned ice-wedge patterns.

beaded esker A sinuous upstanding ridge of sand and gravel (ESKER) in which occasional wider hillocks on the ridge are linked by narrower segments, creating the overall impression of a necklace of beads. Since the esker is known to be of GLACIOFLUVIAL origin it is suggested that the beaded form represents differential input rates of sediments from the meltwater stream. The narrow sections may have been formed during colder periods when melting slowed down, and the 'beads' when meltwater activity increased during warmer phases.

beaded valley A term to describe a valley in which wider parts alternate with narrower linking sections, in the form of a string of beads. It is a rarely used term. *See also* BEADED DRAINAGE, BEADED ESKER.

bearing The direction of a line with reference to the CARDINAL POINTS of a compass. The true bearing is the horizontal angle between a line on the ground and a MERIDIAN of longitude (e.g. N, 45°W). The *magnetic* bearing is the horizontal angle between a line on the ground and the magnetic meridian relating to the MAGNETIC POLE (MAGNETIC DECLINATION). A *back bearing* is the reciprocal of a given bearing.

bearing capacity The average load per unit area that is required before a supporting soil mass ruptures or 'fails'. It is measured in kg/m^2.

Beaufort notation A code of letters to indicate the character of the weather past or present, devised by Admiral F. B. Beaufort in the early 19th cent: *b* blue sky; *c* cloudy; *o* overcast; *g* gloom; *u* ugly threatening sky; *q* squalls; *kq* line squall; *r* rain; *p* passing showers; *d* drizzle; *s* snow; *rs* sleet; *h* hail; *t* thunder; *l* lightning; *tl* thunderstorm; *f* fog; *fe* wet fog; *z* haze; *m* mist; *v* exceptional visibility; *e* wet air but no precipitation falling; *y* dry air; *w* dew; *x* hoar-frost. These were refined by the use of capital letters to indicate high intensity (R = heavy rain) or by repetition of capital letters to indicate continuity (RR = continuous heavy rain). Conversely, other symbols were added to show low intensity (*ro* = slight rain) and intermittence (*ir* = intermittent rain). An international system of ninety-nine symbols for official weather maps (unlike TV forecast charts) has been introduced including: ● = rain, ❟ = drizzle, ▲ = hail, ✱ = snow. *See also* BEAUFORT SCALE.

Beaufort scale A numerical wind-force scale ranging from 0 (calm) to 12 (hurricane) devised by Admiral Beaufort (BEAUFORT NOTATION) in 1805 (modified in 1926), with velocities measured at 10 m above the ground. [*28*]

Beaumont period One of 48 consecutive hours during which the dry-bulb temperature in the screen (STEVENSON SCREEN) has remained at or above 10° C, and the relative humidity has stood at or above 75 per cent on at least 46 of the 48-hourly observation. An important criterion for the issuing of a potato blight warning.

beck A local term in N England for a rapidly flowing stream.

bed **1** The smallest layer (or stratum) of a stratified sedimentary rock which can be divided from its adjacent layers, both above and below, by a planar surface (BEDDING PLANE). It is the smallest of the lithostratigraphic units (LITHOSTRATIGRAPHY). **2** A layer of PYROCLASTIC or VOLCANIC material (e.g. an *ash-bed* or a bed of LAVA). **3** The floor of a water body, e.g. ocean, lake or river.

bedding plane The planar surface which separates one layer (bed) of a sedimentary rock from another. It indicates a break between phases of deposition. *See also* STRATIFICATION.

bedforms Features developed by streamflow over alluvial sediments or by wind on sandy deposits. Their size varies from small-scale RIPPLE MARKS to a large-scale BAR or DUNE. *See also* POINT BAR, RIFFLE.

Figure 28 *Beaufort scale*

Beaufort scale	*Velocity (knots per hour)*	*National Weather Service terminology*	*Specifications for use on land*
0	<1	Calm	Smoke rises vertically
1	1–3	Light air	Wind direction shown by smoke, but not by vanes
2	4–7	Light breeze	Wind felt on face; leaves rustle, vane moves
3	8–12	Gentle breeze	Twigs in constant motion; wind extends light flag
4	13–18	Moderate breeze	Raises dust and loose paper; moves small branches
5	19–24	Fresh breeze	Small, leafy trees sway; wavelets on inland water
6	25–31	Strong breeze	Large branches in motion; whistling in telegraph wires
7	32–38	Near gale	Whole trees in motion; resistance felt in walking
8	39–46	Gale	Breaks twigs off trees; generally impedes progress
9	47–54	Strong gale	Slight structural damage (chimney-pots, roof tiles)
10	55–63	Storm	Trees uprooted; considerable structural damage
11	64–72	Violent storm	Rarely experienced inland; widespread damage
12	73–82	Hurricane	Very rare occurrence except in tropics; catastrophic structural damage; heavy loss of life
13	83–92		
14	93–103		
15	104–14		
16	115–25		
17	126–36		

bedload A term describing the amount of solid material carried along a river-bed by rolling, pushing or bouncing (SALTATION). The bedload, which usually amounts to < 10 per cent of total sediment transported, generally comprises the coarsest fraction of the carried load, in contrast to the finest fraction which is usually carried in SUSPENSION. *See also* COMPETENCE, DISSOLVED LOAD, SIXTH POWER LAW, SUSPENDED LOAD.

bedrock The rock which remains solid and relatively unweathered beneath the superficial layer of soil and weathered rock (REGOLITH). *See also* PARENT MATERIAL.

bedrock meander A term to describe the sinuous, curving course of a stream when it is cut into solid rock, in contrast to unconsolidated superficial materials such as ALLUVIUM or GLACIAL DRIFT. *See also* MEANDER.

bed roughness ROUGHNESS.

Beestonian An early cold stage in the British PLEISTOCENE, the first in which terrestrial conditions can be clearly recognized. At its type-site at Beeston, Norfolk, the deposits include fresh-water sediments, beach gravels and a marl. It precedes the CROMERIAN. [*197*]

beheading of river RIVER CAPTURE.

Behrmann's projection A version of the map projection known as the CYLINDRICAL EQUAL-AREA PROJECTION. The scale is preserved along the 30°N and 30°S standard parallels.

bell pit A cavity created by a shallow mining method in which material excavated and removed through a narrow shaft leads to a bell-shaped underground aperture. Such mining is now obsolete in Britain.

belt 1 A spatial term applied to a distinctive zone of terrain, vegetation or climate. **2** A term applied to a narrow strip of sea-water.

belted-outcrop plain A planar topographic surface of low relief, which may be a PENEPLAIN or a wave-cut platform (ABRASION PLATFORM) which truncates a parallel series of rock outcrops (e.g. the coastal plain of SE USA and the coastal plain of SE Ireland). As rivers adjust to the structure the surface may become dissected into a number of CUESTAS with intervening valleys (SCARP-AND-VALE TERRAIN).

belt of no erosion A term devised by an American, R. E. Horton, in 1945 to describe the uppermost hilltop zone of a slope in which the gradient is insufficient to generate erosion in the form of hillwash. The zone, extending from the crest to a point at which running water attains erosion velocity, is termed the *critical distance*, which varies according to climate, lithology and vegetation cover.

ben A Scottish term for mountain or peak.

bench A narrow, flat or gently sloping ledge or step, bounded above and below by steeper slopes. It may be formed by structural movement (e.g. parallel normal faults STEP-FAULT) or by agencies of erosion (e.g. wave-cut bench ABRASION PLATFORM) or by man-made quarrying.

bench-mark (BM) 1 A virtually permanent point of reference used by a surveyor during a topographical survey (TRIANGULATION survey). A specific point or level on the bench-mark (which may be artificially erected, cut into an existing building or into the solid rock) has a known elevation related to a datum at a fixed level. In the UK this is termed the ORDNANCE DATUM (OD) and in the USA it is the Sea-Level Datum, fixed in 1929. **2** In general, a standard test to enable comparisons to be made between two systems.

Benguela current An ocean current flowing northwards off the coast of SW Africa, generally between Cape Town and 18°S. It is characterized by the upwelling of relatively cold water and has an associated effect on the coastal climate of the region. [*168*]

Benioff zone A seismic zone named after H. Benioff extending at an angle of about 45° from the base of an ocean TRENCH, down through the LITHOSPHERE to the ASTHENOSPHERE. It marks the edge of two plate margins where one is overridden by another during the process of subduction (PLATE TECTONICS). The earthquake focus is shallow near the base of the ocean trench but becomes progressively deeper as the Benioff zone inclines away from the trench under the continent. *See also* ISLAND ARC. [*119*]

benthos The life dwelling on the sea-floor at whatever depth, divided into two: **1** the deep-sea benthos, below 200 m; and **2** the littoral benthos, from 200 m up to high-water springtide level. Benthos occur in numerous forms, either fixed, crawling or burrowing. *Sessile benthos* are fixed to the sea floor while *vagrant benthos* are capable of active movement on or within the sediment. They must be contrasted with the free-floating PLANKTON and the swimming NEKTON. They are important parts of the marine FOOD CHAIN.

bentonite An assemblage of CLAY MINERALS formed by the weathering of volcanic rocks, especially volcanic ash and TUFFS.

berg 1 A term of German and Dutch derivation used locally to describe a single hill or range of mountains, respectively: in Germany, a single hill, e.g. Vogelsberg; in South Africa a range of mountains, e.g. Drakensberg. **2** A colloquial term for an ICEBERG. *See also* BERG WIND.

Bergeron–Findeisen theory of precipitation An explanation, proposed by Bergeron in 1933 and later modified by Findeisen, which suggested that ice crystals present in a cloud will grow at the expense of neighbouring droplets of supercooled water. Since the SATURATION VAPOUR PRESSURE over ice is less than that over water at the same temperature, the ice crystals grow in size rapidly, owing to the flux of water vapour from neighbouring supercooled droplets, ultimately forming snowflakes large enough to fall through the cloud. These flakes melt to raindrops when they reach zones with temperatures between 0° and 4° C. It was discovered that since rain also falls from clouds with no ice crystals the Bergeron theory could not be universally applied. *See also* LANGMUIR THEORY OF RAINDROP GROWTH.

Berghlaup An Icelandic term equivalent to BERGSTURZ.

Bergmann's rule A rule which states that the body size of warm-blooded animals tends to increase in climates with cooler temperatures. *See also* ALLEN'S RULE.

bergschrund A deep, narrow crack near the back of a CIRQUE GLACIER, marking the line along which the glacier ice is moving away from the cirque's BACKWALL. It was once thought that the bergschrund was largely responsible for the retreat of the backwall by virtue of its assistance in the FREEZE–THAW process of rock disintegration. Modern research has demonstrated that such a process is limited to a small zone at the top of the *bergschrund*, for at greater depth the air temperature rarely oscillates across the FREEZING-POINT threshold, thereby obviating rock breakdown.

Berg's index of continentality An empirical attempt to define the spatial limits of a CONTINENTAL CLIMATE. Berg relates the frequency of continental air masses (*C*) to that of all air masses (*N*), in the index of continentality: $K = C/N$ (%). In Europe, the Berg index illustrates that all the areas to the west of a line drawn from Helsinki through Berlin to Belgrade experience non-continental air for more than 50% of the time, whereas all areas to the east of this line are dominated by continental air for the greater part of the year. *See also* CONRAD'S FORMULA OF CONTINENTALITY, CONTINENTAL CLIMATE, GORCZYNSKI'S INDEX OF CONTINENTALITY.

Bergsturz A German term referring to a large-scale (>10^6 m^3) ROCK-FALL.

berg wind Literally, a mountain wind (German and Afrikaans) with FÖHN-like characteristics. Common in S Africa where a hot, dry plateau wind blows down to the east coast, often causing oppressive weather. *See also* FÖHN. [*83*]

bergy bits Small pieces of floating ice less than 1 m in diameter still visible above the sea; remnants of large ICEBERGS, eroded by wave action and melted by increasing temperatures. *See also* GROWLERS.

berm BEACH-RIDGE.

Bermuda high Part of the subtropical high-pressure cell of the W part of the North Atlantic Ocean, frequently linking up with the AZORES ANTICYCLONE. Together they move a few degrees towards the equator in winter but move poleward again in summer.

Bernard convection A system of cellular CONVECTION in which cloudy areas are interspersed with clear areas. When viewed from above by SATELLITE the image is often one of hexagonal patterns.

Bernouilli's theorem A process, outlined at the end of the 17th cent. by J. Bernouilli, relating to the relationship between *kinetic energy* and *potential energy* in a fluid flow, and especially applicable to the hydrology of a stream. In an open-channel flow the slope of the water surface may be equal to the HYDRAULIC GRADIENT. The *energy gradient* stands theoretically above the hydraulic gradient by a distance equal to the *velocity head*. For a given length of channel the decline of the energy gradient downstream owing to friction and other influences represents a loss of energy. The relationship of the energy gradient to the hydraulic gradient reflects this energy change and represents the conversion between potential and kinetic energy. When the two gradients are parallel the flow is termed *uniform flow*; when the energy gradient is steeper than the hydraulic gradient it represents a conversion from kinetic to potential energy and is termed *retarded flow*; when the hydraulic gradient is steeper than the energy gradient potential energy is converted to kinetic energy, giving a state of *accelerated flow*. *See also* HYDROSTATIC PRESSURE.

best units analysis A method referring to a method used in examination of SLOPE profiles in which the profile is divided into units (segments) each with characteristic angles. *See also* SLOPE UNIT.

bevel A flat surface that has been cut across the upper termination of a CUESTA on a hill-top. This is its definition in geomorphology, although strictly speaking a bevel denotes any change of slope-angle which departs from the vertical plane (BEVELLED CLIFF).

bevelled cliff A type of sea cliff in which the junction of the vertical (or near-vertical) face with the horizontal plane at the cliff top is interrupted by an obliquely angled linear slope, frequently mantled by soil and REGOLITH. It is thought that such cliffs, which are common in SW England, were formed in two stages: **1** when a falling Pleistocene sea-level abandoned the sea cliff and led to a reduction in its slope angle by SOLIFLUCTION processes; **2** when a later rise of sea-level, in either interglacial or post-glacial time, led to the renewal of wave attack and the cliffing of the foot of the obliquely angled slope, leaving its upper remnant as a bevel.

B-horizon The layer in a SOIL PROFILE which lies beneath the A-HORIZON and may be enriched by the deposition of material washed down from above. This material includes BASES and COLLOIDS moved from the A-horizon by the process of ELUVIATION and redeposited or precipitated in the underlying zone by the process of ILLUVIATION (hence, zone of illuviation). Some B-horizons may be characterized by the precipitation of iron, thereby leading to the development of a HARDPAN. *See also* SOIL HORIZON. [*235*]

bifurcation ratio The quantitative relationship between the different orders of magnitude of streams within a drainage basin (STREAM ORDER). It is expressed as the number of streams of one order divided by the number of streams of the next highest order, which they create through their confluence. Most values lie between ratios of 2.5 and 3.5, except in long but narrow drainage basins dominated by a master trunk stream. *See also* HORTONIAN ANALYSIS, LAW OF STREAM LENGTHS, LAW OF STREAM NUMBERS.

bight A large-scale indentation of a coastline, generally of continental proportions; e.g. Great Australian Bight, Bight of Benin. The term is also used to denote an indentation in the edge of the ice-sheets of polar latitudes.

bill A beak-like coastal promontory; e.g. Portland Bill.

billabong An Australian Aboriginal term to denote an ephemeral stream, but often confined to describe a cut-off, stagnant pool in an abandoned meander (OXBOW LAKE).

billow clouds A series of parallel regular cloud rolls separated one from another by clear sky bands, generally of similar thickness to the cloud bands. They are thought to result from a strong increase of windspeed with height when airflow remains stable or as waves formed at a surface of discontinuity. They occur usually at 6–8 km above the surface.

bimetallic thermograph An instrument for measuring air temperature in which the temperature sensor is a curved strip formed by welding together two metals with contrasting coefficients of expansion. The changes in strip curvature associated with temperature variations are used to actuate a recording pen through a lever mechanism.

bimodal distribution A statistical term denoting that the frequency curve representing the data DISTRIBUTION has two maxima. A frequency curve with more than two maxima has a *multimodal distribution*. *See also* UNIMODAL DISTRIBUTION.

binary A positional method of counting in which only two digits are used (0 and 1). The fundamental basis of all DIGITIZING.

binodal tidal unit An AMPHIDROMIC tidal system which has two amphidromic points or nodes instead of one.

binomial distribution If a number of identical but independent experiments are carried out in each of which a specified event has the same chance of occurrence, the number of occurrences has a binomial (two terms, e.g. $a + b$) distribution. *See also* BIVARIATE DISTRIBUTION, FREQUENCY DISTRIBUTION,

NORMAL DISTRIBUTION, POISSON DISTRIBUTION, PROBABILITY DISTRIBUTION.

biochemical oxygen demand (BOD) A measure of organic pollution and water cleanliness in which the amount of dissolved oxygen consumed by micro-organisms feeding on organic matter in standing or running water is measured over five days at a temperature of 20° C.

biochore **1** A general term denoting any geographic area supporting a distinctive plant or animal life. **2** More precisely it has been used to describe each of four major vegetation types: forest, grassland, savanna and desert.

biochron A geological time unit equivalent to a biostratigraphic range zone (BIOSTRATIGRAPHY). It was once regarded as synonymous with the term BIOZONE, but this usage is now obsolete.

biochronology **1** A geologic time-scale based on fossils. **2** A study of the relationships between organic evolution and geologic time. **3** The dating of geologic events by evidence from BIOSTRATIGRAPHY.

biocide A toxic substance capable of killing animals and plants.

bioclastic A term referring to rocks composed of broken organic remains. *See also* CLASTIC SEDIMENTS.

bioclimatology The study of climate in relation to living organisms, especially human beings. Matters of health and human comfort are of particular relevance. *See also* WIND-CHILL.

biocoenosis, biocenose A biogeographical term used to describe a mixed biotic community of plants and animals in a given HABITAT. *See also* BIOTOPE, ECOTOPE.

biodegradation The process by which organic substances are broken down by bacteria into basic elements and compounds. Most organic wastes are biodegradable; e.g. certain organisms can consume marine oil spillages.

biodiversity A widely used term that has a variety of definitions: **1** The total number of SPECIES in a region or area. **2** The number of endemic species in an area. **3** The genetic diversity of an individual species. **4** The sub-population of an individual species which embraces the genetic diversity. **5** The distribution of different ECOSYSTEMS.

biofacies **1** Lateral changes or variations in the biologic aspect of a stratigraphic unit. **2** Assemblages of plants or animals that have formed at the same time but under different conditions. *See also* FACIES.

biogenic sediment A sediment that is created by living organisms, either animal or plant; e.g. CORAL limestone.

biogeochemical cycles Mechanisms by which chemicals (e.g. carbon, oxygen, phosphorus, nitrogen and water) are moved through the ECOSPHERE to be renewed over and over again. The three major cycles are gaseous, sedimentary and hydrologic. [*29*]

biogeography The study of plant distributions (PHYTOGEOGRAPHY) and animal distributions (ZOOGEOGRAPHY) together with the geographical relationships with their environments (ECOLOGY), studied over time.

biogeomorphology The role of organisms in the study of geomorphology, either by the influence of flora and fauna on the genesis of landforms or the influence of earth surface processes/landforms on plant and animal distributions.

bioherm A large, unstratified, dome-like deposit of reef limestone (REEF-KNOLL) in which the organisms are in the position of growth.

biokarst A small-scale KARST landform formed largely by organic action. It is common in the coastal regions and is generally in the inter-tidal zone where the rocks are bored by marine organisms.

biological control (biocontrol) The control of pests (fungi, insects) by natural means, e.g. introduction of predators, breeding resistant crops, etc., instead of applying herbicides and pesticides.

biological magnification The increase of toxic materials (pesticides, herbicides, etc.)

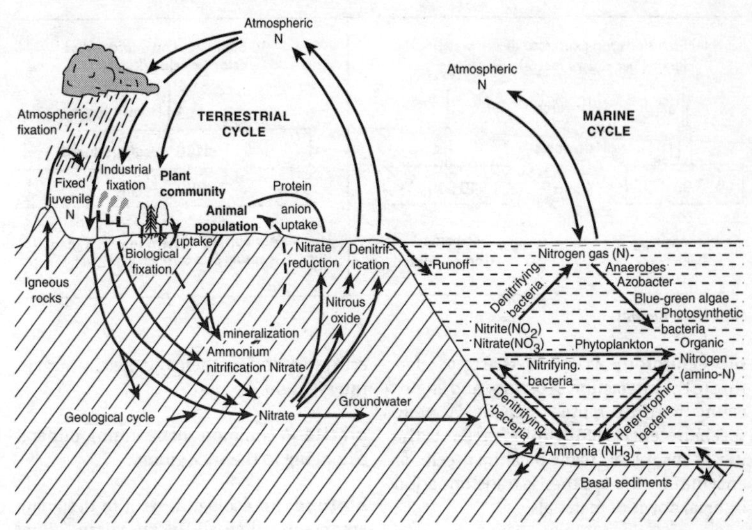

Figure 29 *Biogeochemical cycles*

that are often non-biodegradable (BIODEGRADATION) into an ECOSYSTEM. The toxins accumulate at successive TROPHIC LEVELS as part of the FOOD CHAIN.

biological productivity The rate at which growth processes occur either in an ECOSYSTEM or in an organism. It is normally expressed as: the weight of dry matter/unit area/unit time (kg/ha/year). Occasionally it is expressed as: grams of carbon/unit area/unit time (g C/m^2/day). *See also* NET PRIMARY PRODUCTION.

biomass The total mass of living organisms that can be supported at each TROPHIC LEVEL in a FOOD CHAIN. It is expressed as mass per unit area, measured as dry weight, ash weight or calorific value. The amount of living material present is referred to by ecologists as the *standing crop*, which gives an indication of the pattern of energy flow in the ecosystem. Usually the amount of standing crop in each trophic level decreases with each step on the food chain away from the plants, thus the flow of energy will also decrease. This can be illustrated by means of a *trophic pyramid*, in which energy is lost by conversion of plant substances into animal substances and also by respiration of the organisms within each trophic level. *See also* ECOSYSTEM. [*30*]

biome A major climax community of plants and animals. It generally corresponds to a CLIMATIC REGION. *See also* BIOTA.

biosphere The zone adjacent to the surface of the Earth where all life exists. *See also* ATMOSPHERE, HYDROSPHERE, LITHOSPHERE.

biostasis, biostasy A state of biological equilibrium between soil, vegetation and climate. *See also* HOMOEOSTASIS, RHEXISTASIS.

biostratigraphy That part of STRATIGRAPHY in which rock units are separated and differentiated on the basis of their contained fossil assemblages. *See also* PALAEONTOLOGY, CHRONOSTRATIGRAPHY, LITHOSTRATIGRAPHY.

biostrome A geological term to describe a sheet-like mass of purely organic material derived from organisms *in situ*, in contrast to the dome-like form of a BIOHERM. A thick sheet of limestone is a biostrome.

biota The collective flora and fauna of a region.

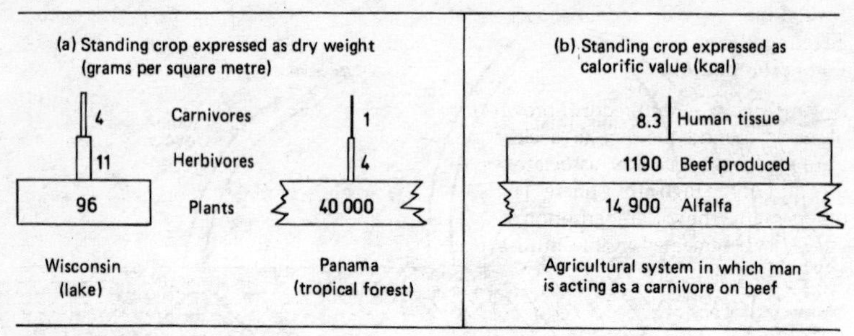

Figure 30 *Biomass*

biotechnology A controversial term referring to the interaction of biology and engineering to their mutual benefit. A more specific recent (1988) definition refers to the use of microbial, plant or animal cells or enzymes in the synthesis, transformation or breakdown of materials.

biotic Pertaining to life. Hence the *biotic factor* relates to the influence of organisms, in contrast to the influence of inorganic factors such as the weather (climatic factor) and the soil (edaphic factor) on the environment. A biotic area is a geographical expanse (e.g. region) which corresponds to the distribution of one or more groups of organisms, usually animals. A biotic succession refers to a shift in the composition and/or structure of the community because of the influence of one particular organism, e.g. the flying beetle *Scolythus scolythus* has introduced Dutch elm disease in Britain and induced a biotic succession.

biotic complex The varied cover extending over much of the Earth's surface, comprising an interacting complex of living organisms which exist in relation to the various environmental conditions.

biotic isolation The isolation of organisms due to genetic incompatibility rather than by geographical or ecological factors. *See also* SYMPATRY.

biotic potential A measure of the reproductive rate of a certain species. The biotic potential of some species (e.g. rats, rabbits) is so high that they threaten to overrun an ecosystem and are thereby culled by human intervention.

biotite A common rock-forming mineral belonging to the MICA group.

biotope An ecological term to designate an area in which all the faunal and floral elements (BIOTA) are uniformly adapted to the environment in which they occur. The ultimate subdivision of the HABITAT.

bioturbation The breakdown and reworking of sediment by the action of its contained organisms, e.g. earthworms.

biozone A biostratigraphic unit (BIOSTRATIGRAPHY) including all strata deposited during the existence of a particular kind of fossil. This term was originally proposed as a geological time unit (BIOCHRON) but this usage is now obsolete.

Birch discontinuity A seismic discontinuity within the Earth's MANTLE, some 900 km below the surface.

bird's-foot delta A type of DELTA formed by the outgrowth of natural river LEVEES into a body of water to form a finger-like pattern, reflecting the number of distributary streams. [*58*]

biscuit-board relief Topography which is characterized by a rolling upland or plateau into which glacial CIRQUES have taken large 'bites' but where the cirques have not yet coalesced.

bise, bize A cold, dry, northerly wind affecting the mountain areas of central Europe during winter.

bishop's ring A dull reddish-brown ring observed around the Sun in a clear sky, attributed to DIFFRACTION associated with fine dust in the high atmosphere. That first observed after the volcanic eruption of Krakatoa (1883) remained visible until spring 1886.

bispectral technique A term used in REMOTE SENSING referring to the use of radiances in two distinct wavelength bands (e.g. visible and thermal infrared) to assist in data collection for satellite climatology. It is adopted, for instance, in the inference of precipitation rates from clouds and also for automated cloud classification (CLOUD).

bit The basic unit of digital BINARY data, capable of taking only two values (conventionally 0 and 1).

bitumen A term applied originally to any of the flammable, viscous, liquid HYDROCARBON mixtures (ASPHALT). Now applied to all hydrocarbons, including gases, mobile liquids, viscous liquids and solids.

bituminous **1** That which is composed in part of BITUMEN, in the form of the tarry hydrocarbons. **2** It is commonly used for certain varieties of COAL which burn freely with flames, although they really contain no bitumen.

bivariate distribution Any FREQUENCY DISTRIBUTION or PROBABILITY DISTRIBUTION in which data are classified according to two variables. *See also* BINOMIAL DISTRIBUTION, NORMAL DISTRIBUTION, POISSON DISTRIBUTION.

blackband ironstone A sedimentary rock comprising a mixture of carbonaceous (coal-like) material and iron carbonate. This combination makes the iron ore virtually self-smelting. The best known is the blackband ironstone of the Coal Measures in Staffordshire, England.

black body An ideal emitter which radiates energy at the maximum possible rate per unit area and wavelength at any temperature. A black body also absorbs all the radiant energy incident upon it. No actual substance behaves as a true black body; however, platinum soot closely approximates the ideal. The spectral distribution of black-body radiation is described by PLANCK'S LAW.

black-body radiation STEFAN'S LAW.

black-box system A system of whose structure nothing is known except that which can be deduced from its behaviour. Thus by manipulating the inputs to a system one can discover statistical relationships between INPUTS AND OUTPUTS. *See also* GREY-BOX SYSTEM, WHITE-BOX SYSTEM.

black-bulb thermometer A type of mercurial maximum thermometer with its bulb blackened and mounted in an outer glass sheath. When exposed in direct sunlight it gives the maximum temperature of the Sun's rays. The use of such a thermometer as a means of measuring solar radiation is declining. *See also* ACTINOMETER.

black cotton soil A type of BASISOL, developed on basic lavas under humid, tropical climates and used extensively for growing cotton in India and E Africa.

black earth CHERNOZEM.

black ice FREEZING DRIZZLE, FREEZING RAIN, GLAZED FROST.

black smoker A plume of HYDROTHERMAL fluid containing finely dispersed particles of black sulphides issuing from a submarine vent on a MID-OCEANIC RIDGE. *See also* WHITE SMOKER.

blanket bog A BOG which drapes all features of lowland terrain, infilling hollows to great depths; it is composed essentially of PEAT upon which rough wet moorland or marshland vegetation prevails, and is formed under conditions of high rainfall incidence and low evapotranspiration; e.g. W Ireland and NW Scotland. *See also* RAISED BOG. [*108*]

blind valley A type of valley in a KARSTIC, limestone terrain. It may be: **1** occupied by a stream which disappears underground at the valley's lower end as it approaches an enclosing rock-wall; **2** a dry valley, formerly occupied by a surface stream, which is closed at the lower end. Blind valleys may have been formed by roof collapse which exposes an underground stream-course.

blizzard An intensely cold and strong wind accompanied by falling snow (WHITE-OUT). *See also* KATABATIC WIND.

block An angular rock fragment, similar in size to a boulder, and more than 256 mm in diameter. It shows little evidence of modification by transportation.

block diagram A three-dimensional perspective representation of geologic and topographic features. It exhibits a surface area and at least two vertical faces or cross-sections (e.g. [*47*]).

block disintegration The process by which well-bedded and jointed rocks are broken up by MECHANICAL WEATHERING. Frost action is one of the most important factors influencing the break-up.

block-faulting The action of faulting on a section of the Earth's crust in which parallel or semi-parallel faults produce a series of individual aligned blocks. Some may be down-faulted (GRABEN), others raised (HORST), while others may simply be tilted (BASIN-AND-RANGE TERRAIN).

blockfield A continuous spread of broken, angular rock fragments (of boulder dimensions) which mantle the surface of a high mountain or plateau. It will only survive on a flat or gently sloping surface; beyond critical slope angles the blocks will tend to move down the mountainside (STRIPES (block stripes)). The blocks are produced by BLOCK DISINTEGRATION, generally by frost action. The German terms *blockmeer* and *felsenmeer* (lit. 'rock-sea') are frequently used synonyms.

block-glide landslide BLOCK-SLUMPING.

blocking anticyclone Also termed a *blocking high*. A system of high atmospheric pressure which remains stationary, thereby preventing the passage of other pressure systems, especially in the westerly zone of mid-latitudes. ANTICYCLONE.

block lava AA LAVA.

blockmeer BLOCKFIELD.

block mountain An area of high relief which is bounded on most sides by faults and which has been either **1** uplifted by earth-movements or **2** left elevated by the sinking of surrounding areas; e.g. Vosges Mts and Black Forest. *See also* BLOCK-FAULTING, HORST.

block, perched **1** A block of rock, or more commonly an ice-transported boulder, left at the lip of a steep slope in glaciated terrain. It is frequently in a delicate state of balance and may be rocked by the application of strong pressure. **2** The term is also used, less commonly, of a rock which is left precariously balanced *in situ* by weathering, as a *pedestal stone* or *capstone*, e.g. on a granite or gritstone tor.

block-slumping A type of MASS-MOVEMENT of large sections of rock material on escarpment faces or sea-cliffs, especially along a curvilinear glide-plane (ROTATIONAL SLIP). It is generally the result of instability in the steep slope due to the undermining of the lower strata by water (SPRING SAPPING), especially of the more easily eroded strata which underlie more massive, well-jointed rock layers. The weakening and removal of the less-resistant basal rocks causes the whole mass to slump bodily downwards and outwards, leaving behind a scar on the escarpment or cliff face. *See also* LANDSLIDE.

blood-rain Raindrops which contain fine red dust, brought by upper winds from neighbouring deserts (e.g. Sahara dust often causes blood-rain over Italy).

blow-hole **1** A vertical or near-vertical cleft in a coastal cliff linking a sea-cave with the cliff-top and through which columns of spray are violently ejected. A blow-hole is formed by wave erosion along a fault or a joint at the back of the sea-cave; when a wave surges into the cave the compressed air forces spray violently up the narrow orifice. In Scotland the term *gloup* is synonymous with blow-hole. **2** A minute crater formed on the surface of thick lava flows.

blow-out A saucer- or trough-shaped hollow usually in a sand-dune terrain. It is formed by wind erosion (DEFLATION) of pre-existing dunes or other loose sand deposits, especially when the protective vegetation cover has been removed or destroyed. The accumulation of sand from the blow-out trough is sometimes referred to as a blow-out dune. The term has also been applied to the wind-eroded hollows which have been recognized among the peat-lands of mountain plateaux. It is suggested that as the peat cover dries out and shrinks it becomes increasingly prone to erosion by the wind.

blow-well, blowing-well A colloquial term in E England for an ARTESIAN WELL, of either natural or artificial derivation.

blue-band A layer of dense bubble-free (hence bluer) ice in a glacier, probably due to the freezing of meltwater in a crevasse. When this layer is exposed at the surface it forms a darker ribbon-like pattern.

bluehole A deep circular hole in coral reefs. There is no agreement over the genesis of this phenomenon: some scientists favour karstic processes (KARST), others point to volcanic action or even meteorite impact.

Blue John An ornamental variety of FLUORITE (CaF_2) mined in N Derbyshire, England.

blue mud A submarine deposit occurring on the CONTINENTAL SLOPE at depths of 228–5,183 m (750–17,000 ft). It contains up to 75% of terrigenous material and derives its colour from organic matter and from iron sulphide. Its individual particles are smaller than 0.03 mm in diameter.

bluff In general usage, a high, bold headland or steep, prominent cliff, but used specifically to denote a river-cut cliff or steep slope on the outside of a MEANDER. A line of bluffs often marks the edge of a former FLOODPLAIN. [*147*]

Blytt–Sernander model The standard sequence and terminology of the HOLOCENE, established by two Scandinavians in the late nineteenth and early twentieth centuries. *See also* FLANDRIAN. [*81*]

BOD DISSOLVED OXYGEN.

bodden A German term for an irregularly shaped coastal inlet along the Baltic coast of E Germany, formed by a rise of sea-level in an undulating terrain. It is especially well illustrated on the island of Rügen.

body waves The transverse or longitudinal seismic waves (EARTHQUAKE) transmitted in the interior of a solid or fluid body, in contrast to surface waves. Body waves include primary (P or longitudinal) and secondary (S or shear) waves.

bog, peat bog **1** A general term for a morass or swamp. **2** A commonly used term in Scotland and Ireland for a stretch of waterlogged, spongy ground, chiefly composed of decaying vegetable matter (PEAT), especially of rushes, cotton grass and sphagnum moss. It is widespread in Russia, Scandinavia and Canada (MUSKEG) and frequently covers layers of glacial TILL which cause extensive waterlogging. Well-preserved tree-stumps and tree-trunks are frequently discovered within the peat-bogs and these are colloquially referred to as *bog-oak*, whatever their species. These woody remains relate to a former vegetation type when drier conditions prevailed (BOREAL period) but the forests became overwhelmed by the increasingly wet and stormy conditions (ATLANTIC and SUB-ATLANTIC PERIODS) which occurred sporadically in post-glacial times (FLANDRIAN). *See also* BLANKET BOG, RAISED BOG, WETLAND.

bogaz A Serbo-Croat term for a deep and narrow chasm in limestone country, especially in KARST. It is formed by carbonate solution following the penetration of water along a joint-plane or a fault-line.

bog-burst The sudden rupture of a BOG and release of floodwater and fluid peat over considerable distances after a period of very heavy rainfall has caused the margins of the dome-like accumulation to collapse through over-saturation.

boghead coal A type of BITUMINOUS coal, similar to CANNEL COAL in appearance, and characterized by a high percentage of algal remains (ALGAE).

bog ore, bog iron ore A spongy variety of hydrated oxide of iron (LIMONITE), occurring as lumps or layers beneath peat bogs and thought to have been formed through precipitation by bacterial organisms. The ore generally has a low sulphur and phosphorus content and can yield very pure iron of some commercial value.

bole A fossil LATERITE, indicative of a PALAEOSOL, now interbedded between two basaltic lava flows. The presence of bole indicates that tropical climates prevailed during its formation.

Bølling Interstadial A Danish term which has been used to describe one of the Late-Glacial INTERSTADIAL stages and which, in NW Europe, has been RADIOCARBON dated to about 12,350–12,750 BP. It was a time when the ice-sheets temporarily waned and a tundra vegetation cover made a very brief appearance in southern Britain and the Low Countries prior to the renewal of a sub-Arctic climate (OLDER DRYAS). This stage is also referred to as Zone Ia–Ib in the LATE-GLACIAL pollen zone chronology. *See also* ALLERØD INTERSTADIAL, BLYTT–SERNANDER MODEL, WINDERMERE INTERSTADIAL.

bolson A closed depression or basin in the mountainous desert environment of SW USA and N Mexico. Material is moved across the low-angle slopes (BAHADA) mainly by sheetwash and the centre of the basin is commonly occupied by a salt-lake (PLAYA) surrounded by sheets of GYPSUM OT ROCK SALT. *See also* BASIN-AND-RANGE TERRAIN. [*189*]

bomb, volcanic A globular mass of lava ejected from a volcano (PYROCLASTIC MATERIAL) in a liquid or plastic form, which solidifies in the air prior to hitting the ground. Ranging in length from 4 mm to over 1 m, volcanic bombs acquire characteristic tear-drop or pear-shaped forms, surface markings and internal structures during their flight through the air or on impact. Bombs less than 4 mm in diameter are classed as VOLCANIC ASH.

bone bed A richly fossiliferous but thin rock stratum containing many fragments of fossil bones, coprolites, scales and teeth of vertebrates. The rapid accumulation and concentration of these fossil remains is thought to result from a catastrophic event, possibly an earthquake.

Bonne's projection An EQUAL-AREA MAP PROJECTION based on a CONICAL PROJECTION, in which the central meridian is straight and truly divided into 10° intervals (lines of latitude). The parallels of latitude are all concentric circles and all the meridians, other than the central meridian, are composite curves drawn through the division points on the correctly divided parallels of latitude. Although the projection has the property of equal area, the distortion of scale and

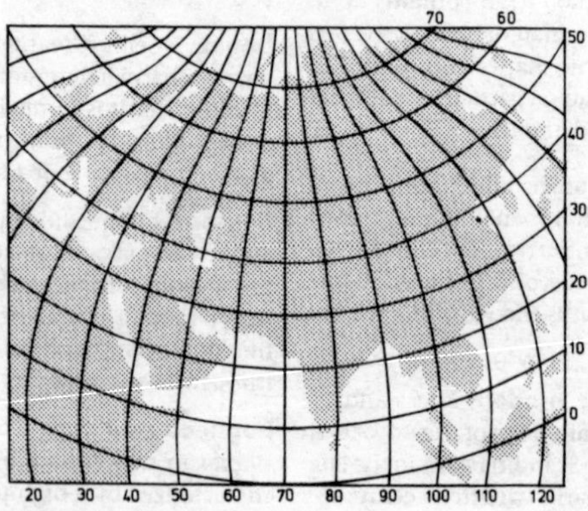

Figure 31 *Bonne's projection*

shape increases rapidly away from the central meridian in an E or W direction. *See also* SANSON–FLAMSTEED PROJECTION. [*31*]

bora A strong, very dry north-easterly wind which blows in cold gusts from the Balkan mountains down to the eastern shore of the Adriatic Sea, most commonly in winter. Occasionally used for a cold north-easterly squall in the Black Sea. It results from the passage of a deep depression over the Mediterranean to the south of a central European high. The mountainous topography helps to funnel the wind, which occasionally reaches speeds of 50 km/h. *See also* MISTRAL.

bore **1** A rapidly advancing tidal wave which moves upstream in any shallow estuary experiencing a large tidal range. Because of the restriction of the estuary shorelines, the retardation by friction of the river-bed and opposition from the river current itself, the advancing 'wall' of broken water may reach heights of >1 m at the outset (e.g. the bore on the R. Severn, England), with exceptional heights of >4 m being reported from N China, but the height gradually diminishes as it progresses upstream. *See also* EAGRE, MASCARET. **2** A borehole or boring. **3** A surge of water downstream. FLASH FLOOD.

Boreal **1** A climatic zone characterized by long, cold, snowy winters and short summers. **2** A term applied specifically to the coniferous forests of the N hemisphere (the Latin term *borealis* means northern). **3** One of the early post-glacial (FLANDRIAN) climatic periods in N Europe, RADIOCARBON dated as ranging from about 9,450 to 7,450 BP in the British Isles. It was a period of cold winters, warm summers and drier conditions than at present, when the birch–pine forests of the preceding PRE-BOREAL gave way progressively to a mixed oak forest with hazel and pine. The Boreal is also referred to as zones V and VI in the Flandrian pollen-zone chronology. [*81*]

borehole A hole drilled into the crust of the Earth for exploratory purposes, especially during prospecting operations for oil, coal and natural gas.

bornhardt A German term, named from a geologist W. Bornhardt, for an isolated residual hill which rises abruptly from a plain. In this usage it is often taken as a synonym for INSELBERG, but some geomorphologists restrict the term bornhardt to describe an inselberg of massive, poorly jointed granite which has resisted reduction by chemical weathering and has been revealed only by subsequent removal of deep layers of weathered REGOLITH. It is sometimes referred to as a *shield inselberg*. *See also* EXFOLIATION, RUWARE, ZONAL INSELBERG. [*282*]

boss A circular-shaped outcrop of intrusive rock, smaller than a BATHOLITH, usually giving rise to a rounded or craggy eminence and descending into the earth with steeply inclined sides. Some geologists regard it as synonymous with STOCK but others believe that the latter exhibits a more irregular outcrop when seen in plan.

bottom **1** The bottom land of the Mississippi FLOODPLAIN and similar alluvial tracts along US river courses. **2** The bed or floor of a body of running water, lake water or ocean water. **3** The floor of an underground passage. **4** The flat land in the floor of a U-shaped valley, formerly occupied by a lake but now infilled by alluvium, in the Lake District of England. **5** The floor of a DRY VALLEY in the chalklands of southern England.

bottom current, bottom flow An underflow or current denser than any part of the surrounding fluid and which flows along the bottom of the water body. *See also* DENSITY CURRENT.

bottom load BEDLOAD.

bottom melting BASAL MELTING.

bottom-set beds The stratified layers of finer material carried out and deposited on the bottom of the sea or a lake in advance of a DELTA. Subsequently, as the delta grows seawards, they are buried by the FORE-SET BEDS. *See also* TOP-SET BEDS. [*85*]

bottom water The lowest layer of water in an ocean, especially in the deeps (ABYSS). It is relatively cold and dense (BOTTOM

CURRENT), with temperatures only a little above freezing-point. The movement of bottom water is influenced by the submarine topography.

boudinage A type of geological structure, caused when a competent bed is set amidst yielding or incompetent beds of rock and is then squeezed or stretched. The resulting structure of detached sausage-shaped segments (from the French *boudin* = sausage) forms because the competent bed is unable to deform plastically to the same extent as the surrounding beds and therefore thins locally before being pinched and fractured into discrete segments.

Bouguer anomaly A gravity anomaly first observed by P. Bouguer in 1735, in which the deviation of the plumb-line in gravitational observations is less than calculated. This deviation (or anomaly) is because of isostatic compensation (ISOSTASY) resulting from a deficiency of mass in the Earth's crust where it underlies mountain ranges. *See also* GRAVITY ANOMALY.

boulder A large, rounded or subangular piece of rock lying on the surface of the ground, embedded in the soil or within a layer of boulder clay (TILL). Its diameter is larger than 200 mm by official British Standards but in the USA it should exceed 256 mm (WENTWORTH SCALE). *See also* PARTICLE SIZE.

boulder clay TILL.

boulder-controlled slope A concept defined by a US geomorphologist, K. Bryan, based on the belief that in arid environments the slope angle will be controlled by the angle of rest (repose) of the boulders which have been broken up by weathering but remain on the slope. Although the size of the boulders will slowly diminish by GRANULAR DISINTEGRATION the slope angle will remain constant since the original size of the boulders resting upon it will remain the same when they are released by weathering, provided that the joint-spacing remains unaltered. *See also* PARALLEL RETREAT OF SLOPES.

boulder-field FELSENMEER.

boulder-pavement **1** A surface of boulder-rich till that has been abraded to flatness by the passage of a glacier. **2** The horizontal alignment of boulders when seen in section in a layer of till.

boulder-train A term for the 'stream' of boulders generated from an identifiable source of bedrock and carried laterally in a more or less straight line by a former glacier or ice-sheet. Such trains can be used to identify former directions of ice movement, but since they rarely lie in exactly parallel lines they have sometimes been referred to as *boulder fans*, with their apexes pointing to the original rock-knobs or exposures from which the ERRATIC rocks were glacially quarried. *See also* GRAVEL TRAIN.

boundary **1** That which fixes a limit or extent of an area; a territory. **2** A plane separating two rock formations.

boundary current A fairly rapid flow of marine water at depth, as part of the general circulation of the oceans, in which the edge of the current marks a sudden change in salinity and temperature. Boundary currents are most marked on the western sides of ocean basins, where cooler waters move from high to low latitudes; in the case of the N Atlantic there is a south-flowing current beneath the Gulf Stream.

boundary layer A term referring to the juxtaposition of a fluid and a solid where relative motion exists. The boundary layer is the zone in the fluid nearest to the solid surface within which a velocity gradient develops because of the retarding frictional effect caused by the solid. The fluid velocity increases with distance away from the solid in a parabolic curve. The velocity gradient of the boundary layer occurs in OVERLAND FLOW, FLOW REGIMES in river channels, BACKWASH and SWASH of beaches and in airflow over a desert DUNE. The boundary layer in rivers becomes fully developed when the velocity profile extends to the surface. *See also* HJULSTRÖM CURVE, REYNOLDS NUMBER, TURBULENT FLOW. The term boundary layer also refers to that layer of the atmosphere in which air movement is governed by the proximity of the ground surface. Two divisions are recognized: **1** the *planetary boundary layer* (100–600 m) where the influence of the ground surface is felt but is not dominant; **2** the *surface boundary layer*, where the influence of the Earth's surface is of paramount importance to a height of 100 m. *See also* MICROMETEOROLOGY.

boundary waves Also known as *internal* waves. These are waves occurring within a body of water but possessing quite different characteristics from waves which develop at the surface of the water body; their amplitude, for example, is considerably greater than that of surface waves. They may have considerable and abrupt changes of density (in which the sharp surfaces of discontinuity are termed *interfaces*), but in other instances the changes may be gradual.

bourne A local term for a stream in chalk country, exhibiting a sporadic or intermittent flow. After periods of heavy rainfall, usually during winter, the surface water percolates (PERCOLATION) through the chalk until it reaches the saturated layers at the WATER-TABLE. After a long period of time the latter rises to a sufficiently high elevation to enable the bourne to flow on the surface of an intermittently DRY VALLEY. The bourne is a widespread feature of chalklands in S England where it has become incorporated in numerous placenames, especially the term 'winterbourne'. Also known as *gypsey, lavant, nailbourne* and *woebourne*.

Bowen–Ludlam process COAGULATION PROCESS.

Bowen ratio An empirical formula devised to determine the ratio of the amount of SENSIBLE HEAT to that of LATENT HEAT lost by a surface to the atmosphere. The Bowen ratio (β) is expressed as $\beta = H/LE$, where H = *sensible heat* flux, LE = *latent heat* flux; β is a variable which ranges from <0.1 for water to >10 for a desert surface. When β exceeds 1.0, sensible heat flux becomes dominant, whereas latent heat flux dominance occurs with values of <1.0. [*32* and *see also 102*]

Bowen's reaction series A series of minerals which crystallize from molten rock of a certain chemical composition (MAGMA). Any mineral formed early in the chain will ultimately react with the melting rock to form a new mineral further down the series.

Boyle's law A law of mechanics which describes the behaviour of atmospheric gases. Boyle's law states that, at a constant temperature, the volume (V) of a mass of gas varies inversely to its pressure (P): $P = k/V$, where k is a constant. *See also* CHARLES'S LAW, DALTON'S LAW.

BP An abbreviated form of the expression 'before (the) present (day)' (as opposed to BC). It is widely used in GEOCHRONOLOGY,

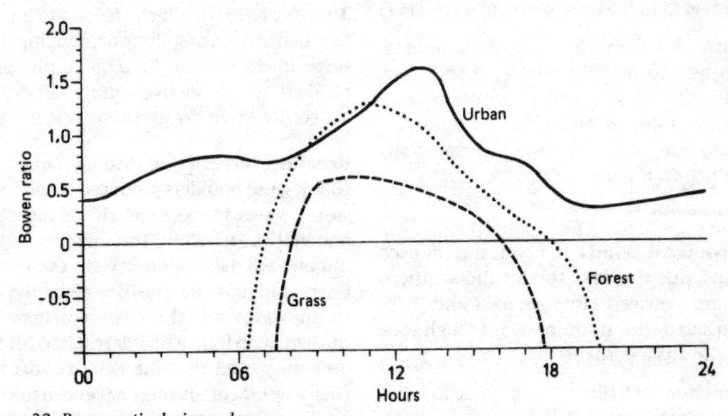

Figure 32 *Bowen ratio during a day*

especially in RADIOCARBON DATING of the later Quaternary period.

brackish A term applied to any water which exhibits salinity intermediate between sea water and fresh water. It is variously defined as containing between 15 and 30 parts of salt per 1,000. Degrees of brackishness in decreasing order of salinity are termed: polyhaline; pliohaline; mesohaline; miohaline; oligohaline.

brae A Scottish term for a hill-slope or BROW of a hill.

braided stream A stream characterized by a network of interconnected converging and diverging channels resembling the strands of a braid. The whole river system is generally shallow, so that the intervening shingle bars and alluvial material are exposed at low water. It is due mainly to the non-coherence of the LOAD, which leads to selective deposition and the subsequent diversion and reunification of the numerous channels. Braiding is commonplace where streams debouch from glaciers and flow across GLACIOFLUVIAL deposits, and also where they cross ALLUVIAL FANS. *See also* ANASTOMOSING.

Bramertonian An interglacial stage of the early PLEISTOCENE in Britain, named from its type site at Bramerton near Norwich, Norfolk. Its pollen assemblage, associated with a marine shelly sand, indicates that it is older than the PASTONIAN interglacial but younger than the BAVENTIAN cold stage. [*197*]

brash **1** A term used by farmers and soil scientists to describe friable, rubbly rock beneath the soil which weathers into 'brashy' soils, excellent for corn-growing (CORNBRASH). **2** Masses of loose ice fragments floating on the sea.

braunerde BROWN EARTHS.

brave west winds A nautical term used to describe the WESTERLIES of the southern oceans, between latitudes 65°S and 40°S. Associated with storminess and rough seas. *See also* ROARING FORTIES.

breached anticline An anticline that has been deeply eroded along its central axis by river incision, thereby exposing older rocks in the central anticlinal valley, bounded by infacing scarps of younger rocks. *See also* INVERSION OF RELIEF.

breached crater, breached cone Any volcanic crater in which the lava breaks through one wall of the crater to create a breach. Similarly, when lava breaks through the wall of a cinder cone it is termed a breached cone.

breadcrust bomb A volcanic bomb (BOMB, VOLCANIC) with a cracked and crusty exterior, created when the skin shrank rapidly on cooling.

breakaway The uppermost part of a slope when capped by a resistant rock layer such as a LATERITE or DURICRUST. Because it is likely to be undermined by the removal of less-resistant underlying strata the slope will retreat (PARALLEL RETREAT OF SLOPES) and the unsupported cap rock will break away and fall to the slopes beneath.

breaker An oversteepened wave of water breaking with a pronounced SWASH on to a shoreline. The oversteepening occurs when the ratio of wave-height (AMPLITUDE) to wavelength exceeds 1 : 7. [*260*] A rough classification of breakers recognizes three types (although there is much overlap): **1** *spilling breakers*, which break gradually over quite a distance; **2** *plunging breakers*, which tend to curl over and break with a turbulent crash; **3** *surging breakers*, which steepen to a peak but then instead of spilling or plunging they surge up the beach. In general, plunging breakers are destructive and the other two are constructive. *See also* BACKWASH, WAVE.

breaker terrace A distinct terrace of cobbles and boulders produced by periodic storm waves in slopes of till or morainic material (KAME) along the margins of ice-impounded lakes (PRO-GLACIAL LAKE). The finer materials are washed away by the storm waves but the coarser material is thrown up to form a bouldery terrace. It has been suggested that the PARALLEL ROADS of Glen Roy in Scotland may have been formed partly in this way.

break of slope An abrupt change of slope due to differences in erosional history of the terrain or lithological contrasts. The change of slope in the long-profile of a river-bed is termed a KNICKPOINT.

breakpoint bar A marine BAR which forms offshore in the zone where waves first break.

breccia An Italian term for a rock composed of angular (CLASTIC) fragments, in contrast to rounded fragments (CONGLOMERATE). There are several ways in which breccias may be formed: **1** a deposit of TALUS or SCREE; **2** a volcanic deposit made up of PYROCLASTIC MATERIAL ejected during an eruption (AGGLOMERATE); **3** a sedimentary rock which has become cemented during consolidation; **4** a *fault breccia* formed by crushing and grinding along a fault-line, sometimes termed a *crush-breccia* (CATACLASIS); **5** a cemented mixture of rocks and bones in a cave deposit, known as a *bone breccia*.

breckland **1** A general term for a heathland. **2** A term used more specifically to denote an area of sandy and gravelly soils in E Anglia, England, into which the Forestry Commission has introduced very extensive coniferous plantations on the poorer soils.

breeze A wind speed between force 2 (light breeze) and force 6 (strong breeze) on the BEAUFORT SCALE. The term is generally applied to winds caused by convection. *See also* LAND BREEZE, SEA BREEZE.

brickearth **1** A general term for any loamy clay that can be used for brick-making. **2** A term used more specifically by soil scientists and geomorphologists to describe the wind-blown, fine-textured soils which have been re-sorted and re-deposited by water, frequently in old river terraces. Many of the brickearths of southern England and N Europe were initially blown southwards by glacier winds which picked up many of the finer deposits of the GLACIOFLUVIAL and morainic materials formed by the melting Pleistocene ice-sheets. Brickearths give fertile, easily worked soils and are generally classified as of high agricultural value. *See also* LOESS.

brickfielder A hot, dry northerly wind which blows from the Australian interior out to the SE coastlands during the summer, generally accompanied by quantities of dust and very high temperatures (>38° C/100° F). Most common in Victoria, normally associated with the advancing edge of a low-pressure trough. *See also* SOUTHERLY BURSTER.

brig A local term for a coastal headland of hard rock; e.g. Filey Brig in E Yorkshire, England.

brigalow scrub A semi-arid scrub vegetation, composed mainly of acacia, in Australia.

Brillouin scale A logarithmic scale of linear measurement produced in 1964 by L. Brillouin. It ranges from 10^{30} down to 10^{50}, with the Earth's equatorial circumference being regarded as 0. *See also* G-SCALE.

brine Any water deposit which is strongly impregnated with salt.

British thermal unit (BTU) A unit of heat equivalent to the amount which is required to raise the temperature of 1 lb of water from 63° F to 64° F. Alternatively, it is equivalent to 1/180th part of the amount of heat required to raise the temperature of 1 lb of water from freezing-point (32° F) to boiling-point (212° F) at sea-level.

brittleness index The potential of a sedimentary material to flow downhill. It is indicated by the equation, *Brittleness index* (Ib) = $[(\tau f - \tau r)/\tau f] \times 100\%$, where τf is the resistance to shear at failure and τr is the residual resistance to shear. A high brittleness index signifies that a large drop in strength occurs in the residual state. In sand the brittleness index increases with the density of the packing because low-porosity materials exhibit a larger drop in frictional strength through failure.

broad An English term for an E Anglian lake of fresh water, usually reed-fringed and connected to a slow-flowing river near its estuary. It is now believed that the Broads of Norfolk and Suffolk were created mainly by the artificial removal of peat in medieval times and subsequent flooding by the neighbouring rivers, into which linking channels have often been cut.

brockenspectre The shadowy effect and surrounding rings of coloured light projected on to a cloud or bank of mist as a diffraction effect when the sun throws the shadow of an observer on to a bank of fog. Called after the summit of the Brocken Mt in Germany.

brockram A type of BRECCIA found in the PERMIAN rocks of N England.

brodel A German term for a highly irregular and contorted INVOLUTION structure in soils which have been churned by a FREEZE–THAW mechanism. The feature often contains masses of silt or clay cut off from their original horizontal continuity by the CRYOTURBATION process.

Bronze Age One of the cultural phases of man's evolution, when an alloy of copper and tin was perfected. This metal (bronze) replaced the stone artefacts of earlier cultures (*see also* MESOLITHIC, NEOLITHIC, PALAEOLITHIC) and the less developed 'Copper Age', in the second half of the 2nd millennium, but was itself superseded by the IRON AGE. In Britain the Bronze Age flourished between about 2000 BC and 500 BC (4,000–2,500 BP).

brook A small stream.

brousse-tigrée A term referring to vegetation banding on arid or semi-arid hillslopes. They are thought to be related to differences in SHEETFLOW conditions in the soil.

brow 1 The top of a hill or mountain slope; the point where a steep slope eases into a gentle slope; the Scottish BRAE. **2** The frontal portion of an overfold in a NAPPE STRUCTURE.

brown alluvial soils One of the subdivisions of the BROWN EARTHS, which are themselves part of the SOIL CLASSIFICATION OF ENGLAND AND WALES. They are typically non-calcareous soils developed on recent ALLUVIUM.

brown calcareous soils Alkaline soils (CALCIMORPHIC SOILS) which develop largely on calcareous rocks under humid, temperate climatic conditions. The soils develop thicknesses of up to 70 cm, particularly on limestones with their base-rich vegetation, and are characterized by a CRUMB STRUCTURE and a high organic content. The A-HORIZON has a dark, reddish-brown colouring and with depth merges into a poorly defined B-HORIZON which is lighter in colour and contains many fragments of the parent material. They have a pH of >7 and support high-quality agriculture. *See also* MOLLISOL.

brown coal A fibrous, subbituminous coal, part way between PEAT and BITUMINOUS COAL. It is high in moisture content but low in heat value. *See also* LIGNITE.

brown earths (brown forest soils) A wide range of brown soils which once supported a thick cover of deciduous forest in humid temperate latitudes to the south of the PODZOL zone, often forming on Pleistocene glacial deposits. Although they form in areas where precipitation exceeds evaporation, they are free-draining, have a pH of 5 to 7 and a well-developed CRUMB STRUCTURE. Because of their rich humus content in the A-HORIZON and their retention of sesquioxides they are classed as good agricultural soils despite their lack of carbonates, which have been removed by LEACHING and/or ILLUVIATION. *See also* BROWN PODZOLIC SOILS, SOL LESSIVÉ. [*37*]

brown podzolic soils One of the subdivisions of the PODZOLIC SOILS, which are themselves part of the SOIL CLASSIFICATION OF ENGLAND AND WALES. They are normally well-drained, loamy or sandy soils with a dark brown or ochreous, friable subsurface horizon and no overlying bleached horizon or peaty topsoil. They are transitional in character and in topographical location between the BROWN EARTHS and the true podzols (*sensu stricto*). Thus, in humid temperate climates, where precipitation exceeds evaporation, they occur on well-drained slopes, below the upland podzols but above the acid brown earths of the valleys and lowlands. They are less rich in bases than a brown earth and therefore exhibit a pH of 5 to 6. [*37*]

brown sands A term used to describe one of the subdivisions of the BROWN EARTHS, which are themselves part of the SOIL CLASSIFICATION OF ENGLAND AND WALES. They develop on fairly free-draining non-alluvial sandy and gravelly deposits.

brown soils One of the seven major groups in the SOIL CLASSIFICATION OF ENGLAND AND WALES. It includes soils which range from well-drained to imperfectly drained but excludes the PELOSOLS. The soils are characterized by an altered, brownish B-HORIZON that has a soil rather than a rock structure and extends to more than 30 cm in depth. The group includes the ARGILLIC BROWN EARTHS, BROWN ALLUVIAL SOILS, BROWN CALCAREOUS SOILS, BROWN EARTHS and BROWN SANDS.

Brückner cycle A cycle of climatic change with a postulated periodicity of thirty-five years, during which there appear to be irregular alternations of cold, wet periods with hot, dry periods. Recognized by E. Brückner in 1890, but the thirty-five-year periodicity has been found to be irregular and the oscillations of temperature and rainfall from the norm of small dimensions (for temperature $<1°$ C; for rainfall from +9% to –8%).

brunizem PRAIRIE SOIL.

Bruun rule An equation used to predict absolute shoreline recession (r) in a horizontal direction, resulting from absolute sea-level rise(s). It is expressed as

$$r = ls/h$$

where l and h are the length and height of the equilibrium cross-profile from offshore to beach crest. The equation works well on open BARRIER BEACHES but less well on shingle beaches of closed or crenellate coastlines.

Bubnoff unit A time-distance unit which assists in the measurement of rates of ground-surface lowering or SLOPE RETREAT. One unit equals 1mm per 1,000 years.

Buchan spell An unseasonable spell of weather, postulated in 1867 by A. Buchan on the basis of some fifty years' observations. The cold summer spells and warm winter spells, which were thought to occur on nine occasions per year, are now thought to be statistically unreliable, so far as their specific dates are concerned: *Cool spells*: 7–14 February, 11–14 April, 9–14 May, 29 June to 4 July, 6–11 August, 6–13 November. *Warm spells*: 12–15 July, 12–15 August, 3–14 December.

bulk density of a soil The mass of dry soil per unit bulk volume. The latter is determined before the soil is dried to a constant weight at 105° C.

bulk specific gravity of a soil The ratio of the BULK DENSITY of a soil to the mass of a unit volume of water.

Bunter The lowest formation of the TRIASSIC system. It consists of thick pebble beds and desert sandstones, ranging in colour from brown, through red, to yellow. The very hard, purple quartzite pebbles of the pebble beds have been incorporated into the glacial drifts of the English Midlands and enter into the GRAVEL TRAINS of the Thames terraces.

buran A strong, north-easterly wind in Russia and central Asia, occurring in any season, but most frequently in winter, when it is very cold and accompanied by BLIZZARDS. The winter blizzards are also called PURGA.

buried topography Any terrain which has been inundated by transgressing seas and buried by sedimentary deposits. Thus, the old land surface becomes a line of UNCONFORMITY which may subsequently become exposed by erosion to reveal the hitherto buried topography. More frequently the buried surface is known only from borehole data and from geophysical information, when buried river channels, terraces and 'raised' shorelines may be identified; e.g. below the carselands (CARSE) of the Firth of Forth, Scotland, a complex suite of formerly raised beaches has been recognized, although these are now buried beneath FLANDRIAN sediments.

burn A Scottish term for a small stream. It is also used in N Ireland and N England.

bush A general term denoting wilderness or uncleared land, as opposed to cultivated and settled land.

bushveld A type of SAVANNA vegetation in Africa in which the ratio of trees to grassland varies from near-forest conditions to open, grassy parkland.

butte A French term for a conspicuous flat-topped hill with steep sides and frequently capped by a resistant layer of rock. It is characteristic of arid and semi-arid regions and is thought to be a remnant of a partly dissected plateau surface. It is smaller in extent than a MESA which it closely resembles in every other way.

butte-témoin A French equivalent of an OUTLIER.

buttress A rugged, protruding rocky ridge or face on a mountainside, analogous to an architectural buttress against the wall of a building; e.g. Milestone Buttress on Tryfan, a mountain in N Wales.

Buys Ballot's law A law, enunciated in 1857 by C. H. D. Buys Ballot, to describe GEOSTROPHIC FLOW. If an observer in the N hemisphere stands with his back to the wind, the atmospheric pressure will be lower to his left than to his right – the reverse being true for the S hemisphere. The law implies that, in the N hemisphere, the winds blow anticlockwise round a low-pressure system and clockwise round an anticyclone; the converse is true of the S hemisphere. *See also* CORIOLIS FORCE, FERREL'S LAW.

bypassing The passage of mass, energy or information around a system component or store.

bysmalith A more or less vertical and cylindrical igneous intrusion which has been injected by pushing up the overlying strata along marginal faults. Its surface expression is similar to that of a PLUG.

byte A unit of computer storage of BINARY data, usually comprising eight BITS, equivalent to a character. *Megabyte* = one million bytes; *gigabyte* = one thousand million bytes.

C

caatinga A Portuguese term which has been widely used to denote the type of thorny, tropical woodland which is found in NE Brazil. It is characterized by xerophilous (XEROPHYTE) species, including acacias, euphorbias and cacti, which intermingle to form a tangled, virtually impenetrable mass of vegetation.

cadastral map A map showing ownership of land, usually drawn at a large scale so that all individual landholdings may be accurately depicted.

Cainozoic (Cenozoic, Kainozoic) A Greek term meaning 'recent life', adapted to describe the third of the eras of geological time. Originally, the term was regarded as being synonymous with the TERTIARY, i.e. succeeding the MESOZOIC and finishing at the QUATERNARY. Such usage is obsolescent and the Cainozoic is now held to include the Quaternary. Thus, it is now divided into two time PERIODS (SYSTEM), the Tertiary and the Quaternary, and therefore includes all the EPOCHS (SERIES) from Palaeocene to Recent. Modern geologists are tending to replace the terms Tertiary and Quaternary with the terms PALAEOGENE and NEOGENE.

cairn A Gaelic word used: **1** to denote an artificial pile of stones erected on a mountain summit or along a mountain path as a guide; **2** as a place-name or part of a place-name in mountainous districts; e.g. Cairngorm. Also *carn*.

calcareous Containing calcium carbonate ($CaCO_3$).

calcareous ooze OOZE.

calcareous rocks Those sedimentary rocks containing high proportions of calcium carbonate ($CaCO_3$); e.g. carboniferous limestone.

calcareous soils Soils containing sufficient calcium carbonate to effervesce visibly when treated with cold 0.1 N hydrochloric acid.

calcicole A plant which requires a lime-rich soil. *See also* CALCIPHYTE.

calcification A soil-forming process in which calcium carbonate accumulates in the B-HORIZON; particularly characteristic of arid and semi-arid climates.

calcimorphic soils Soils which have developed on a calcium-rich parent material. *See also* RENDZINA.

calciphobe, calcifuge A plant which cannot tolerate lime in the soil and flourishes only on ACID SOILS. A calcifuge can just about survive on a lime-rich soil but will not flourish, while a calciphobe (e.g. an azalea) will rapidly die under similar circumstances.

calciphyte A plant that tolerates or requires a large amount of calcium in the soil. *See also* CALCICOLE.

calcite The crystalline form of calcium carbonate ($CaCO_3$) and the principal constituent of limestone. It forms on stream beds and in caves as STALACTITES, STALAGMITES and TUFA. *See also* SPELEOTHEM.

calcrete A sedimentary deposit composed of coarse rock fragments cemented by calcium carbonate. It forms when lime-rich groundwater rises to the surface by capillary action and evaporates into a crumbly powder or into a tough, indurated sheet of variable thickness. *See also* CALICHE, SILCRETE.

caldera A large, circular, basin-shaped volcanic depression created in one of three ways: **1** by destruction of the upper part of the volcanic cone by an eruption of great force, e.g. Katmai in Alaska; **2** by collapse of the volcanic cone inwards (ENGULFMENT); **3** by gradual reduction of an extinct or dormant volcano by erosion. The criterion is that the diameter of the caldera should be many times that of the original volcanic vent; thus, some calderas are wrongly referred to as CRATERS (e.g. Ngorongoro, Tanzania). [*38*]

Caledonian orogeny The mountain-building episode of late-Silurian/early-Devonian times. In W Europe the remnants of the Caledonian DIASTROPHISM can be traced along a NE to SW alignment from Scandinavia, through Scotland (hence Caledonian) into NW Ireland. The worn-down relics are referred to as *Caledonides* and form the chief LINEAMENTS of highland Britain. *See also* ARMORICAN, ALPINE. [*104*]

calf **1** A piece of floating ice which has broken away from a larger piece of land or sea ice. **2** An islet or small island adjacent to a larger island; e.g. Calf of Man, Isle of Man.

calibration **1** The act of graduating any kind of gauge, with due allowance for its irregularities. **2** The act of determining the calibre of an instrument, such as a thermometer. **3** In geology, the determination of rock sequence against a time-scale.

caliche A Spanish term with several meanings: **1** Most commonly, an impure deposit of sodium nitrate ($NaNO_3$) which accumulates as a surface crust in soils of arid lands. It is especially common in the Atacama desert of Chile and Peru where it has been extensively mined as a source of nitrogenous compounds. **2** In Colombia it is used both as a term for a newly discovered mineral vein and as a mining term to describe a clay, sand and gravel deposit. **3** In Mexico it is used to describe a compact, white limestone. **4** In SW USA it is regarded as synonymous with CALCRETE. **5** In Mexico the term is also applied to a white, feldspathic clay.

California current A cold current which flows southwards along the Pacific coastline of the USA, caused by the upwelling of colder water from greater depths due to the southward deflection of the North Pacific Current and the transference of surface water westwards across the Pacific as the North Equatorial current. This cold current causes a marked equatorward bend of the global isotherms and has a marked effect on the coastal climate of Oregon and N California (including the sea fogs off San Francisco). [*168*] *See also* CANARIES CURRENT.

caliper log A technique for recording the varying diameter with depth of an uncased borehole.

calms The absence of appreciable wind, represented by force 0 on the Beaufort scale (velocity <1 knot). Calms can occur at any time in any latitude, especially during anticyclonic spells, but are most common in the HORSE LATITUDES (the subtropical high-pressure belts) and the DOLDRUMS.

calorie (gram calorie) The amount of thermal energy required to raise 1 g of water at sea-level pressure from 14.5° C to 15.5° C. 1 calorie = 3.968×10^{-3} BTU. (*See also* SI UNITS, JOULE (J).) 1 calorie = 4.186 J; 1 calorie = 697.3 W m^{-2}.

camber, cambering A downward bending or draping over of a hard horizontal rock stratum at the edge of its outcrop, especially at the side of a valley (valley-side cambering). It is due to the valleyward flow of an underlying clay layer, the plastic properties of which are particularly susceptible to SOLIFLUCTION and periglacial disturbance. It is particularly common in the Cotswold valleys of S England where the overlying competent oolitic rocks (OOLITES) have draped over the valley edges and collapsed into landslips owing to the incompetence of the underlying Liassic clays (LIAS). [*33*]

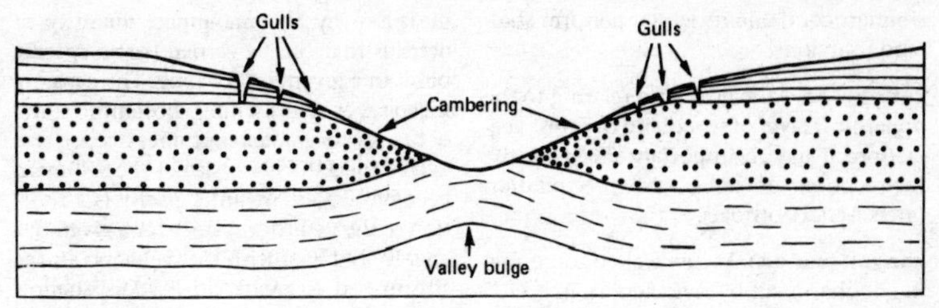

Figure 33 *Camber*

Cambrian The earliest geological period of the PALAEOZOIC ERA. It was named from the ancient name for Wales by A. Sedgwick in 1836 (although his rock system has subsequently been divided into Cambrian and ORDOVICIAN). It includes rocks which were formed between about 570 million years and 510 million years ago. In the British Isles, Cambrian rocks are common in N Wales, the Welsh borders, NW Scotland and E Ireland. Cambrian rocks were the first to show an abundance of fossils, largely primitive members of most of the invertebrate animal phyla known today. The argillaceous rocks have, in places, been converted by metamorphism to SLATE, which has been commercially quarried on a large scale in N Wales.

Campbell's law A law relating to tectonic upwarping and its effect on river drainage systems. It states that, where two streams that head in opposite directions are affected by an even lengthwise tilting movement, the one whose declivity is increased cuts down more vigorously and grows in length headwards at the expense of the other. If the tilting that affects them is part of a general warping, the divide migrates towards the axis of upwarping. This general law that axes of upwarping become drainage divides is inapplicable in the case of an ANTECEDENT river.

Campbell–Stokes recorder SUNSHINE RECORDER.

campo **1** A SAVANNA vegetation in central Brazil where the ratio of trees to grass allows a division into: *campo cerrado*, which has scattered trees, and *campo limpo*, which has tall grass and virtually no trees. **2** An intermontane, fault-guided depression in Argentina.

canal **1** An artificial watercourse cut for purposes of navigation or irrigation. **2** A long narrow arm of the sea connecting two larger stretches of sea. **3** An underground water-filled cave passage.

Canaries current A cold current which flows southwards past Madeira and the Canaries off the Atlantic coast of N Africa. It is caused by the upwelling of colder water from greater depths owing to the southward deflection of the WEST WIND DRIFT in the North Atlantic and the transference of surface water back across the Atlantic by the North Equatorial current. [*168*] *See also* CALIFORNIA CURRENT.

Cancer, Tropic of The tropic of the N hemisphere, at 23°32'N latitude, at which the Sun's rays are vertical at midday about 21 June.

cannel coal A greyish, non-lustrous coal which burns easily with a smoky yellow flame, leaving behind a high amount of ash. It is a fine-grained coal composed of algal rather than woody remains. *See also* BOGHEAD COAL, COAL.

cannon-shot gravel A coarse gravel deposit characterized by large, almost perfectly rounded stones, thought to have been laid down by Pleistocene meltwater streams at ice-sheet margins.

canonical structure The representation of an input/output process in terms of the simplest components which perform standard functions.

canopy An ecological term relating to the uppermost layer of woodland or forest vegetation. It has an important effect on the interception of light and precipitation (INTERCEPTED MOISTURE).

canyon (cañon) A steep-walled gorge, ravine or chasm cut by river action, in which the depth considerably exceeds the width. Such a feature is common in areas of low precipitation, which inhibits local denudation of the valley sides, but where the river is supplied by an external source enabling downcutting to continue. The most striking canyons are produced in areas of horizontally bedded strata where alternating treads and steep risers are characteristic (e.g. the Grand Canyon, USA). Some canyons are produced in limestone terrain by collapse of roofs of underground streams.

capability class, of land A rating that indicates the capability of land for a particular type of use, i.e. forestry, arable farming, etc. A *capability subclass* may indicate lands with similar limitations and hazards. Both ratings are extensively used in land use and conservation planning.

capacity The ability of a current (either water or wind) to transport sediment, as measured by the maximum quantity of detritus that can be carried past a specific point in a given unit of time. The capacity increases as the discharge becomes greater or the stream gradient becomes steeper, and decreases as the grain size of the sediment becomes larger. Stream capacity is a function of the bed width, since for a given discharge and gradient the velocity at the stream bed is greater in a wide, shallow stream than in a narrow, deep one.

cape A headland or promontory of significant size jutting into the sea.

capillarity (capillary movement) The mechanism by which CAPILLARY WATER moves upwards in the soil and subsoil via the hair-like diameter spaces of the capillary tubes. The movement is greatest where evaporation exceeds precipitation and is therefore more typical of arid and semi-arid lands than humid lands. Capillarity is more effective in clays than in sands (a maximum rise of 2.5 m can be expected in the former in contrast to one of 0.7 m in the latter). The zone of movement is termed the *capillary fringe*, and this is the zone in which dissolved salts are deposited, leading to salt accumulation (SOLONCHAK, SOLONETZ) which may be exacerbated by irrigation schemes. *See also* SOIL WATER. [34]

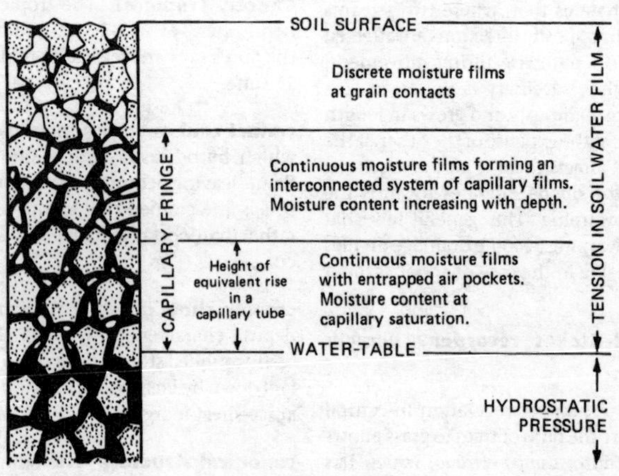

Figure 34 *Capillarity*

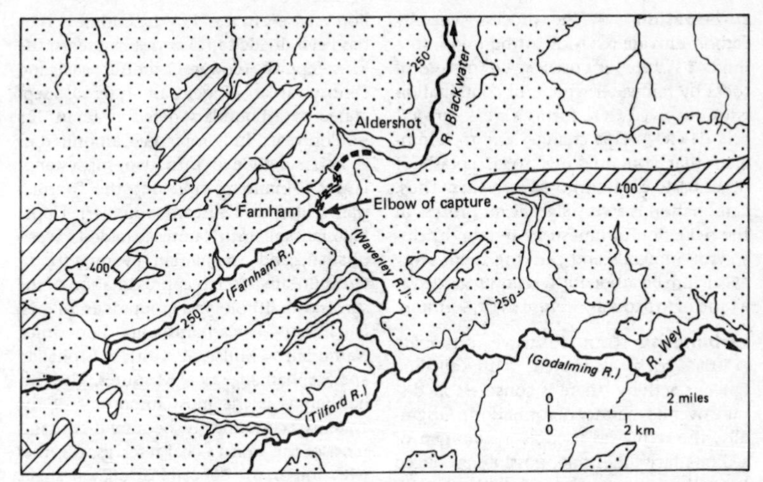

Figure 35 *Capture of the headwaters of the R. Blackwater by the R. Wey*

capillary fringe CAPILLARITY.

capillary water The portion of soil water held by cohesion within the PORES and as a continuous film around the soil particles. This is the water available for plant rootlets, in contrast to gravitational water, which flows down through the soil and is responsible for LEACHING the soil nutrients, and in contrast to HYGROSCOPIC WATER, which remains in the soil even after drying and is not available to plants. The term is now gradually becoming obsolete. *See also* GRAVITATIONAL WATER, SOIL WATER.

capillatus A cloud species (CUMULONIMBUS) characterized in the upper portions by the presence of fibrous or striated cirrus, in the form of an anvil or plume or mass of hair (Latin *capillatus* = having hair). *Cumulonimbus capillatus* is usually accompanied by a thunderstorm with hail and violent squalls.

Capricorn, Tropic of The tropic of the S hemisphere, at 23°32'S latitude, at which the Sun's rays are vertical at midday about 21 December.

cap-rock **1** A layer of hard, resistant rock which forms the flat summit of a hill (BUTTE, MESA) and protects underlying, less-resistant strata. **2** A virtually impermeable stratum overlying an oil or natural gas reservoir, an AQUIFER, or a SALT DOME. **3** A type of DURICRUST. **4** A mass of barren rock overlying an ore-body.

capture, river The diversion of the headwaters of a river system into a neighbouring system, leaving a beheaded stream in the original system. The mechanism which may trigger this act of river piracy is explained in CAMPBELL'S LAW, and the point at which capture takes place may be characterized by an anomalous bend in the beheaded stream, termed an *elbow of capture*. Once the pirate stream has increased its CATCHMENT it will become the master stream, probably becoming more deeply entrenched and therefore perpetuating river capture by virtue of its lower BASE-LEVEL. *See also* ABSTRACTION. [35]

carapace **1** A hard crust at the surface of a soil, especially a CALCRETE or SILCRETE. **2** The upper limb of a RECUMBENT FOLD.

carbon-14 RADIOCARBON DATING.

carbonaceous Pertaining to, or largely composed of, carbon. Among the sedimentary rocks the following can be regarded as carbonaceous: BROWN COAL; LIGNITE; PEAT; and dark-coloured *carbonaceous shale*.

carbonation **1** The process whereby carbon dioxide (CO_2) is introduced into a fluid. **2** A type of CHEMICAL WEATHERING of rocks by rainwater which, in combination with dissolved CO_2, forms a weak *carbonic acid* (H_2CO_3). This changes any rock minerals that contain lime, soda, potash or other basic oxides into *bicarbonates*. Thus, carbonation is a very significant process in the reduction of limestone terrain, by dissolving minerals and carrying them away as calcium bicarbonate. It is now regarded as a form of HYDROLYSIS. *See also* KARST, SOLUTION.

carbonatite An intrusive carbonate (>50%) rock associated with alkaline igneous activity (there is considerable dispute over its manner of formation). Originally, the term was used as a synonym of sedimentary limestone, possibly produced hydrothermally and redeposited as a sedimentary rock. It is now thought that carbonatites were formed either during a magma phase rich in soda and lime or by reaction of a basic magma with existing limestone and dolomite country-rock. The term is *not* synonymous with limestone, since carbonatites occur as central intrusive masses, DYKES, and CONE-SHEETS; they are found in relatively stable continental regions, especially in the East African rift-valley region.

carbon dating RADIOCARBON DATING.

carbon dioxide An atmospheric gas (CO_2), capable of absorbing RADIATION in wavelengths similar to those emitted by the Earth. Thus, it prevents excessive loss of terrestrial radiation and accompanying heat loss. It is thought by some scientists that the rapidly increasing combustion of fossil fuels (coal, oil, etc.) might raise the CO_2 content in the atmosphere to such an extent that it could cause an increase in the mean temperature of the Earth. Other scientists believe that much of the increase is being stored in the oceans. *See also* GREENHOUSE EFFECT, PHOTOSYNTHESIS.

Carboniferous A period of the PALAEOZOIC ERA, following that of the DEVONIAN and preceding that of the PERMIAN. It ranged from about 362 million years to about 290 million years BP. In the USA the Carboniferous has been divided into two subsystems: the Mississippian (Lower Carboniferous) and Pennsylvanian (Upper Carboniferous), separated by a boundary dated at about 325 million years. In Europe the Carboniferous is also subdivided into two subsystems: Lower (Dinantian) split into the Tournaisian (older) and Visean (younger); Upper (Silesian) split into Namurian (eldest), Westphalian, and Stephanian (youngest). In Great Britain a rough approximation of the stages with lithological groups is as follows: Dinantian = Carboniferous Limestone, Namurian = Millstone Grit; Westphalian and Stephanian = Coal Measures. The Carboniferous is the most important of the systems, economically, because it contains most of the world's coal reserves, together with important deposits of oil, oil shale, iron ore and fireclay.

Carboniferous Limestone The lowest lithological division of the CARBONIFEROUS system in Great Britain. Despite its name it includes sandstones and shales. It is divided into the Lower Avonian (= Tournaisian) and the Upper Avonian (= Visean).

carbonization The process of converting to carbon by the removal of other ingredients, and by slow decay or reduction following the death of a plant or animal.

Carbo-Permian The end of the CARBONIFEROUS system and the beginning of the PERMIAN system; a time marked by widespread mountain-building (HERCYNIAN OROGENY) and a glaciation in the S hemisphere (GONDWANALAND).

cardinal points The four main compass directions: north, east, south, west. *See also* POINTS OF THE COMPASS.

carnivora The older of the class MAMMALIA that includes flesh-eating animals, the modern groups of which evolved during the late Eocene and Oligocene. *See also* HERBIVORA.

carr A wooded FEN in a waterlogged terrain where the PH is not too acid and where the soils are not too deficient in mineral

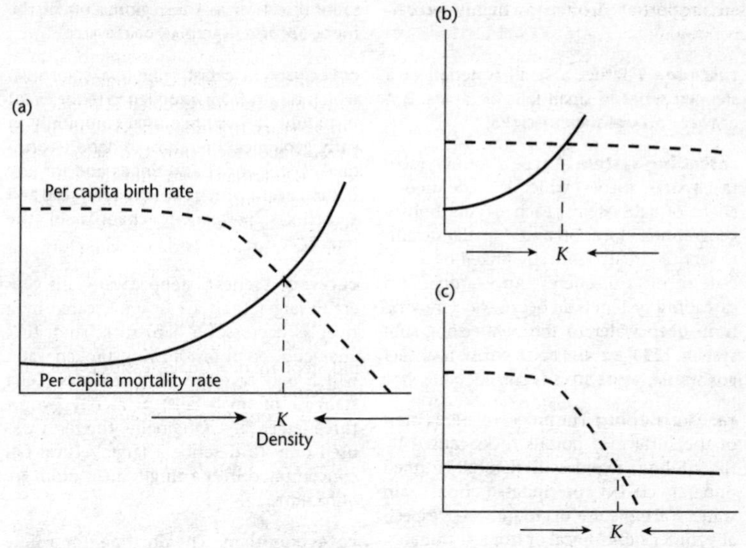

Figure 36 *Examples of carrying capacity*. Source: Begou *et al*. 1986

elements. Characteristic trees include alder, willow and sallow. [*108*]

carrying capacity An ecological term referring to the population size which the resources of a particular environment can maintain without a tendency to decrease or increase. It is usually denoted by the symbol *k*. Three examples are shown in [*36*]. (a) Birth rate and mortality rate are density dependent. (b) Mortality rate is density dependent. (c) Birth rate is density dependent. Below *k* the population increases: Above *k* the population decreases.

carse A Scottish term for the low-lying, alluvial lands along the river estuaries of eastern Scotland. The carse clays are thought to represent deposits in part laid down by a marine transgression in post-glacial times (FLANDRIAN).

carstone A brown sandstone with a LIMONITE cement.

cartographic analysis The scientific examination of maps and charts in order to collect data for subsequent manipulation, statistical testing or graphic representation.

cartography The art and science of map and chart construction. It draws upon TRIANGULATION SURVEYS, AERIAL PHOTOGRAPHS and data from existing maps (COMPILED MAPS). It includes the production of all types of maps and charts, e.g. topographical, geological, pedological, etc. in addition to thematic maps, e.g. world energy reserves, natural-resource maps, etc. HYDROGRAPHY and PHOTOGRAMMETRY are not accepted in Britain as part of cartography. A more recent definition would be: the organization and communication of geographic information (GIS) in either graphic or digital form. *See also* AUTOMATED CARTOGRAPHY, BATHYMETRIC, MAP PROJECTIONS.

cartometric testing A technique to assess the accuracy of the information portrayed by a map or chart in relation to a control. Mathematical formulae are used to test the horizontal and vertical deviations of the cartographic data from their true location or elevation on the ground.

cartouche A type of decorative key or panel on a map or chart in which information such as distance, scale, title, date, etc. are portrayed, often in a highly decorative way.

cascade **1** Either a small waterfall or a stepped series of small falls or rapids. **2** A GRAVITY-COLLAPSE STRUCTURE. [*96*]

cascading system A type of system (GENERAL SYSTEMS THEORY) which is made up of a chain of subsystems, each having both a geographical location and a spatial magnitude, that are dynamically linked by a cascade of mass or energy. An example of a cascading system is an ocean wave moving from deep water to the SWASH zone subsystem. [*236*] *See also* BLACK BOX SYSTEM, GREY BOX SYSTEM, WHITE BOX SYSTEM.

case hardening The process of INDURATION of the surface of porous rocks caused by the infilling of voids with naturally formed mineral CEMENT precipitated from rain water, soil moisture, or groundwater, especially under sub-tropical or tropical climates. If the surrounding case is breached cavernous weathering (TAFONI) may occur.

castellanus (castellatus) clouds Clouds which tower into castle-like profiles, with cumiliform protuberances in the form of turrets.

castle kopje A small-scale INSELBERG in which the joint pattern of the deeply weathered and subsequently exhumed rock has influenced the shape of the summit. Thus, the Afrikaans term *kopje*, which means isolated hill, is described as a castle kopje when its summit blocks and joints create a castellated appearance. The term has also been used to describe a tumbled mass of loose blocks, which may represent a collapsed inselberg or simply a collection of exhumed CORESTONES. *See also* BASAL SURFACE OF WEATHERING, TOR.

cataclasis Rock deformation accomplished by fracture and rotation of mineral grains, as in the production of a CRUSH-BRECCIA (BRECCIA). A rock formed in this way is termed a *cataclasite*.

cataclinal A rarely used geomorphological term relating to rivers which flow in the same direction as the regional DIP of the rocks. *See also* ANACLINAL, DIACLINAL.

cataclysm A catastrophe (CATASTROPHISM) which results from a sudden extreme event in nature. It was once used commonly in early geological literature to denote earthquakes, volcanoes and floods and for any natural event which caused widespread and sometimes permanent changes in the terrain.

cataract A series of stepped waterfalls, generally larger than a CASCADE, created by a river as it crosses a hard rock band. It is characterized by fast-flowing, broken water which may take the form of *rapids* (i.e. a staircase of small falls) or merely two or three larger falls. Originally, the term was used only to describe a large vertical fall concentrated into a single sheer drop. *See also* CHUTE.

catastrophism The doctrine, for a time generally discarded, that past geological changes (often exemplified by contrasts in the FOSSIL record of successive stratigraphic horizons) were caused by sudden catastrophic changes in the operation of natural systems rather than by slow, continuous processes (UNIFORMITARIANISM). Thus, a series of natural disasters was believed to have caused the extinction of certain organisms and their replacement by alternative forms. In the late 20th century a theory of *neo-catastrophism* emerged, challenging the views of slow, almost imperceptible change inherent in uniformitarianism, and highlighting the fact that many geological and geomorphological processes are episodic (EARTHQUAKE, FLOOD, LANDSLIDE, VOLCANO), in which earth processes change from long periods of relative stability to short periods of instability, during which critical limits, boundary conditions, or yield points (termed *thresholds*) are crossed and the strength of materials exceeded. *See also* CATACLYSM, STEP FUNCTIONS.

catchment area In British usage the term refers to the total area from which a single river collects surface RUNOFF. In the USA the term *watershed* is used in this context, and

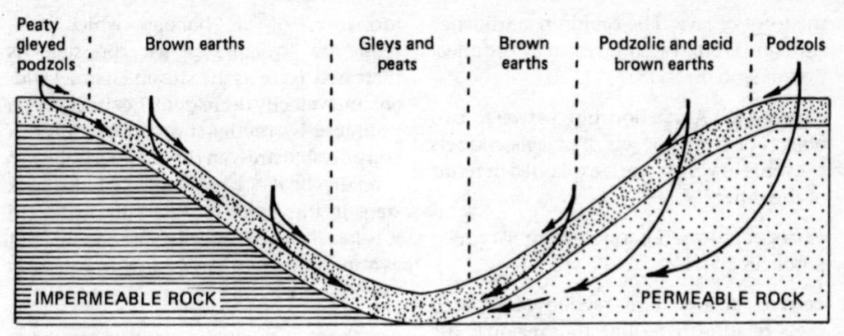

Figure 37 *Catena*

the term *catchment area* is reserved to describe both the intake area and all areas which contribute surface water to the intake area of an AQUIFER. In the UK the term *catchment* has been used loosely by river authorities that are responsible for the water supply, drainage and flood control, since the statutory boundaries of their catchment boards sometimes extend beyond a single river basin. *See also* SOURCE AREA.

catena (toposequence) A repeated sequence of SOIL PROFILES that is geographically related to and associated with relief features. The term is derived from the Latin (= chain), since the soil profiles were found to be linked together in the same fashion as a hanging chain when traced laterally down a slope from the INTERFLUVE to the valley floor. The profiles change character as they are traced along the traverse in accordance with changing slope angle, drainage conditions, etc., so that different degrees of LEACHING and TRANSLOCATION are found, for example. Geomorphic history of the land surface is also involved in the catena concept. [*37*]

catenary curve The curve formed by a chain of uniform density hanging freely from two fixed points not in the same vertical line.

cation An ION carrying a positive charge of electricity (*positive ion*) that moves by electrolysis towards a cathode (negative pole). *See also* ANION.

cation exchange BASE EXCHANGE.

cation exchange capacity The total potential of soils for absorbing CATIONS, expressed in milligram equivalents per 100 g of soil. *See also* BASE EXCHANGE, BASE SATURATION.

cauldron subsidence A geological structure which results from the collapse of a more or less cylindrical block of country-rock into an underlying MAGMA chamber along an encircling ring-fracture of steep inclination. In some cases the ring-fracture does not reach the surface, so that the structure remains entirely subterranean until it is uncovered by subsequent erosion. In other cases the ring-fracture reaches the surface, resulting in the outpouring of magma in

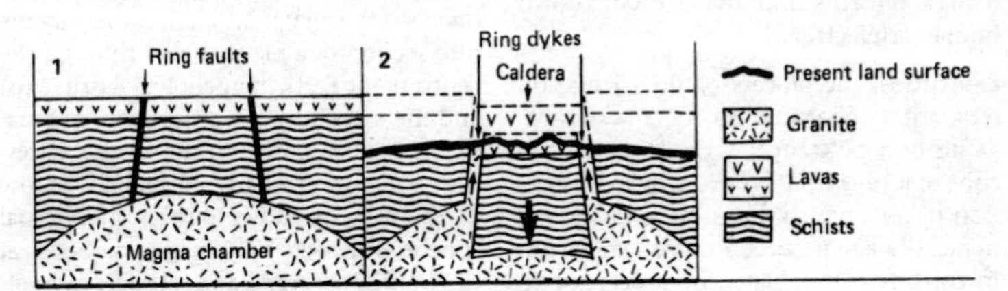

Figure 38 *Cauldron subsidence*

the form of lava. The cauldron subsidence structure is often accompanied by the injection of RING DYKES. [*38*]

causality A relationship between two objects or events or sets of objects and sets of events in which one is explained in terms of the other.

cause A process (i.e. a system input) which produces an EFFECT.

causse A local term, synonymous with KARST, relating to the limestone region of the Central Massif of France.

cave **1** A natural cavity, chamber or recess which leads beneath the surface of the earth, generally in a horizontal or obliquely inclined direction (in contrast to a shaft ABÎME). It may be in the form of a passage or a gallery, its shape depending in part on the joint pattern or structure of the rock and partly on the type of process involved in its excavation. Thus, caves worn by subterranean rivers may be different in character from, and of considerably greater extent than, a sea-cave eroded by marine waves. Cave systems in limestone (KARST) owe very little of their genesis to the SOLUTION process. **2** A cavity or recess in a glacier often created by an underground stream, possibly as the result of enlargement of a CREVASSE (ice-cave). **3** An artificially excavated aperture into a hillside, usually in search of mineral wealth. **4** Used as a verb (*caving*) to indicate the act of cave-exploration (*speleology*). **5** Used as a verb to express the collapse of a gallery or working face in a mine or quarry (*cave in*). *See also* SPELEOTHEM.

cave earth A deposit of clay, silt, sand or gravel infilling or flooring the passage of a CAVE and sometimes containing mammalian remains and the earliest known human artefacts.

cavitation A process of fluvial erosion (CORRASION) characteristic of a WATERFALL, RAPIDS or a CATARACT (i.e. the high-velocity zone of a stream). It is caused by a constriction of flow which raises the velocity and hence the KINETIC ENERGY of a stream. This, in turn, is compensated by a decrease in the pressure of the water, leading to the formation of air bubbles which subsequently collapse when pressure is increased again as the stream channel widens and velocity decreases. The shock waves produced by the bursting bubbles exert a considerable stress on the channel walls and a relatively rapid breakdown of the rock steps in the rapids or waterfalls. Although it is locally a very effective process it is not as important a mechanism in fluvial erosion as corrasion.

cay (key) A small, flat island composed of a bank of sand overlying a coral REEF, just above high tide level. It is particularly common in the Gulf of Mexico, especially off the S coast of Florida (the Florida Keys).

cedar-tree laccolith A type of LACCOLITH in which the igneous intrusion is fed by a single pipe from the underlying magma chamber but branches into a series of laccoliths, one above the other, thus forming a shape which approximates to that of a cedar-tree. [*129*]

ceiling **1** Often refers to the level of the cloud base, in its atmospheric context. **2** In aeronautics the term is used to delimit the maximum altitude to which an aircraft or a balloon can ascend. **3** In a mountaineering context it refers to the physiological limit attainable by a climber without oxygen assistance.

celestial sphere A term used to describe the so-called 'bowl' of the heavens surrounding the planet Earth which lies at the centre of this imaginary sphere. The radius of the sphere is infinite but it is used as a basis for navigation and in astronomy. The point on the sphere directly above the observer is known as the *zenith*. The *celestial equator* is the imaginary circle formed by the intersection of a plane drawn through the centre of the Earth (perpendicular to its axis) and the celestial sphere; thus, in a celestial sense, it corresponds to the Earth's terrestrial equator. Similarly, the plane of the Earth's axis, extended to meet the celestial sphere, intersects this imaginary 'bowl' at the *celestial north pole* and *celestial south pole* respectively.

cell The basic element of information in the RASTER description of spatial entities. *See also* GIS, GRID.

cell, atmospheric ATMOSPHERIC CIRCULATION.

Celsius scale A temperature scale devised by the Swedish astronomer, A. Celsius, in 1742. It divides the interval between the freezing-point and the boiling-point of water into 100 parts, the lower fixed point being marked 100. The present system, in which freezing-point is marked 0 and boiling point 100, was introduced by Christin in 1743. *See also* ABSOLUTE SCALE, CENTIGRADE SCALE, FAHRENHEIT SCALE.

cement **1** In its natural state the term refers to a siliceous, ferruginous or calcareous material that has been chemically precipitated in the interstices of CLASTIC rocks or loose particles of ALLOGENIC deposits, thereby binding them together to form compact rocks, some of which have considerable resistance to weathering and erosion (*see also* DIAGENESIS, LITHIFICATION). **2** A term used for an industrial product (PORTLAND CEMENT) obtained by burning a mixture of pulverized materials containing lime, silica and alumina in varying proportions and by finely grinding the resulting fused clinker.

cementation DIAGENESIS.

cementstone Argillaceous limestone, containing an abundance of calcareous and siliceous minerals together with varying amounts of magnesium carbonate and alumina. It is sometimes quarried for crushing in the CEMENT industry.

Cenozoic CAINOZOIC.

centigrade scale A temperature scale in which 0° C represents the melting-point of ice and 100° C the boiling-point of water. Officially replaced by CELSIUS since 1948. °C = 5/9 (°F – 32°); 100° C = 180° F (212° – 32°). *See also* FAHRENHEIT SCALE, RÉAUMUR SCALE.

central eruption A volcanic outburst or eruption from a single source (VENT) or a closely linked group of vents, as opposed to the volcanic activity of a FISSURE ERUPTION in which material is ejected not from a point but from a linear feature. Hence, a CENTRAL-VENT VOLCANO. *See also* VOLCANO.

central meridian The line of longitude (MERIDIAN) upon which a MAP PROJECTION is constructed. In general it is the central axis of the projection. This term should not be confused with the standard or primary meridian (GREENWICH MERIDIAN).

central tendency The tendency of individual members in a variable population to cluster morphologically around the mean type.

central-vent volcano A VOLCANO which has one single orifice or VENT, around which volcanic debris has accumulated to build up a more or less symmetrical cone. [*271*]

centrifugal force Although this is a hypothetical rather than an actual force, a body moving on a highly curved path is subject to centrifugal force acting radially outwards from the centre, e.g. a HURRICANE, where maximum wind velocities are found in a ring away from the calm central 'EYE'.

centripetal acceleration Acceleration of a body moving on a curved path, equal and opposite to the CENTRIFUGAL FORCE, directed to the centre of curvature.

centripetal drainage pattern A pattern of drainage in which the streams drain radially inwards, either towards a single trunk river which drains the basin, or to a lake which may or may not have an outlet (*see also* ARETIC DRAINAGE, INTERNAL DRAINAGE). In the UK, the rivers which drain into Lough Neagh, N Ireland, are described as part of a centripetal drainage pattern, resulting from the down-sagging of the central part of the Antrim Plateau owing to the considerable weight of the underlying basaltic lavas. *See also* ENDORHEIC DRAINAGE.

centripetal force CENTRIPETAL ACCELERATION.

centroclinal A geological term relating to a structure in which the strata dip inwards to a central low point, sometimes referred to as a *centroclinal structure* or *centrocline*. *See also* PERICLINE.

cerrado A type of SAVANNA found in Brazil, in which trees (9–15 m in height) are intermixed with tall grassland.

CFCs CHLOROFLUOROCARBONS.

chain **1** A linear sequence of interconnected or related natural features, e.g. islands, lakes and particularly mountains. **2** A unit of length equivalent to 66 ft, originally named from a series of links connected by rings, used in land survey, hence the verb, *chaining*. (Ten square chains = 1 acre.) It has generally been replaced by the metal or linen measuring-tape.

chalk A soft, permeable, white to light-grey, amorphous limestone. It is a very pure calcium carbonate ($CaCO_3$) and is composed of the shells or invertebrate skeletal fragments of marine micro-organisms, such as FORAMINIFERA, held in a matrix of finely crystalline calcite. Its genesis is a matter of dispute, for originally it was thought to be entirely of organic origin, possibly a lithified calcareous mud or abyssal OOZE. It is now believed that, although some types of chalk were laid down in shallow seas and are composed mainly of shallow-water marine shells, there are other instances in which the chalk is clearly the product of chemical precipitation and has no organic content. Some chalk deposits are characterized by bands or nodules of very hard siliceous material known as FLINT. Chalk is a very important AQUIFER and much of the domestic water supply of SE England is derived directly or indirectly from the CHALK. [*16*] It forms very distinctive scenery, termed downlands (DOWNS) with low rolling hills, steep scarps, DRY VALLEYS and COMBES. The thin soils support mainly grassland, composed of short turf, except where there is a thick cover of CLAY-WITH-FLINTS, which is capable of supporting natural or introduced woodland. *See also* BOURNE, COOMBE ROCK.

Chalk, the A proper name given to the limestones, marls and flints which form the upper stratigraphic beds of the Upper Cretaceous system, laid down between 100 and 65 million years ago. The following sequence is characteristic of S England: **1** *Upper Chalk* = white chalk, some flints (FLINT) and the *Chalk Rock*, a well-jointed hard limestone; **2** *Middle Chalk* = soft, white chalk, some marl, a few flints and a well-jointed limestone termed the *Melbourn Rock*; **3** *Lower Chalk* = greyish chalk and CHALK MARL.

chalk marl The basal horizon of the Lower Chalk (THE CHALK), comprising an admixture of calcareous (70%) and argillaceous (30%) material known as MARL.

chalk rock A layer of hard limestone at the base of the *Upper Chalk* in the Wessex downs and Chilterns. It is not present in the N. Downs.

chalky boulder-clay A greyish TILL, containing a high proportion of CHALK and many FLINTS, which mantles much of East Anglia in E England. Some authorities regard it as a single till unit but others classify it as indicative of two glacial advances in the Pleistocene. *See also* ANGLIAN.

chalybeate Spring-water which contains a high concentration of iron compounds and is reputedly of therapeutic value. Thus, it is commonly utilized in spa towns.

chamaephytes One of the six major floral life-form classes recognized by the Danish botanist, Raunkiaer, based on the position of the regenerating parts in relation to the exposure of the growing bud to climatic extremes, and totally irrespective of taxonomy. The chamaephytes are herbaceous or woody plants with buds produced close to the soil. *See also* CRYPTOPHYTES, HEMICRYPTOPHYTES, PHANEROPHYTES.

chañaral A type of thorny scrub vegetation found in the south central areas of S America. Derived from the Spanish word *chañar*, a thorny bush.

Chandler wobble The small wobble of the Earth during rotation around its axis, discovered in 1891 and found to have an amplitude of 0°5' and a periodicity of fourteen months. Although it has not been explained it is thought to be linked with earthquake activity. *See also* EARTHQUAKE.

channel **1** An extremely linear three-dimensional object, usually bounded only on the bottom and sides; its usage has been extended to include any means of communication along which materials or information may pass. **2** The trough-like form which contains a river and is shaped by the force of water flowing along it. Its shape is capable of accurate measurement (CHANNEL GEOMETRY, *see also* FORM-RATIO, GRADIENT, HYDRAULIC RADIUS, WETTED PERIMETER). **3** The main routeway or shipping lane, deep enough for navigation in an otherwise shallow estuary or coastal sea area. **4** A very wide STRAIT linking two seas, e.g. the English Channel. **5** An artificially constructed drainage or irrigation ditch. **6** As a verb it is used to denote the action of cutting into an object to produce a trough, or to denote the feeding of information along a path into a system.

channel capacity The maximum volume of flow of a river within its CHANNEL without overtopping its banks (BANK-FULL STAGE). *See also* CHANNEL STORAGE.

channel-fill deposit A sedimentary deposit formed in a river CHANNEL when the transporting capacity of the river has been incapable of removing the detrital material as rapidly as it has been delivered.

channel flow That part of the surface RUNOFF of water which is confined in a river CHANNEL in contrast to that which spreads laterally over wider and less confined areas. *See also* OVERLAND FLOW, THROUGHFLOW, UNIFORM STEADY FLOW.

channel geometry A term used in hydrology and in fluvial geomorphology to describe the spatial properties of a river channel. These include the width, depth slope gradient, bed ROUGHNESS and WETTED PERIMETER of the CHANNEL. *See also* HYDRAULIC GEOMETRY, WIDTH–DEPTH RATIO.

channel resistance The resistance encountered by water flowing in a river channel. Bank vegetation, channel form and ROUGHNESS combine to add to flow resistance. *See also* CHEZY EQUATION, MANNING EQUATION.

channel storage The volume of water within a given section of a river CHANNEL, above a given measuring-point at a given time. If a river is receiving inflow more rapidly than can be moved downstream it will eventually reach CHANNEL CAPACITY, which is equivalent to its maximum channel storage.

chaos theory A mathematical theory referring to the fact that seemingly irregular, unpredictable phenomena can arise from the sensitivity of some well-defined differential equations to small changes in their initial conditions. It is applied particularly in METEOROLOGY and studies of TURBULENCE in hydrological studies.

chaparral A type of evergreen scrub vegetation, common in SW USA and NW Mexico. It comprises poor grasslands interspersed with evergreen oaks and its name is derived from a Spanish term. *See also* MAQUIS.

characteristic angles The most commonly occurring slope gradients under a specific climate or on a particular type of rock. Over a wide range of contrasting environments it has been found that the most commonly occurring angles are (°): 1–4; 5–9; 25–26; 33–35.

characteristic curve A REMOTE-SENSING term referring to a curve which shows the relationship between exposure and the OPTICAL DENSITY or tone of a photographic IMAGE, usually plotted as the optical density (*D*) against the logarithm of the film exposure (log *E*). It is also called the *D* log *E* curve, sensitometric curve and the *H* and *D* curve. The form of the curve determines many photographic characteristics, e.g. tonal

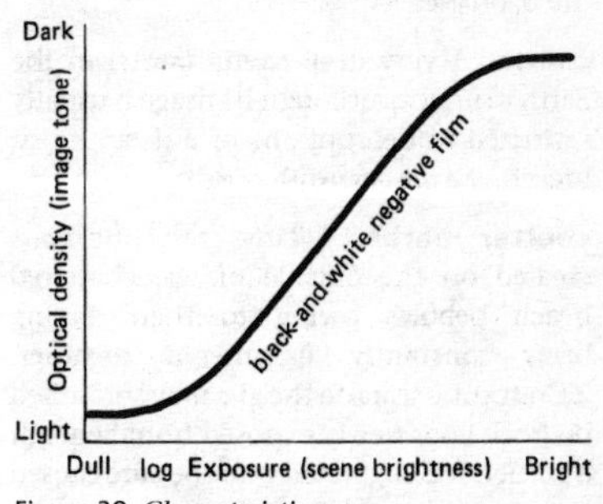

Figure 39 *Characteristic curve*

range and contrast. A manipulation of the curve enables an AERIAL PHOTOGRAPH to be enhanced (IMAGE ENHANCEMENT) in order to facilitate interpretation. *See also* DENSITOMETER. [*39*]

Charles's law A law of mechanics which describes the behaviour of atmospheric gases. Charles's law states that, at a constant pressure, volume (*V*) varies directly with absolute temperature (*T*, measured in °K or degrees Kelvin); $V = kT$. *See also* BOYLE'S LAW, DALTON'S LAW.

Charnian **1** An OROGENY of late Pre-Cambrian times, taking its name from the Pre-Cambrian rocks of Charnwood Forest, Leicestershire, England. **2** A stratigraphic division of the volcanic Pre-Cambrian rocks of the English Midlands.

chart **1** A specialized map of a coastline and its adjoining seas to facilitate marine navigation. It depicts depths of water, lighthouses, TIDAL CURRENTS, etc. *See also* BATHYMETRIC. **2** A specialized map to assist aviation – termed an *aeronautic chart*. There are three varieties: one to assist landing, one to assist radio navigation and one which depicts topography, other hazards (pylons, towers, etc.), restricted zones and flight corridors. **3** A weather map (SYNOPTIC CHART) to aid weather forecasting.

chart datum The level or plane to which all soundings on a marine navigational CHART are related. It is sometimes referred to as the *datum plane* or the *reference plane* and is generally taken as the height of the low-water spring tide. In Britain this lies below the ORDNANCE DATUM.

chasm A very deep cavity (ABYSS) in the Earth's surface, although its usage is usually restricted to descriptions of a deep, linear breach. *See also* CANYON, GORGE.

chatter marks **1** Tiny crescentic scars created on the outside of wave-battered beach pebbles owing to their having been constantly hammered together. **2** Crescentic scars on the glacially smoothed bedrock floor newly exposed from beneath a glacier. It is suggested that they are caused by pounding and 'kneading' of the rock floor by boulders loosely embedded in the BASAL TILL. They should not be confused with STRIATIONS.

cheesewring A curiously shaped rock formation in which the uppermost block of a TOR-like feature is supported on a narrow stem, thereby resembling a mushroom. It was named from an oddly shaped granite tor on Bodmin Moor, Cornwall, England. *See also* GARA.

chelate CHELATION.

chelation The process by which rocks and soils decompose or disintegrate through the action of organisms or organic substances. When water passes through leaf litter it acquires an acidic organic extract capable of combining with the metallic CATIONS of the soil (especially aluminium and iron) to form a *chelate*. The strength of the latter varies according to the different types of organic material: pine needles have a greater effect than broadleaf litter which in turn is stronger than HUMUS and PEAT. Chelation is thought to play a major part in the soil process of *podzolization* (PODZOL).

cheluviation A term derived from the combination of CHELATION and ELUVIATION, whereby water containing organic extracts combines with metallic CATIONS in the soil to form a CHELATE. This sesquioxide-rich solution then moves downward through the soil profile (hence *eluviation*) and moves the aluminium and iron into the lower horizons. *See also* LEACHING, PODZOL.

chemical precipitate A sedimentary deposit formed from the material precipitated by SOLUTION or colloidal suspension. It contrasts, therefore, with deposits formed from transported and detrital material (CLASTIC SEDIMENTS).

chemical weathering The processes which lead to the decomposition or breakdown of solid rocks by means of chemical reactions. These comprise: CARBONATION, HYDROLYSIS, OXIDATION, REDUCTION and SOLUTION. The decomposition is due partly to the removal of the natural cementing agents which hold the grains together, and partly to the formation of secondary minerals which are less resistant to erosion than those of the unweathered rocks. Some scientists regard chemical weathering as synonymous with the term *corrosion*, while others restrict the term *corrosion* to the single process of solution. *See also* MECHANICAL WEATHERING, ORGANIC WEATHERING, REDOX POTENTIAL, SULPHATION.

chemotrophic AUTOTROPHIC.

chenier A beach-ridge which has been formed above swamp deposits in a DELTA as a stationary phase or 'still-stand' of a prograding shoreline. It is typical of the Mississippi delta but a British example exists at Sales Point, Bradwell, Essex. A series of ridges is termed a *chenier plain*, found on coastlines of abundant sediment supply and limited tidal range.

chernozem (tschernosem) A term used in soil science, meaning 'black earth' in Russian, and so derived because of its very widespread existence on the STEPPES of the former USSR. It is a PEDOCAL, similar to a prairie soil, and forms in continental interiors of Eur-Asia, N America, the pampas of Argentina, and SE Australia, where winters are cold, summers warm and annual precipitation averages about 500 mm with a slight summer rainfall maximum. It is a very black, crumbly textured low-humus soil rich in carbonates and exhibits a thick A-HORIZON (0.6 to 1 m) characterized by a layer of MULL humus resulting from the decomposition of the prolific natural grassland which is its climax vegetation. The brown or yellow B-horizon has an accumulation of bases and colloids and an excess of calcium carbonate in the pore fillings of its lower part. Because of their richness in plant foods and excellent crumb-structure the chernozems are among the most agriculturally productive in the world and support large acreages of cereals. They are included in the MOLLISOL order of soils (SEVENTH APPROXIMATION).

chert A hard siliceous rock of opaline and/or chalcedonic SILICA formed either from a crystallization of radiolarian and diatomaceous deposits (OOZE) or by precipitation during HYDROTHERMAL ACTIVITY. It is black or dull in colour, splinters easily and fractures along flat planes in contrast to the conchoidal fracture of FLINT, which is a variety of *chert*. It occurs as bands or layers of nodules in sedimentary rocks, especially in limestones. Lyddite is a dense black variety of chert.

chestnut soil A PEDOCAL found on the steppes of the former USSR, the pampas of Argentina, the Great Plains of the USA and the S African veld, but in drier environments than those of the CHERNOZEM. The horizonation of the chestnut soils is not as clearly marked as that of the chernozems and it does not develop to such depths. It has a brown chestnut colour in its A-HORIZON but its texture is platy unlike the crumb structure of the chernozem. Its B-HORIZON is lighter-coloured owing to the concentration of calcium carbonate with some gypsum. Because of its lower rainfall (200–250 mm) it is not as leached as the chernozem and its grassland is sparser. It is used extensively for grazing rather than for cereal growth, but is subject to wind erosion if the sward is over-grazed. It is included in the MOLLISOL order of soils (SEVENTH APPROXIMATION).

chevron crevasse A CREVASSE or fissure in a glacier which has taken on the form of a chevron owing to differential movement of the glacier ice. *See also* CHEVRON FOLD.

chevron fold A very sharp V-shaped fold in layered rocks. *See also* CHEVRON CREVASSE.

Chezy equation An equation used in HYDRAULICS in order to estimate the average velocity of stream flow in relation to the geometry of a natural channel (CHANNEL GEOMETRY). It is denoted as: $V = C\sqrt{(Rs)}$, where V is the average velocity of the stream flow, C is the Chezy coefficient of channel resistance (due to ROUGHNESS), R is the HYDRAULIC RADIUS and s is the slope (HYDRAULIC GRADIENT). For a channel with a given resistance to flow, velocity increases as a function of increased hydraulic radius and increased hydraulic gradient. *See also* MANNING EQUATION.

chilling A term used in agricultural geography to denote the specific number of hours required by certain crops at low temperatures (but not below freezing) in order to achieve successful growth.

chimney 1 A vertical shaft leading up from the passage of a subterranean cave. **2** A vertical cleft in a rock-wall, utilized by rock climbers. **3** A volcanic VENT. **4** In the USA the term is applied to an erosional rock pillar. *See also* AVEN.

china clay A clay mineral formed by the decomposition of feldspars in solid rocks, especially GRANITE, by the process of HYDROLYSIS combined with ALTERATION (HYDROTHERMAL ACTIVITY) caused by the action of gases rising through the crust from a magmatic reservoir. The alteration of the aluminium silicates creates the common clay mineral kaolinite ($2H_2O.Al_2O_3.SiO_2$), from which the white china clay (kaolin) is derived. Its name is derived from its earliest use in porcelain manufacture in Kiangsi province, China. Much of the china clay in Britain is obtained from the St Austell region of Cornwall. It is used for paper-making as well as in the pottery industry. *See also* CHINA-STONE, KAOLIN.

china-stone A hard rock which produces KAOLIN when it is artificially crushed. It is formed from GRANITE which has been partly altered by kaolinization. *See also* CHINA CLAY, KAOLIN.

chine A local term for the narrow, steep-sided gorges cut into the soft, sandy rocks which form some of the coastal cliffs in Hampshire and the Isle of Wight, England.

chinook 1 A warm, dry south-westerly wind, which blows down the eastern slopes of the Rockies in parts of N America, adiabatically warming and thereby causing a rapid temperature rise and snowmelt in spring. *See also* ADIABATIC, FÖHN. **2** A warm, moist south-westerly wind of the Pacific coastal regions of Oregon and Washington. The first definition is the most universally accepted use of the term, albeit the earliest use of the word (derived from the name of an Indian tribe that formerly lived near the mouth of the Columbia R.) was on the Pacific coast of the continent. [*83*]

chi-square test A statistical procedure to measure the significance of the discrepancy existing between observed and theoretical frequencies in a set of possible events. It is identified as x^2, and measures the probability of randomness in a distribution of data.

chloride A compound of chlorine with one other more positive element.

chlorite group The group of hydrous silicates of aluminium, ferrous iron and magnesium, structurally related to the micas. Most of the chlorites are green-coloured and are often found in low-grade regionally metamorphosed rocks (METAMORPHISM) as the product of the alteration of ferromagnesian minerals.

chlorofluorocarbons (CFCs) A class of artificially produced stable compounds, combining atoms of chlorine, fluorine and carbon; used in refrigerators, aerosols propellants and non-flammable foam. They are effective *greenhouse gases* (GREENHOUSE EFFECT) and they are known to have a deleterious effect on the OZONE layer of the stratosphere.

chop A term describing the surface of the sea when broken by small waves, i.e. a choppy sea.

C-horizon The lowest layer of the soil, underlying the B-HORIZON, the little-altered horizon of the parent material from which the true soil (SOLUM) is derived. *See also* SOIL HORIZON, SOIL PROFILE. [*235*]

chorochromatic map A type of thematic map in which qualitative spatial distributions (especially those relating to resources) are depicted by colour-tinting. Contrast a CHLOROPLETH MAP.

chorographic map Any type of map depicting specific continents or nations, particularly those maps at a small scale which are used in most atlases. Political boundaries are frequently depicted on this category of map.

chorological A term referring to matters of space as opposed to time (CHRONOLOGICAL).

choromorphographic map A type of map which depicts the terrain units and morphological classes of any land area, based on shape rather than landform genesis.

choropleth map A type of thematic map in which quantitative spatial distributions are depicted from data that have been computed from mean values per unit area. Contrast a CHOROCHROMATIC MAP.

chott SHOTT.

chron The smallest interval of geological time (CHRONOSTRATIGRAPHY), equivalent to a *chronozone*, which is the body of rock formed during the time period of one chron. When chrons are grouped together they form an AGE. When the term was originally introduced it was intended to refer to an indefinite division of geological time, but this usage is now obsolete.

chrono-isopleth diagram A graph depicting hourly values of pressure, temperature, etc. plotted against their times of occurrence over a specific time interval; similar values are joined by ISOPLETHS.

chronological A term referring to matters of time, especially in orders of occurrence (chronological order).

chronomere Any interval of geological time (CHRONOSTRATIGRAPHY), of no standard uniform duration.

chronosequence A sequence of related soils that differ from each other in certain properties, primarily as a result of time as a factor in soil formation.

chronostratigraphy One of the branches of STRATIGRAPHY. It is concerned with time rather than spatial distributions and lithology (LITHOSTRATIGRAPHY). It includes **1** a set of terms which are applied to intervals of geological time (the chronomeric standard terms) and **2** a set of terms which are applied to bodies of rock which are laid down during these time intervals (the stratomeric standard terms). The time (chronomeric) hierarchy comprises AEON, ERA, PERIOD, EPOCH, AGE and CHRON (in descending order of magnitude). The lithostratigraphical (stratomeric) hierarchy comprises SYSTEM (= period), SERIES (= epoch), STAGE (= age) and CHRONOZONE (= chron). There are no stratomeric equivalents to aeon and era. The individual time intervals are not of standard duration, nor are the rock divisions of uniform magnitude. Although some geologists regard BIOSTRATIGRAPHY as synonymous with chronostratigraphy, most would regard the former as merely one method of calibrating a chronostratigraphic scale. [*40*]

chronozone CHRON, CHRONOSTRATIGRAPHY.

chute 1 A narrow, sloping CHANNEL of steep gradient through which a stream descends rapidly to a lower level. A CATARACT will probably include a number of chutes. **2** In the USA, a narrow channel with a free current, especially on the lower Mississippi R. **3** Occasionally used to describe a narrow passage of water separating an island from the mainland.

cienaga A Spanish/Mexican term used in SW USA to denote an area in which the WATER-TABLE is at or near the ground surface, thereby giving rise to springs or small marshy patches in surface depressions in a semi-arid environment.

cinder cone A conical-shaped hill surrounding a volcanic VENT, formed by an accumulation of fragmented lava which has solidified as it flew through the air (PYROCLASTIC MATERIAL). The cinders are really glassy and vesicular ejecta (SCORIA), varying in size but mainly between 3 and 4 mm in diameter. *See also* ASH CONE.

circadian rhythm The 24-hour rhythm of activity exhibited by the majority of organisms.

circles A type of PATTERNED GROUND, on nearly horizontal surfaces. *See also* NETS, POLYGONS. [*179*]

circulation index A term used in the measurement of processes in global ATMOSPHERIC CIRCULATION PATTERNS. *See also* INDEX CYCLE, ROSSBY WAVE, ZONAL INDEX.

Table 40 *Chronostratigraphy (British sequence)*

Era	*Period*	*Epoch*		*Absolute time-scale*
CAINOZOIC	QUATERNARY	Holocene		11,000 BP
		Pleistocene		Millions BP c. 2
	TERTIARY	Pliocene		5
		Miocene		23
		Oligocene		35
		Eocene		57
		Palaeocene		65
MESOZOIC	CRETACEOUS	Chalk		100
		Upper Greensand and Gault		105
		Lower Greensand		112
		Wealden		145
	UPPER JURASSIC	Purbeck and Portland		
		Kimmeridge, Corallian, Oxford and Kellaway Beds		162
	MIDDLE JURASSIC	Cornbrash, Great Estuarine Series and Oolites		172
	LOWER JURASSIC	Lias		208
	TRIASSIC	Rhaetic	New Red Sandstone	
		Keuper		
		Bunter		245
PALAEOZOIC	PERMIAN	Magnesian Limestone and Sandstone		290
	CARBONIFEROUS	Upper	Coal Measures	325
			Millstone Grit	
		Lower	Carboniferous Limestone Series	362
			Calciferous Sandstone Series	
	OLD RED SANDSTONE (Devonian)	Upper		359
		Middle		370
		Lower		408
	SILURIAN	Generally referred to as Lower Palaeozoic		439
	ORDOVICIAN			510
	CAMBRIAN			570
PROTEROZOIC & ARCHAEOZOIC	PRE-CAMBRIAN AND ARCHAEAN	Including Dalradian (Upper Dalradian is of Lower Palaeozoic age)		Pre-570

circumference of Earth Because the Earth is an oblate spheroid its circumference at the equator differs from that between the poles. The equatorial circumference is 40,076 km (24,902 miles) while that at the poles is 40,008 km (24,860 miles). *See also* GREAT CIRCLE.

circumferential wave A seismic wave that travels parallel to the Earth's surface.

circum-Pacific belt A seismic belt, which girdles the Pacific Ocean, in which some 75% of the world's earthquakes occur. It is also characterized by many active vol-

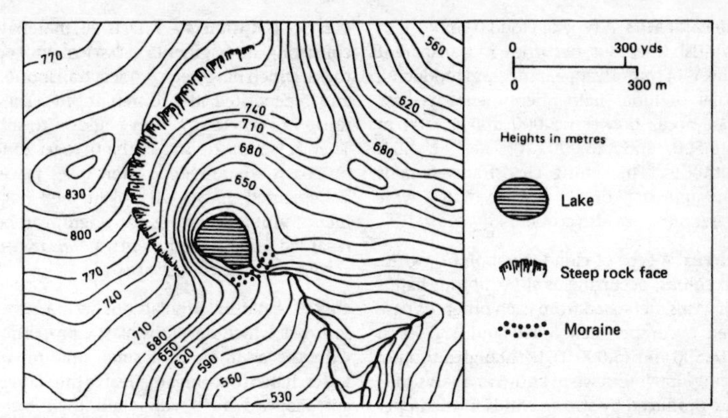

Figure 41 *Cirque*

canoes, hence its popular name, the Ring of Fire. In most places it coincides with the margins of tectonic plates (PLATE TECTONICS).

cirque A French term which has been universally adopted to describe a glacially eroded rock basin with a steep headwall and steep sidewalls, surrounding an armchair-shaped depression. The latter may be occupied by a CIRQUE GLACIER or a small lake (TARN). If the lake is missing the cirque floor is composed of ice-polished rock slabs, although these may be buried by MORAINES. It has been suggested that *cirques* developed mainly from pre-glacial fluvially created hollows in highland terrain. These were slowly enlarged by snowpatch erosion (NIVATION) in which meltwater removed the disintegrated rock. It is suggested that the hollow became occupied by a small glacier, setting in motion the FREEZE–THAW process along the BACKWALL, which progressively retreated and became oversteepened by BASAL SAPPING. The ROTATIONAL SLIP of the glacier deepened the cirque floor by the process of CORRASION, leaving a rock-bar or lip across the mouth of the cirque. Thus, the overall dimensions of the glacially eroded hollows have tended to increase and although they vary greatly in size they generally maintain similar proportions, with a length-to-height ratio of 3 : 1. *See also* COIRE, CORRIE, CWM, GLACIAL EROSION, KAR. [*41*]

cirque glacier A small glacier occupying a glacially eroded armchair-shaped hollow (CIRQUE). It may be contained entirely within the rock basin or it may extend outwards beyond the lip of the cirque as a small glacier tongue. Because of its small size it is claimed that it can form in less than 100 years. It is characterized by a ROTATIONAL SLIP, thereby leading to an overdeepening of the glacial hollow. [*92*]

cirque lake A small water body within a glacially eroded armchair-shaped rock-basin. It may be retained entirely by the lip of the overdeepened rock hollow or it may be dammed by glacial moraines which mark the limits of the former CIRQUE GLACIER. [*41*]

cirque stairway A stepped succession of glacially eroded rock basins, one below another, so that the lip of the upper cirque descends into the headwall of the next cirque in the succession. *See also* GLACIAL STAIRWAY.

cirrocumulus A type of cloud (symbol: Cc), white in colour, occurring in thin sheets or layers without shading but in more-or-less regular ripples or grains. Its base can occur between 5,000 and 13,700 m (16,500 and 45,000 ft), i.e. at high altitudes. Because of its resemblance to fish-scales it is associated with the term 'mackerel sky'. *See also* CLOUD. [*43*]

cirrostratus A type of cloud (symbol: Cs), whitish in colour, occurring in transparent sheets of smooth appearance and producing solar or lunar halo phenomena. Its base can occur between 5,000 and 13,700 m (16,500 and 45,000 ft), i.e. at high altitudes. Thickening cirrostratus heralds the approach of a WARM FRONT in a WAVE DEPRESSION. *See also* CLOUD. [*43*]

cirrus A type of cloud (symbol: Ci), white in colour, occurring as silky, fibrous bands or wisps, detached from each other. Its base can occur between 5,000 and 13,700 m (16,500 and 45,000 ft), i.e. at high altitudes. Long-drawn-out wisps known as mares' tails are produced by strong winds in the upper atmosphere. *See also* CLOUD. [*43*]

CISK CONDITIONAL INSTABILITY OF THE SECOND KIND.

cladistics The study of the evolutionary groups of organisms. Data are plotted on a *cladogram*.

clapotis A type of stationary wave (standing wave) in a water body, created when an ocean wave meets a reflected wave head-on as it returns seaward after impinging upon a vertical cliff, barrier or very steep beach.

clarke A term referring to the average percentage of an element in the Earth's CRUST. The *clarke of concentration* is a measure of the amount of an element within a particular deposit.

class A unit or group in the taxonomic classification of organisms (TAXONOMY). It is composed of one or more orders (ORDER) and itself is a subdivision of a PHYLUM.

classification Any formal arrangement of data into a hierarchy of categories or a distribution of classes. *See also* CHRONOSTRATIGRAPHY, SEVENTH APPROXIMATION, TAXONOMY.

clast **1** An individual part or single constituent of a sedimentary rock, produced by the physical disintegration of a larger mass. The disintegration may take place by FREEZE–THAW processes, DESICCATION, etc. **2** A piece of fragmented rock due to a volcanic explosion (PYROCLASTIC MATERIAL).

clastic sediments **1** Detrital materials consisting of fragments (CLAST) of broken rocks, which have been eroded, transported and redeposited at a different site. They range in PARTICLE SIZE from boulders to silt. They are characteristic of the LITTORAL ZONE on coasts where redeposition takes place. **2** Cemented masses of broken shell fragments, although of organic origin, can be regarded as clastic sediments. *See also* ARENACEOUS, BRECCIA, CONGLOMERATE.

clay **1** A natural argillaceous substance of soft rock which develops plastic properties with the addition of a small amount of water. It then becomes malleable into different shapes which will retain their form when either air- or oven-dried. Clays have different physical properties because of their differing chemical and mineralogical compositions but, in general, they are based on a hydrous aluminium silicate (CLAY MINERALS). **2** A particle-size term in which the size fraction is less than 0.002 mm (1/256 mm), i.e. the smallest particle-size of the British Standard Classification (PARTICLE SIZE). (NB. In the USA <0.005 mm.) **3** A clay soil has more than 30% of its bulk composed of clay. When a clay soil is wet it is almost completely impermeable (PERMEABILITY). *See also* BOULDER CLAY, CLAY PAN, CLAY-WITH-FLINTS.

clayband A clayey rock heavily charged with carbonate of iron, known as siderite, which occurs in uneven beds and nodules, especially among CARBONIFEROUS rocks. Although its iron content is only 20–30% it formed one of the chief sources of iron in Britain during the Industrial Revolution.

clay-dune A DUNE in which about 30% of the material is of clay size. Such dunes commonly occur as LUNETTES on dry lake beds. *See also* SHOTT.

clay–humus complex A mixture of fine clay particles and HUMUS in which the CATION content is high enough to provide sufficient nutrients for plant growth.

clay loam A soil which contains between 27 and 40% CLAY and between 20 and 45% sand. This mixture produces a relatively permeable and friable soil with particles of different sizes. *See also* LOAM, SANDY LOAM, SILTY LOAM, SOIL TEXTURE.

clay minerals Finely crystalline hydrous aluminium silicates and hydrous magnesium silicates, occurring in a fibrous or platy form in a layered structure and possessing an ability to take up and discard water. They are responsible for the plasticity of CLAY and are the main constituents of the ARGILLACEOUS rocks. Exchangeable cations (BASE EXCHANGE) of calcium, sodium, potassium, magnesium, hydrogen, aluminium, etc. usually occur on the surfaces of the silicate layers, in varying amounts. Four main groups of clay minerals may be recognized: the KAOLINITE group; the ILLITE group; the MONTMORILLONITE group; and the VERMICULITE group.

clay pan A layer of compact clay occurring within the lower layers of a soil and responsible for waterlogging owing to its lack of PERMEABILITY. It differs from HARDPAN. *See also* ILLUVIAL HORIZON.

clay skins (clay films) Thin coatings of clay occurring on the surfaces of mineral grains, and soil peds (PED), and as linings of soil pores. *See also* CUTANS.

clay-slate A type of SLATE derived from hard, consolidated SHALE or from a compacted clay.

clay-with-flints One of the superficial deposits overlying the chalklands of S England. It is characteristically a tenacious reddish-orange clay incorporating large unworn, rounded and nodular FLINTS. It varies in thickness from a few metres down to several centimetres, but is rarely thick enough to stop total percolation of water down to the underlying chalk. In some localities it infills solution pipes in the chalk. Its origin is a matter of some dispute, with two alternative theories being proposed. First, that it is a product of chalk SOLUTION. Calculations have nevertheless shown that, to produce 1 m of clay-with-flints, 100 m of chalk would need to be reduced. Because of the time required for this type of dissolution it seems unlikely that this was the sole process involved. The second theory proposes that the clay has been derived from the weathering of Eocene rocks, especially the Reading Beds, which have identical mineralogical assemblages to the clay-with-flints. It is very likely that the majority of the deposit was formed by this latter process but with some clay contribution (together with the flints) by dissolution of the chalk.

Clean Air Acts 1 In the United Kingdom the Clean Air Act was first introduced by Parliament in 1956, hastened by the deaths of 4,000 Londoners in the SMOG of December 1952. It was later strengthened by the Clean Air Act of 1968. In both cases the legislation was directed against the emissions of smoke, grit and dust, but not against sulphur dioxide. The Acts resulted in a spectacular improvement in general air quality and the formation of smokeless zones. **2** In the USA similar legislation was introduced in the 1970 Clean Air Act, in order to combat air pollution. It was amended in 1977. *See also* AIR POLLUTION, SMOG.

clear-air turbulence Known to airline pilots as CAT, this sudden atmospheric turbulence in a usually cloudless sky suggests that it is a mechanism unconnected with the vertical turbulence associated with convection clouds. It is found in the high TROPOSPHERE and especially in the vicinity of a JET STREAM at the interface between two atmospheric layers which exhibit different wind and temperature characteristics (KELVIN–HELMHOLTZ INSTABILITY). *See also* TURBULENCE.

cleavage 1 The tendency for a rock to split or break along closely spaced parallel planes which do not correspond to the BEDDING PLANE and may be highly inclined to it. Cleavage is a secondary structure imposed on the rock as a result of metamorphic pressure (METAMORPHISM) or of deformation. The following types have been recognized: *fracture cleavage* (or shear cleavage); *flow cleavage*, in which recrystallization takes place to produce a foliated rock (SCHIST); *axial-plane cleavage*; *strain-slip cleavage*; and *slaty cleavage*, in which some recrystallization takes place but does not destroy all traces of bedding. The quarryman makes use of cleavage in order to split slates for commercial use. **2** In crystallography the term refers to the tendency of a mineral to split along planes determined by the crystal structure. The perfection of the mineral cleavage depends on the relative strength of the bonds along these planes.

cleavage plane The plane of mechanical fracture in a rock or mineral (CLEAVAGE).

cliff **1** A high, steep, rock face or precipice (FREE FACE). **2** A steep face in unconsolidated deposits such as boulder clay (TILL) or alluvium (when occurring as a river-eroded cliff in a MEANDER). **3** A steep coastal declivity which may or may not be precipitous, the slope angle being dependent partly on the jointing, bedding and hardness of the materials from which the cliff has been formed, and partly on the erosional processes at work. Where wave attack is dominant the cliff-foot will be rapidly eroded and cliff retreat will take place, especially in unconsolidated materials such as clays, sands, etc., frequently leaving behind an ABRASION PLATFORM at the foot of the cliff. Near-vertical cliffs can also be formed in well-jointed rocks, but where wave-attack has diminished and the cliff-foot has been abandoned (possibly in response to tectonic uplift) subaerial processes will become the dominant process and the slope angle will tend to diminish and the cliff become degraded. It is often possible to recognize an abandoned cliff-line behind a RAISED BEACH.

cliff-line CLIFF.

climate The average weather conditions at a specific place over a lengthy period of time (>thirty years), including absolute extremes, means and frequencies of given departures from these means. In modern usage the term is used most commonly of world regions rather than of particular places (CLIMATIC REGION). Climate deals with all the meteorological ELEMENTS (ATMOSPHERIC PRESSURE, HUMIDITY, PRECIPITATION, TEMPERATURE, WIND) and with the way they are influenced by FACTORS such as changes in latitude, altitude, distribution of continents and oceans and location of ocean currents. Today, climate has become the synthesis of weather, representing a profitable fusion of meteorological principles and the facts of geography.

climate-stratigraphic unit GEOCHRON.

climatic change The slow variations of climatic characteristics (SECULAR TREND) over time at a given place. This may be indicated by the geological record in the long term, by changes in the landforms (CLIMATOMORPHOLOGY) in the intermediate term, and by vegetation changes in the short term (e.g. DROUGHT). Small variations in climate can also be observed from the period during which reliable instrumental records have been available, e.g. the increase of CO_2 and the 1° C warming trend witnessed between 1850 and 1940. *See also* MILANKOVITCH RADIATION CURVES.

climatic classification CLIMATIC REGION.

climatic climax CLIMAX.

climatic geomorphology CLIMATOMORPHOLOGY.

Climatic Optimum ALTITHERMAL, ATLANTIC PERIOD, MEDIOCRATIC.

climatic region A specific area in which various combinations of climatic ELEMENTS can be recognized. Numerous attempts have been made to identify and classify climatic regions: **1** Based on latitudinal temperature zones – the TORRID, TEMPERATE and FRIGID ZONES. **2** Based on climatic effects (i.e. vegetation responses) expressed in terms of critical temperature and precipitation characteristics measured seasonally or annually. Those of W. Köppen and A. A. Miller, despite their global coverage, involve complex notation systems and fail to take into account the fact that some vegetation zones do not simply reflect prevailing climates but also differences in geology and soils and the effects of human interference. *See also* KÖPPEN'S CLIMATIC CLASSIFICATION, THORNTHWAITE'S METHOD OF CLIMATIC CLASSIFICATION. **3** Based on climatic causes. *See also* FLÖHN'S CLIMATIC CLASSIFICATION.

climato-genetic geomorphology CLIMATOMORPHOLOGY.

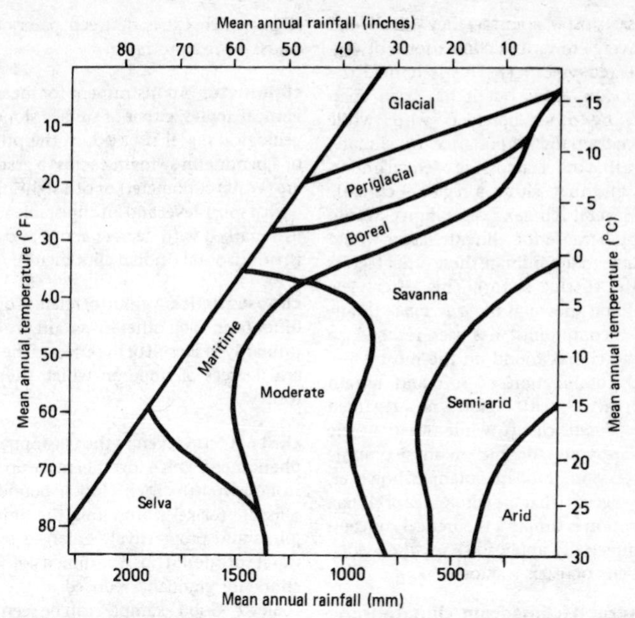

Figure 42 *Climatomorphology*

climatograph A circular graph depicting seasonal mean-temperature changes plotted on a monthly basis. The length and nature of the hot, warm, cool and cold seasons can be identified for any place, based on the assumption that the limiting temperatures of these seasons are >20° C (>68° F), 10°–20° C (50°–68° F), 0°–10° C (32°–50° F), and <0° C (<32° F) respectively.

climatology The scientific study of world climates. It is concerned with accounts and explanations of CLIMATIC REGIONS; their spatial and temporal variations (CLIMATIC CHANGE); and the influence of climates on the environment of life on Earth (BIOCLIMATOLOGY, CLIMATOTHERAPY, LOCAL CLIMATE, MICROCLIMATOLOGY).

climatomorphology (climatic geomorphology) The scientific study of current landform development under different climates. It is concerned with accounts and explanations of the relationships between different climates and geomorphic processes (GEOMORPHOLOGY) and especially with the definition of *morphogenetic regions* (MORPHOGENESIS) on a world basis. Some scientists recognize a separate term – *climato-genetic* geomorphology – which is defined as the scientific study of climatically controlled landform development over time. It is concerned with the explanation of landforms which were created largely under climates no longer operative in the area under investigation. [*42*]

climatotherapy The treatment of disease by a favourable climate.

climax A botanical term referring to the terminal community said to be achieved when a SERE achieves dynamic equilibrium with its environment and in particular with its prevailing climate. Each of the world's major vegetation climaxes is equivalent to a BIOME. Many botanists believe that climate is the master factor in a plant environment and that even if several types of plant succession occur in an area (HYDROSERE, HALO-

SERE, PSAMMOSERE, XEROSERE) they will all tend to converge towards a climax form of vegetation (CONVERGENCE). This is termed the *monoclimax theory*, but it has been challenged by other botanists who, while appreciating the importance of climate, believe that other factors are of equal importance, and that within a regional climatic type several climax vegetations would develop because the climatic factor would be unable to subordinate these other factors (e.g. the EDAPHIC factor). This latter view came to be known as the *polyclimax theory*. Today a compromise has been reached, so that the climax found on the most widespread combination of soil and terrain within each climatic region is now regarded as the *climatic climax*, while others in the same areas are edaphic or physiographic climaxes. Some modern botanists, however, have suggested that the whole idea of climax recognition is simply a theoretical concept, not universally applicable. *See also* ASSOCIATION, PHYTOCLIMAX, PLAGIOCLIMAX.

climograph (climagram, climatogram, climogram) A graphic diagram to illustrate the general climatic character of a particular place. This is an extended usage, for the term was first used by T. Griffith Taylor with reference to human comfort and humidity. The diagram is produced by plotting DRY-BULB and WET-BULB temperatures against one another as ordinates and abscissae respectively.

climosequence A sequence of related soils that differ from each other in certain properties primarily as a result of the effect of climate as a soil-forming factor.

clinographic curve A means of depicting the average slope of an area in graphic form. It is achieved by constructing concentric circles on graph paper, each being equivalent in area to that enclosed by a selected contour traced off the maps. By repeating the exercise for each contour in turn it becomes possible to draw the clinographic curve by (a) plotting each of the contour areas on the horizontal axis against the actual contour intervals on the vertical axis and (b) inserting each section of the average slope which exists between pairs of contours, using a protractor.

clinometer An instrument for measuring vertical angles, either of surface slope or of geological DIP. It is based on the principle of a pendulum swinging across a graduated arc (Watts clinometer) or of a sighting-tube with a spirit-level and an adjustable vertical leaf marked with degrees above and below the horizontal (Indian clinometer).

clinosequence A group of related soils that differ from each other in certain properties primarily as a result of the effect of the different degrees of slope on which they were formed.

clints A term given to the flat-topped rock phenomena which together make up a limestone PAVEMENT. Each clint is bounded by a fissure (GRIKE) worn along the limestone joints and progressively enlarged by rainwater SOLUTION. Thus, the dimensions of the clints are gradually reduced as the grikes coalesce. Good examples can be seen in the Burren of Ireland and at Malham Cove, Yorkshire, England.

clisere A term used by botanists to refer to the slow replacement of one group of plant communities (SERE), in a particular vegetation zone, by another group over a lengthy time period, owing to CLIMATIC CHANGE. For example, in England during post-glacial warming the birch/pine forests of the PRE-BOREAL gave way to the mixed oak forests of the BOREAL.

clitter A local term given to the BLOCKFIELDS of granite boulders which surround the TORS on Dartmoor, England. Clitter is thought to have been formed by frost action during the Pleistocene ice age.

closed system A system (GENERAL SYSTEMS THEORY) in which energy but not matter is exchanged between the system and its environment. A further characteristic of a closed system is its tendency to move towards total homogeneity by the destruction of any heterogeneity which may exist within it. Contrast an OPEN SYSTEM. [*78*]

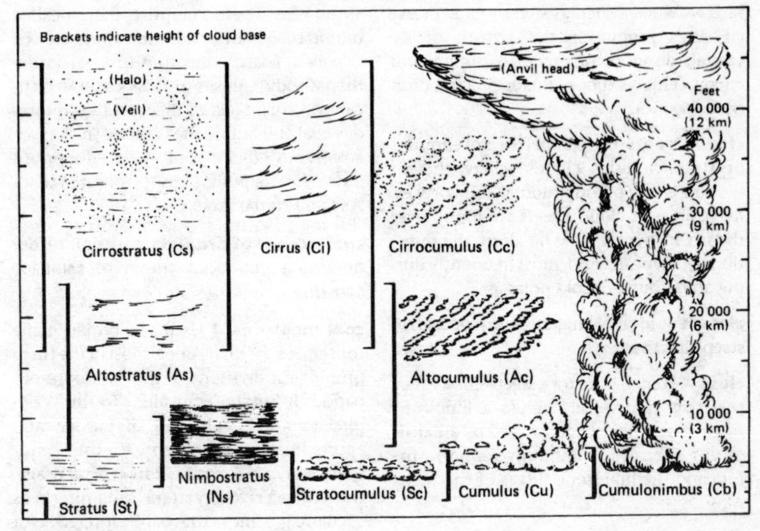

Figure 43 *Cloud types*

cloud A visible mass of tiny particles of water and ice held in suspension by vertical motion of air, formed by condensation around nuclei, such as dust and smoke particles, pollen or salt. A cloud can exist at various elevations between sea-level (FOG, MIST) and 13,700 m (45,000 ft). Clouds are classified according to their height and their shape. *See also* ALTOCUMULUS, ALTOSTRATUS, CASTELLANUS, CIRROCUMULUS, CIRROSTRATUS, CIRRUS, CUMULONIMBUS, CUMULUS, FRACTUS, NACREOUS, NIMBOSTRATUS, NOCTILUCENT, STRATOCUMULUS, STRATUS. The proportion of the sky under cloud cover is expressed in tenths or eighths (OKTA). Cloud is formed as a result of the vertical motion of air associated with CONVECTION and DEPRESSIONS/FRONTS. [*43*]

cloudburst A popular term for a violent torrential shower of rain or hail, generally short-lived but capable of considerable damage because of the large volume of water precipitated. It is associated with violent upward and downward air currents. *See also* FLOOD, THUNDERSTORM.

cloud cluster One of the five categories of tropical weather systems, identified according to their space- and time-scales by satellite photography: **1** An individual cumulus cloud of few hours' duration. **2** A mesoscale convective cell of diurnal time-scale. **3** A cloud cluster (subsynoptic scale) which may persist for 1–3 days. **4** A wave disturbance (2,000–4,000 km wavelength) and a cyclonic vortex of one to two weeks' duration and with a diameter of 650 km. **5** A planetary wave (10,000–40,000 km wavelength) which occurs in the equatorial upper troposphere and also in the equatorial stratosphere. Both **4** and **5** are said to be at 'synoptic' scale. [*117*]

cloud forest The usually broad-leaved evergreen forest that occupies the zone in a mountainous terrain where cloud forms regularly and in sufficient volume to support a substantial vegetation growth, e.g. the laurel forest of the Canary Is. and the montane forest between 2,750 m and 3,050 m in the New Guinea highlands.

cloud-seeding A term used in the experimental procedure of artificial rain-making. It is based on the principles outlined in the BERGERON–FINDEISEN THEORY OF PRECIPITATION in which clouds containing SUPERCOOLED WATER droplets are 'seeded' artificially with

FREEZING NUCLEI, such as silver iodide or DRY ICE, thus promoting the growth of ice crystals alongside supercooled droplets and encouraging vapour fluxes/precipitation (RAIN-MAKING).

cloud streets The parallel rows of cumulus or strato-cumulus clouds where the air motion is one of longitudinal roll vortices. Large areas of cloud streets are found over the oceans where they lie along the direction of the mean wind, most frequently during an outburst of cold polar air.

clough A local N England term for a small steep-sided valley.

cluse A French term for a steep-sided valley that cuts transversely across a limestone ridge. It may have been formed by an ANTECEDENT river. It is commonly found in the Jura and the limestone ridges of the Savoy Alps.

cluster analysis A statistical technique for grouping variables into subsets that in one fashion or another are linearly related.

coagulation process The process in which large rain droplets, possibly resulting from extra-large nuclei, grow more rapidly than those on smaller nuclei. As they fall through the cloud it is postulated that they grow by capturing smaller droplets in their path, particularly in the vortex found on the rear side of the descending droplet. Also known as the BOWEN–LUDLAM PROCESS. This type of process has been observed largely in tropical marine climates. *See also* BERGERON–FINDEISEN THEORY OF PRECIPITATION, COLLISION THEORY OF RAINFALL, LANGMUIR THEORY OF RAINDROP GROWTH.

coal The general name given to stratified, rock-like deposits of carbonaceous material derived from former forest vegetation that accumulated in PEAT beds that later became compacted by burial beneath later sediments. The degree of *coalification* depends on the degree of pressure and associated heating, so that the deposit is classified according to the progressive increase in carbon content allied with the simultaneous decrease in volatiles: PEAT; LIGNITE; subbitiminous; BITUMINOUS; sub-anthracite; ANTHRACITE. These comprise the so-called humic coals, but in addition there are the *sapropelic* coals, consisting of algal rather than woody deposits (*see also* BOGHEAD COAL, CANNEL COAL). Most of the world's bituminous coal and anthracite is found in the CARBONIFEROUS rocks; but very important brown-coal deposits (lignites) occur in MESOZOIC and TERTIARY rocks.

coalescence of droplets COLLISION THEORY OF RAINFALL, LANGMUIR THEORY OF RAINDROP GROWTH.

coal measures **1** Used as a proper name the term refers to the uppermost of the three lithological divisions of the CARBONIFEROUS period in Britain, equivalent to the Westphalian and Stephanian of Europe and approximately equivalent to the Pennsylvanian of N America. **2** In general usage, the term refers to any strata containing beds of coal, e.g. some of the coal-bearing rocks of Scotland occur in the Lower Carboniferous rather than the Upper Carboniferous.

coastal deposits All the materials which have been deposited in the vicinity of a COASTLINE derived from one or more sources: from offshore, brought to the shoreline by waves; from along the coast, moved by LONGSHORE DRIFT; from the coastal cliffs by cliff falls or by COASTAL EROSION; by rivers, moved from inland to coastline by fluvial action. The deposits consist of a wide variety of particle sizes, from finest clays and silts through sands, shingle and cobbles to boulders.

coastal erosion A term which describes the actions of marine waves, through their fourfold processes of HYDRAULIC ACTION, CORRASION, ATTRITION and SOLUTION, when they combine to attack the coastline, thereby causing it to retreat. The rates of erosion and retreat will depend on the rock hardness, joint patterns, lithology, etc.

coastal plain Any gently sloping plain or lowland which borders the landward side of a COASTLINE and which often continues off shore as the submarine CONTINENTAL SHELF. (NB. In the USA the term is restricted to any section of former continental shelf that is now exposed from beneath the ocean by a fall of sea-level or by uplift.) In general usage, however, the term may refer either to an accumulation of ALLUVIUM brought down by rivers, or to a newly emerged sea-floor.

coastal wetlands Shallow coastal shelves and margins that are normally flooded or marshy. They include tidal flats, LAGOONS, MANGROVE swamps and MARSHES.

coastline **1** In general usage the term refers to the line forming the boundary between the land and the water. **2** More specifically the term refers to the highest limit reached by the SWASH of storm waves during high-water spring tides. **3** The ORDNANCE SURVEY restricts the use of the term to the high-water mark of medium tides. *See also* BACKSHORE, FORESHORE, SHORELINE CLASSIFICATION.

cobble A rounded to sub-rounded rock fragment between 64 and 256 mm in diameter. It is smaller than a BOULDER but larger than a PEBBLE. *See also* PARTICLE SIZE, WENTWORTH SCALE.

cockpit karst A type of tropical KARST landform characterized by deeply incised hollows (cockpits), which have developed from alluvial-floored DOLINES due to the fluctuating water-table in the limestone. Between the cockpits are smoothly crested steep-sided hills, often conical in shape. In Jamaica, the Cockpit Country is a good example of this type of karstic terrain which, in German, is termed *kegelkarst*. [*127*]

coefficient A statistical term: **1** of correlation, which refers to an index or measure giving a precise value to the relationship existing between two or more variables; **2** referring to a constant in an equation; **3** referring to an index of reliability. *See also* CORRELATION.

coefficient of variation (Vc) A measure of variability within a sample or population. It is calculated as 100 times the standard deviation divided by the mean.

cohesion The property of a material to stick or adhere together. The cohesion of a rock or a soil is that part of its SHEAR STRENGTH which does not depend upon friction between particles.

coire A Gaelic term for CORRIE.

col **1** In meteorology, a term to describe the area of slack low pressure and low wind-speeds, which has a saddle-like form on a weather map, between two diametrically opposed anticyclones and two areas of even lower pressure (DEPRESSION). It is analogous with a mountain pass between two peaks but leading from one valley to another. **2** In geomorphology, it is the lowest point on a mountain ridge between two peaks.

colatitude The complement of the LATITUDE, i.e. 90° minus the latitude.

cold desert A term generally used as a synonym for TUNDRA and polar regions where plant life is inhibited by low temperatures and physiological drought. Also used occasionally to define enclosed basins of central Asia, cut off by high relief from maritime influences, and having at least one month with a mean temperature below 6° C (43° F), e.g. Tarim Basin. *See also* DESERT.

cold front The front boundary of a mass of advancing cold air where it undercuts and replaces the slower-moving warm air mass, thus forcing it to rise. In this way a wedge of cold air replaces the warm air at ground level, and is accompanied by a marked fall of temperature. Rainfall associated with the passage of a cold front is often heavy and sometimes thundery (THUNDERSTORMS); CUMULONIMBUS and FRACTUS clouds are common and a LINE SQUALL may develop. *See also* ANAFRONT, FRONTOGENESIS, KATAFRONT. [*44 see also 60*]

cold glacier A GLACIER in which the ice temperatures may be as low as –30° C throughout the year, with no appreciable surface melting. Sometimes referred to as a *polar glacier*. Because of the absence of meltwaters the cold glacier may remain static, frozen to the underlying rock. Thus it requires great shear stress before movement occurs, and even then it will be very slow. However, pressure melting-point can bring basal temperatures close to 0° C, when meltwater is trapped between the ice and

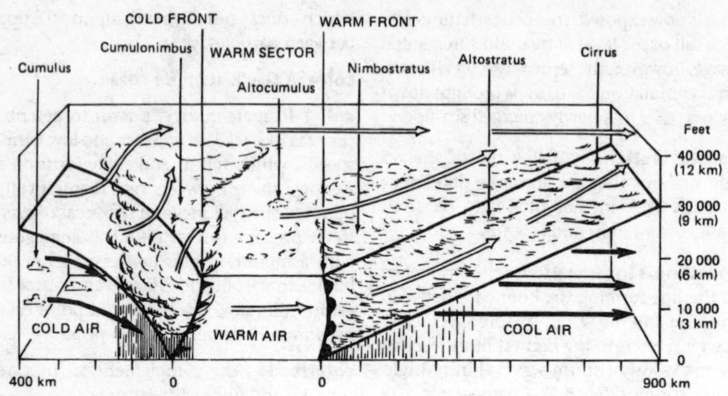

Figure 44 *Section through a depression*

bedrock. *See also* SUBPOLAR GLACIER, TEMPERATE GLACIER.

cold low, cold pool An atmospheric DEPRESSION in which the central low-pressure cell is composed of cold air. Because of the column of cold air at the core of the depression, the low-pressure cell at sea-level will intensify with increasing elevation into the troposphere. Conversely, a high-pressure cell with a cold core (e.g. the Siberian winter anticyclone) weakens with elevation and is replaced aloft by low pressure.

cold occlusion A climatological term referring to an OCCLUSION in which the advancing mass of cold air has lower temperatures than the air mass in front. A secondary DEPRESSION can form at the point where the cold and warm fronts unite to form a cold occlusion. *See also* COLD FRONT, FRONTOGENESIS.

cold pole The point on the Earth's surface with the lowest mean annual temperature. This is commonly claimed to be near to the town of Verkhoyansk in NE Siberia: mean annual temperature –16.3° C (2.7° F), mean January temperature –50° C (–58° F); lowest recorded temperature –70° C (–94° F). It is possible, however, that the Russian base at Vostok, Antarctica, may be even colder, for an extreme temperature of –90° C (–130° F) has been recorded.

cold wall An oceanographic term referring to the boundary, represented by a DISCONTINUITY, between ocean waters of very different temperature characteristics.

cold-water desert A term used to describe regions where the western coastal margins of hot deserts are washed by upwelling cold water associated with equatorward-flowing cold ocean currents. Because cool air moves onshore across the cold current it produces lower summer air temperatures and lower mean annual temperature ranges than are normal for the latitude. Heavy dews and fogs are common on these coastlands. Examples are the Atacama Desert of N Chile and the Namib Desert of SW Africa (DESERT). *See also* INVERSION OF TEMPERATURE.

cold wave **1** The fall of temperature caused by the sudden burst of cold air which flows in after the passage of a low-pressure system (DEPRESSION) in temperate latitudes. In the USA a more specific usage includes a fall of temperature of at least 20° F in 24 hours. **2** The downward movement of the FREEZING POINT into the ground in PERMAFROST environments.

colour display A REMOTE SENSING term relating to the combination of an IMAGE with colour, for easier interpretation.

collapse structures Geological or topographical structures formed by gravity collapse: **1** Folds created when erosion of incompetent strata has allowed slightly more competent beds to move by collapse under gravity (GRAVITY-COLLAPSE STRUCTURES) and thereby become contorted into knee folds, flaps, etc. [*96*]. **2** Sedimentary structures which result from the downward bulging of a competent bed into a softer underlying bed, often leading to the complete break-up of the uppermost bed. **3** The circular collapse structures (CALDERA) caused by the withdrawal of magma from a deep underground chamber. **4** The topographic hollows in a KARST landscape which are formed by the roof collapse of subterranean caves (POLJE).

collision theory of rainfall One of the attempts to explain the generation of *raindrop growth*. Initially it was believed that the collision of cloud particles by turbulence would cause particles to coalesce, thereby creating larger droplets. Experiments showed, however, that collisions due to atmospheric turbulence within a cloud caused not only coalescence but many instances of raindrop disintegration, so that alternative explanations had to be sought. *See also* ACCRETION, BERGERON–FINDEISEN THEORY OF PRECIPITATION, COAGULATION PROCESS, LANGMUIR THEORY OF RAINDROP GROWTH.

colloid A substance in a state of very fine subdivision, midway between a solution and a suspension, in which the particles are so small (1–10 μm) that the forces which tend to support them are greater than the gravitational forces which would otherwise cause them to settle. Colloids become coagulated (*flocculated*) either by heating or by coming into contact with salt water (i.e. when they are carried into the sea by rivers). This latter process leads to the colloids becoming part of the clay fraction (*colloidal clay*) of the sedimentary deposit.

colluvium A heterogeneous mixture of weathered materials (WEATHERING) transported downslope by gravitational forces and deposited at the foot of the slope. Thus, they can be contrasted with ALLUVIUM. *See also* ELUVIAL DEPOSITS, SCREE.

colonization The occupation by an organism of a new HABITAT, thereby extending its ecological or geographical RANGE. *See also* R AND K SELECTION.

colony 1 A biological term referring to a group of closely associated organisms. **2** A macroscopically visible growth of microorganisms on a solid culture medium.

colour composite A REMOTE-SENSING term referring to a colour picture produced by assigning a colour to an IMAGE of the Earth's surface recorded in a particular waveband. For a LANDSAT colour composite, the green waveband is coloured blue, the red waveband is coloured green and the INFRARED waveband is coloured red. This produces an image closely approximating a FALSE COLOUR photograph. Colour composite images are easier to interpret than separate images recording different wavebands. US national experimental crop inventories are based upon visual interpretation of Landsat colour composites.

columnar structure 1 A type of jointing in igneous rocks which results from internal contraction during cooling of lava (BASALT), well illustrated in the vertical columns of the Giant's Causeway, N. Ireland. **2** A type of soil structure characterized by vertical jointing, e.g. BRICKEARTH. [*236*]

combe (coombe) A topographic term used in place-names, mainly in England, but with a variety of meanings: **1** In the chalklands of S England it refers to the head of a dry valley which terminates in a steep-sided amphitheatre. **2** In N Devon it refers to a short valley, usually containing a stream, which descends steeply to the sea. **3** In the Lake District it refers to a glacially eroded hollow in a mountainside (CIRQUE). **4** In the Jura mountains of France the term is used to describe a high-level depression carved along the crestline of an anticlinal fold by river action and which is bounded by in-facing limestone cliffs.

comber A type of BREAKER or water-wave, characterized by a steep crest which is blown forward by powerful wind action and curls over before breaking noisily from a great height on to a beach.

comfort zone, comfort index A term used in BIOCLIMATOLOGY to describe the range of temperature and humidity within which human beings feel comfortable. In general the higher the temperature the lower should be the humidity to obtain comfort. In the temperate zone DRY-BULB temperatures of 20°–25° C (68°–77° F) with relative humidities of 25–75% are regarded as the limits of the comfort zone. In England the figures of 15° C (60° F) with a relative humidity of 60% are commonly accepted as the optimum conditions for comfort. *See also* SENSIBLE TEMPERATURE.

commensalism A relationship between two types of organism living in the same environment without harm to either species and in which one or both may obtain sustenance, protection or other benefits.

comminution The gradual breakdown of rocky materials by weathering and erosion to form progressively smaller particles. *See also* PARTICLE SIZE.

community An organized group of plants or animals, generally of distinctive character and related to a particular set of environmental requirements. The term is used in a more general, collective, sense when it is unnecessary to adopt a more specific designation such as ASSOCIATION OF FORMATION.

compaction A term referring to the decrease in volume of a mass of unconsolidated materials owing to compression. Such consolidation usually results from a closing of pores and the loss of any interstitial water from sediments because of the increasing weight from the overlying deposits (DIAGENESIS). Silts and clays, for example, are converted into mudstones and shales. But drying and shaking of a sediment can also lead to a certain amount of compaction, as can the compressional effect of earth-movements during METAMORPHISM.

compass **1** A navigating instrument designed to indicate direction. It consists of a magnetic needle which is free to swing in a horizontal place, thereby seeking the N and S magnetic poles by following the local MAGNETIC DECLINATION. The needle moves across a graduated card or dial on which the CARDINAL POINTS ARE MARKED. **2** A *dip compass* is a geological instrument, used in the search for magnetic iron ore, in which the needle swings in a vertical plane. **3** An instrument for drawing circles. **4** An instrument for transferring measures from a map or chart to a linear scale (more strictly termed *dividers*).

compensation flow A legally established rate of river discharge below an abstraction point or reservoir supply in order to maintain the water resource of a given region or catchment.

competence **1** A term which is used in fluvial geomorphology and hydrology to indicate the ability of a stream to move particles of a particular size as BEDLOAD. It refers to the largest size of grain that can be carried by a particular stream velocity, a relationship defined by the *sixth-power law* postulated by W. Hopkins in 1842. This states that the largest particle that can be transported increases with the sixth power of the stream velocity. The stream velocity at its bed is controlled by such factors as gradient and the ratio between width and depth (CHANNEL GEOMETRY), e.g. with other things being equal, bed velocity is higher in a wide shallow stream than in a narrow deep one. But a large, slow-moving stream may carry a large quantity of small particles in SUSPENSION and although its load-carrying ability is great (CAPACITY) its competence is small. Conversely, a small but rapidly flowing stream is able to move relatively large particles (i.e. it has a high competence) although its *capacity* is small because it is unable to carry a great volume of material. **2** A term used by geologists to denote the degree to which a bed of rock (*competent bed*) can be folded or raised without appreciable flowage or change of thickness. This is due to its inherent strength or massive character, so that it is able to lift not only its own weight but also that of the overlying strata.

competent bed COMPETENCE.

competition An ecological term referring to the struggle between individual organisms of similar or different species to use greater quantities of available resources or

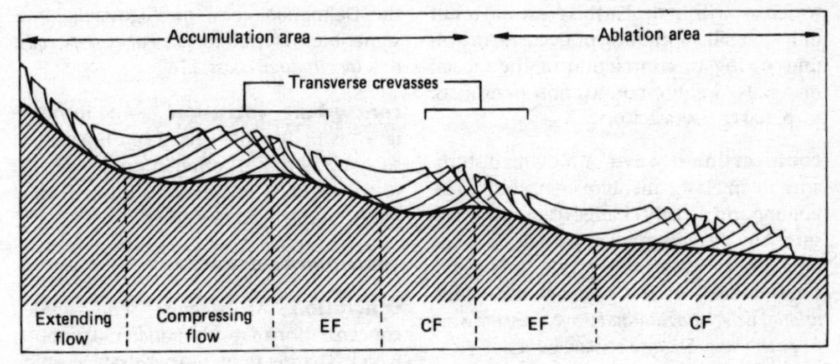

Figure 45 *Compressing and extending flow in a glacier*

use them at a greater rate than their neighbours, e.g. competition for soil moisture by vegetation.

compiled map A map produced by the transfer of information, usually by reduction of scale, from an existing map rather than by original survey work.

complex **1** A situation or state in which a large number of variables have to be considered simultaneously. **2** In soil science, a mapping unit used in soil surveys where two or more soil units are so intermixed geographically that it is cartographically impractical to separate them because of the map scale. **3** A combination of numerous types of ancient, usually metamorphosed rocks in the shield areas of continents (BASAL COMPLEX).

composite coast A coast that has resulted from the alternating spatial subsidence and upwarping of coastal tracts along alignments (often faults) that run transverse to the coastline itself. Thus, the downwarped areas coincide with bays while the upwarped areas produce salients or headlands.

composite map A type of map, usually a COMPILED MAP, in which information from several other maps is brought together and displayed to facilitate comparisons between different sets of data.

composite profile A term used by geomorphologists in topographic map analysis to denote a profile of the relief of any area, constructed as if it was viewed from an infinite distance across a horizontal plane of summit levels. [*195*] *See also* PROJECTED PROFILE, SUPERIMPOSED PROFILE.

composite volcano (strato-volcano) A volcanic CONE, built of alternating layers of LAVA and PYROCLASTIC MATERIAL, over a lengthy time period.

compound shoreline SHORELINE CLASSIFICATION.

compressibility The susceptibility of a material to change its volume and density when subjected to pressure due to loading. It is particularly related to soils and unconsolidated sediments and is an important factor to be taken into account by civil engineers.

compressing flow A glaciological term devised by J. F. Nye to describe the type of glacier flow whereby a reduction in the surface velocity leads to an increase in thickness of a glacier. The phenomenon is characterized by the appearance of SHEAR PLANES (slip planes) along which the ice rides obliquely upwards due to compressive stress. The compression may be caused by the narrowing of a valley, the decreased velocity at the base of an icefall or simply by the termination of the glacier flow at the snout. *See also* EXTENDING FLOW, SHEARING. [*45*]

compressional movement A geological term referring to all the stresses or forces which act within the Earth's crust and result in its overall shortening or decrease in volume owing to contraction of the crustal materials. *See also* CONTRACTION HYPOTHESIS, FOLD, NAPPE, REVERSE FAULT.

compressional wave A moving disturbance in an elastic medium characterized by volume and density changes caused by pressure. Any displacement of particles is in the direction of wave propagation. It is sometimes referred to as a *dilational wave, irrotational wave, longitudinal wave, pressure wave* or *push wave*. *See also* SEISMIC WAVES.

computer A device capable of input, processing of DATA and output (INPUTS AND OUTPUTS). A stored-program computer follows an ALGORITHM specified by the program that it is executing.

concavity One of the morphological characteristics of a slope, in which the gradient becomes progressively gentler as it is traced downwards towards the foot of the slope. *See also* WANING SLOPE.

concentric dyke RING DYKE.

concentric weathering SPHEROIDAL WEATHERING.

conchoidal fracture A fracture in a mineral or rock, e.g. flint, producing a smooth, curved surface, similar to the interior surface of a shell.

concordant Pertaining to the harmonious correspondence between a morphological feature and its underlying geological structure. It is used specifically by geomorphologists to describe the parallel relationship between a drainage pattern and the 'grain' of a region, in contrast to an example where a river course may disregard the underlying structure, in which case the relationship is discordant. *See also* CONCORDANT COAST, CONCORDANT INTRUSION.

concordant coast A coast which lies parallel to the general 'grain' of the regional topography and geological structure. It may take the form of a linear coastline or a series of promontories and narrow elongated islands all lying parallel to each other, e.g. the Dalmatian coast of Yugoslavia. It is sometimes referred to as a *Pacific-type coast* or a *longitudinal coast*. [*56*]

concordant intrusion An intrusive igneous body which has been injected in such a way that its orientation lies parallel to the bedding of the rocks into which it has been intruded. *See also* SILL.

concordant summits ACCORDANT SUMMITS.

concretion An irregular, semi-rounded concentration of resistant material (NODULE) which is harder than the rock strata in which it occurs. It is thought to have been formed around a central nucleus by localized deposition of a cementing material (e.g. calcite, silica) during the consolidation of the bedded rocks. Some ironstone concretions form valuable ores, while calcareous concretions provide a source of cement. *See also* DOGGER.

condensation The process by which vapour changes into a liquid or solid form, either by cooling air below its DEW-POINT or by saturating air. Cooling results from the adiabatic expansion of rising air, from radiation/conduction at the surface on a calm, clear night and advection/conduction when warm moist air crosses over a cold-water body. Cooling beyond the dew-point causes the formation of water droplets as the excess vapour is condensed on existing nuclei. If the dew-point is below 0° C (32° F) vapour will be condensed in a solid form (HOAR-FROST). *See also* CLOUD, COAGULATION PROCESS, DEW, FOG, MIST, RAINFALL, SATURATED ADIABATIC LAPSE-RATE, SNOW.

condensation level, lifting condensation level The level in the atmosphere at which condensation occurs. It is calculated empirically by weather forecasters in order to determine the height of the cloud base. *See also* DEW-POINT, SATURATED ADIABATIC LAPSE-RATE.

condensation nuclei HYGROSCOPIC NUCLEI.

condensation trail, contrail A trail of ice crystals or water droplets left in the atmosphere by the passage of a jet aircraft.

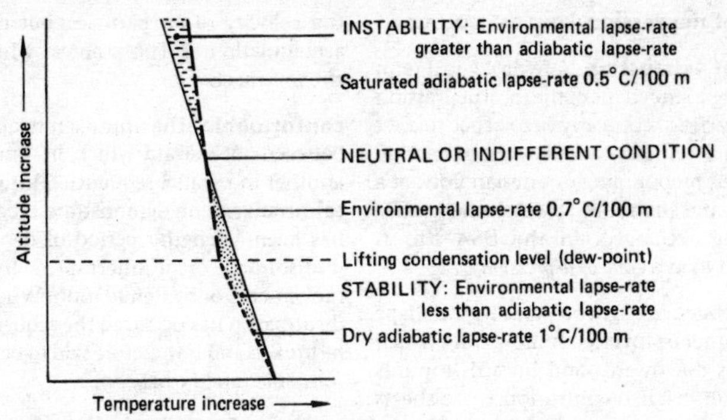

Figure 46 *Conditional instability in the atmosphere*

It is caused by the engine exhaust in which the water vapour condenses and is unable to revaporize. Most contrails initially expand but then slowly evaporate and disappear as mixing with drier air takes place. They will persist only if high-level cloud is beginning to form, but their effect on solar radiation is thought to be minimal.

conditional instability The state of an air parcel when the ENVIRONMENTAL LAPSE-RATE is less than the DRY ADIABATIC LAPSE-RATE but is greater than the SATURATED ADIABATIC LAPSE-RATE. Initially, the rising air parcel (cooling at the dry adiabatic rate) is colder than the environment and stability will exist. However, with continued physical uplift (e.g. OROGRAPHIC) beyond the dew-point, the reduction in cooling rate (at the saturated adiabatic rate) will result in the parcel eventually becoming warmer than the environment above the neutral or indifferent condition. The parcel has now become unstable and will continue to rise well above the mountain summit, leading to the development of towering CUMULUS cloud and THUNDERSTORMS. It must be noted that the change from stability to instability is conditional upon the release of LATENT HEAT of condensation, which slows down the cooling rate of the parcel. The saturated adiabatic lapse-rate is now less than the environmental lapse-rate, so eventually the parcel becomes warmer than the environment and unstable. *See also* ABSOLUTE INSTABILITY, ABSOLUTE STABILITY. [*46*]

conditional instability of the second kind (CISK) A term used in climatology and meteorology, referring to the augmentation of simple cumulus CONVECTION (the 'first kind') by frictional CONVERGENCE (and rotation) in the atmosphere BOUNDARY LAYER. It is considered to be a dominant process in the spin-up of tropical vortices (VORTEX) and some types of polar DEPRESSIONS.

conduction The process in which heat is transferred directly through matter from a point of high temperature to a point of low temperature by molecular impact but without overall movement of the matter itself. Air is a poor conductor of heat. *See also* SENSIBLE HEAT.

conduit **1** The vertical pipe through which magma moves upwards in a volcano towards the VENT. **2** A narrow underground passage that is filled with water.

cone **1** A conical mass of which the base is a circle and the summit a point. The term is used frequently in connection with a VOLCANO (ASH CONE, CINDER CONE) or with reference to the surface morphology of a glacier (DIRT CONE). **2** When used to describe a fan-like deposit the term cone is really a misnomer, although it has widespread usage (ALLUVIAL CONE). **3** The fruit of a conifer (CONIFEROUS FOREST).

cone of dejection ALLUVIAL CONE.

cone of depression CONE OF EXHAUSTION.

cone of exhaustion A hydrological term referring to the shape of the local depression of the WATER-TABLE or any PIEZOMETRIC SURFACE around a well. It is caused by the extraction of water, by pumping or artesian flow, at a rate greater than that at which the AQUIFER is being recharged. In the USA this is referred to as a *cone of depression*.

cone-sheet An igneous DYKE in the shape of a funnel or inverted hollow cone which inclines downwards and inwards towards the top of an igneous intrusion. Cone-sheets usually occur in swarms around a centre since they have formed as a result of the fractures in the country-rock created by the injection of the igneous intrusion. *See also* RING DYKE. [*47*]

confidence limits A term used in statistics when referring to the degrees of confidence that can be placed upon estimates of STANDARD DEVIATIONS or true means when they are based on only a sample of the data. A confidence limit implies the proportion of times that a given outcome may be expected to occur by chance in a statistical analysis. Thus, a 99% confidence limit implies that the given outcome will be expected to occur by chance only once in 100 times.

confluence **1** A term used in climatology to describe the condition during which the streamlines depicting a horizontal flow of air converge upon each other (CONVERGENCE). Confluence causes an increase in the velocity of air particles, but no mass accumulation. **2** The point at which two streams meet.

conformable The unbroken relationship between rock strata which lie one above another in parallel sequence. This geological arrangement demonstrates that there has been a lengthy period of deposition, undisturbed or uninterrupted by earth movements or by denudation. Where such disturbance has occurred the sequence will be broken and is therefore said to be unconformable (UNCONFORMITY).

conformal projection ORTHOMORPHIC PROJECTION.

congelifluction A geomorphological term, proposed in 1951 by J. Dylik, to denote the progressive flow of earth (SOLIFLUCTION) in conditions of a permanently frozen subsoil (PERMAFROST). Its meaning differs from that of GELIFLUCTION, which also includes seasonally frozen ground. *See also* CONGELIFRACTION, CONGELITURBATION.

congelifraction A geomorphological term, proposed in 1946 by K. Bryan, to denote the mechanical weathering of rocks by the freezing of interstitial water, leading to expansion, fracturing and disintegration of the rocks. The effectiveness of frost-splitting depends on the size of the pores. If the pores are very small the water may become supercooled but remain in a liquid state. *See also* CONGELIFLUCTION, CONGELITURBATION, FREEZE–THAW ACTION.

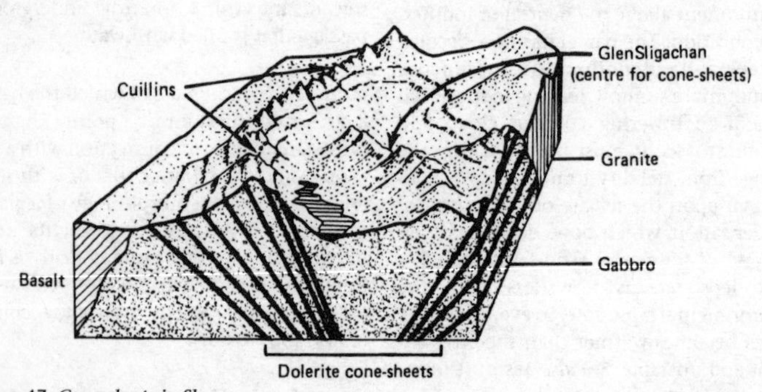

Figure 47 *Cone-sheets in Skye*

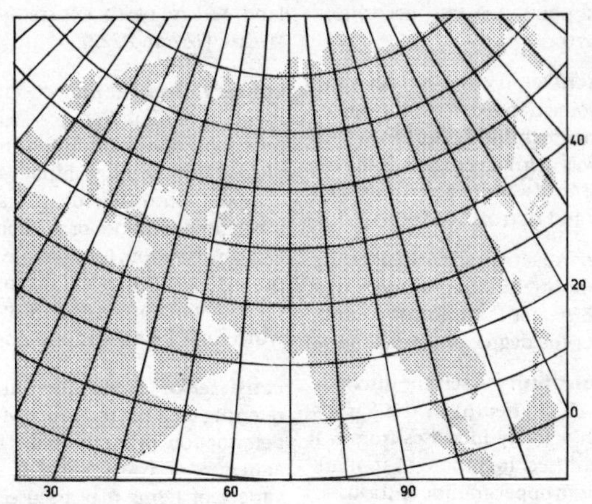

Figure 48 *Conical projection*

congeliturbation A geomorphological term, proposed in 1946 by K. Bryan, to denote the general action of frost disturbance in the ground surface. It involves both large-scale CONGELIFLUCTION as well as local churning and frost-heaving of the soil. *See also* BRODEL, CRYOTURBATION, INVOLUTION, PATTERNED GROUND, SOLIFLUCTION.

conglomerate A sedimentary rock consisting of rounded or sub-rounded fragments, usually water-worn pebbles, cemented together by a matrix of calcium carbonate, silica, etc. The following varieties can be recognized: **1** the *oligomict*, comprising only one type of pebble; **2** the *polymict*, comprising a mixture of pebble types; **3** the TILLITE, or indurated glacial TILL; **4** the *autochthonous conglomerate*, formed by the break-up *in situ* of an existing rock. It is popularly termed a *puddingstone*. *See also* AGGLOMERATE, BRECCIA, CLASTIC SEDIMENTS.

conical projection, conic projection A MAP PROJECTION constructed by projecting the geographic meridians and parallels of latitude on to a cone which is tangential to or intersects the surface of a sphere, with the cone touching the sphere along one or more STANDARD PARALLELS which themselves are their true distance apart. The scale is correct only along the standard parallels, all the other parallels being too long, but it is true along all the meridians. The projection is not ORTHOMORPHIC nor equal-area. The *conical projection with two standard parallels* is often used for maps of a small continent (e.g. Europe) or a particular country. *See also* BONNE'S PROJECTION, LAMBERT'S PROJECTIONS, POLYCONIC PROJECTION. [*48*]

coniferous forest A forest type characterized by cone-bearing, needle-leaved trees. They are generally, but not necessarily, evergreen and relatively shallow-rooted. Since they grow more rapidly than most broad-leaved trees, conifers are extensively planted as a source of softwood timber and pulp. They are tolerant of wide-ranging climatic conditions, of many different types of soil and of considerable differences in terrain. Thus, they are found from the polar latitudes to the tropics, on most types of soils (especially, thin acid soils) and from mountain summits to coastal environments.

coniology KONIOLOGY.

conjunction A term referring to the relative positions of two heavenly bodies in relation to the planet Earth. Thus, when the Sun and the Moon are in a straight line on the same side of the Earth they are said to be in conjunction, resulting in strong tidal forces (SPRING TIDES). *See also* OPPOSITION, QUADRATURE, SYZYGY.

connate water Ocean water or fresh water which has become trapped in the interstices of a sedimentary rock during the time when the material was being deposited. It is synonymous with *fossil water* but contrasts with JUVENILE WATER and METEORIC WATER.

connectivity A statistical term referring to the degree of interlinking of nodes in a network – the greater the number of linkages the higher the degree of connectivity.

Conrad discontinuity A seismic discontinuity (at 10–12 km depth) in the Earth's crust at which velocity increases from 6.1 to 6.4 or 6.7 km/sec. It is thought to mark the junction of an upper granitic and a lower basaltic layer and can occur at a variety of depths. It occurs *only* beneath the continents, since beneath the ocean floors the upper layer is missing.

Conrad's formula of continentality An empirical attempt to define the spatial limits of a CONTINENTAL CLIMATE. Conrad's formula shows continentality (k) as:

$$k = \frac{1{\cdot}7\,A}{\sin\,(\phi + 10)} - 14$$

where A is the average annual temperature range in °C and ϕ is the latitude angle. This is similar to GORCZYNSKI'S INDEX OF CONTINENTALITY, differing mainly in the relative magnitude of k. *See also* BERG'S INDEX OF CONTINENTALITY, CONTINENTAL CLIMATE.

consequent A geomorphological term, introduced in 1875 by J. W. Powell, to describe any surface movement having a course of direction controlled by the original geological fold-structures. Thus, its meaning was intended to refer to external transfer of surface material consequent upon the slope produced by the structure, not simply the downslope movement of water. But modern usage tends to confine the term to drainage direction (CONSEQUENT STREAM).

consequent stream A stream in which the direction of flow is controlled by or dependent upon the original slope of the land. *See also* OBSEQUENT STREAM, SUBSEQUENT STREAM. [*167* and *222*]

conservation The protection and management of natural resources (especially the renewable ones) with a view to a sustained yield. This generally implies careful planning and control in order to maintain either a balanced land use or a stable ecosystem. Soil conservation, for example, includes the protection of the soil against chemical deterioration or physical loss by erosion, which may be natural or man-induced.

consistence A term used in soil science, referring to: **1** the resistance of a material to deformation or rupture; **2** The degree of adhesion or cohesion of a soil mass. A variety of terms may be used to describe the consistence of a soil, e.g. sticky, plastic, friable, compact, loose, indurated.

consociation A biogeographical term referring to any unit of vegetation which is dominated by a single species. *See also* ASSOCIATION.

consolidation 1 In geological use, a term referring to any or all of the processes which cause loose or soft earth materials to become coherent or firm. **2** In soil science, the term refers to the gradual reduction in volume of a soil mass, as a result of increased compression or 'load'. In both uses it involves a decrease in VOID RATIO by the squeezing of water from the pores.

constant A statistical term referring to a value in an equation which does not vary. Thus, *constancy* refers to a state in which values do not change.

constant of channel maintenance A geomorphological and hydrological term referring to the surface area of a river basin that is necessary to support a unit length of stream channel. Although the constant for a given drainage basin will not vary, it will differ from that of another drainage basin if the latter exhibits different terrain, soils, geology, climate and vegetation, all of which influence such factors as PERMEABILITY and RUNOFF.

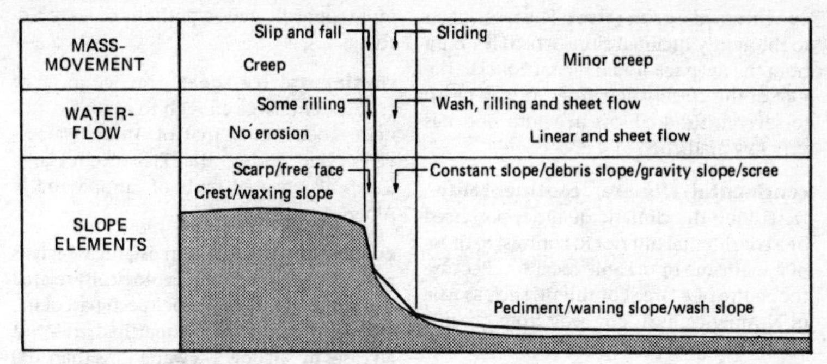

Figure 49 *Standard hill slope*

constant slope The part of a slope profile which stands above the WANING SLOPE and below the FREE FACE and the WAXING SLOPE. It is a slope element produced by the accumulation of rock debris from above and therefore coincides with the *talus slope* or SCREE. Its gradient is determined by the ANGLE OF REPOSE of the fallen debris and is therefore frequently referred to as the *debris slope*. It must be emphasized that the elements shown in [*49*] are based on theoretical deductions relating to a standard hillslope morphology as defined by A. Wood and L. C. King, and that as the slope profile develops over time the form and properties of the slope elements will change. Thus, weathering of the fallen material reduces the debris to finer particles which are subsequently removed farther downslope by wash to produce a concavity, termed the waning slope. As the waning slope rises up the side of the *constant slope* the latter is gradually consumed and disappears. [*49*] *See also* THRESHOLD ANGLE.

constructive wave A water wave breaking on a beach, with a powerful SWASH and a weaker BACKWASH. Thus, it tends to move material up the slope of the beach (BEACH-RIDGE).

contact The interface or boundary between adjacent but dissimilar rock types. Its recognition is an important principle of LITHOSTRATIGRAPHY. The contact may be CONFORMABLE or unconformable (UNCONFORMITY) within a succession of sedimentary rocks; it may be produced by bringing dissimilar rocks together by faulting; or it may result from the injection of an igneous intrusion.

contact (thermal) metamorphism A type of METAMORPHISM related to the intrusion or extrusion of hot magma into or on to existing country-rocks. Where these rocks come into contact with the magma they will undergo a recrystallization of their minerals and a growth of new minerals. Since the metamorphism is by heat alone and is devoid of significant pressure effects, there is little or no deformation of the country-rock and generally no accompanying FOLIATION.

contiguity The state of being in contact or contiguous, or touching without being fused together.

continent A large land mass rising fairly abruptly from a deep ocean floor, although its peripheral areas may be submerged, as a CONTINENTAL SHELF beneath shallow seas. Continents currently occupy almost one-third of the global surface, with the bulk of the continents occurring in the N hemisphere. They are composed essentially of SIAL (CONTINENTAL CRUST).

continental air mass An air mass, usually of low humidity, formed over a continental interior where high atmospheric pressure prevails. It may originate either from a low-latitude source-region (tropical, Tc) or form in a high-latitude source-region (polar, Pc). *See also* AIR MASS.

continental apron (rise) The term given to the gently inclined slope which leads up from the deep sea-floor (ABYSSAL ZONE) to the foot of the CONTINENTAL SLOPE. It may be up to several hundred km in width and has very low gradients, of < 1°.

continental climate, continentality Describing the climatic qualities possessed by a continental interior in contrast to those of a maritime or oceanic location. Because the centre of a large continent, such as Asia or N America, is a long way from oceanic influences, it is characterized by low rainfall amounts with a summer (convectional) maximum and great contrasts of seasonal temperature, and by generally hot summers and cold winters. 'Continentality' indices have been devised by several climatologists, based upon the relationships between mean annual temperature and the mean temperatures of the warmest and coldest months. *See also* AIR MASS, BERG'S INDEX OF CONTINENTALITY, CONRAD'S FORMULA OF CONTINENTALITY, EXTREMES OF CLIMATE, GORCZYNSKI'S INDEX OF CONTINENTALITY.

continental crust That part of the Earth's crust which is composed of SIAL and constitutes the continents. Although its average thickness is 33 km it achieves thicknesses of over 50 km beneath high mountain regions. *See also* ISOSTASY, OCEANIC CRUST.

continental drift A concept, initiated in 1858 by A. Snider but developed and popularized by F. B. Taylor (1908) and A. Wegener (1915), which suggested that continents can move around the Earth's surface because of the weakness of the suboceanic crust. Although it was originally based on the apparent 'jig-saw' fit of the opposing coasts of the Atlantic Ocean, much evidence of matching fossils, geological structures, etc. was accumulated to support the former movements, although no adequate mechanisms were advocated. It was suggested by Wegener that the world's continents had been derived from the breakup of the two super-continents of Gondwanaland and Laurasia, which had themselves been united as Pangaea in pre-Mesozoic times. The entire concept has been replaced by the more sophisticated hypothesis of PLATE TECTONICS.

continental ice-sheet An ice-sheet of great extent, large enough to cover a continent. Today, only that of Antarctica survives, but during the Pleistocene large ice-sheets covered parts of Europe and N America.

continental island An island which is geographically near and geologically related to a continent, having once been part of the mainland but now being detached in almost all cases by shallow sea water (less than 100 fathoms in depth). Thus, the British Isles and Sri Lanka are continental islands, in contrast to the islands (termed *oceanic islands*) which are geographically isolated and rise from the floors of deep ocean basins.

continental margin (terrace) A zone which combines both the CONTINENTAL SHELF and the CONTINENTAL SLOPE and is distinct from the deep sea-floor. It extends from the coastline down to depths of some 2,000 m.

continental platform A term which refers to a CONTINENT together with its CONTINENTAL SHELF but does not include the CONTINENTAL SLOPE. It is composed of SIAL and therefore reflects the true extent of a continent in a crustal sense (CONTINENTAL CRUST).

continental rise CONTINENTAL APRON.

continental sea A partially enclosed sea, linked with the open ocean but lying wholly within the area of a continental shelf; e.g. the Black Sea, the Baltic, Hudson Bay. Synonymous with *epicontinental*. *See also* EPEIRIC SEA.

continental shelf A gently sloping (< 1° gradient) offshore extension of a CONTINENT, submerged by a shallow sea (usually taken as an average depth of 130 m) and which extends to the top of the CONTINENTAL SLOPE. In general, it is structurally similar to the contiguous continent, although it may be blanketed by recent marine or glacial deposits. Some continental shelves represent submerged platforms of marine abrasion (ABRASION PLATFORM) or drowned surfaces of glacial erosion. They may be very

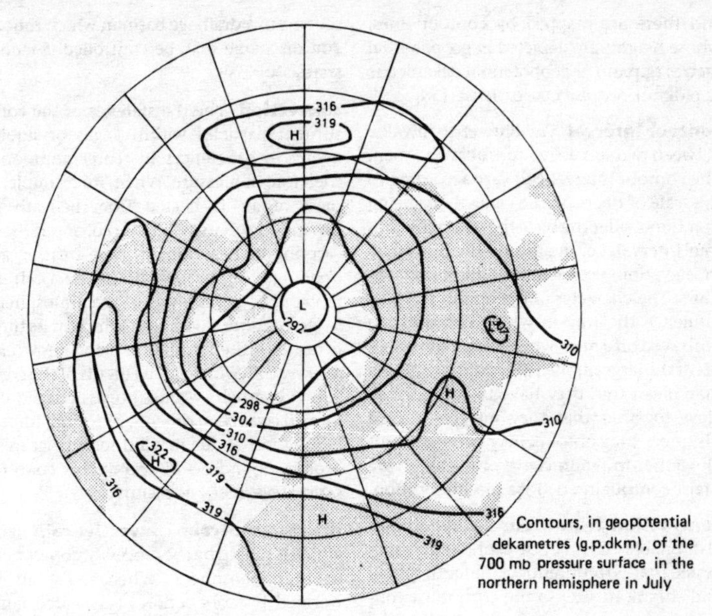

Figure 50 *Contour chart*

wide (560 km off Argentina) or virtually nonexistent (off Chile). Many of them are dissected by valleys and troughs (SUBMARINE CANYON). Continental shelves occupy 7.6% of the ocean floor and are usually of great importance as fishing-grounds. [*162*]

continental slope The continuously sloping portion of the CONTINENTAL MARGIN, seaward of the CONTINENTAL SHELF, and extending down to the deep sea-floor of the ABYSSAL ZONE. It is characterized by gradients between 2° and 5° and by considerable submarine sliding and slumping of the marine sediments. It is cut through by SUBMARINE CANYONS, which debouch on to the deep sea-floor. It constitutes about 8.5% of the ocean floor. [*162*]

continuous variable A variable which, in principle, can take any value within its range, e.g. temperature or volume. *See also* DISCRETE VARIABLE.

contorted drift 1 A term used to describe the crumpled and folded appearance of glacial TILL where it has been deformed by an overriding ice-sheet or by the pressure of an ice-front (PUSH MORAINES). **2** When used as a proper noun, the term *Contorted Drift* refers to the uppermost till in the coastal cliffs at Cromer, East Anglia.

contour 1 A line connecting points on a topographical map or chart, all of which exhibit the same height above or below a given datum (sea-level or ORDNANCE DATUM). Contours below sea-level are termed *submarine contours*. The measurements may be in feet, metres or fathoms (submarine contours). *See also* FORM-LINE, SUPPLEMENTARY CONTOUR. **2** In the context of meteorology, a line used to represent upper air pressure differences on a CONTOUR CHART. [*50*]

contour chart A term used by meteorologists to describe a map showing the pressure patterns at selected heights above the Earth's surface. A particular pressure surface (expressed in millibars) exhibits various undulations within the middle troposphere and these are mapped by contour lines, whose heights are depicted in geopotential metres (g.p.m.) or geopotential dekametres (g.p.dkm.). *See also* GEOPOTENTIAL. [*50*]

contour interval The difference in value between two successive CONTOURS on a map. The contour interval will vary according to the scale of the map, the general rule being that the smaller the scale the wider the contour interval, i.e. the greater the difference in elevation between two adjacent contours. The character of the terrain will also influence the way in which contours are portrayed on a map; where very steep slopes occur the large number of required contours may mean that they have to be drawn so close together that they merge. In such instances the contours may be missed out altogether to retain clarity, thereby giving a false contour interval at a specific location.

contraction hypothesis A hypothesis, fashionable in the 19th cent., which suggested that as the interior of the Earth cooled and shrank in size, so the crust must contract in order to accommodate itself to the smaller bulk. The resulting compression was thought to be responsible for the world's fold mountains and the thrusting mechanisms, all created by crustal shortening.

contrail CONDENSATION TRAIL.

contributing area The area whose soil becomes saturated during a storm, thereby permitting the process of OVERLAND FLOW to occur.

control point, station A point on the ground, and hence on a map, whose exact position and elevation have been accurately determined. Control points are used as constants in a network during a survey which leads to the compilation of a map, so that other details can be plotted at their correct positions and elevations.

control system A PROCESS–RESPONSE SYSTEM in which key components are controlled by some intelligence, thereby causing the system to function in a manner controlled or determined by the intelligence. [*249*] An example of a control system would be the introduction of a river-management scheme in a drainage basin in which runoff and discharge may be controlled. *See also* SYSTEMS ANALYSIS.

convection The transference of the constituent particles within a gas or liquid owing to differences in temperature and therefore in pressure. When, for example, a mass of air is heated from beneath it expands, its density is reduced, and it rises, carrying its heat with it, thus creating an updraught of heated air (THERMAL). It is replaced at lower levels of the atmosphere by descending, cooling air which in its turn is heated – the circulation being known as a *convection cell*. The movements of the constituent particles within the gas or liquid are termed *convection currents*. The term 'forced convection' is due to wind TURBULENCE over broken ground. *See also* CONVECTIVE CONDENSATION LEVEL (CCL), INSTABILITY.

convectional rain, convective rain Rain which is caused by the process of CONVECTION in the atmosphere. When moist air is warmed by CONDUCTION at a heated land surface it expands, its density falls, and it rises convectionally. It is thereby cooled adiabatically and its temperature falls below the DEW-POINT, thereby forming clouds, often of the CUMULONIMBUS form, from which heavy rainfall is generated. Rain of this type is associated with EQUATORIAL CLIMATES and with the cold front of an unstable POLAR AIR-MASS. *See also* CLOUDBURST, THUNDERSTORM.

convective condensation level (CCL) The pressure altitude in the ATMOSPHERE at which condensation (and cumuliform cloud formation) occurs, and in which parcels of air rise by CONVECTION. *See also* LAPSE-RATE and [*132*].

conventional projection A map PROJECTION that is constructed from mathematical formulae, in contrast to a projection based on a geometrical construction (PERSPECTIVE PROJECTION).

convergence **1** In CLIMATOLOGY, the term describes a type of horizontal air-movement in which inflow exceeds outflow, thereby creating a tendency for air to accumulate at

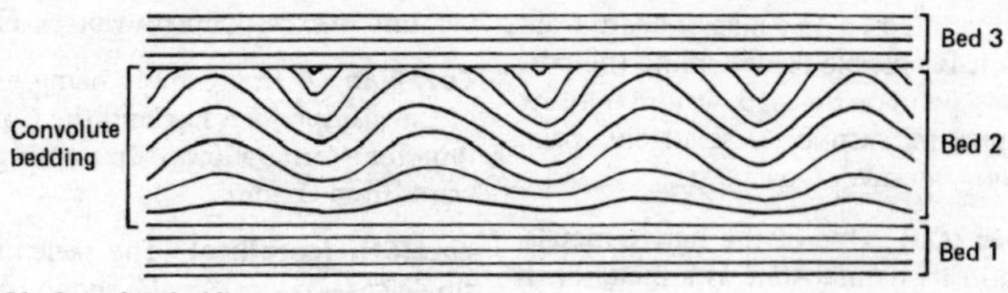

Figure 51 *Convolute bedding*

a particular place in the atmosphere. Two types of air-convergence have been recognized: *streamline convergence* is created when STREAMLINES approach each other although remaining at constant velocity; when a sole air current suffers a progressive reduction of its velocity although its direction remains constant, it is termed *istotach convergence*. If air density remains constant during horizontal convergence a vertical current of ascending air must result. *See also* INTERTROPICAL CONVERGENCE ZONE. **2** In OCEANOGRAPHY the term describes a very defined line at which converging masses of sea water meet. Generally these masses will have different temperature and salinity characteristics, as in the meeting of the GULF STREAM with colder N polar currents. *See also* DISCONTINUITY, DIVERGENCE. **3** In BIOGEOGRAPHY the term refers to the increasing degree of similarity exhibited by different species (plant or animal) which are evolving in such a way that their life-forms become progressively similar. Such convergence can be demonstrated by a study of fossil sequences (PALAEOLOGICAL CONVERGENCE). *See also* CONVERGENT EVOLUTION.

convergent evolution The process whereby organisms that are not closely related produce similar-appearing forms, despite the fact that they are not as closely related genetically as they appear to be. Convergent evolution is thought to result from adaptation to similar environmental conditions; e.g. the similar characters evolved by placental mammals and marsupials which occupied similar habitats but which were isolated on land masses great distances apart.

convexity One of the morphological characteristics of a slope, in which the gradient becomes progressively steeper as it is traced downwards from the top. It is thought to be a result of either MASS-MOVEMENT or structural control (SHEETING). *See also* WAXING SLOPE.

conveyor belt (in depression) A term used to describe a broad airflow (several km deep and a few hundred km wide) which moves parallel to and a short distance ahead of a surface COLD FRONT. In a partly occluded depression the air-flow then rises over the WARM FRONT and turns equatorwards ahead of it. Because of this overriding mechanism convective instability is generated, generally leading to a belt of heavy precipitation developing in the frontal zone. *See also* DEPRESSION.

convolute bedding A term referring to the appearance of a crumpled or highly disturbed layer in a well-bedded sedimentary succession. The distortion is confined to only one bed and is usually abruptly truncated by an overlying undisturbed sedimentary bed. The feature is thought to be the result of: **1** post-depositional slumping and gliding; **2** contemporaneous loading, leading to the expulsion of pore waters; **3** passing eddies from high-velocity currents during deposition. [*51*]

coombe rock A structureless mass of unstratified rubble, consisting of flints, broken chalk fragments, compacted chalk mud and some sand. It was produced by SOLIFLUCTION during cold phases of the Pleistocene period and has accumulated in great thicknesses in the valley bottoms of the chalklands of S England. *See also* DRY VALLEY, HEAD.

co-ordinates Linear or angular quantities, expressed in the form of a two-dimensional grid superimposed on a map or chart, from which it is possible to determine the position of a point on the map. *See also* ABSOLUTE CO-ORDINATES, EASTING, FALSE ORIGIN, GRID, NORTHING, RELATIVE CO-ORDINATES.

copper (Cu) One of the native metals, found in its natural state as a metal or in association with other minerals as a carbonate, oxide or sulphide. It occurs in the cavities of basic igneous rocks and in veins, and is the product of HYDROTHERMAL ACTIVITY and METASOMATISM.

coprolite A geological term referring to the petrified faeces of animals, particularly the Mesozoic dinosaurs. Because they are rich in calcium phosphate coprolites are a valuable source of fertilizer. In Britain they are quarried as nodules in the Cretaceous rocks of E England, especially near Cambridge.

coquina A carbonate rock consisting of poorly/moderately cemented fossil debris (especially shells) in which the interstices often remain unfilled with a matrix.

coral 1 A marine polyp capable of secreting calcium carbonate to build an external skeleton. It flourishes in clear, warm, tropical oceans, usually between 30°N and 30°S of the equator, but is occasionally found in warm seas (>20° C) outside these limits. Corals rarely live at depths below 55 m and will not tolerate fresh or muddy waters. Because of the cold, upwelling currents along the western coasts of continents in the tropical zone, corals are much more common on eastern coasts. **2** The term *coral rock* is often used to describe the coral fossils which build the reef limestones of certain sedimentary formations. They were laid down mainly from the ORDOVICIAN to the CARBONIFEROUS, although some reef limestones extend as far as the Middle JURASSIC.

coral bleaching A term referring to the so far inexplicable loss of coral's colourful symbiotic algae, usually leading to death of the CORAL REEF. It has been suggested that, since greatest bleaching occurs in shallower water, the phenomenon may be explained by CLIMATIC CHANGE, e.g. increased water temperature, or excessive ULTRA-VIOLET RADIATION.

Corallian A stratigraphic name for part of the JURASSIC. In S England the Corallian limestone forms a low scarp running westwards from Oxford.

corallith (corallite) The skeleton of a single CORAL polyp. *See also* RHODOLITH.

coral mud An accumulation of very fine coral fragments which lie on the lower distal zone of the CONTINENTAL SLOPE near to CORAL REEFS.

coral reef A REEF created by the coral polyp along the shallow shores of some tropical seas. It comprises a complex of several subclasses of true coral growing *in situ*, together with debris of littoral shells, algal material and chemically precipitated carbonates of calcium and magnesium. The true corals may comprise only half of the total bulk of the limestone reef, although its existence and its continued incremental growth depend on the successful existence of the coral polyp. *See also* ATOLL, BARRIER REEF, FRINGING REEF, REEF-KNOLL. [*19*]

coral sand An accumulation of fine white sand composed of comminuted coral fragments bordering a CORAL REEF. As it is traced farther away from the reef coral sand grades into an even finer material termed CORAL MUD.

co-range line 1 A line or isopleth drawn on a map joining up those climatic stations which exhibit the same temperature range between January and July. **2** A line joining points with an equal tidal range. [*9*]

cordillera A Spanish term referring to a major system or group of mountains. It includes not only the parallel mountain ranges but also the intervening valleys, plateaux, intermontane basins, etc. It usually refers to an orogenic belt at a continental scale, e.g. the Western Cordillera of the USA, which includes all the ranges between the Pacific and the Great Plains, although individual mountain systems are also called cordilleras.

core The central part of planet Earth, probably consisting of a dense nickel-iron alloy with a temperature estimated at about 2,700° K. The outer perimeter of the core commences at the GUTENBERG DISCONTINUITY, 2,900 km from the surface, where the outer core may be liquid. Within this there appears to be a much denser inner core with a radius of about 1,400 km and which may be solid. [*227*]

core sampling A method whereby relatively undisturbed samples of soil, peat, rock or ice can be withdrawn from the solid material under examination. The samples are obtained by driving a hollow metal tube or cylinder into the material and withdrawing it in the form of a lengthy tubular section or core. There are many types of coring devices, including gravity corers, which penetrate by their own weight, and piston corers which are forced into the sediment by percussion. Core sampling is an important part of sea-floor analysis (OCEANOGRAPHY) and in pollen studies (PALYNOLOGY) of organic sediments. Core drilling with a hollow bit and a core barrel is one of the chief methods of securing geological information, both in mineral-wealth investigation and to ascertain rock structures for load-bearing capacity in civil-engineering studies.

corestone A residual block of hard, undecomposed rock, surrounded by a mass of softer decomposed material. It is formed in well-jointed rocks, particularly granites and arkoses, in which deep subsurface rotting by CHEMICAL WEATHERING has proceeded more rapidly along the joints, thereby leaving the intervening unjoined sections relatively unweathered. Ultimately the corestones themselves will be consumed by decomposition, unless there is a stripping of the weathered incoherent material, thus revealing the corestone complex at the surface. Some authorities believe that a TOR is formed in this way, with the upstanding rock mass being composed essentially of a series of unconsumed corestones (sometimes referred to as woolsacks). *See also* SPHEROIDAL WEATHERING. [*239*]

Coriolis force A deflecting motion or force discussed by G. G. de Coriolis in 1835 and developed by W. Ferrel in 1855 (FERREL'S LAW). The rotation of the Earth causes a body moving across its surface to be deflected to the right in the N hemisphere and to the left in the S hemisphere. The Coriolis force is defined as $2\,vw \sin \phi$, where v = velocity of the object, w = angular velocity of the Earth's rotation and ϕ is the latitude. *See also* PLANETARY VORTICITY. [*90* and *215*] *See also* ATMOSPHERIC CIRCULATION, BUYS BALLOT'S LAW, PLANETARY WINDS.

cornbrash A name given to a thin limestone layer in the Middle and Upper JURASSIC rocks of S England. The name was given to this rock stratum in 1813 by the famous geologist William Smith, who took it from the West Country dialect term for a stony soil (a 'brash'): since its outcrop coincided with excellent cereal crops the name cornbrash was advocated.

cornice An overhanging ledge of compacted snow on the leeward side of a snow-covered mountain ridge or cliff edge. It is formed by a wind EDDY and represents a hazard to climbers and skiers because of its inherent instability and liability to fracture without warning.

corniche A term referring to the inter-tidal organic protrusion that grows out of steep coastal rock surfaces at about mean sea-level. They often consist of calcareous algae.

cornstone A concretionary limestone, usually formed under desert conditions. It is typical of the DEVONIAN and the NEW RED SANDSTONE in Britain and derives its name from its early use in corn-grinding.

corona **1** The coloured rings which surround the Sun or Moon, comprising a range from blue (inner) through green and yellow to red (outer). Often only the bluish inner ring and the outer reddish ring are discernible. Since the corona is produced by DIFFRACTION of light by water-droplets it must be contrasted with the HALO which is due to refraction in ice crystals and in which the colour sequence is opposite. **2** The fringe of radiant light which is discernible around the dark rim of the Moon during a total *lunar eclipse*, or around the dark rim of the Sun during a total *solar eclipse*.

corrasion The process of mechanical erosion of a rock surface caused when materials are being transported across it by running water, by glaciers, by wind, by marine waves, or by gravity movement downslope (MASS-MOVEMENT). The resulting effect on the rock surface is termed ABRASION.

correlation, statistical A statistical term referring to the degree of relationship which exists between two or more variables, such that when one changes the other also changes. The term is used in a number of ways: **1** *Autocorrelation* refers to a relationship between members of a series of samples ordered in space and in time, which is due to a dependence between them. **2** *Coefficient of correlation* (r): an index of interdependence between two variables. **3** *Coefficient of determination* (r^2): the square of the *coefficient of correlation*, expressing the proportion of variation in the dependent variable explained by the association with the independent variable. **4** *Coefficient of multiple correlation* (R): an index or measure akin to r but giving a precise value to the relationship involving more than one independent variable. **5** *Coefficient of multiple determination* (R^2): the square of the coefficient of multiple correlation, expressing the proportion of variation in the dependent variable explained by the joint association with the independent variables. **6** *Multiple correlation*: the correlation between two or more independent variables and one dependent variable. **7** *Negative correlation*: the relationship in which one variable increases as the other decreases. **8** *Partial correlation*: the correlation between two variables when a third, which is related to both, is controlled. **9** *Product-moment*: a parametric test for correlation between two variables on a ratio or interval scale. **10** *Spearman rank*: a non-parametric test for correlation between two variables expressed on a rank or ordinal scale.

correlation, stratigraphical A geological term which refers to the process by which two stratigraphic units or formations may be equated, usually in a time relationship, even though they are spatially separated. Rock strata can be correlated on the basis of: lithological correlation; biostratigraphical correlation (BIOSTRATIGRAPHY); radiometric correlation (RADIOMETRIC DATING); and palaeomagnetic correlation (PALAEOMAGNETISM).

correlation system A system that is identified by virtue of the significant correlations which link its various components.

corrie A Scottish term synonymous with CIRQUE.

corrosion CHEMICAL WEATHERING.

co-seismal line A line which connects those points at the Earth's surface where SEISMIC WAVES arrive at the same time. *See also* ISOSEISMAL line.

cosmic A term applied to phenomena originating beyond the Earth's atmosphere, e.g. *cosmic dust, cosmic radiation. Cosmogeny* is a term referring to theoretical investigations of the origin of the solar system; *cosmography* is a term no longer common but once used by early 'geographers' to describe mapping of the universe; *cosmology* is a modern term which embraces all attempts to study the laws of the universe by current scientific methods. *Cosmos* = an ordered universe (in contrast to chaos).

cosmogeny COSMIC.

cosmography COSMIC.

cosmology COSMIC.

costa A Spanish term referring to a stretch of coastline. *See also* CÔTE.

cost/benefit ratio The comparison of the economic results achieved with the costs of a project or operation, expressed as a ratio. It is a technique that is being increasingly used in planning the use of natural resources.

côte A French term which has two meanings: **1** A stretch of coastline, e.g. Côte d'Azur. **2** An escarpment or steep hill slope, e.g. the numerous *côtes* in the wine-growing country of Burgundy.

co-tidal line A line drawn on certain hydrographic (HYDROGRAPH) charts joining points at which high water (sometimes low water) occurs simultaneously. Co-tidal lines radiate from an AMPHIDROMIC point and are usually expressed as departures from the times of high water (sometimes low water) at a particular port. Alternatively, they may be depicted as time intervals following the lunar transit time. All the British tidal information is revised annually and published in the Admiralty tide tables. [*9*]

coulée (US coulee) A French term with several distinct meanings: **1** A steep-sided, congealed lava flow, generally of glassy rhyolite or obsidian. **2** A glacial MELTWATER CHANNEL, now abandoned, in the northern states of the USA. **3** Occasionally applied to a gorge-like stream valley in the USA. **4** A debris tongue of SOLIFLUCTION material formed by PERIGLACIAL processes.

couloir A French term for a narrow gully with a steep gradient in a mountainous terrain. Its use is generally restricted to gullies which dissect a precipitous mountain slope.

Coulomb's failure law MOHR–COULOMB EQUATION.

country-rock A general term for any type of rock that is penetrated by an igneous intrusion or invaded by a mineral VEIN.

covariance A statistical term referring to the relationship between two variables. It is expressed as $\Sigma xy/n$, where x and y are the differences between each variate and the mean of the set, while n is the number of pairs, and Σ is the sum of all the figures that follow.

cove **1** A small coastal bay, often with a narrow entrance. **2** A steep-sided hollow in the English Lake District. **3** A valley or portion of lowland that penetrates into a plateau or mountain front (RE-ENTRANT) in the USA.

covered karst A limestone terrain (KARST) in which the characteristic karstic features are buried beneath a cover of superficial materials: alluvium, blown sand, glacial till, etc. The drift-covered Carboniferous limestone of the Central Plain of Ireland is a good example.

cover rocks A term referring to the younger sedimentary rocks which overlie the BASEMENT rocks in an unconformable manner (UNCONFORMITY). Their structures are usually less complicated than those of the basement and the unconformity usually represents a lengthy period of erosion prior to the deposition of the cover rocks. The New Red Sandstones of the English Midlands, for example, can be regarded as cover rocks overlying a variety of older folded and faulted rocks of Palaeozoic and Pre-Cambrian age.

coversand A usually thin cover of blown sand, often modified by subsequent fluvial or periglacial reworking. Extensively found peripheral to the European Pleistocene ice-sheets. *See also* LOESS.

crachin A local term to describe the low stratus cloud and its associated drizzle which affects the Gulf of Tonkin and the coast of South China in late winter/early spring.

crag **1** As a geological term, crag refers to sedimentary rock comprising a shelly, marly sand of marine origin. **2** As a proper noun it is used in East Anglia for certain rock formations of Plio-Pleistocene age (e.g. Coralline Crag). **3** As a topographical term it refers to a precipitous, rugged rock face or rocky BUTTRESS on a mountainside.

crag-and-tail A topographical feature in which a resistant mass of rock (the crag) has withstood the passage of an ice-sheet, thereby protecting an elongated ridge (the tail) of more easily eroded rocks on its lee-ward side (e.g. the igneous PLUG of Castle Rock in Edinburgh and the gently sloping tail of sedimentary rocks followed by the Royal Mile). In some instances the tail may be composed of TILL which has survived the erosional powers of the ice-sheet by virtue of its sheltered position.

crater **1** The funnel-shaped basin surrounding the VENT at the summit of a VOLCANO, or on its flanks. It is surrounded by very steep inward facing cliffs and may be several hundred metres in depth. Its floor may contain a LAVA LAKE or may be composed of layers of ejected material. It is formed by either a major ERUPTION or the collapse of a volcanic CONE (ENGULFMENT). **2** The circular depression caused by the impact of a METEORITE. *See also* CALDERA.

crater-lake A fresh-water lake occupying the CRATER of a dormant VOLCANO. It is generally circular in shape and often of a brilliant colour.

craton, kraton A large section of the Earth's crust that has remained relatively stable and unaffected by mountain-building (OROGENY) for a considerable period of geological time. It is made up of ancient rocks (SHIELD) which were once parts of very old mountain ranges but have remained immobile since Pre-Cambrian times. Any feature relating to a craton is termed *cratogenic*. [*91*]

creationism The belief that all matter (organisms) was created by divine means. It is directly opposed by scientists who support DARWINISM and EVOLUTION.

creek **1** A tidal channel in a coastal marsh or between estuarine mud-banks. **2** In parts of the USA any type of small stream. **3** In the arid SW of the USA and in central Australia a shallow stream of intermittent flow (ARROYO). **4** In the eastern coastal states of the USA a wide embayment in a river estuary. *See also* DRAW, NULLAH, WADI.

creep **1** The imperceptible but continuous movement of rock debris (REGOLITH) and soil down a slope (MASS-MOVEMENT) in response to gravity. There is no general agreement on its mechanisms, with opinion being divided between (a) those who believe that the movement is essentially VISCOUS, as a type of flow in which there is internal and permanent deformation and where the stresses are too small to create shear FAILURE (LANDSLIDE); and (b) those who subscribe to the belief that the whole surface mass slips over the underlying rock. Creep occurs under all types of climate where there is a slope gradient, but is thought to be one of the most significant of the geomorphic processes in the humid maritime parts of the PERIGLACIAL environment. It has been suggested that, although the presence of soil water is essential for lubrication, many processes may be operative, singly or in combination: *See also* FREEZE–THAW ACTION, NEEDLE ICE, SOLIFLUCTION. **2** The term creep is also used to describe the gradual deformation of the basal ice of a COLD GLACIER owing to internal pressure and accompanying intergranular motion. **3** In civil engineering, the term *creep* refers to any general slow displacement of soil and subsoil under load. [*146*]

crepuscular rays **1** The rays of sunlight which shine earthwards through small breaks in heavy cloud layers, often of STRATOCUMULUS variety. **2** The rays of sunlight, alternating with 'rays' of shadow, which diverge upwards immediately following the setting of the sun. The light rays represent true sunlight from below the horizon; they are caused by interruptions of the sunlight by hills or by cloud masses.

crest **1** The highest part of any projection on the summit ridge of a mountain. **2** The highest point of an ANTICLINE [*84*]. **3** The highest point of an ocean WAVE. [*169* and *276*] **4** The maximum stage of a FLOOD as it moves down river. **5** The WAXING SLOPE of L. C. King. [*49*] **6** The summit ridge of a sand-dune. [*22*] **7** The highest point of a ripple (RIPPLE MARKS). [*210*]

crêt A French term referring to a limestone cliff or ESCARPMENT in the Jura Mts.

Cretaceous The third and final period of the MESOZOIC era, following the JURASSIC and preceding the PALAEOCENE. It began about 145 million years ago, during which time dinosaurs and other reptiles reached their zenith, prior to almost complete extinction, while ANGIOSPERM plants became widespread, replacing the GYMNOSPERMS as the dominant land plants. In Britain the sedimentary rock succession indicates that a slow marine transgression covered the entire land surface as layers of sands and clays were succeeded by the deposition of the CHALK, which is its most characteristic rock. The Cretaceous is divided into twelve stages which are (in ascending order): the

Ryazanian, Valangian, Hauterivian, Barremian, Aptian, Albian, Cenomanian, Turonian, Coniacian, Santonian, Campanian and Maastrichtian. (NB the Coniacian, Santonian and Campanian together make up the Senonian.) In Britain the first four stages are represented by the WEALDEN rocks (Hastings Sands, Weald Clay); the Aptian and Albian are represented by the LOWER GREENSAND, GAULT CLAY and UPPER GREENSAND; the remaining stages are represented by the Lower, Middle and Upper Chalk.

crevasse A deep fissure of variable width in the surface of a GLACIER, caused by differential movement within the ice resulting from shear stresses. Crevasses can be either *transverse*, when the glacier moves down a steeper gradient, or *longitudinal* (pointing up the glacier) partly because the glacier sides move more slowly than the centre and partly because the glacier spreads laterally whenever the valley broadens. Where these two sets of crevasses intersect a series of ice pinnacles (SÉRAC) will form. Crevasses allow meltwater and morainic debris to penetrate deeply into a glacier with some crevasses becoming partly infilled with meltwater debris (CREVASSE FILLING). [*210*]

crevasse filling A short, linear ridge of GLACIOFLUVIAL sand and gravel lying in an area previously buried beneath a stagnating ice-sheet. It is not as long or as sinuous as an ESKER, which it resembles. It is thought to have been formed by meltwater debris penetrating and partly filling a CREVASSE in the melting ice. *See also* KAME.

crevice **1** In general usage, a narrow crack or aperture. **2** In geology, a mineral-bearing vein, especially one containing gold. **3** An enlarged JOINT, whether mineralized or not. NB Not to be confused with CREVASSE.

crinoidal limestone A limestone composed largely of crystalline remains of one class of marine echinoderms (crinoids), consisting of a stem with rootlets at the base and a cup (calyx) with five branching arms at the top. When quarried and polished it forms a highly decorative ornamental stone.

critical angle **1** The ANGLE OF INCIDENCE of refracted light which, if exceeded, will lead to total internal reflection. **2** The critical angle of slope stability (THRESHOLD ANGLE).

critical density The unit weight of a saturated substance at its critical temperature and under its critical pressure. Below its critical density the substance will lose strength (FAILURE) while above its critical density it will gain strength.

critical distance A term used in geophysics to denote the distance between the source of SEISMIC WAVES and the point in an upper rock horizon where the arrival time of a direct seismic wave is matched by the arrival time of a higher velocity wave refracted from a lower rock horizon.

critical erosion velocity The critical (minimal) velocity required in the process of fluid flow over an erodible river bed that leads to erosion and ENTRAINMENT. It is similar to CRITICAL TRACTIVE FORCE. *See also* CHEZY EQUATION, HJULSTRÖM CURVE.

critical flow FROUDE NUMBER.

critical-path analysis A statistical term referring to a type of geomorphological analysis which establishes the number of sequential steps which have occurred in the creation of the present-day landforms.

critical slope ANGLE OF REPOSE.

critical temperature **1** A 'threshold' temperature of great significance for vegetation growth or survival. For the majority of plants active growth cannot occur below 6° C (43° F), while temperatures below FREEZING-POINT (0° C, 32° F) can severely injure or kill some plants at certain times in their life-cycle. **2** That temperature above which a specified gas cannot be liquefied by pressure alone.

critical tractive force A term used in geology and fluvial geomorphology, referring to the fluid stress, or critical shear stress, necessary to move sediment grains on a river-bed. Because stream velocity is partly dependent on channel depth and slope, fluid stress on the stream bed is correlated with velocity. Equally, since discharge is

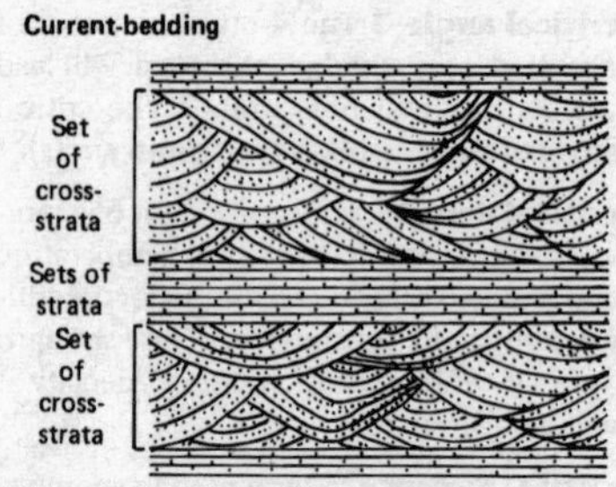

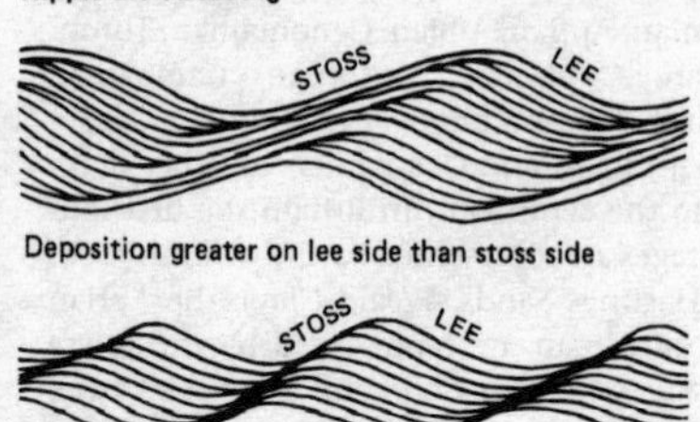

Figure 52 *Cross-bedding*

proportional to velocity and depth, it too is correlated with fluid stress on the stream bed. The critical tractive force (fluid stress) is measured in grams/cm^2 or kg/m^2. It has been demonstrated that the critical tractive force required to move sand grains is less than that required for fine clay particles or for gravel. *See also* CHEZY EQUATION, HJULSTRÖM CURVE.

critical velocity The rate of flow of a fluid in a constrained pipe or channel at which LAMINAR FLOW changes to turbulent flow (TURBULENCE) in which particles begin to move in erratic courses. *See also* CHEZY EQUATION, CRITICAL TRACTIVE FORCE.

Cromerian An interglacial stage of the Middle Pleistocene in Britain, named from its type-site at the foot of the sea-cliffs at West Runton, Norfolk. Here, the sediments exhibit a complete sequence of interglacial marine and freshwater deposits, comprising the Cromer Forest Bed Series, the highest of the periglacial Pleistocene deposits of East Anglia as defined by B. M. Funnell and R. G. West. Reaching a maximum thickness of 8 m, the formation lies on the Chalk near Cromer and is buried by glacial deposits of ANGLIAN age. Farther south-east the formation lies on an earlier pre-glacial Pleistocene deposit known as the *Norwich Crag* (ICENIAN) *Series*. The Cromerian stage commenced about 750,000 BP and experienced a temperate climate in which pine and birch vegetation gave way to oak, elm and lime. [*197*]

cross-bedding The disposition of laminations within a rock stratum transverse or obliquely inclined to the main stratification of the BEDDING PLANES. It is found only in granular sediments, e.g. sandstones, and is caused by changes in the direction of the depositing currents (water or wind). It is also termed *current-bedding*, *cross-lamination*, *cross-stratification* and *false-bedding*. [*52*]

cross-cutting relationships, law of A geological method of determining the relative age (but not the absolute age) of rocks, based on the premise that an igneous rock must be younger than all the rocks through which it cuts. *See also* SUPERPOSITION, LAW OF.

cross-grading A geomorphological process referring to the development of secondary RILLS on a slope. It has been suggested that after the formation of parallel rills on a slope any increase in the discharge of the RUNOFF will cause the water in the individual rills to overtop their divides, thereby initiating *cross flow*. The latter destroys some of the divides and leads eventually to the incremental growth of one dominant rill by the addition of tributary rills which themselves run obliquely across the line of the slope. This is the cross-graded stage. *See also* RILL EROSION.

cross-lamination CROSS-BEDDING.

crossover point A point in a river CHANNEL occurring in a MEANDER belt, where the channel crosses the meander-belt axis either from right to left or from left to right.

cross-profile A diagrammatic means of depicting the transverse profile of a valley or other landform in order to demonstrate the slope gradients, the terrace sequence, etc.

cross-stratification CROSS-BEDDING.

cross-valley moraine WASHBOARD MORAINE.

crumb structure of soil A type of soil structure in which particles accumulate in the form of crumbs, making fine-textured soils more workable by increasing their coarseness. [*236*]

crush-breccia A BRECCIA formed *in situ* by extreme pressure leading to crushing (CATACLASIS).

crushing strength The force required to cause any material to fail by fracture.

crust **1** A hard surface layer, sometimes referred to as *duricrust*, above relatively soft and sometimes unconsolidated sediments, formed by the zonal concentration of aluminous, calcareous, ferruginous and siliceous elements drawn to the surface in solution by CAPILLARITY. *See also* ALCRETE, CALCRETE, FERRICRETE and SILCRETE. **2** A layer of hard surface snow overlying a softer, powdery layer.

crustal movement CYMATOGENY, DIASTROPHISM, EPEIROGENESIS, ISOSTATIC MOVEMENTS, OROGENY.

crust of the Earth The outer layers of the Earth's structure, varying between 6 and 48 km in thickness, and comprising all the material above the MOHOROVIČIĆ DISCONTINUITY. The earlier idea of a cool solid skin (the crust) overlaying a hot molten interior has now been replaced by a concept of a crust composed of two shells: an inner basic unit composed of SIMA (*oceanic crust*) and an outer granitic unit composed of SIAL (*continental crust*). *See also* LITHOSPHERE, PLATE TECTONICS.

cryergic A term referring to FREEZE–THAW processes and their resulting physical phenomena, i.e. frost action.

cryic layer A perennially frozen layer.

cryogenic CRYERGIC.

cryolaccolith HYDROLACCOLITH.

cryology **1** Synonymous with GLACIOLOGY. **2** The study of sea ice. **3** In the USA = the study of refrigeration.

cryopediment A type of PEDIMENT in which the low-angle rock slope is thought to have been produced by CRYERGIC processes (frost action) in a PERIGLACIAL ENVIRONMENT. *See also* ALTIPLANATION, CRYOPLANATION.

cryopedology The scientific study of frost action in soils and the resulting structures (PATTERNED GROUND). It also includes research into the measures required by civil engineers to overcome the problems posed by such frost action.

cryophilous A term in biogeography relating to the responses made by plants and animals to very low temperatures. *See also* CRYOPHYTE.

cryophyte A plant adapted to live in an environment of permanent ice or snow. CRYOPHILOUS.

cryoplanation The slow DENUDATION and reduction of the land surface in a region dominated by frost action. Although cryergic processes are of paramount importance, SOLIFLUCTION and fluvial activity have a part to play in transporting the material produced by frost action.

cryoplankton A term referring to the microscopic plant and animal organisms which inhabit regions permanently blanketed by ice and snow. *See also* CRYOPHYTE.

cryosolic A term used in Canada to denote mineral or organic soils that have perennially frozen material within 1 m of the surface. Thus, they are the dominant soils of the zone of continuous PERMAFROST and are still important, though not as widespread, in the zone of discontinuous permafrost which lies to the south. The active layer (MOLLISOL) of these soils is frequently saturated with water, which leads to extensive GLEYING.

cryostatic pressure A freezing-induced soil pressure in PERMAFROST environments. It occurs in pockets of unfrozen material trapped between the permanently frozen ground and the downward migrating freezing plane of the ACTIVE LAYER. *See also* CRYOTURBATION, MOLLISOL.

cryoturbation Frost action which causes churning, heaving and considerable structural modification of the soil and subsoil in the PERIGLACIAL zone. Since 1946 it has largely been replaced by the term CONGELITURBATION.

cryovegetation Plant communities, especially algae, mosses and lichens, which have adapted to live in environments of permanent snow and ice. *See also* CRYOPHILOUS, CRYOPHYTE.

cryptobatholithic stage One of the six stages leading to the exposure and DENUDATION of a BATHOLITH. In the early stages the batholith is still not exposed, but its presence is indicated by DYKES, SILLS, a METAMORPHIC AUREOLE and by HYDROTHERMAL ACTIVITY in the overlying rocks.

cryptocrystalline A crystalline rock in which the grains are so tiny that their individual character can only be seen with the aid of a powerful microscope.

cryptophytes One of the six classes of plant life-form recognized by Raunkiaer (1934), wherein the perennating buds are submerged in water or below the ground surface. *See also* CHAEMOPHYTES, HEMICRYPTOPHYTES, PHANEROPHYTES.

cryptozoic A geological term derived from the Greek (= hidden life), referring to the primitive life-forms whose traces have been found in the PRE-CAMBRIAN. They are mainly in the form of primitive wormcasts, algal fossils and leaf-like fossils.

crystal, crystalline A three-dimensional body, bounded by plane surfaces which are symmetrically arranged and exhibit constant angular relationships as an outward expression of the inherent regular atomic structure of the substance. Thus, the term crystalline refers to the nature of a crystal, with its regular molecular structure, in contrast to an amorphous body.

crystalline rocks **1** A general term used in geology to distinguish igneous and metamorphic rocks on the one hand from sedimentary rocks on the other. **2** Rocks consisting of minerals in a crystalline state.

crystallography The scientific study of crystals, including the nature and cause of their atomic structure.

cuesta A Spanish term that has been widely adopted to describe an asymmetrical ridge, produced by differential erosion of gently dipping strata, in which the long, gentle slope (BACK SLOPE) is generally accordant with the DIP of the resistant strata which form the cuesta, while the other slope is shorter and generally steeper (ESCARPMENT), except in the case of a HOG'S-BACK RIDGE, where both BACK SLOPE and SCARP may be of the same gradient. It should be noted that in the term's original usage the back slope of the cuesta was related simply to topographic form rather than to structure, so that the *back slope* and the DIP SLOPE were not necessarily the same. In modern usage, however, the term *cuesta* is thought to be a landform in which the back slope and the dip slope are virtually synonymous. [*222*]

cuirass FERRICRETE.

culm **1** A geological formation in SW England, comprising beds of shales and thin layers of impure ANTHRACITE, all of CARBONIFEROUS age. **2** A coal-mining term for carbonaceous shale and fissile varieties of anthracite.

culmination **1** In geology, the term applies to the highest point of a structural fold, especially that of a NAPPE. It is applied particularly to the case where a RECUMBENT FOLD has been refolded, in which case the axes of the two sets of folds will intersect each other at an angle (cross-folding). If two ANTICLINES belonging to the different sets of folds coincide they will create a culmination; if an anticline and a SYNCLINE coincide the result will be a DEPRESSION. **2** In astronomy, the lowest or highest altitude reached by a heavenly body as it crosses the MERIDIAN.

cultural vegetation Any type of vegetation which has been influenced by human activities, either directly or indirectly. *See also* NATURAL VEGETATION, RUDERAL VEGETATION, SEGETAL VEGETATION.

cumec CUSEC.

cumulative frequency A statistical term used to denote the total frequency of all the values falling below the upper class boundary of a given class interval (including the class interval). A table which presents such cumulative frequencies is called a *cumulative-frequency distribution* (*cumulative distribution*). A cumulative-frequency curve is a graph depicting the way in which the values are added successively to each other and finally converted into percentages.

cumulo-dome A dome-shaped accumulation of VISCOUS lava from a volcanic VENT but with no visible CRATER.

cumulonimbus A type of cloud (symbol: Cb), dark grey when viewed from beneath but dazzling white when seen from the side, characterized by its considerable vertical extent, where it rises into towering forms (CASTELLANUS CLOUDS) to heights of over 10 km (16 miles). It is usually associated with THUNDERSTORMS from which torrential rain, HAIL OR SNOW falls. Its top often spreads out in the form of an ANVIL CLOUD. *See also* ABSOLUTE INSTABILITY, CLOUD, CLOUDBURST, COLD FRONT. [*43*]

cumulose deposits Superficial deposits composed largely of organic materials, e.g. peat.

cumulus A type of detached, dense cloud (symbol: Cu), greyish in colour when viewed from beneath but brilliant white where sunlit and seen from the side. Its form is characterized by a relatively horizontal base (at the level of CONDENSATION), varying in elevation between 460 m and 2,000 m (1,500 to 6,500 ft) from which large, white globular masses extend upwards into mounds and domes owing to CONVECTION. It is sometimes associated with good weather, when it is termed *fairweather cumulus*, but at other times may form *congestus* (cauliflower-bulging) and eventually CUMULONIMBUS with associated rain, hail or snow showers. *See also* CLOUD. [*43*]

cupola A narrow dome-like body of igneous rock extending upwards from a larger igneous body at depth (BATHOLITH).

current A term used to define the various movements of air, water or other fluids (also an electric *current*): **1** The vertical movement of air by CONVECTION. **2** The motion of water in a stream or river channel. **3** The movement of surface water in oceans, either permanently or seasonally, *see also* DRIFT, OCEAN CURRENTS, THERMOHALINE. **4** The movement of tidal water through a restricted channel, TIDAL CURRENT. **5** The movement within the ASTHENOSPHERE associated with *convection currents* produced by thermal heating in the upper zone of the Earth's MANTLE below the more rigid crust (PLATE TECTONICS).

current base The maximum depth below which currents of water are incapable of moving sediment.

current-bedding CROSS-BEDDING.

current meter An instrument for measuring the velocity of flowing water. There are several models of which the most common is the rotating variety.

current ripple A ripple mark in sand or mud produced by the action of a current flowing steadily in one direction. It has a long gentle slope facing the direction from which the current comes and a shorter, steeper slope on the leeward side. Ripples migrate with the current in much the same way as sand-DUNES. Where a pre-existing set of ripples is affected by a current from a different direction, two sets of ripples, intersecting at any angle, may form, or a new set of cross-ripples may be formed. *See also* ANTIDUNE, RIPPLE MARKS. [*14* and *210*]

curvature of the Earth (correction for) An adjustment taken from a set of correction tables (during a GEODETIC SURVEY) to compensate for the curvature of the Earth.

curve fitting A statistical term used to denote the fitting of a mathematical curve to any statistical data capable of being plotted against a space or time variable.

curve parallels A set of graphs that illustrate similarities (or disparities) of climatic trends at different stations over the same time period.

cusec A measurement unit (an abbreviation for cubic feet per second) relating to the rate of discharge of a stream. It refers to the volume of water passing through a particular section of the stream (REACH). At a particular station the measurement is obtained by multiplying the cross-section of the channel (sq ft) by the velocity (ft/sec); 1 cusec = 538,000 gallons in 24 hours. Since metrication, cusec has been replaced by cumec ($m^3 s^{-1}$ or m^3/sec, i.e. cubic metres per second); 1 cusec = 0.028 m^3/sec; 1 cumec (m^3/sec) = 35.313 cusecs.

cusp BEACH CUSP.

cuspate delta A symmetrical DELTA, usually formed where a river debouches on a straight coastline, and in which the sedimentary material is deposited evenly on either side of the river mouth.

cuspate foreland An approximately triangular accumulation of shingle and sand jutting out into the sea to produce a point of low elevation at its apex. A cuspate foreland is created by the linking of two BEACH-RIDGES or SPITS that have gradually built out from the coast and approached each other from opposite directions as a result of two major sets of waves (CONSTRUCTIVE WAVE). The foreland is composed of many parallel shingle ridges behind which lies a coastal marsh, e.g. Romney Marsh behind the cuspate foreland of Dungeness.

cut and fill **1** A term referring to the process by which a MEANDER migrates downstream by means of lateral erosion of the river bank on one side of the channel accompanied by deposition of sediment on the other. **2** A civil engineering term.

cutans A term used in soil science to denote a modification of the structure, texture or fabric at natural surfaces in soil materials owing to concentration of particular soil constituents or *in situ* modification of the matrix. Cutans may be composed of any of the component substances of the soil material, but in the majority of cases these thin layers, which coat the aggregates of some soils, are formed by redistribution of the clay fraction. *See also* CLAY SKINS, ELUVIATION.

cut-off A short river CHANNEL formed when the neck of a MEANDER is finally broken through, thereby shortening the length of the active channel and leaving the abandoned meander as an OXBOW. [*147*]

cuvette A French term denoting a large basin of deposition that is not of TECTONIC origin. The Old Red Sandstone of Scotland, for example, was deposited in a number of cuvettes.

cvp index An index used, mainly by foresters, to define the relationships between climate (c), vegetation (v) and productivity (p) of the forest crop. It includes measurements of precipitation, temperature, length of the growing season, evapotranspiration, etc.

cwm The Welsh equivalent of a CIRQUE.

cybernetics The study of guidance and control problems in biological or mechanistic systems. *See also* GENERAL SYSTEMS THEORY, SYSTEM.

cycle **1** In its strictest sense the term refers to a sequence of events which returns to its starting-point, e.g. the HYDROLOGICAL CYCLE. **2** In physical geography, the term has been used to denote a succession of stages which occur repeatedly in the same order: *see also* ARID CYCLE OF EROSION, CYCLE OF EROSION, CYCLE OF SEDIMENTATION, CYCLE OF UNDERGROUND DRAINAGE, GLACIAL CYCLE OF EROSION, MARINE CYCLE OF EROSION, NORMAL CYCLE OF EROSION, PERIGLACIAL CYCLE.

cycle of erosion A concept relating to the evolution and modification of the physical landscape, in which the various stages of erosion are believed to be parts of a cyclic process (CYCLE) which follows an orderly sequence. The concept was developed by W. M. Davis in 1889 and it began ideally with the uplift of a land surface which caused REJUVENATION and incision of the

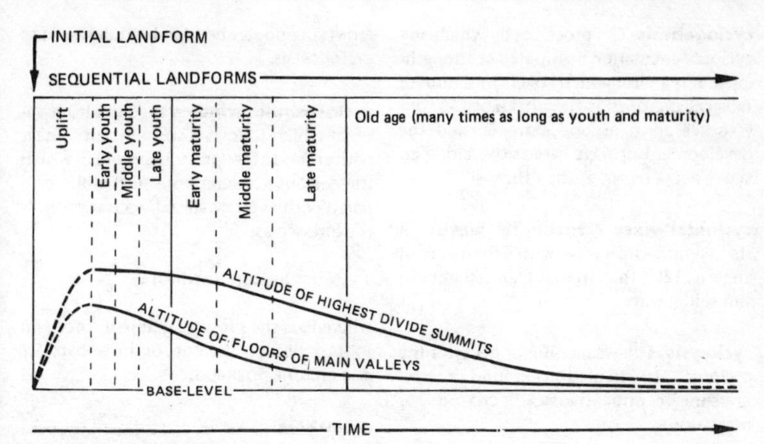

Figure 53 *Cycle of erosion*

drainage pattern; this was called the stage of *youth* or ADOLESCENCE by Davis. Once the relief had become developed its AMPLITUDE at first increased (the stage of *maturity*) but later as the relief degenerated the land surface reached the stage of senility or *old age* when the characteristic landform was a low, featureless plain (PENEPLAIN) with a few residual hills (MONADNOCKS). Further uplift led to the termination of the first *cycle* and the initiation of a second *cycle of erosion*, thereby leaving the uplifted peneplain as a high level EROSION SURFACE. The concept of strict cyclic progression and its associated stages is no longer adhered to as far as landscape development is concerned, partly because complete planation by rivers was probably rarely achieved and partly because CLIMATIC CHANGE means that processes in one region are never constant. The cycle of erosion was termed by Davis the *normal cycle of erosion* in the temperate latitudes, and is also known as the *Davisian cycle*, the *geographical cycle* and the *geomorphic cycle*. [*53*] [*see also 193*]

cycle of sedimentation Sometimes referred to as *cyclic sedimentation*, during which **1** sediment is formed, transported and deposited, or when **2** the depositional phase exhibits a sequential cyclic progression of changing sedimentary character, starting with a fresh-water environment and proceeding through a brackish-water environment to a marine environment (as salinity increases), then back through brackish water to a fresh-water environment again (as salinity decreases). Each of these environments and their associated sediments is regarded as a *cyclic unit*, which tends to be of great thickness and considerable lateral extent. *See also* CYCLOTHEM, RHYTHMIC SEDIMENTATION.

cycle of underground drainage Sometimes referred to as the *karstic cycle*. A concept relating to the evolution of a karstic (KARST) drainage pattern in which the various stages are thought to be part of an orderly sequential *cyclic process*. It is said to commence with a surface drainage which then passes underground during the *youthful* stage. In the *mature* stage the drainage is almost entirely subterranean but cavern collapse is already beginning. The *old age* stage is shown by general collapse of the cavern roofs and the re-exposure of the drainage and a final return to surface drainage. *See also* CYCLE OF EROSION. [*54*]

cyclic time The length of time necessary for the completion of a CYCLE OF EROSION. Cyclic time is one of the three categories in the time-scale of GEOMORPHOLOGY devised by S. A. Schumm and R. W. Lichty. *See also* GRADED TIME, STEADY TIME.

cyclogenesis The processes by which any cyclonic circulation is initiated or strengthened. It is a term which covers initiation of TROPICAL CYCLONES at the INTERTROPICAL CONVERGENCE ZONE, at one extreme, and the development of WAVE DEPRESSIONS along an active POLAR FRONT, at the other.

cycloidal wave (trochoidal wave) A steep symmetrical wave with a sharper crest angle (*c.* 120°) than that of a normal smooth sinusoidal wave.

cyclolysis The weakening or decay of the cyclonic air circulation around a low-pressure centre (TROPICAL CYCLONE or DEPRESSION).

cyclone A system of low atmospheric pressure in which the BAROMETRIC GRADIENT is steep. Winds circulate, blowing inwards in an anticlockwise direction in the N hemisphere and in a clockwise direction in the S hemisphere. Two types are usually distinguished: **1** The TROPICAL CYCLONE of the Indian Ocean, Indonesia and Australasia, associated with very high winds of a destructive nature and with torrential rainfall. **2** The low-pressure system of temperate latitudes. Formerly known as a cyclone, this usage is currently being replaced by the term DEPRESSION, WAVE DEPRESSION or LOW, but the term *cyclonic rain* is still used as a synonym for the precipitation associated with depressions and FRONTS in middle and high latitudes. *See also* CYCLOGENESIS.

cyclostrophic wind A type of air movement which follows a strongly curvilinear path around a low-pressure system in which the CORIOLIS acceleration is negligible compared with the CENTRIPETAL ACCELERATION. It is expressed as:

$$V \text{ (cyclostrophic wind)} = \frac{P_n ½}{R_T}$$

where P_n is the pressure gradient force and R_T the radius of curvature of the isobars. *See also* CYCLONE, TORNADO.

cyclothem A series of sedimentary beds deposited during a single CYCLE OF SEDIMENTATION and therefore indicative of changing environmental conditions. Strictly speaking, the use of the term in connection with cyclic sedimentation illustrates a *symmetrical cyclothem*, i.e. the change is from **1** fresh-water sediments, through **2** brackish-water sediments to **3** marine sediments and back through **2** brackish- to **1** fresh-water sediments (namely 1, 2, 3, 2, 1). In contrast, the term can also be used in connection with RHYTHMIC SEDIMENTATION, in which case it is termed an *asymmetrical cyclothem*, i.e. the change is repetitive: from **1** fresh-water sediments, through **2** brackish-water sediments to **3** marine sediments, followed by **1** fresh-water sediments, **2** brackish-water sediments and **3** marine

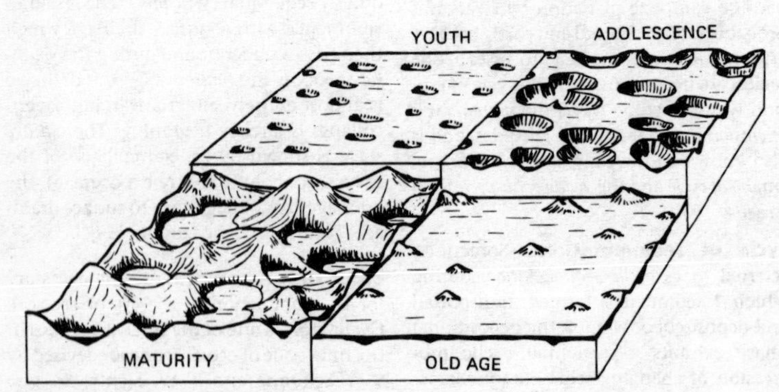

Figure 54 *Cycle of underground drainage*

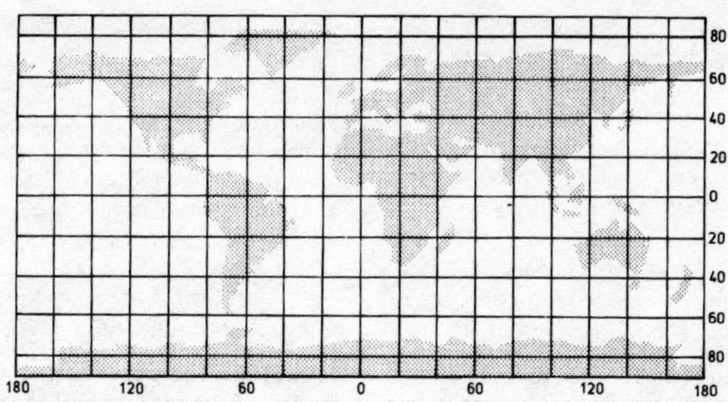

Figure 55 *Cylindrical projection*

sediments (namely 1, 2, 3, 1, 2, 3). A cycle of cyclothems is termed a *megacyclothem.*

cylindrical equal-area (Lambert's) projection A MAP PROJECTION drawn so that the MERIDIANS and PARALLELS are straight lines perpendicular to each other. Although the scale along the equator is correct, the scale on the meridians decreases polewards, causing the parallels to become closer together as they approach the poles. Because the ratio between the meridians and parallels remains constant the projection is one of equal area, but because of the distortion of shape in polar latitudes it is rarely used for areas outside the zones bounded by 40°N and 40°S of the equator.

cylindrical projection A MAP PROJECTION constructed as if the globe is inserted within a cylinder and the global surface projected on to the cylinder at a tangent to the cylindrical surface. MERIDIANS and PARALLELS are drawn as straight lines intersecting each other at right angles. There are three types: **1** *simple cylindrical*; **2** *equal-area* (LAMBERT'S PROJECTIONS); and **3** MERCATOR'S. [*55*]

cymatogeny A term relating to large-scale tectonic warping of the crust to produce a type of *basin-and-swell* structure in which the crests of the swells (domes) collapse to form GRABEN (rift valleys). The term was devised in 1961 by a South African, L. C. King, largely to explain the structural warping of the African land surface, especially the up-doming and rifting of East Africa. The scale of the cymatogenic movement is thought to be some thousands or even tens of thousands of metres in a vertical direction and to be some hundreds of kilometres in breadth.

D

dacite An extrusive igneous rock, of which the principal minerals are QUARTZ, PLAGIOCLASE, HORNBLENDE and/or PYROXENE. It is a fine-grained rhyolite, the extrusive equivalent of DIORITE.

Dalmatian coast A term derived from the Yugoslavian Adriatic coastline in which the coast runs parallel with the LINEAMENT of the topography and probably with the underlying geological structure. A rise of sea-level (EUSTASY) has drowned the coastal area, resulting in a coastline of narrow peninsulas, lengthy gulfs and channels and a number of linear islands. It is an example of a CONCORDANT COAST. [*56*]

Dalradian The uppermost stratigraphic division of the PRE-CAMBRIAN in the British Isles, albeit the youngest strata are now regarded as part of the CAMBRIAN. Its rocks outcrop mainly in the Scottish Highlands and in Ireland.

Dalton's Law A mixture of gases has the same pressure as the sum of the partial pressures of its components. One of the thermodynamic laws applying to perfect gases. *See also* BOYLE'S LAW, CHARLES'S LAW.

dambo A shallow streamless depression at the headwaters of a drainage system, in which no actual channel can be identified.

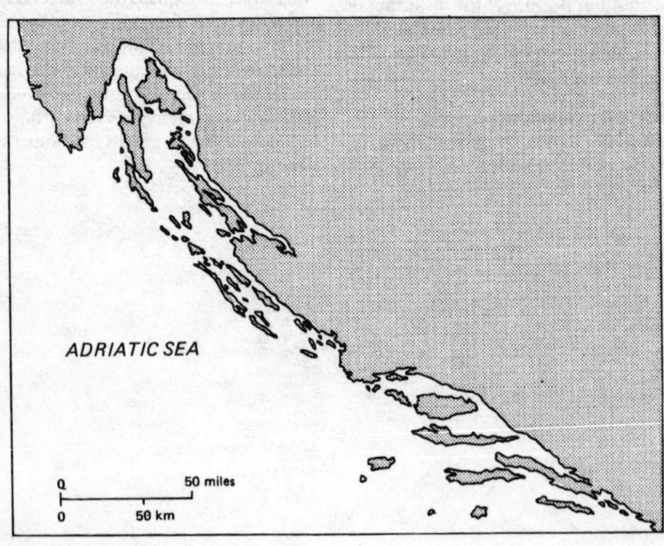

Figure 56 *Dalmatian coast*

The term has been specifically applied to phenomena in tropical Africa but it has been widely recognized (under different names) in other parts of the world outside the tropics.

damping The process by which an effect is diminished in intensity. It is occasionally used to describe the process of self-regulation or negative FEEDBACK.

Daniglacial A cold stage of the Late Pleistocene during which ice-sheets waned in Denmark although Norway and Sweden were still ice-covered. It occurred between 20,000 and 17,000 BP.

darcy A standard measurement unit of PERMEABILITY. It is equivalent to the passage of 1 cm^3 of fluid with a VISCOSITY of 1 centipoise flowing in 1 second under a pressure differential of 1 atmosphere through a porous medium the areal cross-section of which is 1 cm^2 and whose length is 1 cm.

Darcy's law **1** A law formulated in 1865 by a Frenchman, Henri Darcy, to describe the flow of water through a porous medium (in the first instance, of groundwater through the sand beds of an AQUIFER). It was originally defined as: $Q = kS[(H+e)/e]$, where Q is the volume of water passed in unit time, S is the area of the bed, e is the thickness of the bed, H is the height of the water on top of the bed, and k is a coefficient depending on the nature of the sand. **2** The law has been extended and the formula amended to include the flow of fluids, based on an assumption that the flow is laminar and that inertia can be ignored. It is now expressed as: $V = p(h/l)$, where V is the velocity, h is the HEAD of water, l is the length of flow between two given points, and p is the coefficient of PERMEABILITY of the aquifer (DARCY). Thus, the law now states that the rate of viscous flow of homogeneous fluids through isotropic porous media is proportional to, and in the direction of, the driving force.

Darwinism The theory of EVOLUTION by natural selection as stated by Charles Darwin (1809–82). This challenged the biblical notions of CREATIONISM. Darwin's revolutionary theory allowed the disciplines of BIOGEOGRAPHY, ECOLOGY, GEOLOGY, GEOMORPHOLOGY, PEDOLOGY, etc. more time for their processes to operate in sequential progressions, in place of the strictures of Archbishop Ussher, who categorically stated that the world was created in 4004 BC. *See also* CYCLE OF EROSION.

data A representation of numbers, facts, concepts or instructions in a formalized manner suitable for communication, interpretation or processing, usually by a computer.

data base An organized, integrated collection of DATA in a store capable of being accessed by different logical paths.

data capture The encoding of DATA, often by DIGITIZING.

data compression **1** Methods of DATA CAPTURE which reduce the overall data volume. **2** In REMOTE SENSING the term involves the reduction of many images into one image to facilitate interpretation.

data set An organized collection of DATA which bear a common theme.

datum A geographical or numerical quantity or fact which serves as a base or reference point. It is the starting-point in any type of measurement, or in reasoning (*see also* DEDUCTIVE REASONING, INDUCTIVE REASONING). The plural of *datum* is *data*. *See also* DATUM LEVEL.

datum level The zero level (usually sea-level or a mean based on tidal levels) from which land elevations and ocean depths are measured. *See also* DATUM, ORDNANCE DATUM.

Davisian cycle CYCLE OF EROSION.

dead cliff A CLIFF created by marine erosion which has now been abandoned owing to a marine REGRESSION or to an accumulation of beach deposits. Because the wave attack has ceased, the main denuding agent changes from marine erosion to subaerial weathering and the gradient of the cliff declines from near verticality to angles of 45° or less, thus becoming a *degraded cliff*.

dead ground **1** A term used in surveying, referring to the land surface which is invisible to a surveyor when taking a line of sight, owing to the undulating character of the intervening topography. **2** In the USA the term refers to the surplus rock in a mining excavation which has to be removed before the productive ground can be reached.

dead-ice features A hummocky terrain produced by the deposition of **1** GLACIOFLUVIAL sediments and **2** ABLATION TILL, by the melting of a stagnant ice-sheet or glacier (*dead ice*) *in situ*. *See also* KAME, KETTLE HOLE.

debris A French term for 'wreckage', which has been adopted for geological use to describe a superficial collection of broken rocks, earth and other inorganic material that has been moved from an original site by streams of water or ice and redeposited at other localities.

debris avalanche A sudden downslope (gravitational) movement of DEBRIS due to saturation by heavy rain. Strictly, the term AVALANCHE should be restricted to snow and ice movement. A more acceptable term for the phenomenon described above is *debris slide*, in which the mass of material moves downslope without any rotational movement (LANDSLIDE).

debris fan, debris cone A fan-shaped deposit of DEBRIS at the point where a mountain torrent debouches on to a valley floor and where deposition takes place owing to the reduction of stream velocity. It is similar to an ALLUVIAL FAN except that its material is usually much coarser. Sometimes termed a *debris cone*, but this term has the disadvantage of being confused with DIRT CONE, a phenomenon of quite different genesis.

debris flow A rapid flowage of DEBRIS, often the result of additional surface water being added to a DEBRIS AVALANCHE, in which the mixture becomes extremely fluid and flows considerable distances, sometimes with catastrophic results, e.g. the flow which overwhelmed the town of Yungay, Peru, in 1970, following the ice- and rock-fall from the summit of a neighbouring peak, Mt Huascarán, caused by an earthquake. *See also* EARTH-FLOW, MUD-FLOW. [*146*]

debris load BEDLOAD.

debris slide DEBRIS AVALANCHE.

decalcification A term used by soil scientists when referring to the removal of calcium ions from a soil by the process of LEACHING, in which calcium bicarbonate is carried away in solution. Decalcification proceeds downwards from the surface and is thought to be an initial stage of PODZOLIZATION and LESSIVAGE. Many of the chalky sands and gravels deposited by rivers and by solifluction in the chalklands of S England during various stages of the Pleistocene have been largely decalcified throughout their entire depth, since the process acted more quickly at the lower temperatures of the Pleistocene (inasmuch as the solubility of calcium carbonate decreases as temperature rises).

decay RADIOMETRIC DATING.

decibel A unit of measurement related to the intensity of sound. It is used in seismic studies (SEISMOLOGY) and *environmental impact analysis*. It is expressed as: DB (decibel) = $10 \log_{10}(I^1/I^2)$, where I is intensity. *See also* NOISE RATING NUMBER.

deciduous A term referring to the ability of some plants to lose their leaves annually. In the temperate zone the leaf fall occurs during AUTUMN (US = *the fall*) when broadleaf and some coniferous trees (e.g. larch) shed their leaves (or needles). In the tropical forests the fall of leaves may be at any time since there are few restrictions on the growing season. In contrast to the deciduous type of vegetation is the EVERGREEN, which retains its leaves for more than a year.

decile A statistical term referring to a tenth part of a series of values ranked in order of magnitude.

declination 1 *Solar declination* is the angular measurement of the Sun in relation to the *celestial equator* and the *celestial poles* (CELESTIAL SPHERE). The declination of the Sun varies from 23½°N on about 21 June (SOLSTICE) to 23½°S on about 22 December. The solar declination can be a useful aid in marine navigation and can be read from the

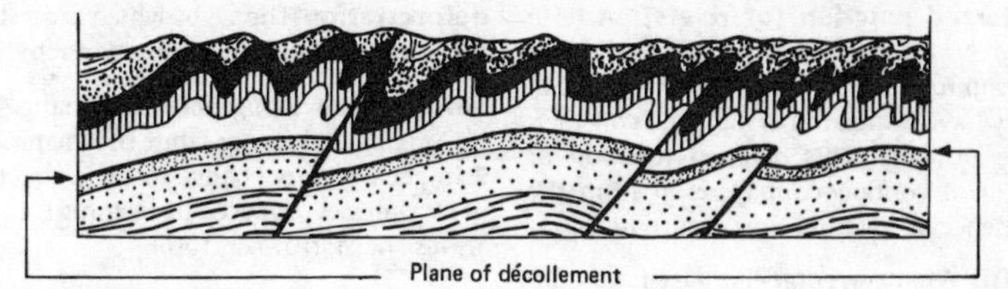

Figure 57 *Décollement*

Nautical Almanac, from which the latitude of the observer can be calculated by adding the *solar declination* to the Sun's zenith distance. **2** *Magnetic declination* is the angle at any point on the Earth's surface between magnetic N and true (geographical) N. *See also* MAGNETIC POLES. MAGNETIC DECLINATION (or *variation*) is expressed in degrees E or W of true N and it is subject to diurnal changes, to irregular short-term changes and to long-term changes. It varies in different parts of the world, e.g. in 1976 the magnetic declination was zero along a meridian passing through Ohio, USA, but in the UK was approximately 8°W. [*120*]

décollement A French term which has been adopted by geologists to describe the deformation of superficial rocks by folding or faulting without the rocks at depth being affected. This is usually achieved when the uppermost rocks slide or glide forward over the underlying rocks along a low-angle planar boundary during folding. The presence of an easily deformed stratum, such as ROCK SALT, often facilitates the movement. [*57*]

decomposition The weakening and breaking-down of rocks by CHEMICAL WEATHERING.

deductive reasoning A type of reasoning which draws upon a theoretical framework to explain a specific phenomenon, i.e. it proceeds from the general to the particular. *See also* INDUCTIVE REASONING.

deepening A term used in meteorology to indicate a decrease in atmospheric pressure at the centre of a DEPRESSION, or system of low pressure. It generally denotes a worsening of the weather. The converse term is '*filling*'.

deep focus A term referring to a SEISMIC FOCUS of an EARTHQUAKE occurring at depths below 300 km. *See also* SHALLOW FOCUS.

deep, ocean A trench of considerable depth on the DEEP-SEA PLAIN of the ocean floor. Its use is generally restricted to trenches which exceed depths of 5,500 m (3,000 fathoms). They occur either in association with ISLAND ARCS or along coasts bounded by high mountain ranges, where they are termed *foredeeps*. The ocean deep is often associated with the SUBDUCTION ZONE at the margins of two tectonic plates (PLATE TECTONICS). *See also* ABYSSAL ZONE.

deep-sea plain A far-spreading, virtually level area on the ocean floor in the ABYSSAL ZONE, from which the ocean deeps (DEEP, OCEAN) descend and the SEAMOUNTS rise.

deep weathering A type of WEATHERING of rocks that extends to great depths (30 m) below the surface in the humid tropics, where the process reaches its maximum efficiency. It is influenced and controlled by temperature, precipitation, rock type and vegetation cover, since the latter can create humic acid during its decay. If EROSION is slight, very thick accumulations of REGOLITH, or chemically rotted rock, will accumulate *in situ*, with the rotted material giving way downwards to a zone of partly weathered material in which corestones (CORESTONE) may be found, until the BASAL SURFACE OF WEATHERING is reached, below which is the unweathered solid rock. Evidence of deep weathering in areas no longer subject to humid, tropical climates is taken to imply that CLIMATIC CHANGE has occurred. *See also* INSELBERG, SHIELD INSELBERG, TOR.

deferred junction (of rivers) A term referring to the inability of a tributary stream to join the main trunk stream until they have flowed in parallel for a considerable distance. This deferred junction or point of confluence may result from the presence of LEVEES on the FLOODPLAIN.

defile A narrow, gorge-like pass in a mountainous region.

deflation The process by which wind removes dry, unconsolidated sand, silt and clay from the land surface, especially in arid and semi-arid regions, but also of importance in shaping coastal dune hollows and in denuding beaches. The finer materials may be carried considerable distances before being redeposited (LOESS), while the coarser material may simply be rolled or bounced along the surface, thereby abrading pebbles which are too large to be moved by the wind (VENTIFACT). Once the fine material has been removed, a surface of deflation remains behind, characterized by gravel-strewn areas interspersed with bare rock exposures or patches of vegetation that have formed protective mats against the windblow. *See also* AEOLIAN.

deflation hollow A large-scale basin or depression formed by the action of the wind (DEFLATION) in arid or semi-arid lands. The removal of the fine superficial material may lower the ground surface sufficiently to cause the WATER-TABLE to be reached, in which case the hollow may be the site of an OASIS. *See also* AEOLIAN.

deflecting force CORIOLIS FORCE.

deflection 1 A term used in GEOPHYSICS, referring to the angle between the GEOID and the SPHEROID. **2** A sudden change in the trend of a LINEAMENT, especially that in a mountain range, as depicted on a map.

deflocculate A term used in soil science referring to the dispersion or break-up of an AGGREGATE. This is achieved by physical or chemical means, e.g. the addition of a monovalent CATION (e.g. sodium) to soil in a colloidal state (COLLOID) causes the particles to disperse and become suspended in the solution.

deforestation The act by which a forest is totally felled and cleared by human activity.

deformation 1 In general, any change in the original volume or fabric of a material. **2** In geology, any change in the character of rock masses produced by tectonic forces. *See also* FAULT, FLOWAGE, FOLD.

de Geer moraines Moraines comprising swarms of small ridges arranged transverse to the glacier flow. Also known as cross-valley moraines and WASHBOARD MORAINES.

deglaciation A term used to denote the slow wastage of an ice-sheet and the uncovering of a formerly ice-covered terrain. It is used specifically to describe an event which took place during an earlier episode of geological time, e.g. the Pleistocene ice-sheets of N Europe. *See also* DEGLACIERIZATION.

deglacierization A term used to describe the slow wastage and shrinkage of an existing ice-sheet or glacier, e.g. the glaciers of the Alps. *See also* DEGLACIATION.

degradation 1 In geomorphology, a term referring to the gradual lowering of a land surface by DENUDATION, in which erosive forces (rivers, ice, etc.) erode material, transport it and deposit it elsewhere. Degradation is, therefore, the opposite of AGGRADATION. **2** In soil science, a term used to describe the process by which a soil becomes weathered and more highly leached (LEACHING), usually accompanied by morphological changes such as the development of an eluviated light-coloured A-HORIZON.

degraded cliff DEAD CLIFF.

degree 1 A unit of angular measurement; one degree is equivalent to 1/360th part of a circle. **2** A unit of angular measurement of LATITUDE and LONGITUDE, in which each degree is divided into 60 minutes (') and each minute into 60 seconds ("). Because of the Earth's oblate spheroidal shape the distance represented by a degree varies slightly according to the location (ELLIPSOID OF REFERENCE): 1° latitude at 45° = 111.132 km (69.057 miles), but 1° latitude at the equator = 110.569 km (68.708 miles); 1°

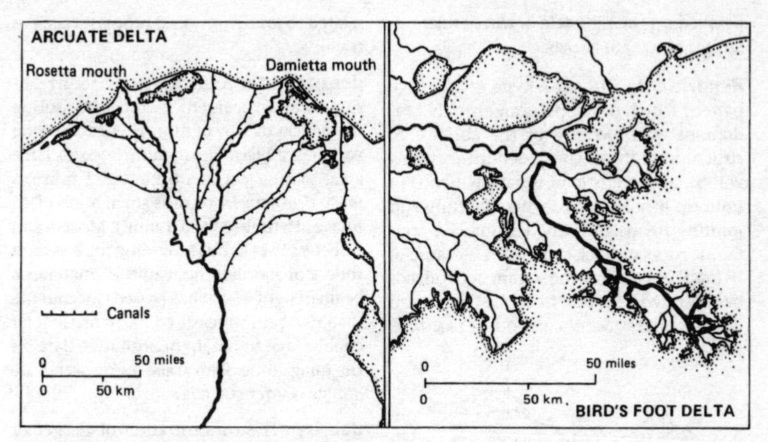

Figure 58 *Deltas: the Nile (left) and the Mississippi*

longitude at 45° = 78.850 km (48.997 miles), but 1° longitude at the equator = 111.322 km (69.175 miles. **3** A unit of temperature measurement *see also* ABSOLUTE TEMPERATURE, CELSIUS SCALE, CENTIGRADE SCALE, FAHRENHEIT SCALE, KELVIN SCALE, RÉAUMUR SCALE.

degree-day A measure of the total daily amount by which the main daily temperature departs from a CRITICAL TEMPERATURE. The departures (either above or below) from this given 'threshold' temperature are summed for a selected period of time and expressed as an ACCUMULATED TEMPERATURE. If, on a particular day, the temperature exceeds the critical value of *h* hours and rises above this given value by *m* degrees the *accumulated temperature* for that day is *hm* degree-hours or hm/24 degree-days.

degree of advancement A term describing the amount of change accomplished by a chemical reaction towards a new state of equilibrium.

degrees of freedom A term derived from systems theory, referring to the capability of variation within a system. The number of degrees of freedom in a particular system is the number of independent variables that can be freely assigned or changed to bring the system into a new state of equilibrium without causing the disappearance or appearance of a phase within the system. It is synonymous with the expression VARIANCE of a system. *See also* GENERAL SYSTEMS THEORY.

delta A fan-shaped alluvial deposit at a river mouth formed by the deposition of successive layers of sediment. As the river current enters the sea (or a lake) the bulk of the coarsest fraction of the LOAD is deposited immediately, but the finer material is carried farther out by the divergent river channels (DISTRIBUTARY). Fine clays, carried in suspension, are deposited by FLOCCULATION when they enter saline water. Deltas will only grow where the amount of river-deposited sediment exceeds that removed by coastal processes. The sequential deposition is characterized by different types of bedding, known as sets (BOTTOM-SET BEDS, FORE-SET BEDS, TOP-SET BEDS). [*85*] The name is derived from the Greek letter Δ, which resembled the form of the Nile delta, but not all deltas have this shape, depending on such variables as tidal currents, wave action, supply of material, etc. *See also* ARCUATE DELTA, BIRD'S-FOOT DELTA, CUSPATE DELTA. [*58*]

demoiselle A term of French derivation used to describe an earth-pillar (usually of TILL) in which a boulder has protected the underlying material from weathering and now survives as a cap-stone precariously perched on the pinnacle of the slender pillar. *See also* EARTH PILLAR.

dendritic drainage A type of drainage pattern which develops as an entirely random network because of the absence of structural controls. Thus a dendritic pattern will be characteristic of terrain which is of uniform lithology, and where faulting and jointing are insignificant, e.g. massive crystalline rocks or thick clay plains. Because of its branching nature the stream network has been termed dendritic from the Greek term *dendron* (= *tree*). *See also* INSEQUENT DRAINAGE. [59]

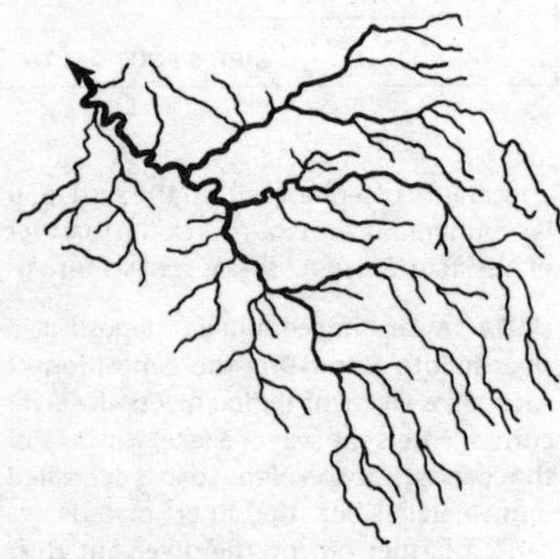

Figure 59 *Dendritic drainage pattern*

dendrochronology, dendroecology The science of interpretation of former environments (climate, vegetation, etc.) by the study of annual growth-rings in certain species of trees. The methodology is based on the premise that the width of the growth-ring will reflect the amount of precipitation and the temperature of the year in which it was formed. In SW USA a study of the bristle-cone pine has traced its history back some 3,600 years in living specimens. The reliability of the technique is limited, however, for the annual tree rings are thought to reflect changes in other environmental factors (soils, forest fires, etc.) so that there is no simple, linear relationship.

denitrification The loss of nitrogen in soils by either biological or chemical mechanisms. This is a gaseous loss and is not related to loss by physical processes such as LEACHING.

densitometer A REMOTE SENSING term referring to an instrument used for measuring the OPTICAL DENSITY of film. There are several varieties: a *macro-densitometer* records large areas of an IMAGE at a time (e.g. 1 mm^2); a *micro-densitometer* records small areas of an image at a time (e.g. 0.01 mm^2). Most densitometers work by providing a constant source of standard light until it matches a beam of light which has passed through the material being measured. A densitometer enables one to obtain quantitative data on the image tone of an AERIAL PHOTOGRAPH. *See also* CHARACTERISTIC CURVE.

density The concentration of matter or mass at a specified temperature and pressure, expressed as an absolute quantity (in kg/m^3) in terms of mass per unit volume. It can refer to the density of air and other gases, to all types of solid materials and to water and other fluids. There is a wide range of differences: e.g. cold water is denser than warm water; the density of salt water varies according to both its temperature and its salinity; the density of ice varies with its temperature; the density of the Earth's CORE differs from that of its MANTLE; the density of air varies according to its temperature and its atmospheric pressure. *See also* SPECIFIC GRAVITY.

density current An ocean current caused by the different temperatures and degrees of salinity exhibited by adjacent water masses, which therefore possess contrasting densities, e.g. cold currents flow south from the Arctic Ocean where they descend beneath the warmer waters of the NORTH ATLANTIC DRIFT as a density current. *See also* BOTTOM CURRENT.

denudation **1** In general, a term used to denote the action of laying bare by the process of washing away of surface materials, such that all surface inequalities would be reduced to a uniform level. **2** In geomorphology, the term is now used in an all-embracing sense to include all processes which cause DEGRADATION of the Earth's surface. In earlier years a distinction was drawn between denudation, which was reserved specifically for the operation of the processes themselves, and degradation, which was used specifically to refer to the end-product, i.e. the resulting form of the land surface. But, in modern usage, denudation is regarded as being synonymous with degradation, and as including ABRASION, CORRASION, EROSION, MASS-MOVEMENT, TRANSPORT, and WEATHERING. Rates of denudation vary in accordance with climate, terrain and lithology. They are lowest in hot, arid lowlands (1.2 mm per 1,000 years) and highest in cold, humid, glacierized uplands (3,000 mm per 1,000 years).

denudation chronology The study of landform evolution. It is based on sequential studies over time in which pieces of evidence are arranged in chronological order, to obtain a series of reconstructions of the land surface at various stages of time. The evidence is usually an amalgam of: **1** surface forms (ACCORDANT SUMMITS, SLOPE, TERRACE etc.); **2** drainage patterns (DISCORDANT DRAINAGE, ELBOW OF CAPTURE etc.); **3** superficial deposits (ALLUVIUM, PALAEOSOL, TILL, etc.). By assembling the various pieces of evidence, studying their spatial relationships and adopting a temporal ordering of the data by means of biostratigraphic and lithostratigraphic principles (BIOSTRATIGRAPHY, LITHOSTRATIGRAPHY) it becomes possible to establish a conceptual framework of stages by which the evolution of the present land surface can be traced. Much of the reconstruction is based on DEDUCTIVE REASONING, so that the reconstruction of former environments is hypothesized and the type of geomorphological processes formerly operating is inferred. Unless methods of absolute dating (e.g. RADIOCARBON DATING) or pollen analysis (PALYNOLOGY) are adopted, denudation chronology must remain a rather subjective and speculative exercise rather than a scientific study.

denudation slope A type of SLOPE from which material is being lost by erosion and weathering, leading to a net loss of ground. The form of the slope is controlled by the rate of denudation. *See also* ACCUMULATION SLOPE, TRANSPORTATION SLOPE.

depleted soil A soil that has lost most of its AVAILABLE NUTRIENTS, either by bad management (over-cropping) or by intense LEACHING.

depletion curve BASE-FLOW RECESSION CURVE.

deposition The laying down of material that has accumulated: **1** after having been eroded and transported (DENUDATION) by various physical processes (wind, ice, running water, marine waves and ocean currents); **2** as the remains of former organisms (coal, peat, coral); **3** by evaporation (EVAPORITE). In the case of deposition by rivers, oceans and the wind, material is dumped in low-energy environments, i.e. when the forward movement in the transporting medium is reduced below the settling velocity of the load (*see also* CAPACITY, COMPETENCE). In the case of ice movement, deposition generally occurs either when the ice-sheet meets an obstacle or when the ice body reaches its maximum spatial extension under the existing climatic regime. A change in this regime towards warmer conditions will cause a marked increase in the rate of deposition as the ice begins to melt. *See also* DEGLACIATION, DEGLACIERIZATION.

depression **1** In meteorology, the term is synonymous with a low-pressure system or '*disturbance*' and is the extratropical CYCLONE, so far as the mid- and high-latitudes are concerned, with its pressure and wind characteristics being similar. It is described as '*deep*' or '*shallow*' according to its intensity, i.e. the number of ISOBARS associated with the system. A depression usually passes through a cycle of 'deepening' to maturity, 'filling' as it becomes less intense and finally losing its identity. The deepest depression recorded in UK was 925 mb in 1884. [*60*; *see also 44*] **2** In physiography, the term is used loosely to describe any hollow in the ground surface, but strictly should be used only for an area surrounded completely by higher land, such that surface drainage cannot escape overland. **3** In geology, the term refers to a complex type of FOLD produced during refolding of an existing fold and leading to the coincidence of an ANTICLINE and a SYNCLINE (CULMINATION).

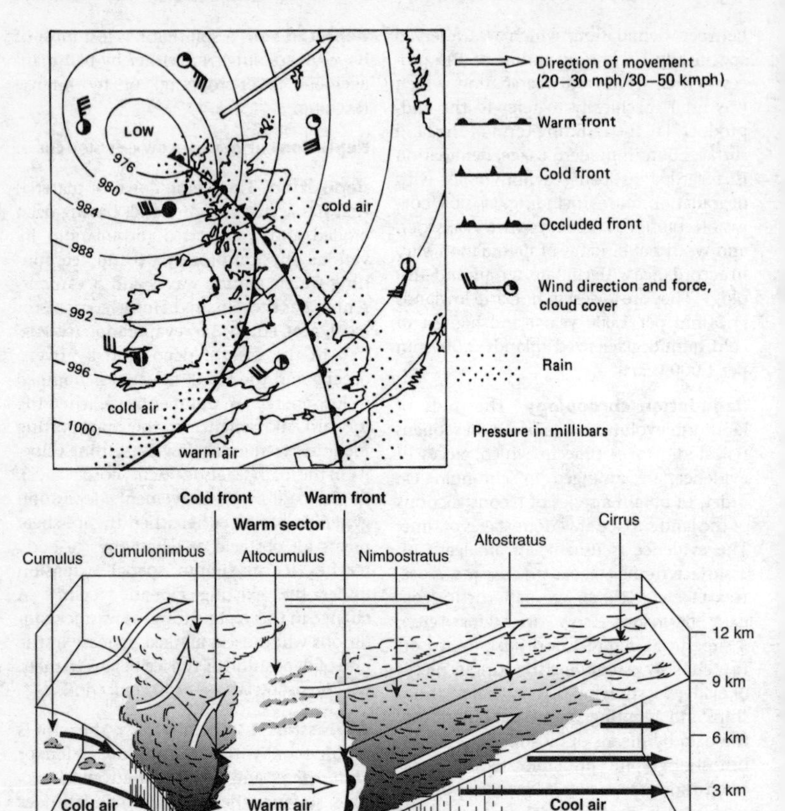

Figure 60 *Characteristics of a depression*

depression storage A term referring to the amount of rainfall that is retained temporarily at the ground surface owing to small irregularities (puddles). It will eventually disappear through PERCOLATION or EVAPORATION, unless the rainfall is persistent enough to exceed the state of depression storage, in which case the puddles amalgamate and OVERLAND FLOW takes place. *See also* SURFACE DETENTION, SURFACE STORAGE.

depth-area curve A meteorological term referring to the amount of rainfall falling over a given area. It is constructed by plotting the area enclosed by each ISOHYET against the mean rainfall within the isohyet.

depth-duration curve A meteorological term referring to the relationship between rainfall magnitude and duration, either at a single site or over a specific area.

depth hoar The upper surface zone of a snowfield which loses heat nocturnally owing to RADIATION. Because of changes in the ice-crystal structure of this surface zone a thin and fragile layer of ice forms above the underlying snow layers.

deranged drainage An uncoordinated pattern of drainage characteristic of a region recently vacated by an ice-sheet. This is probably due to the irregularities produced by the glacially deposited materials (e.g. KAME-AND-KETTLE TERRAIN) and by the fact that there has been insufficient time for the drainage to become adjusted to the structures of the solid rock underlying the glacial drift. The picture is one of numerous watercourses, lakes and marshes, some interconnected and some in local drainage basins of their own (e.g. the Canadian Shield).

Derryvree interstadial A warm phase of the middle stage of the MIDANDIAN glaciation of Ireland, dated at 30,500 BP from organic silts at Derryvree, Co. Fermanagh.

desalinization, desalination **1** The removal of salt from water to make it of a suitable quality for industrial and domestic use. Desalinization plants have been installed in several oil-producing Arab states of the Middle East where there is virtually no natural fresh water and where demands are increasing rapidly as the standard of living rises and industrial activity expands. The process depends largely on solar energy, which is used to evaporate (EVAPORATION) sea water in large pans, after which desalinated water condenses (CONDENSATION) on suitable surfaces. Other methods include that of distillation, which is very energy-intensive. **2** The removal of salt from soil to make it agriculturally productive. This may be necessary after the reclamation of a former sea floor (e.g. the polders of the Netherlands) or to clear the accumulation of surface salts following continuous irrigation (e.g. Imperial Valley, California). The problems may be overcome by such techniques as deep-ploughing, flushing with fresh water, and applications of gypsum.

desert An arid region characterized by little or no rainfall in which vegetation is scanty or absent, unless specially adapted or where groundwater conditions are favourable. The term was once confined to hot tropical and subtropical regions where PRECIPITATION was greatly exceeded by EVAPORATION, but it is now often used to describe mid-latitude low-rainfall areas of the continental interiors and also regions of perennial ice and snow of high latitudes where vegetation cannot exist because of low temperatures and physiological drought rather than a deficiency of precipitation. *See also* ARID, BIOCHORE, COLD DESERT, COLD-WATER DESERT, DROUGHT, THORNTHWAITE'S METHOD OF CLIMATIC CLASSIFICATION.

desertification, desertization The extension of typical desert landscapes, landforms and processes to areas where they did not occur in the recent past. Such changes take place in arid zones bordering the deserts, in areas which experience average annual rainfalls of 100–300 mm. The term is wrongly used by some authors to describe degradation of various vegetation types, including the subhumid and humid forest areas. *See also* ARID, DESERT, SAHEL.

desert pavement An extensive area of bare pebbles or larger stones that occur at the surface as a wind-polished tightly packed crust (sometimes termed *desert crust* or *desert mosaic*) in an arid environment. The pavement of coarse material is exposed at the surface in areas where the finer dust and sand have been blown away, but the hard stony layer protects the underlying finer layers from further wind erosion. The pebbles in the pavement may be cemented by saline deposits drawn to the surface by CAPILLARITY. *See also* REG.

desert varnish A stain or hard glaze that commonly occurs on exposed rock surfaces in the desert. It comprises a coating of dark iron oxide or manganese oxide, drawn to the surface in solution by CAPILLARITY and deposited as a lustrous film on all exposed surfaces by EVAPORATION.

desiccation A progressive increase of aridity, often as a result of climatic change. It may be due to natural changes such as a decrease in precipitation or an interference with a river regime, or it may result from human interference – over-grazing, deforestation, irrigation failure, etc. It is generally accompanied by a falling WATER-TABLE, a drying-out of the soil and a deterioration in the quality and amount of the vegetation cover. *See also* DESERTIFICATION, ARID.

desiccation breccia An accumulation of angular fragments (BRECCIA) derived from the break-up of sun-dried muds and DESICCATION POLYGONS which have subsequently been eroded and transported by floodwaters prior to being re-deposited and compacted. If the broken fragments of the clays and muds become rounded in transport they will be deposited not as a breccia but as a *desiccation conglomerate*.

desiccation conglomerate DESICCATION BRECCIA.

desiccation crack A crack formed in clays and muds by solar heating, caused by shrinkage during the drying-out phase of the bed of a former water body. The cracks may appear in distinctive forms (DESICCATION POLYGONS).

desiccation polygons Polygonal patterns of DESICCATION CRACKS formed in muds during a drying-out phase. They occur when solar heating evaporates a water body and causes the exposed clayey deposits to shrink and crack into irregular shapes.

desilication **1** The removal of SILICA from a soil, generally by LEACHING of the surface material in regions of heavy rainfall. **2** The term is also used to denote the removal of silica either from rocks by CHEMICAL WEATHERING in hot humid climates or from a MAGMA by reaction with the COUNTRY-ROCK (e.g. in limestone this would form lime silicates). *See also* LATOSOLS.

desquamaration EXFOLIATION.

destructive wave A storm wave in the ocean which, when breaking upon a beach, has an almost vertical plunge of water which combs beach material seawards. Thus its BACKWASH is stronger than its SWASH. Contrast a CONSTRUCTIVE WAVE. Destructive waves occur at a high frequency (thirteen to fifteen per minute) so that the backwash of a wave can interfere with the swash of the succeeding wave, thereby reducing its potential for moving beach material landwards. After a period of destructive-wave activity there will be a marked diminution in the size of a beach.

determinism, environmental A concept relating to the inevitability of a set of events or relationships under the influence of physical factors. For example, the great folk-movements of Asian nomads leading to the invasions (by the Mongol hordes) of Europe at the close of the Roman Empire have often been seen solely as a response to drought conditions in the Asian heartland.

deterministic model A MODEL which contains no random elements and for which the future course of the system (GENERAL SYSTEMS THEORY) is determined by its position or character at some fixed point in time.

deterministic process A TIME SERIES or other sequence of numbers whose future values can be predicted with certainty.

detritus Fragmental rocky material produced by the WEATHERING and disintegration of rocks and subsequently moved from its original site. Some writers believe that the term is synonymous with DEBRIS, although others would reserve its use for coarse material only. *See also* REGOLITH.

Devensian The final glacial stage of the Pleistocene in Britain, lasting from about 70,000 to 10,300 BP, and succeeding the IPSWICHIAN INTERGLACIAL. It is named not from a particular type-site but from the fact that its glacial deposits are particularly widespread in the plains of N Shropshire and Cheshire in the vicinity of Chester, whose Roman name was Deva. Its type-site is, in fact, at Four Ashes near Wolverhampton, Staffordshire. The cold stage was characterized by fluctuating ice-advances, interspersed with interstadial periods during which milder climatic conditions prevailed when birch and coniferous forest reappeared temporarily. It also saw the firm establishment of early man in the British Isles (PALAEOLITHIC), although his artefacts date back to the HOXNIAN. Dependent upon their geographical location different parts of Britain remained ice-free for varying lengths of time during the Devensian. Almost all of

S England and East Anglia lay south of the maximum ice-limits but the Western Highlands of Scotland were ice-covered for much of the time. The Devensian has been divided into three: **1** Early (70,000–50,000 BP), including the Chelford Interstadial; **2** Middle (50,000–25,000 BP), including the Upton Warren Interstadial; **3** Late (25,000–10,300 BP) which includes the maximum ice advance (*c.* 20,000 BP), the whole of the LATE-GLACIAL (including the ALLERØD and BØLLING INTERSTADIALS), the DIMLINGTON STADIAL and the LOCH LOMOND STADIAL in Scotland. *See also* WEICHSEL, WÜRM. [*109*]

deviation **1** In general, the variation or departure from the normal or average state. **2** In statistics, the term refers to the difference between an observed value and a calculated value. **3** In geodesy, the angle between the geoid and the spheroid of the Earth, or the angle between the vertical and the normal to the spheroid.

Devonian The fourth geological period of the PALAEOZOIC ERA, extending from 408 to 362 million years, and named from Devon in SW England. It comprises both marine and continental deposits, the latter being referred to as the OLD RED SANDSTONE. In Britain the marine facies, of sandstones, grits, slates and limestones, are confined to the southern counties but owing to a dearth of well-preserved fossils the BIOSTRATIGRAPHY is based on fossils occurring in the Belgian Ardennes. In the Old Red Sandstone, which occurs farther N in Wales and Scotland, the most common fossils are fish, plants and fresh-water molluscs, enclosed in thick layers of brown and red sandstones, marls and limestones, probably deposited in LACUSTRINE environments. CORALS were abundant, especially in the Mid-Devonian. The climax of the CALEDONIAN OROGENY (which began in the Silurian) was reached in the Devonian and was accompanied by widespread vulcanicity and emplacement of granite in northern England, the Southern Uplands and the Grampian Highlands of Scotland.

dew The moisture deposited in the form of water droplets on the surface of vegetation and other objects located near to ground level. It forms when nocturnal terrestrial (long-wave) RADIATION causes heat loss from the Earth's surface thereby cooling the lowest layer of the atmosphere to below the DEW-POINT and leading to CONDENSATION. Prerequisites for dew formation include calm air or low wind-speed (< 1 knot at 2 m), high humidity near the surface, and suitable radiating surfaces. Thus clear skies, calm weather associated with ANTICYCLONES and the cool nocturnal conditions of spring and autumn are most conducive to dew formation.

dew-mound An artificial mound of earth, covered with a layer of flat stones on which DEW condenses and trickles down into the earth itself. It is widely used in the deserts of the Middle East as a means of supplying moisture to fruit-trees.

dew-point The critical temperature at which air, on cooling, becomes saturated with water vapour and below which continued cooling will cause CONDENSATION of water droplets in the form of HYDROMETEORS if atmospheric conditions are favourable. *See also* HYGROMETER, PSYCHROMETER.

dew-pond An artificially constructed hollow in the chalklands of S England. It is lined with clay and is intended to retain moisture for agricultural uses. It is a misnomer, for there is little evidence that any of the water retained in dew-ponds is obtained from DEW by condensation; the majority of the water is rainwater.

dextral fault FAULT.

D-horizon A term relating to a SOIL PROFILE although in fact it refers to the unweathered and undisturbed solid rock occurring beneath the true soil (SOLUM). *See also* SOIL HORIZON. [*235*]

diabatic (the preferred term is **non-adiabatic)** A thermodynamic process in which a mass of gas (air) gains or loses heat from/to external sources, as in ABSORPTION/EMISSIVITY, CONDENSATION, CONDUCTION, EVAPORATION or RADIATION, or mixing by TURBULENCE. Contrast ADIABATIC. *See also* THERMODYNAMIC EQUATION.

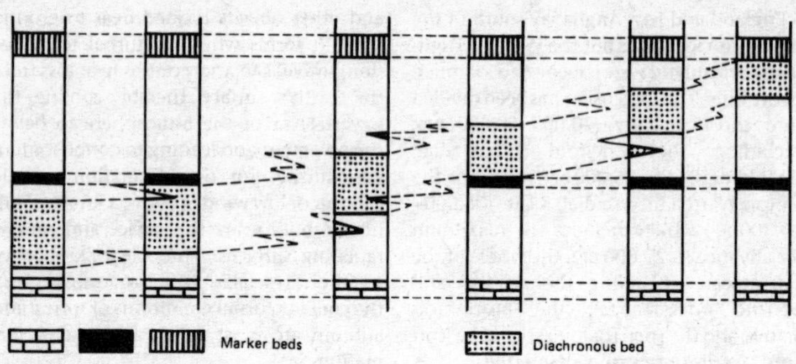

Figure 61 *Diachronous beds*

diachronous A geological term meaning 'time-transgressive' and referring to the cases in which the basal boundary of a sedimentary unit is not a time plane, i.e. its age differs in different lithological successions although the facies are the same and the strata appear to be spatially continuous. Failure to understand diachronism has led to considerable confusion over the recognition of certain geological boundaries, and unless zone fossils are present this argument will probably remain unresolved. [*61*]

diaclinal A geomorphological term introduced in 1875 by J. W. Powell to describe those river courses which cross the strike of the geological structure at right angles.

diagenesis A geological term referring to post-depositional processes of COMPACTION and cementation of sediments when they are at or near to the surface, at relatively low temperatures and pressures. Thus, any changes of volume due to METAMORPHISM are excluded. Compaction occurs owing to the increasing weight of the overlying sediments, while cementation takes place when the individual particles are bonded together by materials precipitated from circulating fluids which may be introduced either by percolating ground-water or be derived from solution of part of the mineral matter of the sediment itself. Calcite, silica and hydrated iron oxides are the most common cementing materials. Most diagenetic changes are controlled by the PH (acidity/alkalinity) and the eH (oxidizing potential) of the circulating fluids. When diagenesis leads on to the formation of massive rock layers it is termed LITHIFICATION.

diamict, diamictite, diamicton Terms referring to rocks and non-sorted terrigenous materials which exhibit a broad range of particle sizes, whatever their genesis. One of the best examples is a TILL.

diamond A cubic crystalline form of carbon found in both ULTRABASIC ROCKS and in alluvial deposits (ALLUVIUM), and initially formed under conditions of intense pressure and heat. It is the hardest known substance (HARDNESS SCALE) and is therefore in demand for cutting and abrasive purposes as well as for its gem qualities.

diapir An updoming structure created by a volume of mobile material rising buoyantly due to its low density relative to its surroundings.

diapirism The mechanism by which a mass of less dense rock domes the overlying layers of denser rock after having pierced the lower layers from beneath. It can be applied to: **1** the intrusion of granitic rocks. **2** the creation of SALT DOMES; **3** the formation of a MUD VOLCANO; **4** the intrusion of diapiric bodies of COOMBE ROCK through overlying solifluction deposits. **5** The creation of MUDLUMPS in deltas.

diastrophism A term referring to the large-scale deformations of the CRUST of the Earth which produce the world's mountain ranges, ocean basins, etc. Diastrophism includes folding, faulting, uplift and depression of the LITHOSPHERE, but not vulcanicity. *See also* CYMATOGENY, EPEIROGENESIS, OROGENY.

diatom A microscopic single-celled marine or freshwater plant: a subdivision of the ALGAE. It is capable of secreting silica and therefore contributes to the formation of some sedimentary deposits (DIATOMACEOUS EARTH, DIATOM OOZE). It is also an important constituent of PLANKTON and therefore important as a source of food for aquatic creatures.

diatomaceous earth A friable organic deposit composed largely of the siliceous remains of DIATOMS. When dried to a white powder form it is known as *diatomite*, used in the manufacture of pottery glaze, dynamite, cosmetics, etc.

diatom ooze A soft, deep-sea, siliceous deposit, composed predominantly of the remains of DIATOMS. It occurs in the ABYSSAL ZONE of oceans where cold waters abound, especially in the N Pacific and in a continuous belt around the Southern Ocean between latitudes 50°S and 60°S. Diatom ooze occupies some 9% of the total floor space of the world's oceans. *See also* GLOBIGERINA OOZE, RADIOLARIAN OOZE.

diatreme A general term for a volcanic VENT that has been drilled through crustal rocks by the highly pressurized gases associated with MAGMA. If it ultimately becomes blocked by PYROCLASTIC MATERIALS it will form a PIPE. The pipes which contain diamond-bearing KIMBERLITE at Kimberley, S Africa, are good examples of diatremes.

die back The first manifestation of stress and subsequent death of vegetation, in which nethermost leaves and shoots wither, followed by cessation of growth. The condition can be engendered by persistent DROUGHT and ACID PRECIPITATION.

dielectric constant A term used in REMOTE SENSING referring to an electrical property of terrestrial surfaces that influences MICROWAVE emissions (passive) and returns (active). Examples include SOIL WATER fluctuations and salinity changes in sea ice, both of which influence the conductivity of electrical energy and hence their microwave signatures.

differential ablation The melting of a glacier surface at different rates to produce a variable degree of relief. The degree of surface melting depends on: **1** the amount of TILL cover; **2** the amount of dust or dirt cover; **3** the colour of the ice. In the case of **1** the thicker the till cover the greater may be the insulative protection. *See also* PERCHED BLOCKS. In **2** and **3** the degree of surface melting is related to the ALBEDO; dirty ice reflects less solar radiation than clean ice (DUST WELL) and dark, bubble-free ice reflects less than whiter, bubbly ice. *See also* DIRT CONE.

differential compaction The relative changes in thickness of sedimentary beds due to reduction of PORE space after loading, depending on the differences in porosity, particle size and rigidity of the individual sediments. COMPACTION also forces water out of the materials, thereby leading to different rates of drying. Clays compact more readily than sandstones or limestones.

differential erosion The relatively greater rate of EROSION in some regions than in others may reflect either climatic differences, contrasting rock hardness, terrain contrasts, or different tectonic histories. Thus, erosion would be at its greatest in a cold, humid climate in a mountain area of relatively soft rocks that had recently been uplifted. Conversely, it would be less effective in an arid climate dominated by hard rocks that had previously been worn down in a tectonically stable area to a plain or basin of low relief. At a local scale a river or glacier will pick out rock weaknesses to create a terrain in which harder rocks remain as eminences and softer rocks will be worn into hollows or vales (SCARP-AND-VALE TERRAIN, KNOCK-AND-LOCHAN). Differences in the degrees of resistance of rocks to erosion are particularly well illustrated on coastlines where marine erosion is particularly selective in its wave attack, thereby creating irregular coasts of bays and headlands.

differential weathering The relatively greater rate of rock disintegration (CHEMICAL WEATHERING, WEATHERING) in certain parts of a rock mass than in others. This may be due to differences in porosity, grain size or mineral composition or simply to the degree of joint development (JOINTS). Rocks with well-developed joint systems will allow water to penetrate more effectively than will massive joint-free rocks. Thus differential weathering will produce an uneven surface (BASAL SURFACE OF WEATHERING, WEATHERING FRONT). Differences in rock colour will also affect the rate of weathering owing to different degrees of reflectivity (ALBEDO). Equally, physical weathering of a rock will vary according to the particular ASPECT of its exposure, i.e. rocks on sunny and shady slopes will undergo different intensities of weathering.

differentiation 1 A geological term referring to the process by which: (a) different parts of a single parent MAGMA assume different compositions and textures as the molten mass cools and solidifies; this is termed *magmatic differentiation*; or (b) certain minerals in a rock are segregated into bands or lenses during METAMORPHISM. This is termed *metamorphic differentiation*, and gives rise to banding in GNEISS. **2** A term relating to GENERAL SYSTEMS THEORY in which FEEDBACK mechanisms ensure that the elements of the system become more differentiated and therefore better adapted to varied environmental conditions. When the feedback mechanism does not function properly, differentiation ceases to occur.

diffluence 1 A term used in climatology to describe the condition during which the streamlines depicting a horizontal flow of air *diverge* from each other (DIVERGENCE). Diffluence causes a decrease in the velocity of air particles. *See also* CONFLUENCE. **2** A term used by geomorphologists in glacial geomorphology (GLACIAL DIFFLUENCE).

diffraction A term used in meteorology to describe the process of radiation spreading, which results when light rays are bent by an obstacle. It is responsible for a wide range of atmospheric optics, e.g. BISHOP'S RING, CORONA.

diffuse radiation Indirect solar radiation received at the Earth's surface after selective SCATTERING (normal <0.4 μm) by atmospheric molecules and minute dust particles. Also termed *skylight* or *sky radiation*. *See also* DUST, INSOLATION, NET RADIATION.

diffuse reflector A REMOTE SENSING term referring to any surface which reflects incident rays in a multiplicity of directions, either because of surface irregularities or because the material is optically not homogeneous. A diffuse reflector is in direct contrast to a SPECULAR REFLECTOR. At visible wavelengths writing-paper is a good example of the former and a mirror is a good example of the latter. Almost all terrestrial surfaces are diffuse reflectors, except for calm water, which is a specular reflector. A perfect diffuse reflector is termed a LAMBERTIAN SURFACE. *See also* ELECTROMAGNETIC RADIATION.

diffusion A term used in: **1** meteorology, to denote random mixing of air bodies either by TURBULENCE (*eddy diffusion*) or by slow molecular diffusion; **2** physics, to describe the spontaneous molecular interpenetration of two fluids without chemical combination. *See also* OSMOSIS.

diffusion coefficient A statistical parameter used in the calculation of rates of DIFFUSION. It varies with the temperature, the nature of the particles being diffused and the nature of the diffusion medium. It is defined as length squared divided by time, expressed by the formula: l^2/t, where l = length and t = time.

diffusion equation A hydrological term referring to an equation to illustrate the transient flow of groundwater through a saturated homogeneous, porous rock possessing identical HYDRAULIC CONDUCTIVITY in each direction. It is expressed as:

$$\frac{d^2h}{dx^2} + \frac{d^2h}{dy^2} + \frac{d^2h}{dz^2} = \frac{pg(\alpha + n\beta)}{K} \frac{dh}{dt}$$

where p is density, g is gravitational acceleration, α is the vertical compressibility of the aquifer, β is the compressibility of water and K is the hydraulic conductivity. *See also* DARCY'S LAW.

diffusion pressure gradient The process by which water is drawn upwards through tree trunks or plant stems, stimulated by the differences in VAPOUR PRESSURE between root-hairs and leaf surfaces. *See also* OSMOSIS.

digital elevation model (DEM) A computer-generated model simulating relief in an array created by a RASTER-format taken from the elevation depicted on a map.

digital image processing A technique used in REMOTE sensing involving the handling and modification of an IMAGE. It includes COLOUR DISPLAY, DATA COMPRESSION, IMAGE ENHANCEMENT, IMAGE RESTORATION AND IMAGE CLASSIFICATION, all of which are used in GEOGRAPHICAL INFORMATION SYSTEMS (GIS).

digitizing The process of converting ANALOGUE maps and any other graphic or pictorial information into a form capable of being read by a computer. It may be *point digitizing* by manually pressing a button when required, or *stream digitizing* where points are recorded automatically at pre-set time or distance intervals.

dike DYKE.

dilatation, dilatation joint A term used to describe the action of *pressure release* within a rock mass by the removal of overlying layers by DENUDATION – a mechanism sometimes referred to as *unloading*. For example, when ice-sheets disappear from a terrain of bare rock, the outward expansion of pressure within the rock may cause it to develop a set of expansion joints, termed dilatation joints, along which the curvilinear rock 'shells' may split in sheet-like concentric layers at right angles to the direction of pressure release. Thus, once loosened, the outer layers may be dislodged by weathering and be moved away from the rock surface by gravitational means. Dilatation is one of the chief mechanisms by which granite domes (INSELBERGS) and TORS develop parallel and broadly horizontal joint-planes which give the appearance of bedding (PSEUDO-BEDDING). *See also* SHEETING.

dilatation joint DILATATION.

dilation **1** In general use, the act of making wider. **2** In geomorphology, the expansion of fissures in a rock by ice, after conversion of water to ice by freezing. **3** In geology, the expansion or widening of an initial aperture by the intrusion of MAGMA.

dilational wave COMPRESSIONAL WAVE.

dilution gauging A means of measuring a river DISCHARGE by putting a tracer, e.g. a coloured dye, into a CHANNEL and timing it over a known river–channel length.

diluvial An obsolete term referring to DILUVIUM.

diluvium An obsolete term used widely by 19th-cent. geologists when referring to SUPERFICIAL DEPOSITS. It was originally given to materials supposedly deposited by the Noachian Flood described in the Bible, then for glacial DRIFT, in contrast to the younger ALLUVIUM, but also used to describe any flood deposit.

dimethylsulphide (DMS) A sulphur compound found in seawater and which is transferred by OXIDATION into the atmosphere where it forms an AEROSOL which in turn becomes an important condensation nucleus (HYDROSCOPIC NUCLEI). It is produced by planktonic ALGAE and the decay of BACTERIA.

Dimlington stadial Dimlington in Holderness, East Yorkshire, has given its name to the *Dimlington Stadial*, which in turn is taken to represent the principal advance of the Late Devensian (DEVENSIAN) ice sheet in eastern England (*c.* 20,000 BP).

diorite A coarse-grained plutonic intermediate igneous rock containing plagioclase FELDSPAR and ferromagnesian minerals. Up to 10% QUARTZ may also be present, but with increasing quartz it grades into a *granodiorite*. If the plagioclase becomes more basic it grades into a GABBRO. Diorite, which is the plutonic equivalent of ANDESITE, is an uncommon rock, forming only a small PLUG or BOSS rather than a large intrusive mass.

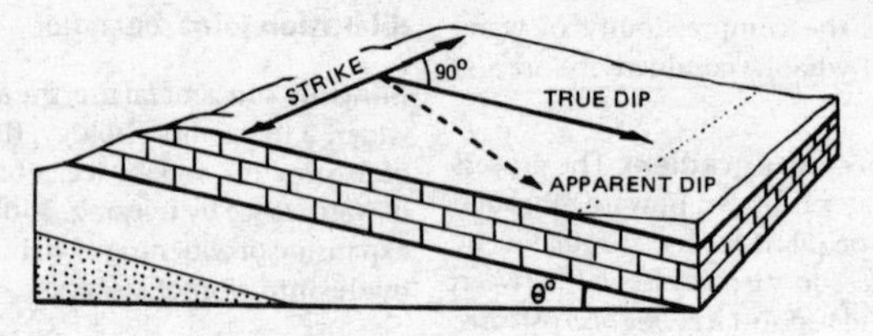

Figure 62 *Dip*

dip 1 In geology, the true dip of a bedding plane is the angle which it makes with a horizontal plane, measured in a direction perpendicular to the STRIKE of the rock strata. The angle is measured by a CLINOMETER and is expressed in degrees. The term *regional dip* is used to describe the general direction of dip over a wide area, by ignoring minor variations. *See also* APPARENT DIP. **2** A geomagnetic term describing the angle made between a horizontal plane and the Earth's MAGNETIC FIELD at any point on the surface. The *magnetic dip inclination* is 90° at the MAGNETIC POLES and 0° at the equator. [*62*]

dip-fault A FAULT that has a STRIKE approximately at right angles (perpendicular) to the strike of the rock bedding planes and is, therefore, parallel with the true DIP of the strata.

diphotic zone The depth in a water body at which sunlight is so faint that virtually no PHOTOSYNTHESIS can occur. *See also* APHOTIC ZONE, DYSPHOTIC ZONE, PHOTIC ZONE.

dipole field The part of the Earth's MAGNETIC FIELD that is inclined at 11° to the Earth's axis of rotation, as if reacting to a dipole magnet (one negative pole, one positive pole) placed at the centre of the Earth.

dip slope A topographic slope whose overall gradient is of the same amount and in the same direction as the true DIP of the rock strata which form the slope. Since this is an infrequent occurrence some authors believe that the term BACK SLOPE should be used to describe the gentler slope of a CUESTA, reserving the term dip slope for those cases where the parallelism of topographic gradient and structural dip is absolute.

dip-stream A stream which flows generally in the direction of the regional DIP of the rock strata.

dirt band A band of dirty or discoloured ice in a GLACIER, variously composed of dust, soot, or debris. It is indicative of a season of ABLATION and is enclosed by layers of cleaner ice which represent the seasonal or annual layers of FIRN accumulation. When exposed at the glacier surface it forms an OGIVE.

dirt cone A conical mound of detritus on a glacier surface. It may be up to 2 m in height and is invariably ice-cored. Its genesis is due to DIFFERENTIAL ABLATION, where a patch of debris protects the ice surface from direct insolation while the surrounding bare ice is lowered. As the AMPLITUDE of RELIEF increases, so the slopes of the dirt cone become increasingly unstable such that patches of the dirt mantle slide down the sides, uncovering the buried ice and leading to its ultimate destruction by ablation.

disaggregation The process by which a group of soil particles, formerly behaving mechanically as a cohesive unit, are broken down into discrete particles.

disappearing stream A stream which changes from a surface course to an underground course when it passes on to an area of limestone rocks (KARST). It usually vanishes down a SWALLOW-HOLE and after a lengthy underground journey reappears at a lower level (RESURGENCE). [*204*]

disasters The culminating events that result from ENVIRONMENTAL HAZARDS, and which are characterized by major losses of life and property. In recent years two schools of thought have developed in disaster studies: **1** the so-called dominant view in which disasters are seen to stem entirely from natural geophysical events of an extreme nature; **2** a structuralist viewpoint in which the causal relationships between natural processes and people (Acts of God) are re-examined. Such a view has led to the so-called *Theory of marginalization*, in which a continual process of human impoverishment, due to a world economy perpetuating technological dependency and unequal exchange, leads to a deteriorating physical environment and hence to an increasing vulnerability to natural hazards.

discharge The rate of flow of a river at a particular moment in time, related to its volume and its velocity. It is usually measured by a current-meter at a gauging-station and is expressed as a CUSEC or a CUMEC. The average velocity is measured at approximately 0.6 m depth below the surface. Other measuring devices include weirs, spillways, flumes or pitot tubes. The global discharge of rivers into oceans is *c.* 30,000 km^3 per annum. *See also* BERNOULLI PROCESS, RATING CURVE.

disclimax PLAGIOCLIMAX.

disconformity A type of UNCONFORMITY where there is a considerable time gap in the sedimentary sequence but where the strata above and below the plane of unconformity have similar DIP and STRIKE characteristics. Thus, unless there is good fossil evidence (PALAEONTOLOGY) it is often difficult to detect an absence of strata in the sequence and, therefore, to recognize a disconformity. [*263*].

discontinuity **1** In geophysics, a term denoting a change in the physical properties of the Earth's interior with increasing depth. A seismic discontinuity is one in which the velocity of SEISMIC WAVES changes as they cross the boundary of discontinuity (*see also* GUTENBERG DISCONTINUITY, MOHOROVIČIĆ DISCONTINUITY). **2** In meteorology, a term denoting a sharp change in the characteristics of temperature, humidity, wind-speed and direction at a marked boundary surface. This is usually a frontal surface (FRONT) between AIR MASSES.

discordance A geological term referring to a lack of parallelism between contiguous rock strata. This is termed an angular UNCONFORMITY.

discordant **1** A general term to describe any phenomenon which does not conform to the normal order of things. **2** In geology, a term used to describe an igneous intrusion (*discordant intrusion*) cutting through the bedding or foliation of the country-rock. **3** In geomorphology, any topographical feature that bears little or no relationship to the geological structure or the 'grain' of the region (*see also* DISCORDANT COAST, DISCORDANT DRAINAGE).

discordant coast A COASTLINE which cuts across the structural 'grain' of the region, e.g. SW Ireland. It is also known as an Atlantic type coast. It is the opposite of a CONCORDANT COAST. [*63*]

discordant drainage A drainage pattern which has not developed a systematic relationship with the underlying structure, in contrast to an ACCORDANT DRAINAGE pattern. The reason for its discordance may be *antecedence* (ANTECEDENT DRAINAGE), CAPTURE, DIVERSION or *superimposition* (SUPERIMPOSED DRAINAGE).

discordant intrusion DISCORDANT.

discordant junction A type of river CONFLUENCE, in which the discordance is due to differences in elevation between the trunk stream and the tributary stream. Thus the junction may be marked by a WATERFALL, possibly where a HANGING VALLEY meets a glacially overdeepened TROUGH. The junction is in contrast to an *accordant junction*.

discrete variable A variable which can take only a restricted set of values (usually whole numbers) within its range, e.g. the number of catastrophic floods at a given site in a decade. *See also* CONTINUOUS VARIABLE.

disharmonic folding Folding in which abrupt changes in the geometry of the folds occur in passing from one bed to another. It is due to differences in COMPETENCE of the various beds, with more intense folding occurring in the least competent beds.

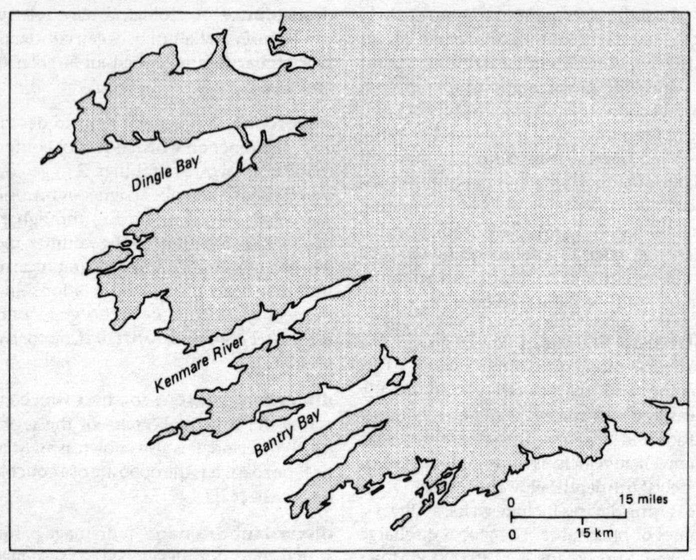

Figure 63 *Discordant coast*

disintegration The process by which rock breaks down into smaller particles by WEATHERING and EROSION.

dislocation A term referring to the displacement of rocks on opposite sides of a FAULT.

dismembered drainage A branching drainage pattern in which the lower reaches have been flooded by a marine submergence. Thus, rivers which were formerly tributaries to the trunk stream now enter the sea directly by separate mouths, possibly in the form of a RIA.

dispersal In biogeography the movement or transport of a vegetative part or seed, or of an egg or larva, eventually leading to the production of another individual.

dispersion **1** In seismology, a division of SEISMIC WAVES of different wavelengths owing to a variation of speed with wavelength. **2** A statistical term referring to the spread of a set of observations or objects about some central point or place. *See also* DISPERSION DIAGRAM.

dispersion diagram A diagram to illustrate graphically the distribution of a set of data, usually over time, e.g. monthly temperatures at a given station. Data plotted in space, or on a surface, are generally referred to as a DISTRIBUTION.

disphotic zone DYSPHOTIC ZONE.

dissection The process by which a land surface is cut up by eroding streams. Thus, the uniformity of a surface is broken up by gullying and stream incision.

dissipative beach One of two basic beach types defined by L. Wright in 1979. It has a wide low gradient between the surfzone and the FORE-DUNE. The waves in the wide surfzone (< 500 m) break and lose much of their energy before they reach the beach face. *See also* REFLECTIVE BEACH, SWASH, BACKWASH.

dissolved load The portion of a river's LOAD that is carried in solution, in contrast to the BEDLOAD. In the Mississippi, for example, some 29% of the load (by weight) is carried in solution. The proportion varies according to climate, the chemical nature of the rocks and the proportions of RUNOFF contributing to the river-flow in relation to the amount of groundwater flow. *See also* SUSPENDED LOAD, WASH.

dissolved oxygen An element present in most water bodies, dissolved from the atmosphere, and having a concentration dependent on the atmospheric pressure and the water temperature. Solubility of oxygen decreases as temperature increases. Serious problems for aquatic fauna occur when the dissolved oxygen deviates from equilibrium. Oxygen is consumed by respiration of organisms, e.g. fish, but also by biochemical OXIDATION of organic material and pollutants. Supersaturation can occur through PHOTOSYNTHESIS during daylight, by algae and macrophytes. Potential oxygen consumption due to organic pollutants is expressed as BOD (biochemical oxygen demand).

dissolved solids (TDS) The total concentration of dissolved material in water, measured by complete evaporation of a given quantity of water. The parameter is of use in any examination of the rate of chemical denudation and of the DISSOLVED LOAD of rivers. Discharge-weighted TDS concentrations range from 5000 mg l^{-1} in arid saline environments to 5 mg l^{-1} in tropical rainforest regions.

distributary A separate river channel created when a river splits, but one which does not rejoin the main channel. Thus, distributaries are common in a DELTA, but the term should not be used to describe a BRAIDED STREAM or an ANASTOMOSING STREAM. [*58*]

distribution **1** A graphic representation to illustrate the interrelationships between a set of objects or observations, usually in a spatial rather than a time dimension. [*200*] **2** In statistics, the term has several different functions: (a) a *normal distribution* is one in which data are grouped symmetrically about the arithmetic mean in such a way that 68.26% of them lie within plus or minus one standard deviation of the mean, 95.44% within plus or minus two standard deviations and 99.73% within plus or minus three standard deviations [*165*]; (b) a *skewed distribution* is one in which data are asymmetrically distributed around the arithmetic mean. More commonly the data exhibit a right-skewed distribution where there is a 'tail' of values to the right of the mean. Less commonly, a left-skewed distribution exhibits a 'tail' of values to the left.

disturbance A low-atmospheric-pressure system of no great intensity. *See also* DEPRESSION.

diurnal range (diurnal variation) A measure of the difference between the maximum and minimum values of temperature (and RELATIVE HUMIDITY), recorded within a 24-hr period. This is usually greatest in desert regions which record high daytime temperatures followed by a rapid heat loss through RADIATION at night, owing to the clear skies. *See also* EXTREMES OF CLIMATE.

diurnal tide A type of tidal pattern found in a few areas of the world where, because of the configuration of the coastline (e.g. the Philippine Is.), only one high tide and one low tide occurs every 24 hours. *See also* TIDE.

divagation A geomorphological term relating to the lateral shifting of a river channel caused by the excessive accumulation of sediments on the stream bed. It is always found in the process of MEANDER formation.

divergence **1** In CLIMATOLOGY, the term describes a type of air movement in which outflow exceeds inflow, thereby creating a tendency for the air at a particular place to decrease. Two types of air-divergence have been recognized: *streamline divergence* is created when STREAMLINES move away from each other, although remaining at constant velocity; when a sole air current undergoes a progressive increase in its velocity although its direction remains constant it is termed *isotach divergence*. If air density remains constant during horizontal divergence a vertical current of descending air must result. *See also* HORSE LATITUDES. **2** In OCEANOGRAPHY the term describes a zone or line from which surface masses of sea water move away as a result of wind-drift, leading to upwelling of deep water. **3** In BIOGEOGRAPHY the term refers to the decreasing degree of similarity exhibited by once similar species (plant or animal) which evolve in such a way that their life-forms become progressively less similar.

divergent erosion The contrast between rates of erosion in sub-tropical and temperate zones. In the latter erosion is strongest on slopes but less effective on flat surfaces, whereas in the sub-tropics chemical weathering is strongest on horizontal surfaces but less effective on slopes.

diversion, river **1** Artificial rerouting of a stream in order to counteract FLOODS, or to utilize its water for IRRIGATION or industrial purposes. **2** The rerouting of a stream by natural causes: glacial blocking by ice-sheets; blocking by a LANDSLIDE or an AVALANCHE; deflection by a coastal BAR or SPIT. *See also* DEFERRED JUNCTION.

diversity The physical or biological complexity of a SYSTEM.

diversivore An animal able to diversify its food intake between animal-eating and plant-eating modes, e.g. humans. Also termed OMNIVORE.

divide The area of high ground which separates two different drainage systems. In the UK, but not in the USA, it is synonymous with WATERSHED. *See also* INTERFLUVE.

division **1** A general term for an individual unit in any splitting or classification. **2** A term relating to a variety of taxonomic studies, e.g. in plant geography divisions are equivalent to the phyla (PHYLUM) of the animal kingdom. *See also* BIOSTRATIGRAPHY, CHRONOSTRATIGRAPHY, LITHOSTRATIGRAPHY.

doab A local term for a low alluvial plain between two converging rivers in the Indian subcontinent.

dogger **1** A large (metric scale) spherical CONCRETION occurring in a sedimentary rock. Ironstone concretions found in Jurassic rocks in Yorkshire are termed doggers, as are the large siliceous concretions in the Dorset CORALLIAN rocks. It has been suggested that the term be restricted to masses with a diameter greater than 256 mm (BOULDER). Any smaller concretionary mass is a NODULE. **2** An obsolete name for the Upper JURASSIC in Europe.

Doldrums The zone of light, mainly westerly winds or calms in equatorial latitudes, applicable largely to the oceans. It moves a few degrees N and S of the equator seasonally, following the passage of the overhead Sun accompanied by violent thunderstorms and squalls. *See also* EQUATORIAL WESTERLIES, INTERTROPICAL CONVERGENCE ZONE.

dolerite A medium-grained basic igneous rock which is intruded in a HYPABYSSAL form. It is finer-grained than GABBRO but is chemically and mineralogically the same. It is sometimes termed *microgabbro*, which in the USA is synonymous with *diabase* (a term which has been largely abandoned in the UK). It is coarser-grained than BASALT, which it resembles mineralogically. It occurs mainly in the form of a DYKE, PLUG or SILL.

doline A term for a circular hollow or depression in the surface of karstic terrain (KARST), in which the funnel-shape may or may not lead down into a vertical shaft descending into the limestone. It varies in size from 10 m to 100 m in diameter and is initially caused by SOLUTION. It is usually the site at which a stream disappears underground (SINK-HOLE, SWALLOW-HOLE). [*64*] *See also* UVALA. [*See also 204*]

dolocrete A type of CALCRETE in which magnesium carbonate is the dominant mineral. *See also* DOLOMITE.

dolomite **1** A common rock-forming mineral of calcium magnesium carbonate $CaMg(CO_3)_2$. **2** A rock in which dolomite is the characteristic mineral. It is synonymous with MAGNESIAN LIMESTONE. The process whereby limestone becomes dolomite by the substitution of magnesium carbonate for part of the original calcium carbonate is termed *dolomitization*. Such a change often results in the destruction of fossils and sedimentary features.

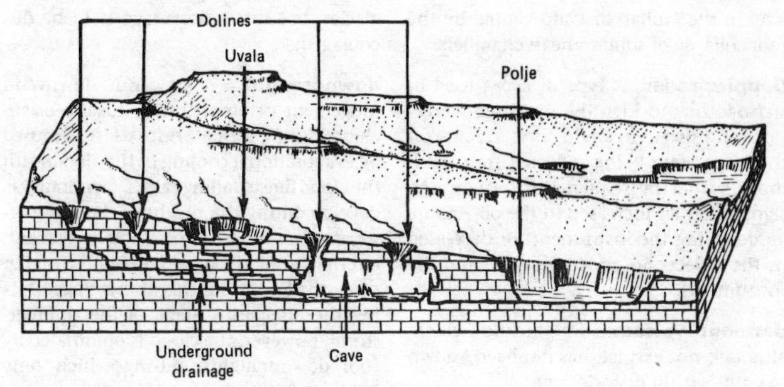

Figure 64 *Dolines, uvala and polje (karstic landforms)*

dome **1** In geology, a fairly symmetrical upfold in which all the beds dip away from a central point. Strictly, it is an anticlinal structure (ANTICLINE) which plunges in all directions (PERICLINE). **2** Topographically, the term refers to a smoothly rounded mountain summit or hill. **3** A rounded snow or ice accumulation on a mountain summit. **4** a low, circular desert dune, generally lacking a SLIP-FACE. **5** An acid volcano, e.g. Puy de Dôme in the French Auvergne. **6** A subterranean crustal upwarp in which oil, salt or natural gas accumulate. **7** A granite hill or knob (INSELBERG). **8** A rounded intrusion of igneous rock (*see also* BATHOLITH, LACCOLITH).

dome-dune DOME.

domed inselberg A term applied to a composite INSELBERG in which a smaller summit inselberg rises from a larger and broader inselberg to give a dome-on-dome effect. Such a landform may have been created by two erosion cycles (CYCLE OF EROSION) in which the earlier cycle has fashioned a residual which has become 'perched' when the land surface was further lowered by the denudation associated with the second cycle.

dominant A term used in biogeography to distinguish that species in a plant COMMUNITY which has an overwhelming influence on the environment of the other species present. It is usually, but not necessarily, the tallest species in the community, although there may be a dominant at each layer or stratum of a forest, i.e. oak may dominate the canopy but hazel may dominate the shrub layer.

dominant discharge A hydrological term to describe the DISCHARGE in relation to the average form of a river channel. It will depend on the CHANNEL GEOMETRY and partly on the river's LOAD.

dominant wave The wave which has the greatest influence on a particular shoreline. Thus, it is the largest wave or series of waves and may approach a coastline from a direction different to that of the PREVAILING WIND.

dominant wind A term introduced to describe the wind which plays the most significant part in a local situation, in contrast to the PREVAILING WIND. Thus a site located in a deep valley may be affected by local valley winds (ANABATIC WIND, KATABATIC WIND), although the prevailing wind may be from a different direction.

Donau The earliest of the Pleistocene glacial stages in the Alps, recognized in 1928–30 by E. Eberl from glaciofluvial outwash around the upper Danube (= Donau).

donga **1** A South African term for a steep-sided gully produced by fluvial erosion (RAVINE) or by floods. It is now used especially for gullies formed as a result of soil erosion. **2** In Australia, the term has been used to describe circular depressions, of varying size, in the Nullarbor Plain caused by the roof collapse of subterranean chambers.

Doppler radar A type of RADAR used in METEOROLOGY to identify air-current directions and speeds in a rainstorm. It relies on the radar echo being reflected from raindrops, so that the frequency of the returning signal is either increased (if the objects are approaching the instrument) or decreased (if the objects are moving away from the instrument).

dormant volcano A VOLCANO which, although not extinct, has not been known to erupt within historic time.

double surface of levelling A geomorphological term introduced by a German, J. Budel, to describe the simultaneous denudation processes being carried out **1** on the ground surface and **2** at depth, in the humid tropics. The *upper surface*, or ground surface, is a surface of erosion and physical weathering, largely by rainwater, in which the products are transported by rivers and by SHEET WASH. The *lower surface* is produced by CHEMICAL WEATHERING down to depths of 25–30 m and has been termed the BASAL SURFACE OF WEATHERING. [25]

double tide A tidal phenomenon caused by the deformation of a PROGRESSIVE WAVE either by the shape of the coastline or by the shallowing of an estuary (or by both). Consequently the *semi-diurnal tide* (TIDE) has a *quarter diurnal tide* imposed upon it, such that the tide rises to a maximum and retreats slightly before rising to a second maximum some two to three hours later. Alternatively, depending on the phasing of the tidal harmonic curves, a *double low tide* may result. It is a characteristic occurrence of the shores along the Solent of southern England (due largely to the location of the Isle of Wight) and gives considerable advantages to the port of Southampton.

downcutting A term normally applied to the action of a stream as it lowers its bed by fluvial EROSION (CORRASION), generally as a result of increased energy due to tectonic uplift. It has been used more loosely to describe the erosion of a glacial trough by a glacier, but this latter usage is to be discouraged.

downdraught A powerful downward movement of air within a CUMULONIMBUS cloud, during a THUNDERSTORM. It is caused by evaporational cooling of the air beneath the cloud due to falling rain. Downdraughts develop during the height of the thunderstorm when surface rainfall is at its heaviest, but the thundercloud also exhibits a higher proportion of *updraughts* at this stage. During the dissipating stage of the thunderstorm, however, the cloud is dominated by cool downdraughts, during which only light rain falls to the ground surface. *See also* UPDRAUGHT.

downs **1** A terrain of gently undulating hills, generally of CHALK, especially in southern England. **2** A plain of temperate grassland in the South Island of New Zealand and in Australia. **3** In the USA an area of coastal sand-dunes (DUNE).

downthrow The side of a FAULT that has moved downwards relatively to the other side which remains unaffected (UPTHROW). [77]

down-valley migration A term referring to the way in which a MEANDER belt moves slowly down-valley owing to lateral shifting of the river channel.

downwarping A slight downward deformation of the Earth's crust caused by sagging under the weight of an overlying burden, leading to a *downwarp* and in some cases to the beginning of a GEOSYNCLINE. The downwarp may be permanent, owing to the accumulation of a considerable mass of sediments (often in the vicinity of a delta, such as the Rhine or the Po); or it may be temporary, as in the case of downwarping beneath the ice-sheets of the PLEISTOCENE. In the latter instance the crust slowly recoils after the melting of the ice-sheets, a process termed GLACIO-ISOSTASY. It has been suggested that the formation of Lough Neagh in Northern Ireland (CENTRIPETAL DRAINAGE PATTERN) resulted from the central sagging of the great thickness of Tertiary igneous lavas which built the Antrim Plateau. Although some writers believe that downwarping is unaccompanied by faulting or folding, it has been suggested that some crustal deformation may take place around the boundaries of the downwarp. *See also* ISOSTASY.

downwasting **1** A term used by geomorphologists to describe the regional process of ABLATION as an ice-sheet diminishes in thickness over a period of time. **2** Occasionally the term is used as a synonym for DOWN-WEARING but this usage should be discouraged.

down-wearing A general term referring to the lowering of a land surface by DENUDATION, in which slopes become progressively gentler as a PENEPLAIN stage is approached. Thus, it contrasts with the term BACK-WEARING.

downwelling The action of sinking of warm coastal waters along a coastline, thereby reducing the nutrient supply near the surface.

draa (dzraa) The largest accumulation of sand in the Sahara, often with small dunes on its summit, and thought to be a coalescence of SEIF-DUNES.

drag fold **1** A minor fold in an incompetent bed (COMPETENCE) formed when the more competent beds on either side move in such a way as to disturb it. **2** A minor fold or pucker in association with a fault, caused by the differential movement of the rocks on either side of the fault.

drainage **1** The process by which water is discharged from an area by a river or by SHEET FLOW. **2** The discharge of water from a soil by PERCOLATION. **3** The removal of water from a marshy area (e.g. the Fens of England) by artificial means, e.g. the introduction of drains.

drainage basin That part of the land surface which is drained by a unitary river system. Its perimeter is marked by a drainage DIVIDE or WATERSHED (English usage only). Geomorphologists use the drainage basin as the unit for research into rates of RUNOFF and DENUDATION and into differences in DRAINAGE DENSITY. The size and shape of the basin is probably controlled by the geology, structure and climate of the region. *See also* DRAINAGE PATTERN.

drainage density A measure of the texture of a DRAINAGE SYSTEM, expressed as the ratio of the total length of all stream channels within a DRAINAGE BASIN to the total area of that basin (i.e. within the DIVIDE or WATERSHED). It is symbolized as D or Dd. Drainage density is influenced by geology, climate and the character of the terrain, with high-relief areas in a humid climate having a high density. Drainage density is also highest on impermeable but easily erodible rocks, e.g. clays. *See also* DRAINAGE PATTERN.

drainage pattern (drainage network) The spatial relationships of all streams within a DRAINAGE SYSTEM. The actual pattern will depend on a number of variables, including soils, geology, structure, present climate, palaeoclimate, tectonic history and human interference. It was once believed that a drainage pattern developed at a constant rate throughout time, but modern studies have demonstrated that, while it evolves very rapidly in the early stages, thereafter it changes very little, having achieved a STEADY STATE. *See also* ACCORDANT DRAINAGE, ANGULATE DRAINAGE PATTERN, ANNULAR DRAINAGE PATTERN, BARBED DRAINAGE, CENTRIPETAL DRAINAGE PATTERN, DENDRITIC DRAINAGE, DERANGED DRAINAGE, DISCORDANT DRAINAGE, PARALLEL DRAINAGE, RADIAL DRAINAGE, RECTANGULAR DRAINAGE, STREAM ORDER, TRELLIS DRAINAGE.

drainage system A river and all its tributaries within a single DRAINAGE BASIN or DRAINAGE PATTERN.

draw **1** A natural linear depression followed by surface drainage. **2** In the USA, a dry watercourse in the shape of a deeply incised RAVINE, occupied seasonally by an ephemeral stream. *See also* ARROYO, CREEK, NULLAH, WADI.

drawdown The act of lowering a WATER-TABLE by artificial means, usually by pumping or by ARTESIAN flow.

dreikanter A German term widely adopted to describe any pebble that has been shaped by aeolian sandblasting into a faceted VENTIFACT displaying plane faces with three (*drei*) sharp angles or edges (*Kanten*) bounding them. *See also* EINKANTER.

drift **1** Any material derived from the process of glacial erosion (DRIFT, GLACIAL). **2** Detrital material washed into an underground cave by rivers. **3** A surface movement of detrital material or of snow by the wind to form sand drifts or snow drifts, respectively. **4** The motion of ocean water generally at a low velocity, as a result of surface friction from the prevailing winds (e.g. North Atlantic Drift). **5** An underground mining passage driven horizontally along a coal seam or a mineral vein. **6** A term initially used with reference to the movement of continents (CONTINENTAL DRIFT) but now superseded by the concept of PLATE TECTONICS. **7** In South Africa a ford across a river.

drift, glacial Any detrital material eroded, transported and deposited by glaciers, ice-sheets or glacial meltwaters (GLACIAL DEPOSIT, GLACIOFLUVIAL). The term *drift* is used by the Geological Survey of Britain to distinguish all superficial deposits (including non-glacial deposits, such as peat and blown sand) from the solid rock. Hence, a *drift map* is a geological map on which all superficial deposits are shown, in contrast to a solid geology map which depicts only the underlying rocks. Drift deposits may be of considerable thickness or merely a veneer. *See also* BOULDER CLAY, DRUMLIN, ERRATIC, ESKER, KAME, MORAINE, SANDUR, TILL.

drift potential A measure of the ability of AEOLIAN forces to move sand in order to determine the resultant drift direction on a SAND-ROSE. *See also* WIND-ROSE.

drift-ice Any type of floating ice that has drifted from its point of origin (ICE-SHELF) and broken into pieces small enough for a ship to sail through. *See also* PACK-ICE.

dripstone A term given to calcite material (TUFA) deposited by dripping water. It is synonymous with both STALACTITE and STALAGMITE.

drizzle Light continuous rainfall, the droplets of which are less than 100 μ in diameter, i.e. like a fine spray. Generally associated with a WARM FRONT and gentle orographic uplift. Because of the small size of the droplets they would be unable to fall to earth if there was strong vertical air motion. Thus drizzle is often associated with stratiform cloud with a low cloud base and a high relative humidity between the latter and the ground surface, in order to compensate for evaporation.

dropstone A CLAST, generally released from floating ice, dropped into soft sediments, thereby causing some deformation.

drought A continuous and lengthy period during which no significant precipitation is recorded. In the UK an official drought is defined as a period of at least fifteen consecutive days on none of which is there more than 0.25 mm (0.01 in) of rain. In the UK a period of twenty-nine days, some of which may experience slight rain but during which the mean daily rainfall does not exceed 0.25 mm, is termed a *partial drought*. *See also* ABSOLUTE DROUGHT, DRY SPELL, PHYSIOLOGICAL DROUGHT.

drowned valley A former valley that has been inundated by a rise of sea-level (*see also* FIORD, RIA).

drumlin An Irish term that has been widely adopted to describe a streamlined elongated hummock or 'whaleback' hillock of glacial drift, generally of TILL. In profile the drumlin has a steeper slope at the 'upstream' end than at the 'downstream' end (in contrast to a ROCHE MOUTONNÉE). Although most drumlins show no internal structures some exhibit a degree of stratification, especially those which contain a very heterogeneous collection of drift. Its long axis is parallel to the direction followed by the former ice-sheet which was responsible for its formation. There is no absolute agreement on its mode of formation, although there are two broad hypotheses – an erosional and depositional genesis. The former suggests that the basal ice moulded a pre-existing landscape of glacial drift (GROUND MORAINE) while the latter indicates that basal till was deposited, possibly around a nucleus of rock or frozen drift, and that ice-movement moulded it contemporaneously. It is further suggested that till deposition takes place beneath the ice-sheet at all places where the ice is overloaded with debris and where friction between the subglacial surface and the basal till is greater than between the latter and the ice itself. Drumlins usually occur in groups termed a *field* or a *swarm*, popularly termed a basket of eggs, because each drumlin is shaped like half an egg. *See also* ROCK DRUMLIN.

druse A cavity in a rock, into which crystals project. The adjectival form is DRUSY. *See also* VUG.

dry In climatology the term refers to a season or climate which is free from or deficient in rainfall. The most important characteristic of a dry climate is that evaporation should exceed precipitation. *See also* ARID.

dry adiabatic lapse-rate (d.a.l.r.) A measure used to describe the change of temperature with height. When a parcel of unsaturated air rises through the atmosphere in equilibrium it expands and cools at a constant rate, termed the *dry adiabatic lapse-rate*, which is 1° C for 100 m (5.4° F in 1,000 ft) of ascent. *See also* ADIABATIC, ENVIRONMENTAL LAPSE-RATE, LAPSE-RATE, SATURATED ADIABATIC LAPSE-RATE. [*132*]

dry-bulb thermometer An instrument for measuring the ordinary air temperature, used in conjunction with the WET-BULB THERMOMETER to derive an expression of the RELATIVE HUMIDITY of the air. *See also* HYGROMETER, PSYCHROMETER, THERMOMETER.

dry delta ALLUVIAL CONE, ALLUVIAL FAN.

dry deposition The process involved when AEROSOLS, gases or particles (particulates) of pollutants are transferred directly from the atmosphere on to liquid or solid surfaces, quite independently of precipitation. The rate of deposition depends on the RELATIVE HUMIDITY, the pH of the surface and its aerodynamic resistance. *See also* ACID PRECIPITATION.

dry ice A term used by meteorologists to describe the solid form of carbon dioxide when used in the process known as 'CLOUD SEEDING' in efforts to produce rainfall. *See also* BERGERON–FINDEISEN THEORY OF PRECIPITATION, HYGROSCOPIC NUCLEI, RAIN-MAKING.

dry snow A type of powdery SNOW with a density of less than 0.1 kg/m^3, in which the very cold crystals do not bond together by REGELATION and therefore remain very small. It is characteristic of continental interiors, away from maritime influences, and forms ideal conditions for skiing. *See also* WET SNOW.

dry spell A period of drought of any length, defined (differently) in the USA and the UK but not internationally accepted. In the USA it describes a period of fourteen days with no measurable rainfall; in the UK it was formerly adopted to define a period of fifteen consecutive days on none of which more than 1.0 mm of precipitation was recorded. *See also* ABSOLUTE DROUGHT, DROUGHT.

dry valley A valley, in chalk or limestone country, that exhibits most of the attributes of a normal river valley with the important exception of the stream itself. This is true for lengthy periods of the year, although some dry valleys may be occupied by ephemeral streams (BOURNE) after periods of prolonged rainfall, especially in the winter. There is no agreement over their mode of formation but the main hypotheses are: **1** A recession of the chalk or limestone escarpment leading to the lowering of the SPRING-LINE and the WATER-TABLE, thereby leaving the valley system perched above the source of groundwater. **2** A fall in the average precipitation owing to climatic change leading to the same result. **3** The formation of the dry valley network could have occurred during an earlier period of PERIGLACIAL conditions during the PLEISTOCENE, when the underlying rock was completely impermeable because of PERMAFROST, so that all water had to run off at the surface, thereby carving out the dry valleys. **4** Some dry valleys have undoubtedly been left abandoned owing to *river capture* (CAPTURE, RIVER). **5** A few owe their formation to glacial diversion of former streams (DERANGED DRAINAGE). **6** In KARST country many valleys become dry because a surface stream disappears underground at a SWALLOW-HOLE. NB In the strictest sense although most glacial MELTWATER CHANNELS are now streamless they should not be described as dry valleys as defined above. *See also* NULLAH, WADI.

dry-weight percentage The ratio of the weight of any constituent of a soil to the oven-dry weight of the soil.

DuBoys's formula A formula proposed by DuBoys in the late 19th cent. to explain sediment transport in fluids. His formula shows that rate of transport is proportional to excess bed stress above a critical bed stress whose value depends upon the grain size of the material being transported. The formula is expressed as: $q_s = \chi\tau_o\{\tau o - \tau crit\}$, where q_s is sediment discharge weight/unit width/unit time; τo is bed stress; τcrit is bed stress required to create movement of particles, and χ is a constant dependent on flow parameters such as velocity and turbulence. *See also* CRITICAL TRACTIVE FORCE.

dujoda A PERIGLACIAL term from Yakutia, Russia, referring to a linear, steep-sided depression with an uneven floor developed in a THERMOKARST environment. [*253*] It forms by the coalescence of individual surface hollows after the BAYDJARAKH stage. *See also* PERMAFROST.

dumpy-level A surveying instrument used for LEVELLING. It consists of a small telescope and spirit-level attached to a supporting base.

dune 1 A mound or ridge of wind-blown sand, rising to various heights up to 50 m, and found in (a) hot deserts and (b) above high-water mark on low-lying coasts where sand is constantly renewed by onshore winds blowing across sandy beaches. Desert dunes are generally free of vegetation whereas coastal dunes are often 'fixed' by marram grass or coniferous trees (PHYTOGENIC DUNES). Dune-building requires: a fairly continuous sand supply; a constant wind strength and direction; an obstacle or series of obstacles to trap the sand. Desert dunes are usually created as wave patterns developed where the air flow interacts with the ground surface to create TURBULENCE, with dunes accumulating between the eddies (EDDY). *See also* BARCHAN, CLAY-DUNE, DOME-DUNE, DRAA, FULJE, HEAD-DUNE, LATERAL-DUNE, LUNETTE, PARABOLIC DUNE, REVERSING DUNE, RHOURD, SEIF-DUNE, STAR-DUNE, SWORD-DUNE, TAIL-DUNE, TRANSVERSE DUNE, WAKE-DUNE, ZIBAR. Coastal dunes are more complex in form than desert dunes owing to plant growth, marine erosion and the presence of groundwater reaching the surface, thereby creating wet hollows in the dune SLACKS. It is often possible to distinguish a development landwards, from the FORE-DUNE, through the main MOBILE DUNE to the STABILIZED DUNE. **2** The term has also been applied to subaqueous dune-like features created by running water (ANTI-DUNE) but these are at a considerably smaller scale and unless the term has a modifying prefix an AEOLIAN derivation is assumed. [*65*]

duricrust CRUST.

duripan A type of HARDPAN formed by siliceous cementation.

dust Tiny particles of solid matter (less than 0.6 mm in diameter) occurring anywhere in the atmosphere and light enough to be carried in suspension, by the wind, sometimes for vast distances around the globe before falling to earth. An average of 600 tonnes of dust are contained in 1 km^3 of air – this can be measured, by an instrument known as a *dust counter*. There are various sources of dust: **1** man-generated industrial and domestic dust, usually from combustion; **2** volcanic dust from an eruption; **3** dust from arid lands, which usually contributes to DUST-STORMS and DUST VEILS; **4** cosmic dust which is of extra-terrestrial origin. Dust particles with an affinity for water (e.g. soot) often act as HYDROSCOPIC NUCLEI for the CONDENSATION of water vapour. They also scatter solar radiation to space and to the Earth's surface as DIFFUSE RADIATION (*see also* SCATTERING). It has been calculated that 30% of atmospheric dust can be

Figure 65 *Classification of dune types controlled largely by vegetation, topographical features or localized sediment sources*

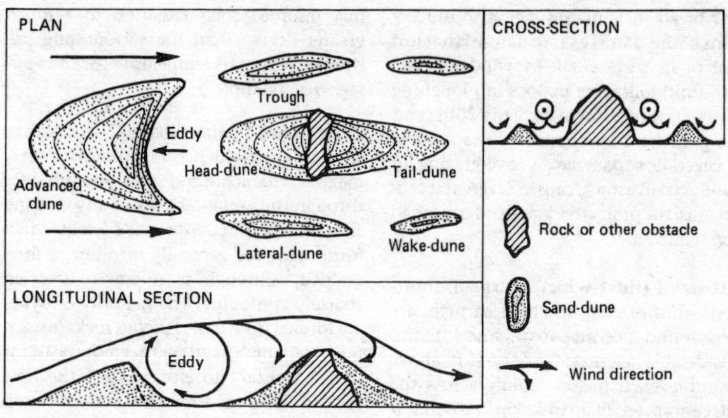

Desert dunes

Type	*Form and position*	*Mode of development*
Blowout	Circular rim around depression	Localized deflation
Parabolic	'U' or 'V' shape in plan view with arms opening upwind to enclose a blowout	Deposition of sand locally deflated upwind: arms are usually fixed by vegetation
Lunette	Crescent-shaped opening upwind	Accumulation downwind of localized sediment sources such as desiccated lake basin or pan
Shrub-coppice dune (nebkha)	Roughly elliptical to irregular in plan, streamlined downwind	Accumulation around and downwind of vegetation clump
Lee dune	Elongated downwind from topographical obstruction	Accumulation on protected lee side of obstacle
Foredune	Roughly arcuate with arms extending downwind either side of obstruction	Accumulation in zone of disrupted airflow immediately windward of obstacle
Climbing dune	Irregular accumulation rising up windward side of large topographical obstruction	Accumulation in zone of disrupted airflow on windward side of obstacle
Falling dune	Irregular accumulation descending leeward side of large topographical obstruction	Accumulation in zone of disrupted airflow on upwind side of obstacle
Echo dune	Elongated ridge roughly parallel to, and separated from, windward side of topographical obstruction	Accumulation in zone of rotating airflow upwind from large obstacle

Source: Summerfield, M. A., 1991 *Global geomorphology*. London and New York: Longman Scientific and Technical and Wiley. Table 10.3

attributed to human activity. *See also* AEROSOLS, POLLUTION.

dust bowl A semi-arid region in the SW plains of the USA (Texas to Kansas) that lost most of its surface soil by wind removal (DEFLATION) following periods of ploughing for cereal cultivation in the early 20th cent. This poor land management, exacerbated by vegetation damage by overgrazing of introduced livestock, caused severe SOIL EROSION once the protective grassland cover had been removed.

dust-devil (dust whirl) A small, short-lived, whirl of dust in which particles are swept round a central VORTEX and lifted to considerable heights (500 m/1,640 ft) above ground-level. It moves rapidly across the surface at speeds of up to 32 km/h (20 miles/h) but dies away without creating a serious hazard. It is caused by intense, local solar heating of the ground surface in arid lands, thereby producing a strong CONVECTION CURRENT. *See also* TORNADO.

dust-storm A storm in arid or semi-arid regions in which the immense volume of dust whipped up by the wind brings a serious reduction of visibility, sometimes almost to zero. Dust is often carried to very great heights (3,048 m/10,000 ft) and the storm is accompanied by excessively high temperatures, low humidity and high electrical tension. Its form may be that of an advancing 'wall' or it may be a VORTEX or whirlwind, either of which is a serious hazard, especially to air navigation. During a dust-storm an average of 2,300 tonnes of suspended matter is contained in 1 km^3 of air. *See also* BRICKFIELDER, DUST-DEVIL, HABOOB, SIMOOM.

dust veil index (DVI) A method devised by H. H. Lamb (1970) to quantify the meteorological significance of various degrees of dust shot into the stratosphere by volcanic ERUPTIONS. The index has a reference of 1,000 (Krakatoa, 1883 eruption). It can give an indication of global cooling by virtue of interference with solar radiation (INSOLATION).

dust well A small pit in the surface of a glacier caused by a patch of dust particles which have melted the ice because the dust has absorbed solar radiation to a much greater degree than the surrounding ice, thereby raising its temperature. *See also* DIFFERENTIAL ABLATION.

dyke (dike) **1** A sheet-like body of intrusive igneous rock which rises upwards from a magma chamber and cuts discordantly through the BEDDING PLANES or any existing structures of the COUNTRY-ROCK. Rocks which form dykes are generally medium to fine-grained, although a RING-DYKE is often coarsely crystalline. The majority of dykes are formed from basic igneous rocks (BASALT, DOLERITE) which may be either more resistant or less resistant to erosion than the host rock itself. Thus a dyke's surface outcrop may form either a wall-like feature or, if considerably eroded, it may create a topographic trench or gully. [*66*] Dykes are very commonplace in Mull and Skye in NW Scotland (TERTIARY VOLCANIC PROVINCE), and some extend over 200 miles south-eastwards into N England. *See also* DYKE-SWARM, CONE-SHEET. **2** An artificial embankment constructed to prevent marine flooding of coastal lowlands, e.g. in the Netherlands. **3** An artificially excavated drainage ditch. **4** An earthwork thrown up by early settlers as a defensive structure, e.g. Offa's Dyke along the Welsh Marches.

dyke-spring A natural outflow of water from an underground source along the line of a DYKE where the permeable sheet of intrusive rock interferes with the normal subsurface flow of water in the COUNTRY-ROCK.

dyke-swarm A multiplicity of DYKES around an igneous INTRUSION, in the form of a radial pattern or parallel to each other. Geologists have calculated that some Tertiary dyke-swarms have caused the crust to stretch some 65 km (40 miles) in the area of occurrence. Dykes that cause this type of stretching are termed *dilation dykes*.

dynamic angle ANGLE OF PLANE-SLIDING FRICTION.

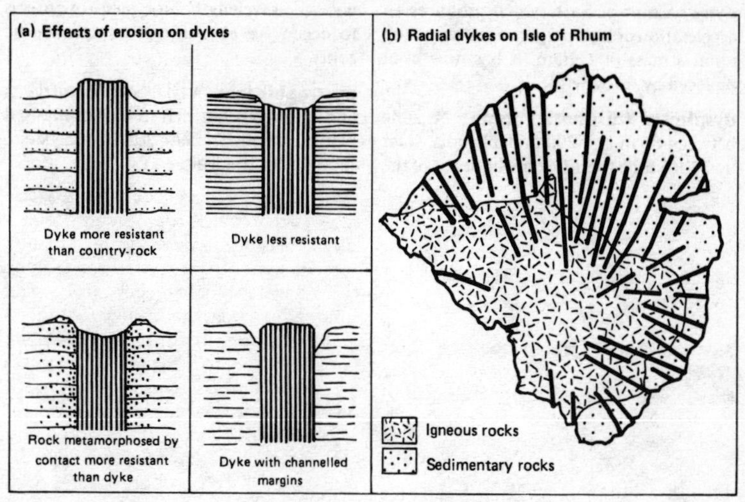

Figure 66 *Dykes*

dynamic climatology A study of the dynamic relationships which govern the general circulation of the atmosphere, in contrast to SYNOPTIC CLIMATOLOGY.

dynamic equilibrium An expression relating to the concept of EQUILIBRIUM within a physical system, in which a moving particle of air, water or solid material is in a state of balance in respect to the forces applied to it. Two possible conditions of motion may occur: the body may be moving at a constant velocity, when it is said to be in a STEADY STATE, because no acceleration or deceleration is required to keep the forces applied to it in balance. Alternatively, the body may be accelerating, subjecting it to unbalanced forces; e.g. when a boulder slides down a slope the reaction provided by frictional resistance (FRICTION) is insufficient to balance the total downslope force. It is an important concept in terms of the rates of rock weathering and debris removal on a hillslope, in which the dynamic equilibrium condition may be disturbed, for example, by climatic change. [*72*] *See also* STATIC EQUILIBRIUM.

dynamic friction The reduced frictional value obtained when a body is in motion. *See also* STATIC FRICTION.

dynamic homoeostasis HOMOEOSTASIS.

dynamic metamorphism A type of METAMORPHISM in which only localized stress is involved, causing rocks to break up, occasionally into a ROCK FLOUR OF MYLONITE. It is sometimes termed *dislocation metamorphism. See also* AUTOMETAMORPHISM, CONTACT METAMORPHISM, REGIONAL METAMORPHISM.

dynamic rejuvenation A term referring to the REJUVENATION of a river due to the uplift of the land or a fall of sea-level. Either of these changes of BASE-LEVEL will initiate a new cycle of erosion in which river incision increases, because of the river's greater energy following rejuvenation. *See also* STATIC REJUVENATION.

dynamic steady state The dynamic state of an OPEN SYSTEM, where the INPUTS AND OUTPUTS of matter and energy of the system are balanced by a steady flow. Any living organism can be described as an open system in a dynamic steady state. *See also* STEADY STATE.

dyne A unit of force which produces an acceleration of 1 cm per second when acting upon a mass of 1 gram. It has now been replaced by the NEWTON.

dysphotic (disphotic) zone A zone between 80 m and 200 m depth in a water body in which there is some light but of the wrong wavelength for PHOTOSYNTHESIS to occur. *See also* APHOTIC ZONE, EUPHOTIC ZONE.

dystrophic lake A lake poor in nutrients and in oxygen but rich in undecomposed plant matter. *See also* EUTROPHIC, MESOTROPHIC, OLIGOTROPHIC.

E

eagre, egre A local term for a tidal BORE on the R. Trent, England.

earth 1 The solid material of the globe that constitutes the land surface, in contrast to the water surface. **2** The loose surface material (including soil) as distinct from solid rock. **3** The fifth in size of the nine major planets which orbit the Sun, and the only planet inhabited by man. It has the shape of an oblate spheroid, a slightly flattened spheroid, but one with the North Pole 45 m farther from the equatorial plane than the South Pole if both were projected on to a theoretical plane at sea-level. Planet Earth has the following constant characteristics: mean density = 5,517 × 10^3 kg m^{-3}; mass = 5,976 × 10^{24} kg; volume = 1,083 × 10^{21} m^3; gravity acceleration = 9.812 m sec^{-2}; total surface area = 510 million km^2 (196.9 million sq miles); land area (29.22%) = 149 million km^2 (57.5 million sq miles); ocean area (70.78%) = 361 million km^2 (139.4 million sq miles); mean radius = 6,371 km (3,956.4 miles); equatorial radius = 6,378.5 km (3,963.5 miles); polar radius = 6,357 km (3,950 miles); equatorial circumference = 40,067 km (24,901.9 miles); meridional circumference = 39,999.7 km (24,859.8 miles); average height = 875 m above sea-level (2,870 ft); greatest height = 8,850 m (29,028 ft) (Mt Everest); lowest point of land surface = 396.2 m below sea-level (−1,299 ft) (shores of Dead Sea); greatest ocean depth = 10,430 m below sea-level (−34,210 ft) (the Marianas trench). *See also* CORE, CRUST, DISCONTINUITY, GEODESY, MANTLE, ROTATION OF THE EARTH.

earth-flow A rapid movement of soil and loose surface material (REGOLITH) downslope when saturated and buoyed up by water (MASS-MOVEMENT). It results from slope instability, perhaps due to human interference (removal of vegetation, undercutting of slope, etc.), but can occur on any steep slope after a period of heavy rainfall, especially when the underlying rocks are impermeable (PERMEABILITY). The gravitational movement commences as a superficial landslip (LANDSLIDE), when a mass of saturated material slumps downwards and outwards, but the amount of contained water may be sufficient to generate a flow, in the form of a tongue, which may extend to the foot of the slope, or beyond, until the velocity of the flow is insufficient to move the plastic mass. The greater the liquid content and the finer the soil particles the further the flow will extend (MUD-FLOW). *See also* SOLIFLUCTION. [*146*]

earth hummock A type of PATTERNED GROUND due to frost heaving in the soil (CONGELITURBATION) under PERIGLACIAL conditions.

earth-movement A differential movement of the Earth's crust caused by diastrophic forces (DIASTROPHISM). It may be a slow movement (CYMATOGENY, EPEIROGENESIS, GLACIO-ISOSTASY), or a rapid movement (EARTHQUAKE). The term should be strictly confined to movements caused by internal forces within the crust (ENDOGENETIC) and should never be used to describe denudation processes acting upon the surface of the Earth (EXOGENETIC).

earth pillar A column or pinnacle of clay (often TILL) or relatively soft earthy material, capped by a boulder, which serves to protect it from erosion by rain. Once the boulder falls from the pinnacle the pillar will rapidly be destroyed. It is found in areas of MORAINE and is also typical of the BADLANDS country of the USA. In France the pillar is termed a DEMOISELLE.

earthquake, tectonic earthquake A shock or series of shocks due to a sudden movement of crustal rocks, generated at a point (FOCUS) within the CRUST or MANTLE. The point where the shock waves (SEISMIC WAVES) reach the surface is termed the *epicentre* [*227*], around which lines of equal seismic intensity can be drawn (ISOSEISMAL LINE). The strength of an earthquake is known as its *magnitude* which can be measured on the RICHTER SCALE. (MERCALLI SCALE, MSK SCALE, ROSSI–FOREL SCALE). Some parts of the world are more earthquake-prone than others, with most of the seismic activity taking place at the margins of the tectonic plates (PLATE TECTONICS), especially around the margins of the Pacific plate. Thus the west coasts of North, Central and South America, the Aleutian Is., Japan, the Philippines, SE Asia and New Zealand are particularly vulnerable, as are the Mediterranean and the Middle East, along the zone of most recent mountain building (ALPINE OROGENY). *See also* AFTERSHOCK, FORESHOCK, SEISMOGRAPH, SEISMOLOGY, VOLCANIC EARTHQUAKE.

Earth Resources Technology Satellite (ERTS) SATELLITE, ARTIFICIAL.

Earth science A collective term for all studies concerned with the physical characteristics of planet Earth, in contrast to its biological characteristics. It includes the following disciplines: CARTOGRAPHY, CLIMATOLOGY, GEOCHEMISTRY, GEODESY, GEOMORPHOLOGY, GEOPHYSICS, HYDROLOGY, METEOROLOGY, MINERALOGY, OCEANOGRAPHY, PALAEONTOLOGY, PETROLOGY, REMOTE SENSING, SEDIMENTOLOGY, SOIL SCIENCE, STRATIGRAPHY, STRUCTURAL GEOLOGY, SURVEYING. *See also* PHYSICAL GEOGRAPHY.

earthslide A downhill movement of a mass of superficial material due to slope FAILURE, often as a result of water reducing the friction along a SHEAR PLANE in the soil mantle. With an increasing addition of water the slide will probably turn into an EARTH-FLOW. Since the movement only affects the soil cover it is a type of superficial landslip (LANDSLIDE). *See also* MASS-MOVEMENT.

earth temperature The temperature of the ground surface, derived largely from long-wave solar RADIATION and partly by CONDUCTION from the earth beneath. The surface temperature fluctuates both seasonally and diurnally, with the latter short-term change leading to the formation of DEW and GROUND FROST.

earth tremor A slight EARTHQUAKE.

Easterly Tropical Jet Stream (ETJ) The high velocity air flow of the JET STREAM over South-East Asia and southern India during the Northern Hemisphere summer. *See also* POLAR FRONT JET STREAM, SUB-TROPICAL JET STREAM.

easterly wave A shallow linear-trough disturbance in the TRADE WIND flow of the tropics, more evident in upper-level winds than in surface pressure patterns. The wave moves westwards transporting bands of convergence with pronounced CUMULUS cloud and heavy showers. It develops from an intense steepening of the LAPSE-RATE associated with the INTERTROPICAL CONVERGENCE ZONE. Some waves (about 10%) act as initiators of a TROPICAL CYCLONE.

easting The first half of a grid reference (GRID), always preceding the NORTHING when map coordinates are being quoted. It represents the distance measured eastwards from the origin of the grid.

ebb channel The channel followed by a tidal current as it moves seawards during the EBB TIDE. It often differs from the *flood channel*, followed by the FLOOD TIDE as it flows landwards. [*67*]

ebb tide The retreating tide, or the outgoing tidal stream, immediately following the period of high tide and preceding the FLOOD TIDE. In coastal areas where the tidal range between high and low water is large, the ebb tide may create a very powerful TIDAL CURRENT.

eccentricity The amount by which a point varies in distance from the centre of a circle

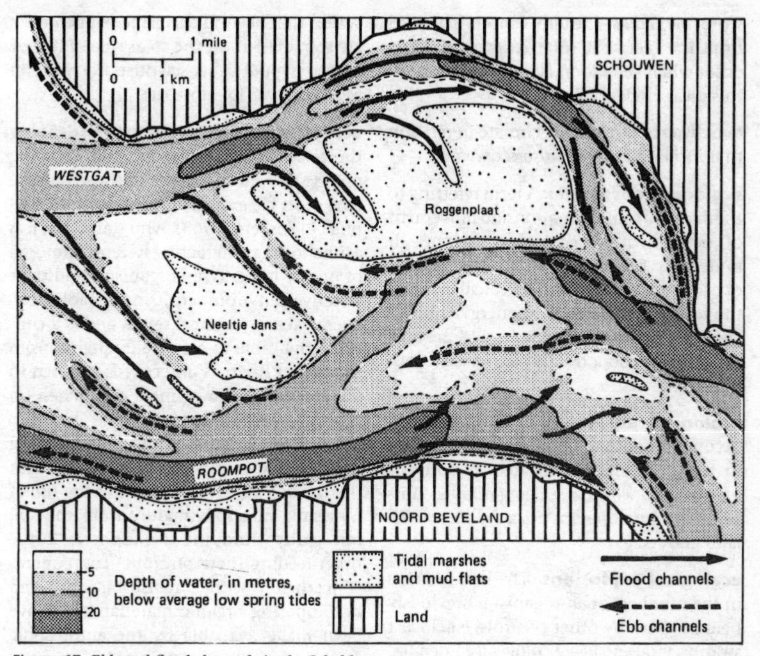

Figure 67 *Ebb and flood channels in the Scheldt estuary*

or orbit. For example, the orbit of the Earth around the Sun is not circular, i.e. the Sun does not lie in the exact centre of the Earth's orbit – the divergence is termed the eccentricity of the orbit.

echo-sounder An instrument designed to measure: **1** the depth of water in a lake or ocean; it measures the time required for a sound wave, generated at the water surface, to travel to the lake bed or sea floor and to return as an echo; and **2** the thickness of an ice-sheet, where the sonic vibration returns from the buried land surface beneath the ice.

Eckert projections A series of world map projections, with many similarities to the MOLLWEIDE PROJECTION, but with the continents having much better shapes. The Eckert projections are characterized by having the poles drawn not as points but as lines half the length of the equator.

eclipse **1** Solar, a situation created when the Moon intervenes in a line between the Sun and the Earth (*conjunction*). **2** Lunar, a situation created when the Earth intervenes in a line between the Sun and the Moon (*opposition*). An eclipse may be total or partial, with the total solar eclipse being most rarely seen at any one place because the narrow cone of lunar shadow is projected on to the Earth's surface as a belt of a mere 145 km (90 miles) width.

ecliptic The apparent path of the Sun in the CELESTIAL SPHERE, along a GREAT CIRCLE, during which it makes an average angle of 23°27' with the *celestial equator* and therefore with the plane of the Earth's equator; such an angle is termed the obliquity, which varies between 22° and 24°45' over 40,000 years. One complete revolution of the Sun along the great-circle path takes one year (strictly 365¼ days), and the ecliptic is divided into 12 sections each of which is symbolized by a *sign of the Zodiac*. The term ecliptic is derived from the fact that solar *eclipses* can only occur when the Moon is on or very near to this great circle. [*73*]

ecoclimate A term for climate in relation to flora and fauna. *See also* ECOSYSTEM.

ecological energetics A term referring to the fixation, transformation and movement of ENERGY within ecological systems. It underpins the concept that the BIOMASS of each TROPHIC LEVEL can be calculated and the results translated into energy equivalents, by using THERMODYNAMIC LAWS. *See also* BIOLOGICAL PRODUCTIVITY, NET PRIMARY PRODUCTIVITY.

ecological equivalents Species which occupy the same or similar ECOLOGICAL NICHES in similar ECOSYSTEMS located in different parts of the world; e.g. kangaroos in Australia and antelopes in Africa are both grassland grazers.

ecological explosions The vast increase in the numbers of an organism, previously held in check by other controls, e.g. locust swarms, viral and bacteriological epidemics.

ecological niche The specific part of an environment to which a particular organism is best suited.

ecological succession The slow change in an ecosystem during which one community of organisms is gradually replaced by a different kind of community. *See also* PRIMARY SUCCESSION, SECONDARY SUCCESSION.

ecology The scientific study of the interrelationships of organisms with each other and with the environment.

economic basement Rock strata below which there is little chance of discovering mineral wealth, especially oil. If mineral resources do exist below this basement, they are unlikely to be capable of exploitation at an economic cost.

ecosphere **1** A term used mainly by astronomers and space explorers to describe the zone of the SOLAR SYSTEM which lies between Venus and Mars, in which life as we know it may be possible. This is based on the observation that beyond this zone temperatures are too low but on the solar side of it they are too high. **2** Other scientists use the term as a synonym for BIOSPHERE.

ecosystem An ecological concept defining the relationships between a set of living objects and the attributes of those objects. One of its clearest definitions is that based on F. R. Fosberg (1963), who states that it is a functioning, interacting system composed of one or more living organisms and their effective environment, in a biological, chemical and physical sense, and is a concept applicable at any scale ranging from the planet Earth as an ecosystem down to the smallest patch of moss and lichen on a rock surface. The ecosystem has two biotic components: **1** the self-feeding (autotrophic) component, concerned with the fixation of light energy and the use of inorganic materials mainly from the soil but partly from the atmosphere; **2** the other-feeding (heterotrophic) component, concerned with redistributing and decomposing organic materials that have been made available by the autotrophic component. *See also* BIOME, CARRYING CAPACITY, GENERAL SYSTEMS THEORY, TROPHIC LEVEL.

ecotone A transition zone marking an overlap rather than a distinct boundary between two plant communities. Although the two communities may appear to blend they may be actively competing for the same territory, so that an ecotone may be a zone of tension.

ecotope **1** The area occupied by an ECOSYSTEM. **2** Sometimes the term is used synonymously with ECOSYSTEM. *See also* BIOCOENOSIS, BIOTOPE.

écoulement The gravitational sliding of large masses of rock, as a result of tectonic deformation. NB It is not synonymous with a ROCK-FALL, ROCKSLIDE OF LANDSLIDE, because all of these are caused by surface DENUDATION.

edaphic A term referring to the soil, particularly with respect to its influence on organisms, in which it is regarded as one of the major factors (FACTOR), together with the climatic factor, affecting plant growth. *See also* EDAPHOLOGY.

edaphic climax CLIMAX.

edaphology The study of the relationships between soil and organisms, including the use made of land by mankind. *See also* EDAPHIC.

eddy A localized movement of a fluid, especially air or water, within a larger overall movement. Wind, for example, will deviate from its non-turbulent LAMINAR FLOW, either in a horizontal or in a vertical plane, in order to circumvent solid obstacles in its path. Such a deviation will create an eddy on the leeward side of the obstacle. The scale of the eddy varies considerably from the vast air circulations of DEPRESSIONS within the general circulation of the atmosphere and the oceanic circulations of water bodies, down to the smallest eddy of air or water behind an obstacle on the downwind or downstream side, respectively. *See also* DIFFUSION, TURBULENCE. [*22* and *65*]

edge 1 A term used in local place-names in England, particularly with reference to the gritstone outcrops of the Pennines (MILLSTONE GRIT), but also to the cuesta of the Wenlock Limestone in Shropshire, Wenlock Edge, and to such sharp ridges (ARÊTE) as Striding Edge in the Lake District. **2** A term used by mathematicians to define a line joining two vertices or nodes.

edge wave A seawater WAVE, either a PROGRESSIVE WAVE or STANDING WAVE most commonly found on beach gradients steeper than 1 in 10. Such a wave represents oscillations moving along the shore due to instability created when energy is trapped against the shore. Short edge waves can form BEACH CUSPS while longer may form crescentic BARS.

Eemian The last interglacial stage of the Pleistocene in NW Europe, equivalent to the IPSWICHIAN of Britain, the Riss/Würm interglacial of the Alps, and the SANGAMON interglacial of N America. It occurred between the SAALE and the WEICHSEL ice advances.

effect The result of an input into a system (CAUSE). In statistical analysis of VARIANCE the term refers to the influence exerted by each separate controlling factor over the average values assumed by the variable.

effective porosity The property of a soil or rock containing interconnecting interstices (PORE), expressed as a percentage of the bulk volume occupied by these spaces. *See also* EFFECTIVE STRESS, POROSITY.

effective precipitation 1 In climatology, that part of the total precipitation which remains after evaporation and which is available for vegetation growth. **2** In hydrology, that part of the precipitation which enters a stream channel. *See also* INTERCEPTED MOISTURE.

effective size (D_{10} size) An expression used in soil mechanics, referring to the grain size which is larger than 10% by weight of the soil particles, as illustrated in the GRADING CURVE.

effective stress A concept relating to a transfer of sediment by rapid MASS-MOVEMENT, and referring to the role played by PORE-WATER PRESSURE in the shear strength of a material. In unsaturated soils or sediments a suction is exerted on the pore space (negative pore-pressure) such that normal stress, and therefore their strength, is increased. In saturated soils or sediments pressure is exerted on the grains (positive pore-pressure), because the pore spaces are now filled. Once a critical pore-pressure has been exceeded the shear strength of the sediment will suddenly fall leading to FAILURE of the slope. *See also* MOHR-COULOMB EQUATION.

effective temperature COMFORT ZONE, SENSIBLE TEMPERATURE, WIND-CHILL.

efficiency of a stream A measure of the capability (CAPACITY, COMPETENCE) of a stream to carry a LOAD. It is expressed by a formula, suggested by G. K. Gilbert in 1914, where stream efficiency is represented by:

$$\frac{\textit{capacity}\ (\text{g/sec})}{\textit{discharge}\ (\text{cusecs})} \times \begin{cases} \textit{percentage slope} \\ \textit{of channel bed.} \end{cases}$$

effluent 1 In general, anything which flows forth. **2** In geology, the flow of a lava from a volcanic fissure. **3** In geomorphology, the exit of a stream from a lake. In humid regions, where the WATER-TABLE is near the surface, groundwater will move gradually towards stream channels to produce perennial streams, termed EFFLUENT STREAMS. *See also* INFLUENT STREAM.

effusion EXTRUSION.

eH OXIDATION, REDUCTION.

E-horizon A-HORIZON.

eigenvalues A statistical term used in factor analysis referring to the characteristic or latent roots of the correlation matrix. Eigenvalues determine the value of the factors and there are as many eigenvalues as variables in the matrix.

einkanter A German term referring to a type of VENTIFACT which has only one (*ein*) facet cut by wind-blown sand. *See also* AEOLIAN, DREIKANTER.

EIS ENVIRONMENTAL IMPACT STATEMENT.

Ekman effect, Ekman flow, Ekman spiral 1 The tendency for upwelling OCEAN CURRENTS to be deflected by atmospheric winds as they approach the ocean surface. The surface flow (the Ekman flow) is complicated by the Earth's rotation and the configuration of the continents. In the N hemisphere the flow is some 45° to the right of the wind at the surface, but with increasing depth the speed of the flow decreases and its direction swings increasingly to the right. Near the equator on the western side of oceans the Ekman effect is virtually absent because there is no poleward deflection and no reverse current at depth. The world's cold currents (Benguela, California, Canary, Humboldt), although upwelling in narrow coastal zones, have their nutritious benefits spread westwards at the surface, thereby increasing supplies to fish populations over wider areas. The *Ekman layer* is the shallow surface layer of the oceans (< 100 m) in which Ekman flow occurs. **2** The Ekman effect in the atmosphere has been termed the Ekman spiral, which refers to the frictional drag of the Earth's surface on circulating air currents. Its frictional effect decreases with height, thereby increasing the windspeed until the GEOSTROPHIC wind becomes dominant between 500 m and 1,000 m. [*68*]

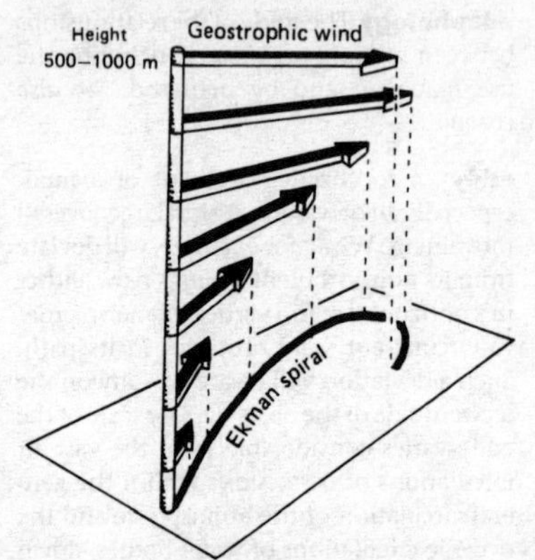

Figure 68 *Ekman effect*

elastic deformation A temporary deformation of a material or body, after which it returns to its former character and shape once the stress has been released.

elastic limit A point marking the maximum amount of stress that a material or a body can withstand before it ruptures or is permanently deformed by solid flow. The term is synonymous with *yield point* or *yield limit.*

elastic rebound The action by which rocks that have been stressed beyond their ELASTIC LIMIT return to a position of zero strain after rupture has taken place at a FAULT. Faulting is a result of the sudden release of elastic energy that has slowly accumulated in the tension zone of the Earth's crust, especially at a plate margin (PLATE TECTONICS) where major TRANSCURRENT FAULTS occur (e.g. the San Andreas fault of California). An EARTHQUAKE is the manifestation of the sudden energy release and the elastic rebound.

E-Layer HEAVISIDE–KENNELLY LAYER.

elbow of capture CAPTURE, RIVER.

electromagnetic energy A form of radiant ENERGY that is transferred by RADIATION.

It travels in a sinusoidal, harmonic wave form and is measured by its WAVELENGTH and frequency. This energy includes heat, visible light and microwaves and is an important data source in REMOTE SENSING. [69]

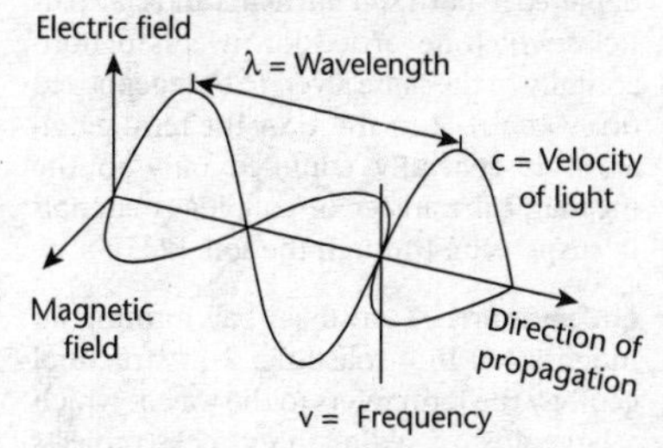

Figure 69 *Electromagnetic energy*

electromagnetic radiation The propagation of radiant energy in a variety of classes which differ only in wavelength, during which electric and magnetic fields vary simultaneously. In order of increasing wavelength (and decreasing frequency) the classes are: gamma radiation; X-rays; ultraviolet radiation; visible radiation; infrared radiation; microwaves; and radio waves. It is an important element of RADAR and INFRARED THERMOGRAPHY. *See also* BLACK BODY, PLANCK'S LAW, RADIATION, STEFAN'S LAW. [70]

electron spin resonance (ESR) An experimental dating technique used to calculate the amount of RADIOACTIVE DECAY occurring in such QUATERNARY materials as corals, molluscs, speleothems, and tooth enamel. *See also* RADIOMETRIC DATING.

electrostatic precipitator A device for removing PARTICULATES from chimney emissions by causing the particles to become electrostatically charged, attracted to a plate of opposite charge, and finally taken out of the air. *See also* POLLUTION.

element **1** In climatology, a term to describe each of the physical constituents whose sum total makes up the science of CLIMATOLOGY. They are 'the elements', namely precipitation, humidity, temperature, atmospheric pressure, and wind. **2** In chemistry and physics, a term to describe a combination of electrons, neutrons and protons. Currently 103 elements are known, ninety-two of which are natural and eleven are laboratory-made.

elevation **1** In general, a term used to describe a topographic eminence. **2** A specific altitude or height above a given level, e.g. above sea-level or ORDNANCE DATUM. **3** In surveying, the term refers to the angle between the horizontal and a point at a higher level. **4** In civil engineering it refers to the drawing of a building, etc. made in projection on a vertical plane.

ellipsoid of reference The shape of planet EARTH expressed as a ratio of ellipticity: ellipticity = $(a - b)/a$ where a is the radius at the equator and b is the radius at the poles. Because the Earth is not a true sphere but an oblate spheroid (or ellipsoid), compressed along the polar axis and bulging slightly around the equator, the radius measurements are slightly different (equatorial radius = 6,378.5 km/3,963.5 miles; polar

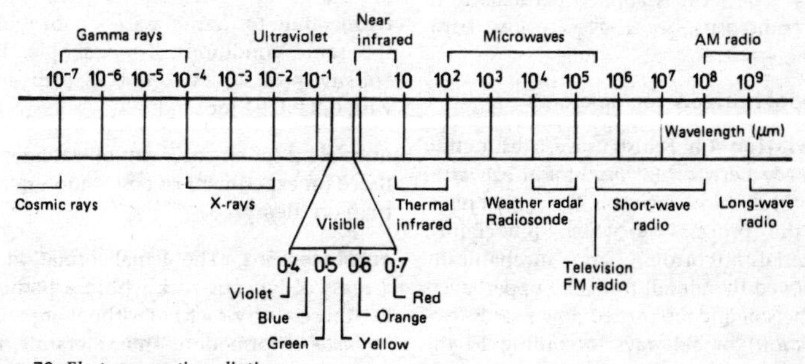

Figure 70 *Electromagnetic radiation*

radius = 6,357 km/3,950 miles). There are several slightly different ellipsoids of reference, depending upon the varying degrees of sophistication used in geodetic measurements. The earliest (1866) is the Clarke spheroid, used for N America (= 1/295); the Airy spheroid (1970) is used by the Ordnance Survey (= 1/299); the Clarke spheroid used by the British Admiralty (1880) (= 1/293.5); the International Ellipsoid of Reference (1924) (= 1/297).

El Niño A name given to the occasional development of a warm ocean current along the coast of Peru as a temporary replacement of the cold PERU CURRENT (HUMBOLDT CURRENT) which normally operates. El Niño is an extension of the *Equatorial current* and leads to an increase in surface-water temperatures of 10° C and a decrease in PLANKTON which thrive in the colder current. As a result of this reduction in their food supply the fish population is seriously depleted. It recurs every seven to fourteen years and results from a weakening of the SE Trades in the Pacific. The cold current oscillation is termed La Niña.

Elster The first of the Pleistocene glacial stages that have been recognized on the North European Plain. It is thought to be equivalent in age to the MINDEL glacial stage of the Alps and the KANSAN of the USA. [*109*]

eluvial deposits Weathered material formed by the break-up of solid rock, components of which have remained at or near to their point of formation, although the finer materials may have been washed away. The term is applied particularly to PLACER DEPOSITS. *See also* COLLUVIUM, ELUVIATION, REGOLITH.

eluvial horizon SOIL HORIZON.

eluviation **1** In British usage, a distinction is made between (a) *mechanical eluviation* and (b) *chemical eluviation*. The former refers to the process by which fine-grained materials (COLLOID) are mechanically removed by rainfall from the upper layers of the soil and are carried downwards (PERCOLATION) or sideways, according to the direction of groundwater movement, in suspension. Chemical eluviation is synonymous with LEACHING, in which nutrients are washed down from the surface layer in solution. As a result the surface layer of the soil, or *eluvial horizon* (A-HORIZON), becomes depleted in both soil nutrients and clay particles, which are carried downwards (or horizontally in the same layer) to be redeposited (ILLUVIATION). **2** In the USA the term eluviation is generally confined only to the mechanical transfer of colloidal materials in suspension through the soil. [*235*]

embayment **1** An open bay forming an indentation in a coastline. **2** In structural geology the term refers to the way in which a large area of sedimentary rocks projects into an exposure of crystalline rocks. **3** In sedimentology, the term describes a basin of sedimentation (GEOSYNCLINE).

emergent shoreline Any type of former coastline that once marked the juxtaposition of land and ocean but which is now either raised above present high-water mark (RAISED BEACH) or located at some distance inland from the present coastline. The emergence of the land/sea margin may be due to either the fall of sea-level (EUSTASY) or to the uplift of the land by earth-movements (ISOSTASY). *See also* SHORELINE CLASSIFICATION.

emission standard The maximum amount of a pollutant (POLLUTION) that is permitted to be discharged from a single source of pollution.

emissivity The ratio of the total radiant energy emitted per unit time per unit area of a surface at a particular wavelength and temperature to that of a BLACK BODY under the same conditions. For example, the emissivity of soil is 0.90–0.98 compared with 0.84–0.91 for sand.

empirical (of an argument or statement) Based on experiment or observation rather than on theory.

emplacement The actual formation of a mass of igneous rock within a body of existing country-rock but without implying its *mode* of formation. Thus, a granite, for example, may be emplaced by the mechanism of INTRUSION, by STOPING or by REPLACEMENT.

endemism The restriction of a particular species to a particular locality (HABITAT) owing to such factors as climate, soil or insularity. Its adjectival form is *endemic* (opposite to EXOTIC).

end moraine A synonym for *terminal moraine*. A mass of ice-transported debris, usually unsorted, marking the maximum limits of a glacial advance.

endogenetic, endogenic A term referring to processes that originate from within the Earth and result in such forces as igneous INTRUSIONS and different types of uplift and depression (*See also* CYMATOGENY, DIASTROPHISM, EPEIROGENESIS, OROGENESIS). *See also* EXOGENETIC.

endorheic (endoreic) drainage A term given to an inward flowing pattern of drainage (CENTRIPETAL DRAINAGE PATTERN) in the world's semi-arid zones. Ephemeral or seasonal rivers, subject to flash floods, flow radially towards large basins, frequently occupied by lakes (PLAYA) and thick infillings of alluvial deposits. Some of the basins may be characterized by internal drainage (ARETIC DRAINAGE). *See also* ARHEIC.

endothermic reaction Any reaction in which absorption of heat occurs. *See also* EXOTHERMIC REACTION.

endotrophic The ability of an organism to receive nourishment from within.

endrumpf A German term, introduced by the geomorphologist W. Penck to describe the almost flat land surface produced as the final or end-product of DENUDATION. It is more comparable with a PEDIPLAIN than with a PENEPLAIN. *See also* PRIMÄRRUMPF. [*193*]

en échelon Pertaining to geological structures which, though remaining parallel, are offset or overlapping (like tiles on a roof viewed from the side).

energy The ability or capacity of doing work possessed by a body or system of bodies. The term includes the following: **1** KINETIC ENERGY (E_k): the 'free' energy which is dissipated continually as heat friction by running water, moving ocean waves, sliding ice, etc. Expressed as $E_k = \frac{1}{2}MV^2$, where M = water mass and V = velocity. **2** *Solar radiant energy*: the energy transferred by ELECTROMAGNETIC RADIATION from the sun. **3** *Potential energy* (E_p): the energy stored in any object prior to its release as 'free' energy, e,g. when water moves downslope or when an ocean wave breaks upon a coastline. In hydrology, $E_p = Wz$, where W = the weight of the water and z = the 'head' of water, or height above base-level. **4** *Geothermal energy*: *kinetic energy* derived from the Earth's interior and which enters into numerous geological phenomena (GEYSER, VOLCANO). **5** *Elastic energy*: *potential energy* associated with a condition of mechanical strain (ELASTIC REBOUND). **6** *Heat energy*: *kinetic energy* generated by the internal random motion of molecules. **7** *Nuclear energy*: *potential energy* contained in the nucleus of an atom. **8** *Chemical energy*: *potential energy* stored in the molecules of compounds.

energy balance HEAT BALANCE.

energy crops Crops which are grown wholly or partly for their energy content, where solar energy can be stored in the form of carbohydrates. Both cellulose and starch can be fermented anaerobically to yield ethanol and methane. Sugar-cane waste is currently being used to provide heat, while other tropical grasses, cereals, algae and seaweed are under investigation.

energy cycle, biochemical A cycle consisting of the absorption of solar energy by plants and its conversion into carbohydrate compounds, which are a food source for animals. After much recycling most of these compounds are oxidized in the respiration process and returned to the atmosphere. A small portion of the compounds is retained in the dead plant and animal bodies where it will form organic layers in the soil, ultimately to be converted to HYDROCARBONS (*See also* COAL, PEAT, PETROLEUM), which are, in effect, storehouses of solar energy.

energy grade line The rate of dissipation of kinetic energy along the length of a stream channel, a rate which varies according to the CHANNEL GEOMETRY. For example, the energy grade line will be steep in a channel with a high velocity and a steep gradient. It always slopes downwards in the direction of flow.

englacial Pertaining to the environment within a glacier. It denotes meltwater moving freely through the ice-body, or debris (*englacial moraine*) which is embedded within the ice, having been derived either by downward movement from the glacier surface or by movement upwards from the bed of the glacier. It contrasts, therefore, with the basal environment (*See also* BASAL MELTING, GROUND MORAINE, SUBGLACIAL, SUPRAGLACIAL).

engulfment The collapse of the CONE of a volcano owing to the extrusion of molten lava laterally through a side fissure, rather than through the central vent. Bereft of its underlying support the volcano may collapse inwards to form a CALDERA.

enthalpy SENSIBLE HEAT.

entisol One of the soil orders of the SEVENTH APPROXIMATION of soil classification. It is a recent soil characterized by the absence of well-formed horizons, owing to the character of the parent material (e.g. blown sand) or to the fact that there has been insufficient time (e.g. on recently extruded lavas). Entisols include thin soils on unconsolidated glacial drift (*regosols*), shallow, stony soils (*lithosols*) and alluvial soils which are periodically buried by incremental layers of alluvium on floodplains (*fluvisols*). Although those entisols which occur in heavily ice-scoured terrain or in sand-dune topography have virtually no agricultural value, those of deltaic regions and alluvial floodplains are highly productive agricultural soils and carry dense populations in Bangladesh, India and China. The entisol is approximately equivalent to the old classification of the AZONAL SOIL.

entrainment A physical process referring to the way in which air or liquid is dragged along in the wake of moving bubble-like bodies. In meteorology the term refers to the process – for example, in a THUNDERSTORM – whereby air rises within a cumulonimbus cloud not as a continuous updraught but in a series of bubble-like air bodies which drag in air from the surrounding regions of the cloud. This has the effect of diluting the warmer rising air, thereby slightly reducing its buoyancy.

entrenched meander A MEANDER that has become incised into its valley floor. This is the US spelling of the British term INTRENCHED MEANDER. *See also* INCISED MEANDER, INGROWN MEANDER. [*112*]

entropy **1** In statistical thermodynamics, the term is used broadly to denote the degree of disorder or randomness in a system, with the greater the disorder the greater the amount of entropy. It refers to the probability of encountering given energy levels, states or events within a system. The condition in which there is a great deal of 'free' energy in a system characterized by heterogeneity, differentiation and hierarchical structuring is termed *minimal entropy* or *negative entropy*. In contrast, *maximum entropy* or *positive entropy* is the condition in which the amount of 'free' energy has declined to zero, and there is an equal probability of encountering given states, events or energy levels anywhere throughout the system because there is now a greater homogeneity. *See also* GENERAL SYSTEMS THEORY. **2** In geology, the term applies to the general degree of uniformity within a sediment. Heterogeneous or mixed sediments have a *low entropy* while uniform or homogeneous sediments have a *high entropy*. Maps showing degrees of entropy of rocks, using *isopleths of entropy*, may be drawn.

environment **1** In general, the sum total of the conditions within which an organism lives. Environmental factors (biotic, climatic, edaphic, etc.) act collectively and simultaneously in a holocoenotic way, so that the action of any one factor may be qualified by the others. **2** More specifically, the term is qualified in numerous ways: e.g. the *natural environment* is that created before the influence of man; the *built environment* refers to the artefacts created by man in the evolution of the cultural landscape. In geology, a major division into marine and continental environments has been recognized, relating to the different types of sedimentary deposition. The *marine environment* includes the following zones: ABYSSAL, APHOTIC (and PHOTIC), BATHYAL, LITTORAL, NERITIC. The *continental environment* includes the following *aquatic environments*: FLUVIAL, LACUSTRINE, LIMNIC, PALUDAL, and PARALIC. It also includes the *lagoonal environment* (LAGOON) which is part continental and part-marine. *See also* TERRESTRIAL ENVIRONMENT, TERRIGENOUS SEDIMENTS, HABITAT.

environmental hazards The risks to which organisms, especially humans, are exposed. They include: **1** *natural hazards* (AVALANCHE, BLIZZARD, EARTHQUAKE, DROUGHT, FLOOD, FOG, HAIL, HURRICANE, LANDSLIDE, TORNADO, TSUNAMI, VOLCANO); **2** *quasi-natural hazards* such as POLLUTION and SMOG; **3** *social hazards*, such as traffic and crime; **4** *human-induced hazards*, such as fire, explosion and industrial accidents; **5** *health hazards*, such as disease and malnutrition. *See also* GEOMORPHOLOGICAL HAZARD, HAZARD.

environmental impact The name given to the negative or positive change imposed upon the stability of an ECOSYSTEM, and especially on the health and well-being of humankind. There are two forms: first, direct impact, which is usually planned and whose effects are generally reversible, e.g. land-use changes, food modification via agricultural practices. Second, indirect impact, generally unplanned and whose social effects are normally undesirable, e.g. many human-produced pollutants. *See also* ACID PRECIPITATION, GREENHOUSE EFFECT.

environmental impact statement (EIS) A legally imposed government directive, related to planning permission, stemming from the US National Environmental Policy Act of 1969. It requires any proposed development to provide: **1** A description of the location and type of the proposed activity; **2** the likely impact on the environment; **3** the adverse or beneficial effects of the proposal; **4** whether the effects are long- or short-term; **5** whether they are reversible or irreversible; **6** the possible range of direct or indirect (knock-on) effects; **7** whether the impacts are of local, national or even international proportions.

environmental lapse-rate The actual rate at which temperature decreases with increasing altitude at a given place at a specific time. It averages about 0.6° C/100 m (3.5° F/1,000ft) of ascent. It is associated with a reduced density of air as elevation increases and with decreasing amounts of water vapour and therefore heat retention because of the remoteness from the surface of the Earth. *See also* ABSOLUTE INSTABILITY, ABSOLUTE STABILITY, CONDITIONAL INSTABILITY, DRY ADIABATIC LAPSE-RATE, INSTABILITY, INVERSION OF TEMPERATURE, PROCESS LAPSE-RATE, SATURATED ADIABATIC LAPSE-RATE. [*132*]

environmental management A term referring to a great variety of techniques and plans that attempt to regularize and rationalize the use of resources, i.e., the flows of energy and materials from their natural states through varying degrees of contact with humankind to their ultimate disposal.

enzyme An organic substance high in protein that is produced within an organism and acts like a catalyst.

Eocene The epoch succeeding the PALAEOCENE and preceding the OLIGOCENE. It was originally regarded as being the oldest part of the TERTIARY. It extends from 56.5 million years (the end of the Palaeocene) to 35.4 million years BP (the beginning of the Oligocene), although some geologists include the Palaeocene as part of the Eocene, thus extending its timespan from 15.5 million years to 27 million years. At this time, mammals were abundant and modern carnivores became well established. In southern Britain a humid subtropical climate prevailed in which rain-forests flourished. The sediments that were formed have survived in two main basins in southern England – the Hampshire Basin and the London Basin. The following stratigraphical successions have been recognized:

	Hampshire Basin	
Youngest	Lower Headon Beds	(Bartonian)
↑	Barton Beds	
	U. Bracklesham Beds	(Auversian)
	L. Bracklesham Beds	(Lutetian)
		(Cuisian)
Oldest	London Clay	(Ypresian)
	London Basin	
Youngest	Barton Beds	
↑	Bracklesham Beds	
	Upper Bagshot Beds	(Auversian)
	Middle Bagshot Beds	(Lutetian)
	Lower Bagshot Beds	(Cuisian)
Oldest	London Clay	(Ypresian)

NB The Thanet Sands and the Woolwich and Reading Beds once regarded as Eocene are now classified as Palaeocene. *See also* PALAEOGENE, TERTIARY VOLCANIC PROVINCE.

Eogene PALAEOGENE.

eolian The US equivalent of AEOLIAN.

eon AEON, CHRONOSTRATIGRAPHY.

Eozoic 1 A synonym for the PRE-CAMBRIAN era, i.e. the earliest of the geological eras, preceding the PALAEOZOIC, MESOZOIC and CAINOZOIC. **2** Some geologists regard the Eozoic as the earliest division of the Pre-Cambrian, followed by the ARCHAEOZOIC and the PROTEROZOIC.

epeiric sea A shallow sea resulting from EPEIROGENESIS. It is always located within a continent or upon a CONTINENTAL SHELF, hence its alternative name of *epicontinental sea*. An example is the Baltic Sea. *See also* CONTINENTAL SEA.

epeirogenesis A broad uplift or depression of the Earth's crust at a continental scale which moves the crustal rocks *en masse* in a vertical or radial direction. It must be contrasted, therefore, with OROGENESIS, in which the tectonic forces create folds, faults and thrusts because they are acting in a direction tangential to the surface of the Earth. *See also* CYMATOGENY.

ephemeral stream 1 In general, a stream which flows at the surface only periodically. In this usage it would be synonymous with BOURNE. **2** Some authors have confined the expression to a stream which is not fed from springs, glaciers or melting snow but only appears as a direct response to a period of heavy rainfall, i.e. in an arid or semi-arid environment.

epicentral angle An expression used in SEISMOLOGY to denote the angular distance between an earthquake's EPICENTRE and a surface recording station.

epicentre EARTHQUAKE.

epicontinental CONTINENTAL SEA, EPEIRIC SEA.

epidiorite A metamorphic rock which exhibits the minerals of a DIORITE but is derived from a thermally metamorphosed basic igneous rock (usually GABBRO or DOLERITE).

epigene EPIGENETIC (EPIGENIC) PROCESS.

epigenetic drainage SUPERIMPOSED DRAINAGE.

epigenetic ice A type of ice lens formed under conditions of PERMAFROST in the subsoil.

epigenetic (epigenic) process Any geological or rock-forming process which takes place upon (*epi* = Greek prefix for 'upon') the surface of the Earth. Its usage has come to include rocks formed at or near to the surface of the Earth and therefore includes both sedimentary and volcanic rocks. It contrasts with *hypogenetic process* (HYPOGENE).

epilimnion The warmer, less dense topmost layer in the water body of a lake or ocean. *See also* HYPOLIMNION, THERMOCLINE. [*252*]

epipedon A surface soil horizon classified in the SEVENTH APPROXIMATION in an attempt to distinguish differences in the history of horizon development (SOIL PROFILE), e.g. those illustrating the influence of man. The six epipedons related to certain classes of soil are: *anthropic*, influenced by farming (>250 ppm acid soluble salt phosphate); *histic*, very high organic content (20–30%);

mollic, very high calcium (>50% base rich); *ochric*, a yellow-brown altered surface horizon; *plaggen*, thick accumulation (>50 cm) of surface material added by human occupation; *umbric*, dark acid surface (>25 cm thick).

epiphyte A plant that grows on another plant, not in a parasitic way (SAPROPHYTE) but using it merely as a supporting base; e.g. orchids that grow in the limbs of trees and lichens and mosses that grow upon the bark of trees.

epithermal A term given by geologists to those HYDROTHERMAL processes which take place at relatively low temperatures in the lithosphere (100°–200° C). It refers to ore deposits formed at shallow depths in the crust by ascending hot solutions.

epoch A term given to the third rank order in a subdivision of geological time (CHRONOSTRATIGRAPHY). Several epochs form a PERIOD. E.g. the Pleistocene and the Holocene are epochs and together form the Quaternary period, which itself is part of the Cainozoic ERA. An epoch is made up of a number of ages (AGE) and its equivalent Stratomeric Standard term is SERIES.

equable A term used in climatology to define a climatic type exhibiting a minimum of variation throughout the year.

equal-area (equiareal) map projection A type of MAP PROJECTION in which areas of the globe exhibit their correct area but are grossly distorted in shape. It is constructed from the centre of a sphere through a point on the sphere's surface to a plane tangent at the South Pole. *See also* AITOFF'S EQUAL-AREA PROJECTION, BONNE'S PROJECTION, CYLINDRICAL EQUAL-AREA PROJECTION, MOLLWEIDE PROJECTION.

equation of time The difference in value between *apparent time* and *mean solar time*. The former is the progression of hours and days governed by the Sun itself and which is changing in value from day to day. The latter is the system of hours and days computed mathematically to give an average (*mean solar day*) of 24 hours. If the Sun arrives overhead at the meridian before 12.00 noon according to mean solar time it is said to be *fast* and the *equation of time* is positive. Conversely, if the Sun arrives late over the meridian the *equation of time* is negative and the Sun is said to be *slow*. On four days of the year apparent time and mean solar time coincide (about 15 April, 15 June, 2 September and 25 December). From January to mid-April the Sun is 'slow' and the equation of time reaches a minimum value of –14 mins. From mid-April to mid-June the Sun is 'fast', reaching a maximum value of +4 mins. From mid-June until the end of August the Sun is again 'slow', reaching a minimum value of –6½ mins. From September until the end of December the Sun is 'fast', achieving a maximum of +16 mins. Values of the equation of time for any day of the year can be seen from a graph termed an *analemma* [*71*], on which the DECLINATION of the Sun is also shown.

equator (of the Earth) The GREAT CIRCLE of the Earth, which coincides with the 0° parallel of latitude. It lies midway between the two poles in a plane at right angles to the axis of the Earth. It is 40,067 km (24,901.92 miles) in length. *See also* CELESTIAL SPHERE.

equatorial bulge The excess radius of the Earth at its EQUATOR (about 20 km) over that at the poles, thereby giving planet EARTH the shape of an oblate spheroid. It results from forces generated by the Earth's rotation and causes the PRECESSION OF THE EQUINOXES.

equatorial climate The type of climate which occurs in low latitudes (approx. 10°N to 10°S) on low ground only. It is characterized by constantly high temperatures and humidity and by an almost equal duration of day and night. Because the midday sun is overhead or nearly overhead at all times the seasonality is reflected largely by differences in patterns of rainfall, as converging AIR MASSES are affected by movements of the INTERTROPICAL CONVERGENCE ZONE. *See also* CONVECTIONAL RAIN, EQUATORIAL TROUGH.

equatorial current A surface current in the oceans of the equatorial regions. In fact

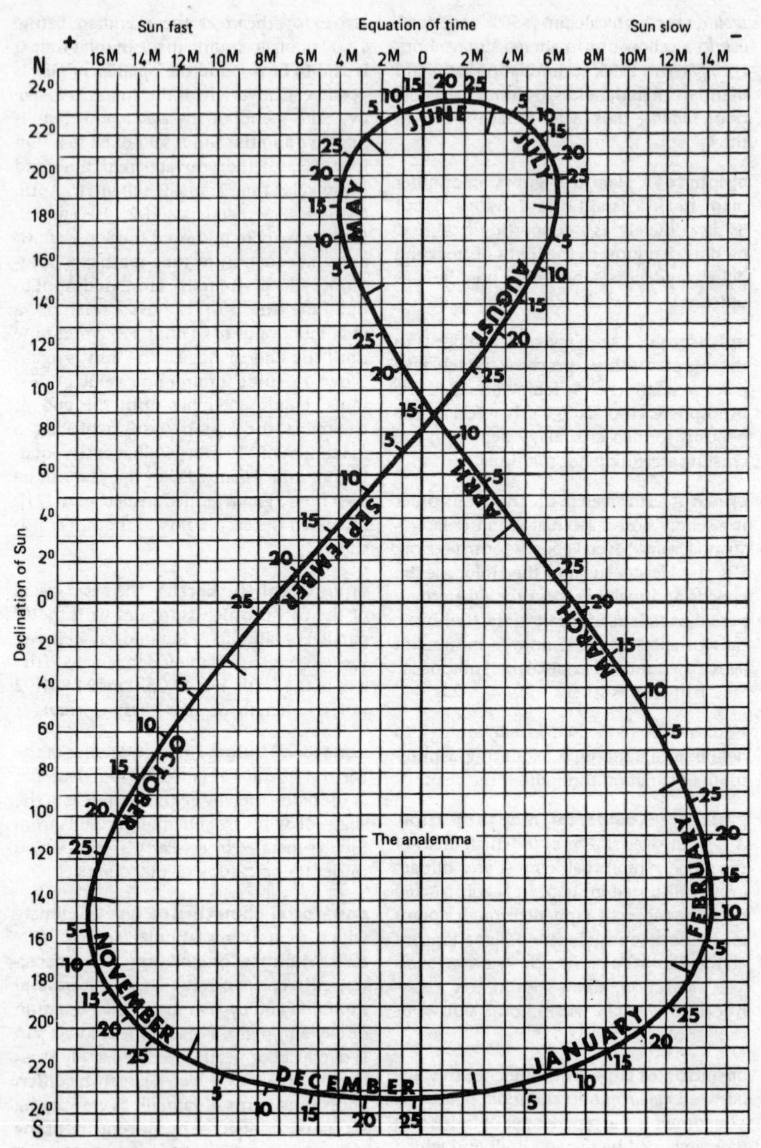

Figure 71 *Equation of time*

there are three *equatorial currents*: the *North Equatorial current*, which flows westwards in the N hemisphere; the *South Equatorial current*, which flows westwards in the S hemisphere; between the two is an eastward-flowing *Equatorial countercurrent*. The last of these is more strongly developed in the Pacific Ocean than in the Atlantic Ocean. The currents flow at speeds of about 2 knots. [*168*]

equatorial rain-forest The forest occurring in the zone of constant rainfall which typifies an EQUATORIAL CLIMATE. In a climatic regime of constantly high temperatures and high humidity there is virtually no seasonal change, with vegetation growing profusely and with great luxuriance all the year round, since there is no cold or dry season. Individual species shed their leaves at different times and since there is a large number of species (almost 2,000 in a square kilometre) there is a constant supply of litter. But somewhat paradoxically there is only a thin cover of leaves on the forest floor and an absence of humus in the soil because of the extremely rapid bacterial decay. The forest is composed of tall, closely spaced trees, the crowns of which form a continuous CANOPY, the trunks of which are commonly buttressed by large roots, and the leaves of which are evergreen. The trees are interlaced with woody vines (LIANA) and frequently covered by EPIPHYTES, although the climax rain-forest has an open floor. Only when secondary vegetation (e.g. bamboo, thorny palms) establishes itself does the forest become impenetrable at ground level – usually where felling or burning has occurred. It is typical of the Amazon basin, the Zaire basin of Africa and parts of Indonesia. TROPICAL RAIN-FOREST has very similar vegetation characteristics but greater seasonality and geographic range.

equatorial trough The belt of low pressure at the equator. *See also* DOLDRUMS, INTERTROPICAL CONVERGENCE ZONE.

equatorial westerlies A name given to westerly winds at low latitudes, caused when the NE and SE TRADE WINDS are deflected on crossing the equator and develop a westerly component, thereby becoming the equatorial westerlies whenever the INTERTROPICAL CONVERGENCE ZONE is more than 5° away from the equator.

equatorial zenithal projection ZENITHAL PROJECTION.

equidistant projection **1** An AZIMUTHAL PROJECTION that maintains its scale by deliberately spacing the meridians and parallels equidistantly outwards from the centre of the map. It is particularly useful as a projection for air navigation maps. **2** When the concept of *equidistance* is introduced on to a CYLINDRICAL PROJECTION (the TRANSVERSE MERCATOR PROJECTION) the map scale remains constant along two straight parallel lines on the map which are equidistant from the central meridian.

equifinality A concept of GENERAL SYSTEMS THEORY indicating that the same final state may be achieved from differing initial conditions and in different ways. In geomorphology, the concept is applicable to landforms, where the overall shape of a particular terrain may be achieved by different processes but where the initial landforms may have been very different.

equilibrium A state of balance in any system, created by a variety of forces in which the state will remain unchanged through time unless the controlling forces change. *See also* DYNAMIC EQUILIBRIUM, METASTABLE EQUILIBRIUM, POISED EQUILIBRIUM, QUASI-EQUILIBRIUM, STABLE EQUILIBRIUM, STATIC EQUILIBRIUM, STATIONARY EQUILIBRIUM, STEADY-STATE EQUILIBRIUM, THERMODYNAMIC EQUILIBRIUM, UNSTABLE EQUILIBRIUM. [*72*]

equilibrium-line **1** A term given to the boundary between the ablation zone and the accumulation zone of a GLACIER. It is similar to but not synonymous with the FIRN-LINE, because the glacier surface of the ablation zone below the elevation of the firn-line is regularly built up by layers of freezing meltwater. **2** A term describing the notional balance between process and form. *See also* EQUILIBRIUM.

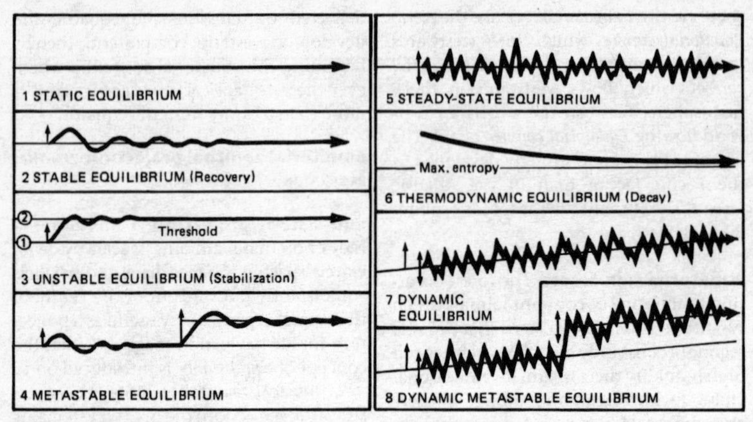

Figure 72 *Equilibrium*

equinoctial A synonym for the celestial equator (CELESTIAL SPHERE).

equinoctial gale A strong wind generally experienced at or around the EQUINOX. Such gales are said to occur in Britain with greater than normal frequency around 20 September and the last week of March (the *equinoxes* are 21 March and 22 September). No physical explanation has yet been offered for this phenomenon and its occurrence is not supported by actual observations. Data from all weather stations in the UK show that the peak frequency of moderate to strong gales is in fact near the winter SOLSTICE.

equinox One of the two periods during the year when the Sun's apparent path (ECLIPTIC) intersects the plane of the Earth's equator. Thus the Sun is directly overhead at the equator at noon on 21 March (the *vernal* or *spring equinox*) and on 22 September (the *autumnal equinox*). The term *equinox* is derived from 'equal night' because at this time the lengths of day and night are approximately equal throughout the world. [73]

equiplanation **1** In general, all geomorphological processes which reduce topography to a plain-like surface. **2** More specifically, the process of landscape levelling under a PERIGLACIAL CLIMATE. The term was introduced by D. D. Cairnes in 1912 but has been replaced by CRYOPLANATION.

equipluve A line joining all those places on a map which exhibit the same PLUVIOMETRIC COEFFICIENT.

equipotential A line drawn on a GROUNDWATER map joining points of HYDRAULIC POTENTIAL. It is in effect a type of contour map of the WATER-TABLE and the distances between the equipotential (contours) gives the gradient of the hydraulic potential. Groundwater flow will be down the gradient, perpendicular to the equipotentials and the mesh formed by a series of equipotentials is known as a FLOW NET. The downhill path of an individual particle of water is termed a STREAMLINE.

equirectangular projection A MAP PROJECTION in which the network of vertical meridians and horizontal parallels are drawn as straight lines, intersecting each other at right angles and producing a GRATICULE of rectangles. If the equator is chosen as the standard parallel the graticule will consist of squares. Although the chosen standard parallel is divided equally and the meridians drawn vertically through these divisions the projection is not conformal and lacks the property of equal area. The projection is useful for large-scale maps of an urban area. *See also* GEOREF SYSTEM.

equivalence The ability of a MAP PROJECTION to represent areas of any size in correct proportion to each other.

equivariable A term given to an interpolated line drawn on a map or dispersion diagram (ISOPLETH) which joins points exhibiting the same COEFFICIENT OF VARIATION.

era A term given to the first interval of a rank-order in the subdivision of geological time (CHRONOSTRATIGRAPHY). There are three principal eras (PALAEOZOIC, MESOZOIC, CAINOZOIC), but older geological texts show the Quaternary as an era, although it is now regarded as a PERIOD. Each era is divided into a number of periods but all three are of different lengths and contain a different number of periods. The Palaeozoic lasted 350 million years and comprised six periods; the Mesozoic lasted 175 million years and comprised three periods; the Cainozoic has lasted 65 million years and contained two periods. In addition to the three principal *eras* the term is also used with reference to the divisions of the PRE-CAMBRIAN, where there is disagreement over the number of eras. Some use the term EOZOIC as a synonym for the entire Pre-Cambrian era; others subdivide the Pre-Cambrian into three separate eras in which the Eozoic is succeeded by the ARCHAEOZOIC and the PROTEROZOIC.

erg **1** An obsolete term referring to a unit of ENERGY, defined as the work done by a force of one DYNE moving through 1 cm in the direction of the force. 10^7 ergs = 1 joule. **2** An Arabic word referring to a region of sand-dunes, generally of very wide extent, particularly in the Sahara.

ergodic hypothesis A proposition that, under particular circumstances, time and space can be considered as interchangeable, e.g. *cycle of erosion*, but it has been pointed out that landforms may be fitted into time sequences simply to fit preconceived DENUDATION CHRONOLOGY.

erodibility The ease with which natural materials break down when they are attacked by agents of DENUDATION. It is dependent upon susceptibility to weathering as well as susceptibility to removal and transportation, and the term is often used in the context of the potential of a soil or sediment to gullying or rill formation. The standard of erodibility of a material depends on the ratio between the proportion of the soil surface which requires binding (i.e.

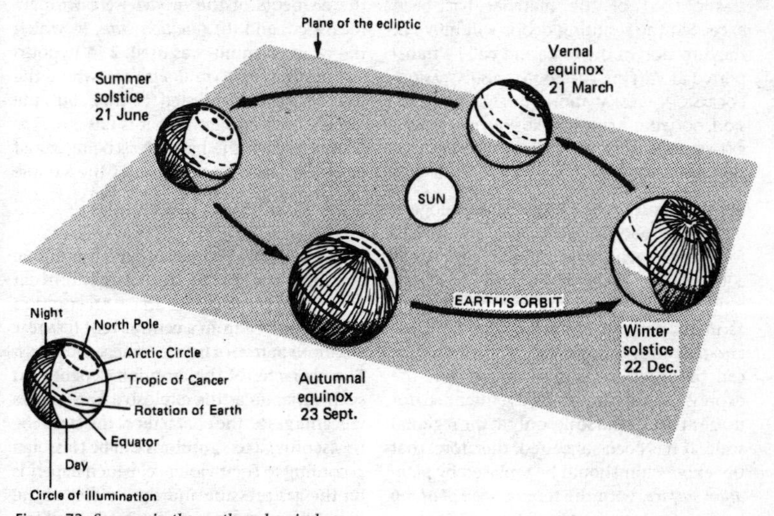

Figure 73 *Seasons in the northern hemisphere*

those particles larger than fine sand) and the quantity of the binding clays. In general, as the percentage of clay or aggregating material increases, soils become less erodible. It is measured in units of volume per unit area per unit time.

erosion A term referring to those processes of DENUDATION which wear away the land surface by the mechanical action of the DEBRIS which is being transported by the various agents of erosion (glaciers, wind, rivers, marine waves, currents). The agents by themselves are incapable of major erosion and would make only a minimal impact on the land surface (e.g. CAVITATION). The processes of erosion must be distinguished from those of WEATHERING in which no transportation is involved. Thus, erosion is not synonymous with denudation as is commonly supposed but merely part of it. Some geomorphologists would not include MASS-MOVEMENT due to gravity as an erosional process but others would include it in so far as they would regard gravity as the agent of erosion and the transported debris capable of wearing away the surface. For erosion to occur the agent must be capable of exerting a force on the land surface greater than its shear strength. Once the ELASTIC LIMIT of the material has been exceeded the resulting debris will move in the direction of the force and will be transported at varying speeds. *See also* ABRASION, CORRASION, DEGRADATION, DIFFERENTIAL EROSION, GULLYING, RAINDROP EROSION, RAINSPLASH, RILL EROSION, ROCK MASS STRENGTH, SHEET EROSION, SOIL EROSION.

erosion surface A geomorphological expression referring to a near-level surface formed by agents of erosion, in contrast to a level surface formed by deposition – i.e. it is degradational not aggradational. In so far as many parts of the Earth's land surface are areas undergoing DEGRADATION then they can be called *surfaces of erosion*, but the expression was never really intended for usage at the global scale, only at the regional scale. It has been suggested, therefore, that the expression should be replaced by *planation surface*, with the term *erosion platform* being introduced to describe an erosion surface of only limited extent. *See also* ABRASION PLATFORM, PEDIPLAIN, PENEPLAIN, DENUDATION CHRONOLOGY.

erosivity A measure of the potential vulnerability of a soil to EROSION by a given geomorphological agent, based on its KINETIC ENERGY.

erratic **1** A large rock fragment (BOULDER, COBBLE) that has been transported, by moving ice, away from its place of origin and deposited in an area of dissimilar rock types. Some erratic boulders are left perched on valley sides or hilltops as PERCHED BLOCKS, after the ice-sheets have melted; others, of great size, can be located along shorelines where they appear to have been deposited from floating icebergs. The majority, however, are incorporated in deposits of TILL or in GLACIOFLUVIAL deposits and are used by geologists to infer the directions of former ice-sheet movements and of the directions of meltwater flow. Erratics should be distinguished from EXOTIC BLOCKS. **2** A statistical term referring to values which seem to vary excessively from the average.

error **1** In a statistical model, the deviation of experiment from theory, possibly from two sources: (a) *measurement error*, in which the elements of the model were wrongly measured; and (b) *equation error*, in which the wrong formula was used. **2** In hypothesis testing. A *Type I error* is when the hypothesis being tested is true, but the sample used implies that it is false. A *Type II error* is when the hypothesis being tested is in fact false but, because of the sample used, it is accepted as true. *See also* STANDARD ERROR.

eruption The process in which volcanic materials are ejected or emitted from an opening in the Earth's surface. An eruption may take place from a central vent (CENTRAL ERUPTION) or from a fissure (FISSURE ERUPTION). The character of the eruption depends on such variables as the explosive force of the volcanic gases, the character of the lava (e.g. its VISCOSITY), etc. Eruptions can be classified according to their violence, which depends on the gas pressure and the fluidity of the lava. The most violent (e.g. Mt Pelée, Mar-

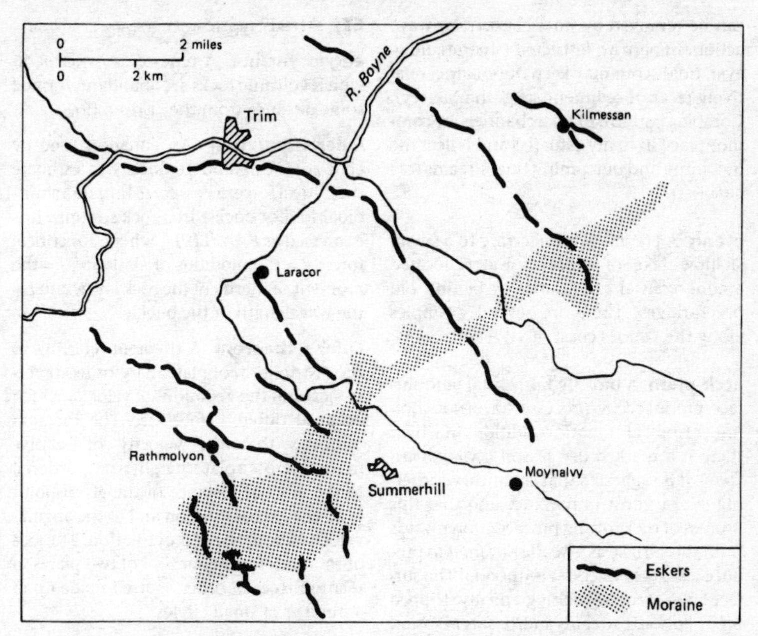

Figure 74 *Eskers in Co. Meath, Ireland*

tinique, 1902) are related to conditions of high gas pressure and low fluidity (i.e. highly viscous). The least explosive (e.g. the Icelandic *fissure eruptions*) result from highly fluid lavas and low gas pressures. *See also* PLINIAN ERUPTION [*271*], DUST-VEIL INDEX.

escarpment 1 The steep slope terminating a plateau or any level upland surface. **2** The steep face which terminates the stratified rocks of a CUESTA. **3** The term is sometimes used as a synonym of cuesta, but this usage should be discouraged. The term *scarp* is synonymous with definitions **1** and **2** but not with **3**. [*49* and *222*]

esker (eskar, osar, asar) A narrow, sinuous ridge of partly stratified coarse sand and gravel of GLACIOFLUVIAL origin. Eskers wind across country and frequently bear no relationship to the modern drainage pattern. Their torrential bedding suggests that they were formed by currents of glacial meltwater flowing under considerable hydrostatic pressure and therefore capable of transporting heavy burdens of glacigenic material at rapid velocities. Some authors claim that this can only have been achieved when the meltstreams followed very restricted subglacial or ENGLACIAL channels and that the esker represents the 'cast' of a former stream tunnel beneath or within an ice-sheet. Others believe that the same structure can be achieved at the edge of a rapidly receding ice-sheet and that an esker is no more than an elongated and linear DELTA which grew at right angles to the ice-front. Good examples of eskers can be found in Ireland, Scotland, Finland, Sweden and in Maine, USA. *See also* BEADED ESKER. [*74*; *see also 219*]

ESR ELECTRON SPIN RESONANCE.

estuary The mouth of a river where it broadens into the sea and within which the tide ebbs and flows, leading to an intermixing of saline and fresh water. Estuaries are usually zones of deposition, especially if the river discharges more sediment than can be removed by TIDAL CURRENTS or wave action. In narrow, restricted estuaries, however, tidal scour may keep the channel relatively clear of sediments (e.g. the Mersey). Complex patterns of tidal channels are commonplace in many estuaries and reflect the incoming and outcoming tidal streams (EBB CHANNEL).

étang A French word referring to a small, shallow lake of brackish water, located among coastal sand-dunes or behind old beach-ridges. There are several examples along the Landes coast of SW France.

etch-plain A broadly horizontal land surface produced by processes of DENUDATION under tropical climatic conditions in which there is a marked dry season (SAVANNA CLIMATE). It is suggested that chemical weathering is of utmost importance and that this process of rock rotting proceeds downwards from the surface as a *weathering front* to produce a BASAL SURFACE OF WEATHERING. The surface layer of rotted rock is gradually stripped off by episodic streams and/or SHEET EROSION to reveal the basal surface of weathering, which is termed an etch-plain. Some geomorphologists have suggested that certain of the horizontal or near-horizontal platforms (EROSION SURFACE) which have been described as uplifted peneplains (PENEPLAIN) in the highland zone of Britain may in fact be etch-plains formed by a process of etchplanation. *See also* PEDIPLAIN.

etch-planation ETCH-PLAIN.

etesian winds Strong north-westerly to north-easterly winds which blow during summer and early autumn in the E Mediterranean, and especially in the Aegean Sea, where they are termed Meltemi. Caused by the steep PRESSURE GRADIENT which exists between Europe and the Saharan low-pressure system, the winds reach their maximum velocity (30 knots) in the late afternoon when convection is at its maximum, generally dying away in the evening. Rough seas associated with the Meltemi characterize the more exposed of the Aegean Is. in summer.

ETJ EASTERLY TROPICAL JET STREAM.

eugeosyncline A type of GEOSYNCLINE in which volcanic rocks are abundant, formed some distance from the CRATON. [*91*]

Euler's equation An equation used by civil engineers and geologists to estimate the critical force necessary to initiate a sinusoidal fold or buckle in a rock stratum. It is expressed as $P = (\pi^2 EI)/L^2$, where: P = critical force; E = the modulus of elasticity; I = the moment of inertia of the rock layer, and L = the wavelength of the buckle.

Euler's theorem A theorem referring to the distribution of plate margins (PLATE TECTONICS) and the variations of velocity which occur at different localities. The theorem indicates that the velocity of relative motion across a plate margin is proportional to the angular distance of the given point from the axis of rotation and to the angular velocity about the axis of rotation. The axis of rotation, for the motion of two plates in relation to each other, is itself made up of a number of small circles.

eulittoral zone That part of the coastal zone which extends seawards of high-water mark down to the limit of attached plants, usually to depths between 40 and 60 m.

eumorphism A property attributed to equal-area MAP PROJECTIONS, which do not show undue distortion of shape.

euphotic zone The surface layer of any body of water through which light can penetrate, thereby leading to PHOTOSYNTHESIS.

European Macroseismic Scale (EMS) MSK SCALE.

eustasy, eustatism A change of sea-level that occurs everywhere throughout the world, due not to a movement of the land (ISOSTATIC MOVEMENTS) but to an actual fall or rise of the ocean itself. Eustatic changes in the last few million years are almost entirely the result of the abstraction of water from the hydrological cycle by the creation of Pleistocene ice-sheets and its subsequent return following their melting (GLACIO-EUSTASY). But it is generally recognized that

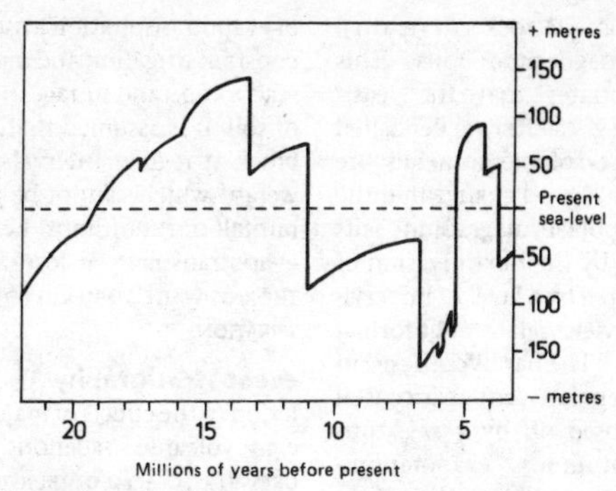

Figure 75 *Eustatic changes in sea-level*

eustatic sea-level changes over the last 20 million years have been controlled by sea-floor spreading and PLATE TECTONICS. The graph [*75*] illustrates the sea-level curve since the Miocene, with the high levels representing marine TRANSGRESSIONS and the low levels coinciding with marine REGRESSIONS. Such calculations of eustatic changes are based partly on OXYGEN ISOTOPE data derived from deep-sea cores.

eutrophic **1** The state of a water body when it has an excess of plant nutrients derived from agricultural fertilizers, from neighbouring farming activities and from other human activities. Thus the water body is originally able to support a large aquatic flora and fauna but this can lead to a depletion of the water's oxygen supply. **2** A solution within a soil which has an optimal concentration of nutrients for growth of plants or animals. *See also* MESOTROPHIC, OLIGOTROPHIC, ALGAL BLOOM.

euxinic environment A geological expression referring to an environment in which there is restricted circulation of sea water owing to the partial or total isolation of a basin (termed a *barred basin*) from the open ocean. Thus, there are large volumes of stagnant, de-oxygenated water, in which the dominating process is one of REDUCTION, giving rise to black, pyritic carbonaceous muds. Euxinic conditions are characteristic of the deeper parts of the Black Sea, hence its name.

evaporation The physical process in which a liquid is changed into a gas by molecular transfer. Evaporation rates depend on such variables as INSOLATION, temperature differences between air and water, humidity of the air (VAPOUR PRESSURE), wind velocity (EDDY), and the nature of the ground surface. Evaporation rates are increased by large insolation values, dry air, high wind-speeds and bare ground. Conversely, rates are decreased by small amounts of insolation, low wind-speeds, a weak vapour flux, high humidity and vegetation-covered ground. *See also* EVAPOTRANSPIRATION, HYDROLOGICAL CYCLE, SUBLIMATION.

evaporimeter (US atmometer) A device used by meteorologists, agronomists, etc. to measure rates and amounts of *evaporation*. The instrument can be in the form of a large, shallow, calibrated tank (the Symons tank) of open water, or a tube of distilled water with an associated piece of porous paper (the *Piche evaporimeter*).

evaporite A sediment formed by the evaporation of saline water, usually sea-water. Some evaporites are formed under special conditions in lakes, e.g. Great Salt Lake, Utah. The type of sediment that is created depends largely on the temperatures at which crystallization occurs together with the concentrations of the various ions in the solution. An evaporite sequence may exhibit the following (in ascending order of deposition): **1** CALCITE and DOLOMITE;

2 GYPSUM or ANHYDRITE; 3 rock salt (HALITE); 4 potash and magnesium salts. This sequence demonstrates that the least-soluble minerals (e.g. calcite) are deposited first and the most soluble minerals are deposited last, with a good chance that the rock salt and the potash/magnesium salts will be redissolved by the next invasion of the sea unless covered by a layer of impervious sediment. Two views relate to the formation of evaporites: **1** The bar hypothesis, in which sea water regularly enters a coastal lagoon, almost closed off by a BAR, thus allowing a continuous evaporation–replenishment system, e.g. Gulf of Kara Boghaz, Caspian Sea. **2** The SEBKHA hypothesis, in which coastal lagoons are only occasionally flooded by the sea, e.g. Abu Dhabi, Persian Gulf. The world's major evaporite fields include SW USA, Chile, Stassfurt in Germany and Cheshire in Britain. *See also* SALT DOME, SALT LAKE.

evapotranspiration The loss of moisture from the Earth's surface by means of direct EVAPORATION allied with TRANSPIRATION from vegetation. Two types are recognized; **1** *potential evapotranspiration* (PE), which assumes an unrestricted supply of water to the surface (e.g. by irrigation) and refers to the water need of a plant and to a theoretical maximum loss; **2** *actual evapotranspiration*, which is the observed or true loss. Actual and potential evapotranspiration will equal each other so long as the soil is above FIELD CAPACITY. PE can be measured more accurately than actual evapotranspiration and is more frequently used. PE is obtained by measuring the evaporation from a large area of actively growing, short-stemmed green crop of uniform height and standard soil moisture maintained by irrigation. PE is also estimated by the use of semi-empirical formulae (e.g. Thornthwaite 1948 and Penman 1952). PENMAN'S FORMULA indicates PE values ranging from about 400 mm/annum in Scotland up to about 510 mm/annum in S England. *See also* EVAPOTRANSPIROMETER, LYSIMETER.

evapotranspirometer An instrument for calculating the rate of potential EVAPOTRANSPIRATION from a vegetated surface. Estimates of evapotranspiration losses are achieved by constant irrigation and measurement of the PERCOLATION and storage in a vegetated block of soil. It is assumed that, by weighing the block at regular intervals, any changes of weight which cannot be accounted for by rainfall or runoff must be a measure of the evapotranspiration loss. *See also* LYSIMETER, THORNTHWAITE'S INDEX OF POTENTIAL EVAPOTRANSPIRATION.

event stratigraphy The study and correlation of the effects of major physical events e.g. volcanic ERUPTIONS, marine TRANSGRESSIONS, GLACIAL OUTBURSTS in the belief that synchronous stratigraphical horizons can be recognized.

everglades **1** In general, a marshy tract of land with tall grasses and occasional trees, seasonally flooded by summer rainfall. **2** The proper name for a low, flat limestone plain in Florida, USA, clothed in coarse grasses and scattered trees. Tiny isolated communities of tropical forest, known as *hammocks*, have survived and on the coastlands extensive MANGROVE swamps can be found. Although the Florida Everglades are covered by shallow floodwaters in summer the region becomes very dry in winter.

evergreen A plant which retains its green foliage throughout the year (in contrast to a DECIDUOUS plant), albeit if there is a dry or cool season the plant may cease growing.

evolution The gradual process by which living ORGANISMS change from more primitive forms and slowly evolve into species different from a common ancestor. The term was first used by the geologist Lyell, in 1832, whose study of FOSSILS led him to the notion of organic transmutation. Various theories of evolution have been proposed, notably by Darwin (DARWINISM) although Lamark was the first to outline the principles of evolution in 1800. Both Darwin and A. R. Wallace independently, developed the theory of evolution by NATURAL SELECTION and they presented a joint paper to the Linnean Society of London in 1858. Darwin's *On the Origin of Species* was published in 1859. The theory has been assimilated by modern genetic research, in which it is now defined as the change in gene frequency in a population over time (neo-Darwinism). *See also* PUNCTUATED EQUILIBRIUM, GRADUALISM.

evorsion A term given to the fluvial process which scours out cavities in the rocky bed of a stream (POT-HOLE), probably owing to the action of eddies.

exaration The action by which glacier ice, unladen with debris, carries out erosion of the bedrock by quarrying or joint-block removal (ICE PLUCKING). This process must be distinguished from that of glacial ABRASION, which is carried out by the transported debris within the glacier. *See also* FRICTION CRACKS, ROBIN EFFECT.

exfoliation A HYDRATION process of breaking, splitting and peeling-off of outer rock layers caused by the expansion of salt crystals within the interstices of the rock surface. Groundwater, containing dissolved salts, is drawn by CAPILLARITY to the surface where the supersaturated water crystallizes as a surface film owing to evaporation. This concept has now replaced that of expansion/contraction arising from heating/cooling of the rock surface, a mechanism which has been found not to operate when simulated in a laboratory. Thus, it has been concluded that rocks do not suffer from 'fatigue' in the same way as metals. The exfoliation process of MECHANICAL WEATHERING is also termed *desquammation, onion-skin weathering, flaking,* and in the USA *spalling*. It should be noted, however, that the spectacular rock domes termed *exfoliation domes* (e.g. Yosemite, California) in the USA have not been created by the simple exfoliation process described above. They are SHEETING structures due to UNLOADING. In contrast to the exfoliation process are the numerous processes of CHEMICAL WEATHERING which take place below ground-level (SPHEROIDAL WEATHERING).

exhumation A geomorphological term referring to the exposing by DENUDATION of a land surface or topographical feature that had previously been buried by deposition. The Pre-Cambrian massif of Charnwood Forest in Leicestershire, England, is often referred to as an example of exhumed relief now partly stripped of its cover of New Red Sandstone. The process of exhumation is thought to result from the combined action of DEEP WEATHERING and subsequent uplift by earth movements leading to accelerated erosion and removal of the weathered rock layers. A fall in sea-level (EUSTASY) will also lead to accelerated erosion by causing rivers to incise into the covering layers of sediment. Some geomorphologists have suggested that many of the EROSION SURFACES of upland Britain represent exhumed landscapes formerly covered by sedimentary rock layers that have now been destroyed. One of the best British examples of exhumation is the landscape of undulating, bare Lewisian gneiss surrounding the monolithic mountainous residuals of younger Torridonian sandstone which rise hundreds of metres above the exhumed surface in NW Scotland.

exogenetic, exogenic Processes of DENUDATION that occur at or very near to the surface of the Earth, in contrast to ENDOGENETIC forces.

exogeosyncline That part of a GEOSYNCLINE which borders a CRATON but receives its sediments from erosion of complementing highlands uplifted from the longer, narrower orthogeosynclinal belt lying away from the craton.

exosphere The outermost zone of the Earth's atmosphere, within the THERMOSPHERE above about 700 km, and beyond which neutral particles of atomic oxygen, ionized oxygen and hydrogen escape into space. *See also* IONOSPHERE.

exothermic reaction Any reaction in which heat is liberated. *See also* ENDOTHERMIC REACTION.

exotic Any plant or animal which is not indigenous and is introduced into an area, usually by man.

exotic blocks Large fragments of rock which occur in sediments and whose mode of deposition is incongruous with the formation of those sediments. Some of these boulders appear to have been transported by submarine slumping, perhaps due to earthquakes, while others may have been moved away from a costal zone by TURBIDITY CURRENTS. Where the exotic blocks have been transported into a zone of unrelated rock types by tectonic movements they are referred to as ALLOCHTHONOUS. They should not be confused with blocks that have been ice-transported (ERRATIC).

exotic river ALLOGENIC.

expanded foot glacier A smaller version of a PIEDMONT GLACIER. It is characterized by the splaying out of the glacier snout into a broad lobe as the glacier debouches on to a plain from the confines of a valley.

expansive soils VERTISOLS.

expert system A GIS package consisting of a data collection of rules, devised by an expert, for use in solving problems by means of a computer.

exponential growth A rate of geometrical growth by doubling.

exposure **1** In medical geography it is the situation of a human being in relation to sun or wind (WIND-CHILL). **2** In climatology, it is the position of a place on the ground surface in relation to the climatic elements (ASPECT). **3** In meteorology, it is the location of a recording instrument in a specific way so that the climatic data are collected uniformly (*see also* ANEMOMETER, RAIN-GAUGE, SCREEN, METEOROLOGICAL). **4** In geology, it is a location where bare rock can be seen at the surface, either naturally or by artificial excavation.

extending flow A concept of differential ice movement throughout the length of a glacier, introduced by J. F. Nye in 1952. It refers to the extension and related thinning of the glacier ice in those zones where the surface velocity increases. In a zone of extending flow the slip-lines within the ice descend to the ice-rock interface. The slip-lines, caused by shear-stress, curve down towards the glacier bed and are always located over the crests of rises in the rock floor and are also found in the steeper gradients of the accumulation zone where velocities are greatest. [*45*] The zones of extending flow alternate with zones of COMPRESSING FLOW and both are responsible for the formation of transverse CREVASSES.

externality A term referring to the link by which a component variable of a system is connected to an external variable.

external perspective projections Any AZIMUTHAL map projection in which the perspective centre lies outside the surface of the GENERATING GLOBE such that the distance of this point from the centre of the globe is greater than the radius of the generating globe. *See also* PERSPECTIVE PROJECTION.

extinction **1** The disappearance of a species in an ECOSYSTEM, often due to the evolution of a new species. It is frequently caused by a rapidly changing environment. Extinction is preceded by the development of a relic COMMUNITY. Most extinctions in the modern world are due to the impact of man, especially in the advancement of agriculture and the extermination of wildlife by hunting and pesticides. **2** The state of a mineral, under investigation with a petrological microscope, when no light reaches the eye of the observer.

extinct volcano A VOLCANO that functioned in the distant geological past and the remains of which occur in an area where there is no longer any active vulcanicity. Its volcanic form (CONE) may have totally disappeared, e.g. Arthur's Seat, Edinburgh, Scotland, or it may be deeply eroded to the stage of a dissected volcano, characterized by the formation of PLANÈZE. It must not be assumed that a volcano which has been inactive for a long period of time is extinct: it may merely be an example of a DORMANT VOLCANO.

extrapolation A statistical term referring to the extension of a plotted line or curve on a graph beyond the limit of observation, by the continuation of the trend exhibited in the plotted portion of the line or curve. Extrapolation can lead to serious error because, although the plotted points may suggest that the line has a steady slope, some new or unsuspected variable may come into play in those areas with no existing data, thereby invalidating the assumption. *See also* INTERPOLATION.

extra-tropical cyclone DEPRESSION.

extremes of climate Climates with considerable differences recorded between the highest and lowest values of atmospheric elements in a specified period (daily, monthly, seasonal, annual or absolute), with particular reference to temperature, pressure, precipitation and wind-speed. *See also* CONTINENTAL CLIMATE, DIURNAL RANGE.

extrusion The thrusting out or emission of magmatic material, in a solid, liquid or gaseous form, from the surface of the Earth during a volcanic eruption. It is the opposite of INTRUSION. Extrusion results in the formation of *extrusive rocks*, which are synonymous with volcanic rocks, characterized by their finely crystalline or glassy form resulting from the rapid cooling of the lava when it appeared at the surface. These rocks contrast with the coarsely crystalline form of PLUTONIC ROCKS. *See also* ANDESITE, BASALT, LAVA, OBSIDIAN, PUMICE, PYROCLASTIC MATERIAL, RHYOLITE.

extrusion flow A theoretical internal movement within an ice-sheet, resulting from differences of pressure, such that maximum rates of movement are postulated for the deeper ice layers which are thought to be more plastic (PLASTIC DEFORMATION). The theory of extrusion flow was expounded by M. Demorest between 1937 and 1943, who contrasted it with gravity flow. It was suggested that this was one of the major mechanisms in large ice-sheets such as those of Antarctica and Greenland. Critics have posed the problem of why upper ice layers are not carried forward on the back of supposedly faster-moving lower layers, and because extrusion flow has not been encountered in boreholes it remains an unproven hypothesis.

extrusive rocks EXTRUSION.

exudation basin A topographic depression that occurs at the head of glaciers as they emerge from the Greenland ice-cap.

'eye' of a storm The central area of an intense revolving TROPICAL CYCLONE, especially one at a HURRICANE strength. It is generally between 16 and 24 km (10 to 15 miles) in diameter and is characterized by light winds and well-broken clouds, around which the high-velocity winds revolve. It exhibits the lowest atmospheric pressure of the tropical cyclone, commonly around 960 mb in intense cyclones, but with a record low of 877 mb recorded near Guam on 24 September 1958. It is caused by cold air descending and warming by the ADIABATIC mechanism at the centre of the system and by the influence of CENTRIFUGAL FORCE in a highly curved circulatory path. [*106*]

eyot AIT.

F

fabric, soil The physical constitution of a soil, expressed in terms of the spatial arrangement of the solid particles and their associated voids. FABRIC, TILL.

fabric, till The structural and textural composition of a TILL, especially the orientation, dip and roundness (ROUNDNESS INDEX) of the included pebbles or boulders. The study of till fabrics is a valuable method of reconstructing former ice directions and till depositional processes.

facet **1** Any plane surface abraded on a fragment of rock by ice or wind (VENTIFACT). **2** The polished surface of a cut gemstone. **3** A face on a crystal. **4** A term used in terrain classification to denote a grouping of land elements of similar genesis.

faceted spur The truncated end of a ridge, where the smooth convex curve of the ridge crest is abruptly terminated by either a steep BEVEL or a cliff. This feature may be due to faulting (FAULT-SCARP), glaciation (TRUNCATED SPUR) or river erosion (BLUFF).

facies **1** In general, the character or appearance of part of a rock by comparison with the other parts. **2** A rock unit or group of units that exhibits lithological, sedimentological and faunal (FOSSIL) characteristics which enable them to be classified as distinct from another rock unit or group. **3** A lateral change of character within a stratigraphic unit, especially in its lithology.

factor **1** A control or cause which leads to an EFFECT. It is often a single variable that causes variations in another set of observations. In ecology, four main classes of environmental factors are recognized: the *biotic factor*; the *climatic factor*; the *edaphic factor*; and the *geomorphic factor*. These factors act simultaneously and affect each other in complex relationships. They function at a global scale (e.g. the climatic factor includes latitudinal and seasonal patterns of insolation, temperature and precipitation) and at a local scale (e.g. the edaphic factor can vary considerably in a small area, thereby providing several habitats). **2** A cluster or family of variables used in FACTOR ANALYSIS.

factor analysis A mathematical technique for examining the interrelationships exhibited by variables or groups of variables. With the advent of the computer considerable quantities of data can now be analysed.

factorial experiment An experiment used in conjunction with variance analysis (VARIANCE) to determine the effect of more than one FACTOR upon a VARIATE, by examining all the possible combinations of their several different levels.

Fahrenheit scale A temperature scale introduced by a German physicist, G. D. Fahrenheit, in about 1709 in which 32° represents the melting point of ice in water and 212° the boiling-point of water. °F = 9/5° C(+32°); 180 °F(212° − 32°) = 100° C. There has been a tendency to replace the Fahrenheit scale by the CELSIUS SCALE (CENTIGRADE SCALE) in recent years.

failure **1** In civil engineering, the conditions at which a structure ceases to fulfil its purpose. **2** In geology and soil mechanics, the deformation of a rock or soil marked by the sudden formation of fractures and loss of cohesion and the inability to resist continued stress. *See also* FRACTURE, SLAB FAILURE, TOPPLING FAILURE, WEDGE FAILURE.

fair-weather cumulus A type of CUMULUS cloud associated with a summer anticyclone during which prolonged spells of dry, sunny weather occur. It is characterized by a limited vertical extent because major convection is prevented by an inversion layer (INVERSION OF TEMPERATURE) linked with the high pressure and subsiding air.

fall face FREE FACE.

falling dune A stationary accumulation of sand on the LEE slope of a hill or cliff. [*65*]

fall-line A line or zone marking the tract where a series of almost parallel rivers descend from a mountain region or plateau edge on to a lowland by means of waterfalls. In the eastern USA the fall-line marks the belt along which some of the Appalachian rivers leave the older folded rocks and debouch on to the younger sedimentary rocks of the coastal plain.

false bedding CROSS-BEDDING.

false colour film A film used for AERIAL PHOTOGRAPHY. This colour film (INFRARED film) consists of three dye layers; infrared radiation is recorded by a red dye layer, red radiation by a green dye layer and green radiation by a blue dye layer. From the resulting photographs it is possible to distinguish vegetation (red) from soil (blue-green), thereby facilitating the mapping of vegetation and land use. It has also been used for monitoring soil-surface moisture and the degree of DESERTIFICATION. *See also* INFRARED PHOTOGRAPHY, REMOTE SENSING.

false drumlin A rocky, drift-covered mound resembling a DRUMLIN.

false origin A zero datum point chosen to the south and west of the true origin of a GRID system, in order to avoid negative quantities in the COORDINATES. The false origin of the NATIONAL GRID in the UK is SW of the Scilly Is, having been transferred 100 km N and 400 km W of the intersection between standard parallel 49°N and the central meridian 2°W. *See also* EASTING, NORTHING. [*160*]

family A collective term used in taxonomic classifications of organisms. A family group is composed of one or more genera (GENUS) and a collection of related families is referred to as an ORDER.

fan ALLUVIAL FAN.

fan cleavage A type of CLEAVAGE in which the planes exhibit a fan-like arrangement (with convergence being either upwards or downwards) instead of the more normal parallel alignment.

fan-folding Folding in which the limbs dip either towards each other (in an anticlinal form) or away from each other (in a synclinal form), because the axial planes of the folds converge towards a centre. It is commonly exhibited in an ANTICLINORIUM and a SYNCLINORIUM.

fanglomerate A type of rock formed by the cementation and compaction of the heterogeneous materials comprising a former ALLUVIAL FAN.

faro A minor elongated REEF enclosing a LAGOON at the edge of an ATOLL or BARRIER REEF.

fast ice An extensive, unbroken sheet of sea-ice formed by the *in situ* freezing of sea water when water temperature falls below –2° C in calm weather. Initially, ICE RIND and pancake structures form which later fuse together to create a sheet of sea-ice, which is subsequently added to by snow accumulation above and further sea water freezing below. In spring, the fast ice is broken up by strong winds and sea swell into ICE-FLOES, which collectively represent the PACK ICE surrounding polar regions.

fathogram The profile of the sea-floor recorded on a chart by an echo sounder in order to ascertain the depth of sea water.

fathom A nautical unit of water-depth measurement used in soundings. One fathom = 6 ft (1.829 m); 100 fathoms = 1 cable; 1,000 fathoms = 1 nautical mile.

fatigue failure **1** A type of FRACTURE in rocks repeatedly exposed to rapid fluctuations of wetting and drying. *See also*

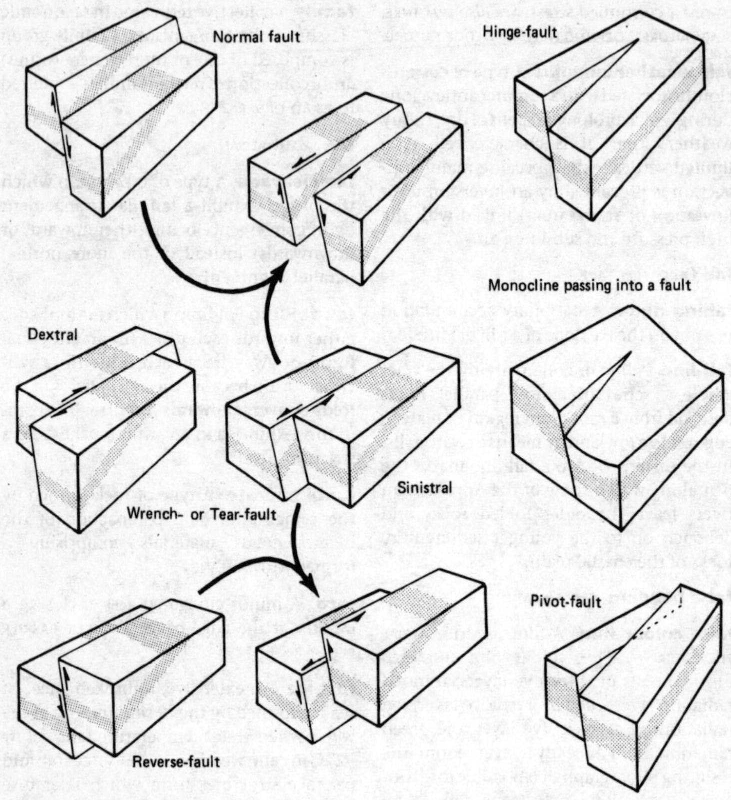

Figure 76 *Types of fault*

THERMAL FRACTURE. **2** Failure of metals due to cyclic stress, e.g. vibration. *See also* FAILURE.

fault A rupture or fracture of rock strata due to strain, in which displacement is observable. Most faults occur in groupings termed a *fault zone* or fracture zone. A *fault-plane* can be vertical or it may exhibit a DIP. The angle which the *fault-plane* makes with the vertical is known as the *hade* (analogous with the angle of dip). If the direction of movement is parallel with the dip of the fault (upwards or downwards) it is termed a DIP-FAULT. If the direction of movement is parallel with the STRIKE of the fault (sideways) it is referred to as a STRIKE-FAULT. The amount of vertical displacement of the *fault-plane* is referred to as the *throw*, and the respective sides of the displacement as the *upthrow block* and the *downthrow block*. The amount of horizontal displacement is known as the *heave*. In a fault which is dipping, the rock surface above the *fault-plane* is termed the *hanging wall* and the rock surface below is termed the *foot wall*. *See also* ELASTIC REBOUND, GRABEN, HINGE-FAULT, HORST, MEGASHEAR, NORMAL FAULT, OBLIQUE-FAULT, PIVOT-FAULT, REVERSE-FAULT, TEAR-FAULT, THRUST-FAULT, TRANSCURRENT-FAULT, TRANSFORM-FAULT. [*76*]

fault-block topography A terrain of sharply defined basins separated by block mountains outlined by BLOCK-FAULTING. [*26*]

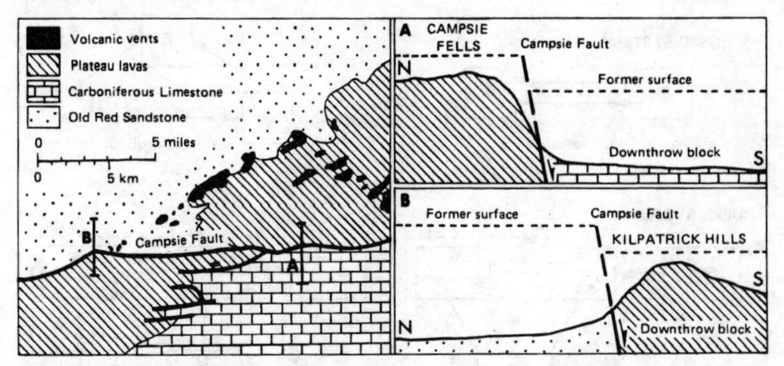

Figure 77 *Fault-line scarp; the Campsie Fault, north of Glasgow*

fault-breccia Broken fragments of rock associated with a fault-plane and created by dislocation *See also* MYLONITE.

fault creep The slow, almost imperceptible displacement of a fault, usually a TEAR-FAULT, which tends to reduce the accumulation of stored strain.

fault-line scarp A scarp resulting from differential erosion that has taken place on either side of a fault-line when rocks of contrasting hardness are brought into juxtaposition. It contrasts, therefore, with a FAULT-SCARP which results from the primary displacement of the fault itself. Because of the different degrees of resistance to denudation the higher ground may occur either on the upthrow or the downthrow side of the fault [*77*]. If the higher ground is on the downthrow side it is termed an *obsequent fault-line scarp*. *See also* RESEQUENT FAULT-LINE SCARP.

fault-scarp A cliff formed directly by the displacement of a recent fault, but usually on a small scale (<10 m in height). It is a relatively transient landform since denudation will soon modify the scarp, turning it into a FAULT-LINE SCARP or obliterating it altogether.

fault-spring A spring formed as a direct result of an impermeable bed of rock being brought into juxtaposition with a permeable bed by virtue of faulting.

fauna A term referring to an association of animals living in a particular place or at a particular time. A *faunal assemblage* may be identified at different spatial scales, ranging from a continental scale (*faunal realm*) through a *faunal province* right down to a local scale. In a geological context (PALAEONTOLOGY) the faunal assemblage will be preserved in the form of FOSSILS. A biostratigraphic unit characterized by the presence of a particular fauna that may have either time or environmental significance is known as a *faunizone*. *See also* WALLACE'S REALMS.

FDSN FEDERATION OF DIGITAL SEISMIC NETWORKS.

feather-edge The thin edge of a wedge-shaped sedimentary rock that tapers out and disappears as it abuts against an area of non-deposition. It is usually overlain by a later bed which exhibits unconformable characteristics (UNCONFORMITY). A feather-edge may be straight but is usually indented.

Federation of Digital Seismic Networks (FDSN) A worldwide network of SEISMOMETERS based on digital recording. It has replaced the Worldwide Standardized Seismograph Network.

feedback A process by which the result of a controlling operation is used as part of the data on which the next controlling operation is based, enabling a system to make a response. Thus, when a change is intro-

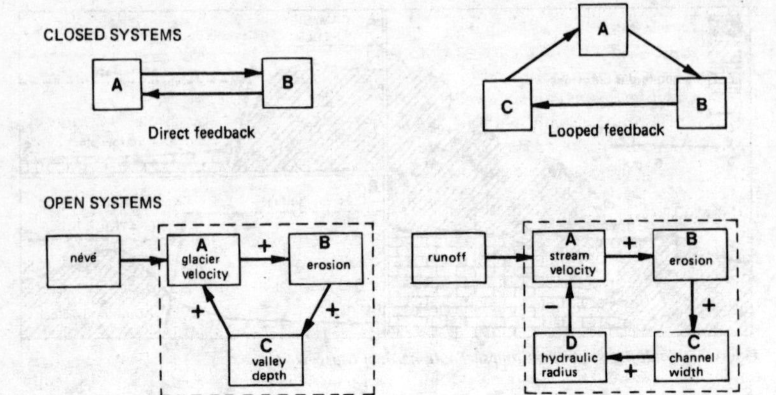

Figure 78 *Feedback*

duced via one of the system's variables, its transmission through the structure leads the effect of the change back to the initial variable, giving a circularity of action. There are two types of feedback: **1** *negative feedback*, in which the effect of the change is to counteract the impact of the initial alteration; and **2** *positive feedback*, when the effect of the change is to cause the system to continue changing in the same direction, i.e. a 'snowballing' effect which may cause the system to go out of control. In nature, *positive feedback* is much less common than *negative feedback*. The latter is very common among most processes in physical geography and is the main factor in promoting self-equilibrium in natural systems. A feedback process in which chance does not operate is known as *deterministic feedback*. A *feedback loop* is the path by which the feedback process is accomplished. *See also* GENERAL SYSTEMS THEORY. [*78*]

feedback control system A system which contains one or more negative FEEDBACK loops and is capable of exerting different orders of self-control.

feldspar, felspar The most important group of rock-forming minerals, comprising silicates of aluminium linked together with those of calcium, potassium or sodium. The sodium and calcium feldspars form a continuous series known as the *plagioclase group*, ranging from *albite* to *anorthite*, which are respectively, important constituents of GRANITE at one extreme and GABBRO at the other. The *alkali-feldspar group* includes various combinations of potassium and sodium feldspars, which are found mainly in alkali igneous rocks, such as granite and syenite, but also in a wide variety of metamorphic rocks. Some plagioclase feldspars are utilized as household abrasives, while certain of the potassium feldspars are important in the glass-making and ceramic industries.

fell A Norwegian term for an open mountainside, which occurs in many place-names of the northern Pennines and the Lake District of England (e.g. Scafell, Crossfell) and is thought to relate to early periods of Viking settlement in northern England.

felsenmeer A German term, literally meaning a 'sea of rock', which has been widely adopted to describe the mountain-top spreads of angular boulders (BLOCKFIELD) due mainly to frost-action on well-jointed rocks, probably during the later cold periods of the Pleistocene.

felsic A term used to describe light-coloured rocks containing an abundance of feldspar and quartz, as compared with MAFIC rocks. It is a mnemonic derived from *fe* (feldspar), *l* (lenads or feldspathoids) and *s* (silica or quartz).

felsite An igneous rock of acid or intermediate composition, characterized by a fine, even graining usually without large isolated crystals (phenocrysts). If phenocrysts do occur among the fine ground mass it is termed a *quartz felsite*. It often occurs in the form of DYKES or veins.

fen An area of wooded, swampy land in which the groundwater is alkaline or neutral but not acid. It occurs, therefore, where the substratum is of limestone, marl or calcareous boulder clay. In the Fenlands of eastern England, around the Wash, most of the waterlogged soils have been artificially drained to form productive arable land (FEN SOILS) but occasional patches of open water fringed by alders, reeds and rushes illustrate the type of environment which once prevailed. The Somerset Levels and the lowland mosses of Lancashire, England, exhibit similar characteristics, while in Yorkshire the wooded fen is termed a CARR. [*108*] *See also* WETLAND.

fen soils A name given to those organic soils which form above *fen peat* in an environment of base-rich groundwater. Thus, fen soils have an alkaline or neutral reaction and when drained give rise to extremely fertile and productive agricultural soils. *See also* PEAT SOILS.

fenster, fenester A German term for 'window', widely adopted to describe the outcrop of a younger rock stratum exposed by an erosional opening cut through a recumbent fold (NAPPE) or overthrust body of older rocks. The term fenster should be used only in this context and should be distinguished, therefore, from the term INLIER, which is a simple exposure of older rocks in the centre of an anticline by denudation.

feral relief A term given to a terrain in which valley sides are deeply dissected by INSEQUENT DRAINAGE related to rapid RUNOFF.

Fermanagh stadial A cold phase of the early MIDLANDIAN glaciation in Ireland (48,000–41,000 BP). Its type site is at Hollymount, Co. Fermanagh.

ferrallitic soils LATOSOLS.

Ferrel cell Part of the global atmospheric circulation pattern, named after W. Ferrel (FERREL'S LAW). It comprises an extensive circulatory system of winds, whereby air rises at about latitude 60°N and S, in association with the inclined plane of convection at the POLAR FRONT, before flowing equatorwards at high altitudes, descending at the HORSE LATITUDES (30°–35°N and S of the equator), and subsequently flowing polewards as the WESTERLIES at the surface [*18*]. Some scientists now question the existence of the Ferrel cell. *See also* ATMOSPHERIC CIRCULATION, HADLEY CELL.

Ferrel's law A law relating to the rotation of the Earth, postulated by an American scientist, W. Ferrel, in 1856. Because of the Earth's rotation a body moving across the surface is subject to an apparent deflection to the right in the N hemisphere, although there is no actual effect at the equator. Its most important effects are shown in oceanic and air movements. *See also* CORIOLIS FORCE.

ferricrete A type of DURICRUST, cemented by iron oxides. *See also* LATERITE.

fer(ri)siallitic soils Tropical soils that form in somewhat drier climates than the LATOSOLS and ones in which weathering and leaching are not as dominant as in the soils of the more humid tropics. Thus the soil profile is not as deep or as well developed, although because of the more limited leaching it is richer in minerals and has a greater cation exchange capacity (BASE EXCHANGE) than the latosols.

ferrisols Tropical soils similar in composition to LATOSOLS and FERRISIALLITIC SOILS. They are more weathered and leached than latosols. They are more fertile than latosols but have a lower cation exchange capacity (BASE EXCHANGE) than the fer(ri)siallitic soils. *See also* ULTISOL.

ferrous Signifying a compound of iron, not saturated by oxygen and formed by the reduction of iron compounds.

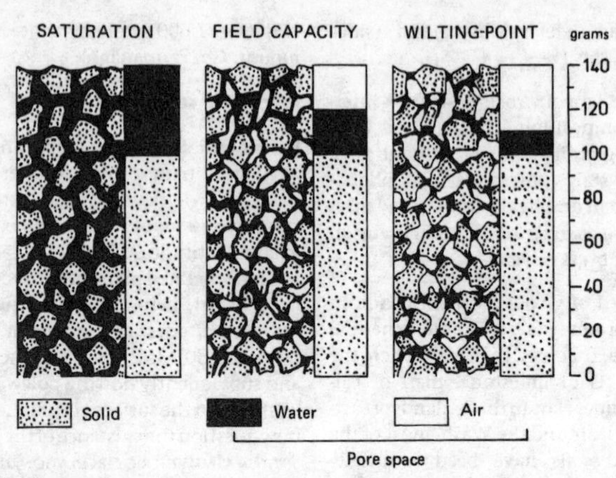

Figure 79 *Field capacity*

ferruginization Staining and/or cementation by iron minerals, usually by LIMONITE.

fertility, soil The status of a soil in relation to the amount and availability of nutrients essential for plant growth.

fetch The distance of open water across which a wind blows or over which a wind-generated water-wave travels, unobstructed by major land obstacles. The amount of fetch helps to determine the magnitude and energy of a wave and therefore its erosional and depositional tendencies on neighbouring shorelines.

fiard (fjard) An indented, drowned coastline fringing a rocky, glaciated lowland but without the deep glacial troughs of a FIORD with which it should not be confused. It is generally deeper than a RIA and is characterized by a large number of islands, similar to the coastal area of SE Sweden. *Fjard* is the Swedish spelling.

fibric layer A layer of fibrous soil material in which the weakly decomposed organic material can be easily recognized and the separate species identified.

field capacity (field moisture capacity) The total amount of water remaining in a freely drained soil after all 'gravity water' has been drained after cessation of rainfall. It is expressed as a percentage of the weight of the oven-dry soil. The remaining CAPILLARY WATER is held on to individual soil particles by surface tension and is usually sufficient for plant growth (AVAILABLE WATER) except in times of drought. [*79*] *See also* SOIL WATER. [*See also 279*]

field intensity The strength of the magnetic field.

field of view A REMOTE SENSING term referring to the solid angle through which an instrument is sensitive to ELECTROMAGNETIC RADIATION. The field of view controls the area of the Earth's surface sensed by the SENSOR. For example, the LANDSAT satellite has a field of view of 11.56° compared to 90–120° for most MULTISPECTRAL SCANNERS. *See also* INSTANTANEOUS FIELD OF VIEW.

filling A term used in meteorology to describe the increasing atmospheric pressure near the centre of a DEPRESSION, which ultimately leads to its disappearance. The converse term is DEEPENING.

film water A layer of water surrounding soil particles, varying in thickness between 1 and 100 molecule layers and not available for plant growth.

filter 1 In general, a contrivance which allows certain magnitudes of mass, energy or information to pass through a system, while retaining or trapping others. The process of treating spatial data in order to separate large-scale regional trends from local variations ('noise') is termed *space filtering*. The process of treating temporal data in order to separate long-term trends from short-term variations is termed *time filtering* and is used, for example, in investigations into climatic change. **2** A REMOTE SENSING term with two uses: (a) As a noun it refers to any object or material which by absorption or reflection selectively modifies the ELECTROMAGNETIC RADIATION transmitted through an optical system, e.g. a camera. Such a filter may be either a physical object or electronic. The filters in a MULTIBAND aerial photographic system are usually placed between the film and the scene being imaged but they can be included within the film itself. (b) As a verb the term refers to the removal of a spatial or frequency component of ELECTROMAGNETIC RADIATION. Spatial filtering enables image details to be enhanced (IMAGE ENHANCEMENT). **3** In a geological sense, the term *filter pressing* refers to the way in which residual liquid may be squeezed out from a crystal mush and rise upwards in the crust during the later stages of crystallization.

fine clay A clay fraction of specified size, always less than 0.2 μm. *See also* SOIL TEXTURE.

fineness modulus A number which indicates the fineness of sedimentary materials etc. during MECHANICAL ANALYSIS. It is synonymous with the term GRADING CURVE.

fines The fine fraction, or smaller particles, identified by MECHANICAL ANALYSIS of soils or sediments. Fine-grained soil or sediment is that which has more than 50% of its bulk by weight comprising particles smaller than 0.075 mm in diameter, according to the US terminology. In Britain a soil which contains more than 35% clay is termed a FINE-TEXTURED SOIL. *See also* SOIL TEXTURE.

fine sand Sand which consists of grains between 0.25 mm and 0.125 mm in diameter, on the International Scale of SOIL TEXTURE. *See also* PARTICLE SIZE, WENTWORTH SCALE.

fine-textured soil A soil which contains more than 35% clay. *See also* CLAY, SOIL TEXTURE.

finger lake A narrow, linear body of water occupying a glacially overdeepened valley and sometimes impounded by a morainic dam. Many of the lakes in the English Lake District exhibit such characteristics, as do numerous Scottish *lochs* and Welsh *llyns*. The lakes of N Italy, in the Piedmont zone of the Alps, are in part finger lakes, while in the USA the water bodies fringing Lake Ontario in New York State are named the Finger Lakes.

Finiglacial stage One of the retreat stages of the Scandinavian ice-sheet during the later part of the Ice Age. It is named from the fact that at this stage, between 10,000 and 8,500 BP, much of Finland was uncovered from beneath the ice-sheets [*109*]. *See also* SALPAUSSELKA.

fiord (fjord) A long, narrow inlet of the sea bounded by steep mountain slopes, which are of great height and extend to considerable depths (in excess of 1,100 m) below sea-level. It is formed by the submergence of glacially overdeepened valleys (TROUGH) due to a rising sea-level after the melting of the Pleistocene ice-sheets. It is characterized by a bottom-profile which carries deep water almost to its head but which shallows near its mouth to form a submerged threshold or 'sill' in solid rock, marking the zone at which the former valley glacier was fanning out as it debouched from the constraining walls of the valley, thereby reducing its erosive ability, and where melting occurred near to its terminal zone. Thus, some of the thresholds have a superficial draping of terminal moraine. Because of the glacial overdeepening of the trunk valley many tributary streams enter the fiord as waterfalls from the HANGING VALLEYS. Examples may be seen in Norway, Greenland, Labrador, Alaska, British Columbia, S Chile, and in South Island, New Zealand. In Britain the best fiords are found in W Scotland, e.g. Loch Etive, although their valley sides are not as steep as those in Norway. NB In Denmark the term *fjord* refers

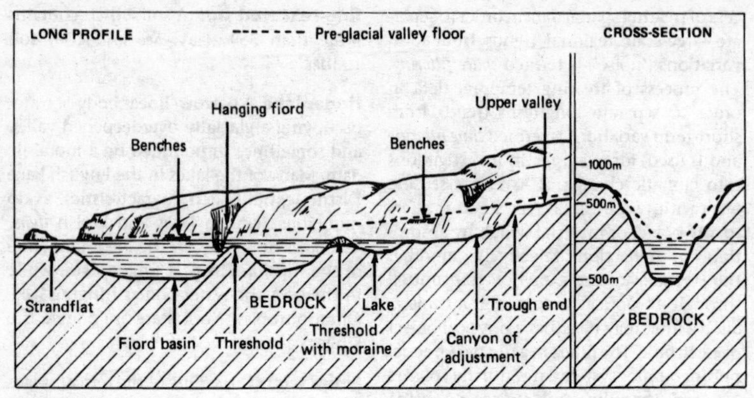

Figure 80 *Fiord*

to a different coastal landform (FÖHRDE), *see also* FIARD. [*80*]

fire clay A clay capable of resisting great heat and therefore in demand in industry for furnace linings. The term was formerly used for any clay which occurred in association with coal seams but in strictest usage it refers only to the clays with refractory properties, many of which occur beneath coal seams of the Lower Coal Measures of the CARBONIFEROUS. It is thought to represent a fossil soil in which the trees of the 'coal forest' became rooted. *See also* SEAT-EARTH. [*224*]

firn A German word meaning old granular snow (density range 0.4 to 0.89 kg m^{-3}) at least in its second accumulation season, in the process of being transformed into GLACIER ICE by FIRNIFICATION. *See also* ALIMENTATION, NÉVÉ.

firnification A combination of processes in which new snow (NÉVÉ) is transformed through FIRN into GLACIER ICE, owing to the release of air and a change in crystallography. The density subsequently increases from 0.1 g cm^{-3} (*névé*) to 0.75 g cm^{-3} (*firn*) to 0.92 g cm^{-3} (*glacier ice*). The processes involved are compaction-settling (when the snow pack exceeds 10 m) (SUBLIMATION), the refreezing of percolating meltwater and crystal diffusion.

firn-line The climatic snow-line or the uppermost line on a glacier to which the previous winter snowfall melts during the ablation season of the summer. It should not be confused with the EQUILIBRIUM-LINE.

firth A Scottish term for a body of coastal or estuarine water. It is most commonly used in the latter sense, e.g. Firth of Clyde, Firth of Forth, where it relates to the drowned valleys of a glacially modified lowland terrain (not to be confused with FIORDS), in which great thicknesses of glacial and post-glacial sediments (CARSE) obscure much of the solid rock.

fissile bedding The bedding of sedimentary rocks in which the individual sheets or laminae are less than 2 mm in thickness.

fissility A property exhibited by certain rocks, especially ARGILLACEOUS sediments, whereby they split easily along closely spaced parallel planes. It is very similar to CLEAVAGE, in which the splitting may take place across the bedding planes. Mudstones, shales, flagstones, and some limestones possess this property, as do slates and schists. *See also* FISSILE BEDDING, LAMINA.

fission-track dating A dating method based on a measurement of the fission tracks generated by the decay of ISOTOPE ^{238}U which occurs in certain VOLCANIC ROCKS. *See also* RADIOMETRIC DATING.

Table 81 *The Flandrian (British sequence)*

					Pollen zones	
Sparks & West (1972)		von Post (1946) Iversen (1958)		Blytt (1876) & Sernander (1908)	Great Britain	Ireland
(Fl. IV)	(Post-temperate)	(Cryocratic)				
Fl. III	Late-temperate	Terminocratic	Decreasing warmth	Sub-Atlantic	VIII	IX
					2,450BP	
				Sub-Boreal	VIIb	VIII
					4,450BP	
Fl. II	Early-temperate	Mediocratic	Climatic Optimum	Atlantic	VIIa	VII
					7,450BP	
Fl. I	Pre-temperate	Protocratic	Increasing warmth	Boreal	VI	VI
						9,000BP
					V	V
					9,450BP	
				Pre-Boreal	IV	IV
					10,300BP	
Late-Glacial					I–III	I–III

fissure A crack or open break in rocks.

fissure eruption A volcanic eruption along a linear crack, fault or other structural weakness in the crust. It is generally characterized by the outpouring of basic and very fluid lavas, especially basalt, leading to the formation of considerable thicknesses of lava which today stand as extensive plateaux, e.g. Iceland and Antrim, Northern Ireland.

fixation The process occurring in a soil whereby certain chemical elements essential for plant growth are converted from an exchangeable form (BASE EXCHANGE) to less soluble and non-exchangeable forms. The conversion of nitrogen to organic combinations, or to forms which can be used in biological processes, is known as nitrogen fixation.

fjeld A Norwegian term for FELL.

flagstone A sandy limestone or micaceous sandstone with very marked FISSILITY. It was formerly extensively used for paving stones (flags), roofing and building stones. The Caithness flags, from the Old Red Sandstone of N Scotland, have also been used (standing on end) as flagstone fences to delimit field boundaries.

flaking EXFOLIATION.

Flandrian A term for the post-glacial stage of the Quaternary in NW Europe, which commenced in 10,300 BP at the end of the LATE-GLACIAL. It was marked by an amelioration of climate, a northward spread of woodland and a submergence of many coastal areas, especially in Flanders and the Low Countries (FLANDRIAN TRANSGRESSION). In the British Isles it has been divided into four zones [81], although we are currently living in the third of these – the fourth (post-temperate) being hypothetical. The first zone (Fl.I) is the pretemperate, or *protocratic*, characterized by the appearance of warmth-loving trees; the second zone (Fl.II) is the early-temperate, or *mediocratic*, which occurred between 7,450 BP and 4,450 BP, during which temperatures were at their highest and warmth-loving trees dominated, at a time popularly known as the

Climatic Optimum. The third zone is the late-temperate, or *terminocratic*, during which temperatures and woodland decline, most markedly after 2,450 BP, when rainfall and storminess increased. [*81*]

Flandrian Transgression A term given to the global rise of sea-level (*c.* 170 m) in post-glacial (Flandrian) times in so far as it affected the coasts of NW Europe. There the rise, which commenced in about 10,000 BP, witnessed, as the ice-sheets waned, the gradual submergence of large tracts of the CONTINENTAL SHELF which had formerly been dry land. The present North Sea was formed and the Straits of Dover created, thereby severing Britain's last link with the continent in about 7,000 BP. It has been suggested that large offshore areas adjoining the British coastline were submerged during the Flandrian, e.g. Lyonnesse off the coast of SW Cornwall, and in some localities the transgression crossed the line of the present coast to leave marine deposits in estuaries and coastal lowlands. Many of these have now been uplifted by ISOSTATIC MOVEMENTS, especially in N Britain and N Ireland (CARSE, RAISED BEACHES).

flap A geological term for a GRAVITY-COLLAPSE STRUCTURE, in which a rock stratum which was originally on the limb of an ANTICLINE has slid down its side, become overturned and finished upside down as it came to rest. [*96*]

flash 1 A very rapid rise of water in a river, leading to a FLASH-FLOOD. **2** The name given to a body of water infilling the small subsidence hollows caused by underground mining. This is particularly true of the salt industry in Cheshire, where the term is found frequently in place-names.

flash-flood A short-lived but rapid rise of water in a river due to snowmelt, heavy rainfall, the collapse of an ice-dam, log-jam or artificial dam. Because of the considerable velocity and discharge achieved in such a short time period the river becomes capable of transporting an exceptionally large load, often with catastrophic consequences (e.g. Lynmouth, N Devon, floods in 1952; the Dolgarrog dam collapse, N Wales, in 1925). Flash-floods sometimes occur in semi-arid areas where violent thunderstorms can turn dry gullies, wadis, etc. into raging torrents in a short space of time. *See also* FLASHINESS.

flashiness The rapidity with which the stage discharge (RATING CURVE) of a stream increases at a given cross-section (AT-A-STATION). It is depicted by means of a HYDROGRAPH and is due to sudden snowmelt, heavy rainfall and accelerated RUNOFF due to land-use changes. The outcome is often a FLASH FLOOD which, in some cases in the semi-arid zone, can be catastrophic.

flat 1 Any smooth, even surface of low relief. **2** A mudbank exposed at low tide, i.e. a *tidal flat*. **3** The low, marshy pasture-land bordering a stream channel in an upland valley, i.e. a *valley flat*. **4** The horizontal portion of a mineral vein in an ore deposit, i.e. a *flat of ore*.

flatiron A triangular-shaped, steeply tilted or inclined mesa-like landform occurring as one of a series of rock formations on the flank of a mountain. A flatiron is commonly associated with the erosion of a dome structure in which sedimentary layers are gradually stripped from the flanks as the older rocks of the 'core' are revealed and an ANNULAR DRAINAGE PATTERN is initiated. The last sedimentary layer clings to the sides of the denuded dome in the form of flatirons which cap the ends of mountain spurs and are separated by V-shaped canyons. Some of the most renowned are the flatirons at Boulder, Colorado, USA, which occur on the flanks of the Rockies.

flatness index A measure of the shape of a rock fragment. It is achieved by summing the long and intermediate diameters of the rock fragment and dividing the result by twice the short diameter. The formula, named from A. Cailleux, is $[(a+b)/2c] \times 100$, where a = the long axis (length); b = the intermediate axis (breadth); c = the short axis (thickness). *See also* ROUNDNESS INDEX, SPHERICITY.

F-layer A zone at a height of about 250 km within the atmosphere that reflects high-frequency radio waves. *See* IONOSPHERE.

fleet **1** A lagoon of brackish water separated from the sea by a BAR or SPIT. The best known is the Fleet, impounded by Chesil Beach, Dorset, England. **2** A small creek or inlet on a coastline.

flexure A type of fold where bedded rocks have slipped over each other owing to tensional stress during deformation in the incompetent rock layers. Sometimes referred to as *flexure-slip* folding.

flint A variety of chalcedonic SILICA, similar to a CHERT. It is a fine-grained black or dark grey material which is very tough but splinters easily when subjected to percussion. Its fracture is curvilinear (conchoidal), in contrast to that of a chert, which fractures along flat planes. Flint occurs in nodules or in bands following the bedding planes of CHALK, almost entirely in the Upper Chalk (CRETACEOUS). It is thought to have been formed when the remnants of silica-bearing organisms of radiolaria and diatoms (OOZE) became incorporated in chalk muds on the sea-floor where the silica was carried downwards in solution and redeposited by precipitation around a core, sometimes a fossil (e.g. a sea-urchin). The newly formed flint nodule grew slowly by accretion within the chalk sediment. The facility with which flint can be fractured to provide a sharp cutting-edge led to its widespread use by prehistoric man, who manufactured many tools and weapons at places such as Grimes Graves, East Anglia, England.

float recorder An instrument for measuring the level of water in a lake, river or well. The float is connected by a cable to a counterweight, the cable passing over a pulley which then transmits vertical float movements to a recording mechanism.

flocculation The process by which soil COLLOIDS concentrate or coagulate into a soil AGGREGATE, thereby coarsening its texture and making 'heavy' soils easier to work. In farming this is usually achieved by 'liming'. The term is also used to describe the formation of muds in estuaries. *See also* CRUMB STRUCTURE OF SOIL, DEFLOCCULATE.

floe ICE FLOE.

Flöhn's climatic classification A classification, outlined in 1950, based on a combination of planetary wind belts and precipitation characteristics. Seven major categories are recognized. *See also* CLIMATIC REGION, KÖPPEN'S CLIMATIC CLASSIFICATION, THORNTHWAITE'S METHOD OF CLIMATIC CLASSIFICATION.

flood The inundation by water of any land area not normally covered with water owing to a relatively rapid change of the level of the particular water body in question. Floods probably account for the greatest loss of life and highest degree of material damage among all the world's natural hazards. Several varieties have been recognized: **1** *River floods due to increase in rainfall*, when the river channel becomes incapable of carrying the greater water volume, thus allowing the floodwater to escape on to the FLOODPLAIN or beyond. **2** *River floods due to snowmelt*, when a sudden thaw of SNOW cover leads to excessive runoff into the river channel (FRESHET), especially where the ground surface is temporarily or permanently frozen (PERMAFROST). **3** *Lake floods*, which result when the water surface rises in response to a sudden influx of river or snowmelt floodwater. **4** *Sea floods*, which occur when high tides coincide with exceptionally high, wind-generated sea-waves and extreme low pressure (STORM SURGE). The monsoonal coasts of India and Bangladesh are subject to periodic cyclonic storm-surge floods of disastrous proportions. **5** *Earthquake-generated sea floods* (TSUNAMI). **6** *Floods due to collapse of dams*. *See also* FLASH-FLOOD, FLOOD FREQUENCY, ANNUAL SERIES.

flood channel EBB CHANNEL.

flood frequency A type of data analysis based on flood records in order to determine the probability of flood magnitudes and the possible recurrence intervals. The most common method is depicted by:

$$T = \frac{N + 1}{m}$$

Where T is the return period, N is the number of years of record from the highest (rank $m = 1$) to the lowest (rank $m = N$). *See also* ANNUAL SERIES.

floodplain That part of a river valley, adjacent to the CHANNEL, over which a river flows in times of flood. It is a zone of low relief and gentle gradients and may incorporate OXBOW LAKES, POINT BARS, abandoned channels, SCROLLS, all indicative of the fact that the river channel has shifted its position continuously during the present regimen of the stream. The floodplain is composed of ALLUVIUM, which generally buries the rock floor of the valley to variable depths. The alluvial deposits were formed either within the channel itself (*c.* 75% of the alluvium) or as overbank deposits (*c.* 25%) in times of flood. *See also* LEVEE, MEANDER.

floodplain scroll SCROLLS.

flood routing A method used to calculate the magnitude and timing of a flood wave at successive stations (AT-A-STATION) as it progresses downriver. The shape of the flood wave will differ according to the CHANNEL GEOMETRY and its particular CHANNEL STORAGE capacity.

flood stage The stage of a river flow which succeeds the BANK-FULL STAGE, commencing at the time when the flow overtops the natural or artificial river banks. *See also* FLOOD.

flood tide **1** A general term referring to the rising tide which occurs between the time of low water and that of high water. **2** The inflowing *tidal stream* in an estuary which flows very strongly for a few hours prior to the slack water at high tide. It is the opposite of EBB TIDE. *See also* TIDAL CURRENT.

flora **1** A term referring to an association or collection of plants living in a particular place or at a particular time. A *floral assemblage* may be identified at different spatial scales from the continental-sized FLORISTIC PROVINCE, down through the regional to the local scale. In geology, a reconstruction of past environments can be made (PALAEO-ECOLOGY) by studying the relationships and character of fossil flora, either at a macro-scale or at a micro-scale (PALYNOLOGY). **2** A scientific publication or treatise listing the species of a particular place or a given time period.

Florida current One of the important OCEAN CURRENTS of the N Atlantic. It commences as a branch of the N Equatorial current, which enters the Caribbean and then returns to the Atlantic Ocean through the Florida Straits as the Florida current, which itself becomes the southern portion of the GULF STREAM. It is a rapidly flowing surface current (felt down to depths of 800 m) which hugs the N American coast as far north as Cape Hatteras, achieving speeds of some 1.5 m/sec.

floristic province A large geographical area containing a distinctive plant assemblage found nowhere else. In 1953 R. Good recognized thirty-seven floristic provinces in the world, ranging in size from the tiny Pacific island of Juan Fernández up to the vast Euro-Siberian province. He classified them into six kingdoms: Boreal; Palaeotropical (including African, Indo-Malaysian and Polynesian subkingdoms); Neotropical; South African; Australian; and Antarctic. [*82*] *See also* FLORISTIC REALM.

floristic realm (kingdom) The largest geographical division into which the world flora have been grouped. Such a scheme completely re-arranges the thirty-seven FLORISTIC PROVINCES of R. Good into six realms: the *Holarctic kingdom*; the *Palaeotropical kingdom*; the *Neotropical kingdom*; the tiny *Cape kingdom*; the *Australian kingdom* and the *Antarctic kingdom*. The boundaries of the thirty-seven provinces differ in detail from those defined by Good. *See also* WALLACE'S REALMS.

flotation **1** The condition of keeping afloat. **2** A method of mineral separation in which some float and others sink.

flow The movement of a substance either as a fluid or by plastic deformation of solids. The following types may be recognized: **1** *Gravitational flow*, including the movement of water, water-saturated earth (EARTH-FLOW) and lava, downslope under the influence of gravity. **2** *Plastic flow*, the movement of solids under stress by rearrange-

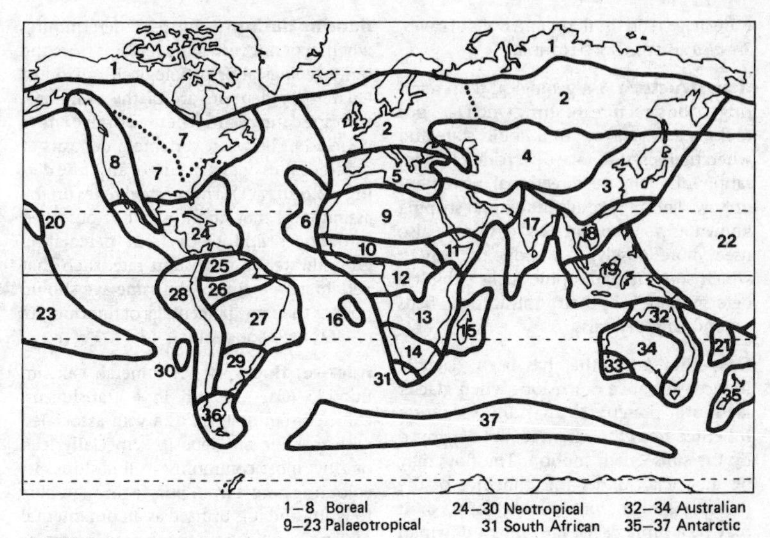

Figure 82 *Floristic provinces*

ment of their particles but without fracturing, whereby the solid material is deformed (PLASTIC DEFORMATION). This type of flow also includes EXTRUSION FLOW of ice. **3** *Atmospheric flow*, in which air moves from an area of high pressure to an area of low pressure (ATMOSPHERIC CIRCULATION) but modified by TURBULENCE owing to friction at the Earth's surface and by the Coriolis force. *See also* LAMINAR FLOW.

flowage The irreversible deformation of rock strata as a result of the stressing of rocks beyond their ELASTIC LIMIT. In a gently folded structure flowage will lead to thinning of strata, particularly by INCOMPETENT BEDS, over the crests of the anticlines and a thickening of beds over the synclinal troughs.

flow-duration curve A graphically plotted curve showing the frequency distribution of the mean daily flow of a river at a particular site, from which it is possible to calculate the percentage of time at which various discharge rates are equalled or exceeded. The slope of the curve indicates the magnitude of the flow (in m^3/sec). In most rivers it will be seen how floods provide very large discharges but occur on so few days that they contribute less water than lower flow rates which occur for a greater percentage of the total time.

flow equations Those equations specifying the relationships between slope, depth, velocity and boundary ROUGHNESS in open CHANNEL FLOW. They include the CHEZY EQUATION, the MANNING EQUATION and BERNOUILLI'S THEOREM. *See also* GRADUALLY VARIED FLOW, UNIFORM STEADY FLOW.

flow net EQUIPOTENTIAL.

flow regimes These four hydrological regimes are defined by specific open CHANNEL FLOW criteria that are closely related to the CHANNEL GEOMETRY, the ROUGHNESS of the stream bed and the sediment transport rate. They are based on a combination of the FROUDE NUMBER (*Fr*) and the REYNOLDS NUMBER (*Re*). The Froude number distinguishes between *sub-critical flow* (Fr < 1) and *supercritical flow* (Fr > 1). The Reynolds number contrasts LAMINAR FLOW (Re < 500) with TURBULENT FLOW (Re > 2000).

flowstone A smooth deposit of calcium carbonate accumulated by flowing karstic water (KARST) on the floors and walls of caves. *See also* RIMSTONE, SPELEOTHEM.

flow structure **1** A geological term referring to the structure resulting when igneous rocks, e.g. a lava, are in a liquid state and when their crystals take up a parallel orientation following the directional movement of flow. This may result in textural striping known as *flow banding*. **2** The term is also used, more loosely, to describe the way in which some metamorphic rocks exhibit a deformation of their micaceous beds around large crystals.

flow till A TILL that has been glacially induced to move downslope when glacier ice is still present, so that any structures inherited from the parent ice are destroyed by the subsequent motion. The flow may be subglacial, under the influence of the pressure-gradient or shearing force, or it may be a supraglacial movement down an ice surface or off the flanks of a sediment-covered ridge of dead ice (DEAD-ICE FEATURES). Any till that becomes remobilized on a slope after the disappearance of the glacier ice should not be classified as a flow till. If a flow till accumulates directly in a water body it is termed a *waterlain flow till*, but if the grains of a flow till become disaggregated it ceases to be a flow till even if resedimentation subsequently takes place. [*255*]

fluidization The process in which finely powdered rock material is invaded by an uprising stream of very hot gas, which welds it together and causes it to flow like a liquid. When such mixtures invade existing rocks during a period of volcanic activity they may cause the rocks to brecciate explosively and contribute to the phenomenon known as a glowing avalanche or glowing cloud (NUÉE ARDENTE).

fluid potential HYDRAULIC POTENTIAL.

flume **1** The channel of an artificially constructed stream built either for industrial uses or for simulation of streamflow characteristics during hydrological or sedimentological research. **2** A deep narrow gorge in the USA, in which the stream flows in a series of rapids and CASCADES.

fluorine dating A means of determining whether or not vertebrate remains from the same geological or pedological horizon of Quaternary deposits are of the same age. Fluorine dating cannot be used to give either correlation between vertebrate deposits at different sites or as a means of absolute dating (ABSOLUTE AGE). The method relies on the assumption that fluorine will be absorbed into bones and teeth from percolating groundwater at a constant rate, such that vertebrate remains of the same age should possess the same percentage of fluorine. *See also* RADIOCARBON DATING.

fluorite, fluorspar A mineral, calcium fluoride (CaF_2), found in a translucent, cubic-crystalline form as a vein associated with metallic ore deposits, especially lead or zinc, most commonly in limestone. Its colouring ranges from purple through blue to white and it is utilized as an ornamental material, particularly the variety known as BLUE JOHN, mined at Castleton in Derbyshire, England.

flush **1** An area of soil enriched by transported materials, either dissolved mineral salts or rock particles. It is sometimes referred to as a *flushed soil*. On most mountain-tops mineral salts will be deficient because of continuous LEACHING of the soils, although base-rich groundwater will reappear further downslope to create flushes in soils of any physical category. A distinction is made between *damp flushes*, around rivulets and spring-heads, marked by a zone of brighter green vegetation, in which there is considerable enrichment by water, and *dry flushes*, composed of freshly weathered rock particles from rock faces, e.g. in SCREES, gullies, etc., where new rock surfaces are constantly being eroded, thereby providing a supply of base-rich material. **2** A sudden rush of water down a stream, perhaps as a result of spring snowmelt (FRESHET).

flushed soil FLUSH.

flute An intermittent subconical groove pattern eroded on the floor of a stream channel as a result of TURBULENT FLOW. It is one example of SOLE markings that are preserved on the base of a bed of sedimentary rock which itself forms a cast of the groove (*flute cast*).

fluted moraine A ground moraine with a distinct lineation parallel to the direction of former ice movement, perhaps formed by a process similar to that which forms DRUMLINS.

fluting 1 A type of corrugation worn on the surface of a rock owing to glacial action, in which a number of glacially eroded furrows lie parallel with each other (GROOVE). **2** A series of smooth vertical grooves on the surface of granites or gneisses, thought to be caused by surface weathering. The term should not be confused with FLUTE.

fluvial, fluviatile Terms pertaining to rivers and river action, respectively. They also refer to organisms which are found in rivers and to anything produced by river action. In recent years attempts have been made to differentiate between the two terms, with *fluvial* being restricted to river flow and its erosive activity, and *fluviatile* being used adjectivally to describe both the products of fluvial action (e.g. *fluviatile deposits*) and the organisms of the freshwater environment.

fluvial cycle One of the CYCLES OF EROSION proposed by W. M. Davis. It is also known as the fluvial geomorphic cycle and the NORMAL CYCLE OF EROSION.

fluvio-glacial GLACIOFLUVIAL.

fluviokarst Landforms of karstic character (KARST) formed by the combined action of fluvial and true karst processes (SOLUTION). It is one of five types of karst landforms recognized by M. M. Sweeting and is much more compartmented than true karst because the limestones are less extensive, both vertically and horizontally. Thus normal river valleys and gorges are common (e.g. R. Dove in the Pennines, England), as are dry valleys. *See also* GLACIOKARST, KEGELKARST.

fluvisol ENTISOL.

flux 1 In general, a state of change. **2** In meteorology, the passage of energy across a physical boundary, e.g. transfer of SENSIBLE HEAT within the earth/atmosphere heat-budget system. **3** A substance that reduces the melting-point of a mixture, e.g. limestone used in iron smelting to lower the fusion temperature of the ore.

flyggberg A Swedish term for an asymmetrical hill, up to 300 m in height and 1–3 km broad, which is shaped by an overriding ice mass. *See also* ROCHE MOUTONÉE.

flysch In its strictest sense, a term referring to sediments associated with the formation of the European Alps, although it has been used universally in a more general usage to describe sediments derived from the erosion of uprising fold structures and which are subsequently deformed by the continuing orogenic movements (OROGENY). In Switzerland the flysch deposits comprise a wide variety of sediments ranging from clays and marls to coarse sandstones and conglomerates, formed during the period between late Cretaceous and early Tertiary times. *See also* MOLASSE.

focal depth A measurement referring to the shortest distance between the FOCUS of an earthquake and the surface of the Earth.

focus The point of origin of an EARTHQUAKE within the Earth's crust. Also known as the *hypocentre*, the *seismic focus* and the *seismic origin*. The source from which SEISMIC WAVES are generated. A deep-focus earthquake has a depth >300 km; a shallow-focus earthquake occurs at depths of < 70 km; an intermediate-focus earthquake occurs at depths between 70 and 300 km.

fog The obscurity exhibited by the surface layers of the atmosphere owing to the presence of water droplets formed by CONDENSATION, amplified by the suspended particles of smoke and dust which may occur. The official meteorological definition of fog is confined to situations where visibility falls below 1 km (0.62 mile). *See also* ADVECTION FOG, ARCTIC SEA-SMOKE, FRONTAL FOG, RADIATION FOG, SMOG, STEAM FOG.

fogbow A virtually colourless arc (with a radius of about 40°) of brighter light, similar in form to a rainbow, seen in a fog, when the sun is behind the observer. The water droplets are too small (0.05 mm) to enable REFRACTION and reflection to create the normal colour spectrum exhibited by a rainbow. The colours overlap and the bow appears white.

fog-drip, fog precipitation A term describing moisture deposited from a bank of fog owing to interception by trees and which in some cases may be sufficient to nurture vegetation in an arid climate, e.g. the Atacama Desert of S America. It is particularly characteristic of a COLD-WATER DESERT coast where humidity is constantly high but other forms of precipitation are scanty. It is commonplace on the central Californian coast.

fog potential index An index devised in Britain to indicate localities where fog frequency on highways (and especially motorways) will be greatest and where visibility will tend to be lowest. The index (I*p*) is expressed by:

I*p* = *f*(*dw*, *tp*, *Sp*, *ep*),

where: *dw* is the distance from and the spatial extent of standing water; *tp* is a function of the local topography at point *p* (including size of the KATABATIC WIND drainage catchment); *Sp* is a function of the road topography at point *p*; *ep* is an expression of any environmental feature likely to help or hinder the formation of RADIATION FOG (e.g. pollution source, woodland, etc.).

föhn (foehn, fön) A warm and relatively dry wind which descends on the leeward side of a mountain range, when moist air is forced to rise over the topographic barrier during the passage of a depression, thereby cooling at the SATURATED ADIABATIC LAPSE-RATE (0.5° C/100 m) before descending on the leeward side and warming at the DRY ADIABATIC LAPSE-RATE (1° C/100 m). One of its main characteristics is the rapid rise of temperatures which it generates, on its descent, generally of more than 10° C in a few hours. Thus it can cause hazardous avalanches and snowmelt in the mountain zones affected by such a phenomenon. The term föhn relates strictly to an example from the European Alps but the *föhn effect* has different names in other regions of the world: *Berg* (S Africa), *Chinook* (Rockies, USA), *Nor-Wester* (New Zealand), *Samun* or *Samoon* (Iran), *Santa Ana* (California, USA) and *Zonda* (Argentina). [*83*]

föhrde A German term for the shallow estuaries of the drowned linear valleys in the boulder-clay lowlands fringing the southern coast of the Baltic, e.g. Kieler Föhrde. It should not be confused with the Norwegian term *fjord* (FIORD) or the Swedish term *fjard* (FIARD) which refer to hard-rock coasts. It should be noted, however, that in Denmark the Baltic föhrdes are referred to as *fjords*. Although the shapes of the föhrdes

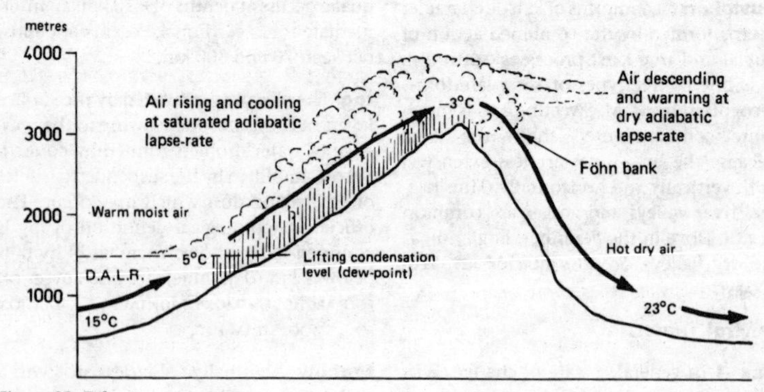

Figure 83 *Föhn*

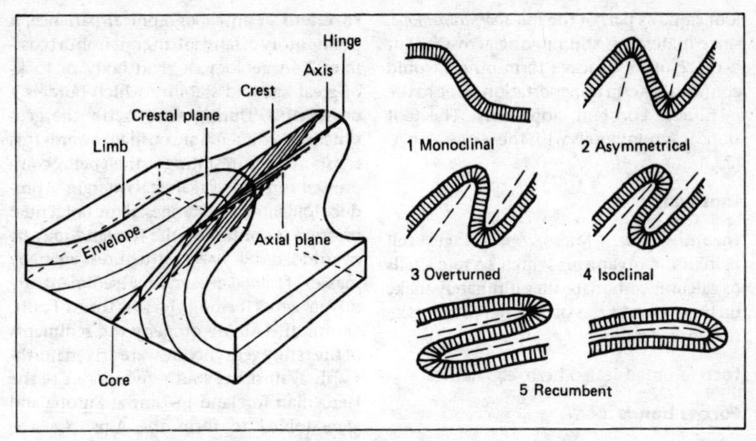

Figure 84 *Types of folding and terminology*

are variable they usually have wide mouths but narrow inland reaches. They are carved entirely in glacial deposits and are thought to have been formed by subglacial streams escaping from beneath the Baltic ice-sheet. They are said to represent the drowned terminations of TUNNEL VALLEYS.

fold, folding A flexure or bending of the Earth's crustal rocks owing to compressional forces. In simple *folding* an upfold is termed an ANTICLINE and a downfold a SYNCLINE. The line along which the DIP of the strata changes direction is termed the *hinge line* and this usually coincides with the position of maximum curvature. The *fold axis* is a line drawn parallel with the hinge line, from which the beds dip away in different directions in the form of limbs. There are many types of folds, depending on the intensity of the compression and on the character of the rock. The simple *monocline* can change into an ASYMMETRICAL FOLD, then to an OVERFOLD or *overturned fold* and finally a RECUMBENT FOLD as the pressure increases from one direction. Folds are generally cylindrical, i.e. semicircular in profile, although *angular folds*, straight limbs and sharp hinges (CHEVRON FOLD) can occur and these give zigzag-shaped patterns instead of the S-shaped curves of *cylindrical folds*. When the axial planes of two sets of folds intersect they may create a CULMINATION and a DEPRESSION. *See also* ANTICLINORIUM, DISHARMONIC FOLDING, NAPPE, PERICLINE, PLUNGING FOLD, SYNCLINORIUM. [*84*]

foliation 1 A laminated or banded structure within a METAMORPHIC ROCK, caused by the segregation of different minerals into parallel layers following the schistosity (SCHIST) of the rock. **2** The stratiform structure of ice in the lower layers of a glacier, probably the result of shearing of one part over another. It differs from the stratification of the ice due to incremental growth during deposition.

food chain The sequence of energy transfer in the form of food from organisms in one TROPHIC LEVEL to those in another, achieved when one organism eats or decomposes another.

foothills The lower line of hills that lie parallel with and at the foot of a higher mountain range.

foot slope The lower part of a SLOPE above the gentler gradient of a valley floor or plain but below the CONSTANT SLOPE. It is part of the ACCUMULATION SLOPE, where colluvial material has been redeposited after having been moved downslope by MASS-MOVEMENT and surface wash. Some authors regard the foot slope as part of the *transportation slope*, since material is still moving across it. But because of its concave form others would confine the term transportation slope to the rectilinear constant slope only. The foot slope is synonymous with the WANING SLOPE. [*233*]

foot wall FAULT.

foraminifera Microscopic single cell planktonic organisms which secrete shells of calcium carbonate that ultimately make up high proportions of OOZE deposits on the sea-floor. An order of PROTOZOA.

forb A broad-leafed herb, e.g. nettle.

Forbes bands OGIVE.

ford A crossing-point on a stream or river that is shallow enough to be crossed without the aid of a bridge.

forecast In a meteorological context this is a statement issued with a fairly high degree of confidence for the anticipated weather trends at certain locations for a given period of time. In the USA the probability is generally stated in percentage terms, but in Europe weather forecasts are generally more definite. They are normally given at three levels of details: **1** short period (24–48 hours); **2** medium range (about one week); **3** long range (about one month). *See also* ANALOGUE WEATHER FORECASTING, SYNOPTIC METEOROLOGY.

foredeep A linear, narrow crustal depression in the ocean floor, on the convex side of an ISLAND ARC, or a coastal range of fold mountains. It is sometimes termed a *foreland basin* because it occurs on the *foreland* to which it is genetically related. *See also* TRENCH. [*119*]

fore-dune A term for part of a coastal DUNE system. It is the dune nearest the sea, occupying the same windward position as does the advanced dune in the desert dune system. The fore-dune is the youngest of the coastal sand-hills and is characteristically colonized by the grass *Agropyron junceum*, which tolerates occasional washing by sea water.

foreland **1** In topographical parlance, a promontory of land jutting out from a coastline. **2** In geology, a rigid body of rocks of great age and stability which borders a GEOSYNCLINE. During an OROGENY the geosynclinal sediments are pushed towards this resistant block, resulting in their being compressed and folded against its margin to produce fold mountain ranges. If the tangential pressure is continued the foreland may be overridden by NAPPE structures whereby masses of folded geosynclinal sediments are driven across it along gigantic thrust-faults. During the Alpine orogeny the sediments of the TETHYS geosyncline were driven northwards against the southern margins of the Hercynian foreland in central Europe and were folded to form the Alps. *See also* HINTERLAND.

fore-set beds Those parts of the bedded layers of a DELTA which are situated between the TOP-SET BEDS and the BOTTOM-SET BEDS. They are inclined at an angle, one above another, as the accumulated sediment descends the steep frontal slope of the delta. [*85*]

foreshock A seismic tremor recorded prior to the main shock of an EARTHQUAKE. It is caused by the initial slip or fracture in the vicinity of the major FOCUS and precedes the main earthquake by a short time interval. *See also* AFTERSHOCK.

foreshore The lower zone of a beach, extending between low-water spring-tide level and high-water spring-tide level, i.e. the part of the beach covered by SPRING TIDE. *See also* BACKSHORE.

forest **1** One of the four main subdivisions of world vegetation (BIOCHORE). **2** A continuous tract of trees over a large area. **3** A royal hunting ground in England where *forest law* prevails over common law, e.g. the Forest of Dean. **4** A treeless moorland in Scotland given over largely to deer-stalking and game-shooting, although it was once a wooded area, e.g. Applecross Forest in Wester Ross, Scotland. *See also* BOREAL, CONIFEROUS FOREST, CLOUD FOREST, DECIDUOUS, MONSOON FOREST, EQUATORIAL RAIN-FOREST, TROPICAL RAIN-FOREST.

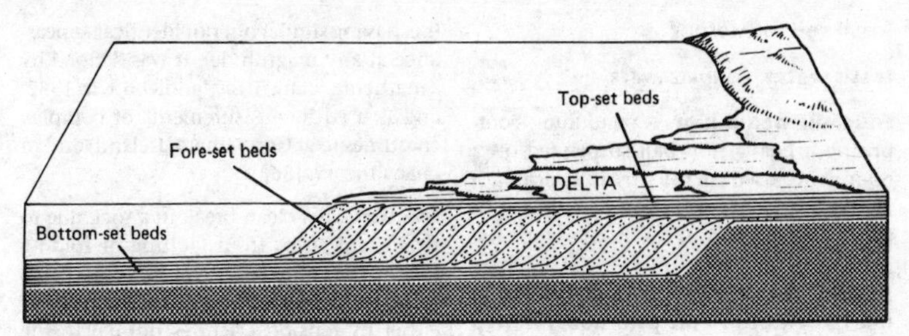

Figure 85 *Formation of a delta*

forest bed An interglacial body of sediment containing macroscopic remains of trees and other vegetation. *See also* CROMERIAN.

forest climate The average weather conditions of a forest region. These differ from non-forested climatic characteristics because the tree masses modify the humid environment, the thermal environment and the air-flow (SHELTER-BELT) in addition to affecting both RADIATION and ENERGY BALANCE because of the tree canopy's reflectivity (ALBEDO) and its mantling or trapping effect. *See also* TAIGA, TROPICAL RAIN-FOREST.

formation **1** In general use, something that has been formed naturally. **2** In biogeography, the term refers to one or more plant communities which exhibit comparable structures within a CLIMATIC REGION. It is synonymous with BIOME. **3** In geology, the term denotes a lithostratigraphic unit (LITHOSTRATIGRAPHY), being the fourth category of a classification of rocks. It is composed of rock MEMBERS and forms part of a lithological GROUP. A rock formation may be composed of one or more strata, the lithological character of which distinguishes the unit from bodies of adjacent rocks. A formation is usually, but not invariably, bounded by a distinct change in the lithological properties of the rock bodies at its upper and lower boundaries.

formation-class A unit of vegetation defined on a geographic basis with reference to its physiognomy. It is a major subdivision of a BIOCHORE; e.g. the tropical forest and the taiga are formation-classes of the forest biochore.

form-line A line drawn on a map where there is insufficient information available to enable a CONTOUR to be constructed. It is not instrumentally surveyed and is intended merely to give an impression of the terrain. It is usually depicted as a dashed rather than a solid line.

form-ratio A relationship between river width and river depth expressed as a ratio.

Fortin barometer A type of mercury barometer in which the level of mercury is adjusted to zero before a reading is made. *See also* ANEROID BAROMETER, BAROMETER.

fosse **1** In general, an artificial trench or ditch around an earthwork. **2** In the USA the term refers to a depression occurring between a glacier and the rock wall of a glacial trough. **3** In France, a fosse is synonymous with a SUBMARINE CANYON.

fossil The remains of the whole or part of any formerly living organism (plant or animal) preserved by natural causes in crustal rocks. It is often chemically altered or replaced or it may be recognized merely by the mould or impression that it has made in the surrounding rocks. The fossilized remnants of the effects of the organism, rather than the fossil itself (e.g. worm burrows or tracks), are termed *trace-fossils*. The study of *fossils* is known as PALAEONTOLOGY. *See also* CONNATE WATER, PALAEOSOL.

fossil fuels All fuels derived from the *fossilization* of formerly living organisms. Examples are COAL, PEAT, PETROLEUM and NATURAL GAS. *See also* HYDROCARBON.

fossil soil PALAEOSOL.

fossil water CONNATE WATER.

Foucault pendulum A pendulum, comprising a heavy iron ball suspended on a 60-m wire, used in 1851 by L. Foucault, a French physicist, during an experiment to prove the rotation of the Earth. Once the pendulum was made to swing it traced a path in an underlying bed of sand, but as time progressed so the path slowly moved round to the right. The implication was that either the plane of the swing had changed or the Earth had rotated, the latter being found to be the case. Under a pendulum suspended over the pole the Earth will make one complete rotation in a day (equivalent to an hourly deviation of 15°) but at the equator the pendulum will exhibit no deviation at all. At intermediate latitudes the rate of deviation varies (e.g. at 45° latitude a complete rotation takes thirty-four hours = a deviation of 10.61 degrees). The amount of turning per hour can be established by $15 \times \sin \theta$, where θ is the latitude.

Fourier analysis A mathematical technique used to analyse a harmonic data series, i.e. a complex curve that exhibits periodic variations. It is used in the analysis of changes in seasonal rainfall and in the examination of wave records. A *Fourier series* consists of terms containing sines and cosines, while a *double Fourier series* develops it so that the dependent variable becomes a function of two independent variables, thereby representing a complex surface instead of a curve.

Fournier's map projections Two projections of 1646, both exhibiting a rectilinear equator and a central meridian, but having the other meridians in elliptical form arranged as in the MOLLWEIDE PROJECTION. Thus, the map of a hemisphere is enclosed by a circle. The first Fournier projection has its parallels as arcs of circles which are not concentric, while the second has rectilinear parallels spaced according to the sine of the latitude.

fractal An object with infinitely detailed dimensions and displaying self-similarity (i.e. having similar but not identical appearance at any magnitude). It was defined by a mathematician, B. B. Mandlebrot, in 1982, and is used in measurements of complex coastlines (e.g. Connemara, Ireland) and in CHAOS THEORY. [*86*]

fracture **1** A clean break in a rock due to strain and stress from faulting or folding (i.e. from extension or compression of the rock). When rocks are extended they fail either by tension fractures normal to the direction of extension or by shearing with fractures oriented so that the direction of extension bisects obtuse angles between fractures. When rocks are compressed they fail either by longitudinal fractures parallel to the principal stress direction or they deform along shear planes inclined at angles of 45° or less to the principal stress direction. **2** The characteristic break pattern of a *mineral*.

fracture zone A zone where large-scale transform faults (TEAR-FAULT) off-set plate structures (e.g. the line of the mid-oceanic ridges) on the floor of the oceans. The fracture zones also displace the palaeomagnetic patterns of the rocks on the sea-floor, indicating that there is differential movement within the tectonic plates themselves (PLATE TECTONICS). They have been recognized in the Atlantic, Indian and Pacific oceans, e.g. the Clipperton and Mendocino Fracture Zones of the E Pacific.

fractus A type of cloud the ragged or torn nature of which indicates storminess in the atmosphere. The term is used as a prefix to other cloud terminology (e.g. fractostratus). *See also* CLOUD.

fragipan A subsurface soil horizon, showing a degree of induration when dry but a weak to moderate degree of brittleness when moist. It is not absolutely impermeable to water, and is often the result of PERIGLACIAL ACTIVITY.

frazil Minute needles of ice which develop into spongy accumulations in moving water below freezing-point.

free-air anomaly A geophysical term relating to the difference at any ground sur-

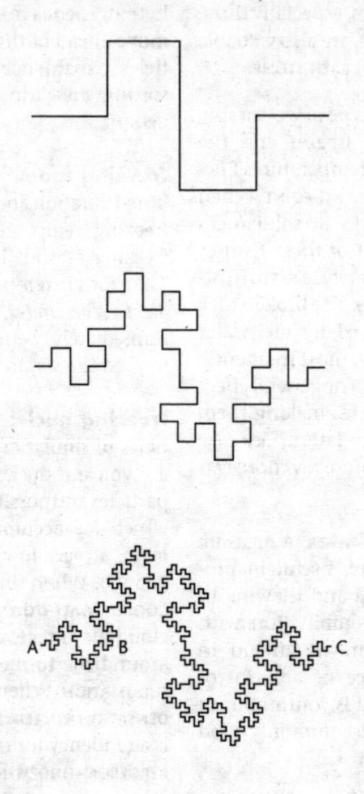

Figure 86 *First three stages in constructing a simple fractal*

face point between the measured gravity and that calculated for the theoretical gravity at sea-level together with a free air coefficient, i.e. a gravity anomaly which takes altitude into account but ignores the attraction associated with isostatic compensation (ISOSTASY).

free energy The capacity of a system (GENERAL SYSTEMS THEORY) to perform work. The maximum work that may be generated from any process is a measure of the increased free energy in the system.

free face The steepest part of a SLOPE profile, with an arbitrary minimum slope angle of 45°, i.e. too steep for any loose material to rest upon it [*49*]. It refers to a rock-cut face and is not generally used to describe vertical or near-vertical cliffs in unconsolidated materials. Because weathered fragments fall from the free face and also from the convex slope above it (WAXING SLOPE) it is sometimes referred to as the *fall face*. Below it in the slope profile is the CONSTANT SLOPE produced by the accumulation of debris from above. *See also* BACK-WEARING, DENUDATION SLOPE. [*233*]

freestone A term used to describe any sedimentary rock which is so evenly textured that it can be freely worked, and in particular can be easily sawn in any direction. The best freestones have long been used as building stone, e.g. two of the most famous in Britain are Bath stone and

Portland stone, both of which are Jurassic limestones. But sandstones, especially those of New Red Sandstone age, are also valuable freestones. *See also* ASHLAR, BATH STONE.

freeze–thaw action A type of WEATHERING process in which water freezes and the resulting ice melts as the temperatures fluctuate below and above FREEZING-POINT. When the water occurs in cracks in solid rocks it causes 9% enlargement of these fissures during freezing and ultimately leads to frost-splitting (CONGELIFRACTION). The broken rock fragments are then moved by meltwater during the period of thaw, most frequently downslope (SOLIFLUCTION). The general effect is a weakening of the rocks, making them more susceptible to denudation. *See also* BLOCK-FIELD, CONGELITURBATION, NIVATION, PATTERNED GROUND.

freezing and thawing index A measure of the severity of climate, useful in projecting depths of freezing and thawing in the soil, and taken in combination with mean annual temperature as an aid in determining the presence or absence of PERMAFROST and whether it is continuous or discontinuous. The most commonly used index (I) is:

$$I = \int_0^t T dt$$

where T is the mean temperature (°C) for a day (d) as represented by (maximum + minimum temperature)/2, and where t is the time period.

freezing drizzle, freezing rain Supercooled droplets, descending from relatively warm air above an INVERSION OF TEMPERATURE, which freeze on impact with the surface (below 0° C) to form GLAZED FROST or ice. *See also* BLACK ICE.

freezing front The limit of freezing in the ground surface debris layer or soil under PERIGLACIAL conditions. The freezing front may extend downwards or laterally into an exposed cliff face, its rate of progress depending upon the character of the material, e.g. differences in grain size and in porosity. Thus, the freezing front is unlikely to be a straight line except where the material is very evenly textured. In heterogeneous materials, the finer particles move ahead of the front while coarser particles remain behind, thereby leading to sorting and the formation of PATTERNED GROUND.

freezing index A measure of the combined duration and magnitude of the below-freezing temperatures which occur in a freezing season, indicated in DEGREE-DAYS. That for air temperatures is known as the *air freezing index*, while that for a layer immediately beneath the soil surface is termed the *surface freezing index*.

freezing nuclei Small atmospheric particles of similar crystalline structure to ice, e.g. volcanic dust, very fine wind-blown soil particles and possibly meteoric dust, around which ice accumulations are thought to form, at very low temperatures (–25° C to –35° C). When tiny ice particles and SUPERCOOLED WATER droplets exist together in a cloud the latter evaporate and are deposited around the former by the process termed SUBLIMATION. When freezing nuclei and CONDENSATION NUCLEI are present in a cloud there is a tendency for raindrops to form. *See also* BERGERON–FINDEISEN THEORY OF PRECIPITATION, CLOUD-SEEDING, NUCLEUS, RAIN-MAKING.

freezing-point The constant temperature at which a liquid or a gas changes from the fluid to the solid state. Some important freezing-points at sea-level are: fresh water 0° C (32° F); sea water about –1.75° C (28.85° F); mercury –39° C (–38.2° F). *See also* SUPERCOOLED WATER.

frequency distribution A statistical method of plotting numerical data to show the frequency with which different values of a variable occur within a sample. An *ungrouped frequency distribution* is simply a list of figures occurring in the raw data together with the frequency of each figure. Once the figures are grouped into classes, it becomes a *grouped frequency distribution*, in which class intervals and class limits can be identified. When the data are plotted graphically on a HISTOGRAM the graph is termed a *frequency curve*. *See also* NORMAL DISTRIBUTION, POISSON DISTRIBUTION.

freshet A surge of floodwater down a small river or stream causing it to overflow its banks, resulting from heavy precipitation or sudden snowmelt around the head-waters. *See also* FLOOD, FLUSH.

fresh water Water with less than 0.2 per cent salinity.

fret A narrow belt of sea-mist along a coastline due to differential surface heating.

friable A term applied to the character of a soil relating to its ability to crumble easily between the fingers when it is wet or dry.

friagem A strong cold wind in southern Brazil caused by a sudden outbreak of cold southerly air associated with a winter anticyclone. Temperatures may fall some 15° C to bring them down to an unseasonal 10° C. Also known as a *surazo*. *See also* POLAR OUTBREAK.

friction **1** In general, the reaction of an equal opposite magnitude to an applied force (FRICTION, COEFFICIENT OF). **2** In meteorology, it refers to the retardation effect of surface roughness on air movement in the atmosphere, acting within 600 m of ground-level. Frictional drag decreases the surface wind-speed (GEOSTROPHIC FLOW). Velocity reductions of about one-third control the influence of the CORIOLIS FORCE, which destroys the geostrophic flow. Thus the dominant force now becomes that of the PRESSURE GRADIENT, causing the winds to blow across the isobars at an average angle of 20°. *See also* EKMAN EFFECT.

friction, coefficient of An equation used by geomorphologists and civil engineers to measure the mechanical force of resistance when two solid objects are brought into contact. The coefficient of friction is defined as the ratio between the frictional stress opposing motion (T) and that of the perpendicular force (P), which is equal to the weight of the object when it rests on a horizontal, supporting surface but is subjected to an outside tangential force. When the object rests on an inclined rather than a horizontal surface it is in a state of incipient motion and the ratio T/P becomes equal to the tangent of the angle of repose ($\tan \propto$).

friction cracks A term proposed by S. E. Harris in 1943 to describe a variety of rock fractures induced by the passage of glacier ice over bedrock (EXARATION). These include *lunate fractures*, *crescentic fractures*, *crescentic gouges*, and *conchoidal fractures*. *See also* PLUCKING.

frigid zone One of the three climatic zones recognized by early geographers (TEMPERATE ZONE, TORRID ZONE). Now used only to describe one of the climatic types defined in KÖPPEN'S CLIMATIC CLASSIFICATION or more loosely as a synonym for permanently snow-covered and PERMAFROST areas.

fringing reef A coral REEF which is attached to the shore, either as a continuous wave-washed erosion platform or separated from the coastline by a shallow lagoon. Beyond its seaward margin the ocean water deepens rapidly. *See also* BARRIER REEF.

frog-earth A term coined by A. P. Currant (1998) to describe those CAVE EARTH deposits which contain large numbers of AMPHIBIAN bones, especially those of the common frog (*Rana Temporaria*). Such accumulations suggest that they were deposited during wetter climatic phases than those experienced in Britain today.

front, frontal surface, frontal zone A term introduced by the Bjerkness School in Norway (1918) to describe a sloping boundary plane or surface separating two AIR MASSES that exhibit different meteorological properties or characteristics. The interface may be narrow and on a small scale, when it is termed a frontal surface (e.g. a *warm front* associated with a DEPRESSION) or it may be broad and on the scale of a frontal zone (e.g. the *Atlantic Polar Front*). It was formerly used in the tropics to describe the INTERTROPICAL CONVERGENCE ZONE (INTERTROPICAL FRONT) but the usage is now restricted to higher latitudes where thermal discontinuities are much more pronounced. *See also* COLD FRONT, FRONTOGENESIS, OCCLUSION, WARM FRONT. [*44* and *60*]

frontal fog The term given to a short-lived period of mist and drizzle associated with the passage of a WARM FRONT in a DEPRESSION. It forms when the cooler layers of air in contact with the ground surface become saturated by the evaporating warmer rain droplets which fall from the advancing FRONT.

frontal rainfall Precipitation which falls along the frontal zones during the passage of a DEPRESSION in middle and high latitudes, although the term *cyclonic rain* is sometimes used as a synonym. A general distinction can be made between the prolonged drizzling rain which falls from the WARM FRONT and the shorter but heavier bursts of rain associated with the squally passage of the COLD FRONT.

frontogenesis The process by which two *air masses* of different physical characteristics are brought together (horizontal CONFLUENCE and/or CONVERGENCE at a *frontal zone*), thus setting in motion the meteorological mechanisms which lead ultimately to the formation of a DEPRESSION with its own frontal systems.

frontolysis The gradual break-up or dissipation of a FRONT or *frontal zone*. This is the antithesis of FRONTOGENESIS, and is effected mainly by horizontal DIVERGENCE of air from the frontal zone, together with SUBSIDENCE.

frost **1** The state of freezing or of becoming frozen, when the air temperature at screen level (1 m) falls to or below the FREEZING-POINT. Qualifying adjectives generally accompany the term in order to indicate its intensity: hard, sharp, killing. **2** A weathering agent which breaks up rocks and soil owing to freezing of the interstitial water. Water expands its volume by 9% when frozen, thereby exerting increased pressure on the surrounding materials; in rocks such pressures induce *frost-shattering* (*frost-riving*) and this *frost-weathering* is one of the most important weathering processes in high mountain regions (FROST ACTION). In unconsolidated sediments of soil, the groundwater near the surface may be periodically frozen, thus producing *frost-heaving* (*frost-stirring*) which in turn creates differential sorting of the soil's constituents. *See also* FROST WEDGES, PATTERNED GROUND, PERMAFROST. **3** Frozen dew, fog or water vapour are sometimes termed *white frost* (in contrast to *black frost* when no RIME is formed). *See also* GLAZED FROST, GROUND FROST, HOAR FROST, RIME.

frost action **1** The MECHANICAL WEATHERING process caused by alternating cycles of freezing and thawing of water in pores, fissures and cracks at the surface. **2** The effects on materials and structures resulting from their mechanical weathering. The term is synonymous with FREEZE–THAW ACTION. *See also* CONGELIFRACTION, CONGELITURBATION, FROST-HEAVING, FROST-THRUSTING.

frost boundary (frost-line) A frost limit, either altitudinally or laterally across the land surface. Also used in a seasonal context. **1** Delimits areas which have never experienced recorded frost – generally the low-altitude tropical zone. **2** Delimits areas with mean minimum air temperatures above FREEZING-POINT (0° C/32° F). **3** Delimits areas which experience a particular number of frost-free days in the growing season. **4** Delimits areas in which the lowest mean monthly temperature remains above *freezing-point*. **5** Delimits areas in which there is no month with a mean temperature above *freezing-point* – the zone of perpetual frost (*polar climate*).

frost crack A fissure opened up in the soil by the development of an incipient ICE WEDGE.

frost-heaving The doming or vertical lifting of the soil surface into frost hillocks owing to pressures caused by the freezing of groundwater under periglacial conditions. Most writers distinguish this term from that of FROST-THRUSTING, although both are included under the term CONGELITURBATION. It causes severe interference with building foundations.

frost hollow, frost pocket A topographic depression susceptible to frosts while the surrounding hill slopes may remain frost-free. The cold dense air, which has lost heat by *terrestrial* radiation at night, drains downslope by gravity as a KATABATIC flow before accumulating in the hollow, wherein the temperature may fall below FREEZING-POINT. *See also* GREENHOUSE EFFECT, GROUND FOG, INVERSION OF TEMPERATURE.

frost-shattering CONGELIFRACTION, FREEZE–THAW ACTION.

frost smoke ARCTIC SEA SMOKE.

frost-thrusting The *lateral* movement of soil and other ground-surface debris caused by the freezing of groundwater under periglacial conditions. Most writers make a distinction between this term and that of *frost-heaving*, which is thought to be a process relating only to *upward* pressures. *See also* CONGELITURBATION.

frost wedging CONGELIFRACTION.

Froude number A hydrological term referring to a dimensionless parameter used to describe the flow conditions of a stream. It is an index of the influence of gravity in stream-flow situations where there is a liquid–gas interface as in an open channel. The Froude number (Fr) is usually defined as:

$$\mathrm{Fr} = \left(\frac{\textit{inertia force}}{\textit{gravity force}}\right)^{1/2}$$

A Froude number of unity is said to indicate *critical flow*; one greater than unity indicates *rapid flow*; and one less than unity indicates *tranquil flow*. *See also* REYNOLDS NUMBER.

frozen ground That part of the Earth's land surface which has temperatures at or below 0° C and generally contains groundwater in the form of ice. The term is usually reserved for **1** seasonally frozen ground and **2** permanently frozen ground (PERMAFROST) in high latitudes. In Sweden it is known as *tjaele*. Seasonally frozen ground is ground frozen by low seasonal temperatures and remaining frozen only through the winter. Some authors restrict its usage to a non-permafrost environment, but most use it to also include the zone of annual freezing and thawing (ACTIVE LAYER) above the permafrost. Ground freezing occurs over 48% of the land mass of the N hemisphere (26% non-permafrost and 22% permafrost).

fula An Arabic word referring to a shallow pool formed during the rainy season in the N Sudan and an important source of water for pastoral tribes.

fulje A deep parabolic depression between closely interlocking BARCHANS.

full A ridge of shingle or sand formed on a beach and more or less parallel to the shoreline. It is formed by CONSTRUCTIVE WAVE action slightly seaward of the line where the waves break. Where there are a series of fulls they are separated one from another by a SWALE or long shallow depression. When a period of DESTRUCTIVE WAVE activity prevails the fulls are destroyed by the combing action of the breakers.

fuller's earth **1** A fine earth similar to a CLAY but lacking its plasticity. It contains the mineral MONTMORILLONITE, which enhances its property of taking up water or other fluids. Because of this ability it was once widely used to remove some of the oil and grease from natural wool prior to weaving, a process known as *fulling*. **2** The term is used stratigraphically to describe a division of the JURASSIC in England.

fumarole A small VENT in a volcanic area from which steam, gases (such as carbon dioxide and chlorine) and various acids are ejected, often with considerable force in the form of jets.

function **1** In general, the purpose of an object. **2** In mathematics, the term describes the relationship between variables, whereby a change in one variable depends on a change in another. The second or independent variable is then a function of the first or dependent variable. An *exponential function* is one in which the dependent variable increases or decreases in geometrical progression while the independent variable increases in arithmetic progression. A relationship is produced in which the rate of change of the dependent variable is always proportional to the corresponding value of the independent variable.

fundamental properties The characteristics of the points or lines of zero distortion of any MAP PROJECTION and the resulting pattern of distortion ISOGRAMS over the map as a whole.

funnel cloud A downward-projecting funnel-shaped dark cloud which descends gradually in the form of a whirling vortex from the base of a thundercloud (CUMULONIMBUS). It may develop into a WATERSPOUT or TORNADO as the long tapering column of spinning air touches the Earth's surface.

furrow **1** A channel or linear depression cut by a plough. **2** A linear depression on the ocean floor. **3** A synonym of a SWALE.

fusain A friable and dusty layer within a bed of coal, high in ash content.

G

gabbro An igneous rock of coarse grain, dark colour and basic composition. It is the plutonic equivalent of the volcanic BASALT and the hypabyssal DOLERITE. It contains low amounts of silica but high percentages of basic plagioclase, pyroxene and usually olivine. Gabbros occur either in the form of a LOPOLITH (e.g. Bush Veld, S Africa) or as a RING COMPLEX (e.g. Ardnamurchan, Scotland). Some of the best exposures in Britain can be seen in the rugged mountains of Rum and the Black Cuillins of Skye, both in Scotland.

gada An Arabic term for a flat-topped table-like hill (BUTTE) in the Atlas mountains of N Africa. It is also used in the Sahara to describe a HAMADA, but its more common usage is related to an elevated landscape.

Gaia hypothesis A recent concept of J. E. Lovelock who, in 1979, described it as 'a complex entity involving the Earth's biosphere, atmosphere, oceans and soils, the totality constituting a feedback or cybernetic system which seeks an optimal physical and chemical environment for life on this planet'.

gal A unit of acceleration named after Galileo. It is equivalent to 1 cm/sec/sec, but since this has been found to be too large to measure the strength of the Earth's magnetic field (GRAVIMETRIC SURVEY) it has been divided into *milligals*. 1 milligal = 0.001 gal (one-thousandth of a gal). 1 milligal = 9.8×10^{-3} Nkg^{-1}.

gale A strong wind, recorded 10 m (32 ft) above ground-level and averaged over a ten-minute period, which has been classified on the BEAUFORT SCALE as follows: force 7, near gale, 32–38 miles/h; force 8, gale, 39–46 miles/h; force 9, strong gale, 47–54 miles/h; force 10, storm, 55–63 miles/h. *See also* HURRICANE.

galena Lead sulphide (PbS), the main ore of LEAD. It occurs in a dull, grey, metallic cubic crystalline form, often in association with barytes, fluorspar and calcite, often in CARBONIFEROUS LIMESTONE after a period of hydrothermal activity. *See also* GANGUE.

gallery forest A dense growth of tropical forest following the course of a river in SAVANNA country. The trees on either bank provide a tunnel-like character for the stream-course owing to the intertwining of their canopies. The term is adapted from the Spanish word *galeria* = an overhanging balcony.

Gall's stereographic projection A map projection of the cylindrical variety (CYLINDRICAL PROJECTION) in which the cylinder intersects the Earth's surface at 45°N and S. The projection is neither orthomorphic nor equal-area because the scale is correct only along the parallels of intersection (45°N and S). The scale increases polewards and decreases equatorwards from these parallels of latitude. Like the parallels the meridians too are straight lines drawn vertically through the equator at true 10° intervals. Although the Gall's projection exaggerates the area and shape of high latitudes rather less than does the MERCATOR PROJECTION, it is infrequently used. [*87*]

gangue The non-valuable metalliferous ores or the relatively worthless non-metallic mineral material in the lode or mineral vein. The term is used in the working of metallic

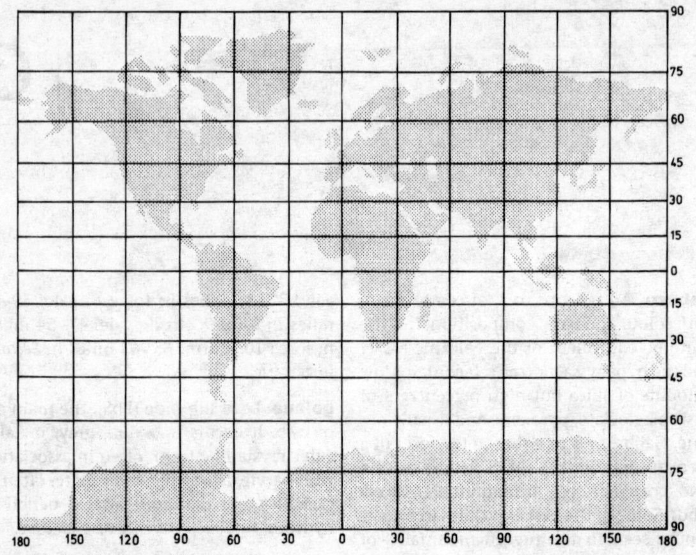

Figure 87 *Gall's stereographic projection*

ores and is frequently applied to such minerals as BARYTES, CALCITE, FLUORITE, PYRITES and vein quartz, some of which may have commercial uses. Because the term has also been used loosely to refer to the waste material from the process of ore separation and concentration it is difficult to give a precise definition. In general, gangue can be defined as that part of an ORE deposit which is of secondary importance during the extraction of one or more metals.

gannister (ganister) An arenaceous SEAT EARTH beneath the coal seams of the COAL MEASURES in Britain. It is a very pure silica sand, the high refractory properties of which make it of great use for furnace linings because of its heat resistance. Its argillaceous equivalent is a FIRE CLAY.

gap A notch or break in a ridge. It may: **1** carry a river (WATER-GAP), e.g. the Thames breaks through the Chalk cuesta of S England at the Goring Gap between the Chilterns and the Berkshire Downs; **2** be a dry gap due to river capture, etc. (WIND-GAP), of which numerous examples exist in the Chalk downlands of S England, many of which have been followed by lines of communication; **3** be a gap cut by glacial erosion where glaciers have overridden a ridge or divide, e.g. Gap of Dunlow, Co. Kerry, Eire.

gara An Arabic word widely used in N Africa and the Middle East to describe an isolated rocky hill with steep slopes and usually with a tabular summit. Its plural form is *gour*. The term should not be confused with *gar'a* which, in Algeria and Tunisia, refers to a vast depression subject to periodic flooding. *See also* BUTTE, GADA.

GARP The *G*lobal *A*tmospheric *R*esearch *P*rogramme, an international project incorporating numerous studies on ATMOSPHERIC CIRCULATION and atmospheric dynamics at a global scale. *See also* GLOSS.

garrigue (garigue) An evergreen scrub with xerophytic characteristics (XEROPHYTE) found on thin limestone soils in the drier parts of regions experiencing a MEDITERRANEAN CLIMATE. It is a French term describing the prickly and thorny shrubs and stunted evergreen oaks which clothe the bare rocky hills of Provence, but it has been widely used to describe all types of dry scrubland and vegetation on uncultivated calcareous soils around the western end of the Mediterranean basin. *See also* MAQUIS.

gas, natural A HYDROCARBON, consisting of ethane and methane, usually found associated with crude-oil accumulations.

gaseous cycle On eof the BIOGEOCHEMICAL CYCLES in which the ATMOSPHERE is the primary reservoir. It includes the NITROGEN CYCLE and the oxygen cycle.

gash breccia A term referring specifically to the broken masses of angular limestone and dolomite occurring in cavities within the Carboniferous Limestone of S Wales. Of uncertain age, the BRECCIA is thought to have been formed either as fault breccia at crossings of major dislocations, or as subterranean collapse structures infilling limestone caverns.

gas laws BOYLE'S LAW, CHARLES'S LAW, DALTON'S LAW.

gat 1 An inshore channel or strait dividing offshore islands from the mainland, e.g. the Frisian Is. of NW Europe. **2** An opening through a line of sea cliffs allowing access to the coast from inland.

gate 1 In topography, an opening, derived from the Anglo-Saxon word *geat*. It appears in many place-names, illustrating: (a) A break in a line of hills (GAP) where a river passes through a relatively narrow and restricted water-gap, e.g. Reigate on the R. Mole where it breaches the North Downs, or the Iron Gate on the R. Danube. (b) An entrance to a natural harbour between two promontories, e.g. the Golden Gate, San Francisco. **2** In hydrology, a barrier across a water-channel built to regulate the flow of water, i.e. a spillway gate of a dam.

gauging station A hydrological term, referring to a point in a stream channel at which a *gauge* has been installed in order to obtain a continuous record of stream-flow and discharge. *See also* STREAM-FLOW.

Gault clay A lithostratigraphic unit of the CRETACEOUS between the Lower and Upper GREENSAND.

Gaussian curve A type of frequency curve, possessing perfect symmetry about the central value and fitting many of the FREQUENCY DISTRIBUTIONS which occur most often. It is synonymous with a normal curve (NORMAL DISTRIBUTION).

Gaussian distribution GAUSSIAN CURVE.

Gauss's map projections 1 The TRANSVERSE MERCATOR PROJECTION. **2** The conformal conical projections with one and two standard parallels which, unbeknown to Gauss, had already been discovered by Lambert (LAMBERT'S PROJECTIONS).

gazetteer A catalogue of place-names and geographical features, usually with a reference to their location by COORDINATES and sometimes with descriptive information.

geanticline (geoanticline) A broad, complex upfold in the Earth's crust, generally the result of compressive folding of the sediments within a GEOSYNCLINE. Its former usage was as a synonym for ANTICLINORIUM but it is now rarely used in this context.

geest 1 Material derived from the weathering of rocks *in situ* (REGOLITH). **1** A landscape of sandy and gravelly soils in N Germany, usually mantled by a heathland vegetation, e.g. Lüneburg Heath. The majority of the sands and gravels are of Pleistocene age, being of GLACIOFLUVIAL origin.

gelifluction A term proposed in 1956 by H. Baulig to define one type of SOLIFLUCTION – that associated with frozen ground, covering both seasonal freezing in addition to PERMAFROST. Thus, it differs from CONGELIFLUCTION, which refers only to permanently frozen ground (permafrost).

gelifluction landforms Soil flow due to GELIFLUCTION can occur on slope gradients as little as 1° or 2° and can be classified as: **1** *Gelifluction sheets*, with a smooth surface and large lateral extent. **2***Gelifluction benches*, with a pronounced terrace form due to different distributions of snow moisture, bedrock and vegetation. **3** *Gelifluction lobes*, characterized by their tongue-like appearance downslope. It has been suggested that *lobes* occur on steeper slopes

(10°–20°) and *benches* on gentler slopes (5°–15°). **4** *Gelifluction streams*, characterized by very pronounced linear form downslope.

gelifraction CONGELIFRACTION.

gendarme A French term now used universally to describe a rocky pinnacle on a narrow mountain ridge (ARÊTE), forming an obstacle to climbers and walkers.

genecology The study of the genetics of populations in relation to HABITAT.

genera The plural form of GENUS.

general circulation ATMOSPHERIC CIRCULATION, FERREL CELL.

general system theory (GST) A logico-mathematical framework developed in biological science and widely adopted in Earth science and physical science to show the interrelationships of objects and ideas with functions and processes as part of an integrated SYSTEM. It outlines the general principles relating to systems, irrespective of the nature of the component elements and of the relations or forces between them. The principles are defined in mathematical language and introduce notions such as wholeness and sum, progressive mechanization, centralization, leading parts, hierarchical order, individuality, finality and equifinality. There is a flow through a system, a flow of information, matter, or energy (INPUTS AND OUTPUTS). In geomorphology and hydrology, for example, energy (which powers the system) continuously enters it from the atmosphere, both as heat and as the KINETIC ENERGY and the POTENTIAL ENERGY of rainfall. Energy entering such a system may be added to from other sources, e.g. the chemically bonded energy resulting from rock formation or the potential energy developed when a land mass is uplifted above base-level. But of all the energy entering or leaving such a system, that of the heat flux is by far the greatest. *See also* BLACK-BOX SYSTEM, CASCADING SYSTEM, CLOSED SYSTEM, CONTROL SYSTEM, CORRELATION SYSTEM, ECOSYSTEM, ENTROPY, EQUILIBRIUM, FEEDBACK CONTROL SYSTEM, GREY-BOX SYSTEM, IRREGULAR-SURFACE SYSTEM, NATURAL-EVENT SYSTEM, OPEN SYSTEM, PHYSICAL SYSTEMS, PLANE-SURFACE SYSTEM, POINT SYSTEM, PROCESS–RESPONSE SYSTEM, REGULATOR, SELF-MAINTAINING SYSTEM, STEADY-STATE EQUILIBRIUM, SYSTEMS ANALYSIS, THROUGHPUT, WHITE-BOX SYSTEM.

generating globe A term used in cartography with reference to a PERSPECTIVE PROJECTION. It defines the spherical surface upon which lie the points that are to be projected to the plane by simple geometry. The radius of the generating globe bears the same relation to that of the Earth as denoted by the representative fraction of the required map.

generic Belonging to a GENUS or class of objects. Not to be confused with GENETIC.

genetic Pertaining to a relationship due to a common origin. Not to be confused with GENERIC.

genus A group in the taxonomic classification of organisms, comprising one or more SPECIES believed to have descended from a common ancestor. The plural form of genus is *genera*, several of which make up a FAMILY. A *generic* name is italicized and always begins with a capital letter, e.g. the genus *Homo sapiens* (modern man) belongs to the family Hominidae.

geo A linear, narrow cleft in a sea cliff worn by marine erosion along a line of weakness in the rock. It originally formed as a narrow cave but various stages of roof collapse can be seen in the many examples found on the hard rock coasts of Britain, especially in Scotland.

geocentric Relating both to the centre of the Earth and to the Earth as a centre.

geochemical cycle A conceptual cycle in GEOCHEMISTRY, demonstrating the way in which an element from a primary MAGMA circulates sequentially through different geochemical environments [*88*].

geochemistry The study of the Earth's elements and its parts in order to determine the composition of the Earth and discover the laws controlling the distribution of its elements. The term was introduced in 1938 and referred initially only to the crystal

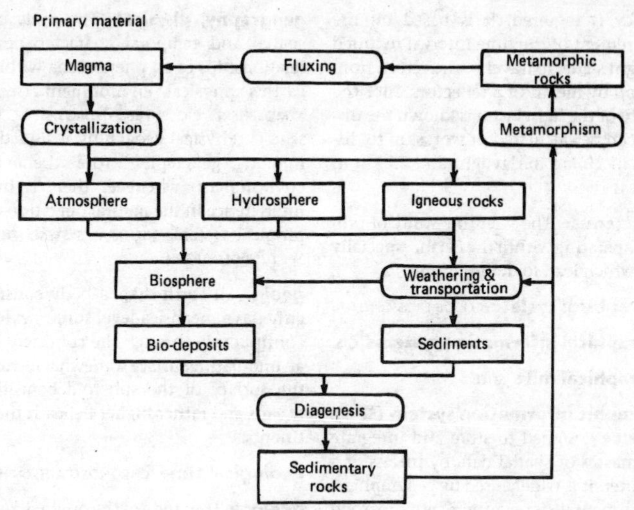

Figure 88 *Geochemical cycle*

rocks. Modern usage has extended it to include the entire Earth and all the planets and moons of the solar system, so that a truer term would be *cosmochemistry*.

geochron A time interval equivalent to a stratigraphic unit in rocks. A geochron may vary between different locations, depending on the age of the rock unit.

geochronology The study of age determination in the history of the Earth. It is divided into *absolute dating* (ABSOLUTE AGE) and *relative dating*. Absolute dating gives a more or less exact date, expressed in years before present (BP) using measurements of RADIOACTIVE DECAY (POTASSIUM–ARGON DATING) and also RADIOCARBON DATING. Relative dating simply orders events or phenomena in age of formation, without being able to put an actual date on them (*see also* CORE SAMPLING, DENDROCHRONOLOGY, FLUORINE DATING, PALAEOMAGNETISM, (counting of) VARVES).

geochronometry The measurement of geological time. This term has been replaced by that of GEOCHRONOLOGY.

geocryology The study of FROZEN GROUND (both seasonally frozen and PERMAFROST) but excluding the study of glaciers (GLACIOLOGY). Instead of using an adjectival form *geocryological*, the term PERIGLACIAL is normally preferred to describe the processes involved.

geode A hollow cavity in a rock, almost spherical in shape, in which inward-pointing crystals line its interior walls. It is also termed a *vugh*.

geodesic line A line drawn on any mathematically defined surface, representing the shortest distance between any two points on that surface. *See also* ORTHODROME.

geodesy The science related to the size and shape of the Earth, together with the determination of the exact position of particular points on its surface (*geodetic surveying*) by taking the Earth's curvature into account. It also studies the Earth's rotation (CHANDLER WOBBLE), its gravitational field (GRAVITY ANOMALY) and its tidal variations (OSCILLATORY WAVE, OSCILLATION THEORY OF TIDES). From geodetic surveying data are obtained which allow the precise location of control points in topographical survey (*see also* SURVEYING, TRIANGULATION).

geodetic surveying GEODESY.

geodimeter An instrument used in SURVEYING in which a considerable degree of accuracy is required. It is based on the measurement of the time interval required for a light signal to travel to and return from a station by means of a reflector. Since the velocity of the light beam is known the distance can be calculated. It works up to distances of 10 km in daylight and 25 km in darkness.

geodynamics The scientific study of processes operating within the Earth, especially those which lead to deformation.

geographical cycle CYCLE OF EROSION.

geographical information systems GIS.

geographical mile MILE.

geographic information system (GIS) A scheme designed to store and integrate large masses of spatial data by means of a computer. It is widely used by geographers, environmentalists, planners, etc., in such fields as environmental planning, LAND CLASSIFICATION and LANDSCAPE EVALUATION.

geographic mean The mean point of a mapped distribution.

geographics A term given to a computer print-out perspective model of the terrain produced by a GRAPH PLOTTER from grid-based map data.

geography The study of the Earth's surface, including all the phenomena which make up its physical environment (GEOGRAPHY, PHYSICAL) and its behavioural environment (*human geography*). These include its topography, oceanography, climatology, biogeography (ATMOSPHERE, BIOSPHERE, HYDROSPHERE, LITHOSPHERE), together with their related processes (*see also* GEOMORPHOLOGY, HYDROLOGY) and a consideration of their spatial and temporal distributions. Geography also includes: the study of natural resources and their exploitation; the recognition and explanation of patterns of human occupance and communication, with the study of ethnic and political distributions and their changes over time; the technology of map-making (CARTOGRAPHY); the scientific development of techniques to assist in the collection of data relating to the surface of the Earth (e.g. REMOTE SENSING).

geography, physical The study of the spatial and temporal characteristics and relationships of all phenomena within the Earth's physical environment (*see also* ATMOSPHERE, BIOSPHERE, HYDROSPHERE, LITHOSPHERE). Physical geography is linked with human geography (the behavioural environment) whenever there is human interference in the natural operation of the physical systems (GENERAL SYSTEMS THEORY). *See also* GEOGRAPHY.

geoid An Earth-shaped body considered either as a mean sea-level surface extended continuously through the continents or as an undulating surface somewhat lower than the surface of the spheroid beneath the oceans and rather higher beneath the continents.

geological time CHRONOSTRATIGRAPHY.

geology The study of the origin, structure, composition and history of the Earth, together with the processes which have led to its present state. It comprises: CRYSTALLOGRAPHY, GEOCHEMISTRY, GEOMORPHOLOGY, GEOPHYSICS, MINERALOGY, PALAEONTOLOGY, PETROLOGY, SEDIMENTOLOGY, STRATIGRAPHY, STRUCTURAL GEOLOGY.

geomagnetic equator The GREAT CIRCLE around the Earth connecting all points of zero geomagnetic latitude. It is equidistant from each of the GEOMAGNETIC POLES and its plane is at right angles to the Earth's magnetic axis. *See also* GEOMAGNETISM.

geomagnetic field The magnetic field of the Earth that causes a compass needle to align North–South.

geomagnetic poles The poles of the Earth's magnetic field, located some 6,400 km above the surface (and not, therefore, corresponding to the surface MAGNETIC POLES). They are situated above 78½°N, 69°W and 78½°S, 11°E respectively.

geomagnetism The study of the Earth's magnetic field. *See also* MAGNETIC POLES, MAGNETIC REVERSALS, MAGNETIC STRIPES.

geometric series A term used in statistics to denote a sequence of numbers in which the ratio of successive members is constant.

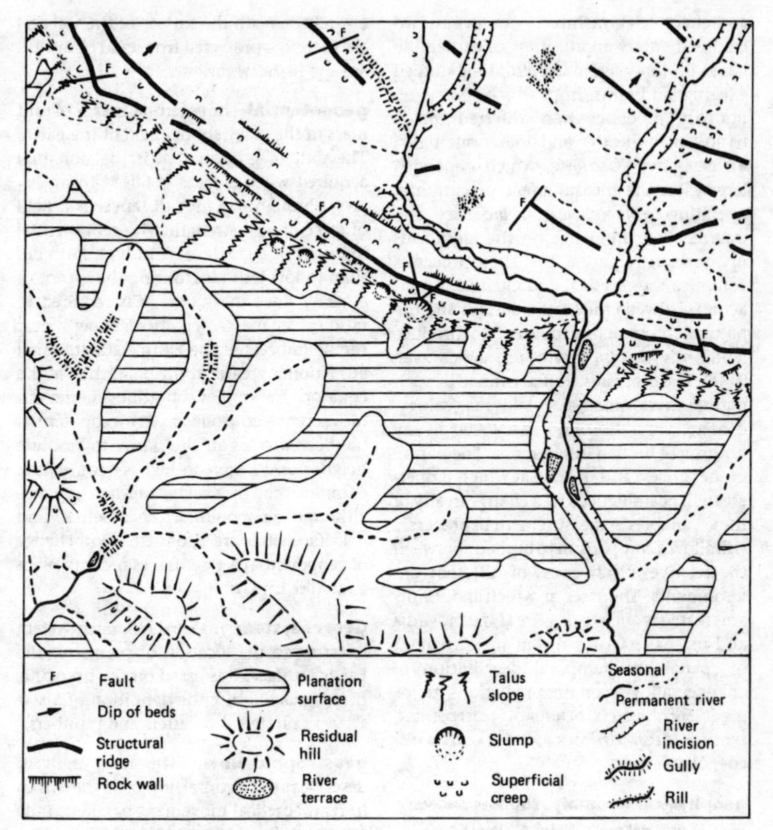

Figure 89 *Geomorphological map*

geometronics A US term used in modern cartographic techniques, involving the use of a computer and electronic devices.

geomorphic cycle CYCLE OF EROSION.

geomorphic region Any region defined solely by the nature of the geomorphic processes operative within it. *See also* GEOMORPHOLOGY.

geomorphological hazard A perceived event, either natural or human-induced, which causes a fluctuation or malfunction in the normal operation of a geomorphological process (GEOMORPHOLOGY) sufficient to pose a threat to life and property. The realization of the hazard usually leads to an extreme event, which may or may not culminate in a disaster. *See also* AVALANCHE, EARTH-FLOW, FLOOD, LANDSLIDE, MUD-FLOW, ROCK-FALL.

geomorphological map A map depicting selected terrain features which assists in understanding the evolutionary history of the landforms of an area (DENUDATION CHRONOLOGY), e.g. a map of river terraces. It should not be confused with a MORPHOLOGICAL MAP or a TOPOGRAPHIC MAP. [89]

geomorphology The scientific study of the origin of landforms based on a cause-

and-effect relationship. It comprises two complementary approaches: an earlier one based on INDUCTIVE REASONING, that studied existing landforms, from which their evolutionary processes were inferred (DENUDATION CHRONOLOGY); and a later one based on DEDUCTIVE REASONING, which depended largely on the measurement of currently operating *geomorphological processes* and inferred their influence on the landforms upon which they are acting. These processes comprise the physical and chemical interactions between the Earth's surface and the natural forces acting upon it to produce landforms (*see also* GRAVITY, ICE, RIVER, WAVES, WIND). The processes are determined by such natural environmental variables as geology, climate, vegetation and base-level, to say nothing of human interference. The nature of the process and the rate at which it operates will be influenced by a change in any of these variables. A measurement of processes will allow a comparison to be made between the relative effectiveness of different environments. The rates at which landforms are created is open to dispute (CATASTROPHISM and UNIFORMITARIANISM), but measurement of spatial and temporal distribution of changes allows certain predictions to be made. *See also* CYCLE, DENUDATION, DEPOSITION, EROSION, GEOMORPHOLOGICAL HAZARD, TRANSPORT, WEATHERING.

geophysical anomaly BOUGUER ANOMALY, GRAVITY ANOMALY, MAGNETIC ANOMALY.

geophysics 1 The study of all the relevant physical phenomena relating to the structure, physical forces and evolutionary history of the Earth, including astrophysics, oceanography, meteorology and climatology. **2** In a more restricted sense, the term applies to the study of the physics of the Earth's crust and interior. This includes GEOMAGNETISM, SEISMOLOGY and VULCANOLOGY.

geophyte One of the six major floral life-form classes recognized by the Danish botanist, Raunkiaer, based on the position of the regenerating parts in relation to the exposure of the growing bud to climatic extremes, and totally irrespective of taxonomy. The geophytes have buds which lie entirely beneath the surface of the soil and are therefore protected from cold or dry air. *See also* HEMICRYPTOPHYTE.

geopotential The POTENTIAL ENERGY of unit mass in the gravitational field of the Earth. The unit of geopotential is the potential acquired when a mass is raised from sea-level through the unit distance in a field of force of unit strength. The geopotential increases at a greater rate in high latitudes than in low latitudes, during the ascent of a body, since more energy is required to raise a given mass to a particular height near the poles because the Earth's gravitational attraction is greater at the poles than at the equator. In studies of atmospheric air-movement (CONTOUR CHART) geopotential has been used as an alternative to absolute height and meteorologists sometimes employ the *geopotential metre* (g.p.m.), although geopotential dekametres and dekakilometres are also used. A knowledge of geopotential is essential in order to place a satellite into orbit.

Georef system A global reference system referring to the location of points on the Earth's surface. It is based on the GRATICULE, thereby facilitating the definition of a MAP REFERENCE in terms of letters and numbers.

geostrophic flow The term used to describe the horizontal flow of air (i.e. wind) in its theoretical movement parallel to the ISOBARS, thus creating a *geostrophic wind*. The wind is a resultant of two opposing forces: the PRESSURE GRADIENT or pressure force in one direction and the deflecting force (CORIOLIS

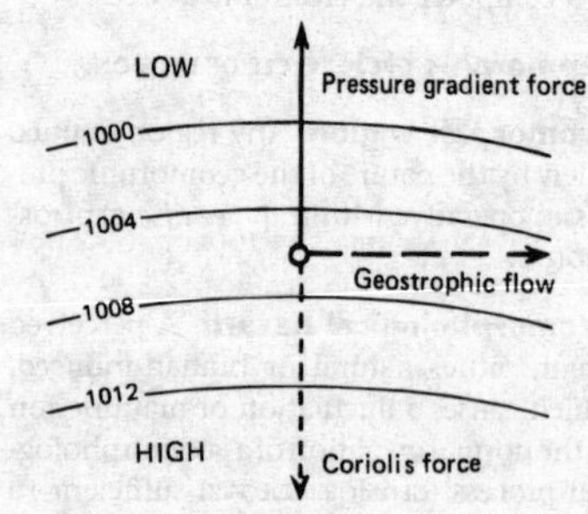

Figure 90 *Geostrophic flow*

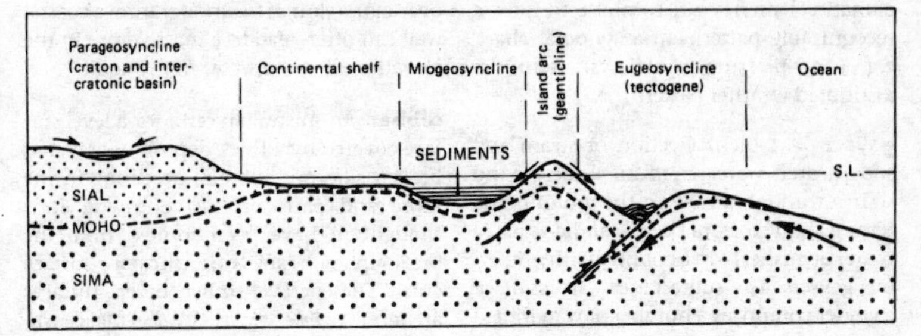

Figure 91 *Types of geosyncline*

FORCE) in the other. [*90*] Geostrophic flow is found only in the upper atmosphere, well clear of the FRICTION created at the Earth's surface, where winds blow obliquely across the isobars, not parallel with them. *See also* FERREL'S LAW, GRADIENT WIND, JET STREAM.

geosyncline (orthogeosyncline) A major structural downfold in the Earth's crust on a subcontinental scale. It comprises a linear basin which serves as a repository for considerable thicknesses of sediment derived from neighbouring land masses (CRATON). The increasing sedimentation causes progressive subsidence of the geosynclinal floor prior to strong deformation of the sediments by OROGENIC movements due to uplift of the fold-mountain mass, e.g. the sediments of the TETHYS geosyncline were subsequently folded and raised during the ALPINE OROGENY. As the fold-mountain chain is being formed there may be granite emplacement at depths and the deeper strata of the sedimentary accumulation often become highly metamorphosed (METAMORPHISM). It is synonymous with the term geotectocline. *See also* EUGEOSYNCLINE, EXOGEOSYNCLINE, MIOGEOSYNCLINE, PARAGEOSYNCLINE. [*91*]

geotechnical processes Processes induced by humans which artificially change the properties of soils.

geotectonics The scientific study of the form, structure and disposition of the rock masses which make up the crust of the Earth.

geothermal energy (heat) The energy derived from *hot rocks* in the Earth's crust, some of which is emitted at the surface as hot water or steam. Geothermal heat originates from the breakdown of radioactive elements, resulting from chemical changes taking place under intense pressure and also from rock movements. Geothermal water provides heating for more than 50% of the population of Iceland. *Dry-steam* fields are generating electricity in Italy and California, while *wet-steam* (hot brine) fields occur in California and New Zealand. *See also* ENERGY.

geothermal gradient The increase in the temperature of the Earth with progressive depth (an increase which does not occur at a constant rate). It has been calculated that the base of the SIAL has a temperature of 986° C and that temperature increases downwards from the ground surface through the sial layer at an average of 1° C for every 28.6 m, or 30.5° C for every km. At the edges of some tectonic plates, however, values of 80° C per km have been discovered and these offer sites for potential extraction of GEOTHERMAL ENERGY.

Gerlach trough An apparatus designed to measure the lowering of a slope by OVERLAND FLOW, in which sediment eroded from the slope is trapped in a trough where it is filtered out.

Gestalt concept A term of German derivation, taken from *Gestalt theory*, in which configurations or structures can be studied in relation to each other. In physical geography the Gestalt concept is generally utilized to denote the ways in which a number of climatic elements can combine to form a recognizable pattern on a synoptic chart, e.g. a low-pressure frontal system and its associated weather pattern.

geyser A violent ejection of steam and superheated water from an underground source through a hole in the ground. The term is derived from the Icelandic *geysir* (= roarer or gusher) – in a country with numerous geysers. The subterranean structure of a geyser comprises a number of water-filled chambers interconnected with a central pipe, the whole system being heated by the increasing pressure head due to the height of the column of water. High pressures force the liquid into a gaseous (steam) phase once a critical temperature threshold has been crossed and the geyser shoots violently into the air, sometimes to heights of 60 m. Some geysers exhibit a remarkably regular periodicity of ejection, e.g. Old Faithful, Yellowstone National Park, USA.

geyserite A deposit of siliceous material around GEYSERS and thermal springs. Not to be confused with TRAVERTINE, which has a different chemical composition. *See also* SINTER.

ghat A Hindi term for a gap or pass through a line of mountains. European settlers later transferred the term to the mountains themselves, e.g. Western Ghats.

Ghyben–Herzberg principle A principle defining the relationship between saltwater and freshwater in a coastal AQUIFER. Being less dense than saline water, freshwater rises above it. Beneath coastlines the interface is oblique with the freshwater near the surface extending towards the sea and the underlying saline water projecting inland. It is expressed as:

$$Zs = \frac{pf}{ps - pf}\ Zw$$

Where *Zs* is the depth below sea level to the interface between fresh and salt water; *pf* and *ps* are, respectively, the density of freshwater and saltwater; and *Zw* is the elevation of the WATER-TABLE above sea level. The principle has been used to demonstrate that over-extraction of freshwater from a coastal well can often lead to a massive rise in the elevation of saline water in the aquifer.

gibber An Australian term for a level surface covered by a thick deposit of gravel or broken siliceous pebbles, occurring in the more arid parts of the continent. It is thought to have been formed from the break-up of a siliceous (SILCRETE) surface crust. These gravel-strewn desert surfaces are termed *gibber plains*. *See also* DURICRUST.

gilgai soil An Australian term for a soil associated with an undulating microrelief of shallow depressions (1 m deep) underlain by non-calcareous soil, interspersed with 1 m high mounds or ridges (termed *puffs*) composed of calcareous soil with lime nodules. It occurs mainly in alluvial soils in regions with an annual rainfall of less than 500 mm (20 in), and is thought to be formed by differential expansion and contraction of the clay-rich soils due to changes in moisture content. Gilgais can be either microbasins and microknolls or microvalleys and microridges, with the latter type parallel to the direction of slope. *See also* VERTISOL.

gill (ghyll) A term derived from Norse and found as a place-name in N England, where it refers to a rapidly flowing mountain stream and its valley, e.g. Dungeon Ghyll in the Lake District and Gaping Gill in the N Pennines (which actually refers to the SWALLOW-HOLE itself).

gipfelflur ACCORDANT SUMMITS.

Gipping till A sheet of boulder clay overlying the LOWESTOFT TILL in parts of East Anglia, England, and thought to have been laid down by ice-sheets during the penultimate glacial stage (WOLSTONIAN) in Britain. It contains much chalky material but also some erratics from N England.

GIS GEOGRAPHIC INFORMATION SYSTEM.

glacial Strictly, the term refers only to a glacier, but apart from its many adjectival and adverbial uses the term is also used as a noun to describe a cold phase during an ice age (GLACIATION). *See also* INTERGLACIAL.

glacial-control theory A theory to explain the existence of CORAL at considerable depths below sea-level. In 1910 R. A Daly postulated that during the Quaternary ice age the level of the oceans fell owing to the abstraction of water to form the ice-sheets. Coral reefs were planed down by marine waves and all living corals were destroyed because of the cooler water temperatures. As the ice-sheets melted the theory asserts that coral began to grow again on the worn-down stumps and as the released waters returned to the oceans so the coral's upward growth kept pace with the rising sea-level (GLACIO-EUSTASY).

glacial cycle of erosion A scheme outlined in 1926 by W. Hobbs to explain the sequential destruction of uplands by the back-wearing of CIRQUE walls. It was based on glacial erosion stages occurring in a cyclic form by analogy with the CYCLE OF EROSION postulated in 1899 by W. M. Davis. He recognized: **1** the *grooved upland* of youth in which much of the pre-glacial surface remained intact; **2** the *early fretted upland* where a smooth-topped ridge is broken by well-proportioned cirques; **3** the *mature fretted upland* in which the residual summits are now HORNS and the intervening ridges ARÊTES; **4** the old-age stage is the *monumented upland* when arêtes are broken through by cols, when diffluent ice (GLACIAL DIFFLUENCE) links up the individual glaciers and the cirque floors coalesce. Some geomorphologists have pointed out that the lowering of mountains by 'glacial peneplanation' is merely a theoretical abstraction not matched in the modern landscape, but others suggest that a surface of glacially induced low relief probably exists beneath thick ice-sheets in Antarctica and Greenland.

glacial deposit An expression which refers to any type of DEBRIS or sediment which has been derived directly or indirectly from a glacier or ice-sheet. The deposit may be stratified or unstratified, homogeneous or heterogeneous. It is synonymous with DRIFT, GLACIAL.

glacial diffluence The local overriding of a pre-glacial subsidiary watershed by glacier ice so that ice flows through a col into a neighbouring valley. The erosive effect of ice on the col is such that it is lowered and the watershed is said to be *breached*. *See also* GLACIAL TRANSFLUENCE.

glacial diversion The blocking of a river drainage system by a glacier or ice-sheet, leading to the redirection of the drainage into a different valley, often by means of an OVERFLOW CHANNEL. A diversion of the lower Thames from a former channel running north-eastwards across Hertfordshire and East Anglia is thought to have taken place during an early stage of the ice age. Ice-sheets advancing southwards are said to have redirected the Thames to its present course.

glacial drainage channel A general expression which covers stream channels associated with glaciers and ice-sheets and including **1** an OVERFLOW CHANNEL and **2** a MELTWATER CHANNEL.

glacial drift DRIFT, GLACIAL.

glacial erosion The destruction of the surface by the passage of a glacier or an ice-sheet and the removal of the eroded material. *See also* ABRASION, ARÊTE, BACKWALL, CIRQUE, EXARATION, FLUTING, FRICTION CRACKS, GLACIAL CYCLE OF EROSION, HORN, ICE, PLUCKING, ROBIN EFFECT, ROCHE MOUTONNÉE, ROCK DRUMLIN, TROUGH, TRUNCATED SPUR.

glacial lake A small body of water impounded between the margin of a glacier and its valley wall or associated moraine, e.g. the Marjelen See at the edge of the Aletsch Glacier, Switzerland. Any larger sheet of water ponded by a regional ice-sheet against a topographic obstacle is termed a PRO-GLACIAL LAKE, e.g. Lake Harrison was impounded in the English Midlands by ice-sheets of Wolstonian age. A lake occupying a glacially excavated rock basin is not regarded as a glacial lake by British geomorphologists, since it is not ice-dammed. This would be a CORRIE lake or TARN. A lake lying within or upon a glacier is termed an englacial lake (ENGLACIAL).

glacial maximum **1** The largest extent achieved by ice-sheets during the PLEISTOCENE ice age. In the N hemisphere this saw the amalgamation of the Scandinavian ice-cap with those of Britain and Ireland while large areas of N America, Greenland and N Russia were also covered. On a global scale some 46 million km^2 were ice-covered, more than three times the size of modern ice-sheets. **2** At a more local level the position achieved by a valley glacier at its greatest extent.

glacial outburst A flood caused by the sudden release of an ice-dammed or englacial lake, often with disastrous consequences. It is also known as a *jökulhlaup*, an Icelandic term, where these outbursts are not uncommon. When large quantities of meltwater are released the glacier or ice-sheet margin may be temporarily lifted by flotation on the ice-dammed lake, thereby allowing the ponded waters to escape at great velocities. Alternatively, in decaying ice, honeycombed with tunnels and cavities, the lake water may find an exit through the glacier itself. Large quantities of glaciofluvial materials are carried down the valley, following the outburst, destroying or burying everything in their path and completely changing the character of the valley or the outwash plain.

glacial protectionism PROTECTIONIST HYPOTHESIS.

glacial recession A phase marked by a reduction in both the area and the thickness of a glacier or ice-sheet. It is marked by downwasting of the ice mass and probably a retreat of the ice front, although the glacier ice itself may still be in a state of forward movement. Great quantities of meltwater are produced at this time and large amounts of debris and sediment (GLACIOFLUVIAL) are transported and subsequently redeposited, often burying the DEAD-ICE FEATURES. *See also* DOWNWASTING.

glacial stairway A glaciated valley the floor of which rises upstream in a series of rock steps (RIEGEL) and alternating, glacially excavated rock basins, often infilled with lakes (PATERNOSTER LAKES). The stairway is often determined by the geological structure, with the steps being formed in the extremely hard rocks (possibly igneous intrusions) and the basins coinciding with bands or zones of more easily eroded rocks. In 1910 E. J. Garwood postulated his so-called PROTECTIONIST THEORY, in which he claimed that each valley step marked the temporary still-stand of the glacier snout, with the step itself protected by ice while downvalley the floor was lowered by the considerable body of meltwater flowing from the glacier. Modern views regard stairways as being formed partly by glacier wave functions, rather than merely by rock structures. [*178*]

glacial striae STRIATION.

glacial transfluence The overriding of a major regional pre-glacial watershed (DIVIDE) by ice-sheets on such a scale that ice overwhelms a mountain range with little regard for the pre-existing terrain. Thus, the ice-shed may not correspond with the original watershed. In Scotland, for example, the greatest ice thickness lay to the east of the main pre-glacial drainage divide, which transfluent ice breached in so many places that erratics were carried from east to west across the highland backbone. The pre-glacial Scottish drainage divide has been considerably displaced eastwards by such transfluence. *See also* GLACIAL DIFFLUENCE.

glaciation **1** In geomorphology, a term used to describe a glacial phase during an ice age when there is a marked extension of GLACIERS and ICE-SHEETS out of the Alpine and polar regions. **2** In meteorology, a term referring to the sudden change which occurs from supercooled droplets to ice crystals at the top of a developing CUMULUS cloud (ANVIL CLOUD, CUMULONIMBUS).

glaciation level (glacial limit) A statistical calculation referring to the altitude above which mountain glaciers should occur. Thus, in this hypothesis, the level rises from near sea level in high latitudes to higher altitudes in low latitudes as a func-

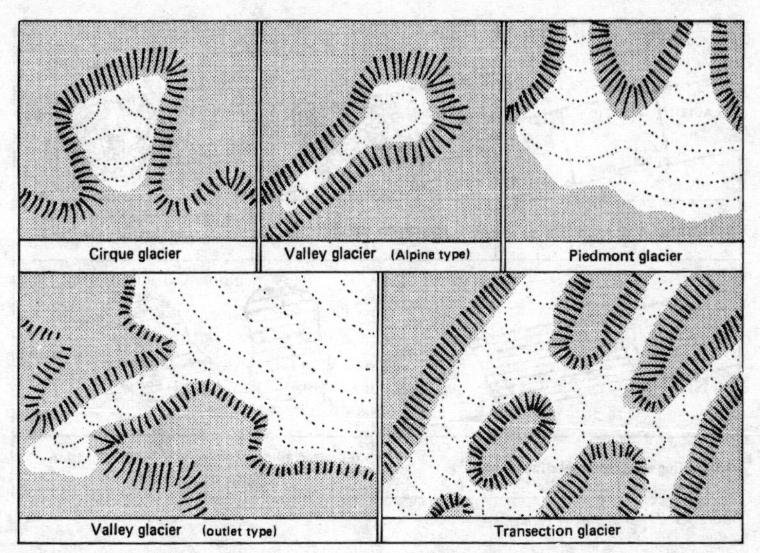

Figure 92 *Types of glacier*

tion of temperature changes. Other factors, however, will influence the actuality on individual mountains. Differences in precipitation, cloud cover, degree of CONTINENTALITY will all have effects on the level.

glacier An extensive body of land ice which exhibits evidence of downslope movement under the influence of gravity and which forms from the recrystallization of névé and firn (FIRNIFICATION). It has a variety of forms but is usually of limited width and is therefore to be distinguished from an *ice-sheet* or an ICE-CAP, both of which have larger dimensions. The term glacier is usually restricted to descriptions of ice masses confined by topographic features, e.g. *cirque glaciers, valley glaciers. See also* ABLATION, ACCUMULATION, COLD GLACIER, COMPRESSING FLOW, CREVASSE, EXTENDING FLOW, FIRN, GLACIERET, HANGING GLACIER, ICE-FALL, NÉVÉ, OGIVE, OUTLET GLACIER, PIEDMONT GLACIER, TEMPERATE GLACIER, TIDAL GLACIER, TRANSECTION GLACIER, URAL GLACIER, WALL-SIDED GLACIER. [*92*]

glacier bands A pattern of alternating colours or textures within a glacier, resulting from dirty layers or from different ice densities and structures (OGIVE). They may be seen either on the surface or in a crevasse section. *See also* DIRT BAND.

glacier burst GLACIAL OUTBURST.

glacieret A term for a tiny glacier in the western mountains of the USA.

glacier flow An expression which covers the complex mechanisms of downslope movement of an ice body. It includes: **1** *Sliding* on the rock floor (BASAL SLIP). **2** *Intergranular adjustment* of ice crystals (as in a bag of lead shot) assisted by local pressure melting and vapour transfer. This mechanism is now thought to be of minimal importance in a *glacier flow*, but a significant factor in the deformation of NÉVÉ. **3** Local and temporary *phase-changes* by pressure melting and vaporization, causing a short-distance transfer of liquid or vapour prior to recrystallization. The *phase-change* mechanism is also thought to be of no great importance. **4** *Gliding* along the glacier's slip-planes (LAMINAR FLOW). Although this mechanism does occur it is probably only on a limited scale. **5** *Intergranular yielding*

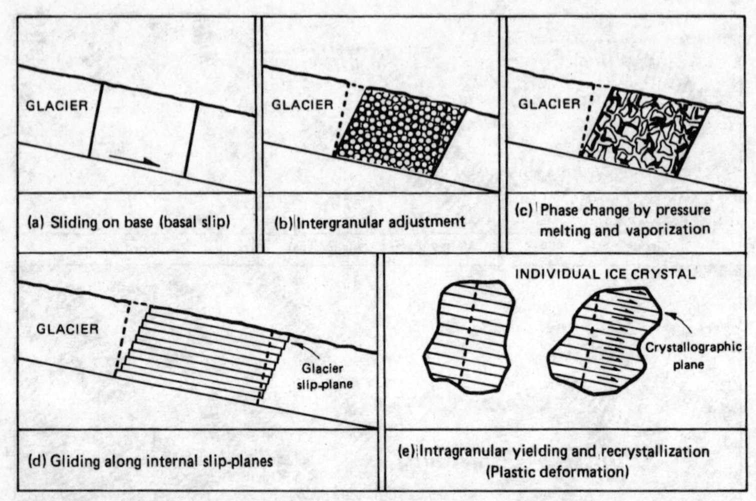

Figure 93 *Types of glacier flow*

(sometimes termed PLASTIC DEFORMATION) within the individual ice crystals by internal gliding along the crystallographic planes (not to be confused with intergranular adjustment **2** noted above). This mechanism of intermolecular movement does not destroy the coherence or solidity of the ice and does not disrupt or alter the crystal's internal atomic arrangement. It is thought that such a process may be the principal mechanism of glacier flow. [*93*] *See also* BASAL SLIP, COMPRESSING FLOW, EXTENDING FLOW, EXTRUSION FLOW, RHEIDITY, STICK SLIP.

glacier ice ICE.

glacierization The gradual advance of an existing ice-sheet or glacier over ice-free terrain. It differs from the term GLACIATION, which refers to a former glacial advance of ice-sheets that are no longer in existence. *See also* DEGLACIERIZATION.

glacier karst The irregular relief and honeycombed character of a stagnant ice mass in which features analogous to those of true limestone KARST are exhibited. It must be emphasized, however, that while the features may look the same, true karst is the result of chemical SOLUTION, while those of glacier karst are produced by meltwater operating beneath a thick cover of moraine and glaciofluvial material which prevents too rapid ablation. The various changes in thickness and character of the glacier's drift-cover lead to DIFFERENTIAL ABLATION, resulting in a variety of pseudo-karst phenomena. These include underground streams, funnel-shaped SINK-HOLES, HUMS, POLJES and UVALAS. They are best seen in the Malaspina glacier, Alaska. Glacier karst should not be confused with GLACIOKARST.

glacier milk A term given to the meltwater stream issuing from a glacier snout, which is characterized by a high percentage of light-coloured, fine clay- or silt-sized particles suspended in the water (ROCK FLOUR).

glacier mill A vertical or steeply inclined shaft or circular hole in a glacier down which meltwater pours. It may extend to the rock floor beneath the ice and it was once postulated that glacier mills (often referred to as *moulins* from the French) were responsible for the growth of POT-HOLES in the solid rock. It is now believed that the character of the glacial pot-hole and that of the plunge pool beneath a waterfall are quite different in detail, and, since the glacier mill is unlikely to stay in the same position for very long, that there is insufficient time for a pot-hole to be formed by this mechanism.

glacier regime The overall state of a glacier in relation to its MASS BALANCE, i.e. its relative gain or loss of bulk. A glacier is said to have a positive regime when *accumulation* exceeds ABLATION, thereby leading to a thickening of the ice and a glacial advance (GLACERIZATION). Conversely, an excess of ablation over accumulation will lead to a negative regime and a GLACIER RETREAT.

glacier retreat The state of a glacier when its front or snout is receding, although the body of the ice itself is still moving forward. This stage is reached when the rate of ABLATION exceeds that of ACCUMULATION, leading to a negative MASS BALANCE. *See also* GLACIER REGIME.

glacier table A large block of rock supported by a pedestal of ice on the surface of a glacier. The block has protected the underlying ice from melting by direct insolation while the surrounding glacier ice is lowered by excessive ablation. Eventually the block will collapse as the underlying pedestal melts.

glacier wind, glacier breeze A cold wind blowing down a valley glacier in daytime, especially in summer, as a result of cooling of the air in contact with the ice. This is a type of KATABATIC WIND, a term which, on a larger scale, is sometimes applied to gravity winds which descend from the high ice-sheets of Antarctica and Greenland. The maximum wind-speed occurs within 2 m of the surface and the peak velocities have been recorded at 1100 hours.

glacio-eustasy A concept referring to world-wide changes of sea-level owing to the growth of ice-sheets during ice ages. It has been calculated, for example, that the abstraction of water by the formation of the Pleistocene ice-sheets caused a fall in sea-level of some 100 metres. Conversely, the melting of ice-sheets during interglacial episodes saw sea-level rises, sometimes to elevations higher than those of present sea-level. Fluctuating sea-levels during the Pleistocene, causing rapid changes in BASE-LEVEL, had profound effects on the rates of fluvial erosion and on rates of deposition in the lower courses of rivers. It must be remembered, however, that base-level changes were not only caused by glacio-eustasy, but also by isostatic readjustments (GLACIO-ISOSTASY) which led to deformation of the Earth's crust.

glaciofluvial (glacifluvial) A term referring to the processes and the landforms related to the action of glacial meltwater. The fluvial transport of material and the associated mechanisms of erosion and deposition are similar to those of a river when the action is occurring beyond the edge of the glacier or ice-sheet. But when the processes occur within or beneath the ice body they are subject to increased pressure and velocity which allow the processes to operate more rapidly and with greater capability, e.g. the carving out of subglacial MELTWATER CHANNELS in solid rock. Because of the great volumes of water available glaciofluvial deposition frequently results in the formation of extensive outwash plains (SANDUR) and associated phenomena. The meltwater deposits are typically stratified, which distinguishes them from those depositions laid down directly by the ice (TILL). *See also* BEADED ESKER, KAME, KAME TERRACE. [*219*, *221* and *267*]

glacio-isostasy An expression referring to the deformation of the Earth's crust owing to the excessive weight of an ice-sheet. The ice-sheet acts in the same way as an accumulation of sediments in a geosyncline, thereby bringing isostatic mechanisms into play (ISOSTASY). Total subsidence of the crust beneath major ice-sheets may amount to 20–30 per cent of the maximum ice thickness and, because of the relatively slow rate of recovery (termed *recoil* of the land surface), following the melting of the ice masses, the downwarped areas may be inundated by marine transgressions in the deglacierization phase. Such inundations will lead to the formation of marine sediments and shorelines that are later to be uplifted by *isostatic* mechanisms (RAISED BEACHES). Because the recovery will be at different rates in different areas the raised shorelines will be differentially uplifted, leading to tilting and warping, e.g. the FLANDRIAN raised shorelines in Scotland.

glaciokarst Landforms of karstic character (KARST) formed by a combination of glacial processes acting upon an area of massive limestone, together with features produced by glacial deposition. The most typical landform is the ice-scoured limestone pavement where the glacier has been selective in its erosion, but greatly enlarged joint planes (producing caves and vertical shafts) and meltwater channels are also characteristics due to the large amounts of meltwater. The two best karstic regions in the British Isles, Ingleborough (England) and the Burren (Ireland), are both predominantly glaciokarstic rather than examples of true karst. *See also* FLUVIOKARST, GLACIER KARST.

glaciology The scientific study of the distribution and behaviour of snow and ice on the Earth's surface. *See also* ABLATION, ACCUMULATION, GLACIER, MASS BALANCE.

glacis A French term for a gentle slope, which when used as a prefix has several different meanings: **1** *glacis d'accumulation* = a BAHADA; **2** *glacis de dénudation* = a PEDIMENT; **3** *glacis d'ennoyage* = detritus slope in a depression; **4** *glacis d'épandage* = an ALLUVIAL CONE; **5** *glacis d'érosion* = a pediment.

glass sand A very pure silica sand extensively used in glass-making and ceramic industries.

glauconite A distinctively bluish-green mineral comprising the hydrous silicate of iron and potassium. It is thought to be indicative of a marine origin for the sedimentary rocks in which it occurs. In Britain it gives a distinctive colouring to the individual grains of the Upper and Lower Greensand in the Cretaceous succession. It should not be confused with the mineral *glaucophane*, an amphibole, commonly found in certain types of metamorphic rocks. *See also* GREEN MUD.

glazed frost, glaze A coating of clear ice which forms when rain falls from an advancing cloud layer on to a ground surface, the temperature of which remains below FREEZING-POINT. On roads this creates a particular hazard known as BLACK ICE, but the thick accumulation of ice on vegetation, telegraph wires, etc. can also cause considerable damage owing to its excessive weight, especially if there is a prolonged period of *freezing rain*. Not to be confused with RIME. *See also* FREEZING DRIZZLE, ICE STORM.

glei soils GLEY SOILS.

glen A Scottish term for a steep-sided valley in the Highlands. It is narrower than a STRATH.

Glenavy stadial An important late PLEISTOCENE cold phase in Ireland (26,000–14,000 BP). It is marked by a series of end-moraines, including the Armoy moraine (Antrim) and the Kells moraine (Meath). The latter part of the MIDLANDIAN.

Glen's law A law outlined by J. W. Glen (1955) referring to the deformation of glacier ice due to SHEAR STRESS, and fundamental to the understanding of GLACIER FLOW. It is shown as:

$$\dot{E} = Ar^n$$

where $\dot{E}$ is the STRAIN-RATE; A is a constant depending on size and orientation of ice crystals, ice temperature and impurity of the ice; r is the shear stress; n is a constant whose normal value is equal to 3. *See also* BASAL SLIP.

gleying, gleyzation, gleyification Soil processes characterized by the REDUCTION of iron from its ferric to its ferrous form, thereby producing the blue-grey (ANAEROBIC) colour of the GLEY SOILS. If the gleying is seasonal rather than permanent the periodic oxidizing conditions will give rise to MOTTLING of the soil profile (GLEY PODZOLS). The general lack of oxygen in the waterlogged soil sometimes leads to the accumulation of a peat, owing to the inability of the bacteria to break down the humus (*see also* HUMIC GLEY SOILS, STAGNOHUMIC GLEY SOILS).

gley podzols One of the subdivisions of the PODZOLIC SOILS, which are themselves part of the SOIL CLASSIFICATION OF ENGLAND AND

WALES. They are characterized by a fluctuating groundwater table or by IMPEDED DRAINAGE, which causes the MOTTLING in the gleyed grey horizon that is overlain by a dark brown or black subsurface horizon. A bleached horizon and/or a peaty topsoil may be present. [*190*]

gley soils (glei soils), gleysols One of the seven major groups in the SOIL CLASSIFICATION OF ENGLAND AND WALES. They are characteristically affected by periodic or permanent saturation by water in the absence of effective artificial drainage. Some varieties exhibit a distinct humose or peaty topsoil which overlies a grey or grey-and-brown mottled (MOTTLING) subsurface horizon altered by reduction of iron caused by the IMPEDED DRAINAGE; these are termed the HUMIC GLEY SOILS and the STAGNOHUMIC GLEY SOILS. Other varieties of gley soils are characterized by an absence of the humose or peaty topsoil; these include the *argillic* (ARGILLACEOUS) *gley soils*, the *sandy gley soils* and the *stagnogley soils* (STAGNOHUMIC GLEY SOILS).

glint-line A Norwegian term (*glint* = boundary) that has been used to describe the edge of the outcrop of the old hard rocks of the Scandinavian Shield where it is overlain by younger rocks, specifically along the Norway–Sweden border where a number of *glint-line lakes* have formed in ice-eroded valleys across the glint-line itself. The term has been adopted universally to describe the edges of ancient SHIELDS (e.g. the Canadian Shield).

global sea-level observing system (GLOSS) A worldwide network of sea-level gauges in order to monitor SEA-LEVEL CHANGE. *See also* GARP.

global warming A claim by the majority of climatologists that the mean temperature of the Earth's atmosphere is rising and will continue to rise as a result of the GREENHOUSE EFFECT. During the last hundred years it has been calculated that global mean temperatures have increased by 0.3–0.6° C and that this rise will accelerate unless emissions from the burning of FOSSIL FUELS is drastically reduced.

globe A map of a heavenly body, usually the Earth, as depicted on a sphere.

globigerina ooze A calcareous marine deposit largely comprising the shells of tiny FORAMINIFERA, especially the *Globigerina species*, in which the calcium carbonate content ranges from 30% to 97%. It occurs over 130 million km^2 of the deep ocean floor at depths of about 4,000 m, but not in the ocean trenches. *See also* PELAGIC, RADIOLARIAN OOZE.

globular projections A term used in cartography to describe those hemispheric GRATICULES which are enclosed by a circle, thereby resembling an AZIMUTHAL PROJECTION, but which do not represent the transverse aspects of the latter projection. The projections are neither conformal nor equal-area. In atlases, the globular projection is commonly used in pairs to illustrate each hemisphere.

Gloger's rule This states that the pigmentation of warm-blooded animals tends to decrease away from equatorial latitudes in line with the decrease in mean annual temperatures towards higher latitudes.

GLOSS GLOBAL SEA-LEVEL OBSERVING SYSTEM.

gloup A Scottish name for a coastal BLOW-HOLE.

glowing avalanche A type of volcanic eruption similar to a NUEÉ ARDENTE but which is propelled more by gravity than by the force of the explosion.

glowing cloud NUEÉ ARDENTE.

gnamma hole An Australian term for a rock hole found at the base or on the flanks of a bare granite dome in an arid environment. The holes vary in depth from a few cm to 2 m and in diameter from 50 cm to 3 m. It is thought that they are formed from differential weathering of the granite (EXFOLIATION), although some of them may have been artificially deepened by human activity, for they provide important tanks in which rainwater survives.

gneiss A coarse-grained, banded, crystalline rock resulting from high-grade regional METAMORPHISM. Most gneisses comprise bands of granular quartz and feldspar, which alternate fairly regularly with thinner schistose bands comprising micas and amphiboles. A distinction is made between *paragneiss*, produced by magmatic alteration of sedimentary rocks, and *orthogneiss* which is formed by the alteration of igneous rocks. In some gneisses the granular bands swell out locally into *eyes* (German, *augen* = eyes) of very coarse porphyritic quartz or feldspar crystals, and these are termed *augengneisses*. The adjectival term *gneissose* is sometimes used for flow-banded granites, but all gradations between the latter and a true banded gneiss can be observed where GRANITIZATION has occurred. The term gneiss is also used in a stratigraphical sense, e.g. the Lewisian Gneiss.

gnomonic projection A MAP PROJECTION of the AZIMUTHAL type produced by projecting from the centre of the generating globe on a tangent plane. It has the characteristic property that the arcs of all the great circles are represented by straight lines, and is therefore useful for navigational purposes and for seismic work (because seismic waves travel more or less in the direction of great circles). But because its scale increases away from the centre the gnomonic projection is of little practical use equatorward of 45° from the poles. [*94*]

gold A rare and therefore precious metal formed by HYDROTHERMAL ACTIVITY. It occurs naturally in veins where it is commonly alloyed with silver, but after erosion and transport it can be redeposited as a PLACER DEPOSIT or a BANKET.

Goletz terraces Terraces formed on hill slopes by a combination of FREEZE–THAW ACTION and SOLIFLUCTION under PERIGLACIAL conditions. They may be several kilometres in breadth or length and may be bounded by stones or by vegetation (STONE-BANKED TERRACE, TURF-BANKED TERRACE). *See also* ALTIPLANATION, CRYOPLANATION.

Gondwanaland The name given by geologists to the super-continent thought to have existed in the S hemisphere over 200 million years ago, and formerly part of the even larger Pre-Cambrian continent of PANGAEA. It was suggested by a German scientist, A. Wegener, that the southern continent of Gondwanaland was separated from the northern continent of LAURASIA by the long narrow ocean of TETHYS. Gondwanaland is said to have comprised Antarctica, Australia, and parts of S America, Africa and India (from which it derives its title from the Gondwana system of India). The hypothetical break-up of this super-continent forms the basis of the theory of CONTINENTAL

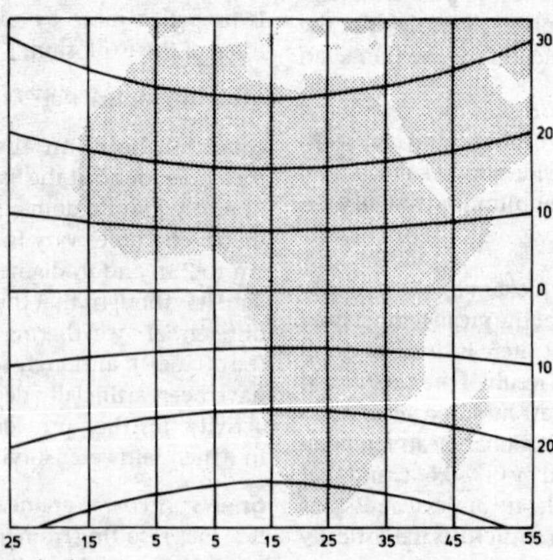

Figure 94 *Gnomonic projection*

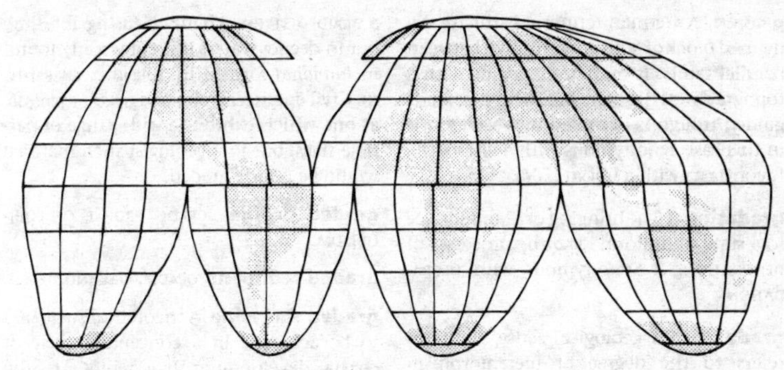

Figure 95 *Goode's projection*

DRIFT, albeit these ideas have largely been superseded by the concept of PLATE TECTONICS.

Goode's interrupted homolosine projection An equal-area MAP PROJECTION in which the oceans are interrupted in order to allow the continents to retain a reasonable shape by recentring them on chosen meridians which are not drawn as smooth curves. The parallels of latitude appear as straight lines but the size of the oceans, especially the Atlantic Ocean, is incorrect. [95]

Gorczynski's index of continentality An empirical attempt to define the spatial limits of a CONTINENTAL CLIMATE. The Gorczynski continentality index $K = 1.7(A/\sin\theta) - 20.4$, where A is the annual temperature range (°C) and θ is the latitude angle. K ranges between –12 at extreme oceanic stations and 100 at extreme continental stations. To illustrate the growing continentality as one progresses eastwards into Europe the following stations can be taken as examples of Gorczynski's index: London, $K = 10$; Berlin, $K = 21$; Moscow, $K = 39$. CONTINENTAL CLIMATE, BERG'S INDEX OF CONTINENTALITY, CONRAD'S FORMULA OF CONTINENTALITY.

gorge A deep and narrow chasm with precipitous rocky walls, currently occupied or formerly occupied by a river (e.g. Gordale Scar, in the Pennines of N England, was once followed by an underground stream prior to the collapse of the cavern roof). It is more steep-sided and enclosed than a *ravine*, but is of the same magnitude. Any river valley of broad floodplain and gentle bluffs may pass into a gorge as it is traced up- or downstream, possibly owing to the presence of a belt of hard rocks, e.g. the Avon Gorge at Bristol. It is noteworthy that the term has been used to describe certain river valleys that do not exhibit very precipitous walls, e.g. the Severn Gorge at Ironbridge, England.

gossan A Cornish term for a hydrated oxide of iron from which the sulphides of copper and the sulphur have been leached away. It is left as a ferruginous residual mass filling the upper part of a mineral vein.

Götiglacial A cold stage later than the DANIGLACIAL in the Pleistocene glacial sequence of Scandinavia, referring to the time period between approximately 17,000 and 10,000 BP, after which the rising sea-level converted the Baltic Ice Lake into the Yoldia Sea – a forerunner of the present-day Baltic Sea.

gouffre A large pipe or shaft in an area of limestone KARST. *See also* ABÎME.

gouge 1 The fine material occurring along the plane of a fault, alongside the FAULT-BRECCIA. **2** The soft, easily worked material lining the walls of a mineral vein. It enables the vein to be exposed fairly rapidly once the material has been gouged out by the miner.

graben A German term for a structurally defined block of land downthrown between parallel faults. It may not coincide with a topographical trough, but where a fault-guided trough is occupied by a valley it is in this case synonymous with a RIFT VALLEY. It contrasts with a HORST. [*208*]

gradation 1 The bringing of a land surface to a state of uniform GRADE by processes of DENUDATION. **2** Synonymous with DEGRADATION.

grade 1 In a geological sense, the term refers to the degree of metamorphism experienced by rocks. *High-grade metamorphism* is achieved by high temperature or high-pressure conditions, while *low-grade metamorphism* is achieved by low temperatures and pressures. **2** An expression referring to the relative quality of the metallic content of an ore. **3** In the USA, a measure of slope steepness, expressed as a percentage (GRADIENT). **4** In geomorphology, a concept of EQUILIBRIUM in relation to a slope or a river, in which they are in a state just capable of maintaining a balance between erosion, transport and deposition. In the NORMAL CYCLE OF EROSION, proposed by W. M. Davis, it was suggested that a smooth, concave long profile (a graded curve) was the manifestation of a state of equilibrium in a river, which was then said to be a *graded river*. It is now known that to produce a state of grade it is not necessary to have a smooth profile and that a graded state may be achieved by a river at any stage of its development, not simply in the later stages as suggested by Davis. A graded river maintains its COMPETENCE by changing its CHANNEL GEOMETRY, the roughness of its bed and the nature of its load, at a given rate of discharge. **5** The division of a circle into 400 equal angles. 400 grades = 360°.

graded bedding A type of sedimentary bedding that exhibits a sorting effect based on a size distribution of particles (PARTICLE SIZE). The coarsest material occurs at the base and the sediment gradually grades upwards to the finest material at the top, e.g. pebbles grading upwards into a coarse sand. This type of sedimentary unit may be formed in a BRAIDED STREAM channel during infilling, but in deeper waters it is commonly found in *turbidites* formed by TURBIDITY CURRENTS. NB Civil engineers refer to a *graded aggregate* as one which exhibits a wide range of particle sizes, but to a geologist such material would be poorly graded.

graded profile GRADE, PROFILE OF EQUILIBRIUM.

graded sediment GRADED BEDDING.

graded shoreline A theoretical stage said to be achieved in a conceptual cycle of coastal development. It is represented by the cutting back of headlands by marine erosion, complemented by the enclosing of bays by the building of BARS across the entrance, thereby leading to the straightening of an irregular coastline.

graded slope A slope exhibiting a continuous cover of REGOLITH, thereby mantling any rocky outcrops. This slope has achieved the ANGLE OF REPOSE, where '. . . the ability of the transporting forces to do work is equal to the work they have to do' (W. M. Davis, 1899). *See also* GRADE, STABLE SLOPE, STATIC SLOPE.

graded stream GRADE.

graded time The time-scale necessary for a stream or a slope to achieve a graded profile (GRADE). It is one of the three categories in the time-scale of GEOMORPHOLOGY, devised by S. A. Schumm and R. W. Lichty. *See also* CYCLIC TIME, STEADY TIME.

gradient 1 The steepness of a slope expressed in degrees, percentages, or as a ratio. The latter is expressed as a proportion between the vertical interval (*VI*) and its horizontal equivalent (*HE*). E.g. a 3° slope = a 5% slope = a 1 in 20 slope. **2** A change or variation in the value of a variable, especially those relating to climate, e.g. temperature gradient. *See also* BAROMETRIC GRADIENT.

gradient wind The wind which results from a balance between the CORIOLIS FORCE and the centrifugal force due to any curvature displayed by the ISOBARS. Strictly speaking a gradient wind can only exist in the upper atmosphere, for nearer the Earth's surface this balance is affected by frictional forces. *See also* GEOSTROPHIC FLOW.

grading The determination of PARTICLE SIZE by sorting of clastic sedimentary material. It can be achieved by MECHANICAL ANALYSIS and plotted on a GRADING CURVE or referred to by means of a GRADING FACTOR.

grading curve A curve on which the PARTICLE SIZE of a sample of soil or sediment is plotted on a horizontal, logarithmic scale and percentages are plotted on a vertical, arithmetic scale, during MECHANICAL ANALYSIS. Any point on the curve indicates what percentage by weight of the particles in each sample is smaller in size than the given point. It is calculated by determining the percentage of the residual material retained on each of a series of standard sieves, each mesh opening being half the size of the preceding one. The sieve series is as follows: 38 mm; 19 mm; 9.5 mm; 4.7 mm; 2.4 mm; 1.2 mm; 0.6 mm; 0.3 mm; 0.15 mm. The grading curve is sometimes referred to as the FINENESS MODULUS.

grading factor The coefficient of sorting of a CLASTIC SEDIMENT. Perfect sorting is equivalent to a grading factor of 1.0.

gradualism A term given to a scientific concept which sees the evolution of species taking place gradually. Its full title is the *theory of phyletic gradualism* which states that: **1** New species arise by the transformation of an ancestral population into modified descendants. **2** The transformation is even and slow. **3** The transformation involves large numbers, usually the entire ancestral population. **4** The transformation occurs over all or large parts of the ancestral species' geographic range. Gradualism contrasts diametrically with the theory of PUNCTUATED EQUILIBRIUM. *See also* UNIFORMITARIANISM.

gradually varied flow A hydrological term, suggesting that in most river channels the stream surface is not parallel to the bed or the ENERGY GRADE LINE because of the variations in the slope of the bed and the cross-sections of the channel. *See also* BERNOUILLI'S THEOREM.

grain 1 A description of the textural coarseness of a rock, i.e. coarse-*grained* or fine-*grained*. **2** A direction of splitting in a rock. **3** The structural or topographical trend of a region, e.g. the trend of the Armorican structures in S Ireland gives an approximate E–W grain to the terrain of the region. **4** A unit of weight equal to 0.0648g – the grain (gr) is the smallest unit of mass in the English system.

grain size The size of mineral particles which make up an IGNEOUS ROCK. The grain size is always determined on the ground mass, with PHENOCRYSTS being ignored. The following classification is generally used: >3 cm = very coarse; 5 mm to 3 cm = coarse; 1–5 mm = medium; <1 mm = fine; zero = glassy. *See also* PARTICLE SIZE.

gram, gramme A unit of weight in the metric system equal to a thousandth part of a cubic decimetre of water. It was replaced by the *kilogram* in the 19th cent. (1,000 g = 1 kg).

granite An igneous rock of coarse grain and PLUTONIC origins, consisting essentially of quartz (20–40%) together with alkaline and plagioclase feldspars, and very commonly a mica. This light-coloured rock is one of the commonest of the igneous rocks but there are many varieties, their classification depending upon grain size and mineral composition. One of the finest-grained is termed *aplite*, while the coarsest is a *pegmatite*. Where mica predominates it becomes a *biotite* or *muscovite* granite. If the content includes hornblende or pyroxene it is a hornblende or pyroxene granite. When alkali and plagioclase feldspars are about equal the rock is termed an *adamellite* (e.g. Shap granite, Cumbria), but if plagioclase dominates it is a *quartz diorite* or *grandiorite*. The largest masses of granite form as BATHOLITHS, but it may occur in a variety of intrusive forms (BOSS, DYKE, PLUG, RING COMPLEX, SILL) as well as in CAULDRON SUBSIDENCE. Granite is the plutonic equivalent of the volcanic rock RHYOLITE. Granites are formed in various ways: **1** by cooling at deep crustal levels of *granodioritic magma*; **2** by more rapid cooling of granitic liquids which rise up into the crust where DIFFERENTIATION occurs and more chemically extreme varieties are produced; **3** by GRANITIZATION. Because of their resistance to erosion granites usually, but not invariably, form hills, mountains and INSELBERGS, while their hardness makes them an excellent building stone. *See also* MICROGRANITE.

granitization A term for the processes which are thought to convert pre-existing rocks into *granite* by the action of *granitic fluids* (ICHOR) rising from deep in the crust, a process termed METASOMATISM. There is no close agreement among geologists about the source of these fluids or about the precise ways in which the mechanisms operate.

granophyre A MICROGRANITE with a GRAPHIC TEXTURE.

granular disintegration One of the main types of MECHANICAL WEATHERING of rocks, caused either by the freezing of pore water and the dislodgement of particles, or by differential expansion and contraction due to INSOLATION, leading to disintegration of the grains. This break-up of a rock can also be caused by CHEMICAL WEATHERING at depth, but there is a school of thought that believes the expression granular disintegration should be confined only to subaerial weathering processes.

granule 1 A spheroidal soil aggregate [*225*]. **2** A fragment of rock, the PARTICLE SIZE of which is between 2 mm and 4 mm diameter. It is smaller than a pebble but larger than a sand grain.

graph 1 A plane surface depicting points and lines in their correct relative positions in relation to COORDINATES. A graph is a visual representation of data in the form of a continuous curve, usually on squared paper (*graph paper*). A line on a graph is always referred to as a curve, even though it may be straight. **2** A term referring to a two-dimensional network in *topology*, in which the arrangement, the proportions and the number of nodes and arcs are studied (*graph theory*). The laws of MORPHOMETRY are concerned with these relationships and are used particularly in the study of river-drainage networks and in studies of the movement of any type of mass (load) across a surface or along a route (channel), e.g. MASS-MOVEMENT. *Graph theory* also assists in the study of networks in other, non-geomorphological, aspects of physical geography, such as circulatory patterns in the oceans and atmosphere (GEOSTROPHIC FLOW) and in the movements of animals in BIOGEOGRAPHY.

graphicacy A term referring to the art of depicting spatial relationships between objects by means of maps, diagrams, photographs and charts. It is analogous with literacy and numeracy. *See also* CARTOGRAPHY.

graphic texture A rock texture, especially in granite, resulting from the regular intergrowth of alkali feldspar and quartz. The textural pattern resembles runic or cuneiform writing or hieroglyphics.

graphite A soft, grey, laminar form of pure carbon, commonly referred to as *black lead*. Its carbon atoms are tightly bonded in sheets but the sheets themselves comprise layers which are free to slide, hence graphite has important lubricating properties. Because of its very high melting-point (3,500° C), excellent mechanical strength and very good thermal and electrical conductivity it has many industrial uses, particularly in thermal nuclear reactors.

graph plotter A standard output device, linked to a computer, drawing lines on a map by moving an automatically guided pen over a flat surface through a number of recorded coordinates which represent GRID REFERENCES. More sophisticated plotters are termed *vector plotters* and those which give the most accurate resolution are *flat-bed electromechanical plotters*. The actual process of recording is known as *digitization*.

grassland One of the four main subdivisions of world vegetation (BIOCHORE). It is characterized by limited precipitation, insufficient for tree growth, and a season of drought. Several types may be recognized: **1** *tropical grassland*; **2** *temperate grassland* (DOWNS, PAMPAS, PRAIRIE, STEPPES, VELD); **3** *montane grassland*. All of these are natural grasslands, but in certain parts of the world grassland has been artificially introduced by human activity as part of a pastoral farming economy in areas that were originally forested, e.g. Britain and New Zealand. More than half of England and over three-quarters of Wales, Scotland and Ireland is under grassland, although these vary between the rich chalk and limestone grasslands of S England and the poorer acid mountain grasslands of Wales and Scotland. *See also* SAVANNA.

graticule An intersecting system of lines drawn on a map to represent MERIDIANS and PARALLELS OF LATITUDE. *Graticule intersections* are the small crosses depicted on the face of a map indicating where the meridians and parallels intersect.

graupel A word of German derivation, meaning soft HAIL, that can also be applied to pellets of partly melted and refrozen snow, each with a soft core, rarely exceeding 5 mm (0.2 in) in diameter.

gravel 1 A general term for an accumulation of loosely compacted coarse, stony material. **2** In mineral resources, a term for unconsolidated deposits of fluvially or glaciOfluvially derived water-worn stones capable of economic exploitation. **3** In the British Standards Classification of PARTICLE SIZE it has a size range between 2 and 60 mm, i.e. larger than coarse sand but smaller than a COBBLE. **4** In the WENTWORTH SCALE it has a size range between 2 and 4 mm, i.e. larger than sand but smaller than a PEBBLE. **5** One definition in the USA regards gravel as all rock fragments retained on a No. 4 sieve, i.e. in the size range 4.76 m to 76 mm. *See also* PEBBLE GRAVEL.

gravel train A synonym for a VALLEY TRAIN. It was a term used to describe the gravel deposits which are found in the higher terraces of the R. Thames where it enters the London Basin in S England. The Thames gravel train, occurring at some 70 to 100 m above present river levels, is thought to be older than the CROMERIAN interglacial. It has been subdivided into a *Higher* and *Lower Gravel Train.*

gravimeter 1 An instrument for determining specific gravity, especially of liquids. **2** A synonym for gravity-meter, an instrument for measuring variations in the gravitational field of the Earth.

gravimetric method A technique to measure SOIL WATER, involving the weighing, oven drying and re-weighing of a soil sample. The moisture content is calculated by expressing the loss of weight as a percentage of the oven-dry weight.

gravimetric survey A geophysical technique (GEOPHYSICS) for determining rock structures at depth. It is based on the variation in the gravitational field of the Earth due to density differences in the crustal rocks. *See also* MAGNETIC SURVEY, SEISMIC SURVEY.

gravitational constant The constant (G) in the law of universal gravitation. It is expressed as: 6.670×10^{-11} N m^2 kg^{-2}.

gravitational gliding The downslope sliding of rock strata resulting from tectonic movements (e.g. nappe formation) that produce uplift. The result of gravity tectonics. *See also* DÉCOLLEMENT.

gravitational water SOIL WATER.

gravity The force exerted by the Earth and its rotation on a unit mass. Newton's law of gravity (1686) states that each body in the universe attracts every other body with a force directly proportional to the product of their masses and inversely proportional to the square of the distance between them, measured from their centres of mass along a line joining these centres. It is the acceleration (g) of a body falling freely in a vacuum in the gravitational field of the Earth. Values differ according to the distance from the centre of mass of the Earth but the standard value is 9.80665 m s^{-2}. It is also expressed in newtons. A NEWTON (N) is the force required to accelerate one kilogram by one metre per second per second.

gravity anomaly An expression used in GEOPHYSICS referring to the difference between a theoretical computed gravity value and an observed gravity value. Excess observed gravity is a *positive anomaly* while

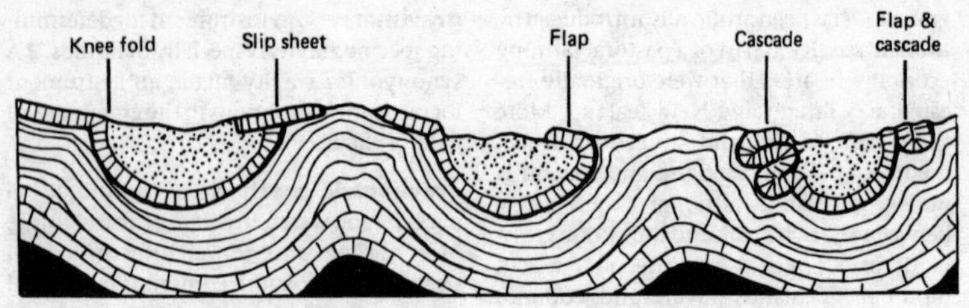

Figure 96 *Gravity-collapse structures*

a deficiency is a *negative anomaly*. *See also* BOUGUER ANOMALY, FREE-AIR ANOMALY, ISOSTASY.

gravity-collapse structure A tectonic structure developed in folded rocks where erosion of INCOMPETENT strata allows more COMPETENT strata to slide or collapse under gravity, thereby creating contorted folds. *See also* CASCADE, FLAP. [*96*]

gravity flow A type of glacier movement first described in 1751 by J. G. Altmann, who realized that gravity was the cause of glacier movement on a sloping surface. *See also* GLACIER FLOW.

gravity movement MASS-MOVEMENT.

gravity slope A slope lying at the angle of rest of the eroded material. Synonymous with CONSTANT SLOPE. [*45*]

Great Basin High A semi-permanent subtropical high-pressure cell located over the Sonoran or SW region of the USA. It is usually replaced by a thermal low-pressure cell in summer.

great circle A circle on the Earth's surface, the plane of which passes through the centre of the Earth. Thus, the shortest distance between any two surface points represents an arc of a great circle. Airlines are conscious of this distance minimization when they plan their long-distance routes,

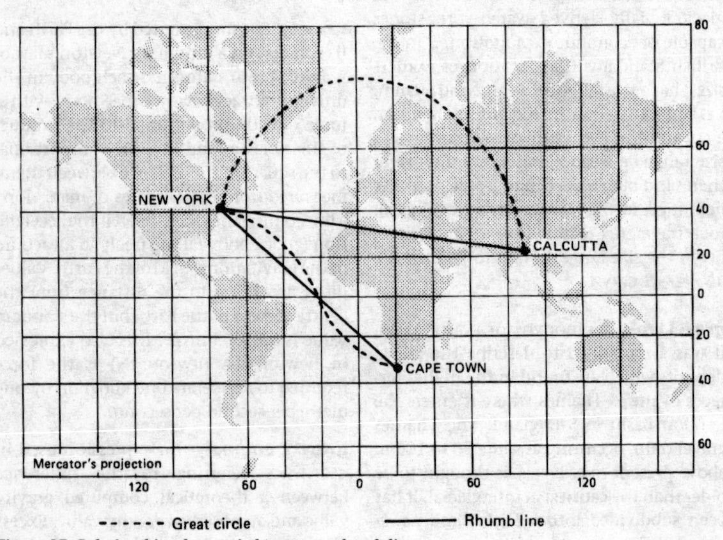

Figure 97 *Relationship of great-circle routes to rhumb lines*

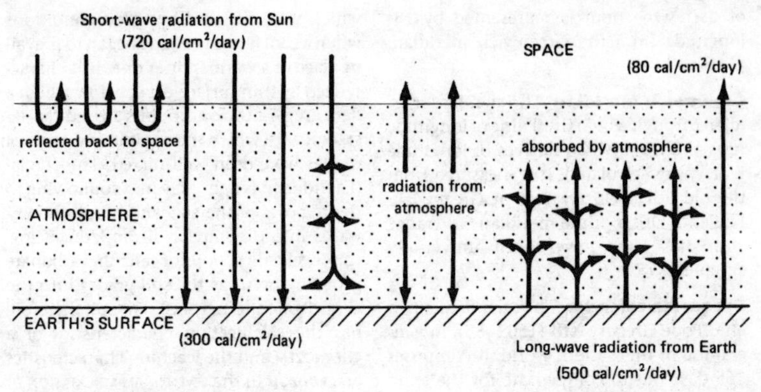

Figure 98 *Greenhouse effect*

so that a flight from London to California passes over Greenland (one of the so-called 'polar' routes), following virtually a *great circle route*. But to avoid a large number of navigational changes of direction long-distance aircraft normally follow RHUMB LINES. [*97*] On a GNOMONIC PROJECTION all great circles are shown as straight lines.

Great Interglacial HOXNIAN.

green flash A momentary optical phenomenon which sometimes accompanies the exact instant of sunrise or sunset. It results from the greater refractive properties of the green waveband (0.5 μm) in the spectrum of light.

greenhouse effect (atmospheric effect) An expression referring to the ability of glass to allow incoming short-wave solar RADIATION to pass easily into a greenhouse, but which blocks some of the reradiated outgoing long-wave terrestrial radiation. The process has been used to describe how short-wave radiation passes easily through the atmosphere to reach the Earth's surface whereas outgoing radiation is absorbed and reradiated by water vapour, droplets and carbon dioxide. It has been shown, however, that since higher temperatures in a greenhouse are due partly to decreased TURBULENCE the analogy is largely erroneous. Nevertheless, it is known that the Earth and the atmosphere retain some of the heat and it has been argued by some environmentalists that by continuing to pollute the atmosphere, largely by releasing carbon dioxide after combustion of fossil fuels, modern man is building an increasingly effective greenhouse effect which will affect global temperatures. For example, since 1900 fossil-fuel combustion has raised the atmospheric carbon dioxide by 10%, thereby increasing global temperature by 0.2° C. The current annual increase of carbon dioxide is 0.4%. *See also* ABSORPTION, ATMOSPHERIC WINDOW. [*98*]

green mud A fine-grained sea-floor deposit, occurring at depths of 150–3,000 m on the CONTINENTAL SLOPE. Its colour is derived from a high proportion of GLAUCONITE, but calcium carbonate is also present in high quantities.

greensand **1** A sand or sandstone with high proportions of GLAUCONITE. **2** A lithostratigraphic unit of the CRETACEOUS (Upper and Lower Greensand).

greenstone A term used by early geologists when mapping slightly metamorphosed basic igneous rocks, e.g. EPIDIORITE.

Greenwich mean time (GMT) The local time at Greenwich, England, located on the 0° meridian (GREENWICH MERIDIAN). It is the basis on which the standard times for virtually all other countries are calculated, at hourly or half-hourly intervals fast or slow of GMT. One hour is represented by 15° longitude from the Greenwich meridian. [*241*]

Greenwich meridian The standard meridian through which the 0° line of longitude runs and from which degrees of longitude E or W are calculated. It is measured from the old Royal Observatory at Greenwich, England. Other prime meridians in France and USA have never been universally accepted. [*176*]

Gregorian calendar The calendar named after Pope Gregory XIII (1502–85), in general use in the western world. It comprises 365 days but to compensate for the extra fraction of a day each year, every fourth year has 366 days (leap year). This allows the calendar to approximate closely to the TROPICAL YEAR.

greisen A mica-rich rock, altered by PNEUMATOLYSIS in which fluorine-rich vapours play an important part, commonly associated with kaolinization (KAOLIN).

grey-box system A SYSTEM the structure of which is only partly known, in contrast to the BLACK-BOX SYSTEM and the WHITE-BOX SYSTEM.

grey-brown podzol A soil type intermediate between BROWN EARTHS and true PODZOLS. It is characterized by a dark surface (MULL), a light-coloured eluvial (Ea) horizon, a brownish illuvial (B) horizon and a basic or calcareous C horizon. It develops in a temperate climate of moderate rainfall but it has a greater organic content and is less leached than a true podzol. It develops largely under woodland but forms the basis of good pastureland where deforestation has taken place. In N America it is sometimes referred to as a *grey-brown luvisol*, but is classified as one of the ALFISOLS in the SEVENTH APPROXIMATION of soil classification.

grey desert soil SIEROZEM (SEROZEM).

grey earth SIEROZEM (SEROZEM).

grey forest soil A soil occurring in a transition zone between forest and steppe (prairie) in Russia and N America, probably originating under grassland conditions which were replaced by forest conditions when a cool humid climate began to prevail in place of a warmer drier one. It is characterized by thin surface LITTER below which a dark grey A-horizon exhibits a good CRUMB-STRUCTURE in which the organic and mineral matter have been well mixed. The eluvial (Ea) horizon is light grey in colour owing to the removal of the clay and humus downwards into the illuvial (B) horizon. Below this level CONCRETIONS of calcium carbonate generally occur as the soil grades into the C-horizon (often a LOESS). The soil therefore has the calcification characteristics of a CHERNOZEM and the leaching characteristics of a PODZOL. In the SEVENTH APPROXIMATION of soil classification it is classified as an ALFISOL.

greywacke An outmoded term for a poorly sorted, fine-to-coarse sandstone or gritstone (ARENACEOUS rocks). It is extremely resistant because of its tough angular quartz particles and its strong argillaceous cement. Greywackes are commonly associated with geosynclinal sedimentation (GEOSYNCLINE). The term TURBIDITE is now more common.

greywether An isolated surface block lying on the downlands of S England and named from its resemblance to a sheep. Synonymous with SARSEN.

grèzes litées A French term for alternating layers of angular stones and finer material occurring in a semi-stratified form on hill slopes. They are thought to be the product of alternating frost-action and solifluction under former PERIGLACIAL conditions during Pleistocene times. *See also* HEAD.

grid **1** A network of parallel lines intersecting at right angles and producing a series of identical squares on a map. *Grid lines* are numbered both east and north of an ORIGIN, thereby giving each square and any point within that square a unique location of EASTINGS and NORTHINGS. Such a location is known as a *grid reference*. Grids differ in various countries. In the UK the NATIONAL GRID is used; in USA each state has one or more grids of 1,000-ft squares based on a LAMBERT'S conformal projection or a TRANSVERSE MERCATOR PROJECTION. **2** A uniformly drawn network on which data are added in order to carry out an exercise in spatial analysis. **3** A colloquial name for the electricity transmission system. [*160*]

grid north The meridian of origin of a GRID system. The direction of all the south–north grid lines on a map are coincident with the meridian of origin, but only the latter coincides with TRUE NORTH, the remainder diverging from it by different amounts, e.g. in the British NATIONAL GRID, grid north and true north coincide at 2°W, but Land's End diverges by 2°50'.

grid reference GRID.

grid square RASTER.

grike (gryke) A deep cleft or fissure formed by solution along a line of weakness, i.e. a joint, in a limestone. It is bounded by a CLINT. In the British Isles its typical measurements are between 15 cm and 60 cm in width and between 0.5 m and 5 m in depth. Where grikes intersect rounded hollows are formed, termed *lapiés wells* or KARRENROHREN. Because the grikes appear as cleft-like ruts across limestone pavements they have been termed *kluftkarren* in German. *See also* KARREN.

gritstone, grit **1** A coarse sandstone in which the particle shape is angular to subangular (ARENACEOUS). **2** A lithostratigraphic name – the *Millstone Grit* – for the middle series of the CARBONIFEROUS system. **3** A quarryman's term in England for a limestone which contains abundant shell fragments.

grivation The angle between GRID NORTH and MAGNETIC NORTH.

groove A large STRIATION or deep scratch cut across a smoothed rock surface by a boulder held in the base of an ice-sheet, or by meltwater (P-FORMS).

ground fog, ground mist A FOG caused by long-wave terrestrial radiation and confined to low-lying areas or valley floors. When the ground surface cools rapidly at night by excessive terrestrial radiation loss, it lowers the temperature of the lowest air layers, which then become denser and flow into topographic hollows owing to gravitational effects. Here condensation occurs as the air is cooled below its DEW-POINT. It may be only a few metres in thickness, so that tall buildings or trees may obtrude from the upper surface of the fog. *See also* FROST HOLLOW, GREENHOUSE EFFECT, INVERSION OF TEMPERATURE, RADIATION.

ground frost A term which signifies that the minimum temperature at the ground surface is below 0° C (32° F), although the air temperature may remain above freezing-point. NB Prior to 1960 a ground frost was recorded only when the ground temperature dropped below –1° C (30.2° F), but this usage has now been abandoned.

ground ice **1** An alternative term for ANCHOR ICE. **2** A collective term for all bodies of clear ice in the ground surface of the PERMAFROST zone. The various forms have been classified as follows: **1** *soil ice*, including NEEDLE ICE (*pipkrake*), *segregated ice* and ice filling pore spaces; **2** *vein ice*, in single veins or ICE WEDGES; **3** *intrusive ice*, including PINGO ice and *sheet ice*; **4** *extrusive ice*, formed subaerially (e.g. AUFEIS); **5** *sublimation ice*, formed in cavities by crystallization from water vapour; **6** *buried ice*, in glacial moraine.

ground information A REMOTE SENSING term referring to information derived from data on the physical state of the Earth. It is obtained from sources other than primary remote-sensing data sources. It typically includes, for example, maps, AERIAL PHOTOGRAPHS, soil-moisture measurements, BIOMASS and temperature measurements. It is collected as simultaneously as possible with that of remotely sensed data. Ground information is sometimes referred to as 'ground truth', but this is a usage which is incorrect and is to be discouraged since ground measurements are only sampled estimates.

ground moraine A thick sheet of TILL forming an undulating surface of low relief and deposited when the morainic debris carried in the base of an ice-sheet or glacier is released during a phase of melting. It tends to obscure former solid rock features. *See also* ABLATION TILL, TILL.

ground swell **1** Waves in deep water exhibiting great height and length. **2** Waves that, on entering water with a depth of less than half the wavelength, decrease in length and increase in height as they are affected by the sea-bed (ground).

ground truth GROUND INFORMATION.

groundwater Water that occupies pores, cavities, cracks and other spaces in the crustal rocks. It includes water precipitated from the atmosphere which has percolated through the soil (METEORIC WATER), water that has risen from deep magmatic sources, liberated during igneous activity (JUVENILE WATER) and *fossil water* retained in sedimentary rocks since their formation (CONNATE WATER). Some authorities exclude VADOSE WATER because it occurs above the WATER-TABLE. The presence of groundwater is necessary for virtually all WEATHERING processes to operate. Groundwater is synonymous with *phreatic water* and is the most important source of any water supply. *See also* ARTESIAN WELL, HYDRAULIC HEAD, SPRING.

group One of the divisions in the lithostratigraphic (LITHOSTRATIGRAPHY) classification of rocks. It is formed from two or more related FORMATIONS and is the first order division in the hierarchy. *See also* BED, MEMBER.

growan A local name on Dartmoor, SW England, for the coarse-grained debris produced by the weathering of granite. *See also* GRUS.

growing season That part of the year conducive to vegetation growth because of the favourable air temperatures. More specifically, the term is confined to the length of time that cultivated crops may be grown in a particular location. In Britain the critical temperature for vegetation growth is 6° C (43° F), but in agricultural terms the number of days between the last 'killing frost' of spring and the first one of the autumn is more critical. Specific crops have different lengths of growing season. In equatorial latitudes the growing season is usually continuous, while in subpolar latitudes it may be a mere eight to ten weeks.

growler A small iceberg, generally awash and therefore a particular hazard to shipping.

groyne, groin An artificial construction built out into the sea from a shoreline to maintain material on a beach, thereby checking LONGSHORE DRIFT *[140]* and inhibiting coastal erosion.

grumusol A modern US name for a BLACK COTTON SOIL or TROPICAL BLACK SOIL. It is included among the VERTISOLS of the SEVENTH APPROXIMATION of soil classification.

grus An accumulation of decomposed granite. GROWAN.

gryke GRIKE.

G-scale A logarithmic scale proposed in 1965 for use in geographical measurements, based on successive subdivisions of the surface area of the EARTH in terms of the power of 10. Thus, the G-scale of a given area of x km^2 is given by $G = \log_{10}(x'/x)$, where x' = the area of the Earth's surface (5.101×10^8 km^2), which is equivalent to a G-value of 0. The G-values rise with decreasing size.

GST GENERAL SYSTEM THEORY.

guano A substance rich in phosphates and nitrogenous material accumulating as a thick deposit of bird-droppings on certain islands and sea coasts of the world.

gulch A deep ravine in the SW of the USA. *See also* ARROYO, CREEK, WASH.

gulf **1** An inlet of the sea of large areal proportions, more indented than a BAY and generally more enclosed. **2** Used to describe, in some regions, an elongated, steep-sided and flat-floored valley in a karstic terrain (UVALA).

Gulf Stream A warm current of sea water, originating in the eastern Gulf of Mexico before flowing past Florida and the eastern seaboard of the USA (at a mean speed of 4–8 km day^{-1}) following the edge of the continental shelf. It leaves the coast at about 40°N and flows north-eastwards as a weaker and broader NORTH ATLANTIC DRIFT (at a mean speed of 5–8 km day^{-1}), towards the British Isles and Norway. The warm water combines with the prevailing south-westerly winds to produce the temperate climate of NW Europe and keeps the Norwegian coast ice-free during the winter months. [*168*]

gull, gulling A fissure which opens up in the surface rocks owing to tension as a result of CAMBER [*33*].

gully A small but deep channel or ravine formed by fluvial erosion but not permanently occupied by a stream.

gullying The process whereby gullies are formed on a land surface owing to the effects of heavy rainstorms. The surface RUNOFF becomes concentrated into shallow channels (RILL) which then combine to form deep gullies which dissect the surface and create BADLANDS. It is often a sign of human interference with the natural processes in a landscape, e.g. removal of vegetation, and is symptomatic of serious SOIL EROSION. One of the remedial measures is to adopt a system of artificial terracing or contour ploughing on hill slopes.

gumbo N American term for a fine-grained alkaline clay soil which becomes very sticky when wet but dries out brick-hard.

gumbotil A weathered, leached, clay-rich TILL in which the clay has been deoxidized. It has properties similar to a GUMBO soil – very sticky when wet, very hard when dry.

Günz One of the earliest of the Pleistocene glacials in central Europe. Distinguished as the earliest phase of glaciofluvial outwash to the north of the Alps by A. Penck and E. Brückner in 1909, it is now thought that there may have been an earlier glacial stage – the DONAU. The Günz glacial stage preceded the MINDEL and is thought to be equivalent to the Pre-ELSTER stage in N Europe and the NEBRASKAN glacial stage of N America. *See also* RISS, WÜRM.

gust A short-lived but rapid increase above the average wind-speed, in contrast to the momentary decrease below average windspeed known as a *lull*. Gustiness increases when the surface is aerodynamically rough, as in cities.

gut A narrow river channel or strait prior to joining an open ocean or estuary.

Gutenberg Channel A layer of plastic, less rigid material occurring at depths between 100 and 200 km below the LITHOSPHERE and at the upper surface of the MANTLE. It is thought to coincide with the ASTHENOSPHERE.

Gutenberg Discontinuity One of the two major discontinuities in the structure of the Earth. The Gutenberg Discontinuity occurs at a depth of 2,900 km and separates the MANTLE from the CORE. It has a considerable effect on SEISMIC WAVES, for at this boundary between a solid and a liquid S-waves cease to be transmitted and P-waves slow down from 16.6 km/sec (in the mantle) to 8.1 km/sec (in the core). *See also* MOHOROVIČIĆ DISCONTINUITY.

guttation The process by which transpired water from vegetation is unable to evaporate into the air because it is already saturated with water vapour. It manifests itself, therefore, as *guttation dew* – single large drops at the tips of grass blades or leaves of ground vegetation. Dew condensed from the air (DEW-POINT) consists of smaller, well-dispersed droplets all over the exposed surface of a grass blade or leaf.

guyot A flat-topped variety of a SEAMOUNT occurring mainly in the Pacific Ocean. Their summits are almost entirely at depths of more than 1,000 m but rise up to 3 km from the ocean floor. The conical shape of the guyot has suggested to some scientists a volcanic origin, with the table-like summit having been planed down by marine erosion and finally submerged by a rise of sea-level. Their summits are covered by sediments of various ages and of different derivation, ranging from Cretaceous sediments to volcanic materials of Tertiary age. [*99*]

Guyou's projection A conformal MAP PROJECTION of the Earth, providing for repetition of the Earth in both directions, north/south or east/west.

gymnosperms A class of seed-bearing plants which do not have their seeds protected by an outer covering, in contrast to

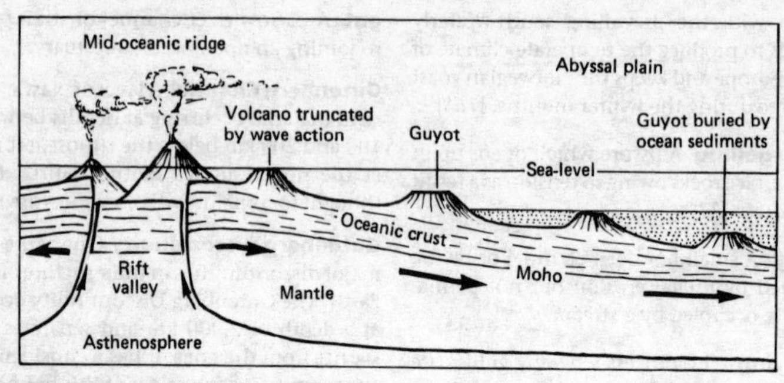

Figure 99 *Formation of a guyot*

the ANGIOSPERMS. They date back to the Carboniferous and flourished during the early Mesozoic, but the majority died out in the late Mesozoic, probably owing to competition with the angiosperms. They include the conifers, cycads and ginkgos.

gypcrete A GYPSUM crust found in deserts.

gypsey BOURNE.

gypsum An EVAPORITE mineral found mainly in clays, shales and limestones, composed of hydrous calcium sulphate ($CaSO_4.2H_2O$), and used in the manufacture of plaster of Paris. In Britain it occurs mainly in the Permian and Triassic deposits. *See also* ANHYDRITE.

gyre A term used in oceanography to describe the circulation of surface water in closed circulatory systems around the subtropical high-pressure systems at latitudes 20°–30°N and S. In the S Atlantic, for example, the gyre consists of a west-flowing South Equatorial current, a southward-flowing Brazilian current, an eastward-flowing West Wind Drift and a north-flowing Benguela current. Double gyres occur in the S Pacific and the N Indian Ocean. *See also* OCEAN CURRENTS.

gyro-compass A type of compass based on a gyroscope, in which a solid rotating wheel mounted in a ring allows its axis to turn freely in any direction. By maintaining the gyro-compass in a constant plane of rotation and with a constant direction of axis it obviates the need for the compass to make use of the Earth's magnetism.

gyttja A Swedish term for a nutrient-rich PEAT or organic mud, consisting mainly of plankton.

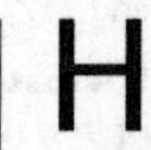

haar A cold, coastal fog which drifts in from the North Sea on to the eastern coast of Scotland and NE England, especially in spring and early summer. Also called *sea-fret* or *sea-roke*. It results from the cooling effect of the North Sea on the lower layers of the atmosphere, which brings the moist air to saturation-point and forms low stratus cloud.

habitat In the broadest sense, the overall ENVIRONMENT, but more specifically the place where an ORGANISM normally lives, characterized by its biotic and physical characteristics, to which the organism must become adapted (ADAPTATION) to survive. Habitats can be examined at different scales: the macroscale (continental), the mesoscale (regional, local) and, of particular importance, the microscale. *See also* COMMENSALISM, NICHE, SYMBIOSIS.

haboob The name (derived from the Arabic *habb* = to blow) for a strong, local wind in the Sudan which raises thick clouds of dust and sand, especially in summer. Most haboobs are thought to result from downdraughts in the large cumulonimbus clouds.

hachures Short lines drawn on a map to run in the direction of maximum slope, which indicate the relief by their thickness and spacing. Hachures give no information on absolute altitude. They are frequently employed to denote precipitous slopes when the contour-spacing becomes too close for clarity. [*100*]

Hacket-Planimeter PLANIMETER.

hade FAULT.

Hadley cell Part of the global atmospheric circulation pattern, based on an idea suggested in 1686 and later modified and developed by G. Hadley in his explanation of the TRADE WINDS in 1735. It comprises an extensive circulatory system of winds, whereby air rises at the INTERTROPICAL CONVERGENCE ZONE near to the equator before flowing polewards at high altitudes, descending at about 30°N and S and subsequently flowing equatorwards as the Trade Winds at the surface. *See also* ANTICYCLONE (warm),

Figure 100 *Hachures*

ATMOSPHERIC CIRCULATION, SUBTROPICAL. [*18* and *117*]

haematite (hematite) The principal ore of iron (Fe_2O_3) and one of the most important sources of iron ore. It occurs in two main forms: the red *botryoidal* form resembling kidneys (hence *kidney ore*), and the *specular* form with bright, black metallic crystals. Haematite is also one of the main cementing agents of sandstones. *See also* RED BEDS.

haff A German term for a shallow coastal lagoon resulting from the growth of a SPIT (in German a *Nehrung*) across a bay or river mouth. The lagoon may be marine or brackish water with virtually no saline content. The haff is best exemplified along the southern coast of the Baltic Sea.

hag (hagg) An eroded cliff or residual of peat due to active dissection of a peat bog.

hail, hailstone A type of precipitation which falls in the form of small pellets of ice (hailstones), normally with a diameter between 5 and 50 mm, although larger hailstones have been recorded (e.g. in USA a hailstone with a 430 mm circumference weighed 0.7 kg, while in China a 4.6 kg hailstone has been recorded). Hail is associated either with the passage of a COLD FRONT in temperate latitudes or with rapidly ascending CONVECTION currents in low latitudes. It is believed that each hailstone 'grows' around an ice nucleus at temperatures well below freezing in a CUMULONIMBUS cloud. After being carried upwards by rising currents within the cloud the nucleus acquires an additional 'shell' or layer of ice, by *collision* and *coalescence* with supercooled water (COLLISION THEORY OF RAINFALL). This creates concentric layering – a clear layer will form when there is a state of so-called *dry-growth* in a part of the cloud with low humidity; an opaque layer will form where there is greater humidity within the cloud and so-called *wet-growth* occurs. The hailstone may acquire several concentric shells or layers before it is large enough to fall through the cloud, overcoming the pronounced updraughts and striking the ground where it can cause considerable damage to crops, greenhouses and glass roofs.

haldenhang A German term to describe the rock-cut slope which occurs as a basal slope, generally covered with TALUS, below the steeper rock face (FREE FACE) above. It is part of a scheme devised by W. Penck in 1924 to explain slope evolution. [*193*] *See also* PEDIMENT.

half-dome A bare rock exposure, generally in coarsely crystalline rocks (especially granite), rising as a prominent 'convex rock spheroid', often at the edge of an escarpment, particularly in tropical latitudes. As a residual hill on the escarpment gets smaller by EXFOLIATION the DOME form extends downwards, destroying the sharp junction between the dome face and the talus slope until its rock face meets the valley floor or plain at an abrupt angle. Since this change usually develops only on one side of the hill or escarpment the resulting landform is a *half-dome*. The world's most famous example is Half Dome in Yosemite, USA, but this has been truncated by glacial erosion.

half-life The length of time taken for one-half of the atoms in a given amount of radioactive substance to decay into another ISOTOPE. *See also* CARBON-14, RADIOCARBON DATING, RADON.

halite Sodium chloride (NaCl) or common salt, formed from sea water or in a salt lake as an EVAPORITE.

halo A ring of light surrounding the sun or moon at a time when the sky is partly obscured by cloud, especially CIRROSTRATUS. It results from a *refraction* of the light through ice-crystals within the cloud. Not to be confused with a CORONA, which is produced by *diffraction*. A halo is often an indication of approaching bad weather, since thickening *cirrostratus* heralds the approach of a WARM FRONT and a WAVE DEPRESSION. *See also* ICE-FOG. [*43*]

haloclasty A process of rock weathering due largely to periodic wetting, leading to the crystallization of salt (HYDRATION) and the creation of stresses in the rock owing to swelling. It is one of the major weathering processes in hot deserts. *See also* CHEMICAL WEATHERING, THERMOCLASTY.

halokinesis The mobilization and flow of subsurface salt. It leads to the formation of SALT PILLOWS and SALT DOMES.

halomorphic soils A general term for those INTRAZONAL SOILS which have developed in areas where salts have accumulated at or near the surface. Three main types have been recognized: **1** the leached SOLOD; **2** the SOLONCHAK, with a predominance of sodium chloride and sodium sulphate; **3** the SOLONETZ, dominated by sodium carbonate.

halons Members of the halogenated fluorocarbon (HF) group of ethane- or methane-based compounds in which hydrogen ions are replaced by chlorine, fluorine and/or bromine. They are thought to be partly responsible for OZONE DEPLETION, where their damage potential is thought to be between three and ten times that of the CHLOROFLUOROCARBONS.

halophyte A term for any vegetation species which is tolerant of salt in the soil or in the air (i.e. sea spray). It is common, therefore, in salt-pan and sea-shore environments, e.g. *Salicornia* sp.

halosere The successional development of a group of plant communities in a saline habitat which has not previously supported vegetation, e.g. a newly exposed coastal mudbank on which a salt-marsh succession becomes established. *See also* HYDROSERE, LITHOSERE, PSAMMOSERE, SERE.

hamada (hammada) An Arabic term for a very flat bare-rock plateau in the desert. Although all the finely divided material has been swept away by the wind which has also polished the rock surface, the ground is characterized by a spread of angular debris created by the splitting of rocks under the influence of extreme temperatures (THERMOCLASTY) and EXFOLIATION, but most probably by HALOCLASTY.

Hammer's projections A number of MAP PROJECTIONS produced by E. Hammer between 1887 and 1910. The best known is an equal-area projection derived from the transverse aspect of the azimuthal equal-area projection (LAMBERT'S PROJECTIONS) which is expanded along the equator until the entire spherical surface can be represented within an ellipse having the major axis (the equator) twice the length of the minor axis (central meridian). Sometimes referred to as the *Hammer–Aitoff projection*. A modification of the latter was made by Wagner in 1949 (the *Hammer–Wagner projection*) in which the poles are represented by curves approximately half the length of the equator.

hanger A local term for a beechwood on a steep chalk slope in S England.

hanging glacier **1** A glacier, usually of small dimensions, on such a precipitous slope that its snout is constantly breaking away as ice avalanches. **2** A tributary or subsidiary glacier occupying a HANGING VALLEY. This latter usage is less common than the former usage above.

hanging valley A tributary valley debouching at an elevation distinctly higher than that of the floor of the glacial trough in either a glacierized or glaciated terrain. A river flowing down the hanging valley will, therefore, descend to the main valley as a waterfall (e.g. the Bridal Veil Falls, Yosemite, USA) or a series of cataracts at the DISCORDANT JUNCTION. It is generally agreed that the greater effectiveness of the main glacier in over-deepening the trunk valley is the main reason for the discordance of the valley floors, but the glacial PROTECTIONIST THEORY once had its adherents. In the cases where marine erosion has truncated normal stream valleys by a series of sea cliffs and where the streams may descend to the sea in the form of waterfalls, the term hanging valley is not really appropriate. [*101*]

hanging wall FAULT.

hardness scale **1** A measure of the resistance of a mineral to scratching. There are several arbitrary scales none of which can be connected by simple conversions. The best known is the Mohs scale, in which ten selected minerals are arranged so that each

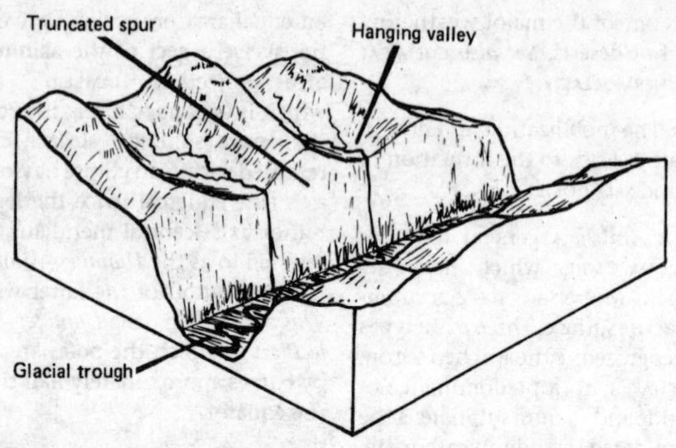

Figure 101 *Features of a glaciated valley*

one can scratch all of those which precede it on the scale. From the softest to the hardest these are: talc, 1; gypsum, 2; calcite, 3; fluorite, 4; apatite, 5; feldspar, 6; quartz, 7; topaz, 8; corundum, 9; diamond, 10. **2** A measure of the hardness of water, now generally expressed as parts of calcium carbonate per million parts of water. In the UK water is said to be soft if it has fewer than 70 ppm and very hard when it has over 2,100 ppm, thereby preventing soap from lathering. This occurs in areas where the water supply is derived from limestone and where there is a high proportion of calcium and magnesium salts present. In the USA the hardness scale differs: slightly hard = 55–100; moderately hard = 101–200; very hard = >200 ppm.

hardpan **1** A hard impervious layer within the soil resulting from: (a) the cementation of relatively insoluble materials, e.g. cementing of sand or gravel by LIMONITE to form DURIPAN, OF IRONPAN; and (b) the leaching and ILLUVIATION of soluble minerals, e.g. calcium carbonate forms *limepan*. *See also* CLAY PAN. **2** In the USA the term is used to denote the cementing of glacial drift.

hard water HARDNESS SCALE.

hardwood The term for the timber of all those broad-leaved trees which comprise the DECIDUOUS woodlands, which is relatively hard and durable, by contrast with that of the coniferous softwoods. It is in great demand for furniture-making, boat-building, etc. The most renowned hardwoods are oak and walnut (temperate varieties), mahogany, teak and rosewood (tropical varieties), in addition to such species as the eucalyptus.

harmattan A desiccating wind of W Africa which blows from the Sahara southwestwards to the Guinea coast. In the interior this NE wind is hot and dusty and can bring unpleasant conditions to the SAHEL zone. Farther S its dryness brings some relief from the high humidity of the coastlands along the Gulf of Guinea, despite the dust carried, hence its local name 'doctor'. During the northern winter its influence is felt as far south as 5°N, but when the planetary wind system moves N in the northern summer the wind is not felt south of 18°N.

harmonic analysis FOURIER ANALYSIS.

hatching A cartographic technique of shading, produced by drawing evenly spaced parallel lines. *Cross-hatching* is a combination of two sets of hatching, crossing each other at a predetermined angle. *See also* HILL-SHADING.

haugh A term used in N Britain to describe a valley floor or floodplain.

Hawaiian eruption An ERUPTION during which great quantities of extremely fluid basic lava flow out from a FISSURE or a central vent to form a typical SHIELD VOLCANO, exemplified by those in Hawaii (e.g. Mauna Loa). The eruption is not usually explosive, so that PYROCLASTIC phenomena are rare. [*271*]

Hawaiian 'high' One of the *high-pressure* cells in the atmospheric circulation. Located in the N Pacific this 'high' is more pronounced in the northern summer than in the northern winter. *See also* ANTICYCLONE (warm), SUBTROPICAL.

hazard A perceived event which threatens the life or the well-being of an organism, especially man. A catastrophe or a disaster is the realization of a hazard.

hazard, natural A natural event which is perceived by mankind as a threat to life and property. It may be generated from within the earth (e.g. earthquakes and volcanoes), upon the surface (landslides, avalanches, floods), or within the atmosphere (high winds, drought, snow, fog). Natural hazards can be contrasted with *human-induced hazards* which, although they kill more frequently, are generally not as catastrophic as natural hazards, which claim more lives in specific disasters. In recent years there has been a recognition of a third type of hazard – the *quasi-natural hazard* – to account for the long-term deterioration of the natural environment by mankind's careless pollution of the world's atmosphere, and its rivers, lakes and oceans. [*161*]

haze A state of obscurity in the lower layers of the atmosphere owing to the presence of large numbers of CONDENSATION NUCLEI. These are tiny particles of dust, smoke or salt spray around which droplets of atmospheric moisture condense. It is officially classified as haze when visibility falls below 2 km but is more than 1 km. In summer the shimmering caused by solar heating of the ground surface causes a *heat haze*. *See also* HAZE FACTOR.

haze factor A term sometimes used in meteorology to define the magnitude of HAZE. It was defined in 1953 as *B*/*B*', where *B*' is the luminance of the object and *B* the luminance of the haze, mist or fog through which the object is viewed.

Haze formula A measure of the permeability of a sedimentary material. It expresses the empirical relationship existing between the saturated hydraulic conductivity (K) of sand and effective grain size. The formula is: $K = 100D_{10}^2$, where D_{10} is the grain size of which 10% is smaller (10th percentile). It demonstrates that smaller grain sizes have a disproportionately large effect on hydraulic conductivity (PERMEABILITY). *See also* DARCY'S LAW.

head **1** A mass of sand, clay and angular stony rubble produced by SOLIFLUCTION under periglacial conditions during Pleistocene times. It occurs in a semi-stratified form as an infill of valley bottoms, on the lower slopes of hillsides and often overlying RAISED BEACHES. When chalky material is involved it is termed COOMBE ROCK. **2** A high coastal promontory or cape, e.g. Beachy Head. **3** A term referring to the pressure exerted by a liquid upon a unit area dependent upon the height of the surface of the liquid above the point where the pressure is determined. Before 1955 all meteorological and laboratory measurements were given in terms of the height of a mercury or water column and expressed in pounds/inch^2 in the height of the column of liquid expressed in feet or cm. Since 1955 the head has been described as force per unit area (kg per m^2). **4** The leading wave of a river BORE. **5** In USA a term denoting the rip current seaward of the line of breakers. **6** In Britain, a local place-name for a valley head, e.g. Wasdale Head, Cumbria. **7** The head of water used in hydroelectric-power generation. **8** HYDRAULIC HEAD.

headcut A term for a step in the bed of a river channel where a gentle gradient suddenly steepens downstream into a vertical drop.

head-dune A variety of DUNE in which sand accumulates on the windward side of an obstacle, in the small space of eddying air due to the uplifting effect of the obstacle on the flow of air [*65*].

heading An adit or horizontal tunnel driven underground to reach an AQUIFER in

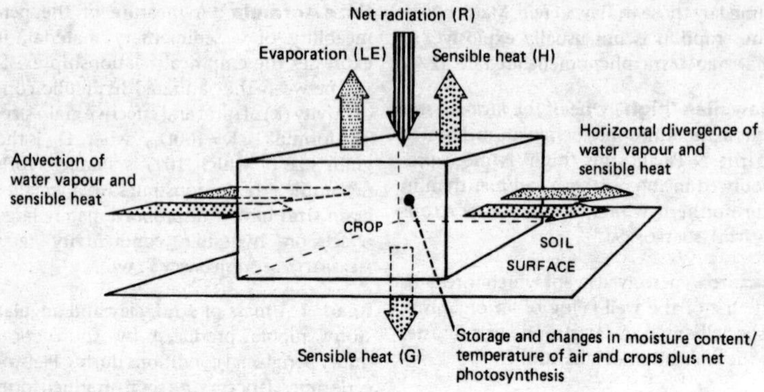

Figure 102 *Heat balance*

order to create or supplement a water supply.

headland **1** A cape or promontory jutting seawards from a coastline, usually with a significant sea cliff. **2** A strip of land left at the end of a furrow in a field in order to facilitate the turning of the plough. In England it often became a field boundary after enclosure.

headwall BACKWALL.

headward erosion The action of a river in cutting back upstream, thereby lengthening its valleys. It is achieved by SPRING-SAPPING and GULLYING and sometimes leads to RIVER CAPTURE.

headwaters The upper parts of a river drainage system; the source or sources of a river at the *head* of its catchment area.

heat Energy flowing between bodies as a result of temperature differences exhibited by the bodies. It can only be measured if it is allowed to flow. The addition of heat to a body can cause rises in temperature, changes in lengths or volume, or changes of state, i.e. melting or boiling. Each of these changes is capable of being measured and a measure of the intensity of heat input is known as TEMPERATURE. Linear or cubic expansion is measured by a coefficient of expansion. In the boiling process the LATENT HEAT of vaporization can be measured. Power production from heat relies on the existence of a *heat* or *thermal gradient* (GEOTHERMAL GRADIENT). Heat is measured in watts.

heat balance The concept of equilibrium which refers to the incoming solar RADIATION received by the atmosphere and the Earth in relation to the outgoing reradiated or reflected heat. It is sometimes referred to as the *energy balance*. The heat balance varies both seasonally and with latitude. In general those areas of the Earth which lie between 35°N and 35°S of the equator have a positive regime, i.e. they receive more radiation than they lose, while those areas which lie poleward of 35°N and 35°S receive less than they lose. As a result of this imbalance heat is transferred from low to high latitudes by AIR MASSES and OCEAN CURRENTS. The *heat-balance equation* for the Earth's surface is: $R + H + LE + G$, where R is NET RADIATION, H is sensible heat, LE the latent heat transfer and G the component transferred into the ground. *See also* ALBEDO, ATMOSPHERIC CIRCULATION, HADLEY CELL, INSOLATION. [*102*]

heat capacity The amount of heat required to raise the temperature of a unit volume of a substance, e.g. soil, by 1° C.

heat-capacity mapping mission A general term, used in REMOTE SENSING, for the first of a series of small, relatively inexpensive 'Application Explorer Missions' conducted by NASA (National Aeronautics and Space Administration) of the USA. The budget

SATELLITE was launched in 1978 with a two-channel RADIOMETER on board: channel 1 to record visible and near-infrared wavelengths. The spatial RESOLUTION is low, averaging around 600 m. Data from this satellite have proved to be of value in energy-balance and soil-moisture studies. *See also* LANDSAT, NOAATIROS SATELLITES, SEASAT I, SKYLAB, SPACE SHUTTLE, SPOT.

heat gradient GEOTHERMAL GRADIENT.

heath 1 A term used originally in a narrow botanical sense to describe a shrubby vegetation dominated by plants of the genus *Calluna* (ling or heath). **2** In a broader sense, the term has now been extended to also include a vegetation comprising *Erica*, gorse, broom, bilberry, and a number of other small, close-growing woody shrubs. **3** A general term for a type of moorland landscape at low altitude, especially on the coast.

heat 'island' A term introduced to describe the zone of slightly increased air temperatures that occur in association with a large urban area, owing to the absorption and storage of solar radiation by the urban fabric and the heat generated by the city's industries, buildings, traffic, etc. This can cause a marked reduction in the number of air frosts experienced annually in the city centre. *See also* URBAN CLIMATE.

heat transport ADVECTION.

heat-wave A popular term to describe a spell of exceptionally hot weather. It is not specified in meteorological circles, either in terms of temperature or in terms of its duration, but in Britain the temperature 'threshold' of 27° C (80° F) is generally taken as a convenient limit.

heave FAULT.

Heaviside–Kennelly layer (formerly **Heaviside layer)** The lower part of the THERMOSPHERE, above the STRATOSPHERE, some 100–120 km above the Earth's surface. It allows short radio-waves to penetrate to the APPLETON LAYER, but reflects medium and long radio-waves back to Earth. It is intensely ionized, and is sometimes known as the E-layer.

heavy metal pollution Pollution of soil, water, and indirectly of flora, fauna and cultivated crops by the artificial introduction of heavy metals (arsenic, cadmium, copper, lead, nickel, silver) into the environment.

heavy minerals The sedimentary minerals of high specific gravity which sink when treated with high-density liquids, such as bromoform (density 2.9 g cm^{-3}), in a laboratory experiment. The 'LIGHT' MINERALS, which float in the liquid, are thereby separated, such separation being an important part of petrological analysis. The small quantities of heavy minerals in arenaceous rocks would otherwise be difficult to identify in the overwhelming bulk of the low-density minerals such as quartz.

heavy soil A soil having a high percentage of fine particles, especially clay, making it difficult to cultivate. *See also* PARTICLE SIZE.

hectare (ha) A metric unit of area equivalent to 10,000 m^2. One hundred hectares = 1 km^2, 1 ha = 2.47106 acres, 1 acre = 0.4047 ha. [*15*]

hekistotherm 1 A plant which can exist in regions with very low temperatures (i.e. the warmest monthly mean <10° C), e.g. lichen. **2** A climate characterized by very low temperatures; the E-climates of KÖPPEN'S CLIMATIC CLASSIFICATION. *See also* MEGATHERM, MESOTHERM.

helicoidal (helical) flow A type of corkscrew flow superimposed on the primary downstream flow of a river. The flow moves across the water surface from the inside (convex) bank of a MEANDER towards the outside (concave) bank where it descends to the channel bed before returning as a reverse flow towards the inside of the meander. The flow tends to erode meanders on the outside (concave) banks and cause deposition of material on the inside (convex) banks of the meander, resulting in a downstream migration of the meander. [*103*]

helictite A type of STALACTITE in which the shape is abnormal, often in the form of a

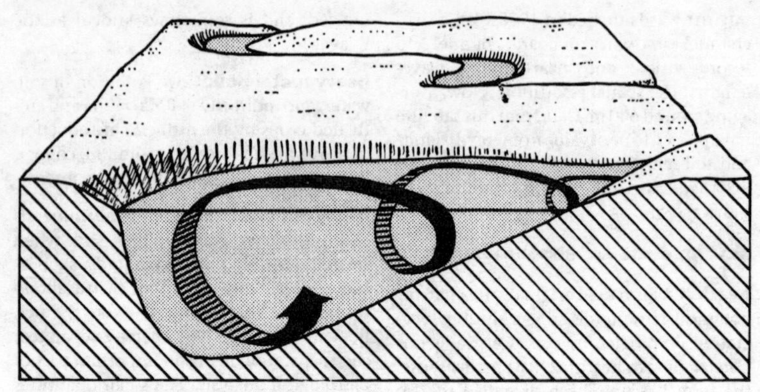

Figure 103 *Helicoidal flow in a meandering stream*

thin column deflected from the vertical by winds.

heliophyte A plant adapted to life in full sunlight. In shade the heliophyte may germinate but would not complete its life-cycle of flowering and fruiting. It contrasts, therefore, with the SCIOPHYTE.

heliotropism The action by which plants turn towards the light.

helm wind A strong and gusty easterly wind in N England that descends intermittently from the summit of the Crossfell escarpment (N Pennines) into the Vale of Eden. It is generally associated with a stationary *helm cloud* which covers the mountain-top (BANNER CLOUD) and a separate parallel line of clouds over the vale itself, resulting from eddies and turbulence (LEE WAVE) on the leeward side of the high escarpment. *See also* LENTICULAR CLOUD.

hemera A period of geological time (CHRONOSTRATIGRAPHY) as determined from an assemblage of FOSSILS.

hemicryptophytes One of the six major floral life-form classes recognized by the Danish botanist, Raunkiaer, based on the position of the regenerating parts in relation to the exposures of the growing bud to climatic extremes, and totally without respect to taxonomy. The hemicryptophytes develop buds which are half buried in the top layer of soil or humus. *See also* CHAMAEPHYTES, CRYPTOPHYTES, PHANEROPHYTES.

hemipelagic deposits Sediments deposited on the CONTINENTAL SHELF and the CONTINENTAL SLOPE.

hemisphere Literally, half a SPHERE. Planet Earth is divided into two hemispheres by the equator. In a less precise usage the continents of S America and N America are referred to as the *Western Hemisphere*.

herbivora The order of the class MAMMALIA that includes plant-eating animals. They range in size from the elephant down to the aphid. *See also* CARNIVORA.

Hercynian orogeny The mountain-building episode of Carboniferous/Permian times. Its name is derived from the Harz Mts in S Germany. Most authorities regard it as including both the ARMORICAN fold trend and the VARISCAN fold trend while others believe it to be synonymous only with the latter. The worn-down remnants of the Hercynian fold mountains occur as faulted and partly metamorphosed tracts in a discontinuous distribution from W to E through central Europe. The Hercynian folds can be seen in S Ireland, Cornwall, Brittany, the Ardennes, the Massif Central, the Vosges, the Black Forest, etc., while further E the Urals and the ALTAIDES are of similar age. In the USA part of the Appalachians is of Hercynian age. [*104*]

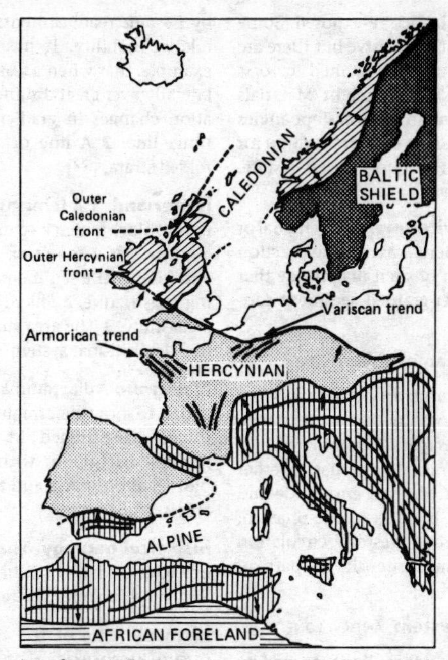

Figure 104 *Tectonic map of western Europe*

heterogeneous Any object which is composed of different properties, in contrast to HOMOGENEOUS.

heterosphere The outer zone of the Earth's atmosphere, distinguished from the underlying HOMOSPHERE by virtue of its gaseous constituents which exhibit very low densities. The following layers have been recognized, separated from each other by transition zones rather than by sharp boundaries: **1** an outermost, hydrogen layer beyond 3,500 km; **2** a helium layer between 3,500 km and 1,100 km; **3** an oxygen layer between 1,100 km and 200 km; and **4** an innermost, molecular nitrogen layer between 200 km and 90 km. *See also* AEROSPHERE, ATMOSPHERE.

heterotrophic Refers to the inability of an organism to manufacture its own food (AUTOGRAPHIC). It must therefore obtain it from other organisms, living or dead, and it is said to depend, therefore, on external nourishment.

heuristic approach An exploratory approach to a problem using successive evaluations of trial and error to arrive at a final approach.

hiatus A gap in a stratigraphic sequence of rocks, where the missing strata either were never deposited or were destroyed by erosion prior to deposition of overlying strata. *See also* UNCONFORMITY.

hierarchy Any number of objects ranked in grades one above another by classes or orders.

high ANTICYCLONE.

high-cohesion slope An expression introduced into geomorphology in 1950 by A. N. Strahler to describe steep slopes formed in cohesive rocks, in which the slope angle is so steep (>40°) that eroded material is rapidly removed and fresh rock re-exposed. Some clays are particularly cohesive but there are examples of vertical cliffs formed in LOESS which rise to over 30 m in height. Materials with high cohesion can stand at slope angles above that of the angle of STATIC FRICTION for the particles when in a non-cohesive state. *See also* REPOSE SLOPE.

high-energy environment Any fluvial or marine environment having wave action or current motion of such magnitude that deposition of finer-grained sediment cannot take place.

high-grade metamorphism GRADE.

high-index circulation An expression used in meteorology to describe the type of atmospheric circulation dominated by a flow of strong surface Westerlies together with ROSSBY WAVES of small amplitude and long wavelength. It refers to the phase of the INDEX CYCLE when westerly circulation achieves its maximum intensity in a particular region. [*113*]

high-pressure system ANTICYCLONE.

high water The highest limit reached by a medium tide. High-water mark in the UK is represented by High-Water-Mark Medium Tides (HWMMT).

hill An area of upland smaller than a MOUNTAIN but with no specific definition of absolute elevation.

hill-shading A type of shading employed to create an impression of relief on a map. Most examples have the light source shining obliquely from the NW, thereby throwing the E-facing and S-facing slopes into shadow. The darker the shading the steeper the slope. *See also* ILLUMINATED RELIEF.

hillslope flow processes *See also* OVERLAND FLOW, THROUGHFLOW.

hinge-fault A fault in which the amount of displacement increases from zero to a maximum amount as it is traced along the strike. It often passes into a FLEXURE. *See also* PIVOT-FAULT. [*76*]

hinge-line **1** A line separating a region of the Earth's crust undergoing uplift caused by *isostatic* readjustments and a region of relative stability. It has been found, for example, that when a RAISED BEACH is traced laterally over great distances its rise in elevation changes in gradient as it crosses a hinge-line. **2** A line of abrupt FLEXURE in folded strata. [*84*]

hinterland **1** A term given to the mobile crustal block which is reputed to move laterally towards a FORELAND, thereby crushing and folding the sediments in the intervening GEOSYNCLINE. **2** The region lying behind a coastline. **3** The area supplying sediment to a depositional system.

histogram A diagrammatic representation of data to show their frequency distribution, usually constructed in the form of a multiple-bar graph, with the frequencies plotted as ordinates and the magnitudes as abscissae.

historical geology That branch of GEOLOGY which deals with the evolution of the Earth and includes STRATIGRAPHY.

histosol One of the soil orders of the SEVENTH APPROXIMATION of soil classification. It is characterized by a very high organic content in its upper horizon and by its high degree of waterlogging, although one of its four sub-orders consists of a relatively free-draining layer of forest litter resting on bedrock. The four sub-orders are based on the degree of decomposition of the organic matter. Histosols occur throughout the world, but especially in cool, humid climatic regions such as Ireland and in Canada around Hudson Bay. They are low in plant nutrients and usually exhibit an acid reaction. They embrace most PEAT soils including the non-acid FEN peats.

Hjulström curve An empirical curve illustrating the critical flow velocity conditions necessary for the EROSION of river channel bed deposits. It was defined by a Swede, F. Hjulström, in 1935 and has subsequently been used in conjunction with the curves of SETTLING VELOCITY to indicate the threshold flow velocities required for the TRANSPORT OF SEDIMENTS on a river bed. The threshold flow velocity is at a minimum for well-graded sand particles (0.2–0.5 mm size range) but higher velocities are needed to move both coarser and finer sediments. *See also* BOUNDARY LAYER.

hoar frost An ice deposit with a feathery and needle-like appearance which forms on most ground-surface objects that have been cooled by terrestrial RADIATION on windless, clear nights so that any condensation (DEW) which occurs will freeze once the temperature drops below freezing-point. Thus, in the USA it is sometimes referred to as *white-dew*, which is also composed in part of ice crystals formed directly from water vapour (SUBLIMATION) below 0° C. It should not be confused with RIME. *See also* DEW, GLAZED FROST.

hodograph A circular diagram based on COMPASS directions to analyse the changes in wind-speed and direction at different heights above a given station. It is constructed by plotting vectors from the origin of winds at different pressure levels.

hoggin A term used in soil mechanics and civil engineering when referring to a well-graded gravel which contains enough clay to bind it together so that it can be used in its natural state to make a road.

hog's-back ridge, hogback **1** A steeply dipping CUESTA, forming a narrow-crested ridge in which both slopes are of comparable steepness, e.g. the Hog's Back near Guildford, Surrey, England. **2** In the USA the term is also applied to a sharply folded anticlinal ridge which decreases in height laterally in both directions.

hohlkarren (mohrkarren) KARREN.

Holarctic **1** In general usage, the whole of the Arctic region. **2** More specifically, in ZOOGEOGRAPHY, the entire northern or Arctic distribution of fauna. *See also* FLORISTIC REALM.

holism The view that the environment can be understood only by looking at it as a complex system of interrelated parts and wholes. Hence, the term *holistic*, which is especially appropriate in the study of GEOGRAPHY, which is said to be based on a holistic approach.

holistic HOLISM.

holm (holme) **1** A small island, e.g. Skokholm, Dyfed, Wales. **2** A floodplain, in N Britain.

Holocene A term meaning 'the whole of recent life' (RECENT) and referring to all of the time which has elapsed since the PLEISTOCENE. It is one of the two epochs which make up the QUATERNARY. The date of its beginning is open to dispute but a general figure of 11,000 BP is often taken as the boundary between the Pleistocene and the Holocene. Some scientists regard the Holocene merely as an interglacial within the Pleistocene. *See also* ALTITHERMAL, BLYTT–SERNANDER MODEL, HYPSITHERMAL, LITTLE ICE AGE, NEOGLACIAL.

holokarst A term introduced by J. Cvijić in 1918 to denote the total entity of karstic landforms (KARST) in a region upon which the KARSTIC CYCLE is said to operate.

Holstein The interglacial which separated the ELSTER and SAALE cold stages in the Pleistocene sequence of NW Europe. It is sometimes referred to as the Great Interglacial and is equivalent to the Mindel–Riss Interglacial of the central European terminology, the HOXNIAN of Britain and the YARMOUTH of N America. It is thought to have lasted from *c.* 230,000 BP to 180,000 BP. *See also* NEEDIAN INTERGLACIAL.

homoclimes, homoclines Places which experience similar climatic regimes as illustrated by their CLIMOGRAPHS.

homoclinal terrain Landforms created from rock strata which are tilted by the same degree of dip in the same direction, resulting in a type of SCARP-AND-VALE topography, especially when referring to one of the eroded flanks of an anticlinal fold. Its use is most common in the USA, whereas in Britain the term UNICLINAL is preferred.

homoeomorphy **1** A similarity in form between organisms of different genetic beginnings due to evolutionary adaptation to a similar HABITAT during CONVERGENT EVOLUTION. **2** In a more restricted usage the term refers to similarity of form in organisms with more closely related ancestries (PARALLEL EVOLUTION).

homoeostasis The concept of self-regulation, in which an organism or a system exhibits an ability to maintain a balance or state of equilibrium by adjusting itself to environmental changes by means of negative-FEEDBACK mechanisms. Such regulation ensures that a system will not go out of control. *See also* BIOSTASIS, RHEXISTASIS.

homoeostat A mechanism for achieving stability in a system which has been disturbed.

homoeostatic system A self-regulating living system in which the control of key variables or responses is maintained by information FEEDBACK, thus allowing the system to accommodate new conditions. HOMOEOSTASIS.

homogeneous Any object which is composed of the same properties, in contrast to HETEROGENEOUS.

homolographic projection EQUAL-AREA PROJECTION.

homolosine projection An EQUAL-AREA MAP PROJECTION combining the MOLLWEIDE and SINUSOIDAL PROJECTIONS. *See also* GOODE'S INTERRUPTED HOMOLOSINE PROJECTION.

homoseismal line, homoseism A line connecting all points on a map that are affected simultaneously by a seismic shock (EARTHQUAKE). It is synonymous with ISOSEIST. *See also* ISOSEISMAL LINE.

homosphere The inner zone of the Earth's atmosphere beneath the HETEROSPHERE and distinguished from it by the higher densities of its gaseous constituents. Its chemical composition is virtually uniform up to an altitude of 90 km, beyond which the heterosphere begins. The homosphere can be subdivided into the TROPOSPHERE, STRATOSPHERE and MESOSPHERE. *See also* AEROSPHERE, ATMOSPHERE.

homotaxis A term referring to those rock strata which occupy the same position in the stratigraphic sequence (CHRONOSTRATIGRAPHY) although they occur in widely separated regions. Such divisions of strata are said to be *homotaxial*.

honeycomb weathering A type of rock WEATHERING in which the rock surface assumes the appearance of honeycomb. It is thought that the CEMENT infilling some joints is harder than the main body of the rock, resulting in a mosaic of recessed hollows surrounded by projecting ridges of more resistant material. Both moisture and wind are believed to play their parts as weathering agents. *See also* ALVEOLE, DIFFERENTIAL WEATHERING.

hoodoo An American term for a weirdly shaped pillar of solid rock in a semi-arid environment, similar in form to an EARTH PILLAR. It is the result of undercutting by the wind.

Hooke's law A law formulated by R. Hooke in the 17th cent. It states that a stressed body deforms to an extent proportional to the force applied, i.e. stress = strain. It refers to the elasticity of a material. *See also* ELASTIC DEFORMATION, ELASTIC LIMIT, ELASTIC REBOUND.

horizon 1 A planar surface separating two beds of rock in a sequence and assumed to be a time plane of no thickness. **2** A soil layer of any thickness (SOIL HORIZON), generally running parallel to the ground surface and exhibiting a relatively uniform character in its material.

horizontal equivalent The distance between any two points reduced to the horizontal plane.

horn 1 A German term for a PYRAMIDAL PEAK. **2** Part of a BARCHAN [*22*].

hornblende A mineral of the AMPHIBOLE group, common in both igneous and metamorphic rocks. It consists essentially of silicates of iron, calcium and magnesium.

hornfels A fine-grained non-schistose metamorphic rock, formed by CONTACT METAMORPHISM in the zone nearest to the igneous intrusion.

hornstone A fine-grained, compact volcanic ash often with irregular but indistinct layering. There is disagreement over the

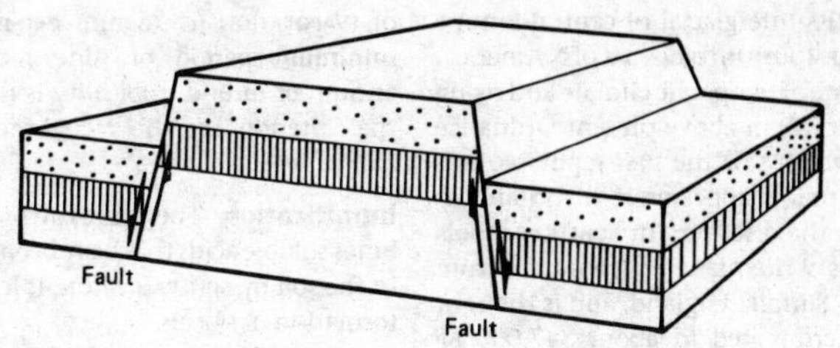

Figure 105 *Horst*

derivation of the term, with some believing that it has the texture of staghorn and others that it is a corruption of honestone, a sharpening stone.

Horse latitudes The belts of high pressure, characterized by calms and light winds, which occur in the sub-tropics at 35°N to 30°N and at 35°S to 30°S. In the pattern of ATMOSPHERIC CIRCULATION they lie between the TRADE WIND belts and the WESTERLIES of higher latitudes. They are zones of mainly descending air at the junction of the FERREL CELL and the HADLEY CELL (in the circulatory system of the Earth's 'climatic engine'), from which surface AIR MASSES move equatorwards and polewards as the Trade Winds and Westerlies respectively. The belts move N and S seasonally.

horst A fault-defined block which has been left upstanding by the sinking of the adjoining land along normal faults or has been uplifted along parallel faults. Although it coincides with a prominent relief form at its inception, the topography may be denuded to leave only the structural formation. *See also* GRABEN. [*105*]

Hortonian analysis A geomorphological technique devised by R. E. Horton in 1945 to show the relationships between STREAM ORDER and a number of variables such as channel length, drainage area and number of streams. This empirical analysis gives a BIFURCATION RATIO expressed as the means of the ratios of the number of streams of any given stream order to the number in the next lower order. In the majority of drainage basins the number of streams in any given order is about three/four times more than in the next higher order. *See also* LAW OF BASIN AREAS, LAW OF STREAM LENGTHS, LAW OF STREAM NUMBERS, LAW OF STREAM SLOPES.

hot spots Localities on the Earth's surface where greater than average thermal activity takes place, often leading to volcanic outbursts (VULCANICITY). Hot spots are thought to be the crustal expression of PLUMES in the underlying mantle and are believed to be closely related to the mechanisms of PLATE TECTONICS.

hot spring A thermal spring with temperatures above 37° C, where water flows out of the ground in a continuous and non-explosive way, in contrast to the flow of a GEYSER. It can occur in non-volcanic areas but generally is more common in areas of current or recently active vulcanicity.

'hot tower' A colloquial term used to describe the vertically rising current of unstable air which rises at the centre of a tropical cumulonimbus cloud owing to intense heating. It also refers to the primary energy cell of a TROPICAL CYCLONE, where rising condensing air releases vast amounts of LATENT HEAT to fuel the storm.

how A Scandinavian-derived term (from the ancient Norse *haugr*) for a low hill in N England.

howe A Scottish term for a low place or depression.

Hoxnian The penultimate INTERGLACIAL stage of the Pleistocene in Britain, and marking the beginning of the Upper Pleistocene. It separated the ANGLIAN from the WOLSTONIAN cold stages and is roughly equivalent in age to the HOLSTEIN of NW Europe, the

Mindel–Riss Interglacial of central Europe and the YARMOUTH INTERGLACIAL of N America. It was a time of temperate climate and rising sea-level (*c.* 30 m above present Ordnance Datum) which saw the first reputable evidence of man's appearance in Britain, as shown by the discovery of ACHEULIAN tools in deposits of Hoxnian age. It is named from Hoxne, in Suffolk, England, and is thought to have terminated in about 347,000 BP when the Wolstonian cold stage began. [*191, 197*]

hum An isolated hill, left as a residual on the floor of a POLJE, named from the village of Hum in the Yugoslavian KARST.

Humboldt current PERU CURRENT.

humic acid A complex organic acid, one of several acids formed from the partial decay of organic matter. It assists in the formation of certain soil types and is thought by some to play a part in CHEMICAL WEATHERING.

humic gley soils One of the subdivisions of the GLEY SOILS, which are themselves part of the SOIL CLASSIFICATION OF ENGLAND AND WALES. They are typically non-alluvial, badly drained loamy or clayey soils with a humose or peaty topsoil, overlying a grey or grey/brown mottled (MOTTLING) horizon which may be calcareous or non-calcareous.

humic layer, humose layer A layer of highly decomposed organic soil in which only a little fibrous material has survived.

humidity The state of the atmosphere in relation to its water-vapour content, and normally referring to RELATIVE HUMIDITY unless otherwise stated. *See also* ABSOLUTE HUMIDITY, DEW-POINT, MIXING RATIO, SATURATION DEFICIT, SPECIFIC HUMIDITY, VAPOUR PRESSURE.

humid tropicality A term introduced in 1960 by B. J. Garnier to describe the climatic condition during a given time period whereby RELATIVE HUMIDITY remains in excess of 65 per cent, mean temperature remains above 20° C (68° F) and atmospheric pressure exceeds 20 mb. At the same time RAINFALL must always equal or exceed the rate of EVAPORATION (*c.* 75 mm per month). A minimum period of nine months per annum of humid tropicality is regarded as the criterion for the recognition of the *humid tropics*.

humification The process which recombines soluble acids that have been produced in the soil by MINERALIZATION. It leads to the formation of HUMUS.

humilis A type of CUMULUS cloud of only limited vertical development, generally flattened into fair-weather cumulus by STABILITY in the troposphere. *See also* CLOUD.

hummock **1** In general, a low mound of earth, rock or ice. **2** More specifically, a mound of earth in a periglacial environment, having either a core of mineral soil (*earth hummock*) or a core of stones (*turf hummock*), up to 70 cm in height and up to 3 m in diameter. Active hummocks (which reappear after artificial levelling) are closely related to the presence of PERMAFROST.

hummocky moraine A type of moraine characterized by considerably undulating terrain owing to the presence of a large number of KETTLE HOLES and CREVASSE FILLINGS. It is often thought to be the result of an ice-sheet stagnating *in situ*.

humus A black or dark-brown organic substance produced by the processes of MINERALIZATION and HUMIFICATION within the soil. *See also* MICRO-ORGANISMS IN THE SOIL, MODER, MOR, MULL.

humus-iron podzol (ferro-humic podzol) One of the subgroups of PODZOL soils. It is characterized by an ILLUVIAL HORIZON, comprising a secondary humus horizon below the ELUVIAL HORIZON and above the iron-stained B-HORIZON. [*190*]

Huronian The stratigraphic name of part of the PRE-CAMBRIAN in Canada, during which there is evidence of an ICE AGE.

hurricane **1** A name of Spanish derivation which refers primarily to the revolving tropical storms of the Caribbean and Gulf of Mexico. It is characterized by a central area of calms – the 'EYE', around which winds of very high velocity (>160 km/h) revolve –

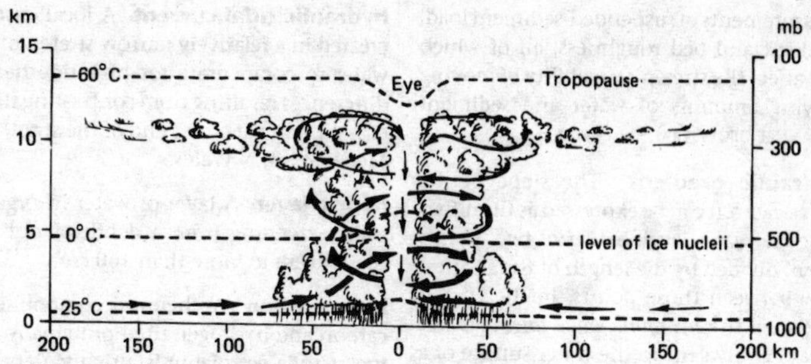

Figure 106 *Section through a hurricane*

and is accompanied by torrential rain and by thunder and lightning. The whole system varies in diameter from 100 km to 1,600 km and moves along a curving track from mid-Atlantic through the West Indies and onshore into the southern states of the USA, generally creating havoc in its path. *See also* LOW, TROPICAL CYCLONE. [*106*] **2** A wind velocity of force 12 (121 km/h or 75 miles/h) on the BEAUFORT SCALE.

hydration One of the major processes of MECHANICAL WEATHERING, involving the addition of water to a mineral, causing it to expand and thereby initiate stress within the rock (GRANULAR DISINTEGRATION). Once minerals have experienced hydration they became more susceptible to the effects of CHEMICAL WEATHERING, especially those of carbonation and oxidation.

hydraulic action The force exerted by moving water on rocks without recourse to the use of its LOAD. The erosive force of fast-flowing water is alone sufficient to remove loose material from river banks, while the air forced into cracks in solid rocks by breaking waves is capable of causing their disintegration by expanding the fissures. In the case of a turbulent flow of water, the suction force of eddies can cause lifting of material from the solid bed of the channel, possibly resulting in removal of fragments. Such a process is termed *hydraulic lift*. The other type of hydraulic action in turbulent stream-flow is termed CAVITATION, whereby the collapse of air bubbles leads to erosion of the channel walls, particularly if the flow is of a subglacial or underground nature. Hydraulic action should be distinguished from ABRASION, in which erosion is caused largely by the load itself.

hydraulic conductivity A measure of the ease with which a fluid flows and the ease with which a porous medium permits its passage through it. Hydraulic conductivity is the parameter K in *Darcy's Law* and its velocity is expressed in ms^{-1}. It differs from PERMEABILITY, which refers only to the character of the porous medium and not to the fluid that passes through it.

hydraulic force CRITICAL TRACTIVE FORCE

hydraulic friction The resistance to the flow of a stream caused by roughness of the channel bed.

hydraulic geometry An expression introduced in 1953 by L. B. Leopold and T. Maddock to describe the hydraulic characteristics of a stream channel. The mean velocity, the mean depth and the width of flowing water are functions of discharge at a given river cross-section (AT-A-STATION), since discharge is the product of mean velocity and the cross-sectional area of flow. It has been shown that with increasing discharge the mean velocity, mean depth and width increase as power functions: $v = kQ^m$, $d = cQ^f$, $w = aQ^b$, where v is mean velocity, d is mean depth, w is width, Q is discharge and k, c, a, m, f and b are numerical coefficients. In addition to the three parameters of velocity, depth and width, the complete hydraulic geometry of a stream channel will include measurements of suspended-sediment load, gradient and bed roughness, all of which will affect the stream's capability of moving varying amounts of water and sediment (CHANNEL GEOMETRY).

hydraulic gradient The slope of the WATER-TABLE. It can be expressed as the difference in water-level between two points (HEAD) divided by the length of the shortest flow between those points. In the USA it is termed the *hydraulic grade line*. It is not always equivalent to the water surface of a stream.

hydraulic head An important part of the equation defining *Darcy's Law*, in groundwater studies. The hydraulic head (h) is the sum of the elevation head (z) and the pressure head (hp), all of which are measured in metres above mean sea-level. The change of the hydraulic head with distance is termed the HYDRAULIC GRADIENT. *See also* HYDRAULIC POTENTIAL.

hydraulic jump A term referring to a sudden change in the discharge along an open river channel. It is expressed as a change in the FROUDE NUMBER, which alters as the flow moves from a state of low velocity in deep water to one of high velocity in shallow water or vice versa.

hydraulic potential (fluid potential) The mechanical energy of water per unit mass. Following DARCY'S LAW groundwater will flow from points of high hydraulic potential towards points of low hydraulic potential, down the HYDRAULIC GRADIENT, perpendicular to the EQUIPOTENTIALS. It is closely related to the HYDRAULIC HEAD.

hydraulic radius (R) An alternative expression for hydraulic mean depth, expressed as the cross-section of the water flowing through a channel divided by the wetted perimeter of the channel. The greater the hydraulic radius the larger the flow for a given channel and gradient. It is shown as $R = A/P$, where A is the cross-sectional area of flow, and P is the WETTED PERIMETER.

hydraulics The study of the flow of fluids, especially water. *See also* HYDRODYNAMICS.

hydraulic tidal current A local current created in a relatively narrow stretch of sea water to compensate for high-tide height differences resulting from contrasting times of high water at either end of the strait, e.g. Menai Strait, N Wales.

hydric layer A layer of water in organic soils, extending from a depth of not less than 40 cm to more than 160 cm.

hydrocarbon A chemical compound of carbon and hydrogen used principally as a fuel. Crude petroleum is a rich variety but it also includes methane, paraffin, kerosene and BITUMEN.

hydrodynamics The branch of HYDRAULICS which deals with the pressures created by water turbulence and its flow through channels, conduits, pipes and over weirs.

hydrogenesis A process of condensation of moisture in the pore spaces of a soil or an unconsolidated sediment.

hydrogenic soil A soil formed in an environment of prolonged waterlogging, mainly in cold humid regions.

hydrogen-ion concentration PH.

hydrogeological map A map of underground water resources. It will depict AQUIFERS, ARTESIAN WELLS, and SPRINGS. Larger scale maps will also show the WATER-TABLE contours and EQUIPOTENTIALS.

hydrogeology The study of GROUNDWATER.

hydrogeomorphology The study of landforms resulting from the action of water, especially FLUVIAL processes.

hydrograph A graph depicting the DISCHARGE of a stream as a function of time at a given station (AT-A-STATION). *See also* HYDRAULIC GEOMETRY.

hydrography **1** The study of all the water bodies of the Earth. **2** More specifically, the term refers to the measurement, collection and plotting of data relating to the ocean floor, the coastline and the tides and currents, all of which can be presented on a *hydrographic chart* as an aid to navigation. *See also* HYDROLOGY, POTAMOLOGY.

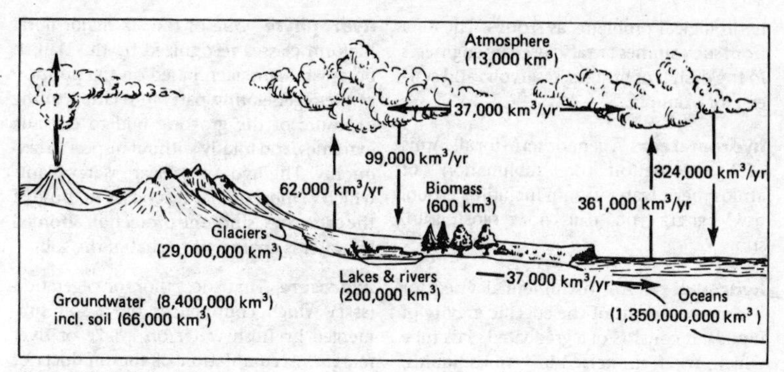

Figure 107 *Hydrological cycle*

hydroisostasy An isostatic (ISOSTASY) reaction to the loading or removal of a large body of water. On a small scale the drying-out and disappearance of lakes can cause former shorelines to become warped. At a larger scale, sea-level changes (EUSTASY) can cause fluctuations in the elevation of CONTINENTAL SHELVES, when a higher sea-level causes WARPING of the Earth's crust downwards and a lower sea-level results in a crustal rebound, upwards.

hydrolaccolith A type of frost-blister in the soil formed under PERMAFROST conditions by the same mechanisms as those operative in PINGO formation. Some writers use the term hydrolaccolith as synonymous with pingo but others regard it merely as a much smaller version of the latter.

hydrological-balance budget A term referring to the relationship between EVAPORATION (E), PRECIPITATION (P), RUNOFF (R) and *ground/surface storage* (ΔS) for a given station over a given period of time. It is expressed as $P = E + R + \Delta S$ where ΔS may be positive or negative. *See also* WATER BALANCE.

hydrological conductivity The rate at which water moves through a soil. *See also* PERMEABILITY, POROSITY.

hydrological cycle The cycle of water movement through the Earth-atmosphere system, initiated through the acquisition of water vapour by EVAPORATION and TRANSPIRATION from water and land surfaces (including vegetation), released into the atmosphere by CONDENSATION (CLOUD) and deposited on land and water surfaces by PRECIPITATION. At the Earth's surface, the precipitation is stored on the surface (lakes, glaciers) or at depth as GROUNDWATER, or is evaporated or transpired (to initiate the 'next' cycle) with the balance returned to the sea through THROUGHFLOW. *See also* WATER BALANCE. [*107*]

hydrological map A map depicting water resources of an area. It is used in the construction of a HYDROLOGICAL-BALANCE BUDGET and a WATER BALANCE.

hydrology The scientific study of the distribution and properties of water (in a liquid and frozen state) within the atmosphere and at the Earth's surface, including PRECIPITATION, EVAPORATION, EVAPOTRANSPIRATION, RUNOFF, soil moisture and GROUNDWATER and the MASS BALANCE of glaciers and snow packs. *See also* GLACIOLOGY, HYDROLOGICAL-BALANCE BUDGET, HYDROLOGICAL CYCLE, WATER BALANCE.

hydrolysis The main type of CHEMICAL WEATHERING, in which water combines with rock minerals to form an insoluble precipitate (CLAY MINERALS) and insoluble components, except in the special case of CARBONATION, where only soluble products are created.

hydrometeorology The study of the interaction between HYDROLOGY and METEOROLOGY, usually to solve or predict such hydrological problems as FLOODS. The data from such studies are also used by engineers in the design of bridges, reservoirs and sewers, for example.

hydrometeors A generic term for all forms of condensation or sublimation of atmospheric water vapour including CLOUD, DEW, DRIZZLE, FOG, HAIL, HOAR FROST, RAIN, SNOW.

hydrometer An instrument devised for the measurement of the specific gravity of liquids. It consists of a graduated glass tube which, when immersed in various liquids, sinks to a particular level.

hydrometry The measurement and analysis of the flow of water, especially by the use of a HYDROMETER.

hydromorphic soils A general term for INTRAZONAL SOILS developed in the presence of excess water. In publications of the Scottish Soil Survey soils are often grouped into *hydromorphic associations*, a classification based on variations in drainage characteristics. *See also* ANAEROBIC, CATENA, HYDROGENIC SOILS, PLANOSOL.

hydrophyte One of the six major floral lifeform classes recognized by the Danish botanist, Raunkiaer, based on the position of the regenerating parts in relation to the exposure of the growing bud to climatic extremes, and totally without respect to taxonomy. The hydrophytes are water plants which exhibit similar responses to those of the GEOPHYTES, since the protection afforded by water is analogous to that of the soil.

hydrosere That succession of vegetation (SERE) which commences on a soil submerged by fresh water on a lake or river margin. Accumulation of the products of decay of a reed-swamp, for example, ultimately raises the soil surface above water-level and enables marsh and fen plants to become established (CARR, FEN). Finally, the surface soil becomes dry enough for the growth of mesophytic trees, which cannot tolerate waterlogged soil, and the hydrosere culminates in climax forest. HALOSERE, LITHOSERE, PSAMMOSERE, XEROSERE. [*108*]

hydrosphere One of the three components of the Earth's planetary system, namely ATMOSPHERE, LITHOSPHERE and HYDROSPHERE, with the latter relating to the surface

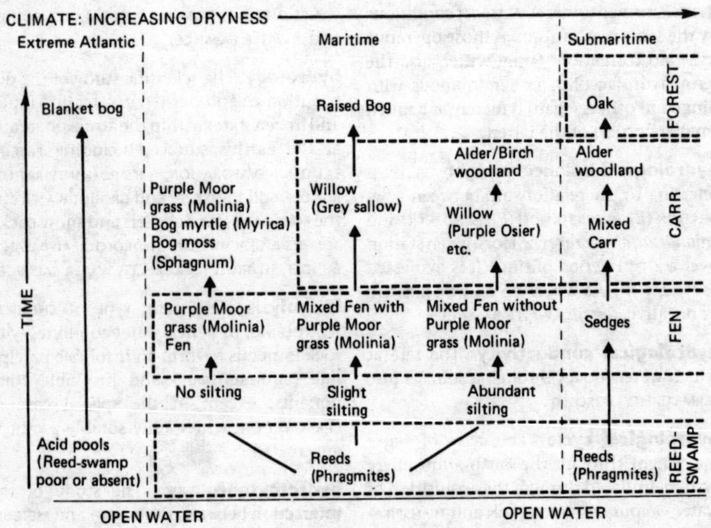

Figure 108 *Development of a hydrosere in different climates of the temperate zone*

waters. The hydrosphere and the atmosphere are interlinked by the HYDROLOGICAL CYCLE. [107]

hydrostatic equation An equation used by meteorologists to define the relationships which exist between altitude, gravity, air pressure and density. A solution to the equation produces a *barometer–height formula*. The hydrostatic equation is $dp/dz = -\varrho g$, where dp/dz is the rate of change of pressure with altitude, ϱ is the density of air, and g is the acceleration of free fall (GRAVITY).

hydrostatic pressure The pressure at any point of a liquid at rest. It is measured by the density of the liquid multiplied by its depth. PASCAL'S LAW.

hydrostatics The branch of HYDRAULICS concerned with the pressures of fluids at rest.

hydrothermal activity Those processes associated with igneous activity which involve heated or superheated water (PNEUMATOLYSIS). The latter is capable of dissolving many substances otherwise insoluble, in addition to breaking down silicates, thereby causing considerable *hydrothermal alteration*, especially in the production of KAOLIN from the breakdown of feldspar minerals in granite (HYDROLYSIS), and in the creation of SERPENTINE from olivine. *Hydrothermal deposition* is responsible for many ore deposits, either in veins or in concentric layers around an igneous body. Many hydrothermal deposits concentrate at the top of the igneous intrusion (CUPOLA).

hydroxide A compound of any element with the radical OH, e.g. calcium hydroxide ($CaOH_2$).

hydrozoa A class of small-sized marine and freshwater organisms.

hyetogram A chart which illustrates the rate at which rain falls. *See also* HYETOGRAPH, NATURAL-SIPHON RAINFALL RECORDER, TIPPING-BUCKET RAIN-GAUGE.

hyetograph 1 A self-recording instrument which measures rainfall amount by means of a recording pen attached to a float which fluctuates according to the amount of water in the RAIN-GAUGE (NATURAL-SIPHON RAINFALL RECORDER, TIPPING-BUCKET RAIN-GAUGE). **2** A diagram which illustrates the mean, maximum and minimum rainfall totals per month for any climatic station, by means of columnar graphics. Standard deviation and probable deviation from the mean are also portrayed.

hygrograph An instrument which records the changing RELATIVE HUMIDITY of the atmosphere. A continuous record is obtained on a graph affixed to a rotating drum, by means of a pen which is linked to a HYGROMETER. The most common type is the hair hygrograph, which uses strands of human hair that increase in length with increasing relative humidity (i.e. a 2½% increase in length occurs as r.h. increases from 0 to 100%).

hygrometer An instrument which measures the RELATIVE HUMIDITY of the atmosphere at any point in time. A continuously recording variety is a HYGROGRAPH. Some hygrometers are based on the shrinking and lengthening conditions of *relative humidity – 'hair hygrometers'*; others are based on the resistance characteristics of a lithium-chloride strip under similarly fluctuating conditions. In both instances the changes are amplified and transferred to a scale. *See also* DRY-BULB THERMOMETER, PSYCHROMETER, WET-BULB THERMOMETER.

hygrophyte A plant species adapted to conditions of a regular and plentiful supply of water (greater than that of the MESOPHYTE). *See also* XEROPHYTE.

hygroscopic coefficient A term used in soil science to denote the amount of moisture (expressed as a percentage of its dry weight) that a dry soil will absorb in saturated air at a given temperature.

hygroscopic nuclei Particles of dust, smoke, sulphur dioxide, salts or similar microscopic substances occurring in the atmosphere around which CONDENSATION occurs. In general, because of their solubility, condensation commences on hygroscopic nuclei before the air is saturated. They range in size from a radius of 0.001 μm to 10 μm and are more prolific over continents than over oceans. *See also* AEROSOLS, DUST, FREEZING NUCLEI, NUCLEUS.

hygroscopic water Water that is adsorbed (ADSORPTION) on to a surface from the atmosphere. It adheres in very thin films to soil particles and is unavailable for plant growth. Under drought conditions the point at which the only water left in the soil is that held tightly by the soil COLLOIDS is termed the hygroscopic coefficient. This is reached when the moisture tension reaches 31 atmospheres. *See also* SOIL MOISTURE.

hypabyssal rocks The smaller-scale igneous intrusions, such as a SILL OR a DYKE, which are intermediate in size between the large deep-seated PLUTONIC ROCKS and the extrusive VOLCANIC ROCKS. The term is rarely used today.

hypocentre FOCUS.

hypogene A term used to denote geological processes that originate at great depths within the Earth. It refers to PLUTONIC ROCKS and some of the rocks which are metamorphosed at depth; it should therefore be contrasted with EPIGENE.

hypolimnion The lowest cold-water layers at the bottom of a lake or ocean. [*252*] *See also* EPILIMNION, THERMOCLINE.

hypothermal An ore deposit formed at high temperatures (300°–500° C) by HYDROTHERMAL ACTIVITY. *See also* EPITHERMAL, MESOTHERMAL.

hypothermia A condition of abnormally low human-body temperature (<32° C/ <90° F) which sometimes results in coma and possibly death, following exposure to severe weather conditions. *See also* WINDCHILL. [*280*]

hypsithermal A term used in the USA to describe those warmer climatic conditions of the HOLOCENE which existed between about 9,000 BP and 2,600 BP, (the dating varies with latitude). In Britain it is sometimes referred to as the Climatic Optimum (MEDIOCRATIC) but is more specifically confined to the ATLANTIC PERIOD between 7,450 BP and 4,450 BP.

hypsographic curve Sometimes termed a *hypsometric curve*. A graph denoting the proportion of a land mass which stands above a given datum. It is plotted in terms of percentages of total area or absolute areas on the horizontal axis and with altitude on the vertical axis.

hypsometer An instrument used in the determination of absolute height above sea-level, which has to be used in conjunction with a set of statistical tables. Its use has largely been superseded by the ANEROID BAROMETER. It is based on the differences exhibited by the boiling-point of water according to the changes of atmospheric pressure with altitude.

hythergraph A type of CLIMOGRAPH, in which two climatic elements (temperature/ humidity or temperature/rainfall) are plotted in diagrammatic form to show broad relationships in relation to human behavioural responses.

I

Iapetus The conjectural ocean thought to have divided North America and Eur-Africa some 500 million years ago, and whose closure led to the CALEDONIAN OROGENY.

ice Water in its solid form, produced by: **1** the normal freezing process of water; **2** the SUBLIMATION process in which ice crystals are formed from water vapour when the air temperature falls below freezing-point; **3** the compaction of snow via the process of FIRNIFICATION into true *glacier ice*; **4** the impregnation of snow layers by running water (possibly from surface melting) and subsequent freezing within the snow mass; **5** the process of REGELATION, in which pressure within an ice mass causes *pressure-melting* (PRESSURE MELTING-POINT), when ice grains are converted into molecules of water, following which, on moving to points where pressure decreases (in a GLACIER this would involve a downhill movement), the water molecules recrystallize and refreeze. The specific gravity of ice (0.9166) is slightly less than that of water, which means that ice floats. *See also* GLAZED FROST, HOAR FROST, RIME, SNOW.

ice age Any period in the Earth's history during which ice-sheets expand considerably and surface temperatures in the temperate latitudes are significantly lowered. The term was first used by K. Schimper in 1837. Although the expression is used most commonly to describe the latest glacial episode (PLEISTOCENE), there is evidence in the world's stratigraphical record to suggest that there were ice ages in Carbo-Permian times, in the Ordovician, and at least three in Pre-Cambrian times (including the Canadian *Huronian* between 2,500 million and 2,000 million years ago, in which three glacial episodes have been recognized). An ice age comprises several glacial maxima interspersed by INTERGLACIAL stages (e.g. HOXNIAN). The glacial–interglacial cycle appears to have a periodicity of some 40,000 years and is thought to result largely from a wobble in the Earth's ECLIPTIC (or tilt) from 21½° to 24½°, modified by two other perturbations of Earth's motion, namely the ECCENTRICITY of the orbit (periodicity of about 90,000 years) and the PRECESSION OF THE EQUINOXES (about 21,000 years periodicity). Longer cyclic changes cannot be recognized with certainty, although tectonic upheavals and changes in levels of atmospheric carbon dioxide are thought to be closely linked with any possible periodicity in ice ages. *See also* LITTLE ICE AGE, MILANKOVITCH HYPOTHESIS, QUATERNARY.

ice apron **1** A thin sheet of ice or snow which adheres to a mountain slope, often at a steep angle. **2** A civil engineer's term for a ramp on the upstream side of a bridge pier which serves to break up floating ice.

ice avalanche AVALANCHE.

ice barchan A crescentic dune or BARCHAN made not by sand but by wind-blown ice crystals at very low temperatures. *See also* SNOW BARCHAN.

ice barrier **1** The outer edge of the Antarctic ice-sheet, namely the *Ross Ice Barrier*. **2** A type of ICE DAM blocking the meltwaters of a glacier. *See also* ICE FRONT.

iceberg A mass of floating ice in a lake or in the sea, which has broken away from an ICE-SHELF or a GLACIER. Arctic icebergs, which are castellated and ragged in character, and Antarctic icebergs, which are mainly flat-topped or tabular, having broken away from the edge of an ICE BARRIER. *See also* BERGY BITS, FAST ICE, GROWLER, ICE-CAKE, ICE CLUSTER, ICE-FLOE, ICE JAM, PACK ICE.

ice blink A whitish glare above the horizon due to the reflection of an ICE SHEET or PACK ICE on to the underside of cloud layers.

ice-cake A flat fragment of floating ice, sometimes classified as an ICE-FLOE less than 10 m in diameter. *See also* ICE PAN.

ice-cap A covering of ice over a tract of land. A distinction has sometimes been made between the term ice-cap (indicating a smaller-sized ice cover) and the term ice-sheet (indicating a very extensive ice cover). Thus ice-cap is frequently used to describe a dome-shaped cover in a highland massif during a mountain glaciation. But accepted usage also includes reference to the 'Greenland ice-cap' and the 'Polar ice-caps' which are of much greater dimensions. *See also* GLACIER, ICE-SHEET.

ice carapace A thin mass of ice occurring on steep rock walls, similar to an ice apron, but flowing like a HANGING GLACIER towards the glacier-filled valley beneath. It lies above the level of the normal glacier TRIM LINE.

ice cascade ICE-FALL.

ice cluster A concentration of floating ice of very wide extent which covers several hundred square km of sea.

ice-contact slope The steeply angled slope of a KAME or other glaciofluvial deposit interpreted as the slope against which the former ice body stood in juxtaposition.

ice-cored moraine Any morainic mound or ridge marginal to an ice-sheet or glacier which encloses buried ice. It is due either to a considerable accumulation of MORAINE in the glacier's terminal or lateral area, particularly in a state of ABLATION (DEAD-ICE FEATURES), or to the burial of ice avalanches or marginal snow banks by morainic debris. Ice-cored moraines are not normally sharp-crested but consist of several round-topped, curved ridges, and research suggests that their formation occurs more readily in regions experiencing a CONTINENTAL CLIMATE, where both rates of *accumulation* and *ablation* are relatively slow and where surface melting in summer is only limited. Ice-cored moraines may be recognized from aerial photographs but have to be confirmed by geophysical techniques or by drilling. *See also* THULE–BAFFIN MORAINES.

ice crystal An individual ice particle which exhibits a regular structure. *See also* SNOW CRYSTAL.

ice dam **1** An obstruction caused by floating ICE-FLOES when they accumulate during a melting phase, either in a river or in a lake; ICE JAM. **2** An ICE BARRIER, formed by a *glacier* or an *ice-sheet* which causes the ponding of meltwaters or natural drainage to create an *ice-marginal lake*, such as the Marjelen See, formed by the Aletsch glacier in Switzerland. *See also* PRO-GLACIAL LAKE.

ice-dammed lake ICE DAM.

ice dome DOME.

ice edge The interface between a mass of floating sea-ice and the open water beyond.

ice-fall **1** A cataract of ice where a GLACIER moves down a steep slope gradient. It is characterized by numerous CREVASSES or ice-chasms, and by unstable ice pinnacles known as SÉRACS. An example is the Khumbu ice-fall beneath Everest. **2** The collapse of a mass of ice from an ICEBERG or a mountain ICE-CAP, which in the latter case leads to an ICE AVALANCHE.

ice-field A term with two distinct usages, only one of which is officially recognized: **1** Both the UK and the USA use the term officially with reference to a widespread and continuous area of floating PACK ICE, generally greater than 10 km in diameter, especially in polar regions. **2** Used more loosely to describe large areas of land ice, especially in Canada, namely the Columbia Ice-field in Jasper National Park, Alberta.

ice-floe A discrete, tabular mass of ice floating on a water body; formed after the break up, usually by spring gales, of the winter FAST ICE. It is flatter and thinner than an ICEBERG and larger than an ICE-CAKE. An accumulation of a large number of individual floes represent the PACK ICE. *See also* BERGY BITS, GROWLER, ICE CLUSTER, ICE JAM.

ice flow GLACIER FLOW.

ice-fog A surface fog associated with ice-sheets and ice-shelves where the lowest layers of the atmosphere are cooled by the ice surface sufficiently for minute ICE CRYSTALS to form in the air. Here they will remain suspended only when calm conditions prevail. Sunlight shining through this 'veil' of ice crystals produces a dazzling effect conducive to snow blindness unless remedial measures are taken. Ice fog also produces optical phenomena such as haloes, owing to the refraction of the Sun's rays. *See also* RADIATION FOG.

ice foot A narrow strip of sea-ice which forms along the shore in Arctic latitudes between high- and low-water marks and is unmoved by the tides.

ice front Strictly speaking the term ice front is applicable only to the vertical cliff which forms the seaward limit of a floating glacier or an ICE-SHELF. The term is also used commonly to describe the distal limit of an ice-sheet on a land surface, although the term *ice wall* has been coined to describe the latter usage.

ice jam **1** A temporary accumulation of small, broken ICE-FLOES which have been carried downstream during the seasonal melting phase and have accumulated against an obstruction to form a type of ICE DAM across a river. **2** A mass of rafted ICE-FLOES which pile up along the shorelines of seasonally frozen lakes.

Icelandic Low One of the cells in the semi-permanent zone of subpolar low pressure, with an average January pressure of 994 mb, located over the North Atlantic between Iceland and Greenland. It is not a stationary low-pressure system but represents the mean pressure resulting from the frequent passage of deep depressions across this area. Since these are more marked in winter the Icelandic Low is more characteristic of the winter months than those of the summer, but even then is only occasionally replaced by high-pressure cells. *See also* ALEUTIAN LOW.

ice-marginal lake PRO-GLACIAL LAKE.

Icenian A term formerly given to the uppermost stratigraphic division of the CRAG deposits of East Anglia. It included the Norwich Crag, the Chillesford Beds and the Weybourn Crag, all of which were thought to represent part of an enormous deltaic system building out into the Pleistocene North Sea. The Icenian Crag sediments of shelly sands and clays overlie the RED CRAG and are succeeded by the CROMERIAN. The term is no longer used and the sediments it described have now been categorized under the following stages of the Pleistocene (oldest to youngest): LUDHAMIAN, THURNIAN, ANTIAN, BAVENTIAN, PASTONIAN, BEESTONIAN.

ice pack PACK ICE.

ice pan A large piece of flat, floating sea-ice, rising less than 1 m above the surface of the sea. *See also* ICE-CAKE.

ice piedmont An ICE-FIELD located on a low-lying strip of land between the ocean and a range of coastal mountains. It frequently terminates in an ICE-SHELF. It is formed when valley GLACIERS coalesce on a coastal plain to form a thick accumulation of virtually stagnant, crevasse-free ice. *See also* PIEDMONT GLACIER. [*92*]

ice pillar, ice pedestal SÉRAC.

ice plucking A process of block removal or quarrying carried out by glacier ice whereby well-jointed blocks are mechanically pulled from the rock floor when the basal ice freezes to them, generally on the downstream side of a rocky protuberance (ROCHE MOUTONNÉE). It is one of the several types of EXARATION. *See also* JOINT-BLOCK REMOVAL, ROBIN EFFECT. [*212*]

ice rafting The action by which rocks and other glacially eroded debris are transported by floating ice, either ice-floes or icebergs. They may be transported great distances, eventually being deposited on beaches or being dropped into soft sediments on the sea-floor as the floating ice melts. This ice-rafted debris is known to be widely deposited on the sea-floor where the Arctic and Antarctic ice-floes meet the warmer currents of temperate latitudes. *See also* ERRATIC.

ice rampart LAKE-RAMPART.

ice rind A thin crust of sea-ice formed by the initial freezing of the surface of a calm body of water. Sometimes termed *slush* or *sludge.*

ice-scoured lowland An area of low relief thought to have reached this condition largely owing to the erosive effects of an ice-sheet.

ice segregation The formation of lenses or bands of clear ice in frost-susceptible materials, especially silt, partly by freezing of existing interstitial soil water but largely by freezing of groundwater drawn upwards by capillary action. Such formation is conducive to uplift of the soil layers, known as frost-heaving or CRYOTURBATION. *See also* CONGELITURBATION, PERMAFROST.

ice-sheet A large, continuous layer of land ice of considerable thickness. The term is generally confined to descriptions of ice masses covering regions on a continental scale (e.g. the Antarctic or Greenland) or to describe the land-ice masses which covered much of the hemisphere during the Quaternary Ice Age. It is larger than an ICE-CAP. [*109*]

ice-shelf A floating ICE-SHEET of great thickness and considerable extent, often the seaward continuation of a land ice-sheet. *See also* ICE BARRIER.

ice storm A period of heavy rainfall in which the rain freezes as soon as it comes into contact with an object the temperature of which is below FREEZING-POINT. *See also* FREEZING DRIZZLE, GLAZED FROST.

ice stream, ice streaming That part of an ICE-SHEET in which there occurs a corridor

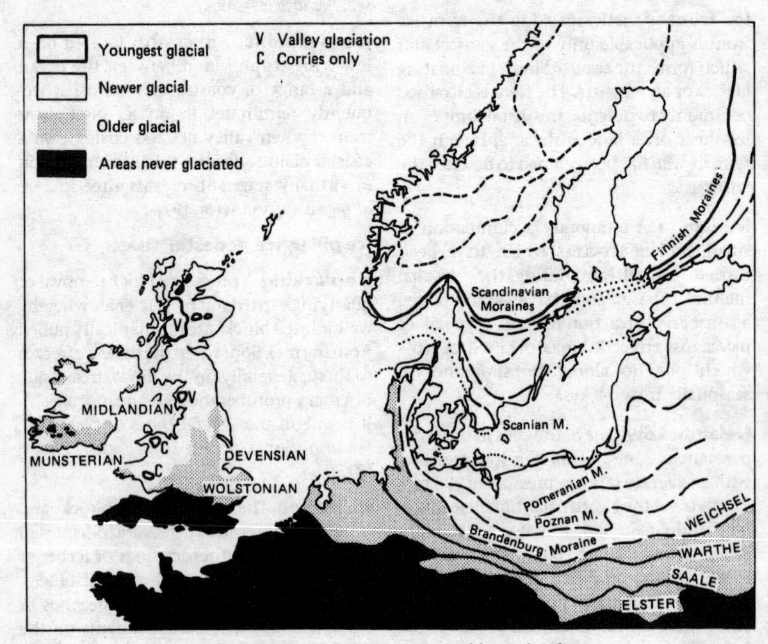

Figure 109 *Glaciation of north-west Europe, showing extent of former ice-sheets*

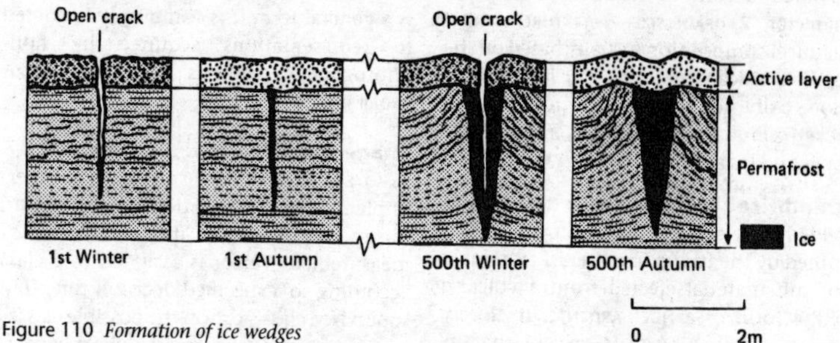

Figure 110 *Formation of ice wedges*

or stream of faster-moving ice surrounded by large areas of slow-moving or stagnant ice. Such differential movement, or streaming, is thought to account for the fact that while some parts of an ice-sheet form a protective cover over summit tors and other rock phenomena, other parts move considerably faster to produce ice-scoured terrain or deep glacial troughs in the same highland massif.

ice streaming ICE STREAM.

ice vein ICE WEDGE.

ice way A linear ice-scoured trough in a lowland terrain, thought to be eroded by an ICE-SHEET, due either to ICE STREAMING or to differences in rock structure and hardness giving rise to DIFFERENTIAL EROSION.

ice wedge A narrow crack or fissure in the ground infilled with ice which may extend below the PERMAFROST level. These V-shaped phenomena, which start life as *ice veins*, have a typically vertical orientation in contrast to the horizontal alignment of an ice lens or bed of ice. Because of the many small air bubbles contained in the wedge the ice has a milky appearance. It is formed by contraction caused by frost-cracking at the surface in a PERIGLACIAL environment, followed by the filling of the initial crack by summer meltwaters to form an ice vein (<5 cm broad) which ultimately widens by addition of frozen water to form a wedge. If there is an amelioration of climate, the ice infilling may disappear and the *ice-wedge cast* become infilled by sediments so that its former location may be seen in section. The largest ice wedges may achieve diameters up to 10 m and they may be buried or reach the surface, where they usually create POLYGONS of frozen ground. The earlier views of S. Taber that both ice veins and ice wedges were not due to contraction cracking but to a downward growth of ground ice from ice lenses due to ICE SEGREGATION are no longer accepted. [*110*]

ichor A term proposed in 1933 to describe the magmatic fluid which penetrates the country-rock around a granitic intrusive body and leads to GRANITIZATION.

icicle A pendant of clear ice, generally in the form of a spike, formed by the freezing of dripping water.

icing The accumulation of a coating of ice on an exposed object. Icing occurs whenever SUPERCOOLED DROPLETS in a cloud come into contact with a surface, such as the leading edges of aircraft wings. Alternatively it can be produced at ground-level by the freezing of droplets on impact, thereby leading to GLAZED FROST. Hence *anti-icing* (prevention) and *de-icing* (removal). *See also* ICE STORM, RIME.

iconic model A MODEL of a real world situation reduced in scale but presenting the same properties. It contrasts with a SIMULATION MODEL.

igneous rocks Rocks that have formed from the crystallization and solidification of MAGMA. They include HYPABYSSAL, PLUTONIC, PYROCLASTIC and VOLCANIC ROCKS, and are classified according to their **1** feldspar character, **2** GRAIN SIZE, **3** TEXTURE, and **4** chemical composition (usually based on the degree of silica saturation). The last of these allows a division to be made into silica-rich rocks (ACID ROCKS BASIC ROCKS, INTERMEDIATE ROCKS and ULTRABASIC ROCKS.

ignimbrite An alternative term for a welded TUFF (one of the PYROCLASTIC rocks) formed by the fusing together of extremely hot tuff material ejected from a volcano in the form of a NUÉE ARDENTE. It closely resembles a flow RHYOLITE, except that its fragments and glass shards are drawn out within the very distinctive banding and appear as flattened fragments.

IGY International Geophysical Year. The most recent took place from July 1957 to December 1958.

Illinoian The penultimate glacial stage of the PLEISTOCENE in N America. It is equivalent to the RISS (Alps), SAALE (N Europe) and the WOLSTONIAN (Britain).

illite One of the groups of the CLAY MINERALS, intermediate in composition between MONTMORILLONITE and MUSCOVITE. It is now known as hydrous mica.

illuminated relief The representation of relief on a map by the use of shading to simulate the shadow thrown by an apparent light source shining from one or more directions across a three-dimensional topography. *See also* HILL-SHADING.

illumination, circle of A term referring to the GREAT CIRCLE on the surface of the Earth which distinguishes day from night.

illuvial horizon A soil horizon (usually the B-HORIZON) into which material is precipitated or redeposited by the process of ILLUVIATION. [*235*]

illuviation The process of depositing soil substances (organic matter, clay minerals and hydrous oxides of aluminium and iron) in the ILLUVIAL HORIZON, usually from an upper horizon. *See also* B-HORIZON.

image A REMOTE SENSING term referring to the representation of a scene as recorded by a remote-sensing system. Although image is a general term it is commonly restricted to representations acquired by non-photographic methods, e.g. a LANDSAT or RADAR image.

image classification: non-manual A REMOTE SENSING term referring to a computer-implemented classification process. In an unsupervised classification each IMAGE measurement vector is assigned to a class according to a specified decision rule. In a supervised classification the possible classes on the ground (e.g. forest, wheat, water, etc.) have already been defined, based on the known data characteristics of each class. Image classification is often an important first step to image interpretation. *See also* PATTERN RECOGNITION.

image enhancement A REMOTE SENSING term referring to the process of improving the visual quality of a remotely sensed IMAGE by optical or digital means. For example: **1** contrast stretching involves the expansion of the original range of digital values in order to use the full contrast range of the recording film or display device; **2** edge enhancement involves the emphasizing of rapid spatial changes within an image. The first use would be of value to most image interpreters while the second would be of value when looking for linear features, e.g. faults. *See also* FILTER.

image restoration A REMOTE SENSING term referring to the removal or correction of effects which mar the quality of the IMAGE.

imbricate structure 1 A geological structure associated with a thrust fault or series of thrust faults in which slices or wedges of crustal rocks are pushed one above another (like overlapping tiles on a roof) [*111a*] without folding. **2** A type of sedimentary fabric in a pebbly rock (CONGLOMERATE) in which the long axes of the pebbles overlap each other as they are inclined in the direction of the former current which deposited the sediment. [*111b*]

immature soil A soil which exhibits poorly developed or indistinct horizons.

Sometimes referred to as a juvenile soil. *See also* AZONAL SOIL.

immigrant A term for any animal or plant newly introduced into an environment.

impeded drainage The restriction of the downward gravitational movement of water through the soil.

imperfectly drained A soil that exhibits a small amount of reduction of iron (GLEYING) owing to short periods of water-logging.

impermeability factor (runoff coefficient) An expression referring to the ratio of the amount of rainfall which runs off a given surface material. It is termed the *coefficient of imperviousness* in the USA. Some typical figures are: slate or tiled roof (0.70–0.95); cobblestones (0.40–0.50); tarmacadam surface (0.25–0.60); gravel road (0.15–0.30); grassland (0.05–0.30); woodland (0.1–0.20).

impermeable **1** The function by which water or other fluids cannot pass through a soil or rock. This may result from the lack of joints or fissures (making the material *impervious*) or because the material is non-porous (POROSITY). NB Clay is porous but becomes impermeable once its pore-spaces become filled with water. **2** Resistant to penetration by plant roots.

impervious IMPERMEABLE.

impoverishment of a soil The action of making a soil less productive by the removal of its nutrients, naturally by LEACHING or artificially by excess cropping.

improvement of a soil The action of making a soil more productive for plant growth, by IRRIGATION, ground drainage, addition of fertilizers, etc.

in-and-out channel An expression used to describe a type of marginal MELTWATER CHANNEL developed in solid rock along lateral ice margins. It may have been formed by superimposition of a meandering SUPRAGLACIAL or ENGLACIAL stream which formerly wandered from the glacier on to a hillside and back again.

inceptisol One of the ten orders defined in the SEVENTH APPROXIMATION of soil classification. They are very youthful soils in which their poorly developed horizons have developed quickly (hence their name, derived from Latin *inceptum* = beginning), usually on newly exposed surfaces recently vacated by an ice or snow cover or on recently accumulated floodplain sediments. They are characterized by a lack of many translocated materials, such as iron or aluminium, in the accumulation horizon, where only carbonate minerals and non-crystalline silica are deposited. In this respect they are more advanced than the ENTISOL but less developed than the ALFISOL. They are widely found in the tundra regions, in montane environments and locally on newly extruded lava flows (ANDOSOL). They include the HUMIC GLEY SOILS, the BROWN EARTHS (BROWN FOREST SOILS) (US usage) and some LATOSOLS.

inch **1** A unit of length (1 in = 2.54 cm). [*136*] **2** A Gaelic (Scottish and Irish) term for a small island, usually rocky.

incised meander A general expression for a river MEANDER which is cut down deeply into bedrock owing to REJUVENATION of the drainage. It includes both an INGROWN MEANDER and an INTRENCHED MEANDER. [*112a* and *b*]

inclination The angle of DIP of a rock stratum, fault or mineral vein.

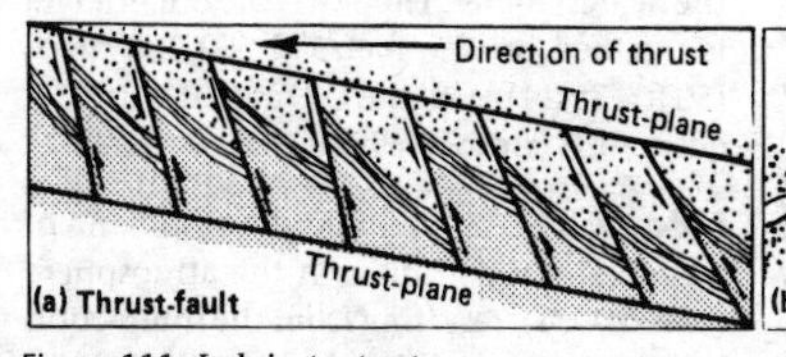

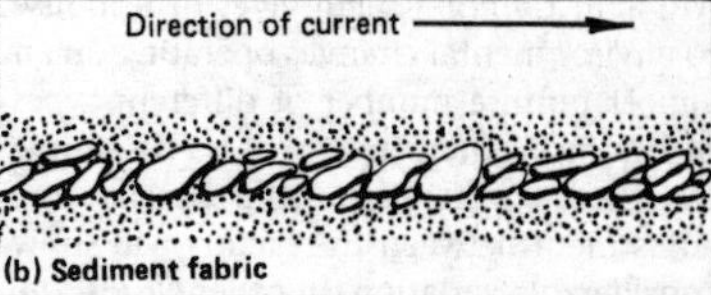

Figure 111 *Imbricate structure*

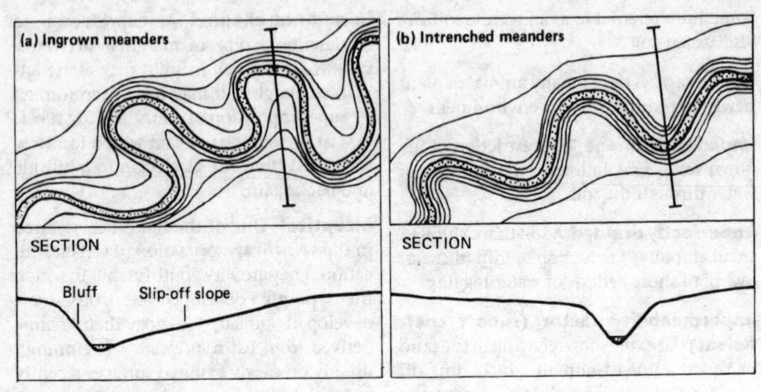

Figure 112 *Incised meanders*

inclinometer CLINOMETER.

inclusion The enclosure of an older rock fragment within a younger rock (XENOLITH) or of crystals of one mineral within another.

incompetent bed A bed of rock which is insufficiently strong to resist stresses incurred during folding, such that it deforms plastically rather than bends, thereby leading to flowage or the development of a slaty CLEAVAGE. It contrasts with a *competent bed* (COMPETENCE).

inconsequent drainage A drainage network that shows no relationship with the underlying structure, i.e. it is not *consequent* upon it, because the drainage either has been SUPERIMPOSED or is ANTECEDENT. It is a type of DISCORDANT DRAINAGE. *See also* INSEQUENT DRAINAGE.

indeterminacy in geomorphic systems A concept introduced in 1964 by L. B. Leopold, M. G. Wolman and J. P. Miller, suggesting that since there are a large number of interrelated factors involved in the creation of landforms and since these factors will adjust among themselves in response to environmental changes operating on an almost infinite number of different types of landform, then there will be no simple cause-and-effect relationship because the adjustment between the factors will show considerable variation. In other words, the relationship will not be a deterministic one (DETERMINISTIC MODEL, DETERMINISTIC PROCESS) but one of indeterminacy. 'This indeterminacy in a given case results from the fact that the physical conditions, being insufficient to specify uniquely the result of the interaction of the dependent variables, are controlled by a series of processes through which any slight adjustment to a change imposed from the environment feeds back into the system. . . . Thus the system tends toward mutual adjustment and equilibrium rather than toward instability.' *See also* EQUILIBRIUM, FEEDBACK CONTROL SYSTEM, GENERAL SYSTEMS THEORY, PROCESS–RESPONSE SYSTEM.

index contour A CONTOUR accentuated in thickness and usually carrying its value in figures in order to assist the reading of elevations on a map. On the maps of the British ORDNANCE SURVEY it occurred at every 250-ft contour on the 1 : 63,360-scale maps and at every 100-ft contour on the 1 : 25,000-scale maps. On the succeeding 1 : 50,000-scale (metric) maps of the Ordnance Survey, the nearest metric equivalent to the 250-ft contour is thickened and its value is given to the nearest metre. The 50-ft contour interval is retained on the metric 1 : 50,000 maps (50 ft = 15.24 m). In the US maps it is usually every fifth contour which is accentuated.

index cycle A periodic development of a general wave motion in the atmosphere (ROSSBY WAVE), resulting from the interaction between polar and tropical air. Over a six-week period, especially in winter, large wave amplitudes develop (termed a low ZONAL INDEX) associated with pronounced oscillations of the JET STREAM, itself occurring along the deformed boundary of the cold polar air stream and the warm tropical air. The cycle terminates when cut-off low-pressure systems (DEPRESSION) develop in the cold air. These destroy the waves and lead to a redevelopment of small-amplitude waves (this state is termed a high ZONAL INDEX). [113]

index fossil A fossil which is restricted to a particular ASSEMBLAGE ZONE and whose name becomes indicative of a particular biostratigraphic zone (BIOSTRATIGRAPHY).

index map A small-scale map portraying the numbers and/or names of larger-scale maps in the same or related series. It is generally in the form of a simple outline map showing the sheetlines of the maps as a means of easy reference.

index of flatness FLATNESS INDEX.

index of roundness ROUNDNESS INDEX.

Indian clinometer CLINOMETER.

Indian summer **1** A period of mild and sunny weather commonly occurring between late October and early November in the USA. Its origin is obscure; some say that the period was used by the American Indians to store their grain in preparation for winter, others that the particular meteorological conditions were first noticed in Indian territory. **2** In the UK the term has been adopted to describe the spell of fine calm weather which sometimes occurs in the late autumn.

indifferent equilibrium NEUTRAL STABILITY.

indigenous An alternative term for *in situ*, referring to any substance (e.g. sediment or ore) which has not been derived from elsewhere or to any organism that has originated in a specific location and remains *in situ*.

inductive reasoning A type of reasoning which attempts to explain phenomena by

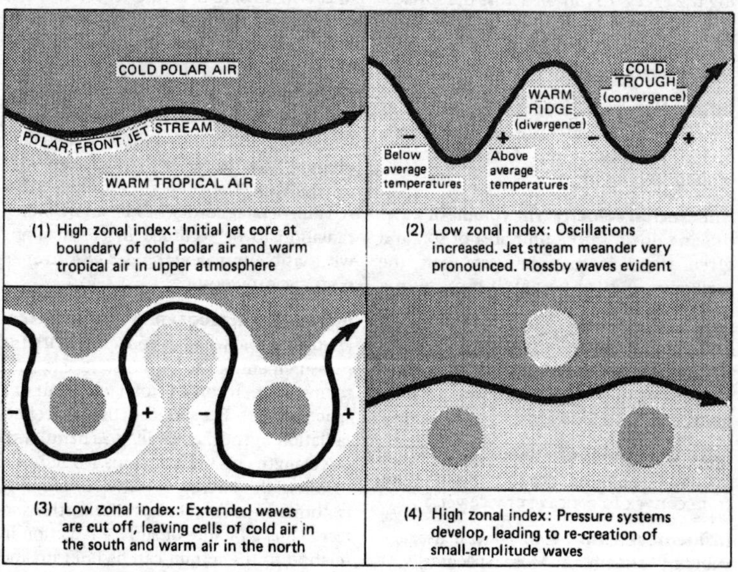

Figure 113 *Index cycle*

drawing on EMPIRICAL rather than theoretical evidence. *See also* DEDUCTIVE REASONING.

induration **1** The process by which sediments are hardened by heating, compression and cementation (DIAGENESIS). **2** The hardening of a soil by chemical processes to form a HARDPAN.

inface The steeper slope of a CUESTA, used especially in a description of coastal lowlands in the eastern USA where the inface is on the landward side of a cuesta.

infiltration The process by which water enters *into* a soil or porous substance in a downward direction from the surface by means of its pores or small openings. Infiltration should be distinguished from PERCOLATION, which denotes flow of water *through* a soil or porous substance. *See also* THROUGHFLOW.

infiltration capacity INFILTRATION-RATE.

infiltration-excess overland flow The type of OVERLAND FLOW occurring when rainfall intensity exceeds the infiltration capacity of the soil (INFILTRATION-RATE), causing the excess to run off across the surface. It is the most important cause of overland flow in the semi-arid zone. *See also* SATURATION OVERLAND FLOW.

infiltration-rate The maximum rate at which water can enter the soil under specified conditions. It was formerly termed *infiltration capacity*.

infiltration velocity The volume of water moving down into a unit area of soil at a given time. It may be less than the maximum (INFILTRATION-RATE) because of a limited supply of water (from rainfall or irrigation).

infiltrometer An instrument for measuring the rate of entry of a fluid into a porous body, e.g. water into a soil.

inflection, point of The position at which a curve changes its direction of curvature, from convex to concave or vice versa.

influent stream **1** In general usage, a tributary stream. **2** More specifically, a stream or reach of a stream from which water is lost by seepage to the WATER-TABLE that lies at a lower level than the stream channel. Such streams are characteristic of arid lands and of limestone and chalk terrain where the surface flow may be seen only in irregular stretches.

information theory The science of the transmission of messages or information by communication systems. Information is a property not intrinsic to any one message but to a set of messages.

infrared Part of the electromagnetic spectrum (RADIATION). In REMOTE SENSING two infrared regions are recognized: **1** The *near infrared*, from a wavelength of 0.7 μm to around 3.0 μm. This is the region in which ELECTROMAGNETIC RADIATION is strongly reflected by plant material. **2** The *thermal infrared*, from about 3.0 μm to around 20.0 μm. Aerial INFRARED THERMOGRAPHY is undertaken within the ATMOSPHERIC WINDOW of 8–14 μm to produce thermal 'maps' of the Earth's surface. *See also* INFRARED PHOTOGRAPHY. [*70*]

infrared photography A technique used in REMOTE SENSING to record reflected visible and INFRARED radiation with wavelengths of up to 0.9 μm. It should not be confused with the recording of temperature (INFRARED THERMOGRAPHY). Black and white infrared film can be filtered into discrete wavebands, e.g. red or infrared, which enables identification to be made of objects with different spectral properties. Colour infrared film or FALSE COLOUR FILM enables infrared radiation to be recorded in colour along with visible wavelengths. *See also* ELECTROMAGNETIC RADIATION.

infrared thermography The branch of REMOTE SENSING which is concerned with the measurement of the Earth's surface radiant temperature, recorded from an aircraft or a space SATELLITE. The surface of the Earth emits radiation in the *thermal infrared* band, with wavelengths of 3–20 microns. To monitor or 'sense' this energy, a THERMAL LINESCANNER is normally utilized to scan the Earth's surface. Thus a picture of surface radiation in relation to the terrain can be built up and recorded electronically on film or magnetic

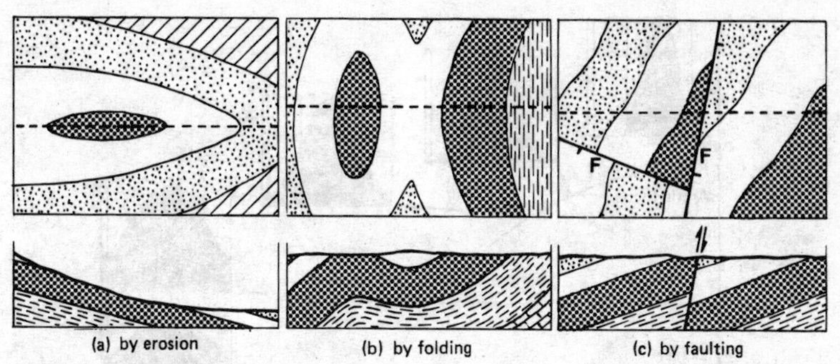

Figure 114 *Formation of an inlier*

tape; the resulting thermal image is termed a *thermogram*. Applications of this technique include urban heat-loss studies, crop surveys and soil-moisture measurement. *See also* ELECTROMAGNETIC RADIATION, INFRARED PHOTOGRAPHY.

infrastructure The zone of complex MIGMATITE and GRANITE structures which exist under conditions of high pressure and temperature deep within the Earth's crust and beneath the less complex folding of the non-migmatic zone.

ingrown meander A type of INCISED MEANDER with asymmetric valley sides. A gentle slope (SLIP-OFF SLOPE) on one flank is faced by a steep slope (BLUFF) on the other, indicating that incision due to REJUVENATION occurred more slowly than the development of the meanders, so that lateral erosion of the river banks for a time kept pace with the vertical erosion of the channel floor, in contrast to the formation of an INTRENCHED MEANDER. *See also* MEANDER. [*112a*]

initial landform The relatively unaltered landform relatively unmodified by geomorphic processes after it has been uplifted and prior to the development of SEQUENTIAL LANDFORMS. In terms of the W. M. Davis concept the initial landform would be the starting-point of the geographic cycle (CYCLE OF EROSION) [*53*].

injection gneiss A gneiss formed very largely by the LIT-PAR-LIT injection of granitic magma.

injection structures Structures formed in certain sediments when a lower layer invades the overlying layer in the form of flame-like streamers or pierces it sufficiently for the lower sediment to flow out on the surface (MUD VOLCANO). Such movements will only occur under particular conditions of water content and under certain types of stress, e.g. earthquake shocks.

inland drainage ARETIC DAMAGE.

inland drainage basin BOLSON, PLAYA.

inland sea A large body of saline or brackish water isolated from and not connected to the open ocean, e.g. the Caspian Sea.

inlet A short narrow opening or waterway which connects a smaller body of water with a larger body, either of fresh water or sea water. *See also* TIDAL INLET.

inlet cave A cave occurring below a SWALLOW-HOLE where surface drainage in a KARST terrain first passes underground. *See also* OUTLET CAVE. [*204*]

inlier An exposure of older rocks which is completely surrounded by younger strata, resulting particularly from the erosion and breaching of an anticlinal structure, but which can also be formed by faulting. It contrasts with an OUTLIER. [*114*]

inner core The probably solid centre of the Earth's CORE, below a depth of 5,000 km.

inputs and outputs Two interrelationships which link an OPEN SYSTEM with its

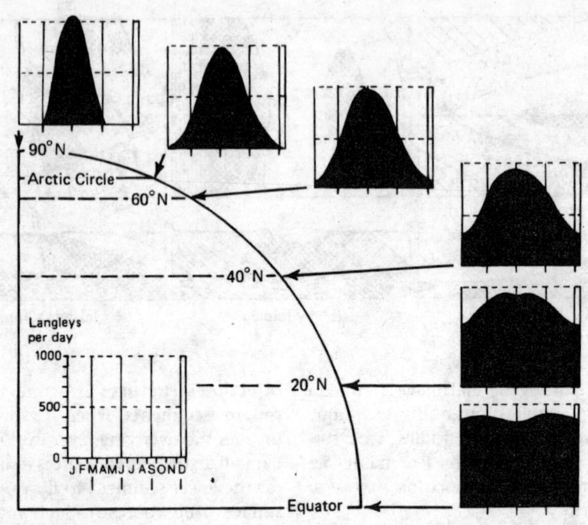

Figure 115 *Insolation*

external environment. Inputs are stimuli which energize the system and provide it with the external material to make it operative. Outputs result from processes and represent the effects of the system itself on the environment. The variable whose value is given as input to a system is termed the independent variable, while that whose value is determined by the system is the dependent variable. *See* VARIABLE. [*249*]

inselberg A German term (literally, 'island mountain') that has been widely adopted to describe a prominent steep-sided hill of solid rock, rising abruptly from a plain of low relief. It is characteristic of tropical landscapes, particularly in the SAVANNA zone, and is generally composed of a resistant rock, such as a granite. Although it may appear to bear a morphological resemblance to a MONADNOCK it is thought to be derived by the process of PARALLEL RETREAT OF SLOPES in which PEDIMENTS encroach into residual uplands during the process of PEDIPLANATION. [*193*]. The inselberg is the end-product of this encroachment and may occur as an isolated hill or in a residual group of hills. Some authors believe that inselbergs are created by the process of deep CHEMICAL WEATHERING, when the buried resistant rock-masses below the BASAL SURFACE OF WEATHERING are gradually revealed by stripping of the overlying regolith (EXHUMATION), but it is preferable to refer to each of these cases as a BORNHARDT (SHIELD INSELBERG). *See also* ROCK DOME, ROCK PAVEMENT, ZONAL INSELBERG. [*180* and *282*]

insequent drainage 1 A drainage pattern which wanders irregularly across the surface and is not influenced by topographic features (DENDRITIC DRAINAGE). This is the normal British usage. **2** In the USA the usage is synonymous with INCONSEQUENT DRAINAGE.

insolation The amount of diffuse and direct solar RADIATION which reaches the Earth's surface. It plays an important role in distinguishing different climatic zones, since its annual amount is 2.5 times greater at the equator than at the poles. It is discharged from the Sun in the form of: **1** very short wavelength X-rays, gamma rays, and ultraviolet rays; **2** visible light rays; **3** short wavelength INFRARED rays. Although the Earth receives only a very tiny fraction of the energy generated by the Sun this is vital to promote terrestrial and atmospheric mechanisms. The amount of insolation reaching the outer limit of the atmosphere is termed the SOLAR CONSTANT but 53% of this is lost during the passage through the atmosphere owing to: **1** ABSORPTION; **2** SCATTERING; **3** reflection back into space from clouds and dust. [*98*] Of the 47% which reaches the Earth's surface about 6% is reflected by the ALBEDO, which varies according to the nature of the surface. Of the remainder, which is not converted into long-wave infrared RADIATION, some heats the Earth's surface and some the lower layer of the atmosphere by CONDUCTION. A considerable percentage is used in the latent heat of vaporization, which is released with condensation in rising air. Insolation is measured in LANGLEYS. Equatorial insolation is fairly uniform at 800 langleys per day while at the poles the daily amount varies from zero in winter to 1,000 langleys in midsummer. [*115*] (NB The langley is now replaced by the *watt* or *megajoule*.)

insolation weathering THERMOCLASTY.

instability A physical condition of the atmosphere in which a body of air rises and expands because it is warmer and therefore less dense than the air immediately above it. As a result of this upward movement the body of air will cool by ADIABATIC mechanisms; if the rising air is moist such cooling usually leads to PRECIPITATION (ENVIRONMENTAL LAPSE-RATE). *See also* ABSOLUTE INSTABILITY, CONDITIONAL INSTABILITY. [*132*]

instantaneous field of view (IFOV) A REMOTE SENSING term referring to the smallest plane angle at which the remote sensor is sensitive to ELECTROMAGNETIC RADIATION and usually expressed in degrees or radians. When expressed in linear or areal units it is an altitude-dependent measurement of spatial RESOLUTION. A small IFOV means that a greater quantity of total energy is being focused onto the sensor, thereby improving radiometric resolution. A typical MULTISPECTRAL SCANNER will have an IFOV of 2.5 milliradians, giving a ground resolution of 10 m at an altitude of 4,000 m. *See also* FIELD OF VIEW.

instrumented catchment The CATCHMENT AREA of a stream in which a number of recording instruments have been set up in order to measure the HYDROLOGY of the stream. *See also* AT-A-STATION.

insular climate A climate experienced by small islands or groups of islands on which the influence of the ocean is of paramount significance. The most important characteristic is the low seasonal-temperature range, which is especially marked at low latitudes. Islands near the equator frequently exhibit annual average temperature ranges of less than 1° C while even in temperate latitudes the annual average temperature ranges are less than 10° C; e.g. Valentia (Ireland), 8° C; Scilly Is., 9° C.

insular shelf The zone of relatively shallow water surrounding a large island, analogous with a CONTINENTAL SHELF.

integer A whole number, which may be positive, zero or negative.

integral The symbol ∫ used in statistics to denote summation or integration.

intensity of an earthquake A measure of the effects (but not of the MAGNITUDE) of an earthquake at a particular place on the Earth's surface. It is measured on either the MERCALLI SCALE or the ROSSI–FOREL SCALE. Its magnitude is measured on the RICHTER SCALE. [*140*]

intensity of rainfall A measure of the rate at which rain falls, expressed in terms of rainfall amount related to a specific time period. The most common usage is that of hourly intensity expressed by the equation:

$$\frac{\textit{total rainfall in mm}}{\textit{number of rainfall hours}}.$$

In Britain the greatest rainfall intensities are usually associated with midsummer THUNDERSTORMS; e.g. Cannington (Somerset) recorded an intensity of 48 mm/h for 4½ hours in August 1924, while Hampstead (London) endured an intensity of 68 mm/h for 2½ hours in August 1975. The world's highest recorded rainfall intensity within 24 hours was recorded on Réunion Is. in the

Indian Ocean in March 1952 when 1,880 mm fell – an intensity of more than 78 mm/h. Intensity of rainfall is classified into: *Heavy rain*, >4 mm/h; *moderate rain*, 0.5–4 mm/h; *light rain*, <0.5 mm/h.

interaction A statistical term used in ANALYSIS OF VARIANCE to indicate the effects of two or more controlling factors upon each other.

interbedded That which occurs between two layers or beds. A sandstone, for example, may be interbedded between two beds of shale or a lava may be interbedded between beds of sedimentary rocks. It is synonymous with *intercalation*.

intercepted moisture That part of the precipitation which is prevented from reaching the soil by any plant surface. Much of the intercepted water may be evaporated directly from the plant surface back into the atmosphere, but if the precipitation is quite heavy the moisture will ultimately overcome the carrying capacity of the leaves and stems and will drip to the ground. *See also* STEM FLOW, THROUGHFALL.

interface A contact surface between two different substances at which their individual characters change suddenly.

interflow Sometimes referred to as SUBSURFACE STORMFLOW, but is usually interchangeable with the term THROUGHFLOW.

interfluve The area of higher ground separating two rivers which flow into the same drainage system. *See also* DIVIDE, WATERSHED.

interglacial The time period between two GLACIAL stages, during which there is an amelioration of climate, with temperatures in the temperate zone rising to about the level of those of the present day. In the Quaternary ice age in Britain at least five interglacials have been recognized. These are (with increasing age): the IPSWICHIAN, the HOXNIAN, the CROMERIAN, the PASTONIAN and the BRAMERTONIAN. Prior to these the ANTIAN and the LUDHAMIAN stages of the Pleistocene exhibited relatively warmer conditions. The evidence for climatic fluctuations is based on fossil pollen, Foraminifera, Mollusca, etc.

intergranular adjustment (translation) GLACIER FLOW.

intergranular yielding GLACIER FLOW.

interior drainage ARETIC DRAINAGE.

interlobate moraine A term given to a terminal MORAINE in which two lobes merge together and curve back at right angles to the ice-frontal margin as a type of medial moraine.

interlocking spurs The interdigitation of the valley slopes as the stream follows a winding course down a V-shaped valley, whereby each projecting spur overlaps its neighbour and obscures the view up- and downstream. These are unlikely to survive the passage of a valley glacier, in which the spurs would be truncated (TRUNCATED SPURS) and the valley converted to a U-SHAPED VALLEY.

intermediate contour A contour that occurs between INDEX CONTOURS.

intermediate rocks Any igneous rock which contains between 52% and 66% silica, by weight, e.g. DIORITE. *See also* ACID ROCKS, BASIC ROCKS, ULTRABASIC ROCKS.

intermittent spring A spring of pulsating flow caused by the character of the cavities and SIPHONS within the underground drainage system of a karstic area (KARST). When the cavity fills the siphon is complete and a pulse of water will emanate from the spring. Because this usually empties the cavity no further water will issue from the spring until the siphon is reactivated.

intermittent stream A stream which flows only part of the time, usually after a period of heavy rainfall when the intermittent spring, related to the height of the WATER-TABLE, is able to function. It may also function seasonally as a result of snowmelt. It contrasts with a PERENNIAL STREAM. *See also* BOURNE, EPHEMERAL STREAM, INTERRUPTED STREAM.

intermontane Lying between mountains or mountain ranges.

internal drainage ARETIC DRAINAGE.

internal waves BOUNDARY WAVES.

International Atomic Time (TAI) An internationally agreed scale of TIME based on the extremely accurate atomic clocks housed in Paris at Le Bureau Internationale de l'heure. *See also* UNIVERSAL TIME.

International Date-Line An internationally agreed time-change line drawn approximately along the 180° meridian but deviating to both W and E to avoid land areas in the Pacific Ocean. A crossing of the date-line entails repeating one day when travelling eastwards and omitting one day when travelling westwards. This is to compensate for the accumulated time-change of 1 hour for every 15° of longitude (TIME ZONE). [*116; see also 241*]

International Map of the World A series of maps at a scale of 1 : 1,000,000 published by a number of countries to an internationally agreed specification, accepted in principle in 1909. It is constructed on a MODIFIED POLYCONIC PROJECTION, with each sheet normally covering 6° of longitude and 4° of latitude, the exceptions being the sheets of high latitudes where the size covers 12° and 4° respectively. The map production has now been centralized at the Cartographic Office of the United Nations.

interpluvial A stage of Quaternary time in low latitudes during which rainfall is thought to have decreased from the maxima recorded during the PLUVIAL stages. It was once believed that there was a direct correlation between the INTERGLACIAL stages of higher latitudes and the interpluvials of low latitudes, but this correlation is no longer accepted.

interpolation A statistical term referring to the insertion of values between measured values. Because the inference relates to unknown areas *within* the scatter of known values it is likely that the degree of error will be small. Interpolation is regularly used, for example, in the drawing of contours on a map, with reference to a series of measured spot heights. *See also* EXTRAPOLATION.

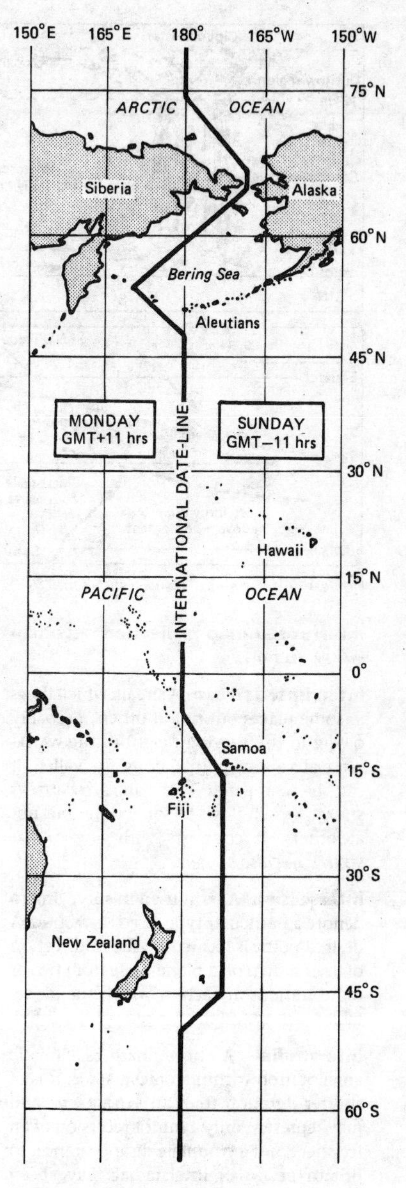

Figure 116 *International Date-Line*

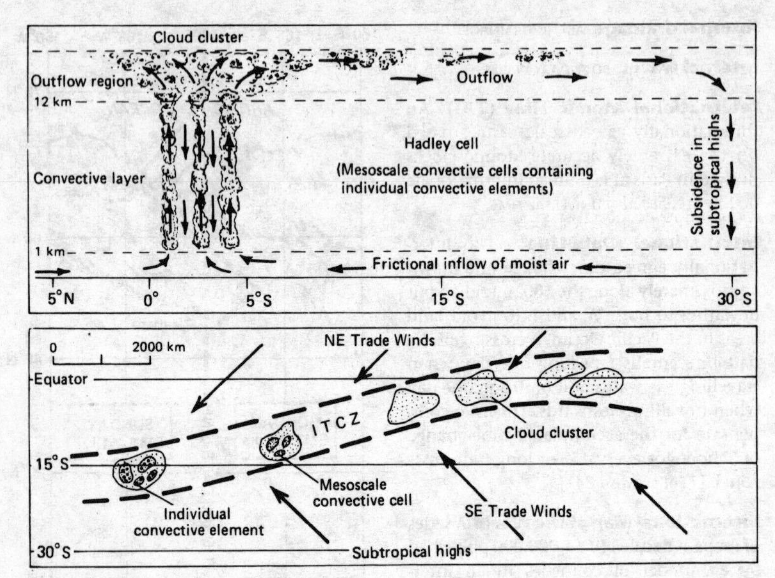

Figure 117 *Intertropical Convergence Zone*

interrupted map projection RECENTRED MAP PROJECTION.

interrupted stream A stream which flows at some places but not at others, probably owing to the varying height of the WATER-TABLE at various points along the valley. It can be compared with an INTERMITTENT STREAM, which flows at some times but not at others. It contrasts with a PERENNIAL STREAM. *See also* BOURNE.

intersection A term used in surveying to denote a particular type of PLANE-TABLING. A desired point is located by the intersection of lines drawn on a plane-table from two or more stations at each of which the plane-table is set up and the desired point sighted.

interstadial A short phase of climatic amelioration within a GLACIAL STAGE. It is of shorter duration than an INTERGLACIAL and may represent only a minor recession of an ice-sheet, not a complete disappearance. In Britain DEVENSIAN interstadials have been recognized at Chelford (Cheshire) and Upton Warren (Worcestershire). *See also* ALLERØD INTERSTADIAL, BØLLING INTERSTADIAL.

interstices Voids occurring as fissures or PORES between the grains or crystals of a rock. The number of INTERSTITIAL spaces will be closely related to the degree of POROSITY of a rock.

Intertropical Convergence Zone (ITCZ) Formerly known as the *Intertropical Front* (ITF), until it was discovered that in general the air masses converged not at a FRONT but at a broad zonal trough of low pressure in equatorial latitudes. The convergence zone lies more or less parallel with the equator but moves N and S with the seasonal passage of the THERMAL EQUATOR. It marks the confluence of the TRADE WINDS, which blow towards the equator, and the meeting-place of northern and southern Tropical Maritime AIR MASSES. Since the zone is weakly defined over the oceans, especially in the areas referred to as the DOLDRUMS, and since there is a discontinuity of convergence both in space and time, modern climatologists have introduced the term *Intertropical Confluence* (ITC). Over some continental areas the converging air masses may exhibit contrasting humidity characteristics (e.g. W

Africa and S Asia) so that mid-latitude frontal conditions may be generated and a slow-moving LOW developed in the *Equatorial Trough*. Occasionally these lows move away from the zone and intensify into severe tropical storms (CYCLONE). The zone is characterized by convection, induced by convergence, which generates cumuliform clouds and CLOUD CLUSTERS. *See also* ATMOSPHERIC CIRCULATION, HADLEY CELL. [*117, see also 18*]

interval 1 The range between two extremes within which a variable can take any value. **2** In statistics, interval data contrast with ordinal data in having real number values rather than INTEGER placings.

interval scale SCALES OF MEASUREMENT.

intraformational Formed by or within an existing geological rock formation. In a newly consolidated sediment, for example, eroded fragments would become instantly incorporated in the continually forming sedimentary formation.

intrazonal soils Soils, the characteristics of which have been influenced by local factors of parent material, terrain or age, rather than of climate or vegetation. They can be subdivided into CALCIMORPHIC SOILS; HALOMORPHIC SOILS; and HYDROMORPHIC SOILS. *See also* ZONAL SOILS.

intrenched meander A type of INCISED MEANDER, with steep and symmetric valley sides indicating that incision of the drainage through REJUVENATION was relatively rapid, so that vertical erosion of the channel floor was more effective than lateral erosion of the river banks, in contrast to the formation of an INGROWN MEANDER. In the USA the intrenched meander is termed an *entrenched meander*. *See also* MEANDER. [*112b*]

intrinsic permeability The ease with which liquids can pass through a porous material. It is measured in DARCYS. For example, the intrinsic permeability of sandstone is between 1 and 2 darcys, that of sand varies between 25 and 140 darcys.

intrusion A body of igneous rock which has been injected in a molten state (MAGMA) into existing crustal rocks (COUNTRY-ROCK) and which, on cooling, becomes an igneous intrusive rock or igneous intrusion.

inversion of relief (inverted relief) Topography created by a prolonged period of denudation in which anticlines form the valleys or low ground and the synclines coincide with the peaks or the high ground. This is because the tensional joints along the anticlinal crest are structurally less resistant to denudation than are the tightly compressed joints along the synclinal axis. Thus, there is a tendency for the anticlines to be eroded more rapidly than the synclines. [*118*]

inversion of structure A reversal of the PRINCIPLE OF SUPERPOSITION, due either to the folding back of strata upon themselves or to overthrusting of older rocks across younger rocks along a THRUST-FAULT. Many examples of this type of structural inversion can be seen in areas of considerable tectonic activity, e.g. the NAPPE structures of the Alps.

inversion of temperature Strictly, *inversion of temperature gradient*. Normally air temperature decreases with increasing elev-

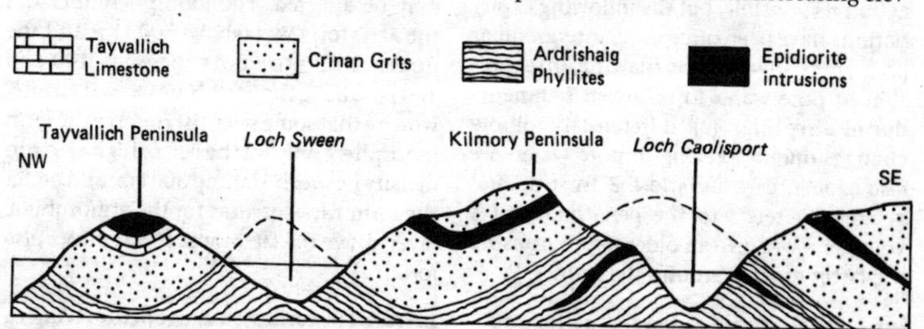

Figure 118 *Inversion of relief (SW Scotland)*

ation but under certain weather conditions the converse may be true. Thus, contrary to the normal ENVIRONMENTAL LAPSE-RATE, over a limited height range, air temperature increases with height so that a layer of warmer air overlies a colder layer. Two types may be recognized: **1** A *high-altitude inversion* due to frontal convergence, when a warm air mass is forced from the ground surface by the undercutting of a cold air mass at a COLD FRONT. Alternatively, a similar inversion can be created when a warm air mass overrides a colder one along a WARM FRONT. Upper-air inversions also develop in the subtropics, associated with the deep subsidence and adiabatic warming in a warm ANTICYCLONE. **2** A *surface inversion* is much more localized and is often dependent on the terrain. It will frequently occur during winter anticyclonic weather when, during calm cloudless nights, there is a rapid heat loss from the ground by *radiation*. Because cold air will flow to a lower level owing to its greater density, temperature inversions are most marked in basins, hollows and narrow valleys. Some mountain settlements have been sited to avoid these cold 'spots', which may become FROST HOLLOWS or frost pockets. [*234*]

Invertebrata All the phyla of animals without a backbone composed of vertebrae.

involution **1** In general, a contortion of superficial rocky material and soils due to slumping, loading and volume changes in sediments. **2** More specifically, the term refers to a deformation of unconsolidated surface materials by PERIGLACIAL activity. It is difficult to determine the exact type of frost action responsible, but the following explanations have been offered: (a) intergranular pressure changes during thawing; (b) expulsion of pore water to unfrozen sediments during freezing; (c) differential volume changes due to freezing of pore water. *See also* BRODEL, CRYOTURBATION. **3** In structural geology, it refers to the penetration of a younger NAPPE into an older nappe, causing the latter to wrap around the former.

ion An electrically charged particle produced when a substance is subjected to electrolysis. In nature ions are thought to be formed by radioactive gases from the Earth or by the action of cosmic and ultraviolet rays. Particles which have an excess of electrons are termed *positive ions* or *cations*, while those having a deficiency are *negative ions* or *anions*. The relationship of positive and negative ions in the atmosphere during a THUNDERSTORM is thought to play a part in RAINDROP formation. Ionic relationships are also important in PEDOLOGY, where the degree to which hydrogen ions are held by the soil colloids determines the PH value of the soil. In addition to hydrogen ions, COLLOIDS are able to attract ions of calcium, potassium and magnesium. *See also* IONOSPHERE.

ion exchange (cation exchange) BASE EXCHANGE.

ionic diffusion The movement of charged ions by the mechanism of DIFFUSION.

ionization The production of charged particles from neutral atoms by the removal of electrons.

ionosphere The layer of the atmosphere above the MESOPAUSE, beyond about 80 km above the Earth's surface, where free ions and electrons occur, having been created by ionization of gas molecules through incoming solar ultraviolet and X-radiation. Several divisions have been recognized (indicated by letters D to G), which reflect radiomagnetic waves back to Earth. The ionosphere plays an important part in radio communications and, because of its frequent disturbances from geomagnetic storms and sunspot activity, transmission may be affected. The ionosphere includes the APPLETON LAYER (above 300 km) and the HEAVISIDE–KENNELLY LAYER (between 100 and 120 km above the Earth's surface). It is noteworthy that some scientists reserve the term ionosphere only for the belt of high electron density between 100 and 300 km, and prefer the term THERMOSPHERE for the atmospheric layer above the MESOPAUSE at 80 km. *See also* ION. [*17*]

Iowan The earliest substage of the N American WISCONSIN glacial, probably equivalent to the early Devensian of Britain, the Brandenburg advance in N Europe and Würm I of central Europe.

Ipswichian The last INTERGLACIAL stage of the Pleistocene in Britain. It separated the WOLSTONIAN from the DEVENSIAN cold stages and is roughly equivalent in age to the EEMIAN of NW Europe, the Riss–Würm interglacial of central Europe and the SANGAMON of N America. It was a time of temperate climate and was characterized by the following fauna: **1** presence of hippopotamus (absent from the HOXNIAN); **2** extreme rarity of horse (present in the Hoxnian); **3** abundance of hyena (apparently absent from the Hoxnian); **4** abundance of a certain species of rhinoceros (rare in the Hoxnian). Its type-site is at Bobbitshole, near Ipswich, England, and it probably terminated about 64,000 BP. [*197*]

Iron Age One of the cultural phases of man's evolution, when the technical improvements in iron-working enabled iron tools and weapons to replace those of the preceding BRONZE AGE. The culture appeared in Europe in about 1500 BC but not until about 500–600 BC in Britain.

ironpan A thin indurated soil horizon in which iron forms the major cementing material in the form of a hydrous iron oxide, a hydrous oxide of iron and manganese, or an iron/organic complex. *See also* HARDPAN.

iron pyrites A common sulphide mineral (FeS_2) occurring in hydrothermal ore veins, in ANAEROBIC sediments and as an accessory mineral in igneous rocks. Because of its glittering appearance it has been termed 'fool's gold'.

ironstone BLACKBAND IRONSTONE, CLAYBAND.

irradiance The amount of radiant power per unit area that falls upon an object or surface. It is measured in watts m^{-2} and carries the symbol E in modern terminology (symbol H in older publications). Spectral *irradiance* refers to the amount of radiant power at a particular wavelength of ELECTROMAGNETIC RADIATION. *See also* RADIANCE, RADIANT EXITANCE, RADIANT FLUX, REMOTE SENSING.

irregular-surface system A spatial PROCESS–response system, such as a land surface, in which flows of energy and mass have spatially variable horizontal and vertical components. *See also* LINE SYSTEM.

irreversibility Any process which proceeds in one direction spontaneously, without external interference.

irrigation The application of water by sprinklers, ditches or canals over a land surface in order to offset aridity and to nurture the growth of plants.

irrotational wave COMPRESSIONAL WAVE.

isallobar A line on a weather chart connecting places where the same atmospheric pressure changes have taken place over a given time period. It is used to give the forecaster a picture of the *pressure tendency*, i.e. how pressure systems, especially *depressions*, are moving and developing. *See also* BAROMETRIC TENDENCY.

isanomal, isanomalous line A line on a map connecting places which exhibit an equal deviation or difference from the mean or normal of any climatic *element*: thus, a line of equal *anomaly*. It is a graphic technique which is frequently used to illustrate temperate differences recorded at places lying along the same line of latitude but at varying distances from the ocean. Isanomalous lines show, for example, the intense winter cold of continental interiors compared with the warmer coastal stations in the same latitude. An area of high positive-temperature anomaly is termed a *thermopleion*, while an area of high negative-temperature anomaly is termed a *thermomeion*.

isarithm ISOPLETH.

isentropic line A line drawn to join up points having equal POTENTIAL TEMPERATURE.

island An area of land completely surrounded by water. Greenland (2.2 million km^2) is the largest island that is not a continent.

island arc A lengthy chain, usually curvilinear, of oceanic islands which frequently exhibit a FOREDEEP or ocean trench (DEEP,

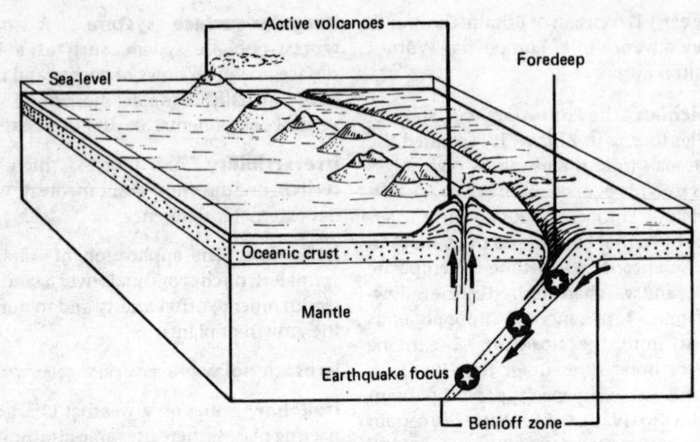

Figure 119 *Island arc*

OCEAN) on the convex side. Many island arcs are volcanic and are generally the location of severe seismic activity (EARTHQUAKE), related to their position along plate margins (PLATE TECTONICS) in the SUBDUCTION ZONE. The trench marks the position of the actual plate junction where one plate is being overridden by the other at a destructive plate margin. The friction generated provides both the thermal energy for the vulcanicity, with the volcanic cones often creating the islands themselves, and also the seismic activity. [*119; see also 91*]

isoazimuthal line RHUMB LINE.

isobar, isobaric surface A line on a weather chart joining places with equal barometric pressure. Because of the effects of altitude on pressure changes the readings at recording stations with different elevations on the Earth's surface have to be corrected to provide a common level, generally that of sea-level. Isobars are usually drawn at 2, 4 or 8 mb intervals on either side of the 1,000 mb isobar, and give a map of pressure 'contours' similar to a topographic contour map. The patterns produced represent the synoptic systems (ANTICYCLONE, DEPRESSION, etc.). The professional forecaster is not only interested in the *isobaric* map of pressure at mean sea-level but also with each of the *isobaric surfaces* at higher levels in the atmosphere. These surfaces are rarely parallel and in fact are constantly deformed by such mechanisms as diurnal heating and nocturnal cooling of land and sea (SEA BREEZE and LAND BREEZE), and seasonal heating and cooling.

isobase A line drawn on a map joining up places having the same amount of uplift or subsidence of land during a specified time period.

isobath A line joining points on the seabed or on a lake-bed, situated at an equal vertical distance beneath the surface. It is sometimes referred to as a *depth contour* and must be distinguished from a submerged or *submarine contour* where the altitude is related to a datum used for mapping adjacent land. This datum may or may not coincide with the water surface. *See also* CONTOUR.

isobathytherm A line joining points having the same temperature at a given depth below sea-level or below the ground surface.

isobront A line on a map connecting places which have thunderstorms at the same time. *See also* ISOCERAUNIC LINE.

isoceraunic line (isokeraunic line) A line joining places having the same percentage frequency of days per annum on which thunder is heard.

isocheim A line on a map connecting places which have the same mean winter temperature.

isochional line 1 A line joining points on the same altitude of snow limit. **2** A line joining places having the same frequency of snowfall, duration of snow cover or depth of snow. *See also* ISOGLACIHYPSE, ISONIF.

isochrones 1 Lines joining points on the Earth's surface which have the same time. **2** Lines joining points at which a SEISMIC WAVE arrives at the same time.

isochronous A process that is coincident in time with another process.

isochronous surface A surface within a body of sediments equivalent to a time plane.

isoclinal fold A fold of any type in which the limbs have parallel dips [*84*]. *See also* FOLD.

isocline A line joining points of equal MAGNETIC INCLINATION.

isoflor A line joining points which exhibit the same floral character.

isofrequency line A line joining places having the same number of tremors following an earthquake (AFTERSHOCK).

isoglacihypse A line joining places exhibiting the same altitudinal limit of the FIRN-LINE.

isogonic line A line drawn on a map joining points of compass declination (DECLINATION). [*120*]

isograd A line joining points where rocks exhibit the same GRADE of metamorphism, and therefore areas of the crust which experienced equal pressures and temperatures during metamorphism.

isogram A line along which values are constant. *See also* ISOPLETH.

isohaline A line joining points in the ocean having the same degree of salinity.

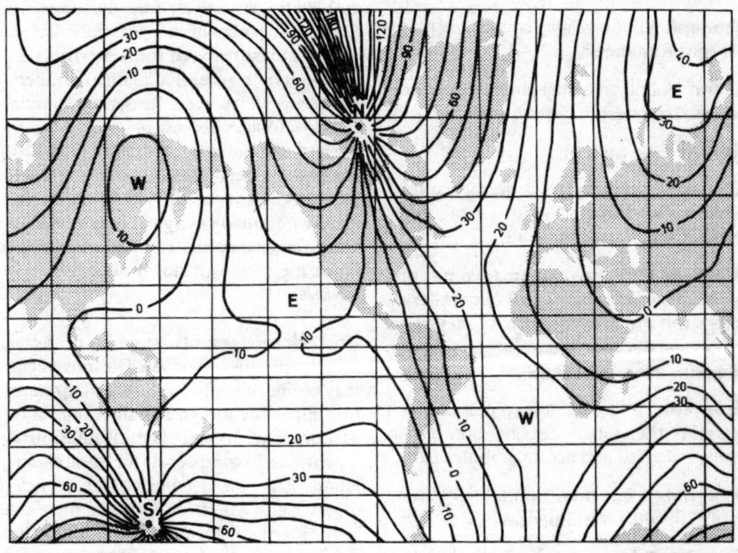

Figure 120 *Isogonic map of the world*

isohel A line on a map joining places which have the same amount of recorded SUNSHINE hours.

isohyet A line on a map joining places of equal rainfall amount. It can be used for any time interval from a single period up to mean annual rainfall totals.

isohypsometric line A line joining points in the crust which exhibit the same height of groundwater table.

isokinetic A line on a map joining places which exhibit equal degrees of wind velocity. Also referred to as an ISOTACH.

isolated system A system in which neither matter nor energy is exchanged between the system and its environment. *See also* CLOSED SYSTEM, OPEN SYSTEM.

isoline ISOPLETH.

isomer A line on a map joining places which have the same mean monthly rainfall expressed as a percentage of the mean annual rainfall.

isometric line ISOPLETH.

isoneph A line on a map connecting places of equal cloudiness.

isonif A line on a map connecting places of equal snow depth. *See also* ISOCHIONAL LINE.

isontic line ISOPLETH.

isopach (isopachyte) A geological term referring to points having the same thickness of a rock stratum.

isopleth A line drawn on a map connecting points assumed to be of equal value. Alternative terms which have been in use at various times are: *isarithm, isobase, isogram, isoline, isometric line, isontic line.*

isopycnic A line joining up points which exhibit the same density on certain meteorological and oceanographic charts.

isoryme A line drawn on a map joining places of equal frost intensity.

isoseismal line, isoseismic line A line joining places having identical degrees of earthquake intensity during a specific earthquake. *See also* ISOSEIST.

isoseist A line joining places at which the earthquake-wave arrival time from a particular epicentre is identical. It is synonymous with HOMOSEISMAL LINE. *See also* ISOSEISMAL LINE.

isostasy The state of balance which the Earth's CRUST tends to maintain or to return to by *isostatic compensation* if anything occurs to upset that balance. This is based on the principle of buoyancy first outlined by Archimedes. It is best illustrated by a high mountain chain which rises above the surface of the Earth but has to be compensated by deep 'roots' of sialic material (SIAL) which penetrate deeply into the underlying SIMA. When denudation lowers the mountains and carries the detritus away compensation causes the mountains to be uplifted to maintain *isostatic equilibrium*. Because of the deficiency of mass in the oceanic sectors of the crust the balance is maintained by an increased thickness of much denser sima. The theories of isostasy proposed by Airey and Pratt, respectively, are illustrated in [*121*]. It is probable that the truth lies in a compromise between these two theories, although both agree that lighter continental masses 'float' on a denser substratum (PLATE TECTONICS). Because of isostatic mechanisms continental masses which are depressed under the load of an ice-sheet will gradually recover as the ice melts (GLACIO-ISOSTASY). Similarly as a great weight of alluvial sediments accumulates at a particular point the crust will slowly sink (ISOSTATIC MOVEMENTS).

isostatic movements A number of crustal movements due to ISOSTASY. The movement may be one of vertical adjustment upwards to compensate for erosion and removal of a land surface, or one of vertical adjustment downwards to compensate for the accumulation of a considerable weight of sediments, e.g. at a delta. In addition, there are horizontal movements of material in the MANTLE as part of a gigantic series of subcrustal convection currents which allow *isostatic*

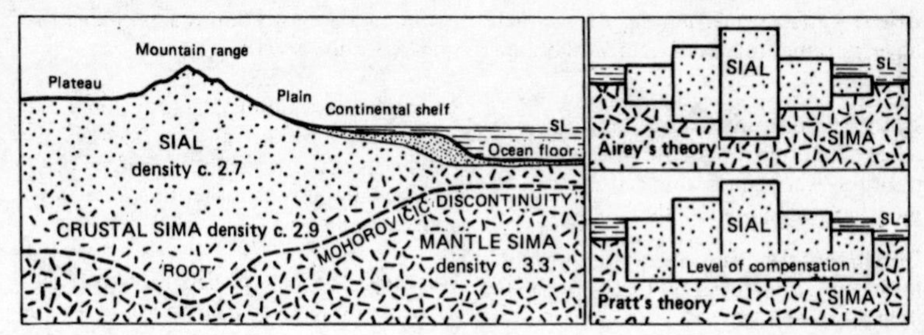

Figure 121 *Isostasy*

equilibrium to be maintained. *See also* GLACIO-ISOSTASY, PLATE TECTONICS.

isosteric surface A surface in the atmosphere which exhibits constant density.

isotach 1 A line on a weather chart joining places of equal wind velocity. Also referred to as ISOKINETIC. **2** A line on a map joining points of equal distance that can be travelled in a given time. **3** A line joining points of equal sound velocity.

isotachyte A line drawn on a map of a glacier joining points which show the same rate of movement.

isothere A line on a map connecting places which exhibit the same mean summer temperature.

isotherm A line on a weather chart or map joining places of equal temperature. Because of the effects of altitude on temperature the readings from stations at different elevations on the Earth's surface have to be corrected to provide a common level, usually that of sea-level.

isothermal layer 1 Formerly a term used for the STRATOSPHERE but this use has now been abandoned. **2** Modern usage refers to any vertical section of the *atmosphere* in which temperature remains constant with height, i.e. exhibits a zero LAPSE-RATE.

isothermobath A line joining points at a given depth in the sea which exhibit the same temperature.

isotopes Elements possessing the same number of protons in their nuclei but having different numbers of neutrons. They have almost the same chemical properties, the same atomic number but different atomic weights. The radioactive carbon isotope (C-14) is used by archaeologists and geomorphologists to date critical organic material (RADIOCARBON DATING) less than 50,000 years old. Lead isotopes are used for dating rocks (RADIOMETRIC DATING), while oxygen isotopes (O^{16}/O^{18}) are used to determine the palaeotemperature in cores of marine sediments and ice-sheets.

isotropic Having the same properties in all directions. If the properties are unequal the term *anisotropic* is used.

isthmus A narrow tract of land separating two bodies of water and connecting two larger bodies of land, e.g. the Isthmus of Panama.

iteration A repetition process of successive approximations in a procedure. It is used in numerical solving of complex equations.

J

jade A green-and-white rock material prized as a gemstone. It comprises two distinct minerals, an AMPHIBOLE (nephrite) and a PYROXENE (jadeite).

Jaeger's projection A 19th-cent. MAP PROJECTION in which the Earth is divided into eight parts by means of a star-shaped projection centred on the N Pole. The parallels are equidistant and parallel straight lines within each of the eight segments but taken in their entirety form octagons about the origin. The meridians are straight lines which meet at true angles at the centre.

jasper A bright-red variety of SILICA, with chert-like properties, formed in association with HYDROTHERMAL ACTIVITY or in a gas-formed lava cavity. It is used as a decorative stone in interior design, e.g. in fireplaces.

jebel An Arabic name for a mountain or hill in a hot desert environment.

jet An extremely compact form of black lignite (CANNEL COAL), utilized as a semi-precious stone in jewellery, etc. because it is capable of being carved and polished. In Britain it was once extensively mined in the sea cliffs near Whitby, E Yorks.

jet stream A narrow belt of high-altitude (above 12,000 m) westerly winds in the TROPOSPHERE. Their speed varies from a mean of 110 km/h in summer to about 184 km/h in winter, although velocities of over 370 km/h have been recorded occasionally. A number of separate streams have been identified: **1** the most constant, that of the subtropics (between 30° and 20°N and S of the equator); **2** that associated with the POLAR FRONT of mid-latitudes, which is greatly variable in its location, partly because of the transitory thermal contrasts of the DEPRESSIONS of these latitudes; **3** an *Arctic jet stream*; **4** a *polar-night jet stream* above the troposphere. Surface depressions and ANTICYCLONES occur in zones which exhibit large surface-temperature gradients. These differences in turn cause marked thermal contrasts in the troposphere, thereby creating DIVERGENCE and CONVERGENCE of air streams in the tropospheric circulation. The *polar-night jet stream*, however, results from a very steep stratospheric thermal gradient in winter (at 20–30 km above the Earth's surface) around the stratospheric cold pole and produces exceptionally strong Westerlies. The velocity and location of the jet stream has had an important effect on travel times of high-flying aircraft, especially in mid-latitude air routes. *See also* WESTERLIES, INDEX CYCLE. [*122*]

JMA scale A scale of EARTHQUAKE intensity (0–VII), rather different from the MERCALLI SCALE. It is used by the Japanese Meteorological Agency because of its greater suitability for local building styles.

joint A fracture in a rock but one which exhibits no differential movement, in contrast to a FAULT. It forms primarily in COMPETENT rock members. Joints can be formed in three ways: **1** By folding and thrusting, which produce *longitudinal joints* parallel to the strike and the axes of folding, *cross joints* parallel with the dip and at right angles to the fold axes, and *oblique joints* at approximately 45° to the direction of tectonic movement. **2** By shrinkage due either to cooling of an igneous rock or to drying out of a sedimentary rock. **3** By unloading of a

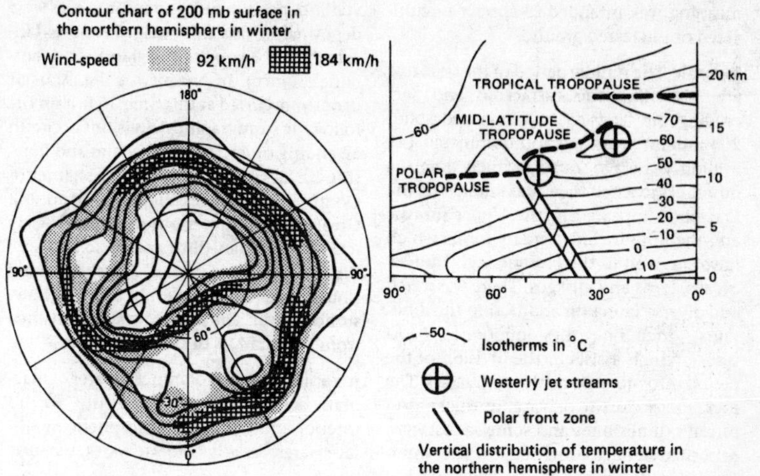

Figure 122 *Jet streams*

rock mass after the removal of its cover rocks, giving rise to *sheet joints* (EXFOLIATION). *See also* GRIKE, TOR. [*123*]

joint-block removal The process of loosening and transporting of jointed blocks of bedrock by the action of a glacier. It is a significant mechanism operating in the steepening of a cirque's BACKWALL (BERGSCHRUND) and in the creation of the steeper end of a ROCHE MOUTONNÉE. *See also* EXARATION, ICE PLUCKING.

jökulhlaup GLACIAL OUTBURST.

jökull An Icelandic term for a small ICE-CAP.

joule An SI (SYSTÈME INTERNATIONAL) UNIT of energy which is equal to the amount of work done by a force of one newton when moving an object one metre in the direction of a force. A joule is equivalent to 0.239 calories or 1 watt second.

junction 1 The contact-plane between two adjacent rock bodies. It may be completely conformable or in the character of an UNCONFORMITY. **2** The meeting-point of two streams.

jungle An imprecise term for a virtually impenetrable tropical forest; its original

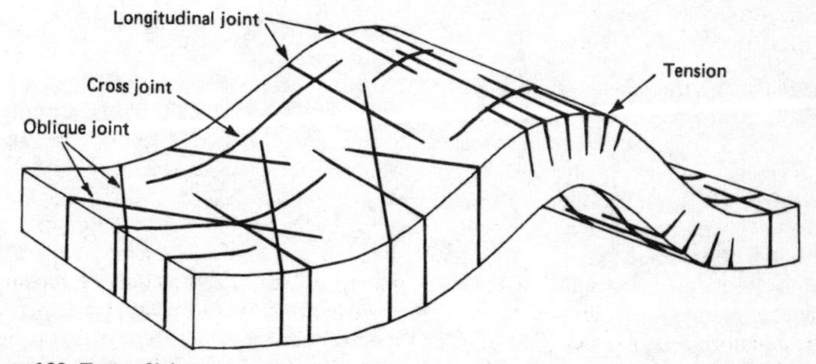

Figure 123 *Types of joint*

meaning was intended to specify uncultivated or uncleared ground.

Jurassic The middle period of the MESOZOIC era, preceding the CRETACEOUS and succeeding the TRIASSIC. It commenced about 208 million years ago and terminated 145 million years ago, during which time the dinosaurs reached their maximum size, the first birds appeared in the Upper Jurassic and the flora included many forms which have survived to the present, e.g. conifers, cycads, ferns and ginkgos. There was a rich and diverse fauna (in addition to the dinosaurs), including the ammonites (MOLLUSCA), which assists in the division of the period into more than 100 fossil zones. The rock strata consist of varying thicknesses of clays, limestones and some sandstones, deposited in fluctuating shallow seas, interspersed with periods of estuarine and fluviatile deposition. Important iron ores were deposited at this time, (e.g. Teesside, Lincolnshire, Northants (England) and Lorraine (France). In N America the LARAMIDE OROGENY occurred at this time. In Britain the following stratigraphical divisions occur (in ascending order): Rhaetic (shales and limestones); Lias (clays and limestones); Inferior Oolite (limestones, clays, sandstones); Great Oolite (limestones); Oxford Clay (clays); Corallian (limestones, some clay); Kimmeridge Clay (shales, clays); Portland (limestones, sandstones); Purbeck (limestones, clays). The Jurassic system is named from the Jura Mts of central Europe.

juvenile water Hot mineral water of magmatic origin derived from the Earth's interior and not from atmospheric or surface water. *See also* CONNATE WATER, METEORIC WATER.

K

Kainozoic CAINOZOIC.

K/Ar dating A type of RADIOMETRIC DATING based on the ratio of potassium (K) 40 to argon (A) 40 isotopes.

kame A steep-sided ridge or conical hill of bedded GLACIOFLUVIAL materials, formed from a crevasse-filling. When part of a series the feature is said to form a *kame moraine*, marking the ice limits of a recessional stage. If the debris-laden meltwater streams emerge from the ice front and deposit their load into an ice-dammed lake or pro-glacial lake the gravels and sands will form a *kame delta* (DELTA). Kames can also result from the accumulation of glaciofluvial sediments in a supraglacial position whence they are lowered by ice melting until they form a tumultuous hummocky terrain after the ice has disappeared (KAME-AND-KETTLE TERRAIN). [*219*]

kame-and-kettle terrain An undulating landscape composed of groups of KAMES and/or KAME TERRACES interspersed or pitted with KETTLE HOLES. This type of landform is sometimes termed a *kame complex* and is created when glaciofluvial sediments are lowered on to the sub-ice surface as the glacier or ice-sheet decays, especially when it can be established that the kame sediments were formed in separate basins of accumulation on the ice surface. This term is preferable to that of *kame moraine*, which tends to imply a simple glacial origin rather than a complex glaciofluvial origin. *See also* KETTLED SANDUR.

kamenitza SOLUTION BASIN.

kame terrace A flat-topped ridge or terrace-feature occurring between a valley glacier and the valley slopes. It is composed of bedded GLACIOFLUVIAL materials deposited by meltwater streams flowing laterally to the glacier. The kame-terrace surface is sometimes carved into by lateral meltwater channels and pitted with hollows (KETTLE HOLES), features which distinguish the kame terrace from a pro-glacial lake shoreline. The terrace is subsequently dissected by post-glacial tributary streams or cliffed by the meandering river after the disappearance of the glacier.

Kansan The name given to the Pleistocene glacial advance which preceded the Yarmouth Interglacial in the USA. It is younger than the NEBRASKAN and older than the ILLINOIAN. In age it is roughly equivalent to the ANGLIAN of Britain, the ELSTER of N Europe and the MINDEL of the Alps. Its deposits in the USA comprise stoneless clays (TILL) with little evidence of glaciofluvial materials.

kaolin, kaolinite, kaolinization Kaolin is a white clay produced by the chemical breakdown of FELDSPARS (especially those in granite) by the process of kaolinization. This may be achieved either by HYDROLYSIS, the main type of CHEMICAL WEATHERING, or by HYDROTHERMAL ACTIVITY (PNEUMATOLYSIS). Kaolin is the clay and kaolinite the two-layered hydrous aluminium silicate ($2H_2O.Al_2O_3.SiO_2$). In Britain most kaolin deposits are associated with the granite outcrops of Bodmin Moor, Dartmoor and St Austell where the china-clay extraction is centred. It is used extensively in pottery and paper manufacture. *See also* BALL CLAY, CHINA CLAY, CHINA-STONE.

kar A German term for CIRQUE.

karren A German term originally used to describe minor SOLUTION furrows or runnels cut by surface water or soil water into limestone, but now referring to all the microforms on KARST limestones, ranging in size from a few millimetres to several metres. The principal factors affecting their formation are the chemical nature of the limestone, the character and degree of precipitation, the texture and dip of the limestone, the presence or absence of peat, soil or vegetation, and the palaeoclimate of the site. The term is synonymous with *lapiés*. The following types have been recognized: *deckenkarren* (very shallow depth by direction action of plants); *hohl-karren* (60 cm to 1 m deep, formed under peat); *kluftkarren* (<4 m wide and <4 m deep, sharp crests when open but smooth when covered), also termed GRIKES; *meanderkarren* of sinuous, winding forms (50 cm deep, <20 cm long); *rillenkarren* (1–2 cm deep, <50 cm long, sharp crests, or inclined surfaces); *rinnenkarren* (50 cm deep, <20 cm long, sharp crests, on inclined surfaces); *rundkarren* (12–50 cm deep, <15 cm long, covered by soil/vegetation, smooth crests on inclined surfaces); *spitzkarren* (50 cm deep, 50 cm wide, very sharp crests, on inclined surfaces and roofs); *trittkarren* (circular, semicircular and step-like hollows on flat surfaces, 3–50 cm deep, 20–100 cm wide). *See also* SOLUTION BASIN.

karrenrohren A German term for a rounded solution hollow formed at the intersection of two GRIKES (*kluftkarren*), often up to 5 m in depth. It is synonymous with *lapiés well*.

karst A German rendering of a Serbo-Croat term referring to the terrain created by limestone solution and characterized by a virtual absence of surface drainage, a series of surface hollows, depressions and fissures, collapse structures, and an extensive subterranean drainage network [*204*]. The classic karst landforms are found in western Yugoslavia. A classification of karstland forms based on climate and process has been devised by M. M. Sweeting (1972): **1** *True karst* (holokarst), produced dominantly by karstic solution. **2** *Fluviokarst*, produced by the combination of karstic and FLUVIAL processes. **3** GLACIOKARST and *nival karst*, produced by glacial scour with differential erosion followed by post-glacial solution, together with solution from glacier and snow meltstreams. (Ingleborough, Yorkshire, England's best-known karstic terrain, is *glaciokarst*, not true karst.) *Glaciokarst* also includes PERMAFROST karst. **4** *Tropical karst*,

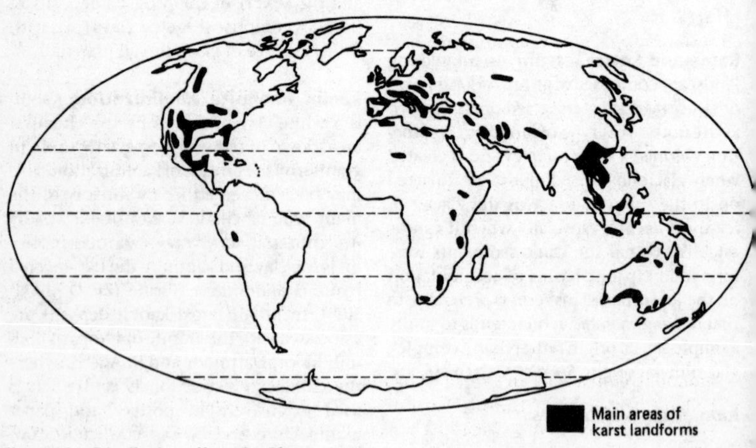

Figure 124 *Distribution of karst landforms*

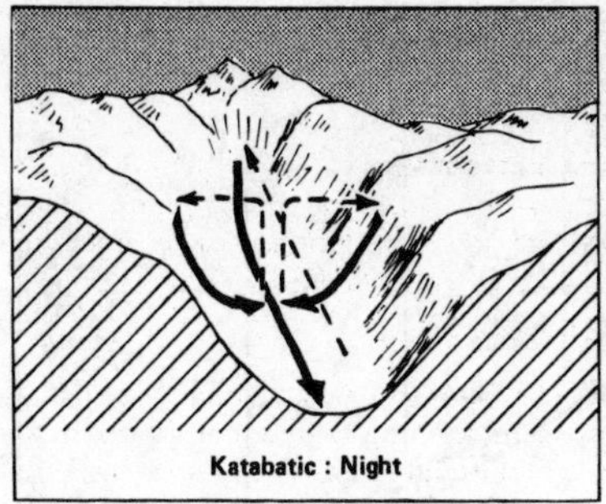

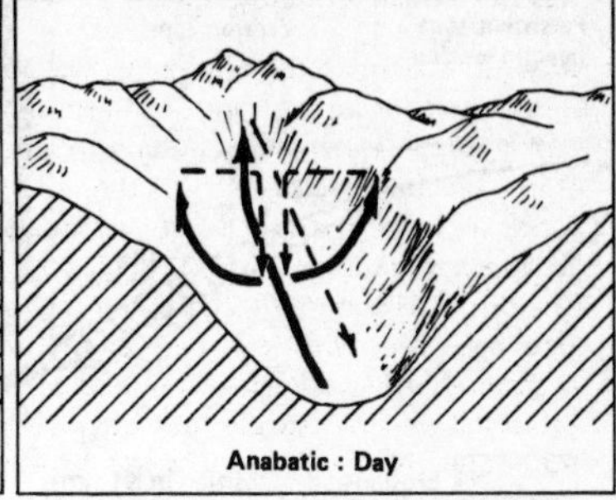

Figure 125 *Katabatic and anabatic winds*

formed under conditions of high rainfall and high temperature and including KEGELKARST and TURMKARST (*tower karst*). **5** *Arid and semi-arid karst*, in which occasional but heavy rainstorms create CALICHE (due to intense evaporation) which forms a tough protective skin of secondary calcium carbonate on the limestone. Desert karst may be due to a formerly wetter climate. *See also* BIOKARST, HOLOKARST, MEROKARST, MESOKARST. [*124; see also 64*]

karstic cycle CYCLE OF UNDERGROUND DRAINAGE.

karst water Water which issues from karst springs and whose chemical composition indicates that it has passed across and through karst limestone. Water which is derived entirely from precipitation over the karst area alone is termed *autochthonous water* and re-emerges at the surface as an *exsurgent* stream (*exsurgence* in contrast to RESURGENCE). Water which is derived in part from precipitation which has fallen on to impermeable rocks outside the karstic area is termed *allogenic water* and this reappears at the ground surface (having passed through the karstic system) as a *resurgent* stream.

karst window That part of an underground river in a karst area revealed by the collapse of a section of the roof, e.g. Cradle Hole in the Marble Arch area of Co. Fermanagh, Ireland.

katabatic wind (stroph) A downhill wind, usually cold, which blows down valleys at night and outwards from large icecaps, such as Antarctica and Greenland. It is caused by ground-surface cooling (often nocturnally) as a result of radiation, which in turn cools the lower air layers. The falling temperature increases the density of the air which then moves downhill by gravity flow. *See also* ANABATIC WIND. [*125*]

katafront A depressed frontal zone in which the warm air is given very little opportunity to rise over the cold air because of the prevailing SUBSIDENCE at high level. It is often associated with the perimeter of a DEPRESSION. *See also* ANAFRONT, FRONTOGENESIS.

kavir A Middle Eastern term for a PLAYA or SEBKHA.

K-cycle A hypothesis relating to the interrelationship which exists between hill-slope soils and erosion and in which it is suggested that soil development exhibits a cyclic periodicity comprising alternating phases of instability and stability. In an unstable phase (u) soils will be eroded from one area, leading to burial of soils lower down the slope by such processes as slope wash, mass movement and loess deposition. Areas of erosion and deposition will vary according to gradient, aspect, exposure, vegetation cover, soil character, etc. During a stable phase(s) there will be a relative pause in both erosion and deposition, in which soils will develop. The expression K-cycle indicates the time unit during which the

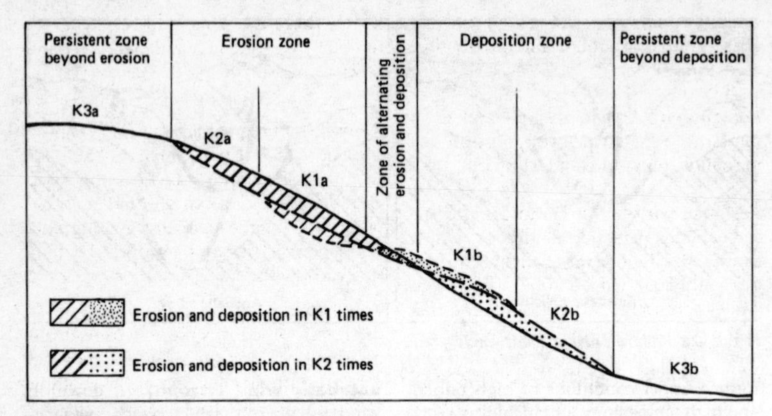

Figure 126 *K-cycle*

unstable period occurs and a number of K-cycles can be distinguished extending back from the present, namely: K1(u), K1(s); K2(u), K2(s); K3(u), K3(s), etc. The concept also recognizes *persistent zones* which lie beyond erosion and deposition and in which stable soil development continues. [*126*]

Keewatin The stratigraphic name for part of the PRE-CAMBRIAN in Canada. Synonymous with Ontarian.

kegelkarst A type of tropical *karst* terrain in which the characteristic landforms are numerous cone-like hills, sometimes separated by cockpits (COCKPIT KARST), which are enclosed depressions interconnected by sinuous corridors between the cones. The cockpits generally occur in lines following joint or fault patterns [*127*]. Kegalkarst is well developed in Indonesia, Jamaica, Puerto Rico, parts of central America, Vietnam, S China and Malaysia. [*262*]

kelvin The SI unit (symbol: K) of thermodynamic temperature. 1 kelvin = 1° C. *See also* ABSOLUTE TEMPERATURE.

Kelvin–Helmholtz instability A term given to the wave development which takes place at the interface of two layers of the atmosphere exhibiting different temperatures and wind-speeds. The waves sometimes increase in amplitude and break up into vortices (CLEAR-AIR TURBULENCE) which very occasionally manifest themselves as clouds, if sufficient moisture is present.

Kelvin scale A temperature scale synonymous with that of ABSOLUTE TEMPERATURE. 0° K = –273.15° C (1 K = 1° C).

Kelvin wave An oceanographic term which relates to a tidal range of a rectangular sea area increases to the right of the direction of the PROGRESSIVE WAVE and decreases to the left. This is exemplified in the English Channel, where the north coast of France

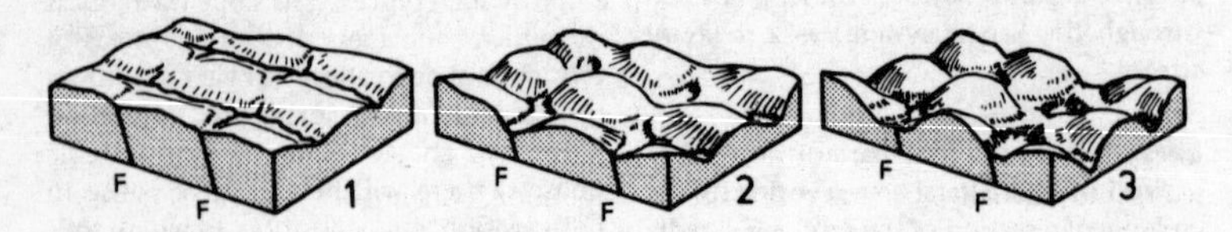

Figure 127 *Kegelkarst*

exhibits a tidal range of 11–13 m while that of the south coast of England has no more than 4 m.

keratophyre A type of trachytic and rhyolitic SPILITE; a fine-grained volcanic rock rich in sodium, potassium and feldspar.

kernlose winter A term to describe the singular temperature conditions which characterize the long winter night of Antarctica. Although temperatures decrease as winter approaches, it has been found that at the end of the winter they are not very much lower than they were at its commencement. This is thought to be the result of the balance achieved between surface radiational cooling and atmospheric counter-radiation from above the strong surface-temperature inversion.

kettled sandur (pitted sandur) An outwash plain of glaciofluvial deposits (SANDUR) the proximal parts of which are pitted with KETTLE HOLES. The sandur was created by ice-melt streams flowing in supraglacial, englacial or subglacial locations all of which, however, have ANASTOMOSING characteristics. Because the proximal part of the sandur was deposited above the decaying ice front, large blocks of ice became buried by the glaciofluvial sediments and these ultimately created *kettle holes* as they melted out. Some examples of kettled sandur are thought to have occurred when the bursting of an ice-dammed lake (*jökulhlaup*) (GLACIAL OUTBURST), carried large blocks of ice far out on to the sandur where they became buried by subsequent deposition. [*219*]

kettle hole An enclosed depression in glacial-drift deposits accumulated in a recently glaciated area. It is formed when a body of ice becomes buried in an area of DEAD ICE FEATURES as an ice-sheet slowly decays. As the buried ice mass finally decays the surface sediments collapse to form a hollow which soon becomes water-filled. As subsequent sediments gradually infill the depression it is rare that a kettle hole survives beyond any but the final glacial stage. *See also* KETTLED SANDUR. [*219*]

kettle moraine A morainic area characterized by an extremely undulating terrain of KAMES and KETTLE HOLES. *See also* KETTLED SANDUR.

Keuper A stratigraphic name for a stage of the Upper Triassic (TRIASSIC) in NW Europe. In Britain it comprises beds of stiff MARL and bright-red sandstones which are good FREESTONES, interbedded with gypsum and salt deposits. The latter are worked extensively in Cheshire, England.

Kew barometer A type of mercury BAROMETER, in which it is possible to read the pressure by noting the top of the mercury column, without adjusting the level of the mercury (FORTIN BAROMETER). *See also* ANEROID BAROMETER.

Keweenanwan The stratigraphic name for part of the PRE-CAMBRIAN in Canada. Synonymous with Algonkian.

key CAY.

khamsin An Arabic term for the very hot, dry, and often dust-laden southerly wind which blows periodically across Egypt from the Sahara desert. Its high temperatures and poor visibility bring unpleasant conditions to the Nile delta. It is most common between April and June when depressions, moving eastwards along the Mediterranean Sea, draw in southerly air masses. *See also* SIROCCO.

kidney ore HAEMATITE.

kieselguhr A German term for DIATOMACEOUS EARTH.

killas A mining term in Cornwall, England, adapted to describe SLATES and low-grade PHYLLITES of Devonian and Carboniferous age in SW England.

kilometre A metric length of 1,000 metres, equivalent to 0.621 of a mile. [*136*] *See also* LENGTH.

kimberlite A brecciated PERIDOTITE, comprising mica and olivine as the most common minerals, infilling a DIATREME or PIPE and often containing diamonds. Its surface is usually weathered ('yellow ground') but at depth becomes less oxidized ('blue ground') above the unweathered kimberlite. It takes its name from Kimberley, S Africa.

Kimmeridgian A stratigraphic term for the middle part of the Upper Jurassic (JURASSIC) in NW Europe. In Britain it comprises thick layers of clays, shales (including oil shales), thin impure limestones and a few beds of calcareous sand. It is named from Kimmeridge in Dorset, England.

kinematic graph A diagram illustrating the transition probabilities related to a Markov process (MARKOV CHAIN).

kinematic wave A concentration of particles or units in a system of unidirectional flow, in which the concentration moves through the particles themselves so that the wave velocity may be greater than the velocity of the individual particles. *Kinematic wave theory* was outlined by M. J. Lighthill and G. B Whitham in 1955 and depends on two principal factors: *flux*, measured by the number of particles or units passing a point at a given time; and *linear concentration*, measured by the mean number of particles or units per unit of distance. The theory was first used to describe traffic flow and river floods but was later applied to the movement of a wave through a glacier, which travels approximately four times faster than the velocity of the ice. When it arrives at the glacier snout the wave gives rise to a sudden forward SURGE, since it represents the mechanism by which the glacier responds to a change in its MASS BALANCE.

kinetic energy The energy of a body derived from its own movement. This energy is continually being dissipated: in the atmosphere through surface friction and turbulence; in the ocean by moving waves; in a glacier by ice sliding on the underlying rock; and by a stream when flowing along a channel. Its magnitude is expressed as: $E_k = MV^2/2$, where M is the mass of the body and V is its velocity. *See also* POTENTIAL ENERGY.

kingdom 1 The largest of the groups in the taxonomic classification of organisms, comprising the two kingdoms of animals and plants. The animal kingdom is made up of several phyla (PHYLUM) while the plant kingdom is composed of a number of DIVISIONS. **2** In 1953 the existing systems of classification were revised by R. Good, who recognized six floristic kingdoms (Boreal, Palaeotropical, Neotropical, S African, Australian and Antarctic). The Palaeotropical he subdivides into three *subkingdoms* (African, Indo-Malaysian, Polynesian) while the remainder are subdivided directly into provinces, which in turn are split into *subprovinces* and *regions*. *See also* TAXONOMY.

Klimamorphologie The German term for CLIMATOMORPHOLOGY.

klint 1 A hill or ridge formed from a reef-limestone (BIOHERM) that has proved more resistant than the surrounding rocks (*see also* REEF-KNOLL). **2** A Scandinavian term for a steep sea cliff in the Baltic, especially in Denmark and Sweden.

klippe (plural **klippen**) A German term referring to an isolated overthrust mass of folded rocks, usually a NAPPE, cut off from the main fold structure by erosion. Many excellent examples can be seen in the Pre-Alps of Switzerland but in Britain some of the isolated mountains in Assynt, Sutherland, NW Scotland, are *klippen*.

kloof An Afrikaans term for a rocky gorge or a mountain pass.

kluftkarren KARREN.

knee fold A type of GRAVITY-COLLAPSE structure resulting from the slipping of competent rocks after the erosion of incompetent beds. [*96*]

knick A term introduced into geomorphology by L. C. King to describe the sharp break of slope between an INSELBERG and its surrounding PEDIMENT. It may be the result of chemical weathering. It should not be confused with KNICKPOINT.

knickpoint An alternative expression for *rejuvenation head*. It refers to the break of slope in the long profile of a stream which results from a fall of BASE-LEVEL due to uplift of the land or a fall of sea-level (REJUVENATION). Because of the renewed downcutting a new

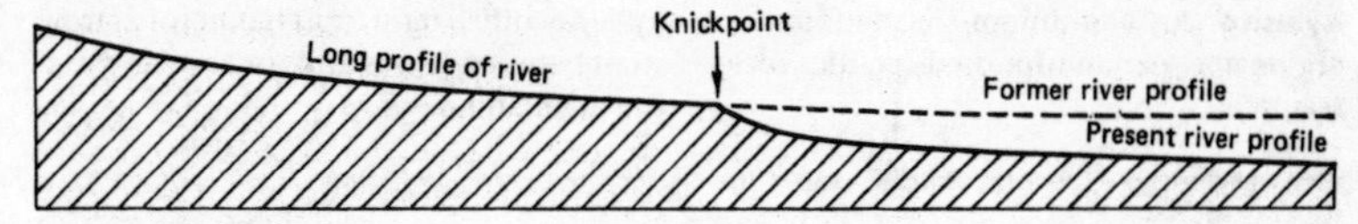

Figure 128 *Knickpoint*

valley long profile is created and where this intersects the long profiles of the former valley a knickpoint will be found. Knickpoints migrate upstream as erosion progresses, but the rate at which they retreat depends on the rock hardness. Where a hard-rock stratum crosses the valley downcutting becomes slower and the knickpoint is held at this feature, often in the form of a waterfall or a stretch of rapids. It is sometimes termed a *nickpoint*. [*128*]

knob-and-kettle An expression used in the United States to describe a kame-and-kettle moraine of undulating relief. *See also* KAME, KETTLE HOLE.

knock-and-lochan An undulating terrain of ice-scoured rocky lowland common in NW Scotland and the Outer Hebrides. It consists of low, rocky knolls (*knock*) and ice-gouged depressions often filled with water (*lochan*). It contrasts with a KNOB-AND-KETTLE landscape which comprises a terrain of glacial drift rather than solid rock.

knoll A rounded hill of no great elevation.

knot **1** A unit of speed equivalent to 1 nautical mile (per hour). The term is derived from the number of knots tied in a length of rope attached to the log-line (LOG). These were arranged at regular intervals. As the rope ran out astern the knots were counted and checked against a timepiece, enabling the ship's speed to be calculated. 1 knot = 0.515 m/sec. **2** The crossing-point or meeting place of two ranges of fold mountains.

koniology (coniology) The scientific study of atmospheric DUST and AEROSOLS. *See also* DUST-VEIL INDEX.

kopi dunes Linear dunes comprising fine-grained gypseous (GYPSUM) sediments, located on the fringes of certain lakes in the semi-arid interior of W Australia. They may have been formed when gypseous crusts were more commonplace around the lake shores than at present.

kopje An Afrikaans term for a small isolated hill in S Africa, usually comprising a prominent pile of rocks (CASTLE KOPJE). *See also* TOR.

Köppen's climatic classification An empirical climatic classification devised by a German biologist, W. Köppen, in 1918. It has been revised many times, especially by Köppen and R. Geiger. It is based essentially on the climatic requirements of certain vegetation types, giving six major categories, A, B, C, D, E and H, each of which is subdivided according to different temperature and rainfall characteristics: A. Tropical zone, 12 months above 20° C; B. Subtropical zone, 4–11 months above 20° C; 1–8 months between 10° C and 20° C; C. Temperate zone, 4–12 months between 10° C and 20° C; D. Cold zone, 1–4 months between 10° C and 20° C; 8–11 months below 10° C; E. Polar zone, 12 months below 10° C; H. Mountain zone.

kratogen The former spelling of CRATON.

***K*-selection** *R* AND *K*-SELECTION.

Kuroshio (Kurosiwo) current A major BOUNDARY CURRENT in the global ocean circulation system. It develops as a deflected part of the North Equatorial current in the Pacific Ocean before flowing northwards as a cold and only partially saline current past the islands of Japan before swinging eastwards and splitting into several branches. It achieves maximum speeds of some 1.5 m/sec. *See also* FLORIDA CURRENT. [*168*]

kurtosis A statistical term for the peakedness of a curve. A high peak is *leptokurtic*, a flat-topped peak is *platykurtic* and a curve intermediate between the two is *mesokurtic*.

kyanite An aluminium silicate found in regionally metamorphosed pelitic rocks (PELITIC).

kyle A Scottish term for a channel or narrow strait between two islands or an island and the mainland.

L

lability A function of the relationship which exists between the nature of a system's components and their responses to disturbance. Those components which are very prone to change are termed *labile,* while those components which are less responsive exhibit low lability or are *non-labile.*

Labrador current A major OCEAN CURRENT in the N Atlantic Ocean. It flows southwards from the Arctic Ocean along the western coast of Greenland parallel with the East Greenland current before combining as a cold surface current near to the Grand Banks off Newfoundland, where they meet the warmer, more saline North Atlantic Drift. The Labrador and East Greenland currents carry icebergs southwards into the N Atlantic shipping routes and are also in part responsible for the high frequency of fogs (ADVECTION FOG) in the Grand Banks area. [*168*]

laccolith, laccolite An intrusive dome-like mass of MAGMA with a pipe-like feeder which causes the overlying sedimentary rocks to be forced up into an arch. When a number of laccoliths are stacked one above another from a single intrusion they are termed a CEDAR-TREE LACCOLITH. [*129*]

lacuna The missing time interval which intervenes between beds above and below an UNCONFORMITY.

lacustrine (lacustral) Pertaining to lakes. *See also* LIMNIC.

lacustrine deposit A sediment accumulated in a lake, generally characterized by depositional layering (LAMINA, VARVE).

lacustrine plain A low-lying tract of land formed from the sediments deposited in a lake which has been completely infilled. The former lake floor is usually marshy and can only be used for farming if artificially drained.

lag 1 A general term meaning a retardation in a movement or transformation of any

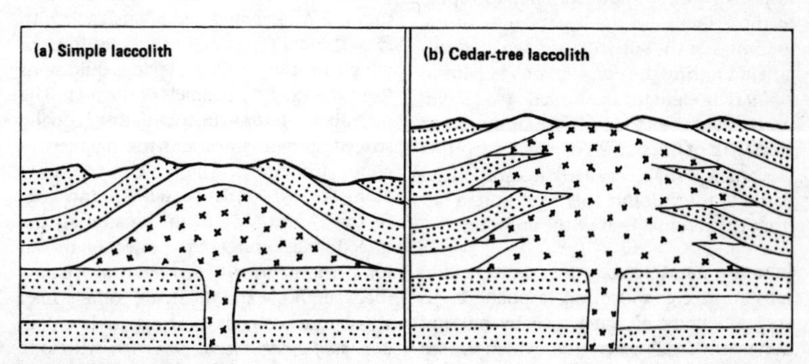

Figure 129 *Types of laccolith*

kind. **2** A type of thrust fault (FAULT) in which the slide or thrust plane has replaced one of the limbs of a RECUMBENT FOLD. If the uninverted limb of the fold is eliminated the thrust is termed a *lag-fault*. **3** The coarsest material (*lag-gravel*) left on a desert surface (DESERT PAVEMENT) after wind has removed the finer material. **4** A coarse deposit left on a stream bed (*lag-deposit*) due to selective transportation of the finer sediments of the stream's load. **5** Any time delay between the arrival of a signal in a measuring instrument and the response of that instrument, especially in meteorological instruments (*time lag*).

lag-deposit LAG.

lag-fault LAG.

lag-gravel LAG.

lagoon A coastal stretch of shallow saltwater virtually cut off from the open sea by a BARRIER-BEACH [*24*], a shingle ridge (FLEET) or an offshore reef, particularly a CORAL REEF. The term is also used for the circular body of water partly enclosed within an ATOLL. [*19*]

lahar A mud-flow composed mainly of volcanic ash lubricated by water derived from the bursting of a CRATER LAKE, from snowmelt (sometimes during an eruption of a snow-capped volcano) or from prolonged torrential rain causing the volcanic ash to flow under gravitational movement. The impact of the mud-flow is often catastrophic to those living on the lower slopes of the volcano, e.g. the burial of the town of Herculaneum during the eruption of Vesuvius in AD 79 is thought to have been due to a lahar. A dry lahar may result from the MASS-MOVEMENT of thick layers of newly deposited volcanic ash shaken by earth tremors on the flanks of the volcano; this type is often hot from the residual heat of the eruption.

lake **1** A standing body of a relatively fluid substance, e.g. water, oil, asphalt. **2** An enclosed body of water, usually but not necessarily fresh water, from which the sea is excluded. *See also* LIMNOLOGY.

lake breeze A local breeze blowing during the afternoon off a large enclosed body of water during periods of calm weather. Its mechanism is very similar to that of a SEA BREEZE, whereby differential heating of land and water during the day produces an area of low pressure and accompanying thermal convection over the land. It is a common feature of the large East African lakes, although its effects are not felt very far inland. *See also* LAND BREEZE.

lake rampart A small ridge along the shore of a lake formed by the shoreward movement of surface ice during the winter freeze. It is sometimes termed an *ice rampart*.

Lambertian surface A REMOTE SENSING term which refers to a perfect DIFFUSE REFLECTOR, capable of reflecting energy equally in all directions. *See also* SPECULAR REFLECTOR.

Lambert's projections A number of MAP PROJECTIONS introduced by J. H. Lambert in the late 18th cent. They comprise: **1** an *azimuthal* (*zenithal*) *equal-area projection*, used mainly for polar areas. The pole is the centre of the projection from which the meridians radiate as straight lines at true angles to each other and with the parallels drawn as concentric circles. All points have their true compass directions from the centre of the map but there are distortions of shape beyond 30° from the map centre. Hence, Lambert introduced an oblique equal-area map to show countries within 40° with little deformation, and also an equatorial equal-area projection for countries at low latitudes. **2** A *conformal conical projection* with two standard parallels, the positions of which are usually taken as lying equidistant from the extreme parallels of the map. The meridians are drawn as radiating, equally spaced straight lines and the parallels as concentric circles. The scale is true only along the two standard parallels; it increases to the N and S of them. Because all the directions are correct this projection is used for international aeronautical charts (between 80°N and 80°S) and for any map where exact compass bearings are essential. **3** A *conformal conical projection* with one standard parallel. **4** A CYLINDRICAL EQUAL-AREA

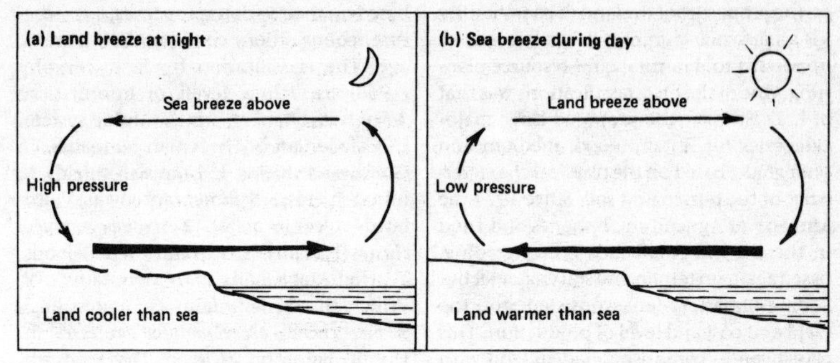

Figure 130 *Land and sea breezes*

PROJECTION. *See also* AZIMUTHAL PROJECTION, CONICAL PROJECTION, CYLINDRICAL PROJECTION, EQUAL-AREA MAP PROJECTION, ORTHOMORPHIC PROJECTION.

lamina, lamination, laminar **1** The thinnest layer of stratification in a layered sequence of soils or sedimentary rocks, defined as less than 1 cm in thickness. Laminae (plural of *lamina*) are generally but not invariably parallel with the bedding, but if they lie at an angle the term *cross-lamination* is used. Each lamina in a sedimentary sequence represents a seasonal or periodic period of deposition by water or wind. **2** The thinnest sheet in a metamorphic rock.

laminar flow **1** In general, a very low-velocity movement of a fluid (water or air) over a smooth surface, in contrast to turbulent flow. **2** In a glacier movement (GLACIER FLOW), the pressure thrust generated by the weight of the ice causes the ice to shear and move forward along thrust- or glide-planes. **3** Non-turbulent flow of air or water with no mixing between the parallel layers (STREAMLINE). Energy is transferred from one layer to the next by the viscosity between the layers. At the ground surface, the stream bed or stream wall (i.e. the boundary), movement is non-existent. The fluid elements follow paths that are straight and parallel to the boundaries. *See also* FROUDE NUMBER, REYNOLDS NUMBER, FLOW REGIMES.

land **1** The solid part of the Earth's surface; the continents. **2** A specified geographical tract of the Earth's surface including all its attributes, comprising its geology, superficial deposits, topography, hydrology, soils, flora and fauna, together with the results of past and present human activity, to the extent that these attributes exert a significant influence on the present and future land utilization. *See also* LAND CLASSIFICATION, LANDSCAPE. •

land breeze A nocturnal air movement from the land, which has been cooled by radiation, towards the sea, which is slightly warmer. Because the temperature differences are small the wind speed is low. Differential cooling of land and sea occurs only under clear skies, in calm weather when the pressure gradient is weak. The slightly higher pressure over the land during the night produces a seaward movement of surface air, which is part of a larger convectional cell, for above the surface land breeze a weaker air movement flows from sea to land. Such mechanisms can be found in most coastal areas, where the land and sea breezes are used by fishing fleets, but they are most notable in the Equatorial zone. *See also* LAKE BREEZE, SEA BREEZE. [*130*]

land bridge A hypothetical land connection between continents, which served as a migratory route for terrestrial organisms and is held to account for the modern distribution of certain flora and fauna. *See also* CONTINENTAL DRIFT.

land classification The arrangement of LAND units into a variety of categories based on the properties of the land or its suitability for a particular purpose. It has become an important tool in rural land-resource planning. One of the first classifications was that of L. D. Stamp, who proposed three major categories for Britain: good, medium and poor grade, based on the physical characteristics of the terrain and soil. Since 1974 the Ministry of Agriculture, Fisheries and Food in the UK has established a new grading, based again on terrain and soil character but adding climatic criteria, consistency of crop yield and cost and level of production. This classification comprises: grade I, land with few limitations; grade II, land with minor limitations on the range of crops that can be grown; grade III, land with moderate limitations; grade IV, land with severe limitations (this land is generally under permanent grass); grade V, land with very severe limitations (given over mainly to rough grazing). Land-classification exercises in the developing world have collected considerable amounts of data, often by the use of aerial photographs and REMOTE SENSING techniques in order to build up a resource inventory for future development plans. The classification is based on the recognition of *land systems*, which are tracts exhibiting HOMOGENEOUS physical attributes, within which a number of distinct *facets* (terraces, alluvial fans, etc.) can be identified and mapped. Each of these latter will possess different degrees of development potential according to the intermixture of terrain, soil, hydrologic variables, etc. *See also* LANDFORM CLASSIFICATION, TERRAIN ANALYSIS.

Landes **1** The proper name for the sand-dunes bordering the Bay of Biscay on the SW coast of France. **2** A term used to describe the heathlands of Belgium.

landform The morphology and character of the land surface that results from the interaction of physical processes (e.g. fluvial action, glacial action, weathering) and crustal movements with the geology of the surface layers of the Earth's crust. Its scientific study is termed GEOMORPHOLOGY.

landform classification The arrangement of LANDFORMS into various categories based on their various properties, e.g. structure, composition, configuration, genesis, age. The classification by R. E. Murphy (1968) uses three levels of information: **1** *Seven structural regions* (A. Alpine system, C. Caledonian or Hercynian remnants, G. Gondwana shields, L. Laurasian shields, R. Rifted shields, S. Sedimentary covers, V. Isolated volcanic areas). **2** *Six topographical classes* (P. Plains, H. Hills and low tablelands, T. High tablelands, M. Mountains, W. Widely spaced mountains, D. Depressions). **3** *Five classes of geomorphic processes* (h. Humid landform areas, d. Dry landform areas, g. Glacial areas, w. Areas glaciated during last glacial stage (Würm, Wisconsin, etc.), i. Ice-caps). By combining the relevant letters for a particular region its landforms can be briefly categorized, e.g. N Scotland is CMw (Caledonian remnant, mountains glaciated by last glacial stage). *See also* LAND CLASSIFICATION.

Landsat One of the unmanned, Earth-orbiting, NASA (National Aeronautics and Space Administration) series of Earth-resource satellites. It represents the first unmanned SATELLITE specifically designed to acquire data on Earth resources on a systematic, repetitive medium-RESOLUTION, MULTI-SPECTRAL basis. Landsat 1 (formerly called ERTS) was launched in 1972, Landsat 2 in 1975 and Landsat 3 in 1978. The satellite orbits at an altitude of approximately 900 km, completes fourteen orbits per day and passes over each point on the Earth's surface every eighteen days. The satellite records IMAGES of the Earth's surface that are relayed to ground receiving-stations located around the world. Images are recorded by means of two REMOTE SENSING SYSTEMS: **1** a three-channel RBV television camera recording in green, red and near INFRARED wavebands; **2** a four-waveband MULTI-SPECTRAL SCANNER recording in green, red and two near-infrared wavebands. Landsat images have proved to be of value in many fields of scientific study. *See also* ELECTROMAGNETIC RADIATION, HEAT-CAPACITY MAPPING MISSION, SEASAT I, SKYLAB, SPACE SHUTTLE, SPOT.

landscape A term derived from the Dutch (*landschap*) which referred simply to rural

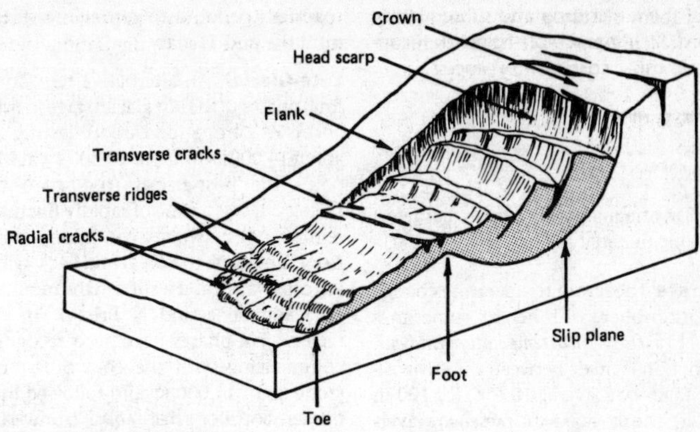

Figure 131 *Features of a landslide*

scenery. Its modern usage relates to the total surface form of any area, rural or urban, and includes both natural and man-made features, i.e. both the natural and the cultural landscapes. Although it is commonly used as a synonym for LANDFORM this is a usage which is to be discouraged, since a landscape will include the landform among its properties. *See also* LANDSCHAFT, NATURAL REGION.

landscape evaluation An exercise in assigning a scenic-quality value to a LANDSCAPE in order to establish its relative importance as a resource in a planning operation. The majority of these attempts have been based on subjective assessments, on biased scaling and weighting and on unacceptable premises, while very few have stood up to statistical testing. The exercises based on simple *character* assessment, referring to those features which can be objectively measured, have been more successful than those based on quality assessment, which depend largely on the personal judgement of each observer.

landschaft A German term which in its broadest sense means LANDSCAPE, but which has been used in a variety of ways. It has been used in the context of scenery, but it can also refer to the regional character of an area.

landslide, landslip The downslope gravitational movement of a body of rock or earth as a unit owing to FAILURE of the material. It may be induced by natural agencies, e.g. heavy rain, earthquake, or it may be caused by human interference with the slope stability. *See also* AVALANCHE, EARTH-FLOW, MASS-MOVEMENT, MUD-FLOW, ROTATIONAL SLIP, TOREVA BLOCKS. [*131; see also 146*]

land system LAND CLASSIFICATION.

langley A unit of energy used as a measure of solar radiation (INSOLATION), suggested in 1942 and named after S. P. Langley, the first director of the Astrophysical Laboratory at the Smithsonian Institution, USA. It is equivalent to 1 calorie per square centimetre per minute. It is now being replaced by the *watt* (W) or the *megajoule* (MJ) (JOULE). The mean total solar radiation is about 392×10^{24} watts and the energy received at the Earth's surface is about 4.95×10^{6} watts^{-2} (or 2 langleys). 1 langley (ly) min^{-1} = 697.6 W m^{-2}; 1 langley day^{-1} = 0.042 MJ m^{-2} day^{-1}. *See also* SI UNITS.

Langmuir theory of raindrop growth A theory to explain the incremental size increase of *raindrops;* a variation of the COLLISION THEORY OF RAINFALL. Experiments by I. Langmuir suggested that because of their greater terminal velocities the larger (>19 μm in radius) raindrops overtook and probably absorbed smaller droplets. Furthermore, some of the latter which escaped initial absorption may be swept into the wake of the larger drop and subsequently absorbed. *See also* BERGERON–FINDEISEN THEORY OF PRECIPITATION, COAGULATION PROCESS.

La Niña EL NIÑO.

lapies KARREN.

lapilli Tiny fragments (4–32 mm diameter) of volcanic material (PYROCLASTIC MATERIAL).

lapse-rate The rate of temperature change in the atmosphere with height, sometimes referred to as the *vertical temperature gradient*. A distinction is made between the ENVIRONMENTAL LAPSE-RATE (average 0.6° C for 100 m of ascent), the DRY ADIABATIC LAPSE-RATE (average 1.0° C for 100 m of ascent), and the SATURATED ADIABATIC LAPSE-RATE (anywhere between 0.2° C and 0.9° C for 100 m of ascent). The lapse-rate is thought to function only as far as the TROPOPAUSE, above which the vertical temperature gradient disappears and the temperature of the lower STRATOSPHERE appears to remain fairly constant with increasing altitude. The lapse-rate is frequently interrupted by an INVERSION OF TEMPERATURE. [*132*]

Laramide orogeny (Laramidian) The mountain-building episode which created the folding of the Rockies, N America. It spanned a period lasting from Jurassic times until the mid-Tertiary.

Late-Glacial An alternative term for the final phases of the last glacial stage in Britain and NW Europe (DEVENSIAN), lasting from about 14,000 BP to about 10,300 BP, at which point the post-glacial (FLANDRIAN) commences. It was a time of rapidly fluctuating temperatures, when conditions shifted from sub-Arctic to Boreal and back again as ice-caps waxed and waned in the mountains of Scandinavia and N Britain. In these regions five phases have been recognized, commencing with the cold OLDEST DRYAS (zone Ia) in 14,000 BP and followed by the milder BØLLING INTERSTADIAL (zone Ia–Ib). Colder conditions returned with the OLDER DRYAS (zone Ic) but in 12,150 BP this was replaced by the milder interlude of the ALLERØD INTERSTADIAL (zone II) during which the ice-sheets waned. The final cold phase of the Pleistocene is represented by the YOUNGER DRYAS (zone III) which lasted from about 11,360 BP to about 10,300 BP. In S Britain the Late-Glacial sequence is represented at Holborough, Kent, where solifluction layers (representing cold zones Ia, Ic and III) have intervening organic horizons and buried soils (representing zones Ib and II), but elsewhere there are very few sites where the climatic oscillation between Ib

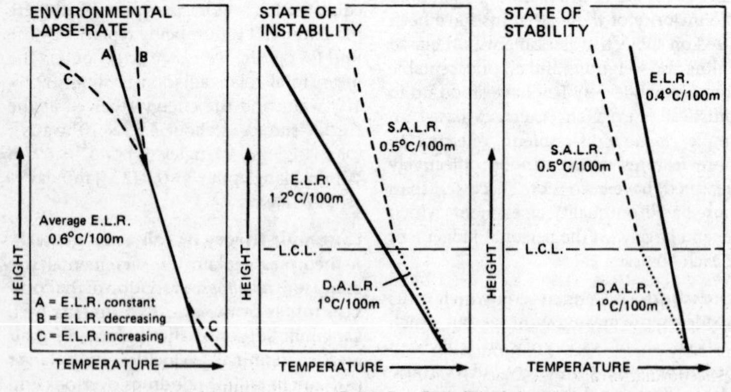

Figure 132 *Lapse-rates in the atmosphere*

and IC can be recognized. The Allerød (zone II) witnessed the appearance of elk and giant Irish deer, with birch and pine spreading across the open tundra landscapes. Zone III saw a dramatic climatic deterioration when the LOCH LOMOND STADIAL ice-cap expanded briefly in the Scottish Highlands and the Cumbrian and Welsh cirque glaciers waxed. Zone III was the last episode of massive periglacial activity in Britain. *See also* WINDERMERE INTERSTADIAL, DIMLINGTON STADIAL..

latent heat The amount of energy emitted or absorbed when a body changes its state or phase, without any change of temperature within that body. It is expressed as calories per gram or megajoules per kilogram, e.g. *latent heat of vaporization* at 20° C = 2.45 MJ kg^{-1} or 590 cal gm^{-1}. Latent heat is very important in atmospheric processes, where water can exist as a vapour, a liquid or a solid, for it is emitted during the processes of CONDENSATION and freezing but is absorbed when EVAPORATION and melting occurs. When a solid is changed to a liquid the heat energy required is termed the *latent heat of fusion* (e.g. ice to liquid: 0.33 MJ kg^{-1} or 80 cal gm^{-1}). When a solid is changed to a gaseous state the heat energy required is termed the *latent heat of sublimation* (e.g. ice to water vapour: 2.83 MJ kg^{-1} or 680 cal gm^{-1}). Thus, latent heat allows energy to be transferred within the general ATMOSPHERIC CIRCULATION, with the atmosphere acting as a heat reservoir, in addition to enabling energy transfer from the ground to the atmosphere. *See also* HEAT BALANCE, INSOLATION, SENSIBLE HEAT.

latent instability An expression referring to the state of that part of a conditionally unstable air mass (CONDITIONAL INSTABILITY) that occurs above the level reached by free CONVECTION in the atmosphere.

lateral accretion The process describing the accumulation of bed sediments at the edge of a stream channel as it shifts laterally, especially during POINT BAR formation. Lateral accumulation also takes place in BRAIDED STREAMS.

lateral-dune A small DUNE alongside a large dune in a desert environment [*65*].

lateral eluviation SUBSURFACE WASH, THROUGHFLOW.

lateral erosion The action performed by a river when it erodes its banks in a horizontal direction. Such action is most marked in a MEANDER zone where the river impinges more strongly on the outside of a meander, thereby undercutting the banks and causing bank-caving (BANK). As this process is repeated the channel migrates downstream in the floodplain tract. In doing so the swinging course of the river is brought into contact with the valley-side BLUFFS which are themselves very gradually cut back, thereby widening the valley. In time the valley sides will be eroded back in this fashion to produce a broad, virtually flat surface at the expense of the INTERFLUVE areas, a process termed *lateral planation*.

lateral flow A term referring to sub-surface flow of water almost horizontally along a permeable soil horizon as distinct from a vertical movement (THROUGHFLOW).

lateral migration The gradual shift of a stream channel across a FLOODPLAIN due to LATERAL EROSION.

lateral moraine A ridge of glacial debris flanking a glacier side or lying along the sides of a valley formerly occupied by a glacier. When ice is present the lateral moraine may bury the glacier edge, in which case the debris may protect the ice from surface melting. In this instance the lateral moraine may become ice-cored. As a valley glacier downwastes, a series of lateral moraines may be deposited at lower and lower levels down the valley sides.

lateral planation LATERAL EROSION.

laterite, laterization A red residual deposit created from the weathering of rocks under humid tropical conditions. The term laterite was suggested in 1807 by F. Buchanan to describe an indurated clay valuable as a building stone (from the Latin *later* = a brick). It consists essentially of oxides of iron and aluminium and is associated with plane surfaces of low relief, suggesting that it is formed on a PENEPLAIN as a FERRICRETE. It is produced by large-scale CHEMICAL WEATHER-

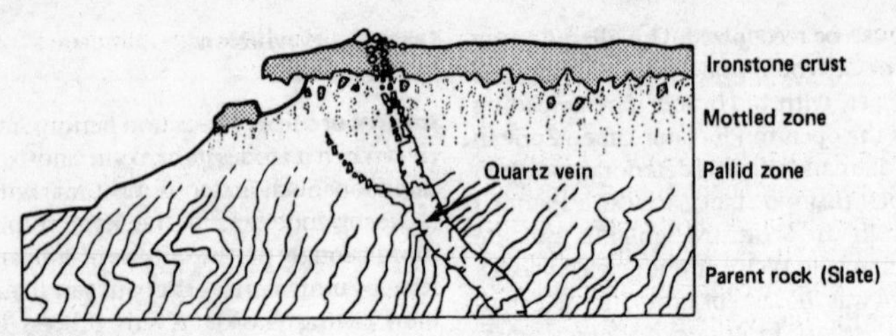

Figure 133 *Laterite*

ING where silica and bases are leached from the parent material, creating a concentration of iron and aluminium sesquioxides, but it is generated by processes of laterization of a much greater magnitude than pedological processes so that laterite must not be regarded simply as a soil (LATOSOLS). It occurs in three forms, either as a soft *lateritic* clay at depth, which hardens on exposure to the air; as a tough, indurated layer forming a HARDPAN at the surface where the superficial material has been stripped off; or as a horizon of nodules or lenses. Laterization is thought to operate over lengthy periods of geological time with many laterites being relics of Tertiary weathering. Although there is no general agreement on their exact mode of formation it is thought that laterites and BAUXITES are formed in similar ways owing to a fluctuating WATER-TABLE due to the seasonal dryness of parts of the humid tropics. Crushed laterite, termed murram, is an important road-building material in tropical countries while *lateritic clays* are extensively used for brick-making. [*133*] The term laterite has now been replaced by *duricrust* (CRUST) or *plinthite*.

latitude The angular distance of a point on the Earth's surface along a MERIDIAN N or S of the equator. it is measured as the angle subtended at the Earth's centre by an arc along a meridian between the point and the equator. The EQUATOR is a circle drawn around the Earth but it has no angle, whereas circles drawn around the Earth PARALLEL to the equator (termed *parallels of latitude*) and at various distances from it will exhibit different degrees of latitude. Every point on a particular *parallel of latitude* will have the same latitude. All *parallels of latitude* intersect all meridians at right angles. Latitude is expressed in degrees, minutes and seconds. The *Tropic of* CANCER and *Tropic of* CAPRICORN *are parallels of latitude* approximately 23½ N and S of the equator, respectively. Because the Earth is not a perfect sphere but an oblate spheroid (EARTH) the distance between the 1° *parallels of latitude* varies slightly. Thus 1° of latitude at the equator = 110.551 km, at 45° it is 111.130 km and near to the poles it is 111.609 km. *See also* LONGITUDE, TROPIC. [*176*]

latosols 1 In general, soils developed on a LATERITE. **2** More specifically, a term synonymous with *ferrallitic soils*, which occur widely in the humid tropics where chemical weathering and leaching are intense, leading to deep soil profiles of free-draining, yellow-to-red acid soils, rich in hydrated oxides of iron, aluminium and manganese. Because they are so heavily leached they are devoid of bases and poor in organic deposits, making them of little agricultural value. Their texture varies from a kaolinitic clay to a loamy sand and five divisions have been recognized in attempts to classify the so-called tropical red and yellow earths (which are now defined as OXISOLS in the SEVENTH APPROXIMATION of soil classification), the divisions being based on differences in climatic regimes which affect the soil character: (a) *aquox*, which are seasonally saturated with groundwater; (b) *humox*, with accumulations of organic carbon, occurring at high altitudes in cooler moist regions of the tropics; (c) *orthox*, occurring in areas with no dry season or only a very short dry season; (d) *torrox*, dry in all horizons for more than half the year; and (e) *ustox*, moist in some horizons for more than six months per year but with a three-month dry season. *See also* FERRISOLS, FER(RI)SIALLITIC SOILS.

Laurasia The northern sector of the hypothetical super-continent thought to have existed after the break-up of PANGAEA by CONTINENTAL DRIFT. It was said to be separated from the southern continent of GONDWANALAND by the intervening ocean of TETHYS. Laurasia is said to have comprised N America, Greenland and all of Eur-Asia north of the Indian subcontinent.

Laurentian The stratigraphic name given to any intrusive granite of Pre-Huronian age in the Canadian PRE-CAMBRIAN.

lava The molten material (MAGMA) extruded from a VOLCANO or by a FISSURE ERUPTION on to the surface of the Earth where it cools and solidifies. Lavas are subdivided into: **1** BASIC LAVAS; **2** ACID LAVAS and **3** intermediate lavas. Their textures may be PORPHYRITIC, non-porphyritic (aphyric), glassy (OBSIDIAN) or VESICULAR. *See also* AA LAVA, BASALT, PAHOEHOE, PILLOW LAVA, RHYOLITE.

lava blister A hollow protuberance on the surface of a *lava-flow* caused by gas bubbles, especially on PAHOEHOE types of lava.

lava cone A volcanic CONE built entirely of *lava-flows* in which explosive activity was minimal. It is usually of low-angled slopes resulting from the mobility of the lava, since most lava cones are built of basaltic lava. In scale it resembles a miniature SHIELD VOLCANO, being of the order of 1 km across and several hundred metres in height. It is generally associated with a FISSURE ERUPTION. *See also* SPATTER CONE.

lava dome A massive protuberance of lava forming a mountain with convex slopes, resulting from the accumulation of numerous *lava-flows* from a CENTRAL-VENT VOLCANO, over a considerable period of time. The Hawaiian volcano of Mauna Loa is a classic example.

lava field A wide expanse of *lava-flows* which have coalesced at the foot of one or more volcanic cones to produce a terrain of continuous lava extending over many square kilometres.

lava-flow A stream of lava issuing from a volcanic VENT in the form of a narrow but lengthy emission. A BASIC LAVA flow is extremely mobile and may extend considerable distances, while a VISCOUS lava-flow will congeal more rapidly and will travel only a short distance from the vent.

lava fountain A forceful vertical expulsion of a jet of molten lava from a VENT or a fissure (FISSURE ERUPTION) during a volcanic eruption. The jet-like expulsion may be rhythmic or periodic and may be in the form of individual '*fire fountains*' or in combination as a '*fire curtain*'.

lavaka The name given to a steep, 30–40 m deep gully cut into plateau edges in Madagascar (Malagasy) due to severe erosion following deforestation by human exploitation.

lava lake A body of molten lava in a volcanic CRATER, e.g. that of Kilauea in Hawaii.

lavant BOURNE.

lava plateau An elevated tableland composed of considerable thickness of *lava-flows* one upon another, built up over a considerable period of geological time, usually from FISSURE ERUPTIONS and covering thousands of square kilometres. Most lava plateaux are composed of BASALT and have been deeply eroded by denudation into prominent cliffs exhibiting COLUMNAR jointing, deep gorges, tabular mesa-like hills (e.g. McLeods Tables, Skye, Scotland) and TRAP LANDSCAPES. In the British Isles the Antrim Plateau of N Ireland and the Is. of Skye and Mull exhibit the most extensive lava plateaux of Tertiary age. Extensive lava plateaux can also be seen in Iceland, the Deccan of India and the Columbia Plateau of the USA. [*134*]

lava tunnel An open-ended tube or passage within a *lava-flow* caused when the lava surface cools sufficiently to form a crust, thereby creating a roof below which the more mobile lava may drain away to leave a lava-enclosed tunnel.

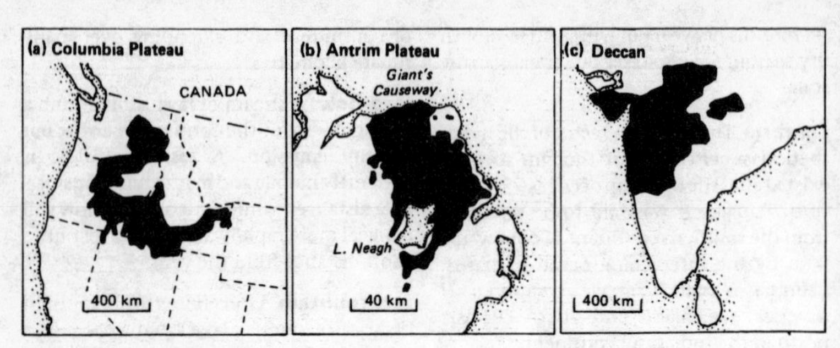

Figure 134 *Lava plateaux*

law A local term for a rocky isolated hill in S Scotland, usually formed from a volcanic PLUG, e.g. N Berwick Law.

law of accordant junctions PLAYFAIR'S LAW.

law of basin areas One of the laws of fluvial MORPHOMETRY formulated by R. E. Horton (HORTONIAN ANALYSIS). It states that: 'The mean basin areas of successive stream orders tend to form a geometric series beginning with the mean area of the first-order basins and increasing according to a constant area ratio.' It is expressed as $\bar{A}u = \bar{A}_1 Ra^{(u-1)}$, where: $\bar{A}u$ is the mean area of basins of order u (STREAM ORDER); $\bar{A}_1$ is the mean area of first-order basins; Ra is the area ratio; and u is the basin order. The law can be given a mathematical expression when plotted on an exponential regression equation in [*135*], which illustrates an example from the Allegheny R., Pennsylvania, USA. It shows how, for example, the area of an entire second-order basin is the sum of the first-order basin areas which it contains plus all the interbasin areas within its own perimeter. Basin area is automatically cumulative because it sums all nested basins of lower orders within it. The closer the plotted points fall to the regression line the greater the validity of Horton's *law of basin areas*. *See also* REGRESSION.

law of cross-cutting relationships A geological law stating that an intruded IGNEOUS ROCK is younger than the rocks through which it cuts.

law of included fragments A geological law stating that if recognizable fragments of one rock are found included within another rock (e.g. a CONGLOMERATE), then the enclosing rock must be younger than the rock material of the included fragments.

Law of the Sea Convention An agreement that all deep-sea mining of the sea floor beyond the 200-mile limit of coastal nations would occur only under the jurisdiction of the INTERNATIONAL SEABED AUTHORITY.

law of stream lengths One of the laws of fluvial MORPHOMETRY formulated by R. E. Horton (HORTONIAN ANALYSIS). It states that: 'The cumulative mean lengths of stream segments of successive orders tend to form a geometric series beginning with the mean length of the first-order segments and increasing according to a constant length ratio.' It is expressed as $\bar{L}u = (\Sigma Lu)/Nu$, where: Lu is the mean length of all stream segments of STREAM ORDER u; Nu is the number of stream segments of stream order u; and ΣLu is the sum of lengths of all stream segments of stream order u. The law can be given a mathematical expression when plotted on an exponential regression equation in [*135*], which illustrates an example from Fern Canyon, California, USA. It is seen how the plotted points of the cumulative mean length (expressed in mls) are summed, e.g. for order 2 the mean lengths of first and second orders are summed; for order 3 the mean lengths of first, second and third orders are summed, and so on. In

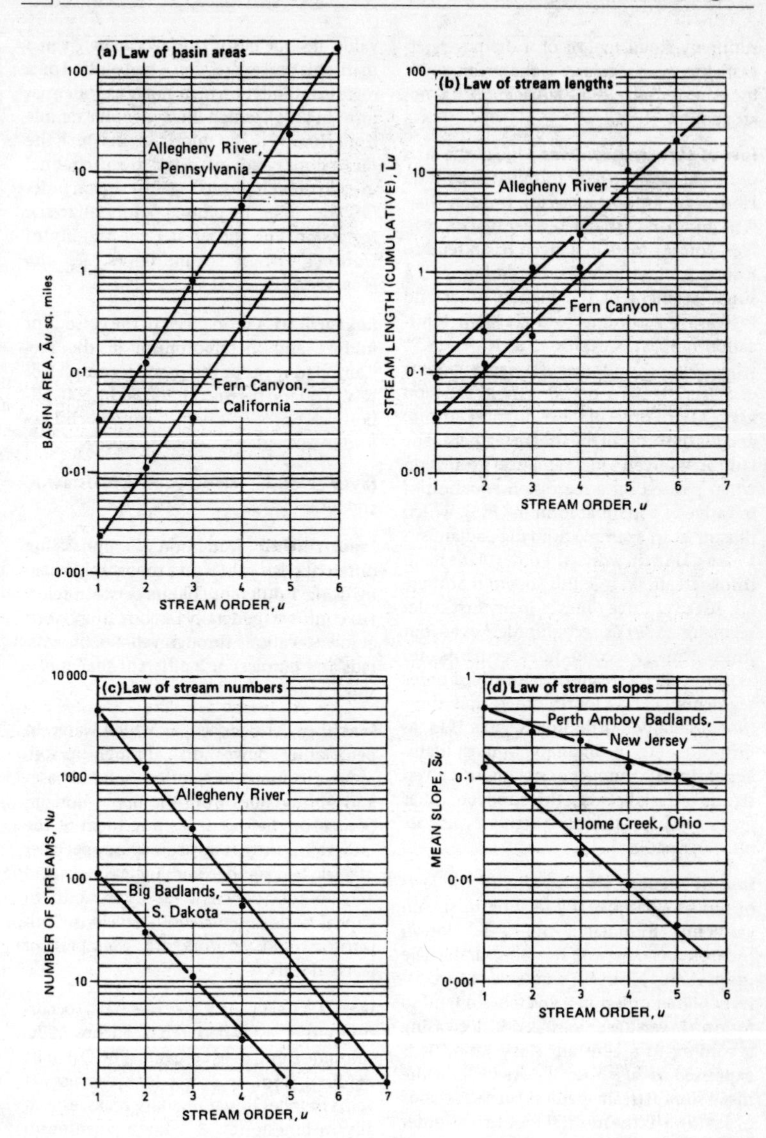

Figure 135 *Laws of fluvial morphometry*

summary, total length of a drainage network increases regularly as the stream order increases. *See also* BIFURCATION RATIO, REGRESSION.

law of stream numbers One of the laws of fluvial MORPHOMETRY formulated by R. E. Horton (HORTONIAN ANALYSIS). It states that: 'The numbers of stream segments of successively lower orders in a given basin tend to form a geometric series, beginning with a single segment of the highest order and increasing according to a constant bifurcation ratio. It is expressed as $Nu = Rb^{(k-u)}$, where: Nu is the number of stream segments of a given STREAM ORDER; Rb is the BIFURCATION RATIO; k is the order of the main trunk stream; and u is the order of the stream segment. The law can be given a mathematical expression when plotted on a regression equation of negative exponential form in [*135*], which illustrates an example from Big Badlands, S Dakota, USA, in which the order of the main trunk stream (k) is 5. It is shown that there are just over three times as many first-order segments (139) as second-order; over four times as many second-order segments (46) as third-order; 3⅔ times as many third-order segments (11) as fourth-order; and three times as many fourth-order segments (3) as fifth-order (1). In summary, in most drainage basins the number of streams in a given stream order is between three and four times greater than in the next higher order. *See also* REGRESSION.

law of stream slopes One of the laws of fluvial MORPHOMETRY, relating to stream gradients, formulated by R. E. Horton (HORTONIAN ANALYSIS). It states that: 'The mean slopes of stream segments of successively higher orders in a given basin tend to form an inverse geometric series, decreasing according to a constant slope ratio.' It is expressed as $\bar{S}u = \bar{S}_1 Rs^{(u-1)}$, where: $\bar{S}u$ is the mean slope (stream gradient) of each STREAM ORDER (u); $\bar{S}_1$ is the mean slope of a first-order stream and Rs is the slope ratio (whose value must be less than 1), which in fact differs from order to order (depending on the resistance of the rock beneath the stream channel), but for which, in the law of stream slopes, Horton assumes a single constant value of slope ratio. The law can be given a mathematical expression when plotted on a regression equation of negative exponential form in [*135*], which illustrates an example from Home Creek, Ohio, USA, in which the mean slope (gradient) of the stream channel in each order is plotted against stream order (u). The closer the plotted points fall to the regression line the greater the validity of Horton's law of stream slopes. *See also* REGRESSION.

Laxfordian A term given to the episode of folding and metamorphism in the Pre-Cambrian rocks of the LEWISIAN complex of NW Scotland between 1,200 and 1,600 million years ago. It is named from Loch Laxford in Sutherland. *See also* SCOURIAN.

layer cloud ALTOSTRATUS, CIRROSTRATUS, STRATUS.

layer-tinting A method of emphasizing different relief values on a topographic map by using **1** different colours between selected contours (generally ranging from green at low elevations through yellows, browns, reds and purples) or **2** different shades of a single colour.

leaching The process by which water, in percolating downwards through a soil, removes humus in solution, soluble bases and sesquioxides from the upper horizon (A-HORIZON) before depositing them in the underlying ILLUVIAL HORIZON. The upper layer of a leached soil becomes increasingly acid and mineral-deficient, as a pronounced eluvial horizon (ELUVIATION) develops. Both LATOSOLS and PODZOLS are examples of leached soils.

lead 1 A heavy, dark grey metal (Pb) occurring either as GALENA (PbS), an ore in its sulphide form, or as cerussite ($PbCO_3$) in its oxidized form. It occurs in HYDROTHERMAL veins or lodes in sedimentary rocks, especially in limestones. **2** A linear opening in sea-ice, either between the PACK-ICE, or between the FAST-ICE and the pack-ice.

lead–lead dating A form of RADIOMETRIC DATING based on the proportion of ^{207}Pb to ^{206}Pb.

lead-ratio (lead–uranium ratio) The ratio of the amount of lead to the amount of uranium in a rock. It is used in RADIOMETRIC DATING of rocks, for the rate of radioactive breakdown of uranium is known. *See also* ISOTOPES.

least-squares method A statistical technique of estimation to avoid individual judgement in constructing lines, parabolas or other approximating curves to fit sets of data. It may also be used to **1** determine the most probable value of a single quantity from a number of measurements of that quantity, and **2** discover the probable error of the mean value of a number of observations.

leat An artificial channel constructed to transport water along the contour of a hillside to increase the catchment of a reservoir or to provide water for an industrial undertaking, e.g. washing ore at a mine.

Le Chatelier's principle A principle proposed by H. L. Le Chatelier which states that if a change of stress is imposed upon a SYSTEM in a state of equilibrium sufficient to induce a change in any of the components of the system, a reaction will occur thereby displacing the state of the system in a direction which tends to absorb the effect of the change, such that if this reaction occurred alone it would produce a change in an opposite sense to that of the original change produced by the stress. In other words the equilibrium will shift in such a direction as will tend to restore the original state of the system. *See also* GENERAL SYSTEMS THEORY.

ledge 1 a rocky shelf on a cliff, above or below sea-level. **2** A projecting outcrop of mineralized rock, especially a quartz vein.

lee, leeward 1 The downwind side or the side protected from the wind. The opposite to WINDWARD. **2** The opposite slope to the STOSS. *See also* ONSET AND LEE.

lee depression, orographic depression A non-frontal low-pressure system which forms to the *leeward* side of a mountain range as a result of an EDDY created in the moving air stream. It is commonly formed in the South Island of New Zealand, to the east of the Southern Alps; also, in Alberta, along the eastern margin of the Canadian Rockies. *See also* HAIL, LENTICULAR CLOUD.

lee-shore Any shoreline towards which the wind is blowing, thereby creating a danger to shipping.

lee wave, lee eddy A stationary wave in an air stream, generated on the *leeward* side of an obstacle such as a hill or a mountain range. A lee wave or eddy will form when a stable layer of air is sandwiched between less stable layers and only when there is a strong and steady wind flow. A gentle wind crossing the obstacle will create only LAMINAR FLOW, but as the wind speed increases a *standing wave* is formed, which in turn becomes a lee wave, as wind velocity increases. Its wavelength varies between 3 and 30 km. A manifestation of the lee wave may be seen in a cloud formation (*altocumulus lenticularis*), when the crest of the wave reaches the air's saturation level. As the air descends on the distal side of the wave the *lenticular* cloud dissipates. *See also* ROTOR STREAMING, TURBULENCE.

lemniscate loop A type of mathematical curve with a streamlined tapering form, frequently used to demonstrate the shape of a DRUMLIN. It is expressed by the formula $p = l \cos k\theta$, where p is the lemniscate loop, l is the length of the long axis and where $k = (l^2\pi)/4A$, which is a dimensionless number expressing the elongation of the loop, in which A is the area of the figure.

length A basic unit of measurement. In Britain the long-established imperial standards (INCH, FOOT, YARD, MILE) have now been replaced by the metric system. [*136*]

lentic environment A term referring to the ecology of standing water, e.g. ponds and lakes. *See also* LOTIC ENVIRONMENT.

lenticular cloud A type of lens- or almond-shaped cloud (*lenticularis*) usually associated with an EDDY over a topographic barrier. The 'Long White Cloud' of the Southern Alps of New Zealand is a good example. *See also* ALTOCUMULUS, LEE WAVE, STANDING WAVE.

Figure 136 *Length*

Unit	*metre*	*inch*	*foot*	*yard*	*trod*	*mile*
1 metre =	1	39.37	3.281	1.093	0.1988	6.214×10^{-4}
1 inch =	2.54×10^{-2}	1	0.083	0.02778	5.050×10^{-3}	1.578×10^{-5}
1 foot =	0.3048	12	1	0.3333	0.0606	1.894×10^{-4}
1 yard =	0.9144	36	3	1	0.1818	5.682×10^{-4}
1 rod =	5.029	198	16.5	5.5	1	3.125×10^{-3}
1 mile =	1,609	63,360	5,280	1,760	320	1

1 Imperial standard yard = 0.91439841 m.
1 yard (scientific) = 0.9144 m.
1 US yard = 0.91440183 m.
1 English nautical mile = 6,080 ft = 1,853.18 m.
1 International nautical mile = 1,852 m = 6,076.12 ft.
1 mile = 1.61 km.

lessivage The process by which CLAY MINERALS are moved mechanically downwards through a soil by percolating water, leading to a degree of decalcification and ILLUVIATION in a lower horizon. Evidence of lessivage is the gradual appearance with increasing depth of CLAY SKINS on the soil PEDS. *See also* ALFISOL, SOL LESSIVÉ.

leucocratic A term describing light-coloured rocks (i.e. rocks with less than 30% of dark-coloured minerals), especially among the igneous rocks. *See also* MELANOCRATIC, MESOCRATIC.

levanter A Spanish term for an occasional easterly wind which blows through the Straits of Gibraltar and between the Balearic Is. and the Spanish mainland, especially in the summer and early autumn, when a depression enters the Western Mediterranean basin. It is associated with storminess, heavy rain and a BANNER CLOUD on the Rock of Gibraltar.

leveche A Spanish term for a hot, dry southerly or south-easterly wind blowing from the Sahara into SE Spain. It is formed in advance of a depression which moves from W to E through the Mediterranean basin. *See also* KHAMSIN, SIROCCO.

levee An elevated bank flanking the channel of a river and standing above the level of the FLOODPLAIN. It is formed naturally by the river in times of flood when the overbank flow causes a decrease in stream velocity after it has left the confining channel walls. Once this ability to transport its LOAD has decreased (COMPETENCE) much of the material is deposited, especially the coarser fraction, along the edges of the channel. The formation of levees tends to raise the level of the river channel above the floodplain, leading to widespread flooding once the levees are overtopped or breached. Consequently a great deal of artificial heightening and levee building has taken place on rivers such as the Hwang-Ho and

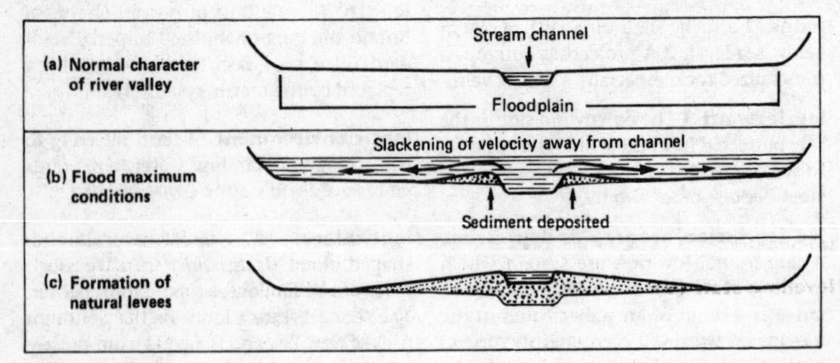

Figure 137 *Formation of levees*

the Mississippi in order to ameliorate the flood danger. [137]

level 1 A synonym for a large tract of alluvial lowland. There are several levels in the Fens of E England. **2** An instrument used in surveying (LEVELLING). It comprises a small telescope on a tripod which can be adjusted to make the level horizontal by using an incorporated spirit-level. The type most generally used is the DUMPY-LEVEL, while a much less sophisticated instrument is the ABNEY-LEVEL. *See also* GEODIMETER, LEVELLING-STAFF.

levelling (US leveling) A surveying method to determine height differences between points in relation to a point of known height (DATUM). It is accomplished by setting up a LEVEL between a point of known height and a point for which an accurate height is required. A LEVELLING-STAFF is held vertically by a surveyor's assistant at the point of known height and a reading is taken on it by use of the level's backsight. Without adjusting or moving the instrument a second reading is obtained by looking through the foresight of the level to another *levelling-staff* held vertically at the point of unknown height. The height difference between the two *levelling-staff* readings gives the difference in height between the two points and the absolute height of the unknown point can then be calculated. The latter can now be used to discover the height of another point as the *levelling traverse* continues. By checking the final traverse reading with that of the first known point the accuracy of the exercise can be determined. If there is a small closing error this may be distributed among the number of surveyed points. An extremely sophisticated type of levelling is termed *geodetic surveying*, which utilizes a GEODIMETER (GEODESY). The determination of height above sea-level by observing values of boiling water is termed *thermometric levelling*, but is little used today (HYPSOMETER).

levelling-staff (US level rod) A graduated pole used in conjunction with a LEVEL in order to determine heights in surveying exercises. The majority of modern staffs comprise three telescopic sections which in total extend to a height of 4 m. The heights are depicted in metric units and can be read through the telescope of the level to the nearest millimetre. *See also* LEVELLING.

Lewisian A group of very old Pre-Cambrian rocks in the British Isles, named from their type-site in the Isle of Lewis, NW Scotland. It is part of the BASAL COMPLEX of NW Britain, comprised largely of GNEISSES and SCHISTS. It is older than the TORRIDONIAN and has been heavily folded, sheared and metamorphosed by two Pre-Cambrian orogenies, the SCOURIAN and the LAXFORDIAN. It makes a barren, ice-scoured SHIELD landscape of KNOCK-AND-LOCHAN terrain in NW Scotland, where it forms the oldest group of British rocks. *See also* CRATON.

liana (liane) A climbing vine-like plant with a woody stem that entwines itself around forest trees in TROPICAL RAIN-FORESTS and EQUATORIAL RAIN-FORESTS, often strong enough to be used as a rope. It should not be confused with an EPIPHYTE.

Lias One of the formations of the JURASSIC in Britain and N Ireland, comprising thick beds of clays, shales and limestones with some sandstones. It is synonymous with the Lower Jurassic. In England it is renowned for its ammonites, fossil fishes and fossil reptiles together with many trace fossils, all of which lived in extensive, muddy and relatively shallow seas. Almost 50% of the British iron production is derived from the Liassic IRONSTONES of Cleveland and Lincolnshire. The rocks of the Lias were laid down between 195 and 172 million years ago and have been divided into the *Lower Lias* (comprising the continental Hettangian, Sinemurian and Lower Pliensbachian); the *Middle Lias* (comprising the continental Upper Pliensbachian); and the *Upper Lias* (comprising the continental Toarcian). The Lias was preceded by the Upper Trias and was succeeded by the Middle Jurassic. The basal calcareous sandstones, marls and shales of SW England (*Lower Lias*) indicate ANAEROBIC conditions and have been termed the *Blue Lias*. In Somerset and Devon/Dorset an impure white calcite mudstone, termed the *White Lias*, is usually included in the RHAETIC, which in Europe is regarded as Upper Triassic, but in Britain has closer fossil affinities with the Lower Jurassic. Nevertheless, the *White Lias* is classified as Triassic.

libeccio A strong westerly wind affecting the western coasts of Corsica, bringing rain or even snow to the Corsican mountains in winter but usually proving invigorating in the heat of summer.

lichenometry A means of estimating the growth-rate of LICHENS in order to determine the approximate length of time that a stone has been lying on the ground surface, e.g. following the withdrawal of an ice-sheet or glacier. Since the growth-rate may be affected by many, especially climatic, variables, the technique has to be used with some care and discretion. Since 1950 some progress has been made in *lichenometric dating*, especially by R. Beschel, who has demonstrated that for various species the growth-rate increases ten times in relation to a fourfold increase in the ratio of precipitation to altitude.

lichens A group of organisms comprising a fungus–alga complex which live together symbiotically (SYMBIOSIS). Lichens grow on bare rock surfaces or on tree trunks, roofs, etc. but the great majority will not tolerate smoke-polluted air. They occur in three forms: *crustose* (as a thin crust); *foliose* (leaf-like); *fruticose* (with stems). One of the best-known is *reindeer moss* (*Cladonia rangerinifera*), a large-branched lichen, up to 7 cm in height, growing abundantly on high Scottish moors and on some sand-dunes at lower levels. One of the most common is *old man's beard* (*Usnea* spp.), which grows in hairy tufts on rocks and trees.

Liebig's law of the minimum A law formulated in 1840 by a German scientist, J. Liebig, referring to the minimal nutrient requirements of certain plants. It has subsequently been expanded to include the effect of other physical factors on plant growth, especially in more recent studies of energy controls in ECOSYSTEMS which examine the photosynthetic efficiency (PHOTOSYNTHESIS) of green plants.

light minerals 1 Those rock-forming minerals which have a low specific gravity (<2.8) compared with the HEAVY MINERALS. The light minerals include quartz, calcite, feldspars, feldspathoids and some micas. **2** Light-coloured (LEUCOCRATIC) minerals, generally the same type of minerals as those in **1** above.

lightning The visible electrical discharge of a THUNDERSTORM, when the increasing electrical charge of the thundercloud overcomes the insulational property of the air, leading to a *flash*. The latter consists of a line of highly ionized air molecules along which electricity flows for a brief moment of time. The first discharge, of low luminosity, is from cloud towards the ground, but this is followed by a highly luminous return stroke which comes up to meet it and continues up to the cloud along the ionized channel. This is the main visible *flash*, transmitting a current of 10,000 amps and travelling at 0.1 times the speed of light. Several varieties of lightning have been defined: *ball, chain, forked, ribbon, sheet, streak*. Of these *sheet lightning* is thought to be the reflection of *lightning flashes* unseen by the observer.

light-year The distance travelled by light in one year. Light travels at approx. 300,000 km (186,000 miles) per sec, or 9.7×10^{12} km (6×10^{12} miles) per year.

lignite A type of low-grade COAL, termed *brown coal*, which is classed midway between PEAT and bituminous coal. Its calorific value, or heat value, is lower than that of bituminous coal.

limb of fold One of the two inclined strata that occur on either side of the axis of a SYNCLINE or an ANTICLINE. [*84*]

lime requirement The amount of agricultural limestone or artificial liming fertilizer required per unit area to a soil depth of 15 cm to raise the pH of a soil to a specific value.

limestone A sedimentary rock formed essentially from carbonate minerals (CALCAREOUS ROCKS), especially CALCITE and DOLOMITE. Limestones have been classified as:

1 *Organic limestones*, comprising the skeletons of former plants and animals. These include algal, coral, crinoidal, foraminiferal, reef and shelly limestones (*see also* ALGAE, BIOHERM, CHALK, CORAL, CRINOIDAL LIMESTONE, FORAMINIFERA, REEF-KNOLL). **2** *Chemically precipitated limestones*, including the EVAPORITES, OOLITIC LIMESTONE and TUFA. **3** *Clastic* (*detrital*) *limestones*, composed of broken fragments of pre-existing calcareous rocks, the classification of which is based upon their GRAIN SIZE. This three-fold classification is currently being replaced by a division based on the particle type and the nature of the matrix (CEMENT). Thus, the prefix of the limestone refers to the particle type, e.g. *Intra*-(intraclasts), *Oo*-(ooliths), *Pel*-(pellets), while the suffix refers to the matrix, e.g. *micrite* (microcrystalline calcite with grain size of less than 0.01 mm) and *sparite* (coarse calcite with grain size greater than 0.01 mm). Consequently, an *oolitic limestone* with a sparite cement would be an *oosparite*. Limestones form characteristic scenery (KARST, CARBONIFEROUS LIMESTONE) and have important uses as building stone, agricultural fertilizer and in the production of cement.

limestone pavement A glacially planed and smoothed surface of bare limestone (GLACIOKARST) which has subsequently been dissected by solution-enlargement of vertical joints (GRIKES) to produce CLINTS.

limiting angles An expression used in geomorphology to describe the range of slope angles within which given processes operate or particular landforms occur. In a humid climate, for example, the *upper limiting angle* for the presence of a REGOLITH cover is normally 40°–45°, while a similar figure refers to the *lower limiting angle* of a FREE FACE, but exceptionally steep regolith-covered slopes of 70°–80° have been discovered in some rain-forest areas. In engineering terms, a stable slope would be defined as one with a gradient below the limiting angle for rapid MASS-MOVEMENT, but this will vary according to the rock type and the climatic conditions. *See also* THRESHOLD ANGLE.

limnic, limnetic Pertaining to an organism which lives in fresh water.

limnology The scientific study of physical, biological and chemical conditions in ponds, lakes and streams.

limnophyte A term for any vegetation species which grows in an environment of constant submergence in fresh water.

limnoplankton A microscopic form of plant and animal organisms (PLANKTON) occurring in fresh water which is relatively static, e.g. a lake.

limon A term given to the AEOLIAN deposits and the reworked, river-deposited, fine-grained LOESS-like materials deposited in Flanders and NW France, from which fertile, brown loamy soils have developed. It is thought that limon represents the fine materials winnowed by outblowing glacier winds from the morainic material of N Europe during the closing episodes of the Quaternary Ice Age.

limonite A term for the yellowish-brown hydrous oxides of iron ($2Fe_2O_3.3H_2O$). It is an important iron ore. *See also* BOG ORE, GOSSAN, LATERITE.

lineament A large-scale linear topographical feature reflecting an underlying structure; a structurally controlled landform. Lineaments include FAULT SCARPS, FAULT-LINE SCARPS, FOLD axes, FISSURE ERUPTION volcanic cones, igneous DYKES and SILLS and structurally controlled coastlines. A lineament at a continental scale is termed a *megalineament* and includes ISLAND ARCS, MID-OCEANIC RIDGES, RIFT VALLEYS and fold mountains at plate margins (PLATE TECTONICS).

linear programming An expression for all techniques which select the optimum solution from a large number of solutions to a problem in which a large number of simple conditions have to be satisfied at the same time and where the relationships involved are linear. *See also* NON-LINEAR PROGRAMMING.

linear theory An expression used in geomorphology referring to the rates of slope recession (BACK-WEARING). It was a theory developed by A. E. Scheidegger in the 1960s as one of his models of slope development. His linear theory assumes that degradation (WEATHERING) acts vertically, while the alternative assumption (the *non-linear theory*) assumes that degradation acts normal to the slope surface. For each of these alternative theories he analysed three cases: case 1 in which degradation (ground loss) is uniform; case 2 in which degradation under all conditions is proportional to altitude; and case 3 in which degradation under all conditions is proportional to slope angle.

lineation A one-dimensional linear structure in a rock, caused by one of the following: **1** orientation of minerals often due to primary flowage of an igneous rock; **2** intersection of the rock surface with a planar surface; **3** intersection of two planar surfaces, e.g. BEDDING with CLEAVAGE; **4** the development of microfolds; **5** orientation of fossils during sedimentation; **6** glacial striae (STRIATION); **7** SLICKENSIDES.

line squall A very stormy phase of weather of short duration, extending along a sharply defined line ahead of a COLD FRONT. It is associated with veering gale-force winds, a thick arch of cloud, hail and thunderstorms, a rapid rise of pressure and fall in temperature.

line system A type of PROCESS-RESPONSE SYSTEM in which flows of mass and energy are considered as linear (e.g. a stream channel).

liquefaction The process by which sediments and soils collapse from a sudden loss of COHESION following a loss of shearing resistance. It involves a temporary transformation of the material into a fluid mass by the sudden increase of interstitial soil water due to an earth tremor, or to suddenly increased loading (e.g. a rock slide) or to a sudden but temporary increase of groundwater from an external source. *See also* QUICKCLAYS.

liquid limit Sometimes referred to as the *Atterberg limit* or the *upper plastic limit*. It refers to the moisture content corresponding to an arbitrary limit between the liquid and plastic states of a clay or soil. *See also* CONSISTENCE, FLOW, PLASTIC LIMIT.

lithification, lithifaction The processes by which newly deposited sediments are converted into a solid rock by COMPACTION and *cementation* (DIAGENESIS). *See also* INDURATION.

lithogenesis The accumulation of sedimentary material in a GEOSYNCLINE prior to DIAGENESIS.

lithology The study of rock characteristics, particularly their GRAIN SIZE, PARTICLE SIZE and their physical and chemical character. Some geologists restrict the term to sedimentary rocks.

lithometeor A mixture of non-aqueous (solid) particles in the atmosphere. *See also* DUST, DUST-STORM, HAZE.

lithomorphic soils One of the seven major groups in the SOIL CLASSIFICATION OF ENGLAND AND WALES. It includes the normally well-drained soils with distinct, humose or organic topsoil formed on bedrock or little altered unconsolidated material which lies at depths of less than 30 cm. The group includes the RANKERS, and RENDZINAS.

lithophyte A term for vegetation which is able to flourish on a bare rock surface, e.g. LICHENS. *See also* LITHOSERE.

lithosere The successional development of a group of plant communities on bare rock faces or blocky scree slopes, i.e. relatively dry, hard and exposed surfaces. It is a type of XEROSERE. The first colonizing plants are usually lichens and terrestrial algae. *See also* HALOSERE, HYDROSERE, PSAMMOSERE, SERE.

lithosol One of the subgroups of the AZONAL SOILS, comprising those immature soils which form on bare rock material. *See also* ENTISOL.

lithosphere The crustal component of the Earth, in contrast to the ATMOSPHERE and the HYDROSPHERE. It includes the SIAL, SIMA and upper MANTLE. Its base lies at 2–3 km under MID-OCEANIC RIDGES, increasing up to 180 km beneath old OCEAN CRUST. Beneath CRATONS it is between 250 and 500 km in thickness.

See also ASTHENOSPHERE, BARYSPHERE, MOHOROVIČIĆ DISCONTINUITY, PLATE TECTONICS.

lithostratigraphy The scientific study of the physical and spatial characteristics and the PETROGRAPHY of stratified rocks in an attempt to determine their STRATIGRAPHY. A *lithostratigraphic unit* is one in which rocks are grouped on the basis of LITHOLOGY rather than on their biological (fossil) characteristics (BIOSTRATIGRAPHY) or on a time basis (CHRONOSTRATIGRAPHY).

lit-par-lit injection A geological term derived from the French (lit. 'bed-by-bed') to describe the forcible intrusion or penetration of igneous (usually granitic) rock in the form of sheets and veinlets into an existing bedded, foliated or schistose rock series. Chemical reaction between the two rock types create a MIGMATITE. *See also* GNEISS.

litre A metric unit of volume. It is defined as the volume occupied by 1 kg of water at its temperature of maximum density (4° C and 760 mm pressure). One litre = 0.22 gallons (UK); 1 gallon (UK) = 4.546 litres; 1 gallon (US) = 3.785 litres. [*272*]

litter The accumulation of leaves, twigs, etc. above the soil, from which HUMUS is ultimately formed.

Little Climatic Optimum An increase in mean annual temperature between AD 750 and AD 1200 in Europe and North America. It allowed more widespread settlement around the shores of the North Atlantic, especially in Iceland and Greenland. *See also* CLIMATIC OPTIMUM.

Little Ice Age 1 A cool period of the SUB-BOREAL during which glaciers re-advanced or were re-established in some of the world's mountain regions (e.g. Sierra Nevada, California). *See also* NEOGLACIAL. **2** A name applied to the climatic cooling in the N hemisphere between the mid-16th cent. and the mid-19th cent. AD, during which Alpine glaciers re-advanced and Britain experienced a preponderance of cold winters. *See also* MAUNDER MINIMUM, SUNSPOTS.

Littletonian The Irish equivalent of the British FLANDRIAN, named from the type site at Littleton, Co. Tipperary.

littoral Pertaining to a shoreline (LITTORAL ZONE).

littoral deposits Materials that accumulate between high- and low-water marks, generally comprising sand, shells or shingle.

littoral zone 1 That area of a shoreline that is in juxtaposition to a body of water. **2** More specifically, the zone between high- and low-water spring-tide marks. **3** Biologists extend the term's usage to include the underwater zone down to 200 m depth (NB – the underwater zone between 60 m and 200 m is referred to as the sublittoral zone). At a global scale the littoral zone covers an area of some 150,000 km^2. *See also* EULITTORAL ZONE. [*162*]

llano A Spanish term referring to the SAVANNA grasslands which occupy the plains and plateaux of the Orinoco region in the northern part of South America.

load 1 The material transported by a river, having been supplied by a variety of means, by BANK-CAVING, CORRASION, CREEP, DENUDATION, LANDSLIDE, LATERAL EROSION, MASS-MOVEMENT, or by GROUNDWATER flow. Once in the river channel the material is carried along in various ways: (a) rivers which have little surface RUNOFF but a steady delayed flow may have a high proportion of DISSOLVED LOAD, in which the material is carried in SOLUTION; (b) in most rivers a considerable amount of the load is carried in SUSPENSION as the SUSPENDED LOAD; (c) a third method of transport is by SALTATION, in which the sandy and fine gravel of the load moves downstream by means of a bouncing or hopping motion; (d) a final method of transport relates to the BEDLOAD, in which the coarsest material moves by rolling and sliding along the channel floor: it is sometimes referred to as the *traction load*. The ability of a river to move different-sized particles (COMPETENCE) is thought to be governed by the SIXTH-POWER LAW which refers to increased current velocity. **2** In a more loose usage the term load is used to describe the materials transported by tidal currents, glaciers, waves and wind.

loading **1** In soil mechanics and civil engineering the term is used to denote the increased weight brought to bear on the ground surface, an increase which might lead to FAILURE or FLOWAGE of the stressed surface material. **2** A term used in FACTOR ANALYSIS denoting the degree to which a given variable is attached to a factor. *See also* DIAPIRISM.

loam An easily worked, permeable soil, much valued by farmers. It comprises an almost equal mix of sand and silt but with less than 30% clay. *See also* CLAY LOAM, MEDIUM-TEXTURED SOIL, SANDY LOAM, SILTY LOAM, SOIL TEXTURE.

lobe **1** A tongue-like extension of a material, as in a glacier or ice lobe extending from an ICE-SHEET. **2** In the USA it refers to the tongue of land enclosed by a meander.

local climate A term used to differentiate the climatic characteristics of a small area from those of the surrounding region. The differences may result from the effects of buildings, terrain, vegetation cover, soil character, amount of water and aspect. *See also* BOUNDARY LAYER, MICROCLIMATOLOGY.

local relief RELATIVE RELIEF.

loch A Scottish term for a body of water, either an enclosed fresh-water lake or a long, narrow penetration of sea-water extending inland from the ocean (e.g. Loch Etive).

lochan A Scottish term for a small, fresh-water loch, generally equivalent to a CIRQUE LAKE. *See also* KNOCK-AND-LOCHAN, TARN.

Loch Lomond Stadial The name given to the last cold stage of the DEVENSIAN in the Pleistocene chronology of Scotland, but subsequently extended to the whole of Britain to describe that part of the LATE-GLACIAL which followed the WINDERMERE INTERSTADIAL. It is approximately equivalent to the YOUNGER DRYAS (zone III), which dates from 11,360 to 10,300 BP, but some authors believe that the Loch Lomond Stadial spanned the period 11,500–10,600 BP. In Scotland, it saw the regeneration of small ice-caps in the Western and Northern Highlands and, less extensively, in the Cairngorms, other parts of the Grampians, the Southern Uplands and some of the Hebridean mountains. In Cumbria small valley glaciers existed, e.g. Upper Langdale, while in Snowdonia only cirque glaciers formed.

lode A Cornish term for a fissure infilled with a metalliferous ore, derived from an alteration of the verb to be led or guided, i.e. a guide to a miner.

lodestone A highly magnetic piece of MAGNETITE.

lodgement till TILL.

loess An unstratified, homogeneous, fine-grained yellow BRICKEARTH. Majority opinion favours a wind-blown genesis, whereby widespread dust clouds were carried outwards from the newly deposited glacial and glaciofluvial deposits of northern Eur-Asia by strong anticyclonic winds blowing from the Pleistocene ice-sheets. It has been suggested that the thick loess sheets of central Europe, Russia and China were deposited largely in interglacial and/or post-glacial times by rainfall washing the dust out of the air, but some American writers believe that some USA loess is of *colluvial* (COLLUVIUM) not AEOLIAN derivation, with that of the Lower Mississippi valley being the result of fluvial reworking of old back-swamp deposits. It was first described in 1821 in the Rhine valley where it took its name from a village in Alsace. The soils derived from loess are of the highest quality and produce excellent crops. They consist of deep, well-drained, easily worked silts, and some fine sands with occasional calcareous concretions, termed *loess dolls*. In the southern Netherlands, fluvially reworked loess deposits mixed with weathered chalk materials are known as *loessoide* and the associated fertile soils as *loessleem*. *See also* LIMON. [*138*]

log **1** A graphic representation of stratigraphic units traversed in a geological BOREHOLE. **2** An apparatus for measuring the speed of a ship (KNOT). **3** An abbreviation of *logarithm*.

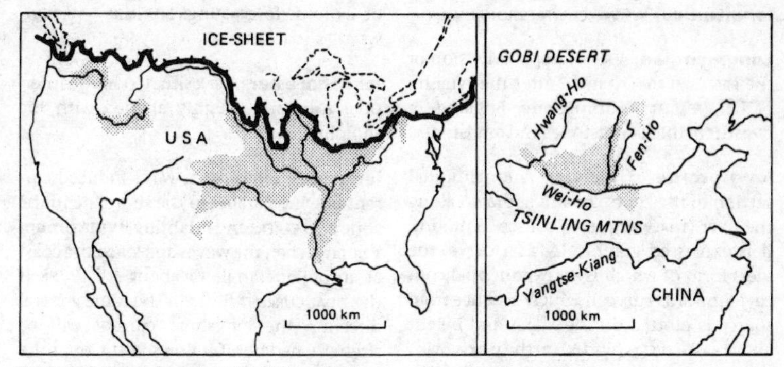

Figure 138 *Loess deposits in the USA and China*

logarithmic scale A constant-ratio scale in which equal distances on the scale represent equal proportions (ratios) of increase. It contrasts with the familiar *arithmetic scale*, which has exactly equal units of distance throughout its entire length. The contrast is illustrated in [*139*]. *See also* SCALES OF MEASUREMENT.

lognormal distribution A distribution of data in which the logarithm of a parameter is normally distributed. *See also* NORMAL DISTRIBUTION.

longitude The angular distance of a point on the Earth's surface east or west of a central MERIDIAN (0° OF GREENWICH MERIDIAN). It is measured by the angle between the plane of the meridian through the point and that of the CENTRAL MERIDIAN. The latter was agreed in 1884 when all readings would henceforth be taken from the meridian which passes through Greenwich, England. A circle drawn around the Earth and passing through each of the poles is a meridian of longitude. It intersects all parallels of latitude at right angles. Every point on a particular meridian will have the same longitude, but as meridians converge towards the poles the length of 1° of longitude decreases from the equator polewards. Thus, 1° of longitude at the equator = 111.320 km; at latitude 45° it is 78.848 km; at 70° it is only 38.187 km; at the poles it is zero. In travelling through 15° of longitude at the equator there is a local time difference of 1 hour. *See also* LATITUDE. [*176*]

longitudinal coast CONCORDANT COAST.

longitudinal-dune SEIF-DUNE.

longitudinal profile of a river LONG PROFILE OF A RIVER.

longitudinal valley A valley which follows the STRIKE of the rock strata and is usually followed by a SUBSEQUENT STREAM. It contrasts with a TRANSVERSE VALLEY.

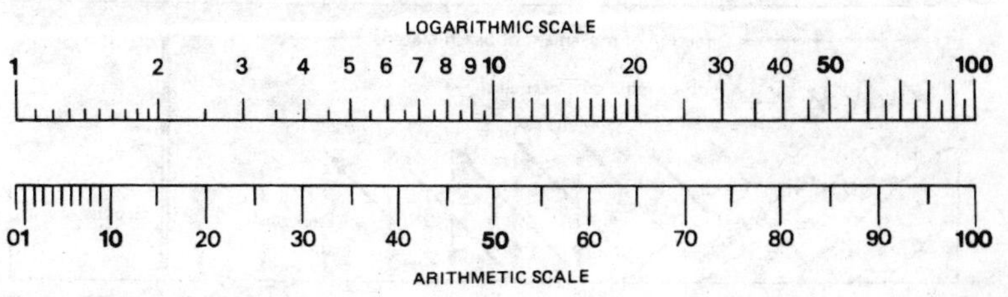

Figure 139 *Logarithmic scale*

longitudinal wave COMPRESSIONAL WAVE.

Longmyndian A stratigraphic division of the PRE-CAMBRIAN, named after the typesite of Longmynd in Shropshire, England. It consists of thick sandstones and mudstones.

long profile of a river A longitudinal section of the course of a river drawn along the river (THALWEG), from source to mouth. It is expressed graphically as a curve, the ideal form of which is an exponential concave-upwards curve in which distance from source is plotted on the *x*-axis and height above BASE-LEVEL is plotted on the *y*-axis, with the ideal curve being referred to as the profile of equilibrium (GRADE) in which a gradual decrease in slope gradient occurs as it is traced downstream. This is due to the fact that discharge increases and hence velocity also increases downstream, so that less gradient is needed to transport the river's LOAD. In fact, few if any *river profiles* fit this ideal curve, but instead are made up of a number of partial curves or different river segments, due to changes of base-level, changes in lithology, together with a number of dependent factors of CHANNEL GEOMETRY. Thus, long profiles are usually irregular rather than smooth and can be considered as a function of the following interrelated variables: DISCHARGE, LOAD, size of debris supplied, flow resistance, velocity, width, depth, slope. The expression *longitudinal profile* should be discouraged. *See also* CROSS-PROFILE, KNICKPOINT, REJUVENATION. [*128*]

long-range forecasting Weather forecasts for a period greater than five days. The accuracy of the forecast is not as good as that of 24-hour forecasting. *See also* ANALOGUE, WEATHER FORECASTING.

longshore bar A BAR situated in the intertidal zone and roughly aligned with the shoreline.

longshore current A wave-induced current flowing parallel to the shoreline in the zone of BREAKERS and attaining its maximum velocity when the waves approach the coast at an oblique angle of about 30° [*209*]. If the TIDAL CURRENT flows in the same general direction the longshore current will be strengthened but if it flows in the opposite direction the current will be obliterated.

longshore drift A general movement of beach material along a shoreline due to the effect of waves breaking obliquely on to the beach. The wave SWASH carries material obliquely up the beach but the returning BACKWASH drags much of the sand and shingle back down the beach although in this case directly down the slope [*140*]. On coastlines dominated by a fairly constant set of waves the movement of material will be unidirectional, with LONGSHORE CURRENTS aiding the longshore drift. If two opposing sets of waves are common the beach material on a given coastline may drift to and fro according to the seasonal wind direction. Drift is checked by GROYNE construction. *See also* BAR, SPIT.

long time-scale A time-scale relating to the length of the PLEISTOCENE glaciation. It is based on RADIOMETRIC DATING of deep-sea sediments obtained by CORE SAMPLING, which suggests that the onset of the Pleistocene glaciation should be moved from

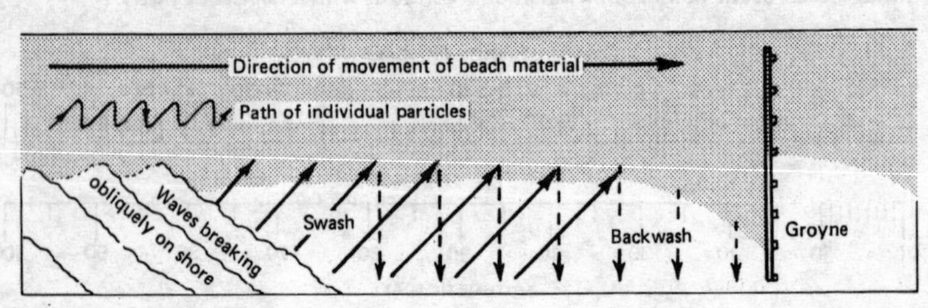

Figure 140 *Longshore drift*

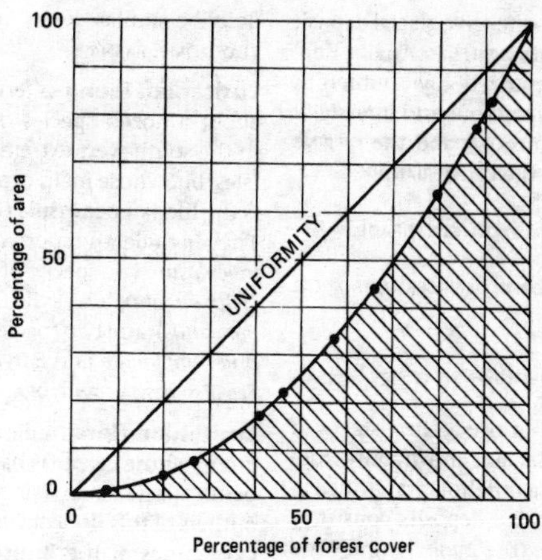

Figure 141 *Lorenz curve*

600,000 BP (SHORT TIME-SCALE) to between 1.64 and 2 million years BP.

long wave ROSSBY WAVE.

long-wave radiation TERRESTRIAL RADIATION.

lopolith A saucer-shaped igneous INTRUSION which, in contrast to a PHACOLITH, is concave upwards. The lopolith is controlled by and concordant with the folding and is related to a SILL. The larger examples (*megalopoliths*) exhibit a banded structure and are thought to be more funnel-shaped than saucer-shaped. The largest of them cover some 200,000 km^2 (e.g. Duluth gabbro, Canada) and make up some of the largest exposures of igneous rocks at the Earth's surface. The *megalopolith* is to the basic rocks what the BATHOLITH is to the granites.

Lorenz curve A mathematical curve used to illustrate the extent to which a given distribution of data is uneven when compared with a uniform distribution. If the density of the data distribution over the area is uniform a straight-line curve will result, running at an angle of 45° across the graph [*141*]. The degree of concavity of the Lorenz curve will show the variations in the density of data distribution for any particular example.

lotic environment A term referring to the ecology of running waters, e.g. rivers. *See also* LENTIC ENVIRONMENT.

lotus projection A recentred MAP PROJECTION of the entire globe introduced in 1958 by Bartholomew. It is designed to provide a continuous map of the world's oceans.

lough An Irish term for a body of water, either an enclosed fresh-water lake or a long, narrow penetration of sea water extending inland from the ocean (e.g. Lough Swilly).

love wave (Q wave) SEISMIC WAVES.

low **1** A low-pressure atmospheric system (DEPRESSION, TROPICAL CYCLONE). **2** A structural low is a structural depression, e.g. a SYNCLINE. **3** A term used in GRAVIMETRIC SURVEY to denote a gravity minimum.

low-energy environment A fluvial, marine or aeolian environment having wave action or current motion of such low magnitude, or in which water movement is lacking, that fine-grained sediment deposition is possible. Contrasted with a HIGH-ENERGY ENVIRONMENT.

Lower Greensand A lithostratigraphic unit of the CRETACEOUS in NW Europe and Britain (GREENSAND).

Lowestoft till An extensive glacial deposit of ANGLIAN age, occurring in East Anglia, England, and characterized by numerous erratics from the Cretaceous and Jurassic. It is older than the HOXNIAN and the GIPPING TILL, but post-dates the CROMERIAN.

low-grade metamorphism GRADE.

low-pressure system DEPRESSION, TROPICAL CYCLONE.

loxodrome RHUMB LINE.

Ludhamian One of the earliest stages of the PLEISTOCENE in Britain, known only from East Anglia, where a borehole was drilled at Ludham, Norfolk. The deposits consist of coarse shelly sands (the *Ludham Crag*) and have been placed as the upper member in the RED CRAG SERIES/Formation. Their included pollen of pine (dominant), birch, elm, oak, alder, etc., indicates that the deposits were formed during the pre-glacial part of the Pleistocene. [*197*]

lunar day The time interval between two successive transits of the Moon across a given meridian. This is longer than the SOLAR DAY (24 hours), in as much as, while the Earth is rotating one complete revolution in 24 hours, the Moon has its own orbital motion around the centre of gravity of the Earth and the Moon. Thus, the latter crosses the particular meridian 50 minutes later than the Earth's transit, and this is the reason why a corresponding high tide is 50 minutes later each successive day (i.e. there is a time interval of 12 hours 25 minutes between successive high tides).

lunette A term referring to an asymmetrical crescentic ridge of AEOLIAN origin on the leeward side of certain Australian lakes and swamps. It is thought to be formed by the wind-blown dust gathered from the shores of the seasonally exposed lake-floor. The term has now been universally adopted to describe similar desert landforms. *See also* CLAY DOME, PARNA.

Lusitanian fauna A term used to describe about a dozen species of fauna surviving in the south-west extremities of the British Isles, but whose main centre of distribution is the Iberian peninsula (Roman Lusitania). They include: the spotted slug (*Geomalacus maculosus*); two species of snail (*Helix pisana, Helix virgata*); a weevil (*Otiorrhynchus auropunctatus*); and two species of beetle (*Eurynebria complanata* and *Barypithes curvimanus*). *See also* LUSITANIAN FLORA.

Lusitanian flora A description of a plant group whose centre is based on the Iberian peninsula (Roman Lusitania). Anomalous outliers of this flora occur in the south-west extremities of the British Isles: the strawberry tree (*Arbutus unedo*); the Cornish Heath (*Erica vagans*); great butterwort (*Pinguicula grandiflora*); pale heath violet (*Viola lactea*); Irish spurge (*Euphorbia hiberna*); St Dabeoc's Heath (*Dabeocia cantabrica*), and several other species. It is thought that the flora spread from Iberia northwards along the Atlantic seaboard during an INTERGLACIAL or early in the POST-GLACIAL, but were eventually isolated by rising sea level. *See also* LUSITANIAN FAUNA.

lutite A sedimentary rock consisting largely of clay or clay-sized particles. *See also* ARGILLACEOUS.

L-wave SEISMIC WAVES.

lynchet An artificial cultivation terrace cut into a hillside more or less parallel with the contours. It is thought to be of Iron Age or Bronze Age date and is widespread on the English chalklands. Some think lynchets are of medieval age.

lysimeter A piece of equipment for measuring the rate and amount of water storage and percolation through a soil. With rainfall recorded, the rate of actual EVAPOTRANSPIRATION can be estimated. *See also* EVAPOTRANSPIROMETER.

M

maar A German term for an explosion-crater created by a violent volcanic eruption but with no extrusion of igneous rocks (EXTRUSION). The crater perimeter is made up only of brecciated COUNTRY-ROCK and the crater itself is usually occupied by a lake.

machair A Scottish term for the highly calcareous shelly sand which fringes many of the Hebridean windward coasts in W Scotland. It forms a low undulating tract of land behind the foreshore, usually clothed in a rich grassland except where it is cultivated by the crofting communities. Seaweed is frequently added to the machair soils to improve their organic content. The soil PH is >7.0.

mackerel sky (cloud) A sky dominated by a cloud type known as mackerel cloud. Convection within a thin cloud sheet often causes it to break up into small cells of CIRROCUMULUS or ALTOCUMULUS, arranged in symmetrical patterns that resemble the skin pattern of a mackerel. *See also* CLOUD.

macroclimate The climate of a large area in contrast to that of a small area (MICROCLIMATE). The climatic characteristics of a *macroclimate* are measured on a standard network of STEVENSON SCREENS. *See also* MACROMETEOROLOGY.

macrofauna Fauna which can be distinguished without the aid of a microscope.

macroflora Flora which can be distinguished without the aid of a microscope.

macrofossils Any FOSSIL, either plant or animal, visible to the naked eye but whose identification and characteristics require a microscope. *See also* MACROSCOPIC.

macrometeorology The study of the large scale atmospheric processes which in combination make up the different types of MACROCLIMATE to be found throughout the world. *See also* ATMOSPHERIC CIRCULATION.

macroscopic A term referring to any physical phenomenon whose scale lies in the range of kilometres to hundreds of kilometres; MACROFOSSILS, therefore, are something of a misnomer. *See also* MESOSCOPIC, MICROSCOPIC.

made land Any part of the land surface which has been accumulated by the actions of mankind. Generally, the term is used to describe infilling of voids and the reclaiming of water bodies by the addition of debris, but it can equally be applied to coastal reclamation (POLDER). Because of its lack of compaction most made land is unsuitable for building development until a lengthy time period has elapsed, and there are numerous cases where settling of foundations owing to surface subsidence has led to building collapse. *See also* ANTHROPOGEOMORPHOLOGY.

mafic A geological term referring to dark-coloured minerals, commonly found in basic igneous rocks and contrasted with FELSIC rocks. It is a term derived from magnesian and iron-rich (*ferromagnesian*) minerals. *See also* MELANOCRATIC.

magma Subterranean molten rock of a highly gaseous and mobile nature, generated under great pressure within the Earth at depths below 16 km, largely as a hot silicate liquid, but capable of being extruded at the surface as a LAVA or intruded into the crustal rocks (INTRUSION). It accumulates locally in *magmatic chambers*, which act as reservoirs for VOLCANOES. Once the magma crystallizes and solidifies at depth it forms IGNEOUS ROCKS, which may be PLUTONIC or HYPABYSSAL, although it is unlikely that any of these rocks are a true representation of the magma from which they were formed since their volatile constituents escape during the solidification phase. When magma separates into two or more parts during its molten phase the process is termed *magmatic differentiation*, which probably results from the following mechanisms: differential crystal formation (fractional crystallization); gas streaming through the molten material; thermal diffusion due to a temperature gradient. A fourth possibility, relating to the impossibility of mixing two different liquids (liquid immiscibility), is considered to be unlikely.

magmatic water Synonymous with JUVENILE WATER, i.e. water arising from a subterranean MAGMA.

magnesian limestone **1** A limestone containing a small proportion (<15%) of magnesium carbonate ($MgCO_3$). *See also* DOLOMITE. **2** A term for one of the lithostratigraphic formations of the PERMIAN in Britain.

magnesite Magnesium carbonate ($MgCO_3$), one of the minerals of the carbonate group. It should not be confused with MAGNETITE.

magnetic anomaly A positive or negative departure from the predicted value of the Earth's MAGNETIC FIELD, measured at a particular point on the ground surface. *See also* ANOMALY.

magnetic declination The acute angle between MAGNETIC NORTH and TRUE NORTH, expressed in degrees W and E of true north. In Britain the magnetic declination in 1980 was about 7½°W, but it differs in various parts of the world. Also termed the *magnetic variation*. It should not be confused with MAGNETIC INCLINATION. *See also* AGONIC LINE, DECLINATION.

magnetic division A stratigraphical unit of rocks used in correlation of lithological sequences. It is based on a grouping of those rocks formed within a MAGNETIC INTERVAL. *See also* MAGNETIC REVERSALS.

magnetic equator (aclinic line) A line joining points on the Earth's surface where the magnetic needle exhibits no MAGNETIC INCLINATION (*dip*), i.e. where it remains horizontal.

magnetic field The space through which the influence (force) created by the presence of a magnet exists. It also refers to the field of force which surrounds a conductor carrying an electric current.

magnetic inclination The acute angle between the vertical and the direction of the Earth's MAGNETIC FIELD in the plane of the MAGNETIC MERIDIAN. It is often referred to as the *magnetic dip*. It should not be confused with MAGNETIC DECLINATION. *See also* ISOCLINE.

magnetic interval The time period which occurs between MAGNETIC REVERSALS in the polarity of the Earth. The rocks formed during this period of time are termed a MAGNETIC DIVISION.

magnetic meridian A line joining the position of the MAGNETIC POLES on the Earth's surface (ISOGONIC LINE). The direction assumed in the horizontal plane by a magnetic needle.

magnetic north MAGNETIC POLES.

magnetic poles The two specific points of the Earth's MAGNETIC FIELD at which the lines of magnetic force are vertical. Their positions slowly change and they do not coincide with the geographical poles. The *magnetic north pole* is located currently on Prince of Wales Is. in N Canada (approx. 70°N latitude and 100°W longitude), while the *magnetic south pole* occurs in South Victoria Land, Antarctica (approx. 68°S latitude and 143°E longitude). A line joining the magnetic poles through the Earth misses the Earth's centre by some 1,200 km.

magnetic reversals The periodic change of polarity (i.e. a 180° change of direction in the Earth's MAGNETIC FIELD) that causes the MAGNETIC POLES to reverse, so that S becomes N and vice-versa. Palaeomagnetic studies of

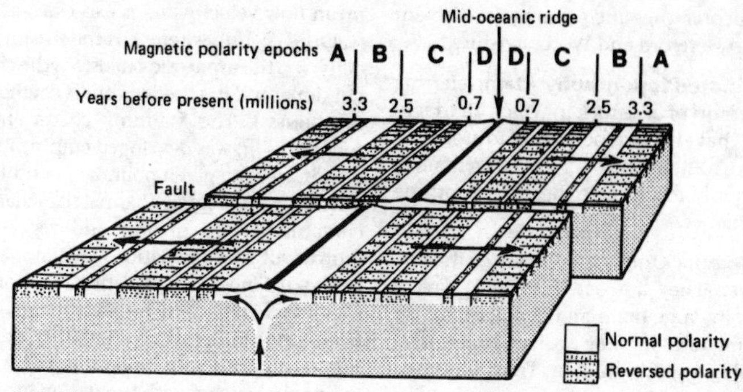

Figure 142 *Magnetic stripes*

rocks exhibiting such reversals in each of the MAGNETIC INTERVALS has assisted in the establishment of a time-scale for the last 4.5 million years. *See also* MAGNETIC STRIPES, PALAEOMAGNETISM.

magnetic storm A temporary but widespread disturbance of the Earth's MAGNETIC FIELD, due possibly to sun-spot activity and solar flares. Such a storm causes serious interference with radio reception and with MAGNETIC SURVEYS.

magnetic stripes The approximately parallel bands of differing polarity which occur among the solid rocks of the ocean floor. The stripes have alternating positive and negative MAGNETIC ANOMALIES owing to the periodic reversals in the Earth's magnetic field (MAGNETIC REVERSALS) being imprinted and 'fossilized' into each stripe of new oceanic crust as it emerges along the MID-OCEANIC RIDGES and becomes incorporated in the mechanism known as SEA-FLOOR SPREADING (PLATE TECTONICS). The stripes have been considerably offset by subsequent faulting. [*142; see also 223*]

magnetic survey A technique for mapping the distortion of the Earth's MAGNETIC FIELD caused by the presence of igneous and metamorphic rocks at depth. Sedimentary rocks have no magnetic properties, so that sediment thicknesses and structures can be deduced since there is a relationship between the source of the distortion and distance. It is a method used in conjunction with GRAVIMETRIC SURVEY and SEISMIC SURVEY to locate reservoirs of oil and natural gas. *See also* MAGNETOMETER.

magnetite A magnetic iron ore (Fe_3O_4) common in the Urals and N Sweden.

magnetometer An instrument for measuring the intensity of the Earth's MAGNETIC FIELD.

magnetosphere The zone around the Earth occupied by the Earth's MAGNETIC FIELD. It extends out from the surface for a much greater distance on the side away from the Sun.

magnitude of an earthquake A measure of the total energy released by an EARTHQUAKE, in contrast to its INTENSITY. It is measured on a scale devised by C. F. Richter (RICHTER SCALE).

Mai-yu rains A Chinese term for the Bai-u rains of Japan (BAI-U SEASON): the '*plum rains*', especially notable in the Yangtze valley in June and July.

mallee A type of dwarf eucalyptus scrub commonly found in the semi-arid parts of W and SE Australia.

malpais A Spanish term meaning 'bad land', which implies a terrain difficult to cross. It usually refers to rough lava surfaces or barren TRAP landscapes.

Malvernian A stratigraphic division of the PRE-CAMBRIAN in England, thought to be of pre-URICONIAN age and forming the Malvern Hills in Hereford and Worcestershire.

mamillated topography A term referring to a terrain of bare rock in which all irregularities have been smoothed off to produce a smooth rolling topography. The agent most commonly responsible for this smoothing is glacier ice.

Mammalia One of the classes of the VERTEBRATES. They are warm-blooded, have a relatively large brain and generally have a covering of hair. They evolved from mammal-like reptiles in late Triassic/Jurassic times but did not achieve supremacy on land until after the extinction of most reptilian groups in the Cainozoic.

manganese A metal (Mn) which commonly occurs as an oxide in a nodular form (NODULE) in certain clay rocks, in the soil and on the deep ocean floor in clays and oozes.

mangrove 1 A vegetation type characterizing a coastal swamp of brackish or saline water, in which specially adapted trees form a dense swamp forest. **2** A term given to those tropical tree species and genera which have adapted themselves to live on saline muds in the tidal zone of certain tropical creeks and estuaries, especially in Indonesia, N Australia, the Amazon delta and the Niger delta. They have developed a dense network of aerial roots which help to aerate the root system and anchor the tree. *Rhizophora* sp. can tolerate prolonged flooding but requires a soft substratum of fine particles; *Avicennia* sp. tolerates shorter periods of tidal flooding and prefers more sandy and less organic soils; *Laguncularia* sp. withstands only periodic immersion and prefers firm soils. *See also* WETLAND.

man-made soils One of the seven major groups in the SOIL CLASSIFICATION OF ENGLAND AND WALES. These include thick (>40 cm) man-made humic soils with or without an artificial topsoil.

Manning equation An equation referring to the relationship between the velocity of a stream and its CHANNEL GEOMETRY. It is a widely used formula to determine a river's mean flow velocity and is expressed as $V = 1.49([R^{2/3}S^{1/2}]/n)$, where: V is the stream velocity; R is the HYDRAULIC RADIUS, S is the channel slope; and n is the roughness coefficient (ROUGHNESS). The Manning coefficient of roughness (n) was developed empirically as a contrast for a given boundary condition, but the channel slope, channel size, channel curvature, depth of flow and vegetation growth all cause variations in the magnitude of n. Some characteristic values for the roughness coefficient (n) are: straight artificial concrete-lined culvert, 0.013; straight natural channel with no bars, 0.030; winding natural stream with vegetation growth, 0.035; winding natural stream with bars, 0.040; mountain stream with boulders and vegetation, 0.040–0.050. *See also* CHEZY EQUATION, HYDRAULIC GEOMETRY.

Mann–Whitney U-test A NON-PARAMETRIC STATISTICAL TEST designed to compare the magnitude of sample measurements of data, without having to make assumptions about the characteristics of the distribution of the populations concerned. It is used to determine whether a difference in the median of two independent samples is statistically significant (i.e. that the samples are drawn from different populations). It can be used: (a) when it is possible to obtain only ordinal data measurement, so long as both data sets are ranked in a single sequence; or (b) when one cannot safely infer the assumptions necessary for a T-TEST. *See also* WILCOXON TEST.

mantle 1 In geology, the layer of the Earth between the CRUST and the CORE, with its upper boundary marked by the MOHOROVIČIĆ DISCONTINUITY and its lower boundary by the GUTENBERG DISCONTINUITY. It consists of ultrabasic rocks with densities of 3.0 to 3.3 g cm^{-3} and the uppermost 100–200 km of its 2,900 km total thickness comprise the ASTHENOSPHERE. **2** In geomorphology, an alternative name for REGOLITH.

mantle plumes PLUME.

map A graphic representation of the geographic features of the Earth's surface on a plane surface. The information displayed may be in the form of symbols or signs positioned by such selected controls as projections (MAP PROJECTION), GRIDS, and scales (MAP SCALE). Among the many types of maps are the following: ANAGLYPH, CADASTRAL MAP, CHOROCHROMATIC MAP, COMPILED MAP, COMPOSITE MAP, INDEX MAP, ORTHOPHOTO MAP, photo map (PHOTOMOSAIC), PLANIMETRIC MAP, PLASTIC RELIEF MAP, RESIDUAL MAP, THEMATIC MAP, TOPOGRAPHIC MAP, TREND SURFACE MAP. *See also* VISUALIZATION.

map projection Any orderly system of MERIDIANS and PARALLELS upon a plane surface in order to portray the curved surface of a terrestrial or celestial sphere or spheroid. A map projection is displayed in the form of a GRATICULE and may be constructed either from mathematical formulae (CONVENTIONAL PROJECTION) or from geometrical data (PERSPECTIVE PROJECTION). *See also* AZIMUTHAL PROJECTION, CONICAL PROJECTION, CYLINDRICAL PROJECTION, EQUAL-AREA MAP PROJECTION, EXTERNAL PERSPECTIVE PROJECTIONS, GLOBULAR PROJECTIONS, ORTHOGRAPHIC PROJECTION, POLAR PROJECTIONS, POLYCONIC PROJECTION, PSEUDO-AZIMUTHAL PROJECTIONS, PSEUDO-CONICAL PROJECTIONS, PSEUDO-CYLINDRICAL PROJECTIONS, RECENTRED MAP PROJECTION, RETRO-AZIMUTHAL PROJECTION, SECANT CONICAL PROJECTION, TRUNCATED CONICAL PROJECTIONS, ZENITHAL PROJECTION. (NB There are a large number of named map projections and these will be found alphabetically under the name of their inventor, e.g. Mercator.)

map reference A means of identifying a point on the Earth's surface by relating it to the GRATICULE or GRID. A map reference is synonymous with *grid reference* and comprises a combined EASTING and NORTHING.

map scale The ratio of the distance measured on a map to that measured on the ground between the same two points. The larger the ratio the smaller the map scale. In Britain most map scales are now metric and are shown, for example, as 1 : 50,000. This latter scale, while not being identical with them, has replaced the former maps of the Ordnance Survey which were at a scale of 1 : 63,360 (1 inch to 1 mile).

map series A set of maps, usually of uniform scale and style, referring to a specific area and identified by a series number.

maquis A low evergreen shrub formation, usually found on siliceous soils in the Mediterranean lands where winter rainfall and summer drought are the characteristic climate features. It consists of a profusion of aromatic species such as lavender, myrtle, oleander and rosemary and often includes abundant spiny shrubs. It has been suggested that the maquis is a secondary vegetation, occupying the lands cleared of their natural evergreen oak forests by human activity. *See also* GARRIGUE.

marble 1 A metamorphic limestone, consisting of CALCITE and/or DOLOMITE, in which a sedimentary limestone has been altered by regional or contact metamorphism. It is widely used as a decorative building stone and in sculpture. **2** A loose term for any decorative stone that will take a high polish, e.g. the so-called Purbeck Marble, which is a sedimentary rock.

mares' tails Feathery and wispy CIRRUS clouds in the upper atmosphere, indicative of strong winds at high levels.

margalitic A term referring to a soil A-HORIZON which has a high BASE status and is dark in colour.

marginal channel A type of MELTWATER CHANNEL cut by glacial meltwaters at the junction of the ice surface and the surrounding rock, i.e. marginal to the glacier or ice-sheet. The channel may be cut: **1** entirely in the hillside, in which case a channel or marginal bench may survive the disappearance of the ice [*7*]; **2** entirely in or on the ice itself (SUBGLACIAL CHANNEL), in which case there will be no surviving evidence after deglaciation; or **3** partly on the hillside and partly on the ice, thereby creating an IN-AND-OUT CHANNEL. An ENGLACIAL meltstream flowing laterally at the ice/rock interface will create a SUBMARGINAL CHANNEL. *See also* OVERFLOW CHANNEL, PRO-GLACIAL CHANNEL. [*143*]

marginal deep 1 A deep-sea trench (DEEP, OCEAN) lying on the seaward side of an ISLAND ARC. **2** An EXOGEOSYNCLINE.

marginal depression A linear hollow surrounding the base of an INSELBERG or at the base of an escarpment where it meets the

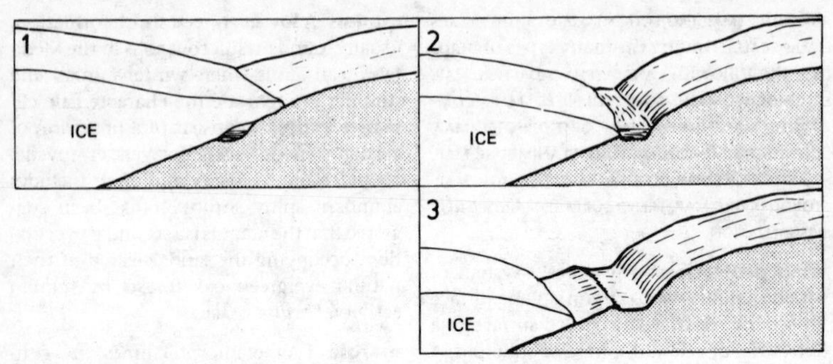

Figure 143 *Formation of a marginal channel*

PEDIMENT (hence the synonym *scarp-foot depression*). It is found almost entirely in tropical regions, suggesting to some geomorphologists that it is due largely to CHEMICAL WEATHERING at the zone where the water-table approaches the ground surface. Other explanations include SPRING-SAPPING and concentrated erosion by surface RUNOFF at the abrupt change of slope.

marginalization, theory of DISASTERS.

marginal sea A partly enclosed section of sea but one that is significantly open to the ocean. It does not lie beyond the CONTINENTAL SHELF but on a downwarped portion of it, e.g. the North Sea.

mariculture Farming of the sea. This includes fishing but also includes experiments on utilizing the thermal gradients of ocean waters which bring nutrients to the surface, thereby encouraging PLANKTON growth which itself is the base of the oceanic FOOD CHAIN.

marigram A graphic representation of the tidal oscillation at a particular coastal station. The tidal curve gives a picture of tidal motion (TIDE) of use to oceanographers and hydrographers.

marin The name given to a warm southeasterly wind blowing onshore in the S of France, induced by the presence of a depression in the Lion Gulf.

marine abrasion The eroding action of sea waves and the shingle and sand which they move (LOAD) whereby the bedrock surface along a coastline is abraded into a *wavecut platform* (ABRASION PLATFORM).

marine-built terrace An accumulation of materials removed during the cutting of a marine ABRASION PLATFORM. The terrace (sometimes termed the *continental terrace*) lies seaward of the abrasion platform. [*144*]

marine cycle of erosion A theoretical scheme outlined in 1919 by D. W. Johnson

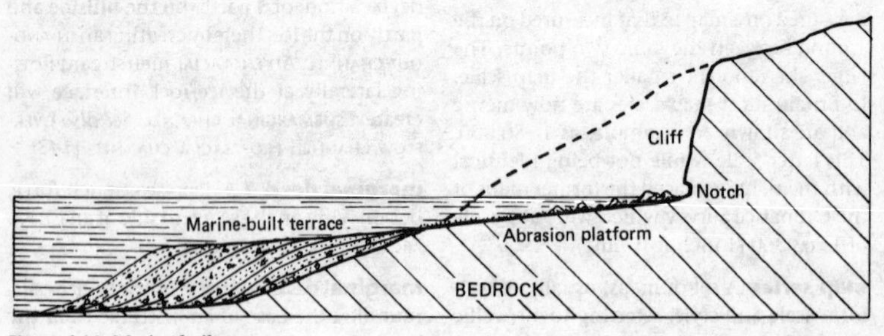

Figure 144 *Marine-built terrace*

to explain the sequential development of EMERGENT SHORELINES and SUBMERGENT SHORELINES, based on the cyclic concept of W. M. Davis (CYCLE OF EROSION). Johnson accepts that the cycle may be interrupted by crustal movements and that the coastal topography will vary according to the nature of the rocks, the configuration of the coastline and the strength of the waves and currents. He suggests that on a *submergent shoreline* the following stages may be recognized: **1** *Initial stage*, with an irregular coastline and numerous islands if there was moderate relief. **2** *Youthful stage*, starting with wave attack concentrated on the promontories. Sea cliffs, stacks, chimneys, arches, and blowholes may abound as an ABRASION PLATFORM is developed. Spits, hooks, and bars extend out from the cliffs and islands, often connecting them. **3** *Mature stage*, seeing the headlands cut back to the level of the intervening bays and the coastline straightened. Connecting bars have been removed. **4** *Old age* is achieved when waves and currents have reduced the shore to a gently sloping plain, while land agents have reduced the relief almost to sea-level. On an *emergent shoreline*: **1** *The Initial stage* is an uplifted marine plain (EROSION SURFACE) into which the waves cut a notch or NIP. Offshore bars are common. **2** The *Youthful stage* sees the growth of the offshore bar and the formation of LAGOONS linked to the open sea by tidal inlets, kept open by tidal scour. The offshore bar is now driven landwards, obliterating the lagoon. **3** The *Mature stage* is the point at which waves have destroyed the protecting bar and are actively attacking the former raised shoreline. **4** *Old age* is a theoretical concept (as in the case of the submergent shoreline) and no examples have been quoted.

marine waves WAVE.

maritime air mass AIR MASS.

maritime climate, marine climate The climate which characterizes those parts of the Earth's temperate zone that are affected by the sea (in contrast to a CONTINENTAL CLIMATE). Because the sea temperatures of the temperate zone do not change very significantly between winter and summer their surrounding land areas experience low annual-temperature ranges and no real dry season. Thus, a typical station in a maritime climate would have mild winters, cool summers and an even distribution of annual rainfall. *See also* OCEANICITY.

maritime polar air mass An AIR MASS originating from a source region in the oceanic tracts of polar latitudes (symbol: mP). It is characteristically cold, moist and unstable and is associated with COLD FRONTS, THUNDERSTORMS, and CONVECTION cloud and brings clear, rain-washed air and very good visibility.

maritime tropical air An AIR MASS originating from a source region in the oceanic tracts of the tropical zone (symbol mT). It is characteristically warm and moist and is associated with WARM FRONTS, DRIZZLE, STRATUS CLOUD, coastal fog, and poor visibility. *See also* ADVECTION FOG, INVERSION OF TEMPERATURE.

marker horizon **1** A bed of rock which has either a distinctive LITHOLOGY or a characteristic assemblage of FOSSILS so that it is instantly recognizable and can be used as a dependable datum in correlation of strata. **2** A bed of rock which yields characteristic seismic reflections in a geographical survey and can be followed over extensive areas. It is synonymous with *marker bed* and *key bed*.

Markov chain A SYSTEM comprising a sequence of possible states such that the conditional probability of transition from one state to another is independent of the way in which the former state was achieved. The statistical model of the Markov chain has been used to describe, but not explain, how in many parts of the world, for example, there is a tendency for dry weather to occur in spells, with the occurrence of a wet day often being independent of the previous weather conditions. Studies have shown that in S England a spell of dry weather has an increasing probability of continuing for up to ten days (termed positive persistence), but that after thirty days there is a strong likelihood of a change to

Figure 145 *Mass*

Unit	*kg*	*lb*	*slug*	*metric slug*	*UK ton*	*US ton*	*amu*
1 kg =	1	2.205	6.852×10^{-2}	0.1020	9.842×10^{-4}	11.02×10^{-4}	6.024×10^{26}
1 lb =	0.4536	1	3.108×10^{-2}	4.625×10^{-2}	4.464×10^{-4}	5.000×10^{-4}	2.732×10^{26}
1 slug =	14.59	32.17	1	1.488	1.436×10^{-2}	1.609×10^{-2}	8.789×10^{27}
1 metric slug =	9.806	21.62	0.6720	1	9.652×10^{-3}	1.081×10^{-2}	5.907×10^{27}
1 UK ton =	1,016	2,240	69.62	103.6	1	1.12	6.121×10^{29}
1 US ton =	907.2	2,000	62.16	92.51	0.8929	1	5.465×10^{29}
1 amu =	1.660×10^{-27}	3.660×10^{-27}	1.137×10^{-28}	1.693×10^{-28}	1.634×10^{-30}	1.829×10^{-30}	1

1 Imperial standard pound = 0.453592338 kg.
1 US pound = 0.4535924277 kg.
1 International pound = 0.45359237 kg.
1 tonne = 10^3 kg.

wetter weather (termed negative persistence).

Markov process A term given to any process which generates a STOCHASTIC series, which will have a statistical influence over any finite-length sequence.

marl **1** A calcareous clay or mudstone which, in British usage, has a minimum of 15% calcium carbonate but which, in US usage, can range between 20 and 90% calcium carbonate. **2** A proper name in stratigraphical nomenclature, e.g. Keuper Marl, Etruria Marl. **3** In agriculture the term is used loosely for any friable clay soil whether calcareous or not.

marram grass One of the commonest species found on coastal dunes: marram grass (*Ammophila arenaria*) nevertheless cannot withstand immersion in sea water. Its rhizomes and roots spread over considerable distances, both laterally and vertically, and therefore marram grass has excellent properties for 'anchoring' mobile sand-dunes. *See also* PSAMMOSERE.

marsh A tract of low, ill-drained ground with patches of open water in which reeds, rushes and sedges abound in temperate latitudes. Some temperate marshes may have alder and willow along the edges of the water bodies, but in the tropics marshes may have a thicker growth of trees and shrubs. Because of their high organic content marsh soils are often very fertile when properly drained (e.g. the Fens) but in general they are left under permanent grassland with occasional hay-cutting. When inundated with tidal water the coastal tracts are termed *salt marshes*, in which a typical HALOSERE will become established. *See also* BOG, CARR, FEN, SWAMP, WETLAND.

marsh gas The alternative name for *methane* (CH_4), resulting from the partial decomposition of plants in a swampy environment.

mascaret A local term for a tidal BORE on the R. Seine, France.

mass The quantity of matter of a body as measured by its weight in the Earth's gravitational field, in contrast to its density and volume. [*145*]

mass balance The relationship between ACCUMULATION and ABLATION on a glacier, from which its budget is calculated. Where accumulation exceeds ablation a *positive mass balance* is said to occur and the ice front will advance. Conversely a *negative mass balance* will cause an ice retreat. Because of variability of snow density, mass balance is expressed in water-equivalent terms. *See also* FIRNIFICATION.

mass extinction The disappearance of a large number of FOSSIL groups in a relatively short time period, such as occurred at the Cretaceous-Tertiary boundary, possibly as a result of an impact of an extra-terrestrial body on the Earth's surface.

massif A French term widely adopted to describe a mountainous mass with fairly uniform characteristics (e.g. geology and structure) and with clearly defined boundaries.

massive **1** An adjectival description of a very thick rock unit without stratification, jointing, foliation, cleavage or flow-banding. **2** The usage to describe an amorphous crystalline structure in a rock should be discouraged.

mass-movement The downhill movement of surface materials (including solid rock) under the influence of gravity but assisted by buoyancy due to rainfall or snowmelt. A distinction is made between *mass-movement processes*, in which large quantities of sediment move together in close grain-to-grain contact, and *flow processes*, in which single grains are dispersed through a fluid transporting medium, although there is a gradual transition between the former and the latter as the fluid input increases. Mass-movement can occur very rapidly (*see also* DEBRIS FLOW, EARTH FLOW, LANDSLIDE, ROCK-FALL) owing to FAILURE, when the critical SHEAR STRENGTH of the material is exceeded; or it can occur more slowly (*see also* CREEP, SCREE, SOLIFLUCTION). Rapid mass-movements along a SHEAR PLANE may be shallow or deep-seated according to whether COHESION is present as a measurable strength factor or not. Cohesive sediments usually have deep-seated failures (COULOMB'S FAILURE LAW). In cohesionless materials slope angle is the only geometric control on slope stability and very dry soils rely on their interparticle frictional strength for stability. In cohesive soils the height of slope is as important as the slope angle. Studies of slow mass-movement have shown that its rate is proportional to the size of the slope angle, to grain size and to temperature, and inversely proportional to the coefficient of friction (FRICTION, COEFFICIENT OF). Slow movement may be due to expansion from insolation (insolation creep), from FREEZE–THAW ACTION, or from undermining of rock particles by flowing water (WASH). *See also* ANGLE OF REPOSE, DEBRIS FLOW, EARTH-FLOW, FRACTURE, MUD-FLOW, LANDSLIDE, QUICKCLAYS, ROCK CREEP, ROCK-FALL, ROCKSLIDE, SHEAR STRENGTH, SOIL CREEP, SOLIFLUCTION, TALUS SHIFT, TOREVA BLOCKS, TRANSLATIONAL SLIDE. [*146*]

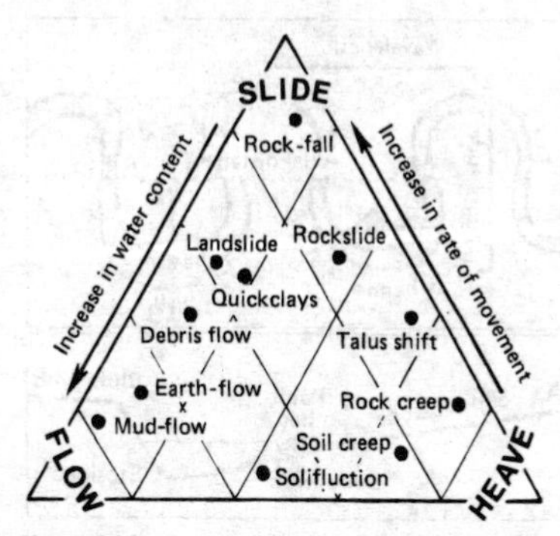

Figure 146 *Mass-movement processes*

mass transport The net transfer of water by wave action in the direction of wave travel. It depends on the mechanism of ORBITAL MOTION, which is asymmetrical and therefore causes a particle of water to move in the direction of wave propagation. *See also* NEARSHORE CURRENT SYSTEM.

mass-wasting MASS-MOVEMENT.

material cycles The paths followed by particular types of material through a SYSTEM. Synonymous with BIOGEOCHEMICAL CYCLES and NUTRIENT CYCLES.

materials, strength characteristics of ANGLE OF REPOSE, COHESION, COULOMB'S FAILURE LAW, FRICTION, COEFFICIENT OF, GRAIN SIZE, PERMEABILITY, PORE-WATER PRESSURE, POROSITY, SHEARING, SHEAR PLANE, SHEAR STRENGTH.

matrix **1** The fine-grained material of a rock in which the coarser components are embedded. In sedimentary rocks the matrix is often a CEMENT binding the coarser fragments together. In igneous rocks it is equivalent to the so-called ground mass. **2** A statistical term referring to an ordered array of numbers on which certain operations may be performed. It may take the form of an *identity matrix*, a *transpose matrix*, an *inverse matrix*, a *square matrix*, a *row matrix*, a *symmetric matrix*, a *column matrix*, a *correlation matrix*, or an *input–output matrix*.

mature soil A soil with well-developed SOIL HORIZONS created by natural soil-forming processes; a soil in equilibrium with its environment.

maturity The advanced stage of land sculpture, shoreline evolution and soil genesis achieved in the cyclic concept of W. M. Davis. *See also* ARID CYCLE OF EROSION, CYCLE OF EROSION, CYCLE OF UNDERGROUND DRAINAGE, GLACIAL CYCLE OF EROSION, MARINE CYCLE OF EROSION, NORMAL CYCLE OF EROSION, SENILITY, YOUTH. [*53* and *181*]

Maunder minimum A time period from AD 1650 to AD 1700 during which there was a marked decrease in the number of SUNSPOTS. It has been suggested that it may correspond in part to the cooler weather of the LITTLE ICE AGE.

maximum and minimum thermometer An instrument for measuring temperature extremes, invented in 1780 by James Six of Cambridge University. It comprises a U-shaped tube part mercury-filled but with a transparent liquid in the upper arms. Above each column of mercury, within the tube, a tiny metal index grips the side of the tube but slides up the arms when pushed by the mercury. With a rise in temperature the transparent liquid in that arm which is totally filled expands and pushes the mercury column down the arm. Consequently, the mercury in the other arm is forced up, together with the index, until a maximum reading is obtained. Conversely, when temperature falls the liquid in the totally filled arm contracts, the mercury in this arm is now forced upwards by the weight of the liquid in the other arm and the second index will register the minimum temperature in the totally filled arm. Each index can be reset daily by means of a magnet.

maximum thermometer Usually this is a mercury-in-glass thermometer, with a constriction in the bore, close to the bulb. On heating, the mercury expands and flows past the constriction up the tube. With cooling, the mercury thread remains in the tube and the end furthest away from the bulb indicates the maximum temperature since the last setting (effected by shaking the mercury past the constriction, until the DRY-BULB equivalent is attained).

M-discontinuity MOHOROVIČIĆ DISCONTINUITY.

meadow soil GLEY SOILS.

mean **1** A statistical term denoting the *average* or arithmetic mean of a set of data obtained by summing the individual values and dividing the total by the number of individuals. *See also* STANDARD DEVIATION. **2** In meteorology, the daily or monthly mean of an element is the average of the maximum and minimum values of that element for a day or a month. *See also* NORMAL.

meander A sinuosity or loop-like bend in a river, characterized by a river-cliff or bluff on the outside of the curve and a gently shelving SLIP-OFF SLOPE on the inner side of the bend. The WAVELENGTH and AMPLITUDE of a meander are related to flow and rate of discharge of the river [*147*] but the exact

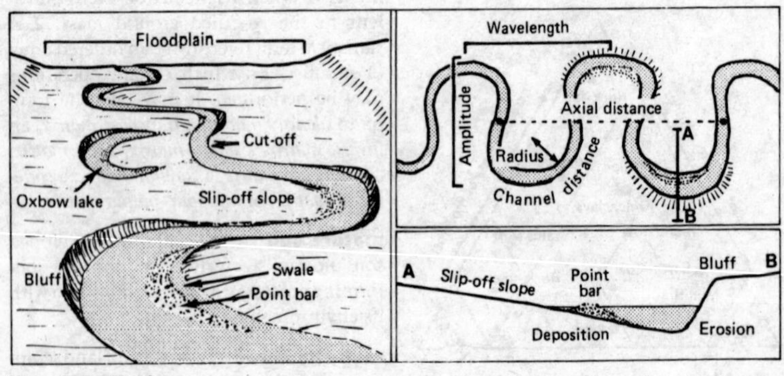

Figure 147 *Meanders*

cause of meandering is not perfectly understood. L. B. Leopold and M. G. Wolman believe that it is a function of frictional drag and that it reflects the smallest energy loss of any curved channel. In their *theory of minimum variance* W. B. Langbein and Leopold suggest that meandering is the most probable form achieved when a river does the least amount of work in turning. Because a meander lengthens the course it reduces the gradient of a river, thereby reducing energy expenditure, despite increasing the volume of water being moved. Some geomorphologists believe that meanders are initiated at the site of riffles (POOL AND RIFFLE) [*192*] or, conversely, at the site of pools. The energy gradient over pools and riffles is almost constant despite the contrasts in bed roughness. Meanders will tend to migrate downstream owing to HELICOIDAL FLOW, with *bank-caving* (BANK) occurring where velocity is greatest on the outside of the bends and POINT-BAR deposition occurring where velocity is least on the inside of the bends. If the load is reduced or the river discharge is increased the tract of the FLOODPLAIN occupied by the meanders increases in width (termed the *meander belt*). *See also* BAR-AND-SWALE, CUT-OFF, INCISED MEANDER, INGROWN MEANDER, INTRENCHED MEANDER, MEANDER CORE, MEANDER SCAR, MEANDER TERRACE, MEANDERING VALLEY, OXBOW LAKE. [*103*]

meander core **1** The hill around which an INCISED MEANDER swings before breaking through the narrow neck of the meander. **2** Sometimes the expression is reserved for the isolated hill left when the river has cut through the meander neck, thereby initiating a CUT-OFF.

meandering valley A sinuous river valley comprising meanders of large magnitude cut into solid rock but on the floodplain of which a set of smaller-scale meanders relating to the present stream are found. The stream is referred to as an UNDERFIT STREAM and it has been suggested that the larger valley meanders were created under a different climatic regime when DISCHARGE at the BANK-FULL STAGE was considerably (perhaps twenty times) greater than at present.

meanderkarren A meander-like form of KARREN.

meander scar A CUT-OFF in which the former river channel or OXBOW LAKE has been partly infilled by slumping of the channel walls and by vegetation growth but where the original MEANDER form is still recognizable.

meander terrace A partly destroyed river terrace caused when a meandering river cuts down into its former floodplain owing to rejuvenation.

mean sea-level The average level of the surface of the sea, determined by averaging all tidal levels recorded at hourly intervals. In the UK the Liverpool datum was abandoned in 1921 and replaced by that at Newlyn, Cornwall, which is 40 mm below that at Liverpool. *See also* BASE-LEVEL, DATUM LEVEL, ORDNANCE DATUM.

mean solar time An expression referring to the 24-hour period of a *mean solar day*. It differs from *apparent time* by the so-called EQUATION OF TIME.

measure of central tendency A statistical parameter which indicates the degree of clustering within the distribution of a sample. *See also* CENTRAL TENDENCY, DISTRIBUTION, MEAN, MEDIAN, MODE.

mechanical analysis A technique for determining the proportions of each PARTICLE SIZE in a dry sediment, soil or till, as a means of classification. Particles larger than 0.06 mm diameter (held by a 200 mesh BS sieve) can be measured by sieving. Particles smaller than this are impossible to sieve; their particle size is determined from their settling velocities in water (i.e. by WET ANALYSIS). *See also* FINENESS MODULUS, GRADING CURVE, GRADING FACTOR.

mechanical eluviation ELUVIATION.

mechanical weathering The physical breakup of rocks owing to internal and external stresses caused by weathering agents. Disintegration is caused by EXFOLIATION, FREEZE–THAW ACTION, GRANULAR DISINTEGRATION, SHEETING, SPALLING, UNLOADING, WETTING-AND-DRYING ACTION. All the processes of mechanical weathering are distinct from those of CHEMICAL WEATHERING. *See also* ORGANIC WEATHERING, WEATHERING.

mechanistic A term referring to the viewpoint that the behaviour of all biological phenomena may be explained in mechanico-chemical terms.

medial moraine, median moraine A linear accumulation of rubbly material extending down the centre of a glacier. It is often ice-cored and varies in width from a narrow ridge to a broader spread of morainic material. The feature is caused by the merging of two LATERAL MORAINES from the point at which two glaciers unite. *See also* ICE-CORED MORAINE, MORAINE.

median The central value in an ordered series of values, with an equal number of values occurring above and below it.

median mass A somewhat dated expression referring to an intermontane plateau or plain lying within a zone of fold mountains. It is thought to represent the relatively underformed central mass of sediments in a GEOSYNCLINE surrounded by the folded and overthrust sediments on either flank, bordering the margins of the approaching FORELANDS.

median ridge MID-OCEANIC RIDGE.

mediocratic A term introduced by von Post (1946) and adopted by Iversen (1958) to describe the climatic phase of NW Europe in post-glacial times (FLANDRIAN), when mean temperatures were some 2°–3° C above those of the present day. This phase was termed the Climatic Optimum and is thought to be synonymous with the ATLANTIC PERIOD (7,450–4,450 BP), and with zone VIIa in the pollen-zone chronology. *See also* PROTOCRATIC, TERMINOCRATIC. [*81*]

Mediterranean climate A climatic type characteristic of the western margins of continents in the world's warm temperate zones, between latitudes 30° and 40°. The Csa and Csb types of the KÖPPEN'S CLIMATIC CLASSIFICATION. It is typified by hot, dry, sunny summers and by warm, moist winters. In the summer the climate is dominated by the subtropical anticyclones and in winter by depressions. Within the Mediterranean basin itself various subdivisions are recognizable, depending on distance from the Atlantic, but elsewhere in the world at this latitude the climate type is restricted to narrower coastal margins, often by topographic barriers, e.g. central Chile, central California, Cape Province of South Africa, SW Australia and the Adelaide region.

Mediterranean front A zone of FRONTOGENESIS active only in the winter, when, at intervals, airstreams from Europe (MARITIME POLAR AIR MASS (*Pm*)) and continental polar air masses (Pc) converge with warmer and drier TROPICAL AIR MASSES (*Tc*) from North Africa. It is associated with winter clouds, rain and gales.

medium-textured soil A soil intermediate in texture between a FINE-TEXTURED SOIL and a coarse-textured soil (SOIL TEXTURE). It includes very fine sandy loam, LOAM, SILT, SILTY LOAM.

megashear **1** Any very large-scale TEAR-FAULT. An alternative name for a TRANSCURRENT-FAULT, i.e. a strike fault of continental proportions, e.g. the Alpine fault of New Zealand. **2** In a strict sense it has been defined as a tear fault whose horizontal displacement significantly exceeds the thickness of the crust.

megatherm **1** The category of plants whose annual temperature requirement is that each month should have a mean no lower than 18° C. The megatherms are found characteristically in the TROPICAL RAINFORESTS. **2** A climate characterized by very high temperatures; the A-climates of KÖPPEN'S CLIMATIC CLASSIFICATION. *See also* HEKISTOTHERM, MESOTHERM, MICROTHERM.

megathermal period ATLANTIC PERIOD.

mekgacha A term given to the streamless valley systems of the Kalahari desert region of South Africa and Namibia. Their origins have never been fully explained but it is thought that they may be related to an earlier climatic episode of heavier rainfall (PLUVIAL).

mélange A jumbled mass of fragments from a wide range of rock types now cemented into a sedimentary rock (BRECCIA). It is thought to have been formed by the slumping of large masses of unstable rock debris at a continental margin.

melanocratic A term used to describe the dark-coloured rocks, especially those igneous rocks which contain between 60% and 100% dark-coloured ferromagnesian minerals (MAFIC). *See also* LEUCOCRATIC, MESOCRATIC.

meltemi ETESIAN WINDS.

melt-out till TILL.

meltwater Water resulting from the melting of glacier ice and/or snow. *See also* GLACIOFLUVIAL, MELTWATER CHANNEL.

meltwater channel A form of channel,

Figure 148 *Mercalli scale, modified*

I	Not felt except by a very few under especially favourable circumstances.
II	Felt only by a few persons at rest, especially on upper floors of buildings. Delicately suspended objects may swing.
III	Felt quite noticeably indoors, especially on upper floors of buildings, but many people do not recognize it as an earthquake. Standing motor-cars may rock slightly. Vibration like passing of truck. Duration estimated.
IV	During the day felt indoors by many, outdoors by few. At night some awakened. Dishes, windows, doors disturbed; walls make cracking sound. Sensation like heavy truck striking building. Standing motor-cars rocked noticeably.
V	Felt by nearly everyone, many awakened. Some dishes, windows, etc., broken; a few instances of cracked plaster; unstable objects overturned. Disturbances of trees, poles, and other tall objects sometimes noticed. Pendulum clocks may stop.
VI	Felt by all, many frightened and run outdoors. Some heavy furniture moved; a few instances of fallen plaster or damaged chimneys. Damage slight.
VII	Everybody runs outdoors. Damage negligible in buildings of good design and construction; slight to moderate in well-built ordinary structures; considerable in poorly built or badly designed structures; some chimneys broken. Noticed by persons driving motor-cars.
VIII	Damage slight in specially designed structures; considerable in ordinary substantial buildings, with partial collapse; great in poorly built structures. Panel walls thrown out of frame structures. Fall of chimneys, factory stacks, columns, monuments, walls. Heavy furniture overturned. Sand and mud ejected in small amounts. Changes in well water. Persons driving motor-cars disturbed.
IX	Damage considerable in specially designed structures; well-designed frame structures thrown out of plumb; great in substantial buildings, with partial collapse. Buildings shifted off foundations. Ground cracked conspicuously. Underground pipes broken.
X	Some well-built wooden structures destroyed; most masonry and frame structures destroyed with foundations; ground badly cracked. Rails bent. Landslides considerable from river banks and steep slopes. Shifted sand and mud. Water splashed (slopped) over banks.
XI	Few, if any, (masonry) structures remain standing. Bridges destroyed. Broad fissures in ground. Underground pipelines completely out of service. Earth slumps and land slips in soft ground. Rails bent greatly.
XII	Damage total. Practically all works of construction are damaged greatly or destroyed. Waves seen on ground surface. Lines of sight and level are distorted. Objects are thrown upwards into the air.

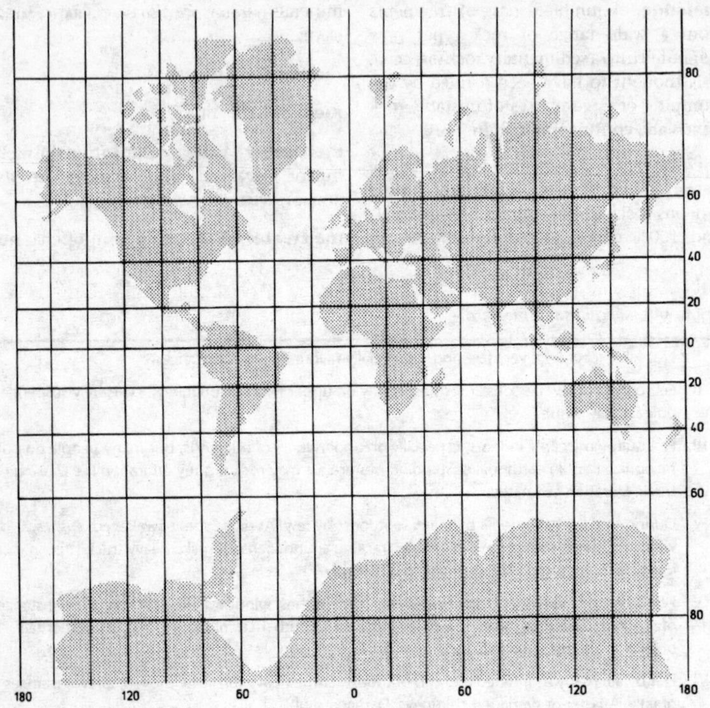

Figure 149 *Mercator projection*

perhaps no longer carrying a stream, cut into solid rock or drift deposits in areas of former glaciation, but unrelated to the present drainage system. It frequently cuts across current drainage divides, has very steep sides, remains of fairly constant width and is often associated with DELTAS, KAME TERRACES, OF ESKERS at its intake end, and with outwash fans, eskers and KAMES at its distal end. Several classifications have been attempted based on morphology or on genesis, of which the latter is usually more satisfactory. In 1973 R. J. Price suggested the following classification: **1** *Marginal/submarginal channels* (MARGINAL CHANNEL), including lateral, frontal, col channels and IN-AND-OUT CHANNELS. **2** *Subglacial channels*, including marginal (chutes), basal (normal and those with UP–DOWN PROFILES), englacial (superimposed varieties), and col channels. **3** *Open ice-walled channels*. **4** OVERFLOW CHANNELS (including marginal, pro-glacial and col channels). **5** PRO-GLACIAL CHANNELS.

member A division of rock bodies in LITHOSTRATIGRAPHY. It is made up from a number of BEDS and is part of a FORMATION. It is distinguished by a lithological characteristic which contrasts with that of adjacent members.

Menapian A stratigraphic name for the top stage of the Lower Pleistocene in Europe.

Mercalli scale, modified A subjective measure of EARTHQUAKE intensity devised by an Italian scientist in 1902, replacing an earlier seismic-intensity scale of de Rossi and Forel (ROSSI–FOREL SCALE). The scale was modified in 1931 to take into account the urban life-style of Western culture but it still depends upon a descriptive list of phenomena which become increasingly more catastrophic [*148*]. It should be contrasted with the RICHTER SCALE of earthquake magnitude.

Mercator projection A MAP PROJECTION of the CYLINDRICAL type with ORTHOMORPHIC properties. It is named from G. Mercator, who used it for his world map of 1569, albeit the projection was used in 1511 by Etzlaub. Its parallels are straight lines, drawn at the same length as the equator and intersecting the meridians at right angles. It has the valuable property that its RHUMB LINES are straight, making its directional information (BEARING) correct. Thus the Mercator projection has been used extensively for navigation purposes. Its greatest disadvantage is the increasing amount of distortion of area as high latitudes are approached, thereby depicting some of the islands of the Canadian Arctic as considerably larger than those in Indonesia. Its orthomorphic (conformal) properties are achieved by increasing the spacing of the parallels with increasing distance from the equator, to conform with the expanding scale along the parallels as they are traced polewards, but the equator is the only one that is true to scale. *See also* TRANSVERSE MERCATOR PROJECTION. [*149*]

mercury (Hg) The only metal which remains liquid at normal temperatures. Its most important source is cinnabar (mercuric sulphide). Used extensively in THERMOMETERS.

mere A small lake: found in some English place-names, especially in the counties of Cheshire and Shropshire, where it is used to describe both water-filled KETTLE HOLES (e.g. Ellesmere) and subsidence hollows (FLASH) due to underground salt extraction.

meridian A line of LONGITUDE that joins the North and South Poles as part of a GREAT CIRCLE (semi-great circle), but that terminates (a single meridian) at the geographic poles. Meridians are numbered from 0° to 180° both east and west from the GREENWICH MERIDIAN, the currently accepted PRIME MERIDIAN. *See also* CENTRAL MERIDIAN. [*176*]

meridian day The term used for the day on which the INTERNATIONAL DATE-LINE is crossed by a traveller.

meridional flow, circulation Any large-scale air movement from north to south or vice versa, i.e. along the MERIDIANS or lines of longitude. *See also* INDEX CYCLE, ROSSBY WAVE, ZONAL CIRCULATION.

merokarst An imperfect or partially developed KARST, in contrast to HOLOKARST. It occurs where limestones are relatively thin and in areas of impure limestone, often interbedded with marly bands. Although SWALLOW-HOLES and caves are present there are no POLJES or KARREN, although some DOLINES may develop. A surface soil cover is more widespread than in true *karst* and surface drainage more commonplace. Thus the hydrology is less complex. Examples are found in the Jura and in Galicia (Poland), and Cvijić also included the English and French chalklands (CHALK).

mesa A Spanish term widely adopted to describe a tableland or isolated flat-topped hill, bounded on at least one side by steep slopes or cliffs. It develops most easily in terrains with horizontally bedded strata where a resistant layer of sedimentary rock, a tough lava flow or an igneous SILL may form a cap rock to the eminence. Alternatively, a hard surface CRUST may form the capping. A mesa represents a stage in the denudation of a plateau in a region experiencing an arid or semi-arid climate, and is commonly found in the SW of the USA and in Mexico. It is of greater lateral extent than a BUTTE, which it resembles in every other way.

mesic layer A layer of organic material within a soil which has reached a stage of decomposition between that of the FIBRIC layer and the HUMIC layer.

mesocratic A term used to describe rocks, particularly igneous rocks, which contain an intermediate mix of dark-coloured minerals (MELANOCRATIC) and light-coloured minerals (LEUCOCRATIC). Mesocratic rocks are usually defined as having between 30% and 60% of dark-coloured minerals (MAFIC).

mesokarst An alternative term for *transitional karst* midway between MEROKARST and HOLOKARST in its development. It is characterized by limestone plateaux separated by gorge-like valleys, with the plateau surfaces occupied frequently by DOLINES and POT-HOLES, but in a terrain in which POLJES are absent although UVALAS may be present. The limestone is thicker and purer than that of the merokarst but not so deep nor so pure as that of the holokarst. It is exemplified by the Causses, Vercours and Chartreuse of France. *See also* KARST.

Mesolithic An archaeological term adapted from the Greek, meaning *'middle stone' age*, and used to describe the culture reached in the early POST-GLACIAL when mankind had moved from the herd-hunting practices of the upper PALAEOLITHIC but had not yet discovered or adopted the use of agriculture (NEOLITHIC). In Britain the culture spans the approximate time period from 10,300 BP to 4,450 BP (FLANDRIAN), during which there was a general climatic amelioration leading up to the Climatic Optimum, when England became separated from the continent as sea-level rose. It coincided with the entire PROTOCRATIC and with part of the MEDIOCRATIC, spanning pollen zones IV, V, VI and part of VIIa (POLLEN ANALYSIS). The earliest Mesolithic inhabitants, living frequently in lakeside habitats, were part of the Maglesmosian culture, in which microliths (small sharp flints), bone and antler tools were extensively used, while chipped-flint axes were introduced, capable of felling the widespread woodlands in which red deer, roe deer and elk roamed. The dog became the first domesticated animal. Later Mesolithic peoples, classified as being of Sauveterrian and Tardenoisian cultures, had already made considerable inroads into the forests, possibly by the use of fire, in such places as Dartmoor and the N Yorkshire moors, in order to provide clearings for semi-permanent villages. In Scotland and Ireland the so-called 'Strand-loopers' lived on a diet of shellfish along the sea coasts and lacked both axes and microliths. *See also* BOREAL, PRE-BOREAL.

Figure 150 *Mesometeorology: three scales of atmospheric motion*

	Macro	*Meso*	*Local*
Characteristic dimension (km)	>483	16–160	<8
	Synoptic	*Meso*	*Micro*
Period (h)	>48	1–48	<1
Wavelength (km)	>500	20–500	<20

Source: Atkinson, 1981

mesometeorology The study of the atmospheric processes which combine to produce those weather phenomena the patterns of which are intermediate in size between those of MACROMETEOROLOGY and MICROMETEOROLOGY. A denser network of weather stations is needed than in *macrometeorology* in order to obtain data from *mesosystems* or MESOSCALE SYSTEMS, which are often topographically induced. *See also* BORA, FÖHN, KATABATIC WIND, LEE WAVE. [*150*]

mesopause The upper limit of the MESOSPHERE at an elevation in the atmosphere of some 80 km above the Earth's surface. Above this lies the THERMOSPHERE. *See also* STRATOPAUSE, TROPOPAUSE. [*17*]

mesophyte A plant which grows under average but not excessive water-supply conditions, demanding less than the HYGROPHYTE. *See also* XEROPHYTE.

mesoscale systems, mesosystems A term used by meteorologists to describe medium-sized atmospheric SYSTEMS, in contrast to macro- and microsystems. The mesosystem has a horizontal diameter between 15 and 150 km, i.e. the order of a THUNDERSTORM or a LEE DEPRESSION.

mesoscopic A term referring to any physical phenomenon whose scale lies in the range of METRES. *See also* MACROSCOPIC, MICROSCOPIC.

mesosphere One of the concentric layers of the ATMOSPHERE, occurring between the STRATOSPHERE and the IONOSPHERE, and bounded by the STRATOPAUSE (below) and the MESOPAUSE (above). It lies between 50 and 80 km above the Earth's surface. Within

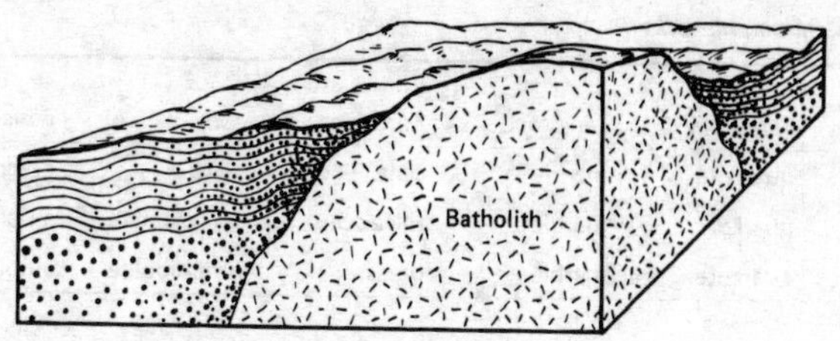

Figure 151 *Metamorphic aureole*

it temperatures decrease from 0° C at the stratopause to –80° C at the mesopause [*17*].

mesosystems MESOSCALE SYSTEMS.

mesotherm **1** The category of plants whose temperature requirements are that: (a) the warmest month should have a mean temperature above 22° C; and (b) the coldest monthly mean should not fall below 6° C. Many of the shrubs found in regions experiencing a MEDITERRANEAN CLIMATE fall into this category. **2** A climate characterized by warm–temperate conditions; the C-climates of KÖPPEN'S CLIMATIC CLASSIFICATION. *See also* HEKISTOTHERM, MEGATHERM, MICROTHERM.

mesothermal An ore deposit formed at intermediate depths and intermediate temperatures (200°–300° C) by HYDROTHERMAL ACTIVITY.

mesotrophic A state of a water body or of a hydrous solution within a soil which is intermediate in character between a EUTROPHIC and an OLIGOTROPHIC state, so far as its nutrients are concerned.

Mesozoic A Greek term meaning 'middle life', adapted to describe the second of the *eras* of geological time. It succeeds the PALAEOZOIC and precedes the CAINOZOIC. It is divided into three time PERIODS (SYSTEM) known as the TRIASSIC, JURASSIC and CRETACEOUS, which together lasted from about 245 million years to 65 million years BP. The lower boundary of the Mesozoic is placed between two systems (the Permian and the Triassic).

mesquite A drought-resistant shrub (XEROPHYTE) which forms dense thickets in the arid regions of Mexico and the SW USA.

Messinian The latest stage of the MIOCENE epoch, during which many EVAPORITE basins developed in the region now occupied by the Mediterranean Sea. Between 6.5 and 5.2 million years ago there was a marked fall in world sea-level, probably due to the extraction of water during the growth of the Antarctic ice-sheet. As a result the western end of the TETHYS ocean became cut off from the Atlantic, leading to a widespread increase in the SALINITY of the proto-Mediterranean Sea. This is known as the *Messinian salinity crisis*, which resulted in large deposits of GYPSUM and HALITE being formed on the former sea floor and a severe incision of surrounding rivers, responding to the lower BASE-LEVEL.

metacartography The portrayal of spatial properties on maps as opposed to those properties expressed in graphs, photographs, etc.

metalliferous deposits Concentrations of metallic compounds in the Earth's crustal rocks. When the degree of metal concentration is economically profitable enough to exploit, the deposit is termed an *ore*. When judged on their abundance (taken as percentage by weight), aluminium and iron are relatively abundant, but mercury, silver, platinum and gold are comparatively rare.

metamorphic aureole The zone of rocks surrounding an intruded mass of igneous rock (INTRUSION) and in which considerable metamorphic changes have occurred (METAMORPHISM). The aureole is usually widest in

Figure 152 *Metamorphic rocks*

Metamorphic grade	*Sandstone*	*Limestone*	*Mudstone*	*Clay*	*Basic lava*
Low	quartzite	marble	slate	slate	greenstone
Medium	quartzite	marble	mica schist	amphibolite	amphibolite
High	quartzite	marble	gneiss	amphibolite	amphibolite

ARGILLACEOUS rocks and narrowest in ARENACEOUS ROCKS. [*151*]

metamorphic differentiation The process by which, in *high-grade metamorphism* (GRADE), some mineral constituents migrate within the metamorphic rock, thereby creating a FOLIATION parallel to the *schistosity* (SCHIST) and leaving layers of contrasting chemical and mineralogical composition.

metamorphic rocks Rocks which have been altered from their original state by various metamorphic processes (METAMORPHISM), generally as a result of mountain-building (OROGENY) and the intrusion and extrusion of MAGMA. The term *metasediments* includes all those rocks altered from original sedimentary rocks and the table [*152*] illustrates how different rocks can be altered by different intensities of metamorphism, termed *metamorphic grade* (GRADE). *See also* EPIDIORITE, GNEISS, GREENSTONE, MARBLE, METAQUARTZITE, PHYLLITE, SCHIST, SLATE.

metamorphism A term covering all the processes by which rocks are altered in their mineralogy, texture and internal structure owing to external sources of heat, pressure and the introduction of new chemical substances rather than changes induced merely by burial, i.e. DIAGENESIS. The different intensity of the processes is described in terms of the *metamorphic grade* (GRADE), referred to as *rank* in US usage. *See also* AUTOMETAMORPHISM, CONTACT METAMORPHISM, DYNAMIC METAMORPHISM, REGIONAL METAMORPHISM.

metamorphosis The biological process in which there is a complete transformation of the bodily form of an animal.

metapedogenesis A term referring to the modification of soil properties as a result of human interference. *See also* OXISOL, ULTISOL.

metaquartzite A term indicating a *metamorphic quartzite*, formed when a highly quartzose sandstone (ARENACEOUS) is altered by METAMORPHISM, partly through pressure and partly through the addition of silica (SiO_2) which completely infills the granular interstices. When broken the fracture passes through the quartz grains, unlike the fracture surface of an unaltered sedimentary quartzite (ORTHOQUARTZITE), which follows the grain boundaries.

metasomatism A process of METAMORPHISM whereby changes in the mineralogy of a rock take place as a result of chemical interactions induced by migrating fluids from external sources. Metasomatism often results in the complete replacement of one mineral by another but without loss of the original texture. Examples include GRANITIZATION and PETRIFACTION. It should be distinguished from PNEUMATOLYSIS.

metastable equilibrium The condition in which STABLE EQUILIBRIUM will prevail only until a suitable catalyst or trigger carries the system state over a threshold into a completely new equilibrium regime. [*72*]

meteor **1** A body of matter travelling at great speed through space which becomes luminous when it enters the atmosphere, at about 150 km above the Earth's surface, because it is heated by friction. Generally, this latter process dissipates the material into METEORIC DUST, although the unconsumed relic may reach the Earth's surface as a METEORITE. A meteor is popularly termed a 'shooting' or 'falling' star. **2** The World Meteorological Organization refers to any phenomenon in the atmosphere (other than cloud) with the suffix *-meteor*: *electrometeor* = any manifestation of atmospheric electricity; *hydrometeor* = atmospheric water; *lithometeor* = solid phenomena; and *photometeor* = luminous effects.

meteoric dust (cosmic dust) Fine dust in the atmosphere which is derived from extraterrestrial sources (METEOR) but becomes trapped within the Earth's gravitational field. It has been estimated that in one year some 5 million tonnes of meteoric dust fall on to the Earth's surface where they contribute small but significant amounts to deep-sea sediments. *See also* DUST.

meteoric water Any water on the surface of the Earth which is derived from atmospheric precipitation, in contrast to JUVENILE WATER. *See also* CONNATE WATER.

meteorite A solid body that manages to reach the Earth's surface from an extraterrestrial source, despite the frictional effect of the atmosphere, which consumes the majority of meteorites. It is composed of various proportions of a nickel–iron alloy (typically 10% nickel and 90% iron) and silicate minerals.

meteorite crater A depression in the Earth's surface created by the impact of a METEORITE. The best-known is that in Arizona, USA (termed the Great Meteor Crater), which is 5 km in circumference and 174 m in depth, below which lies a 3-million-tonne remnant of the meteorite, which has buried itself to a depth of about 260 m. ASTROBLEME.

meteorology The scientific study of the atmosphere. The term is usually confined to a study of the tropospheric and stratospheric processes only, for these are the layers in which surface weather is generated. The atmospheric processes in the MESOSPHERE and the IONOSPHERE are generally regarded as part of GEOPHYSICS, where different methods and techniques of data collection are employed. The most important outcome of meteorological study is the production of a SYNOPTIC CHART for use in weather forecasting. *See also* MACROMETEOROLOGY, MESOMETEOROLOGY, MICROMETEOROLOGY, SYNOPTIC METEOROLOGY.

meter **1** An instrument for measuring a specific quantity. **2** A US spelling of METRE.

methane A colourless, odourless gas, often referred to as MARSH GAS. It is a major constituent (84.6% by weight) of NATURAL GAS, burns with a clean flame, and has a high calorific value.

metre A unit of LENGTH in the metric system. 1,000 m = 1 km, 100 cm = 1 m, 1 m = 39.3701 in = 3.281 ft [*136*].

metrication SI UNITS.

meulière A SARSEN-like stone occurring in the TERTIARY of France.

mica A mineral group with sheet-like structures characterized by excellent CLEAVAGE. They include *biotite* (dark brown to green); *lepidolite* (lilac to grey); *muscovite* (light brown, transparent, green or red) which, like biotite, is common in granitic rocks, gneiss and schist (e.g. *mica schist*); and *phlogopite* (yellow) which occurs in marbles and peridotites.

mica schist SCHIST.

micrite LIMESTONE.

microclimate The climate within a few metres of the ground and in a relatively small area, in contrast, to that of a MACROCLIMATE. *See also* BOUNDARY LAYER, HEAT BALANCE.

microclimatology The scientific study of local climatic conditions, especially in the few metres of the atmosphere in contact with the ground surface. *See also* MICROCLIMATE.

microcracks Tiny apertures in rocks of all varieties that provide one of the main means for CHEMICAL WEATHERING and MECHANICAL WEATHERING to operate. They have been classified as: *macrofractures*: greater than 0.1 mm in width and several metres in length; *microfractures*: less than 0.1 mm in width and a few centimetres in length; *microfissures*; less than 1 μm in width and length.

micro-erosion Small-scale processes of denudation which occur to depths of only a few centimetres, e.g. PIPKRAKE, *rillenkarren* (KARREN).

microfauna Animal organisms too small to be identified without the aid of a microscope, e.g. FORAMINIFERA, PROTOZOA. *See also* MICRO-ORGANISMS IN THE SOIL.

microflora Plants that are too small to be identified without the aid of a microscope, e.g. DIATOMS, BACTERIA. *See also* MICRO-ORGANISMS IN THE SOIL.

microfossil A FOSSIL too small to be identified without the aid of a microscope. The scientific study of microfossils is termed *micropalaeontology*.

microgranite An acid IGNEOUS ROCK with the mineralogical and chemical properties of a GRANITE but having a medium-grained texture. Porphyric varieties are termed quartz porphyry and occur in larger bodies than the ordinary microgranites, which most commonly occur in the form of DYKES, SILLS and small PLUGS.

micrometeorology The study of small-scale atmospheric processes, particularly those operating near the ground surface, where large varieties of MICROCLIMATE are found, because of the different soils, crops, vegetation, surface textures and terrain characteristics. Very sophisticated instruments are required to monitor small changes of moisture, temperature, etc. in very localized conditions. *See also* MACROMETEOROLOGY, MESOMETEOROLOGY.

micron **1** A unit of length (μ) equivalent to one-millionth of a metre (10^{-6}). Since 1968 it has been officially replaced by the term micrometre (μm). **2** A unit of pressure equal to 10^{-6} m or 10^{-3} mm expressed as the height of a column of mercury.

micro-organisms in the soil The tiny organisms which help to incorporate plant residues into the soil in the form of HUMUS. The plant residues have fulfilled their functions and are virtually devoid of starches, sugars and fats, leaving only the old structural elements with small amounts of protein, some ash and traces of waxes and resins to be converted into humus. The most important micro-organisms in the soil are: **1** *microfauna*, including PROTOZOA, nematodes, earthworms and insects; **2** *microflora*, including BACTERIA, fungi, ALGAE and DIATOMS.

micro-relief The slight undulations of the surface created by changes in the form of the REGOLITH rather than by irregularities in the underlying solid rock. These may be due to the erosion of PEAT, to burrowing by wild animals, to trampling by domestic animals (TERRACETTES) or due to minor SOLIFLUCTION activity or fluvial activity which creates landforms less than 1 m in height, e.g. some types of SCROLL bars on a floodplain. *See also* GILGAI, HUMMOCK, MIMA MOUND.

microscopic A term referring to any physical phenomena whose scale lies in the range 5 μm to 2 mm, i.e., it is only visible under an optical microscope. *See also* MACROSCOPIC, MESOSCOPIC.

microseism A small earth tremor generated not so much by EARTHQUAKE activity as by ROCK-FALLS, storm waves pounding on a coastline, artificial explosions and the passage of trains and heavy road vehicles. Microseisms are measured on a very sensitive *microseismometer*.

microtherm **1** The category of plants whose temperature requirements are that: (a) the warmest month should have temperatures between 10° and 22° C; and (b) the coldest monthly mean should not fall below 6° C. Most of the deciduous trees in Britain come into this category. **2** A climate characterized by cool temperate to cold conditions; the D-climates of KÖPPEN'S CLIMATIC CLASSIFICATION. *See also* HEKISTOTHERM, MEGATHERM, MESOTHERM.

microwave A REMOTE SENSING term referring to the region of ELECTROMAGNETIC RADIATION in the wavelength range 0.1 mm to beyond 1 m [*70*]. An ACTIVE REMOTE-SENSING SYSTEM operated at microwave wavelengths is RADAR and a PASSIVE REMOTE-SENSING SYSTEM operated at microwave wavelengths is *passive microwave*. Radar remote sensing has proved to be of value in geological, geomorphological and land-use studies and *passive microwave* has proved to be of value in the determination of soil surface moisture.

mictite A clastic rock (CLAST) with a wide range of GRAIN SIZES. *See also* DIAMICTITE.

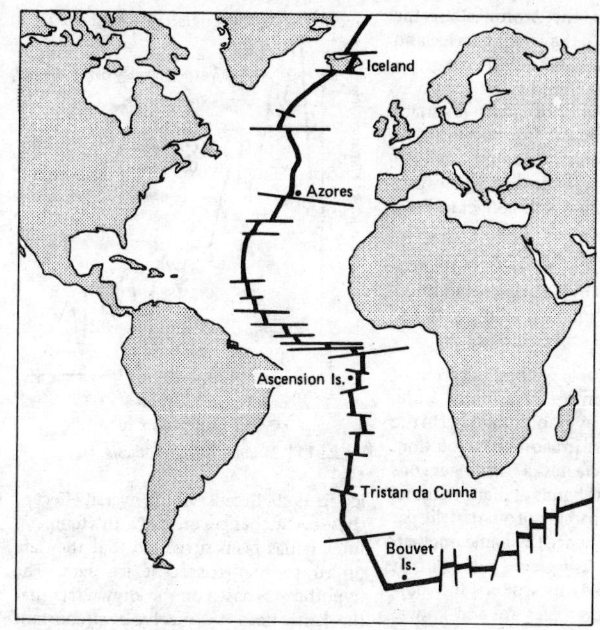

Figure 153 *Mid-oceanic ridge*

Midlandian The name given to the final glacial stage of the PLEISTOCENE in Ireland. It is equivalent in age to the DEVENSIAN of Britain. [*109*] *See also* GLENARG STADIAL.

mid-latitude zone A synonym for the TEMPERATE climatic zone of the world. It refers specifically to the zone lying between 23½° and 66½° N and S of the equator. It coincides broadly with the C-climates in KÖPPEN'S CLIMATIC CLASSIFICATION. *See also* MESOTHERM.

Midnight Sun The term given to the phenomenon which occurs between May and July in the N hemisphere and between November and January in the S hemisphere, during which the Sun remains above the horizon when viewed between latitudes 63½°–90° N and S of the equator. The regions of northern Scandinavia (Lapland) are often referred to as the *Lands of the Midnight Sun*.

mid-oceanic ridge (median ridge) A large-scale linear elevation rising 1–3 km from the ocean floor, approximately in the middle of the N and S Atlantic and the Indian oceans. It marks the boundary of adjoining constructive plate margins (PLATE TECTONICS), where upwelling basaltic magma from the MANTLE arrives at the Earth's surface and adds to the oceanic crust, which spreads laterally away from the line of the ridge (SEA-FLOOR SPREADING). [*188* and *223*] The ridge, which is generally more than 1,000 km in width, is thought to include an extensive RIFT VALLEY (termed a median rift or median valley) along its crest, although the line of the rift has been considerably offset by wrench-faults known as TRANSCURRENT-FAULTS. The submarine volcanic activity reflects the relative youthfulness of the rocks being formed along the ridge and this is also a zone of high seismic activity (EARTHQUAKE). Although the majority of the ridges lie beneath 3,000–3,500 m of sea water several small islands, of volcanic origin, rise at intervals (e.g. Tristan da

Cunha, Ascension, the Azores) along the mid-oceanic ridge in the Atlantic. In Iceland the median rift runs through the centre of this larger island which spans the ridge, here characterized by a zone of extensive surface vulcanicity. The mid-Atlantic ridge is linked with that of the Indian ocean by a median ridge situated part way between S Africa and Antarctica and from which Bouvet Is. rises. [*153*]

mie scattering The SCATTERING of ELECTROMAGNETIC RADIATION by particles in the atmosphere which are comparable in size to the wavelength of the scattered energy.

migmatite A coarse-grained injection of GNEISS formed from the invasion of a pre-existing host-rock by granitic material in the form of a MAGMA or a HYDROTHERMAL solution. This admixture creates a heterogeneous rock with schistose bands of mafic material interspersed with patches of quartz-feldspar granitic material (hence the name migmatite from the Greek *migma* = mixed). It is generally associated with REGIONAL METAMORPHISM, although examples are also known from CONTACT METAMORPHISM, but usually formed under the influence of high-grade metamorphism (GRADE). Different stages of alteration produce different types of migmatite, ranging from the least-altered *nebulite* to the completely altered *anatexite*. When a network of granitic veins invades the host-rock during the LIT-PAR-LIT INJECTION process the migmatite is known as an *agmatite*. *See also* GRANITIZATION, METASOMATISM.

migration The movement of floral and/or fauna from one habitat to another. *See also* DISPERSAL.

migration of divide A change in position of an INTERFLUVE owing to the lateral recession of slopes by erosional processes (PARALLEL RETREAT OF SLOPE). *See also* RIVER CAPTURE, UNEQUAL SLOPES, LAW OF.

mil A unit of length equal to one thousandth of an inch (also called a *thou*).

Milankovitch hypothesis Graphic representations of possible variations in solar RADIATION received at different latitudes over a period of time, calculated by M. Milankovich in 1940. Because of the overall effect of these cyclic changes on the Earth's temperatures it has been suggested that they are linked to the PLEISTOCENE ice ages. The hypothesis is based on the known fact that the Earth does not revolve at a constant velocity in a circular orbit around the Sun. The three variables in the Earth/Sun geometrical relationship are: **1** the ECCENTRICITY of the Earth's orbit, with a periodicity of 91,800 years; **2** the obliquity of the plane of the ECLIPTIC which varies between 21°58' and 24°36', with a periodicity of 40,400 years; and **3** the PRECESSION OF THE EQUINOXES, i.e. whether the EQUINOX occurs in APHELION or PERIHELION, with a periodicity of about 21,000 years. The effect of these variations, either singly or in combination, is to give cyclic changes in the amount of distribution of solar radiation at various localities on the Earth's surface, although it is now thought that the changes are likely to be so small (1° or 2° C) that they will have little effect on global temperatures. [*154*]

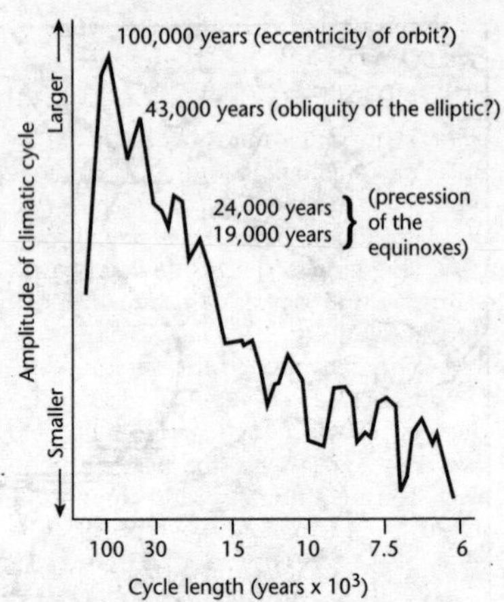

Figure 154 *Milankovitch hypothesis*

Milazzian One of the raised shorelines created by a former sea-level of the Mediterranean during the PLEISTOCENE. It stands at elevations of 55–60 m above present sea-level and the date of its formation has tentatively been correlated with one of the earlier INTERGLACIALS. The name was introduced by

C. Depéret in 1918, from its type-site of Milazzo in Sicily. *See also* MONASTIRIAN, SICILIAN, TYRRHENIAN.

mile 1 The *statute* mile is a British and US linear measurement (LENGTH of 1,760 yards, possibly based on the Roman marching unit of 1,000 double paces, each double pace being five feet. **2** The *Irish mile*, of 2,240 yards, survived in parts of Ireland until well into the 20th cent. **3** The *nautical mile* is the average meridian length of one minute of latitude or 1/21,600th of a mean GREAT CIRCLE. It equals 1.1516 statute miles and one KNOT equals 1 *nautical mile* per hour. The British Admiralty and mercantile marine use the 6,080 ft *nautical mile* (one minute of arc at 48° latitude), whereas all other nations use a *nautical mile* of 1,852 m (one minute of arc at 45°), the distance recommended by the International Hydrographic Conference in 1929. 1 English *nautical mile* = 1.00064 international *nautical miles*. [*136*] 1 mile = 1.61 km.

military grid 1 A GRID system used on British War Office maps prior to the introduction of the NATIONAL GRID. **2** A grid using the metre as the basic unit of length and with a network of squares each of which is 1,000 m wide. The maps of the USA are based on a military grid. *See also* UPS GRID, UTM GRID.

Miller's projections A group of MAP PROJECTIONS introduced by O. M. Miller in the mid 20th cent. They include: **1** A bipolar oblique conformal conical projection used to represent North and South America separately. The vertex for the N American projection is centred in the Atlantic Ocean, and that for S America is centred in the Pacific Ocean, but the two parts fit together systematically along the line joining the two vertices. **2** A projection based upon the conformal transformation of an oblique-aspect STEREOGRAPHIC PROJECTION. **3** A CYLINDRICAL PROJECTION in which the spacing of the parallels is obtained from the MERCATOR PROJECTION by multiplying the usual equation by two constants.

millibar A unit of pressure (mb) equivalent to 1,000 dynes per cm^2. Atmospheric pressure is measured by a BAROMETER, 1,000 mb = 1 bar, or on the SI SCALE 1 mb = 100 N m^{-2}. Many barometers are calibrated to indicate inches of mercury, so that a conversion to millibars may be necessary. There is no universally accepted conversion formula, but at a constant temperature (0° C) and latitude (45°) the following is applicable: 1,049 mb = 31 in of mercury; 1,016 mb = 30 in; 982 mb = 29 in.

milligal GAL.

millimetre A metric unit of LENGTH equal to one-thousandth of a metre and 0.0394 of an inch. [*136*]

Millstone Grit The name given to the coarse-grained sandstone which divides the underlying CARBONIFEROUS LIMESTONE from the overlying COAL MEASURES in the stratigraphical division of the CARBONIFEROUS system in Britain. It is equivalent to the NAMURIAN stage of the European nomenclature of the Carboniferous. It is thought to have been deposited as part of a deltaic series in N Britain, where its locally thick beds form prominent scarps (EDGE) and moorlands in the Peak District and elsewhere in the Pennines. Its name is derived from its use in grinding cereals in the production of flour.

mima mound Sometimes referred to as a *pimple mound*, hence the term *pimple plain*, on which a flat land surface is dotted with mima mounds. The mound is a type of earth HUMMOCK, up to 10 m in diameter and up to 2 m in height, and is characteristic of western USA. The mounds are probably due to PERIGLACIAL activity in the soil, although some may have been due to burrowing animals. *See also* MICRO-RELIEF.

Mindel One of the earlier Pleistocene glacials recognized in 1909 by A. Penck and E. Brückner in the valley of the R. Mindel on the Swiss–German border to the N of the Alps. It is older than the RISS and WÜRM but younger than the DONAU and GÜNZ glacial stages. The Mindel glacial is believed to be equivalent in age to the ELSTER of N Europe, the ANGLIAN of Britain and the KANSAN of N America.

mineral **1** A natural inorganic substance possessing a definite chemical composition and almost always in crystalline form. Minerals are structurally homogeneous (i.e. their fundamental atomic structures are constant and continuous throughout the mineral unit). Ice is regarded as a mineral but coal, oil and natural gas are not because they are of organic origin. Nevertheless, some bedded phosphates, some limestones and some siliceous rocks have organically derived constituents but are still treated as minerals. **2** A loose term for an economic MINERAL DEPOSIT exploited by mining.

mineral deposit A body of naturally occurring MINERALS of commercial value. It includes: **1** *magmatic concentrates*, e.g. diamonds, magnetite; **2** *products of* METASOMATISM, e.g. some iron ores and pyrites; **3** *hydrothermal deposits* in veins, lodes, reefs, etc. and those formed by REPLACEMENT, e.g. most copper, lead and zinc deposits; **4** *products of* PNEUMATOLYSIS, e.g. tin and tungsten; **5** *sedimentary deposits*, e.g. sedimentary iron ore (primary sediments), gypsum, salt and potash (EVAPORITE), those diamonds, gold and tin ores found as PLACER DEPOSITS; **6** *metamorphic deposits*, e.g. asbestos, graphite and talc.

mineralization **1** The process by which minerals of ORE and GANGUE are introduced into pre-existing rocks by REPLACEMENT or in the form of VEINS. **2** The conversion of an element in the soil from an organic form to an inorganic state as a result of microbial decomposition.

mineralogy The scientific study of MINERALS, in which the following properties are determined: the CRYSTAL form; colour; CLEAVAGE; FRACTURE; hardness (HARDNESS SCALE); lustre; specific gravity; streak; and transparency.

mineral soil A soil composed predominantly of MINERAL material and low in HUMUS content. It represents all the soil material below the A-HORIZON.

minette **1** A term given to any sedimentary iron ore which consists primarily of LIMONITE. **2** More specifically, the sedimentary Jurassic IRONSTONE of Lorraine (France) and Luxembourg, with an iron content of between 24% and 40%.

minimum thermometer An instrument to record the lowest temperature attained during a given time period. The most common type is filled with alcohol and includes a sliding metal index within the tube, kept just below the meniscus of the alcohol by surface tension. The index is drawn down by the liquid as the level drops in response to a fall in temperature and it remains at the minimum reading. After recording, the index can be returned to its former position by the use of a magnet. *See also* MAXIMUM AND MINIMUM THERMOMETER.

minimum variance theory A term referring to the concept within GENERAL SYSTEM THEORY (GST) that once a condition of equilibrium has been reached there will then be minimal change in the future.

minute **1** A unit of time; one-sixtieth of an hour. **2** A unit of angular measurement; one-sixtieth of a degree. Both are survivals from the Babylonian system of working in units of sixty. Attempts to decimalize them have been unsuccessful.

Miocene An epoch of the TERTIARY, occurring before the PLIOCENE and after the OLIGOCENE. It extended from about 23.5 million years ago to about 5.2 million years ago, and together with the Pliocene and Pleistocene it comprises the NEOGENE. It was a time when mammals became of great importance among the world's fauna and hominids first made their appearance in the fossil record. The ALPINE orogeny reached its climax during the Miocene and although deposits of this age occur elsewhere in Europe (FLYSCH, MOLASSE) there are no known Miocene deposits exposed at the surface in Britain. *See also* CAINOZOIC.

miogeosyncline A GEOSYNCLINE in which the products of vulcanicity are virtually absent from its sedimentary sequence. *See also* EUGEOSYNCLINE. [*91*]

mirage An optical illusion which results from the refraction of light through layers of air that have developed large temperature gradients due either to the heating of the lowest layer by CONDUCTION from the ground surface or to cooling from a cold surface. Because of its marked temperature gradient the refraction index of the air also changes in a vertical direction, so that light appears to travel in curved paths. Two types of mirage can be distinguished: the more common is the *inferior mirage*, such that on a hot day light rays near the ground surface are strongly refracted upwards, thus producing refractions of the sky which give the impression of shimmering pools of water on the ground; the second is the *superior mirage*, when light rays are bent down from a warm layer of air resting on a colder layer (INVERSION OF TEMPERATURE) which has been cooled by the underlying ground surface. Thus in polar latitudes a double inverted image of a distant object may appear to the observer to be suspended above the flat snow or ice surface.

mire An area of marshy or waterlogged ground. *See also* OMBROTROPHIC.

misfit stream A term given to streams that are not proportionate in size to the valleys they occupy. It was originally suggested that they included both *overfit* and *underfit* streams but since no examples of the former have been discovered it is doubtful if they exist. Thus the term has now become synonymous with UNDERFIT STREAM.

Mississippian A period of geological time in the stratigraphical nomenclature adopted by the USA. It lasted for some 45 million years between 370 million years ago (the end of the DEVONIAN) to 325 million years ago (the beginning of the PENNSYLVANIAN), i.e. it corresponds approximately to the Lower Carboniferous (CARBONIFEROUS).

mist The degree of atmospheric obscurity midway between HAZE and FOG, with visibility officially recorded as being within 1 and 2 km. It results from condensation within the lower layers of the atmosphere when RELATIVE HUMIDITY rises to over 95 per cent, manifesting itself as a veil of suspended microscopic water droplets. *See also* RADIATION FOG, SCOTCH MIST.

mistral A French term for a strong, cold, dry wind which blows from the Massif Central of France southwards to the Mediterranean Sea. It is usually funnelled down the Rhône valley, thereby increasing its wind speed to over 60 km/h. It is most common in winter, especially in association with cold fronts when pressure is low over the Mediterranean and high over central Europe. It brings cold bursts of air to the Rhône delta and crop damage can occur if precautions, in the form of hedges and wind-breaks, for example, are not taken. *See also* BORA.

mixing corrosion The increased degree of corroding power (CORROSION) of KARST water due to the liberation of CO_2 during the mixing of two saturated waters of differing concentrations of calcium carbonate.

mixing ratio The ratio of the weight of a parcel of atmospheric gas to the total weight of air with which the gas is mixed. Its most common use is for water vapour (the *humidity mixing ratio*) where it is stated as g of water vapour per kg of dry air (expressed as g kg^{-1}). *See also* SPECIFIC HUMIDITY.

mizzle A misty drizzle.

mobile belt A lengthy linear portion of the Earth's crust in which intense deformation takes place during an OROGENY. It comprises thick sedimentary rock layers which have infilled a major GEOSYNCLINE and which are compressed when two continental FORELANDS approach each other (PLATE TECTONICS). It is the site of a future range of fold mountains, e.g. the mobile belt of TETHYS eventually became folded and uplifted into the ALPINE mountain ranges.

mobile dune A dune, in a coastal environment, representing the transition phase between a FORE-DUNE and a STABILIZED DUNE. At this stage about half the surface has been fixed by vegetation (e.g. by marram grass) but sufficient bare sand is exposed for winds to cause the formation of BLOW-OUTS by the action of DEFLATION. Thus many coastal dune areas exhibit a mixture of sandy depressions interspersed with mobile dunes in an apparently confused pattern.

mock sun PARHELION.

mode **1** A statistical term referring to the most commonly occurring value in a series of grouped data. **2** In geology, the percentage of each of the component minerals contained in a metamorphic or an igneous rock.

model A formalized expression of a theory, event, object, process or system used for prediction or control; an experimental design based on a causal situation which generates observed data. A model can be viewed as a 'selective approximation which, by the elimination of incidental detail, allows some fundamental relevant or interesting aspects of the real world to appear in some generalized form' (R. Chorley and P. Haggett). Several classifications have been made in the Earth sciences, one of which defines: **1** *scale models*, in which a prototype object is scaled up or down to a size convenient for study, e.g. to emphasize MICRORELIEF; **2** *conceptual* (*diagrammatic*) *models*, which are based on observation and express a segment of the real world in idealized form, retaining essential features but omitting extraneous details (PROCESS–RESPONSE MODEL), e.g. a volcanic eruption; (c) *mathematical models*, which are abstractions of physical models in that they replace events, forces, objects, etc. by using expressions that contain mathematical variables, constants and parameters. These models can be further subdivided into: *deterministic models*; *statistical models* involving variables, constants and parameters and one or more random components to represent unpredictable fluctuations in the experimental data due to measurement-error, equation-error or inherent variability of the objects being measured (sometimes referred to as a *random model*); and *stochastic-process models*, which are statistical models but with a specific random process built in to describe the phenomenon on a probability rather than a determinist basis. *See also* ICONIC MODEL, RHEOLOGICAL MODEL, SIMULATION MODEL.

model-building The creation of a MODEL to simplify a real world problem while maintaining a balance between accurate representation of reality and mathematical manageability. It has been suggested that this aim may be achieved by: **1** omitting certain variables which have only a small effect on the performance of the system but which add great mathematical complexity; **2** changing the nature of the variables, e.g. by treating some as constants; **3** changing the relationship between variables, e.g. by assuming that a function is continuous rather than discontinuous; **4** modifying the constraints by adding to or subtracting from them as appropriate.

moder A type of HUMUS, intermediate in character between MULL and MOR. Decomposition is greater than in mor but has not proceeded as far as in mull and although moder occurs as a distinct organic layer above the MINERAL SOIL it is permeated by loose mineral particles, especially in its lower layer, due to the activity of soil fauna.

modified polyconic projection A type of POLYCONIC PROJECTION introduced in 1909 as the map projection on which the INTERNATIONAL MAP OF THE WORLD was based (scale 1 : 1 million). The scale is preserved along the two extreme parallels of each map sheet and along two meridians each of which lies 2° from the central meridian. The meridians are straight and not curved.

mofette A fissure in the ground surface of a region of dwindling volcanic activity, from which carbon dioxide, water vapour and some other gases are emitted. It is a type of FUMAROLE and represents a late stage in a period of vulcanicity, e.g. the Phlegraean Fields, Naples, Italy. *See also* SOLFATARA.

mogote A residual mass of KARST limestone that rises precipitously from a flat valley floor or plain as an example of tower karst (TURMKARST).

Mohorovičić Discontinuity A seismic continuity occurring between the CRUST OF THE EARTH and the underlying MANTLE, across which the velocities of P-waves and S-waves are significantly modified (SEISMIC WAVES), owing to the different densities of the crust and mantle. The discontinuity (sometimes referred to as the 'moho') occurs at an aver-

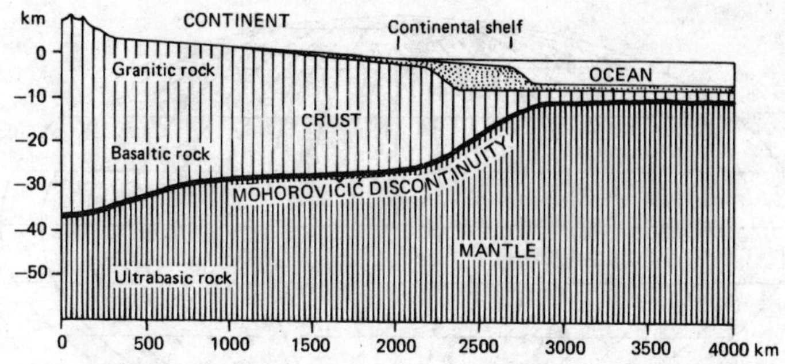

Figure 155 *Mohorovičić Discontinuity*

age depth of 35 km below the continents and at about 10 km below the oceans. It was named after a Yugoslav seismologist who discovered it in 1909. *See also* GUTENBERG DISCONTINUITY. [*155*]

Mohr–Coulomb equation Also known as *Coulomb's failure law*. An important law relating to the strength of materials and the frictional resistance of soil particles to imposed stress. It is expressed as:

$$s = c + \sigma \tan \phi$$

where s is the SHEAR STRENGTH at failure; c is the COHESION component and ϕ is the FRICTION of the soil; σ is the normal stress. The equation is modified if PORE-WATER PRESSURE is taken into account to give an EFFECTIVE STRESS analysis. [*156*] *See also* SHEARBOX, TRIAXIAL TEST.

Mohs scale HARDNESS SCALE.

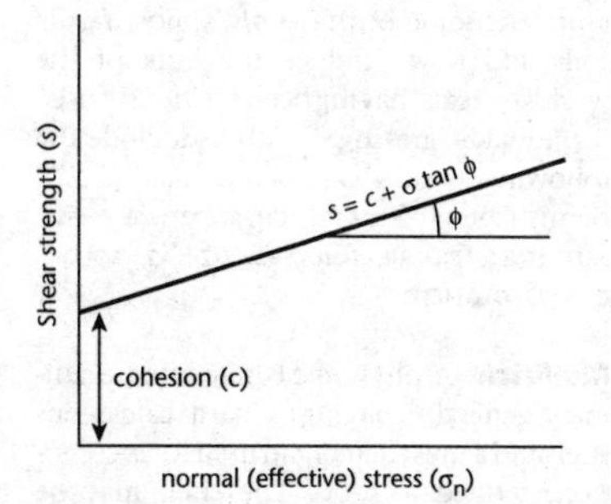

Figure 156 *Mohr–Coulomb equation*

Moinian A stratigraphic division of the PRE-CAMBRIAN in Scotland. It comprises a highly folded and faulted complex of metamorphic rocks (equivalent to the unmetamorphosed TORRIDONIAN). The Moinian is composed essentially of the Moine Schists, a group of rocks occupying most of the Northern Highlands between the *Moine Thrust* in the NW and the Great Glen Fault in the SW, but is also found in the north-eastern part of the Grampians of Scotland. The rocks are a series of MICA SCHISTS and quartz-feldspar granulites, altered by high-grade metamorphism on several occasions prior to and during the CALEDONIAN OROGENY. It was this prolonged period of deformation and folding that saw the Moinian rocks carried bodily north-westwards across the underlying Cambrian rocks and the older Pre-Cambrian rocks of the foreland along the 320 km length of the Moine Thrust (THRUST FAULT).

moisture index An index relating POTENTIAL EVAPOTRANSPIRATION to PRECIPITATION amount, as a basis for climatic classification. It was devised in 1948 by C. W. Thornthwaite in an attempt to indicate whether a station has a positive or negative water balance, i.e. a surplus or a deficiency of precipitation. It is expressed by the formula $MI = 100\ (P - PE)/PE$, where: P = precipitation; and PE = potential evapotranspiration. The moisture index will exceed +100 when precipitation greatly exceeds potential evapotranspiration but will fall to −100 when precipitation is zero. The climatic

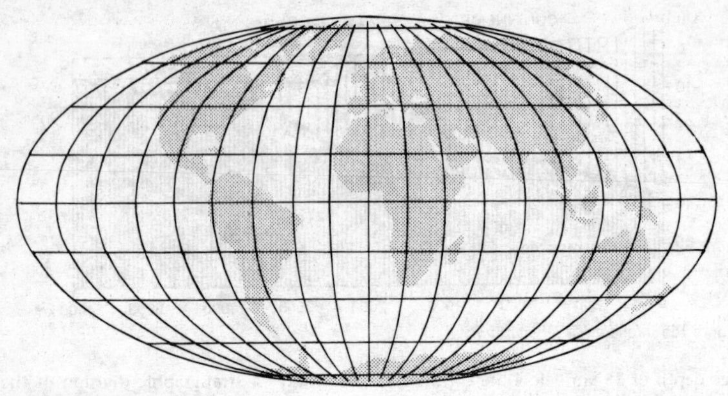

Figure 157 *Mollweide projection*

classification ranges from type A-prehumid to type E-arid. *See also* THORNTHWAITE'S METHOD OF CLIMATIC CLASSIFICATION.

moisture-retention curve A graph used in soil science to show the relationship between soil-water percentage (expressed by weight or volume) and the applied pressure or tension. Points on the graph can be obtained by varying the applied pressure or tension over a specified range. *See also* SOIL WATER.

moisture tension HYGROSCOPIC WATER, SOIL WATER.

molasse A term originally used in Switzerland to describe the soft glauconitic sandstones, marls and conglomerates of MIOCENE age (but younger than the FLYSCH) found in the lower plains and plateaux surrounding the Alps. It is thought to have been deposited mainly in fresh-water lakes, having been derived from materials worn from the recently uplifted mountain ranges. The term is now used to describe all detrital sediments of similar genesis laid down in FORE-DEEPS developed during the later stages of a major DIASTROPHISM. *See also* ALPINE OROGENY.

molecular attraction The attraction of rock surfaces for water molecules and the attraction by each water molecule for another. Much groundwater is held in fine-grained rocks by this phenomenon.

mollisol **1** A term used to describe the seasonally active layer of ground above the PERMAFROST, which thaws in the summer and freezes again in the winter. Environmental factors (climate, terrain, vegetation cover, soil character, snow cover, soil moisture, etc.) create considerable variations in the thickness of the active layer. **2** One of the orders of the SEVENTH APPROXIMATION of soil classification. It is a soft CALCIMORPHIC SOIL characterized by a very dark surface horizon of great thickness (25–100 cm), with a high proportion of calcium among the exchangeable cations (BASE EXCHANGE), and with a dominance of clay minerals of a high base-exchange capacity. Mollisols form mainly under grassland in climates with a marked seasonal moisture deficiency, e.g. the steppes of Eur-Asia, the Great Plains of N America and the pampas of S America. They represent some of the world's most fertile soils and now produce the bulk of the world's cereals, having been previously used for nomadic grazing. Mollisols include the following: BROWN CALCAREOUS SOILS, BROWN EARTHS (some of them), CHERNOZEMS, CHESTNUT SOILS, PRAIRIE SOILS, RENDZINAS, SOLONCHAKS, SOLONETZ.

Mollusca A phylum of invertebrate animals generally having a hard calcareous shell and a muscular organ used for locomotion, termed a 'foot'. The shell may be bivalve or univalve. The phylum is divided into a number of classes: Amphineura, Scaphoda, Gastropoda, Lamellibranchiata, Cephalopoda.

Mollweide projection A MAP PROJECTION of pseudo-cylindrical type, showing the entire Earth enclosed within an ellipse. The major axis (equator) is twice the length of the minor axis (CENTRAL MERIDIAN). The parallels are straight lines but the meridians are curved, with the exception of the central meridian, the curvature becoming greater away from the central meridian. Thus although it is an equal-area projection, there is considerable distortion of shape and direction. The Mollweide projection is often used in an interrupted form, with the interruptions in the oceans and with each of the continental areas having its own central meridian. Named after K. B. Mollweide who introduced it in 1805, the projection is useful for illustrating world distributions of raw materials, for example. [*157*]

molybdenum A metal associated with molybdenite ores produced by HYDROTHERMAL ACTIVITY in proximity to acid igneous rocks.

moment measures An expression used in statistics to describe the character of a distribution curve. They comprise: **1** the MEAN, **2** the STANDARD DEVIATION, **3** the SKEWNESS, **4** the KURTOSIS.

monadnock An isolated hill or type of RESIDUAL due to denudation which has left it rising conspicuously above a gently rolling plain (PENEPLAIN). It is usually, but not necessarily, related to an outcrop of more resistant rocks, although the term refers to the morphology not the structure. It is named after Mt Monadnock in New Hampshire, USA. *See also* NORMAL CYCLE OF EROSION.

Monastirian One of the composite raised shorelines created by former sea-levels of the Mediterranean during the PLEISTOCENE. They stand at elevations of 18 m (Main Monastirian) and 7.5 m (Late Monastirian) above present sea-level and the dates of their formation have tentatively been correlated with the last INTERGLACIAL. The name was introduced by C. Depéret in 1918, from its type-site of Monastir in Tunisia. *See also* MILAZZIAN, SICILIAN, TYRRHENIAN.

monoclimax theory CLIMAX.

monoclinal shifting US equivalent of UNICLINAL SHIFTING.

monocline, monoclinal fold A gentle bend or flexure with a step-like form in horizontally bedded or gently dipping rock strata. *See also* FOLD. [*84*]

monocyclic landscape A topography which has been created by a single CYCLE OF EROSION, in contrast to a POLYCYCLIC LANDSCAPE.

monoglaciation A glacial episode during which a single major ice advance occurred, although the limits of the ice-sheet may have fluctuated sufficiently to lay down more than one TILL deposit. Earliest writers on the Ice Age tended to be *monoglacialists* but more recent writers have emphasized the multiplicity of glacial advances during the Pleistocene.

monolith A vertical section of a SOIL PROFILE removed from the ground and mounted for display or research purposes.

monsoon A term derived from an Arabic word, *mausim,* meaning season, and originally restricted to the winds of the Arabian Sea, which blow some six months from the NE and some six months from the SW. It is currently extended to include any wind which has periodic alternations of direction and velocity related to the seasonal reversal of the SUBTROPICAL jet stream and pressure over continents and neighbouring oceans. The term has been further extended to cover the rains which accompany the winds. It is caused primarily by atmospheric pressure changes due to differential heating of land and water but is complicated by the poleward shift of the hemispheric pressure and wind belts during the summer. Because of topographic differences and contrasts in configuration of coastlines the character of the monsoon varies from place to place. The strongest monsoonal effects are felt in India, Pakistan, SE Asia and China but monsoons are also found in E Africa, the Guinea coast of W Africa and in N Australia.

monsoon forest A type of TROPICAL RAIN-FOREST restricted to regions which are affected by a monsoonal climatic regime (MONSOON). It contains many valuable hardwood trees (e.g. the teak forests of Burma), the majority of which are of deciduous character, losing their leaves in the dry season.

Monte Carlo technique A technique used to create a type of SIMULATION MODEL that deals with a discontinuous distribution of data in which elements of probability and 'chance' are involved. Both spatial and time aspects are included as variables of the model and every stage of development in the spatial pattern may be dependent on the preceding stage, a dependence which, to some extent, may be governed by 'chance'.

montmorillonite The name given to a group of CLAY MINERALS, produced during the alteration of basic rocks under alkaline conditions. Such clay minerals are composed of complex hydrous aluminosilicates of iron, magnesium and sodium. These minerals can absorb water and exhibit considerable properties of expansion, thereby causing soils to swell when wet.

monzonite A granular plutonic rock intermediate in character between DIORITE and SYENITE.

Moon The only satellite of planet Earth, around which it revolves in one *sidereal month* (27 days 7 h 43.25 min). It follows an orbital path which brings it to within 348,292 km of the Earth at its nearest point. During each orbit the Moon rotates once on its axis, thereby presenting the same face towards the Earth at all times. When viewed from the Earth the Moon passes through four phases as follows: **1** first quarter (QUADRATURE), during which time it is said to be *waxing*; **2** full Moon (OPPOSITION); **3** third quarter (*quadrature*), during which time it is said to be *waning*; **4** new Moon, during which time it is invisible (CONJUNCTION). The time period elapsing between one new Moon and the next is termed a *lunar month* or *synodic month* (of 29 days 12 h 44 min) which represents one complete revolution of the Moon in relation to the Sun. *See also* LUNAR DAY, SYZYGY, TIDE.

moon milk ROCK MILK.

moorland, moor **1** In general, a tract of unenclosed wasteland, usually covered in heather. **2** In more strict English usage the term is used primarily to describe uplands covered with heather and other Ericaceous dwarf scrubs, especially *Vaccinium myrtillus* (bilberry, whortleberry, whimberry, blaeberry), but it is also used more widely to include acidic grasslands, peat bogs (BOG) and mosses (MOSS). **3** In German usage the term *moor* refers to any area of deep peat, whether acid or alkaline. Thus, a EUTROPHIC fen-peat in Germany is a *niedermoor*, which may pass through a stage of MESOTROPHIC peat (termed transition moor or *übergangsmoor*) to a stage of OLIGOTROPHIC peat, known as *hochmoor*. It is important to note that the literal translations of *niedermoor* and *hochmoor* are 'low *moor*' and 'high *moor*', respectively, while the term moor should not be used to describe a true FEN in the English sense. But the term has been used to describe a lowland marsh or unenclosed pasture in England, e.g. Otmoor near Oxford and Sedgemoor in Somerset. *See also* HEATH.

moorpan HARDPAN.

mor An organic-matter distribution (including HUMUS) forming a ground-surface layer, sharply delineated from the A-horizon, of the underlying MINERAL SOIL, which often takes the form of a PODZOL. Little intermixing of the humus layer with the mineral soil occurs because of the absence of earthworms and the paucity of soil microfauna owing to the lack of oxygen in the soil (MICRO-ORGANISMS IN THE SOIL). It is derived either beneath conifer forest or on open heath and moorland in cool, moist climatic conditions. It has a very acid character. *See also* MODER, MULL.

moraine An accumulation of heterogeneous rubbly material, including angular blocks of rock, boulders, pebbles and clay, that has been transported and deposited by a glacier or ice-sheet. Classifications of moraines are based usually on the mode of their formation, i.e. in a SUPRAGLACIAL, ENGLACIAL OF SUBGLACIAL environment. An accumulation of GROUND MORAINE, for example, is made up of a combination of both ablation till and lodgement till (TILL). Contrasting with the poorly sorted and widespread character of the *ground moraine* are the great varieties of linear moraines, some of which exhibit sorting by meltwater on their outer slopes where GLACIOFLUVIAL processes have been active. Among the linear moraines are: DE GEER MORAINES, ICE-CORED MORAINES, LATERAL MORAINES, MEDIAL MORAINES, PUSH MORAINES, ROGEN MORAINES, RECESSIONAL MORAINES, STADIAL MORAINES, TERMINAL MORAINES, THULE–BAFFIN MORAINES, WASHBOARD MORAINES.

morass An area of waterlogged, marshy ground, thought by some to be a term derived from MOOR. *See also* BOG, MARSH, MOSS, SWAMP.

morphochronology A term referring to the dating of morphological features or landforms.

morphoclimatic zone CLIMATOMORPHOLOGY.

morphogenesis The origin of forms, especially landforms (GEOMORPHOLOGY). The concept of morphogenetic control is based on the belief that climate controls geomorphic processes, which in turn control the character of the landforms (CLIMATOMORPHOLOGY). *See also* MORPHOGENETIC REGIONS.

morphogenetic regions A concept developed by L. Peltier in 1950 in which it is claimed that under a certain set of climatic conditions particular geomorphic processes (GEOMORPHOLOGY) will predominate thereby giving different regional landscapes different characteristics. He postulated nine morphogenetic regions, defined in terms of temperature and moisture conditions and each one characterized by a dominant climatically controlled geomorphological process: **1** glacial; **2** periglacial; **3** boreal; **4** maritime; **5** selva; **6** moderate; **7** savanna; **8** semi-arid; **9** arid [*42*]. All but the boreal and maritime morphogenetic regions have been recognized to some extent. *See also* CLIMATOMORPHOLOGY.

morphographic map A map illustrating the terrain by pictorial symbols as if the landscape was being viewed obliquely from the air.

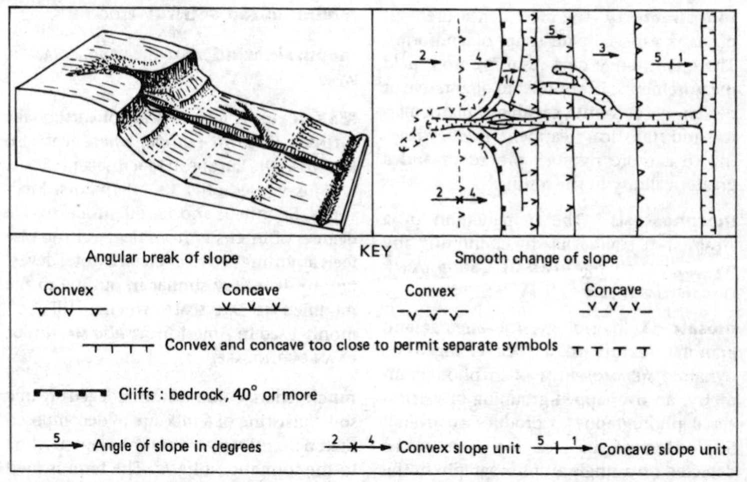

Figure 158 *Morphological map*

morphological map A map showing the detailed location of the MICRO-RELIEF of a small area by means of standardized symbols. Thus, breaks of slope and changes of slope are depicted by a type of FORM-LINE which also shows whether these variations are of a convex or a concave nature. Slope gradients may also be included but no attempt is made to include information on landform genesis (GEOMORPHOLOGICAL MAP). [*158*]

morphological region An area of land demarcated on the basis of its morphology, its structure and its geomorphological evolution. The following hierarchy (increasing in size and complexity) was suggested by D. L. Linton: **1** site; **2** stow; **3** tract; **4** section; **5** province; **6** continental subdivision. It should not be confused with a MORPHOGENETIC REGION.

morphological system A type of SYSTEM based on the existence of significant correlations between its morphological properties (MORPHOLOGY). [*249*] *See also* SYSTEMS ANALYSIS.

morphology **1** The scientific study of landforms. *See also* CLIMATOMORPHOLOGY, GEOMORPHOLOGY. **2** The scientific study of the structure of organisms.

morphometry The precise measurement of shape, especially the shape of landforms. The quantitative data resulting from field measurements, in contrast to qualitative or descriptive accounts, capable of mathematical and statistical treatment, thereby leading to a more rigorous procedure and a greater validity in the results.

morphostasis The perpetuation of a STEADY-STATE EQUILIBRIUM by countering any tendency to change in a SYSTEM. *See also* GENERAL SYSTEMS THEORY.

mosaic **1** In geology, the angular and granular texture of a rock changed by dynamic METAMORPHISM. **2** In photogrammetry, an overlapped grouping of vertical aerial photographs to produce an overall impression of an area too large to be depicted on a single air photograph. In the USA referred to as a *print lay-down*.

moss **1** A class of lower green plants of the phylum *bryophytes*. **2** A general term for marshy or boggy land. **3** In a stricter sense, a term for an accumulation of wet acid peaty vegetation (BOG), as used in the Pennines of N England, in which it corresponds to the German *moos*, the Danish and Norwegian *mose* and the Swedish *mosse*. **4** It is also applied to an accumulation of basic fen-peat in the Lancashire lowlands of NW England, e.g. Chat Moss, and to some areas of coastal marshes around the Solway Firth.

mother-of-pearl cloud NACREOUS CLOUD.

mottling Patches or spots of different colours. It is a term used to describe the patchiness of some soil colouring at depth, often due to the reduction of iron in an ANAEROBIC environment (GLEYING).

moulin GLACIER MILL.

mountain A portion of the land surface rising considerably above the surrounding country either as a single eminence or in a range or chain. Some authorities regard eminences above 600 m (2,000 ft) as mountains, those below being referred to as HILLS. *See also* SEAMOUNT.

mountain-building OROGENY.

mountain-top detritus BLOCKFIELD.

mountain wind ANABATIC WIND, KATABATIC WIND.

MSK scale A revised scale of earthquake INTENSITY, used in Europe, where it is also known as the European Macroseismic Scale. MSK is named after its originators, Medvedev, Sponheur and Karnik; it has twelve degrees of intensity from degree 1 (no one feels anything) up to degree 12 (total devastation). It is very similar in outline to the modified MERCALLI SCALE, which is still commonly used in America. *See also* MAGNITUDE OF AN EARTHQUAKE.

muck A dark-coloured waterlogged organic soil consisting of a mixture of decomposed material and MINERAL SOIL. A stage leading to the formation of PEAT. The term is used mainly in N America.

mud **1** A mixture of clay and/or silt with water to form a plastic mass with a particle size preponderantly below 0.06 mm diameter. It is deposited in low-energy environments in lakes, estuaries and lagoons. It may also be deposited in deep-sea environments. **2** An artificial slurry created for use in drilling boreholes.

mud-flat A tract of low-lying shoreline submerged at high tide and composed of silt or clay to varying depths. It often forms behind a BAR or SPIT and is carved into channels by tidal ebb and flow [*24*]. In the tropics it may carry a MANGROVE vegetation but in more temperate latitudes it supports a SALT-MARSH community. The mud of tidal flats is deposited by FLOCCULATION.

mud-flow A type of MASS-MOVEMENT on an unstable slope, similar to an EARTH-FLOW but of greater velocity because of the high percentage of water present in the mixture. It constitutes a considerable hazard in some parts of the world (LAHAR), but generally comes to a halt before moving very far down-valley. It has an abrupt, well-defined margin and decreases in thickness towards the tongue or 'toe'. It is usually poorly sorted owing to the presence of large boulders and many small pebbles in the clay/silt matrix, although some mud-flows exhibit rudimentary GRADED BEDDING. It is difficult to distinguish from a TILL when it has solidified. [*146*]

mudlump A small-scale (MESOSCOPIC) landform in a DELTA. It is probably formed as a result of LOADING of deltaic sandy sediments on top of an unstable plastic clay. This creates DIAPIRISM and the updoming of the underlying clay through the overlying sandy sediments to form mounds known as mudlumps. They are common in the Mississippi delta.

mud-pot A small pool of hot mud which occurs in areas of minor volcanic activity, e.g. Rotorua, New Zealand. It may simply bubble quietly or occasionally erupt as a GEYSER of sulphurous mud.

mudslide A LANDSLIDE of essentially homogeneous clay-silt material, but without the velocity or linear extent of a MUD-FLOW. *See also* SLUMP.

mudstone A term for an ARGILLACEOUS sedimentary rock, originally used by Sir R. Murchison to describe fine-grained shales in the SILURIAN rocks of Wales. Mudstone is now regarded as a rather imprecise term to cover both claystone and siltstone but it is not as fissile (FISSILITY) as shale.

mud volcano A small (< 50 m) conical mound which simulates a true volcano but from which mud is ejected rather than lava. It is a type of INJECTION STRUCTURE, often created when liquid mud is forced upwards through fissures during an earthquake or a volcanic eruption. The cone thus created is usually short-lived.

mulch A loose layer of organic or inorganic materials that forms naturally or is spread artificially on the ground surface to protect the soil and the plant roots from the effect of raindrops, evaporation and freezing.

mulga A scrubby vegetation type in central Australia, composed largely of the spiny acacia (*Acacia aneura*).

mull An organic-matter distribution (including HUMUS) forming a ground-surface layer in deciduous forests and lying directly on the mineral soil with which it becomes intermixed owing to the activity of the soil fauna (mostly earthworms). It is derived from leaf-mould which, in general, has a neutral pH, but it does not accumulate as a distinct matted layer as in MOR and MODER. It is typical of the BROWN EARTH type of soil.

multiband camera A REMOTE SENSING term referring to a camera that exposes different areas of one or more films either through a single lens and a beam splitter, or through two or more lenses equipped with different filters, to provide two or more photographs of the same area in different wavebands. This type of camera is used for taking an AERIAL PHOTOGRAPH because it allows objects with different REFLECTANCE properties to be distinguished from each other. *See also* INFRARED PHOTOGRAPHY.

multiple correlation systems CORRELATION, STATISTICAL.

multispectral A general REMOTE SENSING term referring to a sensor that records IMAGES of the Earth's surface in a number of wavebands. For example, the MULTISPECTRAL SCANNER aboard the LANDSAT SATELLITE records in green, red and infrared wavebands to enable objects with different REFLECTANCE properties to be distinguished. *See also* ELECTROMAGNETIC RADIATION.

multispectral scanner (MSS) A REMOTE SENSING term referring to a scanning RADIOMETER that simultaneously acquires images in various wavebands at the same time. A multispectral scanner can be carried aboard an aircraft or satellite. The LANDSAT multispectral scanner records images in four wavebands of visible and near INFRARED ELECTROMAGNETIC RADIATION to enable objects with different REFLECTANCE properties to be distinguished.

multivariate analysis The study of the relationships between a number of variables greater than two. The data can be analysed by multiple CORRELATION or by multiple REGRESSION. *See also* FACTOR ANALYSIS.

Munsell colour system A colour code used in pedology to describe a soil's characteristic *hue* (spectral colour), *value* (departure from absolute whiteness) and *chroma* (departure from greyness) as the basis of a colour matching system. A Munsell colour notation, for example, of 5 YR 5/6 denotes a hue/value/chroma combination which describes a yellowish-red soil.

Munsterian The name given to the penultimate glacial stage of the PLEISTOCENE in Ireland. It is equivalent in age to the WOLSTONIAN of Britain and was preceded by the Gortian interglacial (= Hoxnian of Britain). [*109*]

murram A term used for crushed lateritic material (LATERITE) used extensively for road surfacing in tropical Africa. It is a mixture of reddish clay and ironstone nodules.

Muschelkalk A German term for the middle series of the TRIASSIC in western Europe. It consists largely of fossiliferous limestones.

muscovite MICA.

muskeg A Canadian Indian term for a sphagnum moss-covered MUCK soil or BOG in the sub-Arctic latitudes of Canada and Alaska. It is a region of marshy depressions, stagnant mosquito-infested pools and slow, meandering rivers.

mutualism An association between several species in which all derive benefit and none can normally survive without the other species.

mylonite A fine-grained type of BRECCIA formed by the pulverizing and milling of rocks along a FAULT. If the shearing and dislocation is prolonged a foliation may develop. *See also* CATACLASIS, CRUSH-BRECCIA.

N

nacreous cloud A rare cloud-form, resembling mother-of-pearl because of its iridescence, which forms in the STRATOSPHERE, at about 20–25 km elevation, and is visible in high latitudes during low-pressure weather conditions. It is particularly noticeable at night when illuminated against a dark sky by the sun shining from below the horizon. Nacreous clouds are thought to result from a wave motion over a topographic barrier (LEE WAVE) being transmitted through the TROPOSPHERE and into the stratosphere during a period of consistent wind at all heights within the troposphere. *See also* NOCTILUCENT CLOUD.

nadir **1** In general usage, a term referring to the lowest point. **2** The point opposite to the ZENITH on the CELESTIAL SPHERE.

Nahangan stadial A cold phase (10,500–10,000 BP) of the late PLEISTOCENE in Ireland. Named from Nahangan corrie in Co. Wicklow.

nailbourne BOURNE.

Namurian The stratigraphic name for the lower stage of the Upper CARBONIFEROUS in Europe. In Britain it includes the MILLSTONE GRIT, but it also comprises shales, mudstones, sandstones and thin coal seams.

nanism A term describing dwarfism in animals and plants.

nappe **1** A large-scale tectonic overfold in the Earth's crustal rocks, which has moved forward as a RECUMBENT FOLD sometimes for tens of kilometres along a thrust plane (THRUST-FAULT), thereby detaching itself from the so-called zone of 'roots'. The tangential pressures are produced during a mountain-building episode (OROGENY) when the sediments within the former geosyncline are considerably compressed by converging plate margins (PLATE TECTONICS), leading to crustal shortening by nappe formation. Nappes were first described in the Alps

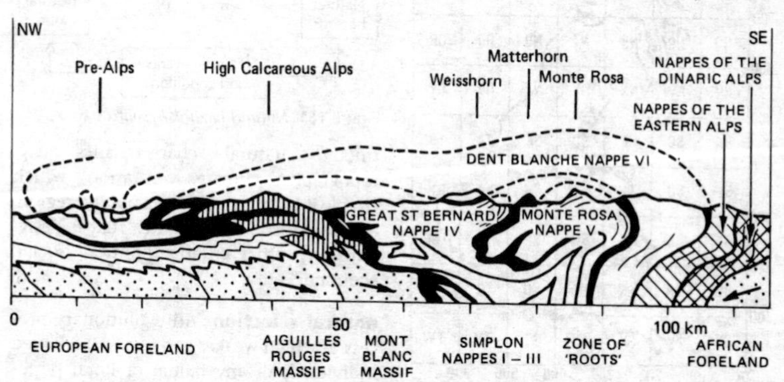

Figure 159 *Tectonic section through the Western Alps*

where a distinction is made between the AUTOCHTHONOUS nappe, which can still be traced to its root zone, and the PARAUTOCHTHONOUS nappe, which has been moved some distance. [*159*] **2** In hydraulics, the term refers to a sheet of water flowing over the crest of a weir or dam. **3** In some countries the term is used to describe an overlying sheet of rocks, especially a basaltic lava flow.

narrows A term used as a place-name to denote a constriction in a waterway or valley.

National Grid The metric GRID currently used by the British ORDNANCE SURVEY in the production of its maps. It consists of 500-km squares (each designated by a letter) divided into 100-km squares (each designated by a second letter) which are themselves subdivided into 1-km squares. It is based on the TRANSVERSE MERCATOR PROJECTION, with the axes at 2°W and 49°N intersecting at the true origin (ORIGIN), while to avoid the use of negative grid references the latter has been transferred to a FALSE ORIGIN. Any point within the UK can be given its own unique *grid reference* by using intersecting COORDINATES. [*160*]

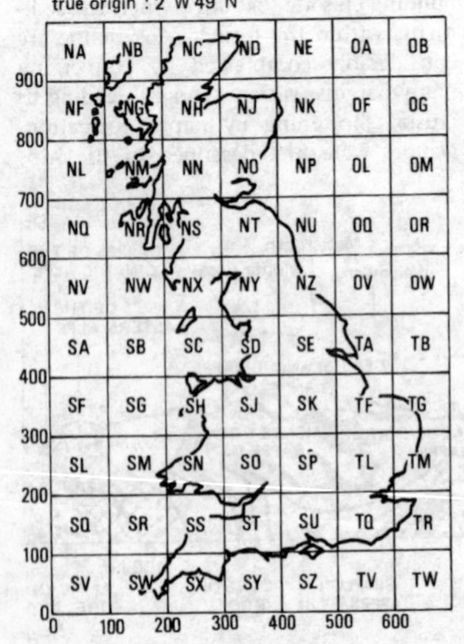

Figure 160 *National Grid*

natural bridge 1 A natural rock exposure that has been converted into a bridge spanning an erosional valley, perhaps owing to collapse of adjacent rock strata up and downstream. Thus it is not uncommon in some KARST regions where subterranean drainage is revealed with some bridges still intact. **2** The term is sometimes used as a synonym for a coastal arch created by marine erosion, e.g. the Green Bridge, Pembrokeshire, S. Wales. *See also* ARCH.

natural-event system A SYSTEM which incorporates the magnitude, frequency, duration and temporal spacing of a natural event.

natural gas A mixture of HYDROCARBON gases occurring in subterranean rock reservoirs, often in association with PETROLEUM deposits. It is composed (by weight) of METHANE (84.6%), ethane (6.4%), propane (5.3%) and several other minor hydrocarbons. The reservoirs are not hollow caverns but solid rocks with pore spaces sufficiently large to accommodate the accumulation of the gas.

natural hazards HAZARD, NATURAL.

natural region An area of the Earth's surface containing relatively uniform and dis-

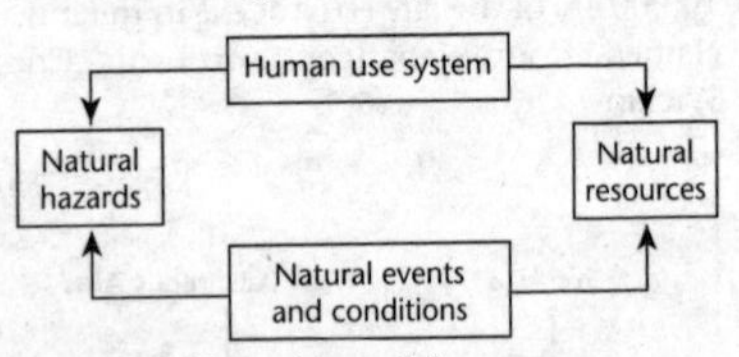

Figure 161 *Natural hazards/resources*

tinctive natural characteristics (e.g. topography, climate, vegetation) which enable it to be classified as a natural region, by comparison with an area that is designated by other criteria (political, social, economic, etc.).

natural selection An evolutionary process in which weaker and less well-adapted individuals of any faunal or floral population tend to be eliminated in competition with more vigorous and better-adapted individuals. Thus, the inferior specimens will not be perpetuated. The idea was proposed by C. Darwin and A. R. Wallace in 1858 in order to account for the origin and diversity of organisms. *See also* CATASTROPHISM, EVOLUTION, DARWINISM.

natural-siphon rainfall recorder A self-recording rain-gauge (HYETOGRAPH) which records both duration and intensity of a rain event. The rainwater enters a float chamber via a funnel, causing the float to rise as the water-level increases. The change in float-level is recorded by a pen on a calibrated clock-driven chart. When full, the chamber is emptied by an automatic siphoning device but it is noteworthy that during periods of torrential rainfall the 12-second interval of siphoning time may fail to record some of the falling rain. The data obtained is of value in FLOOD forecasting. *See also* TIPPING-BUCKET RAIN-GAUGE.

natural vegetation **1** Used loosely to describe any plant life that is not organized or influenced by mankind, in contrast to CULTURAL VEGETATION. **2** More strictly, the term refers to the world's major vegetation climaxes (CLIMAX). *See also* SEMI-NATURAL VEGETATION.

nautical mile MILE.

neap tide A tide occurring near the lunar QUADRATURE during which the tidal-producing forces do not supplement each other, thereby causing relatively small tidal ranges. Thus high tides are somewhat lower and low tides are somewhat higher (by some 10–30%) than average, all of which slows down the velocity of TIDAL CURRENTS at this time (about every 14.75 days). *See also* SPRING TIDE.

nearest-neighbour index A statistical method, originally devised by botanists to describe plant distributions, now widely used by geographers as a test for non-randomness of any spatial data distribution, in addition to allowing, on a continuous scale, comparisons to be made between two or more spatial distributions. The index ranges from 0 (when all points are closely clustered) through 1 (a random distribution of points) to 2.15 (indicating that all points are uniformly distributed throughout the area, and therefore as far away as possible from each other). The index (R) is expressed as $R = \bar{D}obs/\bar{D}ran$, where $\bar{D}obs$ is the measured mean-distance between nearest-neighbour points in a given area, and $\bar{D}ran$ is the mean distance to be expected from a similar number of points randomly distributed in the same area. *See also* RANDOM POINT DISTRIBUTION.

nearshore current system A circulatory system of marine currents in the vicinity of the BREAKER zone along a shoreline, and which are induced directly or indirectly by wave activity, as compared with TIDAL CURRENTS, local wind-induced currents and DENSITY CURRENTS. The system includes LONGSHORE CURRENTS, MASS TRANSPORT currents and RIP CURRENTS.

neat line **1** The boundary of a map, usually coincident with a GRID line. **2** In civil engineering, the line defining the sides of an excavation from within which material is removed.

nebka (nebkha) An Arabic term for a small arrow-shaped hummock of sand which has collected leeward of a rocky obstacle or clump of vegetation. It is generally less than 1 m high and 1.5 m long.

Nebraskan One of the earliest of the N American glacial stages (PLEISTOCENE). It is older than the KANSAN, from which it is separated by the *Aftonian* interglacial. It is thought to be approximately equivalent in age to the pre-ELSTER stage of N Europe and to the GÜNZ of central Europe. The Nebraskan ice-sheet reached as far south as the present Mississippi–Missouri junction.

nebular hypothesis A hypothesis suggesting that the matter which forms the Sun and the planets originated as a disc-shaped cloud of gas or *nebula* which eventually contracted into discrete bodies. It was proposed in 1796 by the Marquis de Laplace.

neck **1** An eroded remnant of solidified lava and/or AGGLOMERATE which formerly filled the pipe of a VOLCANO but which has now been exposed by denudation of the surrounding CONE. **2** A narrow strip of land similar to an ISTHMUS or TOMBOLO.

Needian interglacial A stratigraphical stage name for the HOLSTEIN interglacial of the Pleistocene. It separates the ELSTER from the SAALE glacials in the N European chronology.

needle A pinnacle or pointed mass of rock detached from a sea cliff or a mountain rock face, e.g. the chalk sea stacks of The Needles, Isle of Wight, and Napes Needle, Great Gable, Cumbria, England. In the Alps the Chamonix needles (French *aiguilles*) are celebrated rock-climbing phenomena.

needle ice An accumulation of bristle-like ice crystals at, or immediately below, the ground surface, ranging in length from a few mm to over 30 cm, depending on soil moisture and temperature conditions. The needles grow parallel to the soil's temperature gradient, i.e. perpendicular to the cooling surface, with several layers sometimes occurring, each separated by a thin veneer of mineral soil. Needle ice assists in the process of FROST-HEAVING and in the formation of some types of PATTERNED GROUND, especially small-scale forms such as NUBBINS. Synonyms: *kammeis, pipkrake*.

negative feedback FEEDBACK.

negentropy A synonym for negative entropy (ENTROPY), used in cybernetics as a synonym for 'information'.

negro head A term for broken blocks of CORAL, subsequently darkened by lichen growth, detached by wave action from the outer reef (CORAL REEF) and left on the surface of the reef. It has been suggested by some writers that they may have been torn off by seismically generated sea waves (TSUNAMI).

nehrung A German term for a lengthy coastal stand SPIT created by LONGSHORE DRIFT on the southern shores of the Baltic Sea. The coastal water bodies which the nehrungs almost cut off from the open sea are termed HAFFS.

nekton The actively swimming marine organisms near the ocean surface which have the ability to move in opposition to tides and currents, in contrast to those organisms which float and are powerless to direct their own movements. *See also* BENTHOS, PLANKTON.

neo-catastrophism CATASTROPHISM.

Neogene The later of the two stratigraphical divisions of the CAINOZOIC and now generally accepted as including the epochs of the MIOCENE, PLIOCENE, and PLEISTOCENE. Some geologists, however, continue to use the term to refer only to the Miocene and Pliocene rocks (i.e. later TERTIARY) as distinct from the PALAEOGENE (the earlier division of the Cainozoic), thereby excluding the Pleistocene from the Neogene. It is generally regarded as having spanned the time period between 23.3 and 1.64 million years.

neoglacial A term used increasingly in N America to describe the increased glacial activity in Alaska and in the western cordillera of Canada and USA following the Climatic Optimum (HYPSITHERMAL). In Europe this post-glacial waxing of the glaciers has been termed the LITTLE ICE AGE.

Neolithic An archaeological term adapted from the Greek, meaning '*new stone*' *age*, and used to describe the culture achieved in the middle POST-GLACIAL when mankind had extended the ability to make stone tools from percussion and flaking (PALAEOLITHIC and MESOLITHIC) to grinding and polishing. In addition, the Neolithic saw the introduction of agriculture (sometimes referred to as the Agricultural Revolution) by crop cultivation and domestication of animals. In Britain the culture flourished around the period known as the *Climatic Optimum* (MEDIOCRATIC) during that part of the FLANDRIAN which spanned part of pollen zones VIIA and VIIB (POLLEN ANALYSIS) from approximately 5,200 BP until the advent of the BRONZE AGE in 4,000 BP. Although it succeeded the Mesolithic, some communities still continued to practise this earlier culture in the time period officially designated as Neolithic. Forest clearance became more widespread as farming activities developed

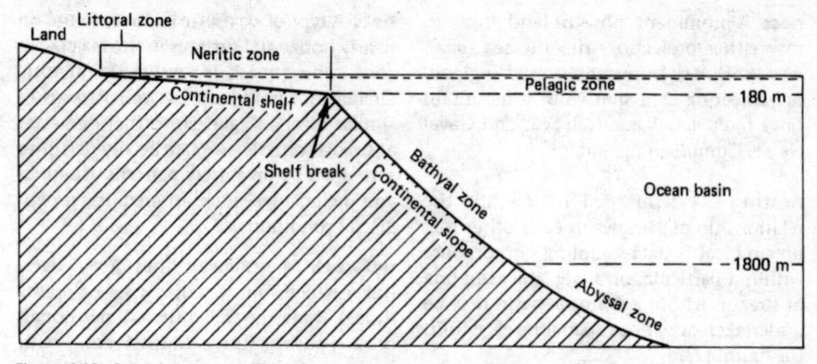

Figure 162 *Oceanic zones*

with the selective felling of the elm (perhaps for the use of the leaves as cattle fodder), at a time when Britain's grasslands were not as extensive as today, but possibly because the elm favoured the more fertile soils now required for cultivation. The chalklands of Britain became heavily colonized by Neolithic farmers, especially in Wiltshire, where early megalithic monuments (e.g. the earliest part of Stonehenge) and long barrows abound. The first British evidence of ploughing occurs near Avebury, and with the increasing demand for high-quality flint tools, flint-mining appeared, e.g. at Grimes Graves, Norfolk, while stone-axe factories utilized hard igneous and metamorphic rock exposures in Wales, Cumbria, Scotland and N Ireland. *See also* ATLANTIC PERIOD, SUB-BOREAL PERIOD.

neotectonics The study relating to the crustal movements (TECTOGENESIS) during the NEOGENE and their effect on the processes and landforms. *See also* ANTECEDENT DRAINAGE, CYMATOGENY, EPEIROGENESIS, WARPING.

nephanalysis The analysis of cloud patterns depicted on weather-satellite (SATELLITE, ARTIFICIAL) photographs prior to the data being graphically displayed on a cloud chart. Because of the much greater coverage obtained by satellite photography than by the ground observer it is now possible to decipher many more cloud characteristics for use in weather forecasting.

nephoscope An instrument used by meteorologists for measuring direction and speed of cloud movements.

neptunean (neptunian) dyke An approximately vertical sheet of sediment, usually sandstone or breccia, that cuts through a contrasting rock type. It is thought to have been formed by the infilling of a fissure in the existing rock body by a sediment, either on the sea-floor or on the land surface. It should be distinguished from an igneous DYKE and from sedimentary dykes injected from below (INJECTION STRUCTURES).

Neptunist A devotee of the Neptunean (Neptunian) theory, expounded by A. G. Werner in the late 18th cent., which proposed that (in contrast to the theory of Plutonism – PLUTONIST) all rocks were formed by water or in aqueous environments.

neritic zone That part of the ocean which includes both the estuarine zone and the CONTINENTAL SHELF. Thus, neritic deposits are of shallow-water origin, occurring below low-tide level down to depths of about 200 m, and are composed of a mixture of terrestrially derived sediments and some calcareous organic remains. The neritic zone coves about 10% of the total ocean floor and is affected by TIDAL CURRENTS and in part by the NEARSHORE CURRENT SYSTEM. It is part of the PELAGIC division of the oceans and, in the strictest sense, lies below the LITTORAL ZONE. [*162*]

ness A prominent 'nose' or land promontory either projecting into the sea (*naze, naes, nab*), e.g. Dungeness, Kent, England, or projecting as a spur from a mountain ridge (*nab, neese*), e.g. Nab Scar and Gavel Neese, Cumbria, England.

nesting A term used to describe the relationship of samples to each other in a hierarchical data-sampling procedure within a particular area, e.g. the sampling of streams within a drainage basin may be undertaken at various spatial levels within the basin. [*218*]

net primary productivity (NPP) The rate at which plants store energy as organic matter in excess of that used in respiration. It is usually expressed as: dry weight of organic matter/unit area/unit time (kg/ha/year). *See also* BIOLOGICAL PRODUCTIVITY.

net radiation The net exchange between all wave outgoing and incoming radiation. It is expressed by the formula: net radiation = $(Q + q)(1 - \propto) + I\downarrow - I\uparrow$, where: Q is the direct SOLAR RADIATION; q is the DIFFUSE RADIATION of indirect solar radiation; $(1 - \propto)$ is the fraction of the shortwave radiation reflected by surface ALBEDO; $I\downarrow$ is the infrared long-wave radiation, counter radiation from the GREENHOUSE EFFECT; and $I\uparrow$ is the outgoing TERRESTRIAL RADIATION (long-wave radiation) lost to space via the ATMOSPHERIC WINDOW. *See also* HEAT BALANCE.

net radiometer A meteorological instrument which measures NET RADIATION by upward and downward-facing blackened sensors (*thermopiles*) which absorb all wave RADIATION, thus raising their temperatures. The temperature difference will indicate the *radiation balance* in negative or positive terms. The sensors, which are enclosed in thin polyethylene spheres (inflated with dry air) to eliminate the disturbing influence of SENSIBLE HEAT FLUX and CONDENSATION, are transparent to both shortwave and long-wave radiation (between 0.3–100 μm). The voltage output is continuously monitored on volt-time integrators and printers. *See also* RADIOMETER.

nets A type of PATTERNED GROUND, found on nearly horizontal surfaces in the PERIGLACIAL zone, whose mesh is neither dominantly circular nor polygonal but otherwise is of similar form and genesis to that of CIRCLES and POLYGONS. These can be divided into sorted and non-sorted patterns (SORTING) and include HUMMOCKS formed under periglacial conditions. [*179*]

network A system of lines (arcs, edges, links) which join together a set of points (NODE). It may be *planar* (two-dimensional) or *non-planar* (three-dimensional), with most physical-geography networks being of the former variety. The technique which plans a number of steps or operations within a SYSTEM is known as *network analysis*.

neutral shoreline A term devised by D. W. Johnson in 1919 to describe coastlines whose essential characteristics depend on causes independent of either submergence or emergence (EMERGENT SHORELINE, SUBMERGENT SHORELINE). Typical examples are shorelines of coral reefs, dunes, deltas, marshes, mud-flats, sandur, volcanoes and those produced by faulting. *See also* SHORELINE CLASSIFICATION.

neutral soil A soil with a PH value of around 7.0.

neutral stability A term used by meteorologists to describe the state of an unsaturated column of air when its ENVIRONMENTAL LAPSE-RATE equals the DRY ADIABATIC LAPSE-RATE, or when the environmental lapse-rate of a *saturated* column of air equals the SATURATED ADIABATIC LAPSE-RATE. It is sometimes called *indifferent equilibrium*, which illustrates the fact that during a state of neutral stability a given parcel of air will move neither up nor down but will be in equilibrium with its surroundings.

neutron probe An instrument for measuring SOIL WATER content.

neutrophyte An organism adapted to life in a near-neutral medium.

Nevadian The mountain-building episode (DIASTROPHISM) of late Jurassic/early Cretaceous times in western N America. The late Palaeozoic and Mesozoic sediments of the EUGEOSYNCLINE were uplifted, folded, faulted and invaded by gigantic granitic plutons during the formation of the Sierra Nevada, the Canadian Coast Ranges, the Klamath Mountains, the Transverse Ranges of Nevada and Arizona and Baja California. It pre-dated the LARAMIDE OROGENY, which formed the Rockies, and its structures were later to be broken up by wrench-faulting, block-faulting (e.g. BASIN-AND-RANGE TERRAIN) and later uplift of Cainozoic age.

nevados The name given to a type of KATABATIC WIND, characterized by its extreme coldness, blowing from the Andean snow peaks down into the valleys of Ecuador, S America.

névé A French term, widely adopted to describe freshly deposited snow (density range 0.4 to 0.89 kg/m^3) in the accumulation area of a glacier during the accumulation season of a MASS BALANCE year. Most glaciologists regard névé as synonymous with FIRN. *See* FIRNIFICATION.

New American Comprehensive Soil Classification SEVENTH APPROXIMATION.

Newbournian A stratigraphic stage name for the base of the British Pleistocene (RED CRAG SERIES). The name has now been abandoned and its continued use is to be discouraged.

Newer Drift A term originally used by the Geological Survey of Great Britain to describe the younger glacigenic deposits of the PLEISTOCENE, equivalent in age to the drift of DEVENSIAN times. The landforms of the Newer Drift have a relatively fresh appearance, steeper slope gradients, and are less deeply weathered than those of OLDER DRIFT age.

Newlyn datum The DATUM LEVEL currently used by the ORDNANCE SURVEY in Britain. *See also* MEAN SEA-LEVEL.

New Red Sandstone The European name given to the red sediments, mainly but not exclusively sandstones, occurring above the CARBONIFEROUS and below the JURASSIC. It includes both the PERMIAN and the TRIASSIC, so that the boundary between the Palaeozoic and the Mesozoic eras is drawn midway through the New Red Sandstone.

newton (N) The unit of force in the SI UNITS system. It is the force required to accelerate a mass of 1 kg by 1 m per second per second. It is equivalent to 10^5 dynes or 100 grams weight. It is named after Sir Isaac Newton and its use was officially authorized in 1938. One newton per m^2 (N m^{-2}) is now internationally recognized as the basic SI unit of pressure (PASCAL) equal to 1.4504×10^{-4} pounds per sq in.

niche **1** A term used in ECOLOGY to describe the optimum HABITAT in an environment for a specific organism. **2** A recess or small shelf in a rock face.

niche glacier A term used to describe a small glacier occurring in a small hollow or gully on a steeply sloping mountain wall. *See also* HANGING GLACIER.

nickpoint The US spelling of KNICKPOINT.

nife A mnemonic term for the material forming the CORE of the Earth. It is based on the symbols for nickel–iron (Ni–Fe).

nightglow A very faint radiance in the night sky resulting from radiation emitted by the ionized products of ultraviolet waves during the day.

nimbostratus A type of cloud (symbol: Ns), dark grey in colour, occurring in sheets thick enough to blot out the Sun. Low ragged FRACTUS clouds frequently occur beneath its base, which occurs between 900 and 3,000 m (3,000–10,000 ft), i.e. at low altitudes. It is associated with more or less continuously falling rain or snow. *See also* CLOUD. [*43*]

nimbus **1** A term for any CLOUD from which rain is falling. *See also* CUMULONIMBUS, NIMBOSTRATUS. **2** The name given to a weather SATELLITE launched in 1964 (it had a life of only twenty-six days).

nip A notch or low cliff cut at the edge of a coastal plain during the initial stage of wave attack on an EMERGENT SHORELINE.

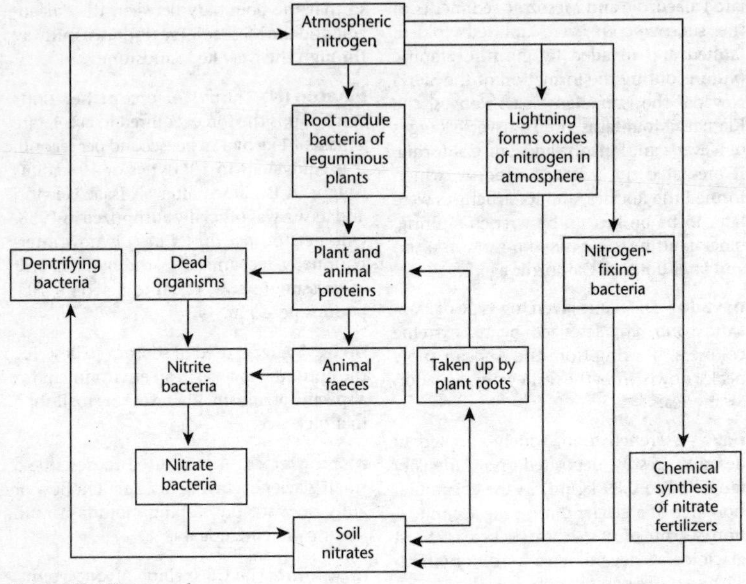

Figure 163 *Nitrogen cycle*

nitrification The biochemical oxidation of ammonium to nitrite and thence to nitrate by the action of BACTERIA, thereby making it available in a form usable by plants. When soils become acid and poorly aerated (ANAEROBIC), nitrates may be converted back to unusable nitrogen gas, a process termed *denitrification*.

nitrogen The most abundant gas of the ATMOSPHERE and a very important constituent of soil, from which it is utilized by certain bacteria (NITROGEN FIXATION). *See also* NITRIFICATION, NITROGEN CYCLE.

nitrogen cycle The sequence of biochemical changes by which nitrogen is taken from the atmosphere into the soil by nitrogen-fixing bacteria (NITROGEN FIXATION), making it readily available for assimilation by living organisms which at the end of their life-span die and are decomposed, thereby returning to the atmosphere. [*163*]

nitrogen fixation The conversion of elemental nitrogen to organic combinations or to forms which can be easily utilized in biological process. It is performed by certain *nitrogen-fixing* BACTERIA.

nival A term referring to any environment of snowy character.

nival karst KARST.

nivation A term introduced in 1900 by F. E. Matthes to describe the localized erosion of a hill slope by frost-action, mass-wasting, and erosion by meltwaters at the edges of, and beneath, melting snowdrifts. The degree of nivation depends in part on the presence or absence of PERMAFROST in the underlying material, for in permafrost-free areas snow-patch erosion can continue in the snow-free areas as well as at the feather edges of the snowdrift. In addition to the sheet-flow and rillwork of the meltwater, mass-wasting is also accomplished by chemical weathering. The combination of processes causes lingering snowdrifts to countersink into hillsides, thereby producing *nivation hollows*, *nivation benches* or terraces (CRYOPLANATION), leading in some instances to the formation of so-called NIVATION CIRQUES. [*196*]

nivation beach (terrace) NIVATION.

nivation cirque A large NIVATION hollow, approaching the scale of a true CIRQUE and possessing its semi-circular plan, but lacking its basin- or armchair-shaped form or its rocky 'threshold'. It may be associated with a PROTALUS RAMPART at the front of the lingering snowdrift. It should not be confused with the landform produced by a landslide. Many authorities prefer to use the term *nivation hollow*, reserving the term cirque for the stage at which glacier ice replaces the snow patch. [*196*]

niveo-aeolian A term first used in 1971 to describe mixed deposits of snow and wind-driven sand and other detritus in addition to the forms and microforms generated by them. The microforms are seasonal features which rarely persist, but the larger forms of stratified snow and sand-dunes are common in subpolar environments and persist in some polar environments.

niveo-fluvial A term introduced in 1964 to describe a combination of PERIGLACIAL processes that are thought to have been responsible for the formation of COMBES in a chalk escarpment in S England, by means of selective erosion. They include SOLIFLUCTION, MASS-MOVEMENT and NIVATION.

NOAA/TIROS satellites A number of REMOTE SENSING SATELLITES designed for meteorological applications and operated by the National Oceanic and Atmospheric Administration (NOAA) of the USA. They obtain low RESOLUTION IMAGES twice daily by means of a MULTISPECTRAL SCANNER operating in visible (0.6–0.7 μm) and near INFRARED (10.5–12.5 μm) wavebands. Images from this series of satellites have proved useful for synoptic and dynamic climatological and meteorological research and for weather forecasting. *See also* HEAT-CAPACITY MAPPING MISSION, LANDSAT, SEASAT I, SKYLAB, SPACE SHUTTLE, SPOT.

noctilucent cloud A rare cloud-form of bluish or silver colour and of CIRROSTRATUS shape which, unlike the latter type however, forms at very high altitudes (50–80 km) in the STRATOSPHERE and at latitudes greater than 50°. It is thought to consist of either ice crystals or METEORIC DUST. *See also* NACREOUS CLOUD.

nodal point NODE.

node **1** A point or dot of no dimensions, representing the position at which two or more lines of a NETWORK meet. **2** A point, line or surface in a vibrating medium at which a stationary wave is produced owing to the amplitude of the vibration being reduced to zero by the interference of wave trains operating in opposing directions. This is the usage which describes the nodal point of a tidal unit (AMPHIDROMIC SYSTEM). **3** In botany, a point on a plant stem from which a bud grows.

nodule A small, rounded CONCRETION harder than the sedimentary rock in which it is contained.

noise **1** A term used in statistics to explain VARIANCE after the effects of the known variables have been accounted for. Any unexplained variance exhibiting no discernible pattern in terms of any other variable, or in space and time, is referred to as *random noise*. Unexplained spatial variance is termed *spatial noise*. **2** A term used in geomagnetic surveying and geophysical prospecting to describe the more or less random and extraneous data obtained.

noise-rating number An index of noise level used in *environmental impact analysis*. It is an arbitrary number based on the fact that under different conditions different amounts of noise can be tolerated, e.g. noise tolerated in an office. It varies with frequency but at 1,000 hertz it is equal to the noise level expressed in DECIBELS with respect to a noise level of 2×10^{-5} Nm^{-2} (the smallest pressure difference detectable by the average human ear). The noise-rating number of a workshop is 65 Db and of a bedroom 25 Db.

nominal scale SCALES OF MEASUREMENT.

nonconformity A type of UNCONFORMITY in which the underlying older rocks are of

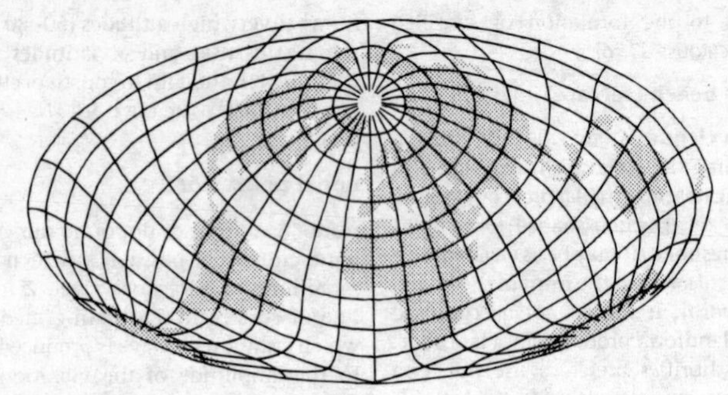

Figure 164 *Nordic projection*

igneous or metamorphic rather than sedimentary origin. [*263*]

non-dipole field The small part of the Earth's MAGNETIC FIELD which causes its irregularities. Because its distribution differs in each hemisphere the Earth's magnetic dip poles are not antipodal. *See also* DIPOLE FIELD.

non-linear function A function whose variables do not increase in a linear fashion.

non-linear programming A type of PROGRAMMING involving the use of an optimizing procedure which enables the maximum and minimum values of a combination of non-linearly related variables to be identified subject to a number of constraints on the values which they may take. *See also* LINEAR PROGRAMMING.

non-linear regression Any type of REGRESSION equation in which the exponent of the independent variables is greater than unity. The associated regression curves are usually sketched in by eye to give a much better idea of how the variables are related than would a more precisely constructed least-squared straight regression line (linear regression) in addition to giving a better prediction.

non-parametric statistical tests Statistics used in distribution-free tests which make no assumptions about the distribution of the background data population. The non-parametric tests become more important when the samples being tested are small. They can be applied to data on nominal and ordinal scales only (SCALES OF MEASUREMENT) and include the following: CHI-SQUARE TEST, MANN–WHITNEY U-TEST, WILCOXON TEST. *See also* PARAMETRIC STATISTICAL TESTS.

non-relict slopes Slopes whose form is related entirely to processes that are similar to those of today, implying that there has not been a CLIMATIC CHANGE over a sufficiently long period to obliterate any earlier forms. A slope with TIME-INDEPENDENT FORM is non-relict, but the reverse is not true. *See also* SLOPE EQUILIBRIUM, CONCEPTS OF.

non-sequence A break in a conformable sequence of rocks caused by a period of non-deposition.

Nordic projection A MAP PROJECTION designed to give optimum representation to Europe and to routes across the Atlantic, Arctic and Indian oceans. It is an oblique, equal-area projection with an origin on the GREENWICH MERIDIAN at 45°N latitude, and with the South Pole shown twice on the bounding ellipse. It was introduced in 1951 by J. Bartholomew. [*164*]

norite A variety of GABBRO.

normal **1** A term used in climatology to denote the average value of any climatological element over a given period of time, usually thirty years (MEAN). **2** A term used in statistics to denote a data distribution that corresponds to an accepted frequency distribution about a population mean. *See* *also* MEAN, NORMAL DISTRIBUTION. **3** A general term meaning at right angles to or perpendicular to (ORTHOGONAL).

normal cure NORMAL DISTRIBUTION.

normal cycle of erosion A synonym for the FLUVIAL CYCLE, the DAVISIAN CYCLE and the CYCLE OF EROSION by rivers in mid-latitudes (in contrast to the ARID CYCLE OF EROSION and the GLACIAL CYCLE OF EROSION). It was an expression introduced by W. M. Davis but is currently being little used, since it was realized that mid-latitude processes were not typical or normal. [*53*]

normal distribution A statistical term referring to a data frequency distribution which, when plotted, shows a symmetrical grouping about the arithmetic mean (DISTRIBUTION) in the form of a GAUSSIAN CURVE. The equation of the curve is expressed as

$$y = \frac{N}{\sigma\sqrt{2\pi}} e - (x - m)^2/2\sigma^2$$

where m = mean; σ = standard deviation; and y is the frequency with which any x occurs. *See also* FREQUENCY DISTRIBUTION, POISSON DISTRIBUTION. [*165*]

normal fault A gravity fault at which the HANGING WALL has been depressed relative to the FOOT WALL. *See also* REVERSE-FAULT, THRUST-FAULT. [*76*]

normal solution A term used in volumetric analysis of liquids. It contains one gram equivalent of replaceable hydrogen per litre of solution, e.g. a normal solution of hydrochloric acid (HCl) contains 1 gram molecular weight of HCl per litre of solution, while a normal solution of sulphuric acid (H_2SO_4) contains half a gram molecular weight of H_2SO_4 per litre of solution.

normal stress The STRESS applied to the surface of an object, which produces a STRAIN or deformation. It is perpendicular to the shear stress.

north GRID NORTH, MAGNETIC NORTH, TRUE NORTH.

North Atlantic Drift The name given to the north-eastward flowing extension of the GULF STREAM which swings away from the eastern coast of the USA and flows across the Atlantic ocean as a number of shifting bands of water which bifurcate as they progress eastwards. The southernmost branch runs eastwards to link with the cool Canaries current, while the more important northern branches push to the NE sectors of the North Atlantic bringing warmer waters to the British Isles and Norway, thereby ameliorating their climates. [*168*]

North-East Trades TRADE WINDS.

norther (norte) A cold wave of air sometimes experienced in the southern USA, when temperatures may fall some 17° to 22° C (30°–40° F) in 24 hours. It is caused by a sudden southern penetration of polar

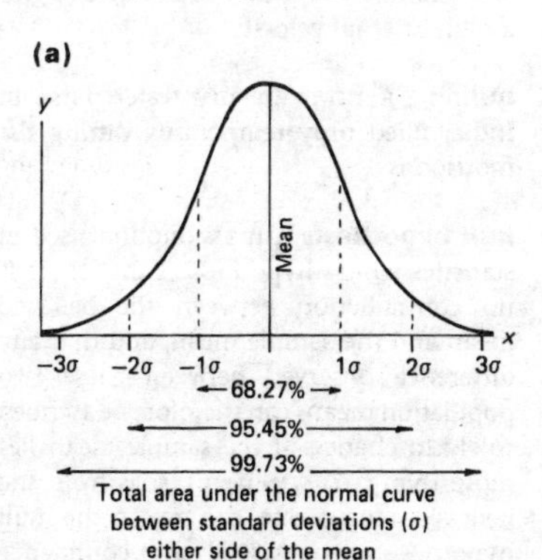

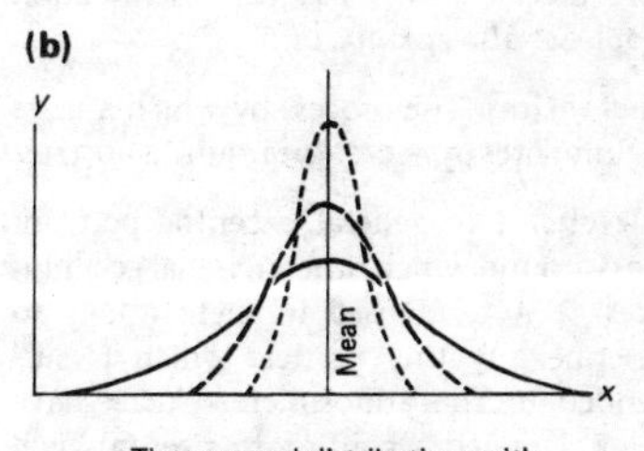

Figure 165 *Normal distribution*

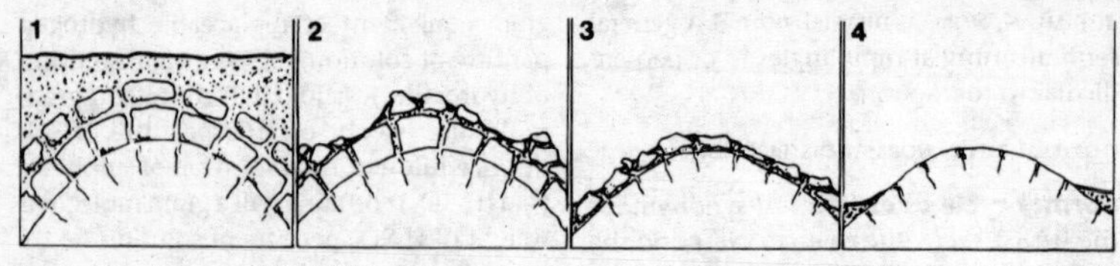

Figure 166 *Formation of a nubbin*

air at the rear of a depression. It can be a serious hazard to fruit crops. *See also* FRIAGEM, POLAR OUTBREAK, SOUTHERLY BURSTER.

northing Any one of the east–west GRID lines on a map, measured from the grid's origin in a northerly direction. It forms the second half of a grid reference (after the EASTING) when coordinates are being quoted.

nor'-wester A strong wind blowing from the New Zealand Alps, across South Island, with certain FÖHN-like characteristics.

Norwich Crag ICENIAN.

notch **1** A wave-cut recess at the base of a marine cliff. **2** A name given to a break in a line of hills, e.g. in the Appalachians of the USA.

nubbin **1** A small rounded or elongated lump of earth (a few centimetres in diameter) thought to be created by NEEDLE ICE action at the ground surface in PERIGLACIAL climates. **2** A small granite dome, created by subsurface weathering of joints in the outer rock 'shells', with its unweathered core revealed by subsequent denudation [*166*]. *See also* BORNHARDT.

nucleation The process by which a mass accumulates by ACCRETION around a NUCLEUS.

nucleus **1** In general, a central point or mass around which other material accumulates. **2** A term used in meteorology to describe any tiny particle which is suspended in the atmosphere. Nuclei have been classified according to size: (a) *giant nuclei* – radius greater than 1×10^{-4} cm; (b) *large nuclei* – radius between 0.2 and 1×10^{-4} cm; (c) *minute* (*Aitken*) *nuclei* – radius smaller than 0.2×10^{-4} cm. The Aitken nuclei are most abundant (average 10^6 m^{-3}), but are more numerous over oceans and industrial areas (AEROSOLS). Those which take up water (HYGROSCOPIC NUCLEI) are very important constituents of the precipitation mechanism. *See also* FREEZING NUCLEI.

nuée ardente A French term (literally 'glowing cloud') for an incandescent mass of volcanic gas, dust and superheated steam together with some PYROCLASTIC MATERIAL, all of which is ejected violently in a horizontal direction from a side vent of a particular type of volcano. The *glowing avalanche* travels at a considerable speed down the slopes of the volcanic cone and overwhelms the surrounding area with devastating effects, e.g. St Pierre on the island of Martinique, West Indies, was totally destroyed by a *nuée ardente* in 1902 when Mt Pelée erupted. The horizontal force of the explosion is usually the result of a blockage of very VISCOUS lava in the vent, while the mobility of the 'avalanche' is caused partly by the material being carried on a body of air (in much the same way as a hovercraft) and partly by the internal particles (IGNIMBRITE) sliding easily over each other to give a high internal velocity.

nullah A normally dry watercourse in India, filled only temporarily during the monsoons.

null hypothesis An assumption used in statistics which hypothesizes that there is no contradiction between the believed mean and the sample mean, and that any difference observed between these two population means can therefore be ascribed solely to chance. If the sample mean lies more than two STANDARD ERRORS from the believed mean, one can reject the null hypothesis at the 95% level of confidence (CONFIDENCE LIMITS).

numerical weather forecasting A means of objective weather forecasting based on the solving of sets of equations which relate to observed data of atmospheric phenomena. With the advent of the computer it has become possible to deal rapidly with great masses of data, for after a period of 48 hours differences between the observed and the computed values become unacceptable. A variety of forecasting models are used which compute the expected values of selected isobaric surfaces, thereby cutting the subjective bias of the forecaster to a minimum. *See also* ANALOGUE WEATHER FORECASTING.

nummulitic limestone A limestone of EOCENE age used for the construction of the Pyramids. It is characterized by many FORAMINIFERA (*nummulites*).

nunatak An Eskimo term that has been widely adopted to describe an isolated peak which protrudes above an ice-sheet and may never have been glaciated.

nutation A periodic minor shift in the position of the Earth's axis, resulting from the gravitational attraction by the Sun and Moon on the Earth's equatorial bulge. It has a periodicity of about nineteen years.

nutrient cycles BIOGEOCHEMICAL CYCLES.

nutrient status A term referring to the degrees of nutrient present in soils, lakes and peatlands. *See also* EUTROPHIC, MESOTROPHIC, OLIGOTROPHIC, OMBROTROPHIC.

nye channel A small-scale channel incised into bedrock underlying a GLACIER. It acts as a conduit for the discharge of glacial MELTWATER. At a larger scale, i.e. beneath an ICE-SHEET, such a channel would be termed a MELTWATER CHANNEL. *See also* RÖTHLISBERGER CHANNEL.

O

oasis An area of any size in the midst of a desert which has sufficient water to support plant growth. The water may reach the surface naturally (e.g. at an ARTESIAN WELL) or may have to be raised by deep-well boring. In some instances the WATER-TABLE may be near enough to the surface for surface springs to occur, especially in erosional hollows created by DEFLATION.

objective analysis A method of weather interpretation in which meteorologists examine a particular ELEMENT in a totally objective way so that the results are independent of human subjectivity. This is usually achieved by adopting the statistical technique known as the LEAST-SQUARES METHOD.

oblate The flattening of opposing parts of a sphere, e.g. the Earth.

oblique aerial photograph AERIAL PHOTOGRAPH.

oblique-fault, oblique-slip fault A FAULT whose STRIKE is oblique to the strike of the strata through which it passes. An oblique-slip fault (which can include *normal*, *reverse* and *thrust faults*) is one in which the net slip lies between the direction of DIP and the direction of strike.

obsequent fault-line scarp FAULT-LINE SCARP.

obsequent stream **1** A stream flowing in a direction opposite to that of the DIP of the rock strata; an anti-dip stream. This definition is favoured by most British geomorphologists. **2** When the term was introduced by W. M. Davis it was intended simply to refer to a stream direction opposite to that of a CONSEQUENT STREAM, which is dependent on the original slope of the land surface and which may or may not coincide with the dip of the rocks. Some US geomorphologists use the term in both contexts. *See also* SUBSEQUENT STREAM. [*167; see also 222*]

obsequent valley A valley carved by an OBSEQUENT STREAM, e.g. the valley of the R. Avon in the section between Bradford-on-Avon and Bath, England.

obsidian A dark-coloured type of RHYOLITE with a high proportion (>65%) of silica. It is so fine-grained as to be glassy (sometimes referred to as volcanic glass) and it has a conchoidal fracture pattern, in contrast to a pitchstone, which has a flatter fracture pattern, but which it otherwise resembles. Because of its higher water content pitchstone is less lustrous and rather more waxy (i.e. like pitch) than the glassy obsidian. Both are extrusive igneous rocks, occurring as veins or dykes in the country-rock.

obsidian hydration dating (OHD) A type of dating based on the rate of formation of a hydration layer on a newly exposed surface of OBSIDIAN. It is used by geologists and archaeologists to date events ranging in age from a few hundred years to a few million years.

occlusion An atmospheric process in which the COLD FRONT overtakes the WARM FRONT of a DEPRESSION, thereby lifting the warm sector off the ground surface, by its undercutting motion, to form an occluded front. Such a process occurs because a cold front normally travels more quickly than a warm front. There are several types of occlu-

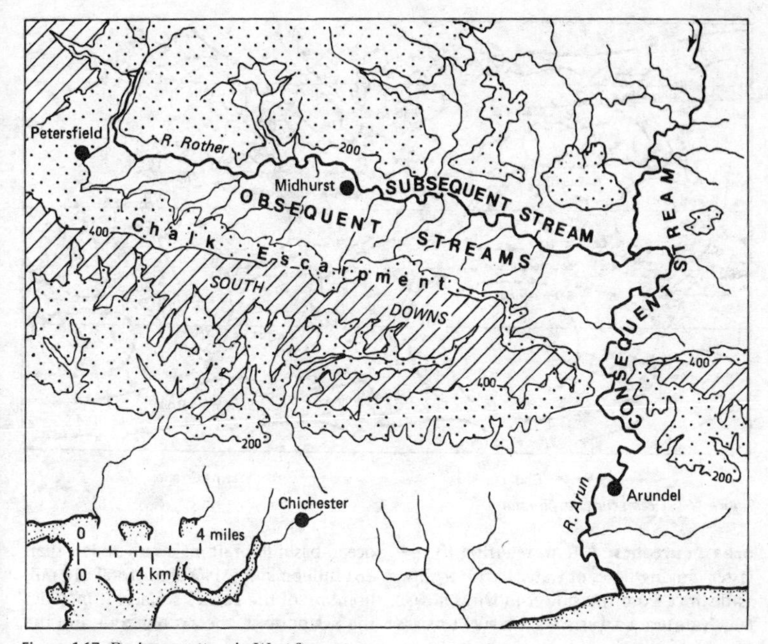

Figure 167 *Drainage pattern in West Sussex*

sion: **1** A *neutral occlusion*, when there is no marked temperature difference between the air ahead of the original warm front and that behind the original cold front. **2** A *cold occlusion*, when the air behind the cold front is much colder than the air ahead of the warm front. **3** A *warm occlusion*, when the overtaking air behind the cold front is not as cold as the air ahead. [*60*]

occult deposition The wet deposition of acidic pollution particles (ACID DEPOSITION) by the impaction of cloud droplets and fog on to surfaces. *See also* DRY DEPOSITION, POLLUTION.

ocean 1 The mass of water occupying all of the Earth's surface not occupied by LAND, but excluding all lakes and inland seas. **2** More specifically, the mass of water which fills the OCEAN BASINS, thereby excluding MARGINAL SEAS such as the Mediterranean, Caribbean and Baltic. The oceans comprise: the N Atlantic, S Atlantic, N Pacific, S Pacific, Indian and Arctic, although some authorities refer to the oceanic areas south of 40°S as the Southern Ocean. *See also* HYDROSPHERE.

ocean basins Those areas of the sea-floor which lie seaward of the CONTINENTAL SHELF but which include the CONTINENTAL SLOPE. Thus, the ocean basins comprise only the BATHYAL ZONE and the ABYSSAL ZONE but exclude the NERITIC ZONE. It is difficult to give an absolute figure for the depth at which MARGINAL SEAS give way to ocean basins, but an approximate figure of 200 m is usually taken. The floor of the ocean basins has been divided into: **1** the continental slope; **2** the DEEP-SEA PLAIN, from which the SEAMOUNTS rise; **3** the very deep parts of the abyssal zone (below 5,500 m), known as ocean deeps (DEEP, OCEAN); **4** the MID-OCEAN RIDGES. NB It is important to distinguish the ocean-basin floor from the OCEAN FLOOR, which is more extensive. [*162*]

ocean climate MARITIME CLIMATE.

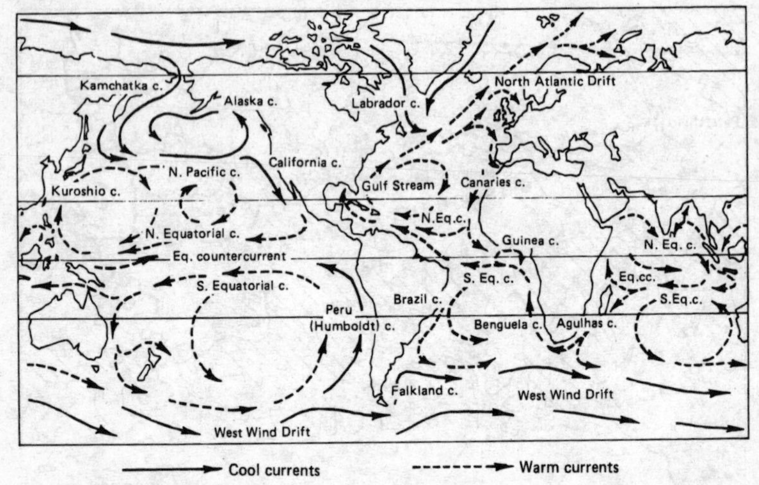

Figure 168 *Ocean currents (January)*

ocean currents A term referring to the faster-moving flows of water in the *oceans*, as distinct from the slower-moving flows, which are termed DRIFTS, e.g. WEST WIND DRIFT. At a global scale, ocean currents assist in maintaining the Earth's HEAT BALANCE by transferring heat from low to high latitudes. They are set in motion by PREVAILING WINDS, which cause a frictional drag on the surface, but are also generated by differences in sea-water densities from one location to another (DENSITY CURRENT). Once the water has started to move it will be affected by the CORIOLIS FORCE and the configuration of the oceans. Thus, ocean water will tend to accumulate on certain continental coasts but gravity will equalize the general ocean level by setting up other currents. The major ocean currents are shown in [*168*]. *See also* BENGUELA CURRENT, EQUATORIAL CURRENT, GULF STREAM, GYRE, KUROSHIO CURRENT, LABRADOR CURRENT, NORTH ATLANTIC DRIFT, PERU CURRENT, WEST WIND DRIFT.

ocean floor A term which refers to the entire sea-bed below low-water mark and which must therefore be distinguished from the ocean-basin floor (OCEAN BASIN). The ocean floor includes the CONTINENTAL SHELF, in addition to all the subdivisions of the ocean-basin floor. It makes up no less than 361 million km^2 (71%) of the total 510 million km^2 of the Earth's surface. It includes the NERITIC ZONE, the BATHYAL ZONE and the ABYSSAL ZONE. *See also* BENTHOS. [*162*]

ocean-floor spreading SEA-FLOOR SPREADING.

oceanic crust The section of the Earth's crust (CRUST OF THE EARTH) beneath the ocean-basin floor (OCEAN BASINS). It comprises layers of young sediments overlying basaltic lavas and intrusive gabbro, in total some 10–16 km in thickness. It is suggested that oceanic crust is constantly moving, by ocean-floor spreading, away from the MID-OCEAN RIDGES, or constructive plate margins, towards the destructive plate margins where it is destroyed by subduction (PLATE TECTONICS). *See also* CONTINENTAL CRUST. [*223*]

oceanicity A measure of the degree to which a MARITIME CLIMATE prevails at a selected point on the Earth's surface, e.g. Hawaii has a more oceanic climate than Cyprus, in so far as the former has greater annual precipitation and a much lower annual temperature range than the latter.

oceanography The scientific study of all things relating to the oceans, including the structure of the ocean floor, the character of the sea water, the oceanic flora and fauna, the ocean currents and drifts and the impact of human interference.

ocean waves The ocean is affected by different types of WAVES, generated in a variety of ways: *see also* BOUNDARY WAVES, BREAKER, SURF, SWELL, TSUNAMI.

offlap The disposition of bedded sediments deposited in a regressing sea, in which the youngest sediments occur farthest from the shoreline marking the limit of the former transgression. Thus, the oldest beds extend further than the youngest, in direct contrast to the deposition pattern of OVERLAP.

offshore A term applied to any phenomenon that moves away from or is located in an area away from the shoreline. It contrasts with the terms onshore, inshore and nearshore.

offshore bar A ridgelike accumulation of sand lying in a semi-submerged position seaward of the BREAKER zone but not elevated enough to be termed a BARRIER-BEACH. *See also* BAR.

offshore zone The sea area extending from the outer limit of the BREAKER zone to the zone of the SHELF EDGE.

ogive 1 A CUMULATIVE FREQUENCY curve, used to depict a numerical distribution of data. **2** A curved band across the surface of a glacier, with the convexity of this transverse phenomenon facing down-glacier. The bands (sometimes referred to as Forbes bands or Alaskan bands) are characterized by their alternating dark and light coloration. The darker bands contain dirt and are devoid of ice bubbles, having been formed by partial melting and refreezing during summer as the particular tract moved down-glacier. Because of their darker colouring the darker bands have greater conductivity to solar radiation (INSOLATION) and have therefore been slightly lowered by melting (ABLATION) to form very shallow transverse troughs. The lighter bands are dirt-free and their whiteness is due partly to the freshness of the newly accumulated winter snow and partly to the presence of air bubbles. Because the whiter ice reflects insolation more effectively than the darker ice, the lighter bands stand out as minor transverse ridges up to 1 m above the darker troughs.

ohm The practical unit of electrical resistance, named after G. S. Ohm and established officially in 1881. The *international ohm* (1908) was defined as the resistance of a column of mercury 106.300 cm long and 14.4521 g in mass at a temperature of melting ice (0° C). It was replaced by the *absolute ohm* in 1948, which involved a slight adjustment in value (1 international ohm = 1.00049 absolute ohms).

oil-pool An accumulation of PETROLEUM in the pores of a sedimentary rock. It is not a pool in the normal sense of the word.

oil-sand Porous sandstone or sand in which PETROLEUM occurs.

oil-shale A fine-grained argillaceous rock containing HYDROCARBONS in the form of a waxy substance termed kerogen. After distillation the hydrocarbons produce crude PETROLEUM. Extensive oil-shale deposits occur in USA, Russia, western Europe and Brazil, mainly in rocks of Carboniferous age, e.g. in West Lothian in Scotland (Strathclyde), often in association with coal-bearing strata.

okta A meteorological term to express that proportion of the sky which is cloud-covered, measured in eighths of the total. 8 oktas = total cloud cover. 0 oktas = clear sky.

old age SENILITY.

Older Drift A term originally used by the Geological Survey of Great Britain to describe the older glacigenic deposits of the PLEISTOCENE, equivalent in age to all the glacial drifts of pre-DEVENSIAN times (*see also* ANGLIAN, BEESTONIAN, WOLSTONIAN). The landforms of the Older Drift are more subdued and more deeply weathered than those of the NEW DRIFT. Because of their greater age the older glacial deposits have been deeply eroded and carried away to be re-deposited as river gravels, etc., except where they have survived on interfluves or in deep basins.

Older Dryas One of the five phases of the LATE-GLACIAL episode of the DEVENSIAN stage in Britain and NW Europe. It is named from the Alpine plant mountain avens (*Dryas octopetala*) which flourished at lower elevations during the short period of climatic deterioration which followed the BØLLING INTERSTADIAL (zone Ia–Ib) when glaciers waxed and solifluction increased during the period of colder temperatures. The Older Dryas (zone Ic in the pollen zonation) was terminated in 12,150 BP by the return of warmer conditions in the ALLERØD INTERSTADIAL (zone II). *See also* BLYTT–SERNANDER MODEL, WINDERMERE INTERSTADIAL.

Oldest Dryas The oldest of the five phases of the LATE-GLACIAL episode of the DEVENSIAN stage in Britain and NW Europe. It is named from the Alpine plant mountain avens (*Dryas octopetala*) which flourished at lower elevations as the ice-sheets of Devensian age disappeared and the glaciers rapidly waned from about 14,000 BP. It was characterized by tundra conditions and continued solifluction as the cold climate ameliorated before giving way to a brief period of climatic warming during the BØLLING INTERSTADIAL (zone Ia–Ib). The Oldest Dryas is placed in zone ia in the Late-Glacial pollen zonation. *See also* BLYTT–SERNANDER MODEL.

Old Red Sandstone An expression given to the sedimentary rocks which make up the continental facies of the DEVONIAN system in the British Isles. It comprises an extensive series of brown, dark-red and white sandstones, red shales, marls, conglomerates and some limestones.

Oligocene The epoch of the TERTIARY following the EOCENE and preceding the MIOCENE. It extended from about 35.4 million years BP to about 23.3 million years BP, a time when the warmer conditions of the Eocene were replaced by a general reduction in average temperatures. In Britain only the lower part of the Oligocene is represented and the deposits are found mainly in the Hampshire Basin and in the Bovey Tracey beds of Devon. In Europe the Oligocene rocks occur more extensively in N Germany, Belgium and the Paris Basin. In Britain the sedimentary rocks are composed of clays, marls, limestones and a few lignites, indicative of deposition in shallow fresh-water lakes or brackish lagoons. The following stratigraphical succession has been recognized in the Isle of Wight:

Youngest	Upper Hampstead Beds	(Rupelian)
	Lower Hampstead	
	Bembridge Marls	
	Bembridge Limestone	(Lattorfian)
	Osbourne Beds	
	Upper Headon Beds	
Oldest	Middle Headon Beds	

The Oligocene is now regarded as the uppermost epoch of the PALAEOGENE.

oligomict A detrital sedimentary rock comprising fragments of a single type of material, as distinct from a POLYMICT.

oligotrophic The state of a water body when it has a low nutrient content and is therefore unable to support a large aquatic flora and fauna. *See also* EUTROPHIC, MESOTROPHIC, NUTRIENT STATUS.

olivine The chief member of a group of rock-forming silicate minerals. The majority of the latter occur in BASIC ROCKS and ULTRABASIC ROCKS but occasionally in acid rocks (some kinds of PITCHSTONE). It is thought to be a major constituent of SIMA and of the Earth's MANTLE. It is usually dark green in colour, exhibits no cleavage and contributes to a continuous series of solid solutions ranging from a silicate of magnesium (Mg_2SiO_4) to a silicate of iron (Fe_2SiO_4).

ombrotrophic A term referring to plant communities which are dependent on precipitation-fed soils that have low NUTRIENT STATUS, e.g. bogs and mires. *See also* EUTROPHIC, MESOTROPHIC, OLIGOTROPHIC.

omnivore An organism which eats both plants and animals (also termed *diversivore*).

onion-skin weathering EXFOLIATION.

onlap OVERLAP.

onset and lee An expression referring to the form of a solid rock mass which has been overridden and abraded by moving ice, with smoothing on the upvalley (onset) side and plucking on the downvalley (lee) side. A ROCHE MOUTONNÉE is a good example of such a landform. [*212*]

Ontarian The stratigraphic name for the oldest of the PRE-CAMBRIAN rocks in Canada.

onyx A banded variety of chalcedonic SILICA, used as a semi-precious stone in the manufacture of jewellery and *objets d'art*.

oolite OOLITIC LIMESTONE.

oolith, ooid A spherical rock particle formed by the gradual accretion of material around an inorganic (e.g. a sand grain) or organic (e.g. a piece of shell) nucleus, possibly by means of the action of ALGAE. Ooliths are of small diameter (0.25–2 mm) and in appearance have been likened to fish roe (from which their name is derived), but larger varieties (3–6 mm diameter) are termed *pisoliths*. Very many SEDIMENTARY IRON ORES are made up of ooliths, and are termed *oolitic ironstones*, but ooliths are most commonly seen when they are cemented together in the form of OOLITIC LIMESTONE.

oolitic limestone, oolite A sedimentary rock made up essentially of OOLITHS and synonymous with the term oolite. At one time the latter term was given to the sedimentary rocks of Upper JURASSIC age in Britain and Europe (*Great Oolite* and *Inferior Oolite*). Oolites are usually calcareous, but non-calcareous oolites are known, e.g. those formed from iron minerals which replaced calcareous ooliths.

ooze A fine-textured sediment formed in ocean basins at depths greater than 2,000 m in the ABYSSAL ZONE and the deeper parts of the BATHYAL ZONE. A distinction is made between the *biogenic oozes* (calcareous and siliceous) and the *non-biogenic sediments* (red clay). The *calcareous oozes* form between 2,000 m and 3,900 m and consist largely of the skeletal remains of Foraminifera (PROTOZOA) and pteropods (MOLLUSCA). They comprise the GLOBIGERINA OOZE and the *pteropod ooze*. The *siliceous oozes* occur at depths greater than 3,900 m, where calcium carbonate is too soluble to survive. Their main constituent is RADIOLARIAN OOZE. At even greater depths much of the silica is dissolved, so that below 5,000 m only the non-biogenic sediment known as RED CLAY will be found. All the foregoing deep-sea sediments are termed pelagic deposits.

open-cast mining A method of mining without recourse to shafts and tunnels, in which overburden is stripped from the surface (hence the synonym *strip mining*) in order to expose the economic mineral to be worked by large mechanical excavators and drag lines.

open-channel flow The characteristic movement of water in a river channel, influenced by the following physical properties: mass; density; weight; specific weight (weight/unit volume); viscosity; shear; temperature – all of which affect the CHANNEL GEOMETRY and help to explain sediment transport. *See also* FROUDE NUMBER, LAMINAR FLOW, REYNOLDS NUMBER, TURBULENT FLOW.

open fold A FOLD the interlimb angle of which exceeds 70°, i.e. the limbs diverge at a large angle.

open system A SYSTEM in which energy and matter are exchanged between the system and its environment; e.g. a living organism. *See also* CLOSED SYSTEM, ISOLATED SYSTEM. [*78*]

opisometer An instrument for measuring distances on maps. It comprises a small wheel the number of revolutions of which are recorded on a dial thereby recording the total distance traversed by the wheel.

opposition A situation existing when three heavenly bodies are in line with each other. If, for example, the Earth lies directly between the Sun and the Moon, the two latter bodies are said to be in opposition to the Earth and to each other. *See also* CONJUNCTION, QUADRATURE, SYZYGY.

optical density A REMOTE SENSING term referring to the measurement of image tone, ranging from 0 = white to 4 = black. It is also used as a general remote-sensing term

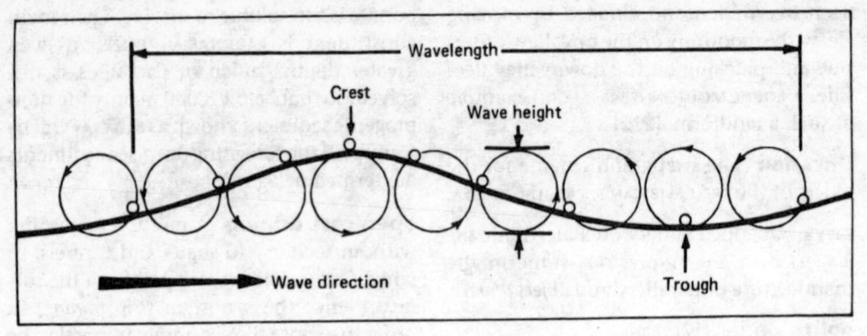

Figure 169 *Orbital motion*

in relation to the relative intensities of ELECTROMAGNETIC RADIATION recorded on an IMAGE. *See also* CHARACTERISTIC CURVE, DENSITOMETER.

orbital motion The circular or near-circular movement of water particles beneath a WAVE. The orbital path is not a completely closed loop because a water particle moves slightly in the direction of wave propagation during each WAVE PERIOD, thereby setting up an *orbital current*. As the wave enters shallower water the *wave orbit* becomes increasingly elliptical in shape. [*169*]

orbit of the Earth The counterclockwise path followed by planet Earth as it revolves at about 106,000 km/h around the Sun in one sidereal year. The orbit is not circular but elliptical, although the degree of ellipticity is only slight. *See also* APHELION, PERIHELION.

orbit of the Moon The counterclockwise path followed by the Moon as it revolves around the Earth. The orbit is not circular but elliptical, considerably more flattened than the ORBIT OF THE EARTH. The Moon's speed of revolution is somewhat slower near to APOGEE and faster near PERIGEE. It completes one revolution during a sidereal month.

order **1** A systematic ranking of data or objects. **2** A unit or group in the taxonomic classification of organisms (TAXONOMY). It is composed of one or more families (FAMILY) and itself is a subdivision of a CLASS. **3** A term used in geomorphology to denote a hierarchy of streams in a drainage network (STREAM ORDER). *See also* BASIN ORDER, SOIL ORDER.

ordinal scale SCALES OF MEASUREMENT.

ordinate A form for the vertical axis (or *y* axis) in a graph, in contrast to the horizontal axis (or *x* axis). *See also* ABSCISSA.

Ordnance Datum (OD) The mean sea-level used as a DATUM for calculating absolute height of land on official British maps. It was calculated from tidal observations taken at hourly intervals for a six-year period at Newlyn, Cornwall.

Ordnance Survey The government body in Great Britain responsible for the survey, production and publication of topographical maps. It was founded in 1791, based upon the needs of military surveys (hence the original link with ordnance) and until the 1970s the principal officials were commissioned army officers. *See also* GEODESY, SURVEYING.

Ordovician The second geological period of the PALAEOZOIC ERA. It was named from an ancient tribal name in N Wales and the Welsh borders by C. Lapworth in 1879. Rocks of Ordovician age were originally included as part of the CAMBRIAN in 1836 and later became known as Lower SILURIAN. The period includes rocks that were formed between about 510 million and 439 million years ago. In the British Isles Ordovician rocks are exposed mainly in N Wales, the borders of N Wales, Cumbria, the Southern Uplands of Scotland and parts of NE Ireland

and SE Ireland. The thick layers of sedimentary rocks are mainly argillaceous, including shales, mudstones and flagstones, but contain grits. Some of the argillaceous sediments have been converted to slates by metamorphism. In some areas thick layers of volcanic rocks are interbedded with the sediments, indicative of widespread volcanic activity (especially in Snowdonia and the central Lake District). The Ordovician period is preceded by the Cambrian and succeeded by the Silurian. The *Lower Ordovician* consists of the Arenigian (Skiddavian) (the oldest), the Llanvirnian and Llandeilian; the *Upper Ordovician* consists of the Caradocian and Ashgillian (youngest).

ore A metalliferous material or aggregate of minerals in which the metal content is of sufficient economic value to justify working. *See also* GANGUE, PROTORE.

organic Pertaining to organisms (either plants or animals) which are characterized by the possession of organs.

organic acids Acids produced during the first stage of ORGANIC MATTER decomposition (termed MINERALIZATION) and before reaggregation into HUMUS by HUMIFICATION. If the latter process is incomplete acids remain in the soil and add to the LEACHING power. They include acetic, lactic and oxalic acids.

organic deposits Sediments formed by organisms and their remains. *See also* CHALK, COAL, CORAL, DIATOMACEOUS EARTH, LIGNITE, OOZE, ORGANIC SOIL, PEAT.

organic horizons Certain of the layers in which humified matter occurs above the MINERAL SOIL in a SOIL PROFILE.

organic matter That portion of the soil which contains material of organic origin, broken down to varying degrees and providing an important source of nutrients in addition to improving the structure of the soil.

organic soil A soil composed predominantly of ORGANIC MATTER, in contrast to a MINERAL SOIL. *See also* BOG, PEAT.

organic weathering The breakdown of rocks by faunal and floral activity. The burrowing action of animals and the probing action of tree roots can cause mechanical disintegration, while the effects of chemical action are even more marked. The decomposition of plant materials creates ORGANIC ACIDS which act upon the rock minerals, and the increased carbon dioxide content of the soil, resulting from plant respiration, has an important influence on the rate of WEATHERING of the underlying rocks. *See also* CHEMICAL WEATHERING, MECHANICAL WEATHERING.

orientation 1 The assignment of a definite spatial direction. **2** The action of aligning a surveying instrument or a map into a north–south direction so that the compass direction agrees with the alignment of the grid lines on the map. The term stems from the fact that early maps were constructed with the east, rather than the north, at the top of the map. **3** In geology, the spatial arrangement of particles in a rock or till (FABRIC, TILL).

orientation of stones PATTERNED GROUND.

oriented lakes A term referring to groupings of enclosed water bodies exhibiting a parallel alignment. They have been described from a variety of environments and some may occupy structural hollows picked out by selective glacial erosion or deflation by wind. Many have been described in Arctic latitudes: these are due, perhaps, to differential thawing of the PERMAFROST under the influence of predominant winds. But their genesis is complex, with littoral transport, thermal activity of permafrost, microclimatic effects on vegetation growth and the effects of lake ice all being suggested as factors in their formation. [*170*]

origin 1 An arbitrary starting-point on a scale. **2** The point defined by the intersection of two axes in a system of coordinates. In cartography, the *true origin* of a GRID is the point defined by the intersection of the central meridian with selected line drawn at right angles to it. But in order to avoid negative grid values the origin is removed beyond the boundary of the gridded area, the new point being termed the FALSE ORIGIN. [*160*]

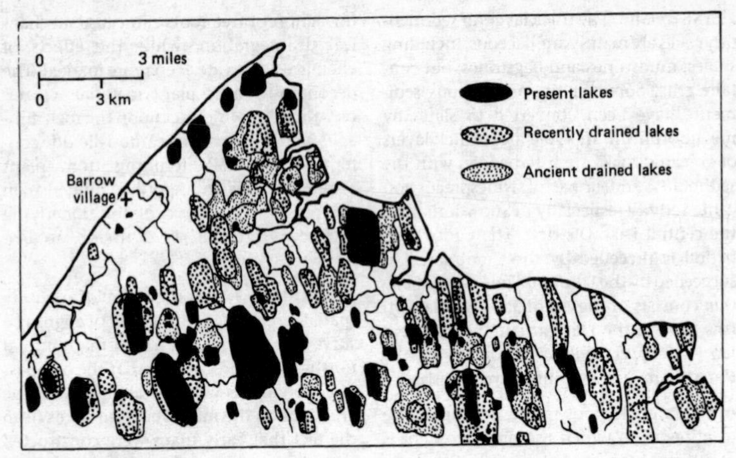

Figure 170 *Oriented lakes in Alaska*

origin, seismic FOCUS.

orocline The name given to any structural mountain chain or orogenic belt, the line of which has been sharply displaced by subsequent horizontal movement to create a marked change in trend.

orocratic period A term introduced by A. C. Ramsay to describe a period of maximum crustal movement (DIASTROPHISM) and/or vulcanicity. *See also* OROGENY.

orogen MOBILE BELT.

orogenesis OROGENY.

orogeny A major period of fold-mountain formation, during which the process of *orogenesis* occurs. Such a process includes folding, faulting and thrusting, often as a result of PLATE TECTONICS, during which sediments within GEOSYNCLINES are buckled and deformed as they are compressed into long, linear mountain chains (*orogens*). Since the crustal movements are tangential to the Earth's surface they must be contrasted with the essentially radial movement of EPEIROGENESIS. Orogenies are often accompanied by igneous activity and the emplacement of PLUTONS. In Britain the following orogenic phases have been recognized: SCOURIAN, LAXFORDIAN, CHARNIAN (all Pre-Cambrian); CALEDONIAN; HERCYNIAN; ALPINE. in N America the following have been recognized: Older Laurentian, HURONIAN, Hudsonian, Great Bear Lake, Beltian, Grenville, Wichita (all Pre-Cambrian); Taconic; Acadian; Appalachian; NEVADIAN (mainly granite emplacement); LARAMIDE; Cascadian (mainly vulcanicity); PASADENIAN.

orographic cloud A cloud-form that results from the uplift of an air stream by a topographic barrier. If the uplift of moist air reaches the CONDENSATION LEVEL a layer of orographic cloud will form, often associated with the generation of OROGRAPHIC RAINFALL. *See also* BANNER CLOUD, LEE WAVE, LENTICULAR CLOUD.

orographic depression LEE DEPRESSION.

orographic rainfall The type of precipitation resulting from a vertical uplift of an airstream by a topographic barrier (OROGRAPHIC CLOUD). In fact, for heavy rainfall to occur it is necessary for cyclonic or convective processes to be operative because the orographic component is normally weak and acts merely as a triggering mechanism.

orography A rarely used word referring to the study of mountain systems and the depiction of their relief.

orophyte A plant that will grow only in a mountainous environment.

orthoclase A mineral of the FELDSPAR group, a common constituent of granitic rocks.

orthodrome GREAT CIRCLE.

orthogeosyncline GEOSYNCLINE.

orthogneiss GNEISS.

orthogonal A line drawn at right angles to or perpendicular to another line. It is often used in WAVE REFRACTION diagrams (where it is termed a wave ray) whereby a line is constructed perpendicular to a series of wave crests in the vicinity of a shoreline. When a series of orthogonals come close to each other near a coastline it indicates a zone of CONVERGENCE and a zone where wave energy is concentrated. Conversely, where orthogonals spread out in a zone of DIVERGENCE this indicates a zone of energy dissipation. It is a term also used in a description of GRATICULES in the construction of MAP PROJECTIONS.

orthographic projection A type of MAP PROJECTION of very great antiquity, achieved by projecting a global hemisphere on to a perpendicular plane surface when the perspective centre is considered to be situated at an indefinite distance from the centre of the generating globe. Its scale is accurate only at the centre of the projection and inaccuracy of scale increases rapidly towards the perimeter. It is used mainly for pictorial world maps and for astronomical charts but its inaccurate scale renders it of little use for most cartographic purposes. [*171*]

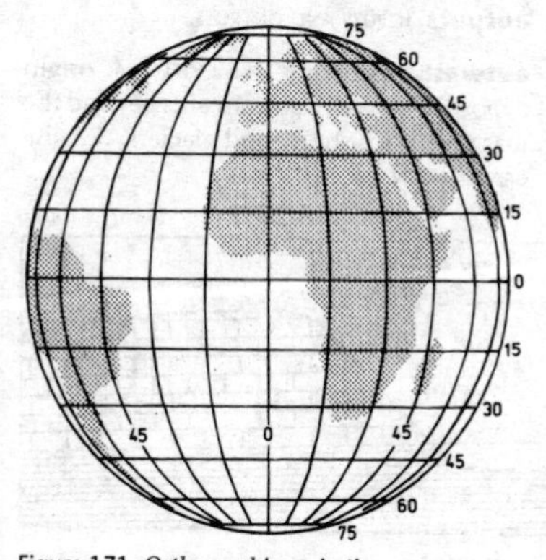

Figure 171 *Orthographic projection*

orthomorphic projection A type of MAP PROJECTION in which the true shape has been preserved over a small area. The scale at any point is the same in all directions and around any point the angles are truly represented. It is also referred to as a *conformal projection*. Examples of the orthomorphic projection are MERCATOR, TRANSVERSE MERCATOR, the conformal LAMBERT'S PROJECTIONS and the STEREOGRAPHIC PROJECTION.

orthophoto map A type of PHOTOMAP which eliminates the distortion of perspective by transforming the central projection of the air photographs to an ORTHOGONAL type of projection.

orthoquartzite A clastic sedimentary rock in which more than 95% of the grains are of quartz firmly cemented by silica. It is generally white or greyish in colour and when broken the line of fracture characteristically follows the grain boundaries, in contrast to the METAQUARTZITE. It is sometimes termed quartzarenite. *See also* ARENACEOUS.

osar ESKER.

oscillation **1** The act of swinging backwards and forwards in the motion of a pendulum. **2** A vibration above and below a mean value. **3** In meteorology, it refers to a cyclic variation of an ELEMENT around its mean value.

oscillation current A type of wave-induced movement of sea water on the floor of a shallow sea. It is generated by the increasing ellipticity of the path of water particles in their ORBITAL MOTION as waves move onshore into shallowing water. The shoreward movement of the water particle beneath the wave crest and the counteracting seaward movement of the particle at depth sets up an oscillatory current on the sea-bed. The net movement may cause the bed deposits to move shorewards.

oscillation theory of tides A modern concept explaining tidal behaviour, replacing an earlier concept known as the PROGRESSIVE-WAVE THEORY OF TIDES. The oscillation theory postulates that each ocean can be divided into tidal units each centred around a nodal point (AMPHIDROMIC SYSTEM) and each governed by the depth, size and shape of the particular water body within the tidal unit. The body of water is set into rhythmic motion, or OSCILLATION, by tide-producing forces (TIDE) but the periodic rises and falls of level will vary according to the size and shape of the different ocean bodies. Tidal development is greatest where the particular size and shape of the ocean body produces a period of oscillation approximately equal to that of the tide-producing forces. If the two are completely different, however, the water body will have no marked tides.

oscillatory wave A wave advancing across the ocean surface notwithstanding the fact that the water particles beneath the wave circulate by ORBITAL MOTION in almost closed loops. The speed of advance of the wave may be quite rapid (depending largely on WAVELENGTH) but the movement of water particles in the direction of wave travel is usually very slow.

oshana An Afrikaans term referring to a periodic river channel in the more arid regions of S Africa. The channel only appears during floods and disappears soon afterwards owing to the extreme flatness of the land, on which no clearly defined river courses survive.

osmosis, osmotic pressure The movement of solvent molecules (normally water) from a dilute to a more concentrated solution through a selectively permeable membrane. The extra pressure, which must be applied to the concentrated solution to prevent the movement of solvent molecules by osmosis, is termed osmotic pressure.

outcrop That part of the Earth's surface on which a particular type of rock occurs whether or not it is covered by an overlying superficial cover of drift or detritus. In the English usage this definition contrasts with that of a rock EXPOSURE, but in the USA the two terms are used synonymously.

outer core The layer of the Earth lying between depths of 2,900 km and 5,000 km below the surface. It is separated from the overlying MANTLE by the *Gutenberg Discontinuity*. *See also* CORE.

outlet cave A hollow cavity in a cliff or hillside marking the point at which an underground KARST watercourse re-emerges. *See also* INLET CAVE, RESURGENCE. [*204*]

outlet glacier A lobe of glacier ice issuing from a plateau ICE-CAP and flowing down a peripheral valley. [*92*]

outlier An outcrop of stratified rocks occurring in a detached location away from the main body of similar rocks, the intervening portions having been removed by denudation. In plan the younger rocks of the outlier will be surrounded by older rocks, e.g. Bredon Hill beyond the scarp of the Cotswolds in S England. *See also* INLIER. [*172*]

outputs INPUTS AND OUTPUTS.

outwash Material of GLACIOFLUVIAL origin deposited by meltwater streams beyond the margins of ice-sheets and glaciers. *See also* GRAVEL TRAIN, VALLEY TRAIN.

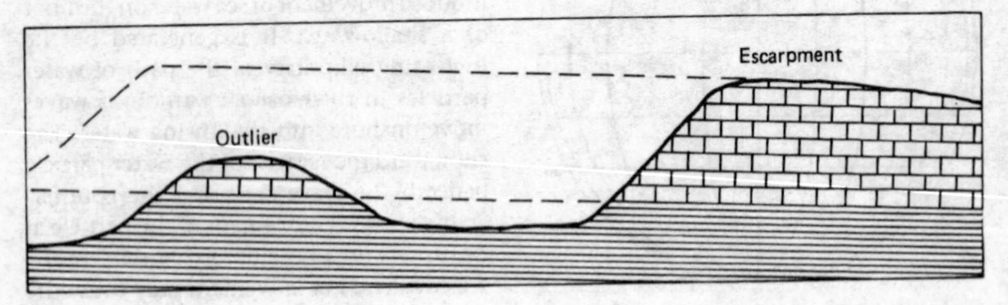

Figure 172 *Outlier*

outwash plain SANDUR.

outwash terrace An OUTWASH deposit that has been incised by glacial MELTWATER to form a terrace. Such terraces are often pitted with KETTLE HOLES and show evidence of former BRAIDED STREAM networks.

overbank flow stage of a river The stage at which the flow of a river cannot be contained within its normal channel, leading to overflow and flooding of the FLOODPLAIN. Any fluvial material deposited beyond the confines of the channel during this stage is termed an *overbank deposit.*

overcast A term used by weather forecasters and pilots when there is a complete cloud cover.

overdeepening The process by which valley floors are excavated by glaciers, sometimes below BASE-LEVEL, so that post-glacial rivers will be forced to infill the rock basins with sediment (AGGRADATION) to restore the valley's long profile, albeit at a level considerably below that of the pre-glacial valley. *See also* FIORD, TROUGH.

overflow channel A channel, often streamless, cut in solid rock or in drift, having been carved out by the overflow of an ice-dammed lake. It can only be recognized with certainty where it is associated with deltas, lake shorelines and lake-bottom deposits formed in the formerly impounded PRO-GLACIAL LAKE. The channels associated with the Pleistocene extensions of the Great Lakes in N America and that of Ironbridge Gorge, Shropshire, England, are good examples. Overflow channels should be distinguished from glacial MELTWATER CHANNELS.

overfold, overturned fold A fold in which the beds in one limb have been rotated through more than 90° so that they are now inverted or overturned. [*84*]

overland flow The surface movement of water derived from precipitation which is not intercepted by vegetation and which runs as a shallow unchannelled sheet across the soil. This type of flow occurs during periods of intense rainfall once the INFILTRATION-RATE of the soil has been exceeded. The depth of the sheet of water increases with distance down a hill slope until near the foot slope the sheet may become channelled into rills, at which point overland flow ceases to exist. The depth of overland flow together with its velocity determines the size of the soil particles which can be moved, but on bare hill slopes and/or relatively impermeable soils, soil erosion is usually severe, especially in tropical and semi-arid environments where convectional rainfall is intense. It is also termed storm runoff and quickflow [*216*]. *See also* INFILTRATION-EXCESS OVERLAND FLOW, SATURATION OVERLAND FLOW, SHEET EROSION, SHEET FLOOD, SURFACE WASH, THROUGHFLOW.

overlap (onlap) A term referring to a particular relationship of rock strata at an UNCONFORMITY, when the beds above the unconformity were deposited by a transgressing sea. Thus, each successively younger bed extends further on to the older underlying rock series than does its predecessor. It contrasts with OFFLAP. *See also* OVERSTEP. [*173*]

overlay A map or graphic display of data depicted on a transparent medium (e.g. plastic), which may be superimposed

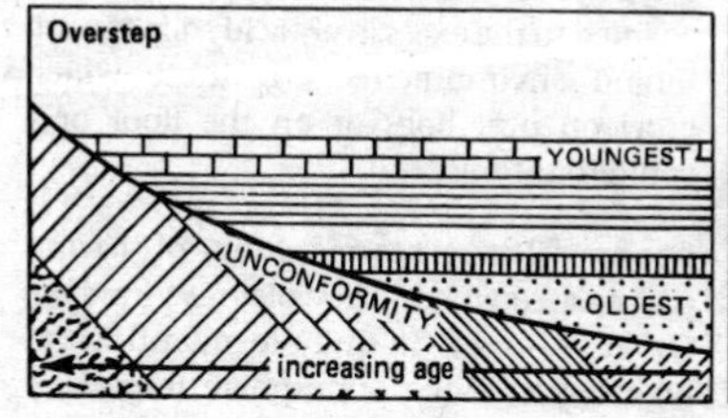

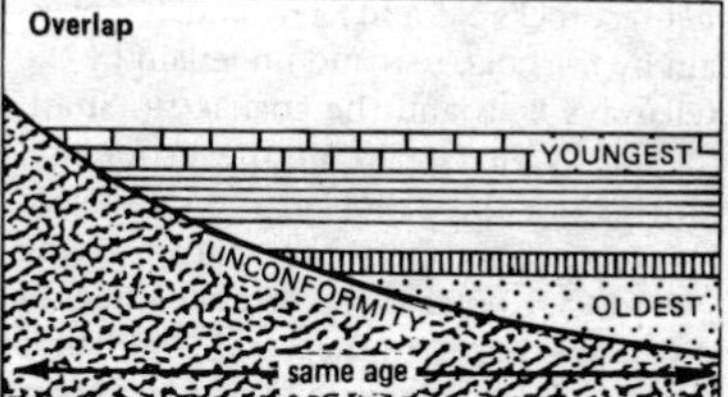

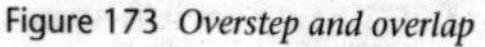
Figure 173 *Overstep and overlap*

on another map, photograph or graphic display.

overstep A term referring to the disposition of rock strata at an UNCONFORMITY when the younger series deposited during a transgression rest upon progressively older strata of the underlying rock series. *See also* OVERLAP. [*173*]

overthrust A low-angle THRUST-FAULT produced by crustal shearing during intense folding, especially during NAPPE formation. [*254*]

overtopping The process whereby the crests of coastal BARRIER BEACHES are built up by SWASH waves that have insufficient force to cross the crest and form OVERWASHING.

overwashing The process whereby storm-generated SWASH waves carry beach-face sediment over the crest of a BARRIER BEACH, and deposit it on the landward-facing slope. Storm waves may breach the crest in several places, creating *overwash throats*, behind which *overwash fans* may be created on the backslope. Overwashing will gradually reduce the crest height and may cause the beach to move landwards as wave erosion causes material to be transported from the seaward to the landward face. *See also* OVERTOPPING.

oxbow lake A crescent-shaped lake occurring on a river FLOODPLAIN, having once been part of a river MEANDER that has been cut through and abandoned (hence the alternative term *cut-off*) by lateral erosion of the banks at the meander neck. The oxbow lake is variously referred to as a BAYOU, BILLABONG or mortlake. [*147*]

Oxford Clay The lowest of the clay strata which dominate the Upper Jurassic sedimentary rocks of S and E England. It is overlain by the CORALLIAN and underlain by the Kellaways Beds and the CORNBRASH. Stratigraphically, it is part of the Oxfordian, which spans the Callovian and KIMMERIDGIAN stages. The unresistant nature of the Oxford Clay outcrop has created a tract of low-lying country in the S and E Midlands of England, where it has long been an important source of clay for brick-making.

oxidation One of the varieties of CHEMICAL WEATHERING, in which OXYGEN dissolved in water reacts with certain rock minerals, especially iron, to form oxides and hydroxides. This manifests itself in a brownish or yellowish staining of the rock surface, which ultimately disintegrates. It is often shown as oxidation/reduction to indicate the reversibility of the two processes and their close relationship. (*See also* REDUCTION.)

oxisol One of the soil orders of the SEVENTH APPROXIMATION of soil classification. It has a loamy or clayey texture, a very low BASE EXCHANGE capacity and is characterized by an extreme degree of weathering of its minerals to form free oxides and kaolin. The oxisols occur over vast areas of the humid tropics where LATERITES abound and iron and aluminium oxides dominate the soil minerals of the LATOSOLS. Because they are heavily leached and have been exhausted by hand tillage for lengthy periods of time the oxisols are low in nutrients and have little fertility.

oxygen One of the main constituents of the ATMOSPHERE, second only to nitrogen in amount (20.95% dry air). At high levels of the atmosphere it breaks down into monatomic oxygen (O) which combines with oxygen molecules (O_2) to produce OZONE (O_3), Oxygen is an important absorber of RADIATION in the ultraviolet wavelength (0.13–0.17 μm). Combined with other elements oxygen makes up nearly 50% by weight of the Earth's CRUST. It is an essential ingredient of all life-enhancing processes, i.e. PHOTOSYNTHESIS in plants, respiration in animals.

oxygen isotope ISOTOPE.

oxyphytes All plant species which are adapted to the excessively acid soils of cool, humid environments, e.g. those which grow on peat bogs or on the floor of a coniferous forest.

ozone The triatomic form of OXYGEN (O_3) created when ultraviolet rays irradiate O_2. It is found in minute quantities at high levels in the atmosphere (especially between 20 and 25 km) where it is sometimes referred to as the *ozonosphere*. Since ozone readily absorbs most of the shorter waves of the Sun's radiation (0.23–0.32 μm) it prevents the shorter ultraviolet waves from reaching the Earth's surface but allows through the more beneficial ultraviolet radiation with longer wavelengths. Environmentalists are particularly concerned about damage or interference with the *ozone layer* which may result from nitrogen oxides released by nuclear explosions and supersonic aircraft in the atmosphere. Ozone is also depleted by the propellant gas freon in fluorocarbon aerosol sprays, which were banned in the USA in 1978. *See also* CHLOROFLUOROCARBONS.

ozone depletion In 1984 scientists in Antarctica confirmed earlier beliefs that high-altitude ozone depletion had accelerated since the late 1970s. A springtime ozone hole was identified by SATELLITES between 10 km and 24 km above the Antarctic. World governments were asked to reduce production of CFCs. In 1987 forty countries adopted a CFC reduction protocol (50 per cent reduction by 1999), known as the *Montreal Protocol*. By 1997 the Antarctic ozone 'hole' was shown to be complete when 97.5 per cent of ozone was recorded as destroyed. Ozone depletion has serious effects on ultraviolet (UV-B) radiation levels which have shown a marked increase in recent decades, with a simultaneous increase in skin cancer cases. For these reasons the *Copenhagen Protocol* (1992) agreed a total phase-out of CFCs by 1996. *See also* CHLOROFLUOROCARBONS, HALONS.

P

Pacific-type coast CONCORDANT COAST.

pack ice Any area of floating sea ice in contrast to FAST ICE, which remains attached to the shore. Its classification takes into account the amount of water exposed between the ICE-FLOES and is expressed in tenths: 10/10 represents consolidated pack ice in which the floes are frozen together. *See also* LEAD.

padang A grass and shrub vegetation type, characteristic of certain poor soils in South-East Asia.

pahoehoe A Hawaiian name for a fluid basaltic type of lava that solidifies into corded or rope-like forms (hence *ropy lava*). It contrasts with AA LAVA.

paired terrace, non-paired terrace A river terrace (ALLUVIAL TERRACE) whose remnants on opposite sides of a valley will correspond in altitude along a particular stretch of the valley. A paired terrace may imply a partial erosion cycle related to intermittent uplift, or it may represent alternations between periods of aggradation and degradation related to static changes of BASE-LEVEL when valley deepening has ceased and lateral erosion is dominant [*211*]. Non-paired terraces on either side of a valley do not correspond in altitude and were formed when there was continued downcutting (perhaps due to slow uplift) during which the channel shifted laterally back and forth. By the time the river had moved from one side of the valley to the other the valley floor had been lowered, leaving the opposing terraces at different altitudes. In 1940 it was suggested by C. A. Cotton that these terraces should be called *cyclic* and *non-cyclic terraces*, respectively. [*174*]

palaeobotany The scientific study of plant life throughout geological time based on the fossil record. It is a branch of PALAEONTOLOGY.

Palaeocene The first epoch of the TERTIARY, preceding the EOCENE. It began some 65 million years ago (at the end of the Cretaceous), terminating 57 million years ago. This time period was once included within the Eocene epoch. It was the beginning of the age of mammals. The sediments deposited in Britain are represented by estuarine sands and mottled clays (the Reading and Woolwich Beds) in both the Hampshire and London Basins, and by marine sands (the Thanet Sands) in the London Basin only. The following stratigraphic successions have been recognized. *See also* PALAEOGENE, TERTIARY VOLCANIC PROVINCE.

	Hampshire Basin:	
Youngest ↑	*Woolwich and Reading Beds*	(Sparnacian)
	Absent	(Thanetian)
	Absent	(Montian)
	London Basin:	
Youngest ↑	*Woolwich and Reading Beds*	(Sparnacian)
	Thanet Sands	(Thanetian)
	Absent	(Montian)

Palaeochannels The abandoned river or stream channels that no longer carry water in a fluvial system (except in times of flood). They may occur as topographical features on a FLOODPLAIN or recognized only in

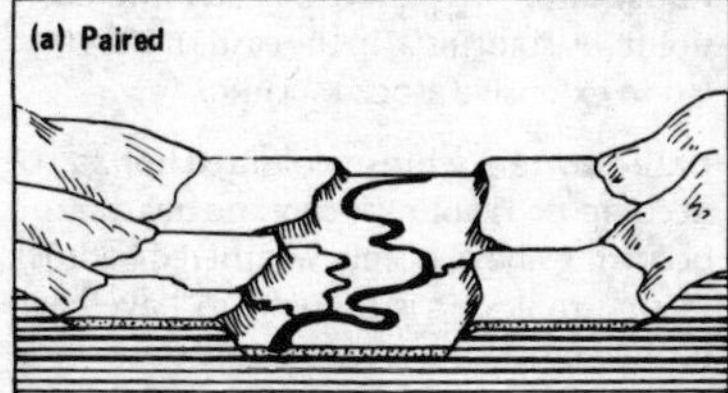

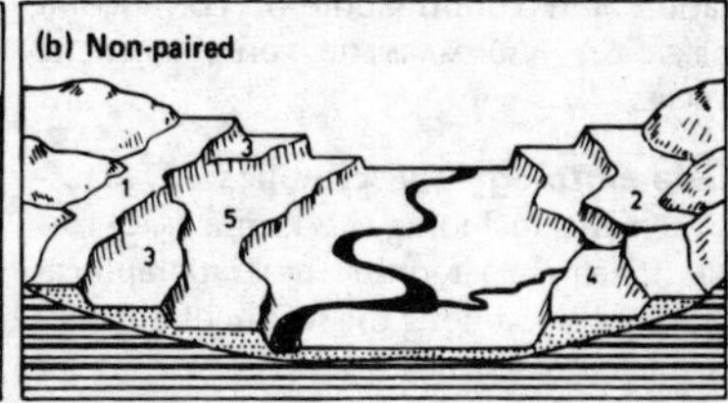

Figure 174 *River terraces*

sediment sections where they have been infilled.

palaeoclimatology The scientific study of former climates in the geological past, based on investigations of sedimentary rocks, fossils, deep-sea cores, etc.

palaeoecology The scientific study of the relationship between fossil organisms and their palaeoenvironments, achieved by an examination of the geological record using such techniques as POLLEN ANALYSIS and RADIOCARBON DATING. [*191*]

Palaeogene The lower part of the TERTIARY. It comprises the three epochs of the PALAEOCENE, EOCENE and OLIGOCENE, which in total spanned some 42.5 million years between 65 million years and 23.3 million years ago. *See also* NEOGENE, TERTIARY VOLCANIC PROVINCE.

palaeogeography A term referring to the spatial reconstruction of former land masses, oceans, etc. during specific periods of geological time. *See also* PLATE TECTONICS.

palaeohydrology A term referring to the spatial reconstruction of former stream patterns, water bodies and groundwater during specific periods of geological time.

Palaeolithic An archaeological term adapted from the Greek, meaning *'old stone' age*, and used to describe the earliest human culture, when mankind fashioned some of the earliest known artefacts (*palaeoliths*), especially from flint. The earliest toolmakers lived during the PLEISTOCENE and the following cultures, named after sites in France, have been recognized: Pre-Chellean, Chellean (Abbevillian), Acheulian, Mousterian, Aurignacian, Solutrean, Magdalenian and Azilian. In Britain the *Lower Palaeolithic* period has evidence of three distinct flint industries, the Clactonian, Acheulian and Levalloisian, which began in the HOXNIAN interglacial and lasted through the WOLSTONIAN and (with the exception of the Clactonian) into the IPSWICHIAN. The early artefacts are found only in lowland Britain, with no finds reported from Ireland, Scotland or N England. By the beginning of the DEVENSIAN glacial the Mousterian culture had appeared in Britain (*Middle Palaeolithic*), at which time Neanderthal Man is known to have lived in caves during the earliest interstadials, e.g. Kent's Cavern, Torquay, and Wookey Hole, Mendips. During the main Devensian ice advance, however, man seems to have been absent from the British Isles (25,000–14,000 BP) only reappearing as *Homo sapiens* (Modern Man) in the Creswellian and Cheddarian cultures of the *Upper Palaeolithic*, which flourished during the LATE-GLACIAL. Upper Palaeolithic man left sophisticated cave paintings and finely carved bone and antler ornaments, particulary in SW France, and it is possible that the horse and reindeer may have become semi-domesticated. *See also* MESOLITHIC, NEOLITHIC.

palaeomagnetism The faint traces of polarization surviving in igneous rocks and those sedimentary rocks containing magnetite, having been preserved since the rocks accumulated. Since the direction of polarization changes periodically (MAGNETIC FIELD) and these MAGNETIC REVERSALS are 'frozen' into the rocks it becomes possible to establish a palaeomagnetic sequence which can be calibrated by RADIOMETRIC DATING, thereby assisting in stratigraphic correlation and confirmation of CONTINENTAL DRIFT. *See also* MAGNETIC POLES, MAGNETIC STRIPES.

palaeontology The scientific study of fossil remains including FOSSILS and TRACE FOSSILS. It helps to determine the stratigraphical relationships among the sedimentary rocks by means of BIOSTRATIGRAPHY.

palaeopedology The scientific study of fossil soils (PALAEOSOL).

palaeosol An ancient soil horizon or buried fossil soil in which organic remains provide possibilities of dating or reconstruction of palaeoenvironments by such means as RADIOCARBON DATING and POLLEN ANALYSIS. Interglacial and interstadial palaeosols of the Pleistocene are covered by younger glacigenic or solifluction deposits.

Palaeozoic A Greek term meaning 'ancient life', adapted to describe the first of the eras of geological time after the Pre-Cambrian and prior to the MESOZOIC. It lasted from about 600 million years to 235 million years BP. In Europe it comprises the CAMBRIAN (oldest), ORDOVICIAN and SILURIAN (which together constitute the *Lower Palaeozoic*), together with the DEVONIAN, CARBONIFEROUS and PERMIAN (which form the *Upper Palaeozoic*). In the USA the Carboniferous is replaced by the MISSISSIPPIAN and PENNSYLVANIAN, thereby giving seven rather than six periods. The Palaeozoic included the CALEDONIAN OROGENY and was terminated by the HERCYNIAN OROGENY. It also saw the appearance of fish and amphibians among the fauna, but many of the invertebrate types (e.g. graptolites, trilobites and rugose corals) had become extinct by the end of the era.

palaeozoology The scientific study of fossil fauna, both vertebrate and invertebrate.

palimpsest A surface bearing superimposed inscriptions of differing date, e.g. the remnants of the original rock texture surviving in a metamorphic rock.

pali ridge A sharp-crested ridge separating two river valleys in deeply dissected terrain of a volcanic cone. *See also* PLANÈZE.

Palisadian A period of early Mesozoic uplift (DIASTROPHISM) in the eastern USA that led to extensive BLOCK-FAULTING.

pallid zone A whitish-coloured horizon of decomposed kaolinitic clay and quartz sand occurring above deeply weathered bedrock in the tropics. It is thought to have been formed by CHEMICAL WEATHERING under hot, humid conditions but is now found in some arid areas of the tropics as a result of CLIMATIC CHANGE. It is characterized by the virtual absence of iron minerals, which may have been transferred upwards through the *mottled zone* to help form surface LATERITE. *See also* CRUST.

palsa A mound or ridge of peat or peaty earth containing perennial ice lenses and a core of PERMAFROST. Its surface is usually criss-crossed with open fissures caused by frost-cracking, dilation-cracking (due to doming) or desiccation. It is thought by most geomorphologists that palsas are formed by differential frost-heaving linked to the thermal conductivity of the peat or peaty earth. They are characteristic of boggy areas in sub-Arctic latitudes in the zone of discontinuous permafrost where snow cover is relatively thin and winters are long. It differs from a PINGO by the characteristic presence of peat, which is comparatively rare in pingos, that also tend to have cores of clear ice rather than the separate ice lenses of the palsa.

paludal A term referring to a marshy or swampy environment and to the sediments deposited therein.

paludification The term given to the gradual lateral expansion of a bog or marsh as peat growth inhibits water drainage, leading to a rise in the WATER-TABLE.

palynology POLLEN ANALYSIS.

pampa (pampas) A Spanish name for the extensive temperate grasslands of Argentina and Uruguay around the Plate estuary. Originally the more humid eastern part was covered in coarse tufted grass (*pampas grass*) but much of this has now been replaced by farmland or used for cattle ranching. The western part, nearer the Andes, is more arid

Figure 175 *Formation of a panplain*

with a semi-desert vegetation. *See also* PRAIRIE, STEPPES, VELD.

pampero A Spanish term for an outbreak of cold, dry polar air blowing northwards or north-eastwards across the PAMPA of S America, owing to the passage of a depression and its associated COLD FRONT. *See also* SOUTHERLY BURSTER, POLAR OUTBREAK.

pan **1** A strongly compacted indurated soil horizon or one that is very high in clay content. *See also* CALICHE, CLAY PAN, DURIPAN, FRAGIPAN, HARDPAN, INDURATION, IRONPAN, PRESSURE PAN. **2** A dried-up deposit in a natural basin or depression. *See also* SALT PAN.

pancake ice Small, thin discs of ice formed during the initial stages of freezing of the sea's surface. It later contributes to the formation of FAST ICE.

pan fan **1** A landform produced during the ultimate stage (SENILITY) in the ARID CYCLE OF EROSION when extensive pediments have developed, intervening ridges have been almost totally denuded and basins virtually infilled. **2** A name given to smooth domelike granitic hills in the Mojave Desert, California, by A. C. Lawson. They are characterized by gentle slopes (<5°) and lengthy detritus-covered pediments and may extend between 5 and 20 km in length.

Pangaea The name given by A. Wegener to an enormous primeval land mass of Pre-Cambrian age in his concept of CONTINENTAL DRIFT. He hypothesized that the supercontinent of Pangaea broke up to form GONDWANALAND and LAURASIA. It was surrounded by a hypothetical sea known as Panthalassa.

panplain (panplane) An almost flat land surface created by the coalescence of floodplains owing to the lateral erosion by rivers of the intervening divides. The process of *panplanation* (*lateral planation*) by laterally eroding rivers should be contrasted with that of PENEPLANATION. [*175*] The concept was proposed by C. H. Crickmay in 1933.

pantanal A Brazilian term referring to the SAVANNA-type vegetation occurring along some Brazilian rivers where lengthy dry periods are interrupted by occasional floods.

Panthalassa The ancestral Pacific Ocean which surrounded PANGAEA.

pantograph An instrument used in map enlargement or reduction.

papagayo A Spanish term for a dry, cold, north-easterly wind in eastern Mexico.

parabola A mathematical curve ($y = x^2$) which passes through the origin of the graph and increases steadily in gradient away from the axis on both sides, with the value on the y-axis increasing as the square of the value on the x-axis.

parabolic dune A type of curved sand DUNE, similar in shape to a BARCHAN but with the horns pointing upwind instead of downwind. It is usually formed by the process known as a BLOW-OUT in which the centre of the dune is partly removed and carried downwind, leaving the horns behind and drawn out in an elongated form.

parabolic projection An EQUAL-AREA MAP PROJECTION of a pseudo-cylindrical type in which each MERIDIAN is represented by a PARABOLA. A large number of variants are theoretically possible, ranging in between the MOLLWEIDE PROJECTION and the SINUSOIDAL PROJECTION.

paradigm A pattern, example or MODEL constructed by a researcher to test a hypothesis or to formalize a set of data so that they may be systematically analysed and synthesized.

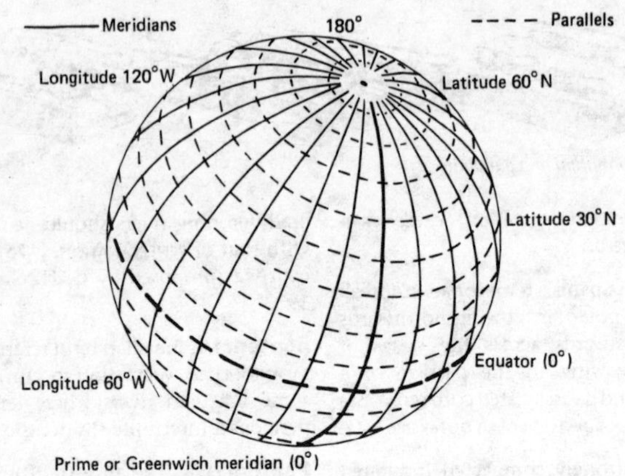

Figure 176 *Parallels and meridians*

parageosyncline A GEOSYNCLINE lying within a CRATON. [*91*]

paragneiss GNEISS.

paralic A term referring to a LITTORAL environment where shallow waters predominate, e.g. a lagoonal environment.

parallel A line of LATITUDE running round the Earth parallel to the equator. The parallels decrease in length away from the equator until they are replaced at each pole by a point. [*176*]

parallel drainage A drainage pattern characterized by streams which run in virtually parallel courses, owing to structural control by the bedrock, parallel folds or a uniform slope gradient.

parallel evolution The process by which similar organisms evolve over time in almost identical fashion (CONVERGENT EVOLUTION).

parallel retreat of slopes A hypothesis referring to a type of slope evolution in which ground loss measured in a horizontal direction occurs at the same rate over the whole length of the slope. During retreat the maximum slope angle remains constant, the absolute length of all parts of the slope except the concavity also remains constant, the concavity increases in length and the angle at any point on the concavity decreases. It has been suggested that parallel retreat may be applicable to the upper part of a slope (i.e. excluding the pediment and concavity) in regions with structurally controlled landforms such as MESAS and ESCARPMENTS. Parallel retreat of slopes may operate preferentially on certain lithologies, favouring harder rocks rather than less resistant rocks such as clays, where slope declines. Parallel retreat occurs with a FREE FACE or without a free face. [*177*] If debris accumulates, because its removal does not keep pace with debris formation, parallel retreat is impossible. *See also* HALDENHANG [*193*], PEDIMENT [*180*], SLOPE DECLINE, SLOPE EQUILIBRIUM, CONCEPTS OF, SLOPE REPLACEMENT.

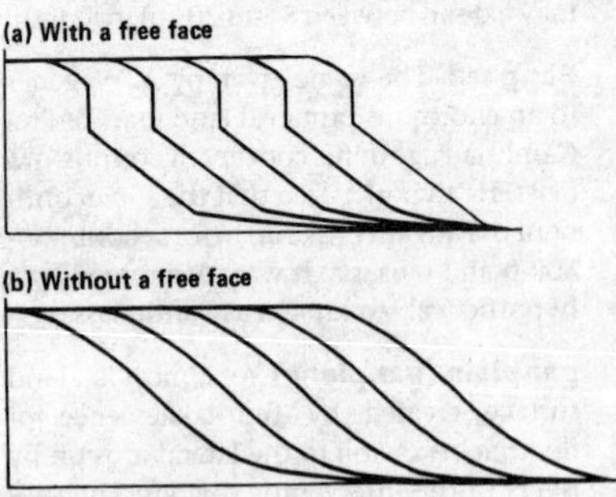

Figure 177 *Parallel retreat of slopes*

parallel roads Terrace-like phenomena that run horizontally at exactly the same levels on both flanks of a highland valley. Those in the Glen Roy area of Scotland are thought to mark the position of former PROGLACIAL LAKES of late-Pleistocene age.

parameter **1** A term used in statistics to denote an invariant numerical value of a population, e.g. mean, standard deviation, etc. **2** It is used more loosely as being synonymous with a VARIABLE or measurable quantity, e.g. volume, height, etc.

parametric statistical tests Statistics used in tests which make certain assumptions about the distribution of the background population data from which the samples are drawn, assumptions which are not always warranted, e.g. that the data are approximately normally distributed (NORMAL DISTRIBUTION). In general, the smaller the sample to be tested the less likely the validity of a *parametric test*. Thus, in the latter case and where normality cannot be assumed, NON-PARAMETRIC STATISTICAL TESTS should be used. The T-TEST is the best-known example of a parametric test. The latter type of test can only be applied to data on an interval scale (SCALES OF MEASUREMENT).

paramos The bleak and barren plateaux of the high Andes, between the snowline and the PUNA. They carry an Arctic–Alpine vegetation in which mosses and lichens are common.

paraselene Synonymous with *mock Moon*. It is an image of the Moon which is formed in the same way as the PARHELION of the Sun but because of its weaker light is less bright than the parhelion.

parasitic cone A small cone occurring on the flanks of a large VOLCANO. Also termed an ADVENTIVE CONE.

paratectonics The tectonics of the relatively stable areas of the present crust of the Earth, which may have been deformed in earlier OROGENIES and DIASTROPHISMS but which now form CRATONS.

parautochthonous Used of folded and transported bodies of rock, generally in NAPPE structures, that have been displaced a long distance from their root zone, in contrast to AUTOCHTHONOUS nappes.

parent material The material from which soil is formed. The soil may form *in situ* on a weathered rock surface or it may be derived from the action of soil-forming processes on SUPERFICIAL DEPOSITS which bear no resemblance to the underlying bedrock. Certain characteristics of the parent material (e.g. mineral composition, permeability, texture) will have an important effect on the soil-forming processes, but they are also affected by climate, vegetation and human interference. The uppermost part of the parent material is referred to as the C-HORIZON when describing a SOIL PROFILE.

parhelion (mock Sun) An image of the Sun which appears as an intensely bright spot on or near to the solar halo. It is caused by the refraction of sunlight by a particular type of ice crystal, especially in CIRRUS clouds. It is much brighter than the mock Moon (PARASELENE).

park savanna An expression referring to the type of SAVANNA in which trees are scattered relatively thinly among the grasslands, in the manner of a European park.

parna A term for a type of loessic clay (LOESS) found in the arid interior of Australia. Much of the deposit comprises clay pellets but with a certain amount of sand. It occurs as a thin, discontinuous sheet, perhaps formed by AEOLIAN processes on dry lake floors. *See also* CLAY DUNE, LUNETTE.

paroxysmal eruption A very violent volcanic ERUPTION of the PLINIAN ERUPTION type, usually resulting in the top of the volcanic cone being blown off, e.g. Mt St Helens, USA, in 1980.

Parshall measuring-flume A type of FLUME used by hydrologists to measure the flow velocity of water in open channels. It has both an expanding length and a contracting length separated by a throat and sill at which a VENTURI EFFECT operates. It was developed by the US Department of Agriculture. *See also* VENTURI FLUME.

partial area model A type of hydrological model that forecasts the saturated area around a stream head as a means of flood HYDROGRAPH prediction. It contrasts directly with the OVERLAND FLOW model proposed earlier by R. E. Horton. *See also* SOURCE AREA.

partial drought A climatological term used in Britain to distinguish a period of twenty-nine consecutive days during which the precipitation of any single day does not exceed 0.25 mm, in contrast to an ABSOLUTE DROUGHT. *See also* DROUGHT, DRY SPELL.

partial duration series A series of events used in FLOOD FREQUENCY analysis, referring to all the flood peak discharges above a given threshold discharge.

particle form A composite term used to describe the shape (morphology) of sedimentary particles, as opposed to the PARTICLE SIZE. The functional expression (F) is shown as:

$$F = f(Sh, A, R, T, F, Sp)$$

where *Sh* denotes the particle shape, *A* its angularity, *R* its ROUNDNESS INDEX, *T* its surface texture, *F* its FLATNESS INDEX and *Sp* its SPHERICITY.

particle size The diameter of a particle, usually determined by comparison with that of other particles by a technique known as *particle-size analysis*, and of considerable importance in geomorphology, soil science and sedimentology. Particle size may be determined either by sieving through sieves of different mesh, plotting the data on a GRADING CURVE, or by noting their settling time when placed in a body of water. The British Standard classification is as follows: >200 mm (BOULDER); 60–200 mm (COBBLE); 2–60 mm (GRAVEL); 0.6–2 mm (coarse sand); 0.2–0.6 mm (medium sand); 0.06–0.2 mm (fine sand); 0.002–0.06 mm (SILT); <0.002 (CLAY). NB Other classifications differ in detail. *See also* GRAIN SIZE, SOIL TEXTURE, WENTWORTH SCALE.

particle-size distribution GRADING CURVE.

particulates Solid particles or liquid droplets that move through or are suspended in the atmosphere.

Pasadenian A phase of mid-Pleistocene folding (OROGENY) in N America, named from a suburb of Los Angeles, California.

pascal (Pa) A unit of STRESS equal to a force of one NEWTON per square metre. It is named after B. Pascal, a 17th-cent. scientist, and is now recognized as the SI unit of pressure. 1 mb = 10^2 Pa; 1 lb force per sq in = 6894.8 Pa.

Pascal's law A law of hydrostatics formulated in 1646 by B. Pascal. It states that in a perfect fluid the pressure exerted on it at any point is transmitted undiminished in all directions.

passive glacier One of the three types of glacier in a dynamic classification, namely active, passive and dead. A passive glacier continues to receive nourishment by ACCUMULATION, but it is only of low magnitude. Since summer melting is also small the glacier displays little activity and its rate of flow is extremely slow in contrast to that of a fast-moving active glacier with its positive mass balance. Passive glaciers occur mainly in the centre of continents or major ice-sheets, e.g. Greenland, where precipitation is low. *See also* COLD GLACIER, SUBPOLAR GLACIER, TEMPERATE GLACIER.

passive remote-sensing system A REMOTE SENSING system that relies on ELECTROMAGNETIC RADIATION reflected or emitted from the Earth's surface as its source. A camera and a RADIOMETER are both examples of passive remote-sensing systems inasmuch as the Sun provides the electromagnetic radiation that they record.

Pastonian A name given to the youngest stage of the Lower Pleistocene in Britain. It is represented by a deposit of estuarine silts and fresh-water peat, indicating a temperate climate prior to the colder conditions of the succeeding BEESTONIAN stage, which marks the beginning of the Middle Pleistocene. It is named from its type site at Paston, on the Norfolk coast (although the most complete sequence of this interglacial stage has been recorded in the cliffs at Beeston). [*197*]

paternoster lakes A series of lakes in a glaciated valley, impounded behind moraines or in rock basins closed at their

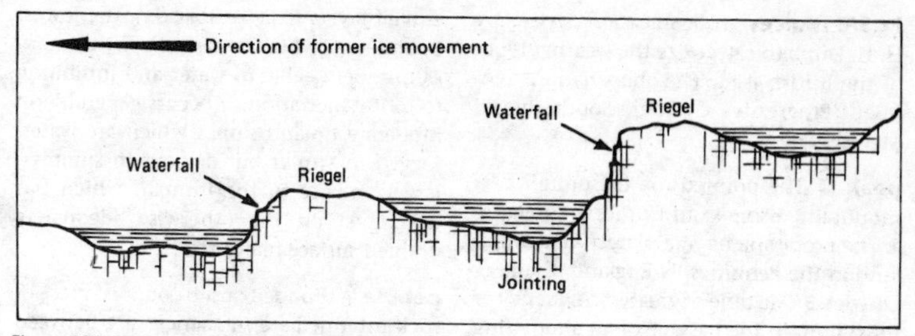

Figure 178 *Paternoster lakes on a glacial stairway*

lower ends by rock bars. Their name comes from their likeness to a paternoster or string of beads. When the lakes occupy a valley comprising a descending series of rock steps the feature is termed a GLACIAL STAIRWAY. [*178*]

patina A discolouring or surface film occurring on rock exposures owing to CHEMICAL WEATHERING.

patterned ground A collective term for the approximately symmetrical forms that are characteristic of, but not necessarily confined to, that part of the ground surface which is subject to intensive frost action. For its optimum development patterned ground requires a combination of moderate moisture and frequent cycles of FREEZE–THAW ACTION but it also depends on the susceptibility of the soil to frost SORTING and GELIFLUCTION and the presence or absence of vegetation. It has been claimed that a vegetation cover deters frost sorting more than gelifluction and often controls the lower altitudinal limit of sorted patterns, while the upper limit is controlled by perennial snow or ice cover and by lack of thawing. Most classifications of patterned ground are based on geometric form and presence or absence of sorting. *See also* CIRCLES, NETS, POLYGONS, STEPS, STRIPES. [*179*]

pattern recognition A REMOTE SENSING term referring to an automated process through which unidentified patterns can be classified into a limited number of discrete classes through comparison with other class-defining patterns or characteristics. Pattern recognition is an essential part of the classification of remotely sensed IMAGES and is used as an aid to image interpretation. *See also* IMAGE CLASSIFICATION: NON-MANUAL.

pavement 1 In general, a bare rock surface fashioned by glacial erosion (LIMESTONE PAVEMENT), aeolian scour (DESERT PAVEMENT) or simply by weathering processes. **2** In different parts of the world local names refer to bare or stony surfaces: *see also* CRUST, GIBBER, HAMADA, PAN FAN, REG, SERIR.

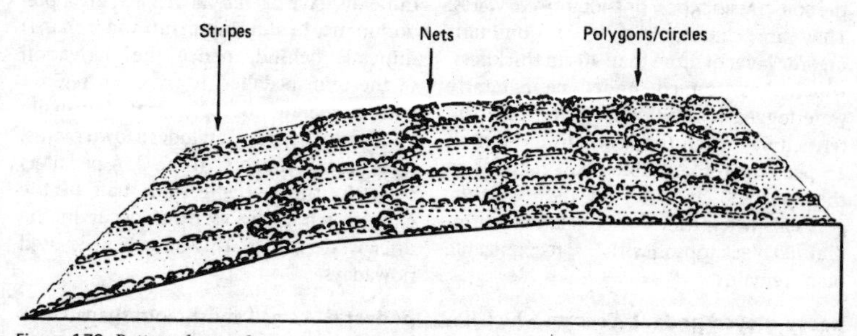

Figure 179 *Patterned ground*

PCSM indices A classification devised by H. H. Lamb to categorize the weather types of the British Isles. The abbreviations represent: Progressive, Cyclonic, Southerly and Meridional.

peak **1** The pointed top or summit of a mountain; also *pike* in Cumbria, England. **2** The proper name of a plateau-like moorland in the Pennines, N England: the Peak District. **3** The time of greatest frequency or maximum in the passage of an alternating cycle, e.g. the peak flow of a river or flood.

peat Unconsolidated black or dark-brown soil material consisting largely of slightly decomposed or undecomposed fibrous vegetable matter that has accumulated in a waterlogged environment (ANAEROBIC). It is formed under cool, humid climatic conditions and in post-glacial times there were lengthy phases of peat growth during the ATLANTIC PERIOD, the SUB-ATLANTIC PERIOD and the TURBARIAN periods in Scotland. The majority of peat vegetation creates acid conditions with SPHAGNUM moss as the dominant species (BLANKET BOG, RAISED BOG), but alkaline or neutral peats can also be formed in such base-rich environments as a FEN or CARR. Peat is a very important source of fuel in some northern countries (e.g. Russia, Ireland), but it has low calorific value – about 75 therms per tonne, compared with 275 therms per tonne for bituminous coal. Diagenesis of peat leads to COAL and LIGNITE formation. *See also* WETLAND.

peat bog BOG.

peat soils One of the seven major groups in the SOIL CLASSIFICATION OF ENGLAND AND WALES. They are characterized by a dominant organic layer of more than 40 cm thickness, which has formed under permanently waterlogged conditions. Peat soils are termed *raw* when they contain more than 15 per cent recognizable plant remains in the upper 20 cm. They are termed *earthy peat soils* when they have a relatively firm, drained black topsoil with few recognizable plant remains.

peaty-gleyed podzol A PODZOL which has an acid organic layer of PEAT overlying the eluvial layers (ELUVIAL HORIZON), the lowest boundary of which is marked by a thin IRONPAN, impermeable to water and inhibiting to root penetration. It occurs generally on moderate upland slopes which are waterlogged in winter but dry out in summer, largely owing to the ironpan which has developed on the otherwise adequately drained surface materials. [*190*]

pebble A rounded or sub-rounded piece of rock intermediate in PARTICLE SIZE between the smaller GRAVEL and the larger COBBLE. It is not recognized on the British Standard Classification of Particle Size but is generally thought of as clastic stony material with a diameter between 10 and 50 mm. On the WENTWORTH SCALE of grade sizes, *pebbles* are classified in the 2–64 mm diameter division. Pebbles can be rounded by fluvial, glaciofluvial or aeolian action.

pebble gravel A term given to one of the higher terrace deposits of Lower Pleistocene age in the Thames valley, England. It lies on an eroded land surface at an elevation of some 130 m in Buckinghamshire and Hertfordshire and is thought to represent one of the earliest stages of the R. Thames drainage sequence when it flowed in a north-easterly direction across East Anglia.

ped An aggregate of soil particles formed by natural processes. If it is formed by artificial means, such as ploughing, it is termed a *clod*. Not to be confused with *pedon* (EPIPEDON). *See also* SOIL STRUCTURE. [*236*]

pedalfer A major division of soils in the USA. It is a leached variety in which base minerals have been removed leaving a predominance of aluminium (*al*) and iron (*fer*) minerals behind, hence the derivation of the term pedalfer. It occurs in regions with moderate to high rainfall (usually >600 mm/25 in) and includes BROWN EARTHS, LATOSOLS and PODZOLS. In the USA pedalfers occur in the wetter eastern half of the country while the PEDOCALS occur in the drier western half. This term is little used nowadays.

pedestal A small, mushroom-shaped rock pinnacle, with a narrow base and stem and

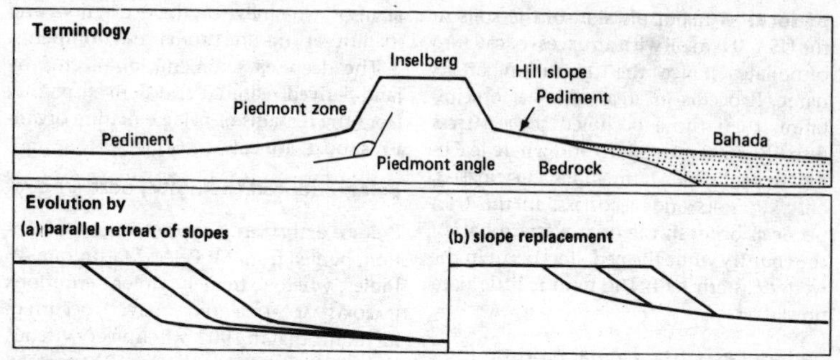

Figure 180 *Formation of a pediment*

a large head or cap rock. It is thought to have been formed, in an arid climate, by the action of sand-blasting (AEOLIAN) upon less-resistant rock strata. *See also* ZEUGEN.

pediment A gentle slope, cut in bedrock, occurring below a markedly steeper slope and extending at a low gradient down towards a river or alluvial plain. The pediment is separated from the steeper upper slope (CONSTANT SLOPE, DEBRIS SLOPE) by a relatively rapid change of slope angle in a transitional zone termed the *piedmont zone*, in which the change of gradient is termed the *piedmont angle*. The pediment is generally concave in profile, albeit a gently concave profile, and it may be bare rock or carry a thin veneer of debris [*49*]. A gentle slope that resembles a *pediment* but which is formed by deposition is termed a BAHADA. Pediments have gradients of less than 10°, with most of them below 6°. A slope of slightly more than 10° which superficially resembles a pediment can be called a FOOT SLOPE. Pediment genesis has been explained by: **1** SURFACE WASH (SLOPE WASH) [*25*]; **2** RILL WASH; **3** lateral planation by PARALLEL DRAINAGE; **4** slope retreat by PARALLEL RETREAT OF SLOPES or SLOPE REPLACEMENT. All the processes can be grouped as the process of *pedimentation*, which some geomorphologists believe is instrumental in the formation of a PEDIPLAIN. It is possible that pediments are formed by more than one type of pedimentation process and although they are said to be typical of the arid and semi-arid regions they are also common in the humid tropics, while some authorities have recognized them in humid temperate climatic zones. Studies have shown that surface flow increases downslope on the pediment, enabling the same amount of debris to be transported across a progressively gentler angle. Thus over a short period it is virtually a TRANSPORTATION SLOPE. [*180*] *See also* PERIPEDIMENT.

pedimentation PEDIMENT.

pediplain (pediplane) A surface of low relief, broken by occasional residual hills (INSELBERG), thought to be produced by the coalescence of several PEDIMENTS. It is a relatively smooth surface with a gently concave profile and with relatively few streams. Some geomorphologists, notably W. Penck, believe that it is the final stage of a landscape in which PARALLEL RETREAT OF SLOPES is the dominant process. It can be contrasted with a PENEPLAIN, in which down-wearing is more dominant than back-wearing and where SLOPE DECLINE is most characteristic. It is claimed by L. C. King that the majority of the interior erosion surfaces of Africa are ancient *pediplains*. *See also* PANPLAIN. [*193*]

pediplanation The action of processes which taken together result in the formation of a PEDIPLAIN. Since it is currently believed that the latter is produced by the coalescence of PEDIMENTS, those processes which are operative in *pedimentation* must also play major roles in pediplanation, i.e. SLOPE WASH, RILL WASH, PARALLEL RETREAT OF SLOPES, lateral planation by PARALLEL DRAINAGE and SLOPE REPLACEMENT.

pedocal A major division of the soils in the USA. It is a soil with an excess of calcium carbonate, hence the derivation of its name. It occurs in areas of lower precipitation than those occupied by PEDALFERS (usually < 600 mm/25 in) and where leaching is negligible. It includes CHERNOZEMS, CHESTNUT SOILS and SIEROZEMS. In the USA pedocals occur in the drier western half of the country while the pedalfers occur in the wetter eastern half. The term is little used nowadays.

pedogenesis The natural process of SOIL FORMATION.

pedology The scientific study of soils as phenomena which occur naturally, taking into account their method of formation, their composition and their distribution, and to some extent the ways in which they are utilized (AGRONOMY).

pedon A long column of soil which includes both the PARENT MATERIAL and the SOLUM, and ranges between 1 and 10 m^2 in size. It has replaced the two-dimensional SOIL PROFILE in the USA, where the SEVENTH APPROXIMATION has been based on the pedon concept, linking contiguous pedons into *polypedons* and thence into *soil series*.

pegmatite A very coarse-grained igneous rock, whose PHENOCRYSTS are over 250 mm in length and with some crystals reaching lengths of over 1 m. It occurs in the form of veins and DYKES emanating from GRANITE, but the name is also linked with other coarse-grained rocks (e.g. *gabbro-pegmatite*) which are termed *pegmatitic rocks*. The pegmatites contain abundant accessory minerals produced by HYDROTHERMAL ACTIVITY and PNEUMATOLYSIS (e.g. tourmaline, topaz, fluorite and apatite) and are important economically because of their association with many rare elements in addition to tin and tungsten.

pelagic **1** The environment of the open ocean as distinct from the ocean floor. **2** The marine organisms which flourish independent of the ocean floor and shoreline environments. These include both the floating PLANKTON and the free-swimming NEKTON. Thus, they may be contrasted with BENTHOS, which live on the ocean floor and in littoral (NERITIC ZONE) environments. **3** The deep-sea sediments unaffected by land-derived material and derived mainly from the remains of pelagic marine organisms (OOZE, RED CLAY). [*162*]

pelagic deposit (pelagite) OOZE, RED CLAY.

Peléan eruption A type of volcanic eruption, named from Mt Pelée, Martinique, W Indies, where extremely violent eruptions (PAROXYSMAL ERUPTION) have occurred, including that in 1902, when a very viscous lava plug solidified in the vent. Ultimately, the cone and its lava spine were pulverized by the ejection of a NUÉE ARDENTE which devastated a wide area.

Pelé's hair Fine threads or filaments of volcanic glass produced during a volcanic eruption or the bursting of bubbles in a lava lake. It is named from Pelé, the Hawaiian goddess of fire, but since most of the Pelé's hair is produced from fluid basaltic lava it should not be confused with PELÉAN ERUPTION, in which the lava is viscous and acid.

Pelé's tears Tiny droplets (6–13 mm in length) of volcanic glass (LAPILLI) rejected during a volcanic eruption, usually of a basaltic origin. They are named from Pelé, the Hawaiian goddess of fire.

pelitic A term, once used synonymously with all ARGILLACEOUS rocks, which in current usage is confined only to metamorphosed aluminium-rich argillaceous rocks (*pelites*). *See also* PSAMMITIC, PSEPHITIC.

pelosols One of the seven major groups in the SOIL CLASSIFICATION OF ENGLAND AND WALES. It includes slowly permeable non-alluvial clayey soils that crack deeply in dry seasons. They are characterized by a brown, greyish or reddish, slightly mottled (MOTTLING) subsurface horizon which dries out into a blocky or prismatic structure. They can be argillic (ARGILLACEOUS), calcareous or non-calcareous in composition.

peneplain An undulating surface of low relief, interspersed with occasional residual hills, known as MONADNOCKS, and claimed to have been formed by the widening of

Figure 181 *Stages in the cycle of erosion (according to Davis)*

floodplains and the wearing down of interfluves (SLOPE DECLINE) by subaerial DENUDATION. It is regarded as the end-product of the NORMAL CYCLE OF EROSION, as defined by W. M. Davis in 1889 to describe a stage when RELATIVE RELIEF has been considerably reduced and rivers are, in theory, completely graded (GRADE). This theoretical concept of peneplain formation has been criticized by other geomorphologists on the grounds that, first, it was designed primarily to explain surfaces of low relief only in the humid temperate zone, thereby ignoring the arid and semi-arid lands; and second, that the length of time required for peneplain formation is so great that it is unlikely to have occurred during one cycle without a change in BASE-LEVEL leading to REJUVENATION. The peneplain can be contrasted with the PEDIPLAIN and the PANPLAIN, both of which are produced by the back-wearing rather than the down-wearing of slopes (PARALLEL RETREAT OF SLOPES). Some writers have used the term *peneplane* to describe a platform of marine erosion (ABRASION PLATFORM) but this usage should be discouraged. [*181*] *See also* SENILITY.

peneplanation The lowering of a land surface by subaerial denudation to form a PENEPLAIN. The spelling peneplaination is incorrect.

penetrometer An instrument designed to measure the vertical resistance of soil or snow to penetration by a rod or cone to a particular depth at a specified rate. It is used in the determination of the strength of materials (MATERIALS, STRENGTH CHARACTERISTICS OF). *See also* SOIL MECHANICS.

peninsula A narrow neck of land projecting into a body of water. The adjectival form is *peninsular*.

penitent 1 A spike of ice formed by melting of a glacier surface. The term is also used to describe a *snow-penitent*, on the surface of old compact snow. **2** An isolated, slightly inclined rock outcrop due to DIFFERENTIAL WEATHERING of joint planes in rocks with a joint pattern slightly removed from the vertical.

Penman's formula A climatological method of defining potential evapotranspiration (EVAPOTRANSPIRATION) devised by H. L. Penman between 1952 and 1963, in which he expressed evaporation loss in terms of sunshine duration (related to RADIATION), mean air temperature, mean air humidity and mean wind speed (which limit heat and vapour loss from the surface). It is used to determine when IRRIGATION is theoretically necessary for crop growth. *See also* EVAPOTRANSPIROMETER, THORNTHWAITE'S INDEX OF POTENTIAL EVAPOTRANSPIRATION.

Pennsylvania A period of geological time in the stratigraphical nomenclature adopted by the USA. It lasted for about 45 million years from the end of the MISSISSIPPIAN (about 325 million years ago) to the beginning of the PERMIAN (about 290 million years ago). It corresponds approximately to the Upper Carboniferous (CARBONIFEROUS).

pentad A name given to a period of five successive days and used in climatology because a pentad divides into the number of days in a year (365) more effectively than does a week.

percentile A value of a ranked data series expressed as a percentage division, e.g. the upper ten percentile indicates the value exceeded by 10% of the data.

perception A stimulus–response psychological process in which mankind makes subjective judgements in response to physical environmental stimuli. In 1955 F. H. Allport defined perception as: 'Our awareness of the objects or conditions about us. It depends to a large extent on the impressions that these make on our senses. It is the way things look to us, or the way they sound, feel, taste or smell. . . . Perception covers the awareness of complex environmental situations as well as single objects – things can and do look differently from the way they are.'

perched block **1** A large boulder left by a glacier in a precarious position on a formerly glaciated hillside or standing in a delicate state of balance, sometimes as a rocking stone, on a glacial PAVEMENT. **2** The expression is also used to describe blocks which have been left *in situ* by weathering of surrounding strata (PSEUDO-ERRATIC) in contrast to an eroded PEDESTAL. **3** An isolated boulder which may have fallen by gravity from a cliff face and which now stands in an unstable state of balance on the top of other rocks.

perched karst spring The emergence of underground water at the surface some height above the basement of a limestone massif owing to the presence of a relatively impermeable bed among the calcareous strata. By restricting the vertical movement of water this creates a PERCHED WATER-TABLE.

perched water-table A WATER-TABLE caused by the retention of water on an isolated relatively impermeable layer (PERMEABILITY) within the rock strata some height above the normal water-table, from which it is separated by layers of unsaturated rock. The top surface of a perched AQUIFER.

percolation In general, the process by which surface water moves downwards through cracks, joints and pores in soil and rocks. More specifically, the downward flow of water in saturated or near-saturated soil at HYDRAULIC GRADIENTS of 1.0 or less. *See also* LEACHING, LYSIMETER.

percoline A horizon along which soil water moves laterally by seepage in the soil. It is part of the process of THROUGHFLOW.

percussion marks CHATTER MARKS.

pereletok A Russian term for a frozen layer at the base of the ACTIVE LAYER in the soil, which remains unthawed for one or two summers. Most definitions of PERMAFROST would exclude such a temporarily frozen layer if it survived for only one summer, but some scientists include pereletok with permafrost. *See also* TALIK.

perennial stream A stream that flows at all times of the year. It contrasts with an INTERMITTENT STREAM.

perfect elasticity A property possessed by a material which is able to return to its original form after the removal of an applied force. *See also* ELASTIC LIMIT.

perfect plasticity A property possessed by a material which is unable to return to its original form after an applied force has been removed but retains a new form instead.

perforation deposit A conical mound of glaciofluvial material of KAME-like form thought to have been created by the so-called perforation hypothesis of kame formation. This suggests that SUPRAGLACIAL debris is accumulated in surface pools of a decaying ice-sheet and that the meltwater, by releasing frictional heat as it flows, melts through to the glacier base, thereby lowering the deposit and leaving it as a mound after the ice-sheet has disappeared.

pergelation A term describing the formation of PERMAFROST.

pergelisol A term introduced by K. Bryan in 1946 to describe perennially frozen ground or PERMAFROST.

perhumid MOISTURE INDEX.

pericline **1** A crustal fold structure in the form of a dome or a basin in which the beds dip around a central point (outwards in the case of a dome, inwards in the case of a basin). **2** Some geologists restrict the term to describe only a dome structure.

peridotite One of the ULTRABASIC ROCKS, consisting largely of olivine (40–90 per cent), hence its predominantly dark green colour. It is coarsely crystalline and may be with or without other ferromagnesian minerals.

perigean tide A tide occurring when the Moon is in PERIGEE. It results in an increased high tide and a decreased low tide.

perigee **1** The point in the orbit of the MOON when it is closest to the Earth (i.e. 348,292 km). It contrasts with APOGEE. **2** The point in the orbit of any planet when it is nearest to the Earth.

periglacial A term introduced in 1909 by W. Loziński to describe the type of climate and the climatically controlled surface features adjacent to the Pleistocene ice-sheets. Subsequently the meaning has been extended to designate non-glacial processes and features of cold climates regardless of age and of any proximity to ice-sheets. Purists, however, restrict its usage to describe the climate, processes and features created by FREEZE–THAW ACTION in a zone bordering ice-sheets, whether or not it coincides with the zone of PERMAFROST. There is considerable overlap between periglacial and permafrost environments but they are not synonymous, since GELIFLUCTION and some types of PATTERNED GROUND are periglacial but are not necessarily associated with permafrost. Although the periglacial concept is somewhat imprecise and broad enough to defy quantification, a periglacial MORPHOGENETIC REGION has been defined by L. Peltier. It is characterized by an average annual temperature ranging from –15° C to –1° C and an average annual rainfall (excluding snow) ranging between 127 mm and 1,397 mm. Apart from frost activity the other essential criterion of a periglacial environment is seasonally snow-free ground. *See also* ACTIVE LAYER, CONGELIFRACTION, CONGELITURBATION, CRYOPLANATION, FROST ACTION, FROST-HEAVING, FROST-THRUSTING, GEOCRYOLOGY, ICE SEGREGATION, ICE WEDGE, NEEDLE ICE, NIVATION, NUBBIN, PALSA, PERELETOK, PINGO, ROCK GLACIER, SAND WEDGE, SOLIFLUCTION, SORTING, TALIK, THERMOKARST.

periglacial cycle A cycle of erosion dominated by FROST ACTION and SOLIFLUCTION, proposed in 1950 by L. Peltier as a counterpart to the cycles proposed by W. M. Davis (e.g. CYCLE OF EROSION). It has been criticized as too idealized and as having failed to take the action of running water adequately into account. *See also* CYCLE.

perihelion The nearest point to the Sun reached by a heavenly body during its orbit. In the case of the Earth this is 147.3 million km (92 million miles) on 3 January. *See also* APHELION.

period A geological time interval in the standard chronostratigraphic classification (CHRONOSTRATIGRAPHY). A period comprises a number of EPOCHS and several periods make up an ERA. The body of rock formed in this time period is known as a SYSTEM.

periodicity An oscillation in a time series of events which recurs at approximately equal intervals of time. It is frequently used in climatological research (CLIMATOLOGY) in attempts to recognize recurrent weather patterns over long periods in order to improve long-range weather forecasting (ANALOGUE WEATHER FORECASTING). The climatic data are plotted on a deviational graph known as a *periodogram*.

peripediment A rarely used term describing a low-angle alluvial surface adjacent to a true PEDIMENT. A preferable term is BAHADA.

permafrost A condition existing below the ground surface, irrespective of its texture, water content, or geological character, in which the temperature in the material has remained below 0° C continuously for more than two years and, if pore water is present in the material, a sufficiently high percentage is frozen to 'cement' the mineral and organic particles. The term was introduced by S. W. Muller in 1947 to describe permanently FROZEN GROUND, but permafrost has been subdivided into *continuous and discontinuous permafrost*, while *sporadic permafrost* is confined to alpine environments. *Continuous permafrost* covers about 20.85 km^2 × 10^6 of the Earth's surface, but where it breaks up into patches, as it merges with seasonally frozen ground, it is known as *discontinuous permafrost*, which covers 17.30 km^2 × 10^6 of the land surface. It has

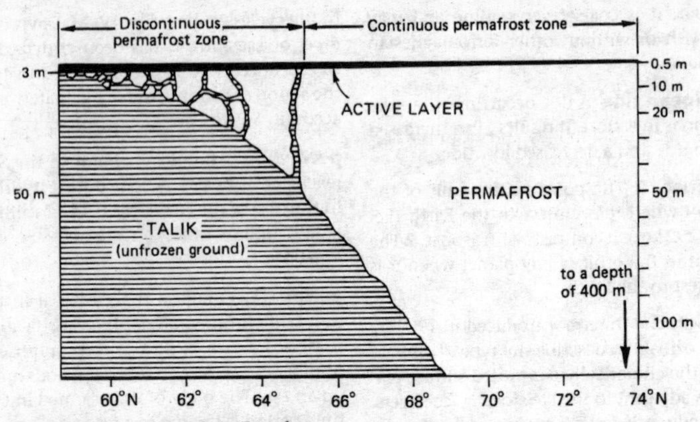

Figure 182 *Permafrost in northern Canada*

been claimed that permafrost underlies some 26% of the world's land surface (including glaciers), but it also occurs offshore in some Arctic and Antarctic latitudes. Its aggradation and degradation are controlled by the thermal regime, which in turn is affected by climatic, geomorphic and vegetation factors, but it is suggested that most permafrost may have originated during the Pleistocene. Today the upper part of most *continuous permafrost* is in balance with the present climate, while most *discontinuous permafrost*, both at surface and base, is either out of balance with the present climate or merely in very delicate equilibrium. *See also* ACTIVE LAYER, ICE WEDGE, PERELETOK, STEFAN'S METHOD, TALIK, THERMOKARST. [*182*]

permafrost table The more or less irregular surface which marks the upper limit of PERMAFROST and separates it from the overlying ACTIVE LAYER in the soil. It is usually a rigid surface capable of bearing heavy loads without deforming.

permanent hardness of water Water hardness which cannot be removed by boiling because of the very high concentration of calcium and magnesium sulphate. *See also* HARDNESS SCALE.

permeability The ease with which liquids (or gases) can pass through rocks or a layer of soil. The permeability of a material will depend on its GRAIN SIZE or PARTICLE SIZE and on the shape and packing of the grains or particles. Permeability is expressed as a rate of discharge per unit area (m^3/day) and some typical examples are as follows: least permeable, clay; most permeable, sand and gravel. Most writers regard *permeable rocks* as being synonymous with *pervious rocks*. *See also* DARCY'S LAW, HYDRAULIC CONDUCTIVITY, POROSITY.

permeameter An instrument used in the laboratory for measuring the coefficient of PERMEABILITY of a soil sample. For poorly permeable materials (clay, silt) a *falling-head permeameter* is used, while for more permeable materials (sand, gravel) a *constant-head permeameter* is used.

Permian The final geological period of the PALAEOZOIC era, extending from about 290 million years ago to 245 million years ago. It succeeded the CARBONIFEROUS and preceded the TRIASSIC, with which it is linked to form the NEW RED SANDSTONE of Permo-Triassic age because of the difficulty in Britain of recognizing a boundary between the two stratigraphic systems. Notwithstanding this difficulty the Palaeozoic–Mesozoic boundary is placed above the Permian. It is named from Perm in Russia, where marine rocks of this age exhibit fairly complete stratigraphic sequences. In Britain, however, continental conditions prevailed for much of the time, during which thick layers of bright red sandstone were formed. In N England there is evidence of marine conditions, exemplified by the marls and limestones (MAGNESIAN LIMESTONE). The Permian saw the extinction of many of the corals and some brachiopods and trilobites, but the increasing importance of reptiles. Among the flora the PTERIDOPHYTA were superseded by the GYMNOSPERMS.

persistence 1 An ecological term referring to the continued occupation of a site by a plant or animal which would no longer be able to colonize that site if necessary. **2** A meteorological term describing the type of weather pattern which exists for a longer than average period of time.

persistence effect The tendency for a number of values or events of similar magnitude to be grouped together in time.

perspective projection A category of MAP PROJECTION which includes the CONICAL PROJECTION, the CYLINDRICAL PROJECTION and the AZIMUTHAL PROJECTION.

perturbation A meteorological term to describe any disturbance in the STEADY STATE of a system, especially when referring to changes or breaks in the ZONAL CIRCULATION of atmosphere.

Peru current A cold current which flows northwards along the western coastline of S America, caused by the upward welling of colder water in response to the northward deflection of the West Wind Drift and the transference of surface water westwards across the Pacific as the South Equatorial current. Also called the Humboldt current. This cold current causes a marked equatorward bend of the global isotherms and has a pronounced effect on the coastal climate of Chile and Peru. *See also* BENGUELA CURRENT. [*168*]

pervious PERMEABILITY.

Petermann's projection A MAP PROJECTION comprising an eight-pointed star shape based on a recentred AZIMUTHAL EQUIDISTANT PROJECTION.

Peters's projection A type of MAP PROJECTION introduced by A. Peters in 1973 in an attempt to escape from the prevailing Eurocentric concept of the world. It is a modified CYLINDRICAL EQUAL-AREA PROJECTION based on the standard parallels of 46°N and S in which

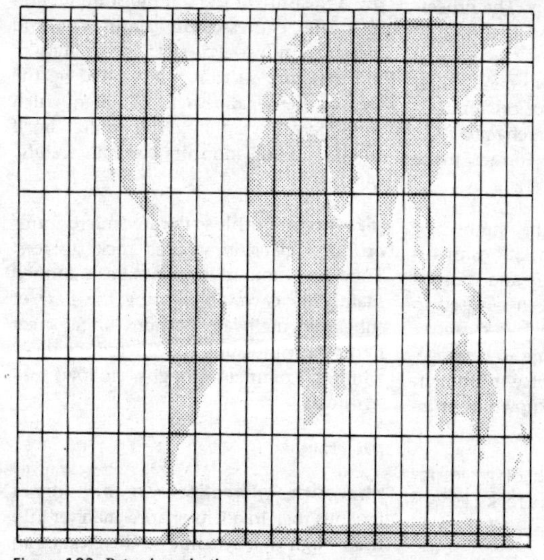

Figure 183 *Peters's projection*

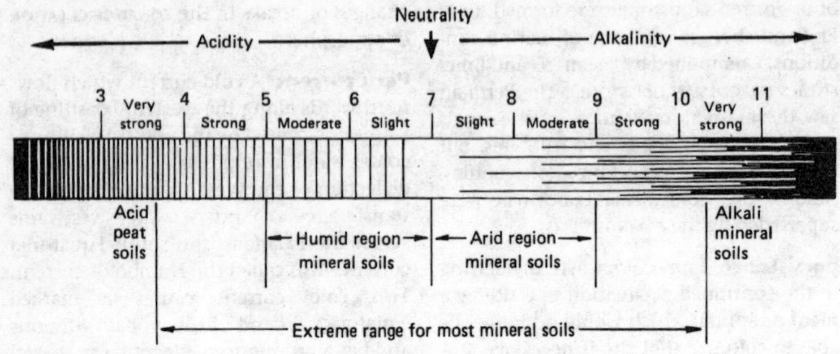

Figure 184 *pH index*

the properties of shape and equal area are lost owing to distortion. The usual GRID of 180 meridians from W to E and 180 parallels from N to S is replaced by a decimal degree network which divides the globe both E and W and N and S into 100 fields each without losing accuracy of direction. It is claimed that the distortions are distributed at the poles and the equator, leaving the Earth's more densely populated zones depicted in proper proportion to each other. It is sometimes referred to as the Orthogonal Map of the World. [*183*]

petrifaction, petrification The process by which any organic object is turned to stone. This is achieved by the gradual replacement of the tissues by mineral matter from percolating solutions of calcium carbonate, silica, etc., which thereby preserves the smallest detail of the former tissue structure.

petrofabric analysis The study and measurement of the spatial relationships of the rock fragments, particles and mineral grains of a rock or till. It can be used to determine the direction of palaeo-currents, the direction of stress leading to cleavage, or the direction of a former ice-movement (in the case of a till). Also known as *petrofabrics*.

petrogenesis The branch of PETROLOGY which examines the origins of rocks, especially igneous rocks.

petrography LITHOLOGY.

petroleum A complex mixture of HYDROCARBONS, derived from crude oil, with a carbon content of 83–87%, a hydrogen content of 11–14% and minor amounts of oxygen, nitrogen, sulphur and traces of metals (e.g. lead). It occurs in sedimentary rocks of marine origin in suitable reservoirs. Its literal meaning is rock-oil, indicating that it differs from animal and vegetable oils.

petrology The scientific study of rocks. It includes GEOCHEMISTRY, LITHOLOGY, MINERALOGY and PETROGENESIS.

pF A measure of the soil moisture SUCTION at a site, expressed in centimetres and measured by a TENSIOMETER. The strength of soil moisture suction is indicated by the $\log_{10}$ of the negative head of water, which will increase rapidly with quite small changes of soil moisture content. *See also* CAPILLARITY.

p-forms Smoothly sculptured micro-forms on bare, glacially scoured rock surfaces. They are thought to have been formed mainly by CAVITATION during the flow of subglacial meltwater streams under great hydrostatic pressure. They include GROOVES, winding channels, troughs, hollows and POT-HOLES.

pH PH INDEX.

phacolith, phacolite An intrusion of igneous rock in a lens-shaped manner into the trough of a syncline or the crest of an anticline without too much disturbance of the strata themselves, i.e. it is a concordant intrusion.

phanerophytes One of the six major floral life-form classes recognized by the Danish botanist, Raunkiaer, based on the position of the regenerating parts in relation to the exposure of the growing bud to climatic extremes, and without respect to taxonomy. The phanerophytes bear buds well above ground-level (five subdivisions are made based on height above soil-level) and are therefore fully exposed to climatic extremes. Consequently, they are most common in regions with warm, moist conditions.

Phanerozoic A Greek term meaning 'obvious life' which has been applied by geologists to the sedimentary rocks that have been formed since actual remains of plant and animal life (FOSSIL) first appeared in the geological stratigraphic record. It refers, therefore, to all rocks formed from CAMBRIAN times onwards, in contrast to those which have only primitive life-forms (CRYPTOZOIC). The Phanerozoic has given its name to the geological time-scale (BIOSTRATIGRAPHY, LITHOSTRATIGRAPHY).

phase **1** In geology, a variety differing slightly from the normal type; a facies. **2** In geophysics, an event on a seismogram marking the arrival of a group of SEISMIC WAVES and indicated by a change of amplitude and/or period. **3** In GENERAL SYSTEMS THEORY, a distinct type, region or economy of systems operation, commonly demarcated by thresholds. The *phase rule*, which governs chemical equilibrium relationships, the number of DEGREES OF FREEDOM is two greater than the difference between the number of components and the number of phases ($F = C - P + 2$).

phenoclast A CLAST in a sedimentary rock larger than 4 mm in diameter.

phenocrysts The large, very conspicuous crystals set in the matrix or fine-grained ground mass of PORPHYRITIC igneous rocks, suggestive of slow cooling.

phenology The study of periodicity in the life-cycle of plants and animals in relation to climatic events, e.g. the time of leafing, flowering, fruiting in various localities. *See also* BIOLOGICAL PRODUCTIVITY, ECOSYSTEM.

pH index A number used to express the hydrogen-ion concentration of a solution and hence its degree of acidity. It is widely used to denote the acidity or alkalinity of soils. It was introduced by S. P. Sorensen in 1909 and is measured in terms of the negative index of the logarithm of the hydrogen-ion concentration expressed in gram ions per litre of solution. The scale is from 0 to 14, with a *neutral* solution having a pH of 7 (i.e. one part in 10 million or (10^{-7}). An *alkaline* solution will range from pH 7 to a strongly alkaline solution of pH 14 (10^{-14}), while an acid solution will range from pH 7 down towards zero, with a very acid solution having a pH of 3 (i.e. 10^{-3}). Most British soils have pH values between 5 and 7. [*184*]

phi scale A scale used in sedimentary PETROLOGY to describe and delimit the range of particle sizes (PARTICLE SIZE) into a number of size classes to indicate degrees of SORTING. It is based on a negative logarithm, unlike the millimetre scale of particle size.

phonolite A fine-grained lava with PHENOCRYSTS of feldspar and nepheline and the extrusive equivalent of nepheline (SYENITE).

phosphate A compound salt of phosphoric acid with a complex mineralogy. *Phosphatic deposits* include: **1** *marine phosphates*, which are rare as sediments but occasionally form phosphatic limestones; **2** BONE BEDS and COPROLITES; **3** GUANO. Phosphate is one of the three main elements needed for successful plant growth. A *superphosphate* is one that has been treated with sulphuric acid to form an important agricultural fertilizer.

phosphorite Rock phosphate, containing various calcium phosphates most of which are derived from apatite. *See also* PHOSPHATE.

phosphorus compounds in the soil PHOSPHATE.

photic zone The shallow water layer of the ocean, lying above a level of 150 m depth, in which light can easily penetrate and make PHOTOSYNTHESIS possible. It contrasts with the APHOTIC ZONE, DIPHOTIC ZONE.

photochemical fog A state of poor visibility caused by the chemical reaction of sunlight on HYDROCARBONS in the atmosphere. Water droplets are not present and thus it must be distinguished from other types of FOG because of the absence of moist air. In the strictest sense it should also be distinguished from SMOG, but the constant interchange of the terms, especially in the USA, has made them virtually synonymous. Photochemical fog is really a HAZE produced when sunlight reacts with hydrocarbons and nitric oxides present in vehicle exhausts. Many world cities suffer varying degrees of photochemical fog, but the most notorious is Los Angeles, California, where large quantities of nitrogen dioxide and ozone can be generated, largely by road traffic during daylight hours, to which hazard industrial emissions are added. Owing to the commonly developed phenomenon of an INVERSION OF TEMPERATURE, Los Angeles, in its mountain-girt basin, regularly suffers from photochemical fog which endangers human health when it reaches particular concentrations. *See also* AEROSOLS. [*185*]

photocontour map A TOPOGRAPHIC MAP produced from information derived from aerial photographs using STEREOSCOPES.

photogrammetry The science of obtaining reliable measurements by means of photography. The principles of photogrammetry are applicable to ground photographs (*terrestrial photogrammetry*) as well as to AERIAL PHOTOGRAPHS (*aerial photogrammetry*). For geographical applications, as in topographical mapping or in height or distance measurement, aerial photographs are usually preferred.

photomap A quick method of map production by adding map data (e.g. settlement names, boundary lines) to a PHOTOMOSAIC. It normally does not contain contours (PHOTOCONTOUR MAP).

photometeor Any luminous phenomenon in the atmosphere that is produced by diffraction, refraction or reflection of light, e.g. BROCKEN SPECTRE. *See also* FOGBOW, HALO, RAINBOW.

photomosaic A grouping of overlapping vertical AERIAL PHOTOGRAPHS (MOSAIC) or parts of aerial photographs joined together to leave minimal scale alterations. It can be controlled or uncontrolled (CONTROL POINT, STATION).

photoperiodism The growth and flowering response of flora in relation to latitudinal and seasonal changes in the length of daylight hours.

photo-relief A type of map-shading in order to give the impression of a photograph of a relief model.

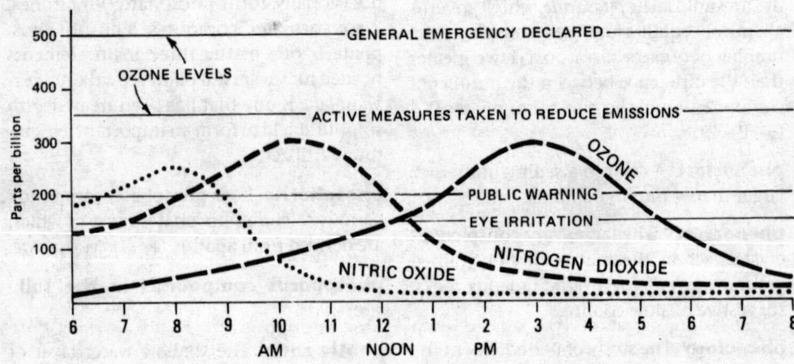

Figure 185 *Photochemical fog during a typical pollution day in Los Angeles*

photosynthesis A complex process occurring within the cells of green plants where sunlight is utilized in combination with carbon dioxide and water to produce oxygen and simple sugar or food molecules.

photothermograph An elaborate piece of meteorological equipment, devised in the 19th cent., to produce a continuous record of DRY-BULB and WET-BULB THERMOMETER readings in graphic form.

phototropism The response of a plant or animal to a source of bright light, e.g. the orientation of a flowering plant towards the sun, or the flight of an insect towards a light at night.

phreatic activity The violent reaction between a hot LAVA flow when it enters a lake, river or the sea and quickly chills at the surface. *See also* PILLOW LAVA.

phreatic divide A hydrological term referring to an underground WATERSHED. Such divides of GROUNDWATER basins need not coincide with surface topographic watersheds.

phreatic eruption A type of volcanic ERUPTION produced when underground lava comes into contact with groundwater, thereby generating a sudden, violent emission of steam and mud at the ground surface.

phreatic water GROUNDWATER.

phreatophyte A plant whose root system has the ability to descend to such great depths that it can tap GROUNDWATER reservoirs. It is found along dry water-courses in arid and semi-arid regions.

phyllite A clay sediment altered by low-grade regional METAMORPHISM into a metamorphic rock midway between a SLATE and a SCHIST. It is coarser-grained and less perfectly cleaved than a slate and comprises a foliated (FOLIATION) pelitic structure.

phylum A unit or group in the taxonomic classification of fauna (TAXONOMY). Each phylum is composed of one or more classes (CLASS). Its corresponding category in the plant kingdom is a DIVISION. Its plural form is *phyla*.

physical geography GEOGRAPHY, PHYSICAL.

physical geology The study which combines both GEOLOGY and aspects of GEOMORPHOLOGY.

physical systems A research methodology based on scientific method and analogous to GENERAL SYSTEMS THEORY, but differing from it in being characterized by a greater dissection of the specific problem into its component parts, such that the operation of each part and the interactions between the parts can be conveniently examined. In this way it becomes possible to synthesize the components into a working whole. A general-systems approach is more concerned with the operation of a SYSTEM as a whole rather than with detailed study of individual elements of the physical systems. *See also* SYSTEMS ANALYSIS.

physical weathering MECHANICAL WEATHERING.

physiographic climax CLIMAX.

physiography **1** An outmoded term for the study of landforms, now replaced by the term GEOMORPHOLOGY. **2** A term for the combined scientific study of GEOMORPHOLOGY, PEDOLOGY and BIOGEOGRAPHY.

physiological drought A temporary state of DROUGHT which affects plants during daytime, often causing them to wilt, owing to the water losses by TRANSPIRATION being more rapid than the uptake by roots, although the soil may have an adequate supply of water. During the night, when transpiration slows down, the plants normally recover.

phytoclimax The climax of vegetation in relation to climate (CLIMAX).

phytogenic dune A type of sand-dune which is almost completely covered by vegetation, having originally been 'fixed' by marram grass or other vegetation.

phytogeography A study of the geographical distributions of plants on the Earth's surface.

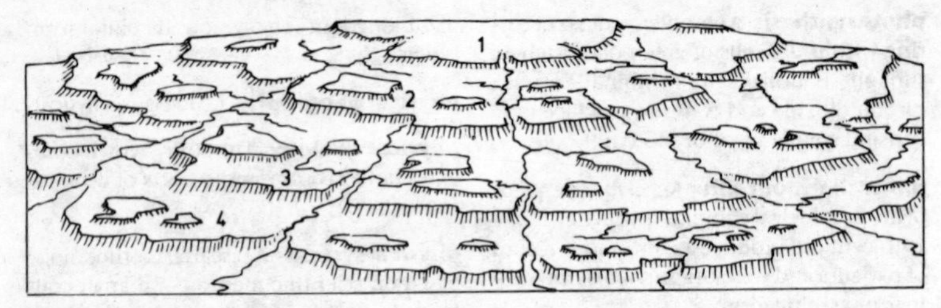

Figure 186 *Piedmonttreppen*

phytogeomorphology The study of the relationships between landforms and vegetation.

phytokarst BIOKARST.

phytomorphic soils The well-drained ZONAL SOILS that have developed naturally in an area.

phytoplankton Tiny, sometimes microscopic, PLANKTON in a water body. They consist of floating plant life, including DIATOMS and sargassum weed. They are the vegetable counterparts of ZOOPLANKTON.

piedmont Literally (in French), the foot of a mountain. Used to describe the gentle slope leading down from the steep mountain slopes to the plains and including both the PEDIMENT and the accumulation of COLLUVIAL (COLLUVIUM) and ALLUVIAL material which forms a low-angle slope beyond the pediment (BAHADA). The term is used as a proper name in N Italy and in the Appalachians of the USA. [*180* and *189*]

piedmont angle PEDIMENT.

piedmont benches PIEDMONTTREPPEN.

piedmont glacier The name given to the coalescence of a number of valley glaciers at the base of a mountain range. The combined glacier is generally of great breadth in relation to its length, e.g. the Malaspina glacier, Alaska. During the Quaternary most of the valley glaciers of the Alps contributed to broad piedmont glaciers. [*92*]

piedmonttreppen A German term, introduced by the geomorphologist W. Penck, to describe the succession of step-like benches (*treppen*) which occur around the flanks of the Black Forest and other mountains. Penck believed that they resulted from continuous accelerated uplift of an expanding dome, the summit area of which would be a PRIMÄRRUMPF. The piedmonttreppen are similar landforms to PEDIMENTS but differ in so far as the latter are bounded at their upper ends by steep risers while the piedmonttreppen risers are very indistinct. The *treppen concept* of Penck has not met with universal support by other geomorphologists. [*186*]

pie-graph An alternative name for a *wheelgraph* or *divided-circle graph*, in which a circle is divided into sectors, each proportional to the value it represents. In the strictest sense this type of visual representation of quantitative data is no more than a diagram and should be compared with the true *line-graphs*, in which a series of points is plotted by means of COORDINATES and then joined by a line (GRAPH).

piezometer An instrument for measuring the pressure head of liquids. It comprises an open-topped tube containing mercury for high pressures and water or kerosene for low pressures.

piezometric surface A subterranean surface marking the level to which water will rise within an AQUIFER.

pike PEAK.

pileus cloud A cloud which develops into a smooth white cap at some distance above a CUMULUS cloud. It is caused by a temporary updraught of moist air from the vigorous convectional activity of the cumulus cloud.

pillow lava A lava which has cooled and solidified into an agglomeration of rounded pillow-like masses (< 2 m diameter) each of which exhibits a VESICULAR interior, a concentric banded structure and fine-grained or 'glassy' skin. The morphology is thought to result from the extrusion of lava under water leading to a sudden cooling. *See also* PHREATIC ACTIVITY.

pimple mound MIMA MOUND.

pinch and swell BOUDINAGE.

pinch-out A geological term referring to the way in which a rock stratum, especially a seam of coal, thins out and disappears.

pingo An Eskimo term for a domed, perennial ice-cored mound of earth formed as a HYDROLACCOLITH in a PERIGLACIAL environment. The largest pingos rise to 60–70 m in height but the smaller ones are difficult to distinguish from PALSAS. Two methods of pingo formation have been suggested: **1** The freezing of a body of entrapped water within a silted or vegetation-filled lake in a PERMAFROST zone to create a massive ice core which domes the lake sediments as it expands. This is the cryostatic process, thought to have been responsible for the formation of the so-called *closed-system* pingos of the *Mackenzie Delta type* (N Canada) [*253*]. **2** The second method, known as the artesian concept, refers to groundwater flowing under artesian pressure below thin permafrost or in TALIKS within permafrost. The water eventually freezes as it forces its way upwards, thereby forming an ice core that heaves the surface into a dome. Such pingos are of the *open system* variety or the *East Greenland type*. Ultimately the doming leads to dilation cracking of the overlying material, the exposure of the ice core to thawing and eventually to collapse of the mound's centre. The circular rampart pattern of a collapsed pingo has been used to identify fossil pingos and therefore former areas of permafrost during the Pleistocene. [*187*]

pinnate drainage A feather-like drainage pattern containing a large number of closely spaced tributaries.

pioneer A term given by biologists to a plant or a plant community that occupies a newly available site, such as a cooled lava-flow, a newly emerged shoreline, a freshly deposited flood sediment or a new sand-dune.

pipe **1** In general, a tubular opening in a solid body. **2** In geology, an alternative

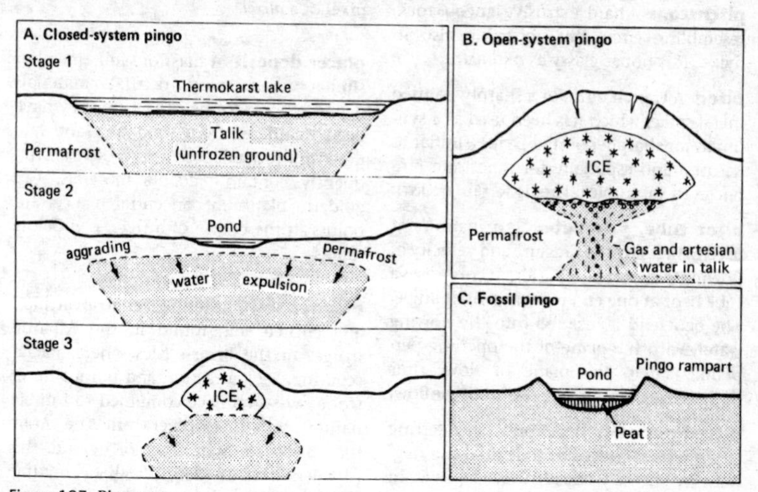

Figure 187 *Pingos*

name for the feeder or VENT which allows the lava to rise into a crater of a volcano. **3** A cylindrical pit, usually filled with sandy or gravelly sediments, in the chalklands, formed by SOLUTION enlargement of a joint. **4** A deep circular void now filled with a mass of mineral ore (LIMONITE) or diamond-filled clay (KIMBERLITE). **5** A sloping tube-like feature in the soil created by the process of THROUGHFLOW (SEEPAGE), usually along the boundary of distinctive horizons.

pipe clay BALL CLAY.

pipkrake NEEDLE ICE.

piracy of streams An alternative expression for river capture (CAPTURE, RIVER).

pisolith OOLITH.

pitch 1 The angle formed by the dipping axis of a FOLD away from the horizontal plane. If the axis of the fold departs from the horizontal the fold is said to plunge; the amount of *plunge* is the pitch of the fold. **2** An alternative name for ASPHALT or bitumen.

pitchblende A black, lustrous oxide of URANIUM (uraninite); the chief ore of uranium, occurring in sulphide-bearing veins.

pitchstone A hard, extrusive igneous rock, resembling PITCH, but occurring also in dykes. It is not as glassy as OBSIDIAN.

piton A French term for a sharply pointed metal spike, which has been used as a synonym for sharply pointed peaks (similar to NEEDLES) and especially for limestone pinnacles in subtropical karstic terrain (KARST).

pitot tube, pitometer An instrument used to measure the pressure and velocity of flowing water (or moving air). It comprises a tube bent at one end to form a right angle. The bent end is inserted into the flowing water, with the plane of the opening perpendicular to the plane of flow, thus determining the impact pressure of the flow.

pitted-outwash An expression referring to the terrain created by the burial of a stagnant ice-sheet by GLACIOFLUVIAL deposits, in which a large number of KETTLE HOLES have been formed. A SANDUR with considerable kettle-hole development is termed a KETTLED SANDUR or PITTED SANDUR. [*219*]

pivot-fault (scissor-fault) A fault which has allowed one block to rotate around a point on the fault-plane. Thus as the initial NORMAL FAULT is traced towards its rotational point its amount of THROW decreases. Beyond the rotational point (at which there is no displacement) the fault becomes a REVERSE-FAULT for the remainder of its length. [*76*]

pixel A REMOTE SENSING term, derived from 'picture element', referring to the component units from which a remotely sensed IMAGE is constructed. Pixels have both spatial and spectral aspects; the spatial variable defines the apparent size of the ground represented by the data values, while the spectral variable defines the intensity of reflected ELECTROMAGNETIC RADIATION from that specific area on the Earth's surface. The LANDSAT SATELLITE produces images with seven million pixels per waveband. The size of the pixel determines the spatial RESOLUTION of the remote-sensing system. Where a pixel covers more than one ground feature, e.g. water and land, it is referred to as a mixed pixel or a *mixel*.

placer deposit A mass of sand, gravel or similar sedimentary detrital materials deposited near to the site where they have been weathered from solid rock and laid down as ILLUVIAL deposits. Such deposits frequently contain valuable minerals, e.g. gold, tin, platinum, concentrated at certain points in the course of a river, e.g. below rapids.

plaggen soil The term given to an artificially created soil, found in the Atlantic fringes of the British Isles where glacial scouring, aeolian action and porous limestone PAVEMENTS have combined to inhibit natural soil development. In the Aran islands, off the west coast of Ireland, the islanders have created a 'soil' by mixing sand, seaweed and manure.

plagioclase One of the commonest rock-forming minerals, comprising a series of sodium and calcium FELDSPARS.

plagioclimax A type of CLIMAX in which a plant community is permanently checked in its seral development (SERE) by the interference of human activity. Most of the British heather moorlands can be regarded as examples of a plagioclimax and it has been argued that in the strictest sense much of the world's vegetation today is plagioclimax.

plagiosere The stage in the seral development of a vegetation community (SERE) when the plant SUCCESSION has been temporarily arrested by non-climatic controls in the form of human interference. If the human interference becomes permanent the plagiosere is a PLAGIOCLIMAX.

plain **1** An extensive tract of flat land or a gently undulating terrain without prominent hills or depressions. *See also* BELTED-OUTCROP PLAIN, COASTAL PLAIN, FLOODPLAIN, PANPLAIN, PEDIPLAIN, PENEPLAIN, SANDUR. **2** A proper name for a region of extensive grassland, e.g. Salisbury Plain, Great Plains of N America.

Plaisancian A stratigraphic name for the Upper PLIOCENE stage in Europe.

planation The process by which an area of the Earth's surface is denuded by SUBAERIAL or marine agencies to produce a flat surface or surface of low relief known as a *planation surface*. This term is now generally preferred to that of EROSION SURFACE. *See also* ABRASION PLATFORM, PANPLAIN, PEDIPLAIN, PENEPLAIN.

Planck's law A law of RADIATION which describes the SPECTRAL relationships between the temperature and the radiate properties of a BLACK BODY. It is expressed as:

$$W\lambda = c_1\lambda^{-5}/(e^{c_2/T} - 1)$$

where: T is temperature (°K) of the black body; $W\lambda$ is the energy emitted in unit time for an area of interest within a unit range of wavelengths centred on λ; c_1 and c_2 are universal constants and e is the base of natural logarithms (2.718). Planck's law allows an assessment to be made of the proportion of emitted ELECTROMAGNETIC RADIATION that occurs between selected wavelengths. Such knowledge is of great value in the interpretation of REMOTE SENSING observations. *See also* STEFAN'S LAW, WIEN'S LAW.

plane-bed A term referring to the bed of a stream on which there are no discernible BEDFORMS. This type of flat bed can occur when the sediment is virtually stationary and the FROUDE NUMBER approaches unity (critical flow).

plane-surface system A type of PROCESS–RESPONSE SYSTEM in which flows of mass and energy possess only spatially variable horizontal components.

planet One of the heavenly bodies, including the EARTH, which revolve around the Sun. There are nine major planets, which shine only by reflected sunlight.

plane-table A surveying instrument used for drawing maps of small areas. It consists of a small drawing-board fixed on a tripod which itself is set up over one end of a measured BASELINE. Upon the attached sheet of paper, points of detail are established when rays are constructed as the directions of the various topographic features are sighted by use of an ALIDADE. Once a sufficient number of points has been determined, the plane-table is transferred to the other end of the baseline and, after orientation, rays are constructed to the same points of detail. The exact location of the latter can be determined by the intersection of the two rays. For greater accuracy three rays may be drawn from known fixed points, although this may lead to the formation of a TRIANGLE OF ERROR. *See also* RESECTION, TRIANGULATION.

planetary vorticity The vorticity of air associated with the spin of the Earth (CORIOLIS FORCE), which is zero at the equator and increases to a maximum at the poles. It is balanced by RELATIVE VORTICITY in order to maintain conservation of ABSOLUTE VORTICITY. [*1*]

planetary winds A common name for the air movements within the Earth's ATMOSPHERIC CIRCULATION. Two main components are recognized: first, the latitudinal meridional component due to the CORIOLIS force; and second, the longitudinal component and the vertical movement, resulting largely from varying pressure distributions due to differential heating and cooling of the Earth's surface (HADLEY CELL). *See also* JET STREAM, MONSOON, TRADE WINDS, WESTERLIES.

planetismal hypothesis A hypothesis proposing that the planets of the solar system were created by the coalescence of numerous tiny planets (*planetismals*) owing to gravitational attraction of collision. It was proposed by F. R. Moulton and T. C. Chamberlin in 1904, and contrasts with the NEBULA HYPOTHESIS.

planèze A French term referring to the triangular or wedge-shaped landform which results from the dissection of a volcanic CONE by denudation.

planimeter An instrument used in measuring distances on a map. The most common type is a wheeled device which is linked to a recording dial, but others incorporate tracer-arms in which a pointer is carefully moved across the map surface.

planimetric map A map in which no contours are shown and there is no other indication of vertical relief, merely the horizontal relationships of surface features.

planina The name given to each of the flat-topped, lengthy limestone ridges which follow the structural GRAIN of the Dalmatian karst country in Yugoslavia.

plankton **1** A collective term for the minute plant and animal organisms that float and drift in the seas and oceans, especially where upwelling currents occur. Thus, they contrast with BENTHOS and with NEKTON. The plant forms are known as *phytoplankton* (including DIATOMS), while the animal forms are referred to as *zooplankton*. Plankton also contain PROTOZOA some of which (e.g. FORAMINIFERA) contribute also to benthos. Plankton are of considerable importance in the maintenance of the FOOD CHAIN of marine environments. Their fossil forms are important in palaeontology where they assist in stratigraphical research (BIOSTRATIGRAPHY). **2** A separate use of the term plankton is that referring to CRYOPLANKTON. **3** The variety occurring in slow-flowing rivers is termed *potamoplankton*.

planosol An INTRAZONAL SOIL of the hydromorphic variety (HYDROMORPHIC SOILS). It is a strongly leached soil which develops on flat or gently sloping upland surfaces under humid climates and where there is a high degree of waterlogging. The leached A-horizon passes abruptly into a clay-rich B-horizon represented by a CLAYPAN, owing to the soil process of LESSIVAGE. It is this horizon which causes waterlogging and the development of GLEYING. In the soil classification known as the SEVENTH APPROXIMATION the planosols are subdivided into ALFISOLS, MOLLISOLS and ULTISOLS.

plant geography PHYTOGEOGRAPHY.

plant productivity, index of An empirical method devised to estimate the speed and volume of vegetation growth in relation to a number of climatic variables, e.g. length of growing season.

plastic deformation A permanent change in the shape of a solid without FAILURE or rupture occurring. It involves some recrystallization, gliding within individual grains and some rotational movement of grains. *See also* GLACIER FLOW.

plasticity index The numerical difference between the LIQUID LIMIT and the PLASTIC LIMIT of a soil.

plastic limit The water content of a soil corresponding to an arbitrary limit at which it passes from a plastic state to a more or less rigid solid state; the state where a plastic soil begins to crumble. *See also* LIQUID LIMIT.

plastic relief map A three-dimensional map produced by printing a map on to a plastic sheet and moulding the latter in a press to create a relief model.

plastic soil A soil that is capable of being deformed or moulded by moderate pressure into a different shape.

plate PLATE TECTONICS.

plateau An elevated tract of relatively flat land, usually limited on at least one side by a steep slope falling abruptly to lower land. It may also be delimited in places by abrupt slopes rising to residual mountains or mountain ranges, as in the Tibetan Plateau, where it occurs as an intermontane plateau. The term is also used to refer to a structural surface such as the Meseta of Spain, in which case it is a tectonic plateau. It is also used to describe areas of extensive lava-flows (LAVA PLATEAU). (Plural: *plateaux*.)

plateau basalt The name given to those basaltic lavas which occur as horizontal surface flows (LAVA-FLOWS) and which accumulate one upon another to form volcanic plateaux. It is sometimes termed *flood basalt*. *See also* BASALT, TRAP LANDSCAPE.

plateau gravel **1** In general a deposit of sandy gravel capping ridges, hills and plateaux. **2** A term used by the Geological Survey in Britain to denote deposits of different ages and varying genesis which cap some of the chalk hills of SE England. The gravels may include river-terrace deposits, glacigenic deposits and solifluction deposits.

Plate Carrée projection An ancient MAP PROJECTION of the equidistant cylindrical type (CYLINDRICAL PROJECTION). It is also known as the EQUIRECTANGULAR PROJECTION.

plate tectonics A scientific concept developed in the 1960s to explain the pattern of the Earth's structural components and the mechanisms by which they were formed. Seven major plates and twelve smaller plates have been recognized, each of which extends down into the upper MANTLE. The plates, which are thought to be moved by large-scale thermal convection currents, are composed of continental or oceanic crust or a combination of both (SIAL and SIMA). They are bounded by *plate margins*, which may be subdivided into three types: **1** *constructive*, where new ocean floor is accumulated on either side of a MID-OCEANIC RIDGE. **2** *destructive*, where ocean floor is lost by engulfment along a SUBDUCTION ZONE; **3** a neutral or *conservative* margin where plates move past each other laterally without adding or destroying ocean floor. Such margins are marked by TRANSFORM-FAULTS or

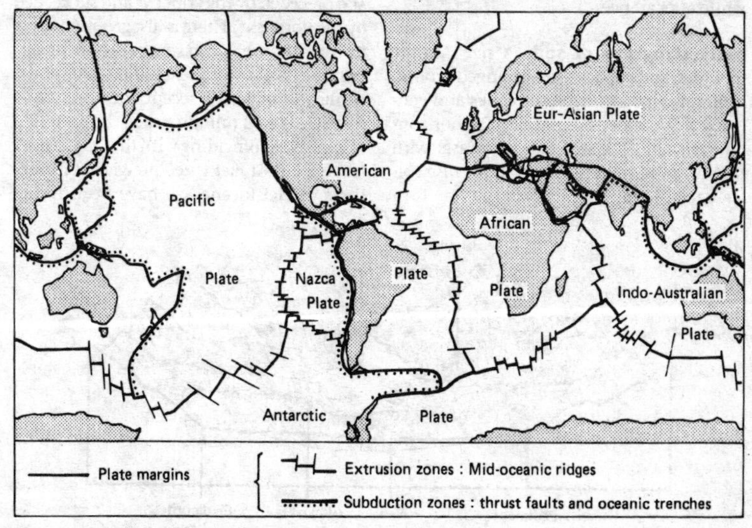

Figure 188 *Tectonic plates of the world*

TRANSCURRENT-FAULTS. It has been claimed that many of the Earth's structures can be explained by plate tectonics, especially MOBILE BELTS, ISLAND ARCS and ocean deeps (DEEP, OCEAN), while global patterns of EARTHQUAKES and VOLCANOES can also be explained. The concept includes the hypothesis of SEA-FLOOR SPREADING, in which magma rises at the mid-oceanic ridges to form new crust which is moved laterally towards the plate margin, as if on a conveyor belt, where it is destroyed by subduction or causes the ocean to increase in width (e.g. the Atlantic, the Red Sea). When the plate movements lead to the rate of construction being exceeded by that of destruction the ocean will begin to close (e.g. the Mediterranean). *See also* CONTINENTAL CRUST, CONTINENTAL DRIFT, MAGNETIC STRIPES, OCEANIC CRUST. [*188*]

platform 1 A relatively flat, low-angle surface cut in solid or in drift by marine erosion (ABRASION PLATFORM). It is not produced by subaerial agencies (*see also* EROSION SURFACE, PLANATION). **2** A geological term referring to an area of thinner sediments adjoining a geosynclinal zone of thicker beds.

platinum (Pt) A rare metal of lustrous white appearance.

platy structure of soil A *soil structure* in which soil particles are arranged around well-developed horizontal planes and generally are bounded by relatively flat horizontal surfaces. This structure contrasts with the PRISMATIC STRUCTURE OF SOIL. *See also* SOIL STRUCTURE. [*236*]

playa 1 A Spanish term referring to a level or almost level area occupying the centre of an enclosed basin in which a temporary lake forms periodically. It is generally composed of stratified beds of clay or silt, deposited within the lake, that usually contain large amounts of soluble salts (ALKALI FLAT, SALAR, SALINA, SALT FLAT). The gentle slopes running down to the playa are known as BAHADA. [*189*] **2** More strictly it is the Spanish word for a beach.

Playfair's law A 'law' of river development expounded in 1803 by a Scottish mathematician, J. Playfair (1748–1819). It can be summarized by three major conclusions: **1** Rivers cut their own valleys and the latter are proportional in size to the streams which they contain. **2** The angle of slope of each river shows an adjustment of equilibrium with the velocity and discharge of water and the amount of material carried, i.e. the modern concept of GRADE. **3** A whole river system is integrated by the mutual adjustment of the constituent parts, i.e. the modern concept of accordant-stream junctions.

Pleistocene The first epoch of the QUATERNARY, preceded by the PLIOCENE and succeeded by the HOLOCENE. There is disagreement over its time of onset, with some authors retaining the SHORT TIME-SCALE (600,000 years) while the majority accept the LONG TIME-SCALE (1.64 to 2 million years). It is referred to loosely as coinciding with the Quaternary Ice Age but in fact three pre-glacial formations of Pleistocene age have been recog-

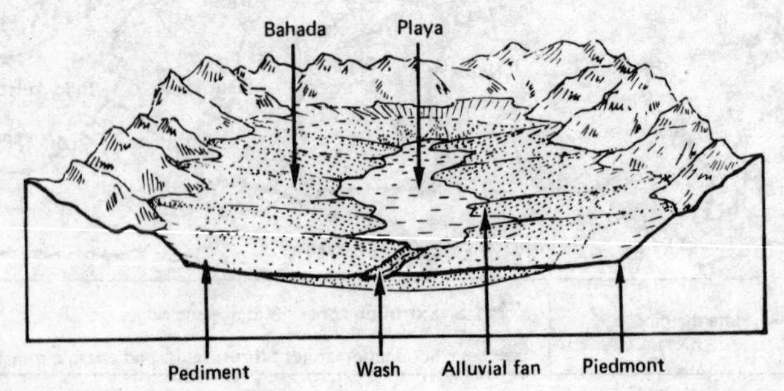

Figure 189 *Desert landforms in a bolson*

nized in East Anglia, Britain, namely the RED CRAG SERIES, the Norwich Crag Series (ICENIAN), and the Cromer Forest Bed Series (CROMERIAN), all of which indicate climatic deterioration during the onset of the Ice Age. In the glacial part of the Pleistocene there were considerable fluctuations of temperature, resulting in the waxing and waning of ice-sheets. In Britain the following divisions have been recognized: **1** *Lower Pleistocene* (oldest to youngest) PRE-LUDHAMIAN, LUDHAMIAN, THURNIAN, ANTIAN, BAVENTIAN, BRAMERTONIAN, PRE-PASTONIAN, PASTONIAN. **2** *Middle Pleistocene* (oldest to youngest) BEESTONIAN, CROMERIAN, ANGLIAN. **3** *Upper Pleistocene* (oldest to youngest) HOXNIAN, WOLSTONIAN, IPSWICHIAN, DEVENSIAN. These glacial and interglacial stages have been correlated with stages in N Europe, the Alps, the Mediterranean and N America, but exact chronostratigraphy of Pleistocene deposits is not always possible. *See also* ICE AGE, LATE-GLACIAL, MILANKOVITCH RADIATION CURVES, PALAEOLITHIC. [*197*]

plicate A term meaning folded.

Plinian eruption A volcanic eruption of great violence (PAROXYSMAL ERUPTION) in which the explosion cloud towers to considerable heights (up to 50 km), in the shape of a pino (Latin = pine tree). It is named after Pliny the Younger, who observed the eruption of Vesuvius in AD 79 when more than 50% of the volcanic cone was destroyed. It has been suggested that volcanic eruptions of this magnitude are due to hydro-explosions when groundwater invades the magma chamber and is rapidly transformed into steam. [*271*]

plinthite CRUST, LATERITE.

Pliocene The last epoch of the NEOGENE, occurring after the MIOCENE and before the PLEISTOCENE. It began about 5.2 million years ago and lasted for about 3.56 million years, during which time there was a considerable cooling of climate in northern latitudes, a net fall of sea-level and widespread denudation as rivers were constantly rejuvenated. The falling temperatures caused the migration of fauna to warmer regions and witnessed the extinction of many groups of mammals, although hominids continued to develop, especially in Africa where remains of man-like creatures have been discovered, e.g. *Australopithecus* and *Homo*. In Britain sedimentary deposition was limited almost entirely to East Anglia where shelly limestones and sands were laid down, including the Coralline Crag (CRAG). The so-called Lenham Beds of Kent are also of this age. Considerable discussion has ensued about the position of the Plio-Pleistocene boundary in Britain, where there is no clear break in the biostratigraphy, but it has been placed within the lower part of the RED CRAG SERIES.

plottable error An expression referring to the smallest distance on the ground that can be depicted on a map, according to the scale. This is due to the minimum thickness (about 0.25 mm) attainable when drawing lines on a map.

plotter A drawing device for maps and graphic figures. An *electrostatic plotter* works by placing small electrical charges on the paper, so that a dark powder (toner) will adhere to them. A FLATBED SCANNER works when the device traverses the paper, which is fixed to a flat surface. *See also* SCANNER.

ploughing block A boulder that moves downslope by frost creep and/or SOLIFLUCTION in a PERIGLACIAL environment. Usually the long axis of the boulder lies in the direction of the slope gradient and its track is marked by a linear depression. Such boulders are termed *Wanderblöcke* in German. *See also* BLOCKFIELD, GELIFLUCTION, ROCK GLACIER.

plucking ICE PLUCKING.

plug 1 A cylindrical body of volcanic rocks, representing the denuded remnant of an ancient volcanic VENT, in which PYROCLASTIC MATERIAL and/or solidified lava have accumulated. This usage is synonymous with volcanic NECK. **2** The surface expression of any more or less cylindrical body of intrusive igneous rock forming a BYSMALITH. *See also* PUY.

plume A term describing the mechanism in which rocks in the Earth's MANTLE begin to melt, thus becoming less dense and rising towards the surface as the vertical component of a thermal convection current. This is thought to cause doming and finally rupturing of the CRUST at certain points, which are referred to as HOT SPOTS (e.g. Hawaii), where continuous volcanic activity occurs. *See also* POLLUTION PLUME.

plum rains MAI-YU RAINS.

plunge pool A hollow at the base of a waterfall, thought to have been formed largely by CAVITATION.

plunging fold PITCH.

pluton **1** Formerly used to describe any large-scale deep-seated body of igneous rock. **2** Modern usage restricts the term to a description of a cylindrical mass of granitic rocks emplaced at high level and at low temperatures in a near-solid state.

plutonic rocks **1** In general, igneous rocks of deep-seated origin. **2** Rocks with coarse GRAIN SIZE, which results from their having been formed at depths where cooling and crystallization have occurred slowly. *See also* DIORITE, GABBRO, GRANITE, PERIDOTITE, SYENITE. **3** Used by H. H. Read to describe a division of rocks comprising granites, MIGMATITES and regionally metamorphosed rocks (METAMORPHISM).

Plutonist (Vulcanist) A devotee of the theory of Plutonism, expounded by J. Hutton in the late 18th cent. It proposed that not all rocks were formed from mechanical and chemical deposits on the ocean floor, as suggested by the NEPTUNIST school, but that certain rocks (now termed IGNEOUS ROCKS) were formed by upwellings from the interior of the Earth, reaching the surface as volcanic materials. Hutton has been proved wrong in his belief that SEDIMENTARY ROCKS were consolidated by heat alone, since it is known that sediments are converted into solid rock by LITHIFICATION.

pluvial A term introduced by A. Taylor (1868) to refer to all surface processes but which is now used only to describe a lengthy period of time when rainfall was considerably heavier than in preceding and succeeding stages. It has been most commonly used about Africa where it was once suggested that the tropics experienced PLEISTOCENE pluvial stages which coincided with glacial maxima at higher latitudes, owing to the equatorward shift of pressure belts and depression tracks during the Quaternary. This so-called *pluvial hypothesis* has been largely refuted, so far as central Africa is concerned, and the term pluvial is now restricted to any post-glacial climatic stage which was wetter than its predecessor or successor (e.g. the ATLANTIC PERIOD). *See also* INTERPLUVIAL, PLUVIAL TERRACE.

pluvial terrace A TERRACE of depositional or erosional derivation, now left abandoned at elevations above the current levels of certain lakes, following a change in climatic and hence hydrological conditions. They are most common in arid and semi-arid regions where lake basins have no outlet, but can occur even in humid regions. The terraces around the former Lake Bonneville in western USA are some of the best known, but there are also examples around Lake Victoria and other East African lakes. *See also* PLUVIAL.

pluviometric coefficient An expression used by climatologists to define the ratio between the mean rainfall total of a particular month and the hypothetical amount equivalent to each month's rainfall were the total rainfall to be equally distributed throughout the year. If, for example, the December mean rainfall for a station is 125 mm and its mean annual total rainfall is 1,000 mm, the pluviometric coefficient may be derived as follows: December rainfall evenly distributed is 31/365 of 1,000 mm = 84.9, thereby giving a pluviometric coefficient: 125/84.9 = 1.47.

pneumatolysis The process in which chemical changes occur in rocks owing to the action of hot gaseous substances during igneous activity. Such a process excludes the action of heated water, which is termed HYDROTHERMAL ACTIVITY. Pneumatolysis is a process which affects both the igneous mass and the COUNTRY-ROCK into which it is intruded, particularly during the late stage of cooling. It includes three main processes: **1** *Tourmalinization*, during which the mineral tourmaline replaces feldspar and biotite. **2** *Greisening*, when the margins of granite intrusions are altered to GREISEN by fluorine-rich vapours. **3** *Ore mineralization*, when certain ORES, such as tin and wolfram, are concentrated and deposited. Kaolinization (KAOLIN) is usually restricted to hydrothermal activity.

pocket beach An accumulation of coastal beach sediment at the head of a bay (BAY-HEAD BEACH).

pocket valley A flat, steep-sided valley enclosing a KARST stream below its RESURGENCE. It extends headwards into a limestone massif and is usually terminated by a cliff, e.g. Malham Cove, Yorkshire, England. It is the opposite of a BLIND VALLEY.

podzol, podzolization A Russian term, meaning 'ash soil', adopted to describe a type of soil formed in cool, seasonally humid climatic regions where LEACHING is a dominant process. The true podzol has a characteristically thin layer of raw HUMUS at the top of the A-horizon overlying an ashen-coloured, eluvial layer in the remainder of the A-horizon. The soluble bases and sesquioxides are carried down by ELUVIATION into the underlying B-horizon, which is subdivided: the Bh horizon is dominated by redeposited humus, while the Bfe and Bs horizons contain the re-deposited iron, clay and some aluminium. If Bh and Bfe horizons are present it is an *iron–humus podzol*; if only the Bh horizon, it is a *humus podzol*; if only the Bfe horizon, it is an *iron podzol*. The true podzols usually carry a vegetation cover of heathland or coniferous forest. *See also* CHELUVIATION, LESSIVAGE, PODZOLIC SOILS. [*190*]

podzolic soils One of the seven major groups in the SOIL CLASSIFICATION OF ENGLAND AND WALES. It includes soils ranging from well-drained to poorly drained, but typified by a black, dark brown or ochreous B-HORIZON in which aluminium and/or iron have accumulated in amorphous forms in association with organic matter. The overly-

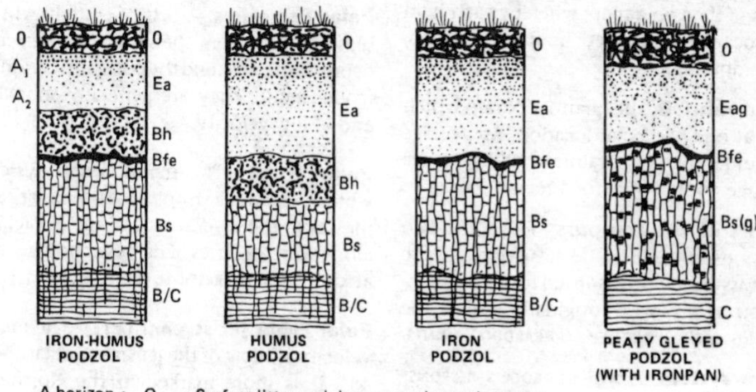

A-horizon : O – Surface litter and decomposed organic matter
Ea – Dark mineral soil (A_1) underlain by light-coloured leached layer (A_2)
Eag – Gleyed horizon
B-horizon : Bfe – Iron pan
Bh – Horizon enriched by humus
Bs – Horizon enriched by clay minerals and sesquioxides (g=gleying)
B/C – Transitional horizon
C-horizon : C – Weathered parent material

Figure 190 *Podzols*

ing bleached horizon, typical of the true PODZOLS (*sensu stricto*), and the peaty topsoil, or both, may or may not be present. The group includes the BROWN PODZOLIC SOILS, the GLEY PODZOLS, and the STAGNOPODZOLS, in addition to the true *podzols*.

poikilotherm Any animal whose body temperature stabilizes at the level of its habitat and changes as the temperature of the habitat fluctuates.

point 1 A small headland or cape projecting into a water body. **2** A location without dimensions. **3** A position on a reference system determined by a survey. **4** A decimal point. *See also* POINT DISTRIBUTION.

point bar A depositional feature composed of sand and gravel that accumulates on the inside of a river MEANDER below the SLIP-OFF SLOPE, from which it is usually separated by a trough (BAR-AND-SWALE). The trough or swale is eventually infilled by deposition of finer alluvial sediments. [*147* and *192*] *See also* POOL AND RIFFLE.

point distribution A statistical method of summarizing the locational characteristics of data on a map, whereby each item is allocated to a discrete point on the map, e.g. a dot map.

point rainfall The quantity of rain which falls at one particular location. An average of numerous point rainfalls will give the average rainfall total for a region.

points of the compass The thirty-two divisions into which the 360 degrees of the COMPASS are divided as an aid to marine navigation. They were originally called the *rhumbs of the wind*. *See also* CARDINAL POINTS.

point system A type of PROCESS–RESPONSE SYSTEM in which the inputs of mass and energy are treated as being essentially vertical, relating to one point in space (e.g. a single plant).

poised equilibrium The condition occurring when opposing forces are balanced. *See also* EQUILIBRIUM, STATIONARY EQUILIBRIUM.

Poisson distribution A type of statistical data DISTRIBUTION, named after Poisson, who discovered it in the early 19th cent. It is used to describe the case in which the PROBABILITY of an event occurring (p) is very small compared with the probability that it will not occur (q). Whereas in a NORMAL DISTRIBUTION $p = q = ½$, in the Poisson distribution p is very much smaller than q.

Poisson's ratio The ratio between the fractional transverse contraction and the fractional longitudinal extension of a body under tensile STRESS caused by a force from one direction. *See also* ELASTIC LIMIT, ELASTIC REBOUND, SHEAR STRENGTH, STRENGTH OF SEDIMENTS.

polar air mass An airmass, characterized by cool temperatures, which originates between latitudes 40°–60°, either over the ocean (MARITIME POLAR AIR MASS, symbol: mP) or over a continental interior (*continental polar*, symbol: cP). *See also* AIR MASS. [*5*]

polar distance (PD) The complement of the DECLINATION of a heavenly body, measured from the North Pole (North PD) or the South Pole (South PD).

Polar Easterlies A belt of easterly winds blowing in a zone between the westerly depression tracks and the polar anticyclones (POLAR HIGH). They are generally sporadic and of low velocity.

Polar Front The main frontal system which separates tropical and polar air masses in the N Atlantic and the N Pacific, along which a series of depressions is generated. *See also* DEPRESSION, FRONTOGENESIS. [*18*]

Polar Front Jet stream (PFJ) The high-velocity air flow of the JET STREAM at latitude 45°. It is more marked in the Northern Hemisphere where the ROSSBY WAVES are more strongly developed.

polar glacier COLD GLACIER.

polar high The semi-permanent ANTICYCLONE which forms periodically in polar latitudes owing to the cold air of the Arctic and Antarctic ice masses. The Antarctic high is more permanent than that of the Arctic.

polar-night jet stream JET STREAM.

polar outbreak The occasional incursion of a cold air mass, bringing strong winds and cool weather, from middle latitudes into lower latitudes. *See also* FRIAGEM, NORTHER, PAMPERO, SOUTHERLY BURSTER.

polar projections **1** A general term referring to any circular MAP PROJECTION which is centred on one of the geographic poles. **2** The term is used more specifically in association with particular map projections (AZIMUTHAL EQUAL-AREA PROJECTION, AZIMUTHAL EQUIDISTANT PROJECTION). The *polar equal-area projection* is a recentred case in which the continental areas are represented by lobes radiating from the North Pole. The parallels are concentric circular arcs and the meridians are curves. It was introduced in 1929 by J. P. Goode.

polar vortex A low-pressure system (DEPRESSION) occurring in high latitudes in winter when the polar-night jet stream (JET STREAM) is at its strongest.

polar wandering curve A theoretical line drawn across the Earth's surface joining up the successive former positions of the MAGNETIC POLES throughout geological time.

polder A Dutch term referring to flat tracts of coastal land reclaimed from the sea and protected from inundation by embankments (DYKE), since it usually remains below sea-level.

pole **1** One of the geographical extremities at the northern and southern ends of the Earth's AXIS. **2** One of the two MAGNETIC POLES. **3** One of the two poles on the CELESTIAL SPHERE.

poles of rotation Points around which the lithospheric plates are supposed to have rotated during PLATE TECTONICS.

polje A depression in KARST terrain, in which the long axis is developed over a lengthy distance parallel to the structural GRAIN. If it is drained by surface watercourses it is termed an *open polje*, but if it drains by means of SWALLOW-HOLES it is a *closed polje*. The depression is thought to have been formed by the coalescence of collapsed cave systems. Its floor is generally covered by alluvium. *See also* UVALA. [*64*]

pollen The fine powdery substance generated by the higher flowering plants (ANGIOSPERMS, GYMNOSPERMS). It consists of numerous microscopic spores which are easily windborne for varying distances before descending as a *pollen rain*. Since it is very resistant to destruction, pollen assists in the deciphering of palaeoenvironments (POLLEN ANALYSIS), especially those of the Quaternary.

pollen analysis A technique used by palaeontologists, botanists and biogeographers to assist in the reconstruction of palaeoenvironments. It is based on the identification and counting of the various POLLEN types which have been preserved in peat, organic soils and lake muds. Careful analysis of each horizon of the organic sample allows the periodic changes of the former vegetational assemblages to be understood, thereby enabling reconstructions to be made of changing climatic conditions. Well-defined sequences can be identified for certain INTERGLACIAL and INTERSTADIAL environments, for example, thus allowing spatial correlations to be made. Although pollen analysis cannot give an ABSOLUTE AGE for a deposit it can assist in relative dating. The study of pollen and other spores is termed *palynology*. [*191*]

pollen count An expression originally used by botanists to indicate the number of pollen grains in 1 cm^2 of a microscope slide. More recently it has been used to indicate the frequency of pollen and other spores occurring in the atmosphere during the flowering season, as a means of forecasting the impact expected by asthma sufferers.

pollution The condition of being physically unclean or impure. **1** *Air pollution* is caused by the emission of AEROSOLS, most of which are created by the combustion of fossil fuels (coal, oil, etc.). It is at its worst in heavily industrialized regions and can lead to health hazards (SMOG) and to serious damage to crops and vegetation (ACID RAIN) [*185*]. **2** *Fresh-water pollution* is caused by careless disposal of effluents from industry,

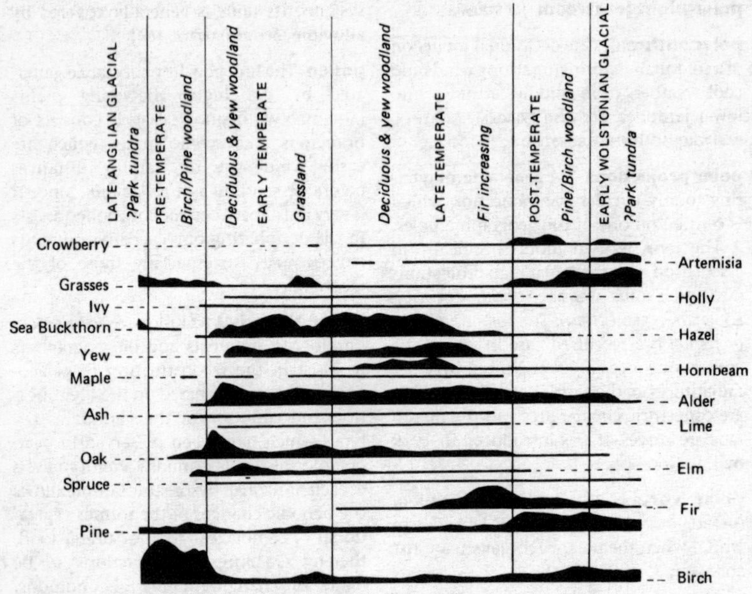

Figure 191 *Pollen analysis of Hoxnian interglacial lake sediments*

domestic buildings and even farming. It can destroy virtually all organic life in lakes and rivers and lead to human health hazards from drinking-water impurities, and to stock losses from concentration of toxic fluids in livestock watering supplies. *See also* THERMAL POLLUTION. **3** *Saltwater* (*marine*) *pollution* is caused largely by oil spillages from ocean shipping, but also by sewage outfalls and industrial wastes. **4** *Noise pollution*, from machinery, aircraft, road vehicles, etc., is becoming an important health hazard in parts of the heavily urbanized regions of the world (DECIBEL, NOISE-RATING NUMBER). **5** *Visual pollution* is a term recently introduced to denote the desecration of a picturesque view by the introduction of a foreign element into the landscape, but this is not as easily measured as in types 1–4 and is regarded as an unsatisfactory usage. *See also* GREENHOUSE EFFECT, NUCLEUS, PHOTOCHEMICAL FOG, SMOG.

pollution dome The dome-shaped layer of polluted air occurring within the warm air layer trapped beneath an INVERSION OF TEMPERATURE. It is usually generated by a large city in association with the HEAT 'ISLAND' effect, when winds are very light or calm. When winds increase in strength the pollutants are carried downwind to produce a POLLUTION PLUME.

pollution plume The narrow streamer of polluted air carried downwind from a city or other major source of pollution during periods of moderate to strong air movement.

polyclimax theory CLIMAX.

polyconic projection A MAP PROJECTION of the modified conical variety (CONICAL PROJECTION) in which each parallel is treated as a standard parallel on a normal conical projection and in which each parallel is drawn on its own radius as a non-concentric circle. All the parallels and the central meridian (which is drawn as a straight line) are truly divided. All the other meridians are curves on which scale increases with distance from the centre. It is neither an EQUAL-AREA PROJECTION nor an ORTHOMORPHIC PROJECTION, but is suitable for maps of countries of great meridional extent. *See also* MODIFIED POLYCONIC PROJECTION.

polycyclic landscape Any landscape or tract of country the geomorphic features of which have been developed under several cycles or part-cycles of erosion (CYCLE OF EROSION), which may be initiated by changes of BASE-LEVEL or by CLIMATIC CHANGE. Different terrains exhibit elements from several different cycles, not simply the current cycle. Thus, a region of stepped EROSION SURFACES with KNICKPOINTS on the rivers is an example of a polycyclic landscape.

polygenetic landscape Any landscape or tract of country the geomorphological features of which have been developed under different types of climates and different types of processes at various times during its formation. Thus, a landscape which has been fashioned primarily by FLUVIAL erosion and deposition will be modified by GLACIATION, for example, and further modified by a PERIGLACIAL phase, thereby creating a polygenetic landscape. It is probable that the landscape will also be a POLYCYCLIC LANDSCAPE.

polygenic (polygenetic) soil A soil formed by two or more different processes so that none of the horizons are genetically related.

polygnomonic projections A group of variants of the GNOMONIC PROJECTION which may be used to extend a gnomonic map beyond the hemisphere or to reduce the excessive radial scale towards the edges of smaller maps.

polygons A collection of many-sided geometric figures, occurring as a type of PATTERNED GROUND, and usually formed by FROST ACTION in the *periglacial* environment. They occur as part of a mesh of sorted/non-sorted materials on nearly horizontal surfaces but tend to merge into STRIPES as the slope gradient increases above some 2°–5°. They can be subdivided into: *sorted polygons*, which have a border of stones surrounding an area of finer material; *non-sorted polygons*, where the border of stones is lacking. Their borders may be raised above a low centre but more commonly occupy a perimeter furrow around a slightly domed centre of finer material, with the topographic differences often picked out by differences in vegetation. As in the case of similar mesh-like phenomena (CIRCLES, NETS) the origin of polygons may be in a single process or a combination of processes, including DESICCATION-cracking, DILATION-cracking, frost-cracking, frost-heaving, salt-cracking or salt-heaving, in addition to PERMAFROST-cracking. [*179*]

polyhedric projection A type of MAP PROJECTION employed for large-scale topographical map series. It is constructed by projecting a small quadrangle from the spheroid on to a plane trapezoid, and retains true scale on both the central meridian and along the sides. It was popular in the 19th and early 20th cent.

polykarst (polygonal karst) A term first used to describe KARST landforms in New Guinea characterized by a soil-covered polygonal network when viewed from the air.

polymict A term referring to a detrital rock which is composed of fragments of many different materials, in contrast to an OLIGOMICT. *Polymictic rocks* are characteristic of geosynclinal regions and include GREYWACKES, but also include such rocks as a TILLITE.

polymorphism A term meaning existing in more than one physical form. For example, minerals may have similar chemical compositions but different physical forms.

polynya Any non-linear opening in PACK ICE or between FAST ICE and pack ice. *See also* LEAD.

polyphase deformation An expression used by geologists to describe the several phases of faulting and folding which occur during an OROGENY.

ponor The Serbo-Croat name for a SWALLOW-HOLE, albeit some authors restrict its use

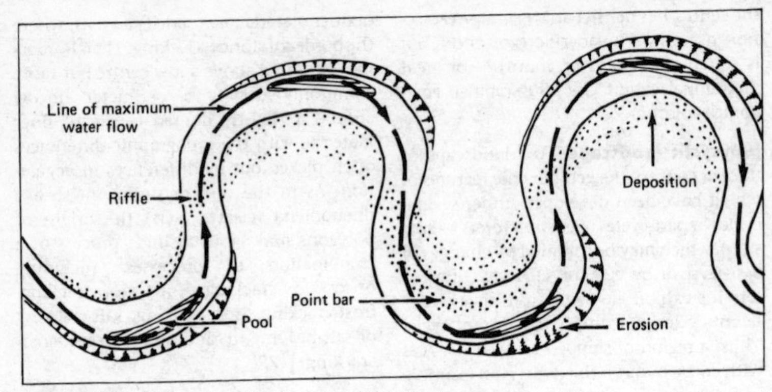

Figure 192 *Erosion and deposition in a meandering stream*

to a deep swallow-hole in a POLJE. *See also* AVEN.

pool and riffle An expression used by hydrologists and geomorphologists to describe the alternating pattern of deep pools and shallower reaches of a river where it flows across gravel bars (riffles). Their distribution is regularly spaced, with the distance between the pools being between five and seven times the channel width. Riffles tend to migrate to the inner side of a bend in a sinuous stream where they become linked with POINT BARS. Conversely, pools become deepest on the outer side of the bend. Because of these developments and because MEANDER lengths are about twice the distance between pools, it has been suggested that there may be a relationship between the pool-and-riffle formation and the initiation of meanders. [*192*] *See also* CHANNEL GEOMETRY.

population 1 In general, a term for people. **2** In biology, all individuals of a species living in a fairly intimate association with each other. **3** In statistics, all objects or values which make up a related group.

pore A void or space in a soil or rock not occupied by solid mineral material.

pore-water pressure The pressure exerted by water contained in the voids and interstices of a soil or rock, which under saturated conditions will force particles apart and possibly cause FAILURE. It is measured by inserting tubes into the soil linked to a Bourdon pressure-gauge.

pororoca The name given to the tidal BORE at the mouth of the R. Amazon.

porosity The volume of water which can be held within a rock or soil, expressed as the ratio of the volume of the voids (PORE) to the total volume of the material, e.g. a material containing pores equal to half its total volume would have a porosity of 50%. Most porous rocks are permeable (PERMEABILITY) with the exception of clay, in which the pore spaces are so small that they are often sealed with groundwater held by surface tension. Different rocks and soils vary considerably in their porosity: alluvial gravel, 25–35%; till, 20–40%; conglomerate, 5–25%; slate, 0.001–1%. Where there is a high organic content in soils the porosity is high (40–60%), but organic matter is reduced by agricultural cropping, thereby lowering the soil porosity.

porphyritic A term referring to an igneous rock texture in which large crystals are set in a finer-grained ground mass which may be crystalline and/or glassy. *See also* PORPHYROBLASTIC.

porphyroblastic A term referring to a rock texture that looks like a PORPHYRITIC rock but in which the large grains or crystals are pseudo-PHENOCRYSTS produced by thermodynamic metamorphism. Thus it refers to metamorphic rather than igneous rocks.

porphyry A geological term for any medium- to fine-grained igneous rock (GRAIN) which contains PHENOCRYSTS. There is an increasing tendency to abandon the use of porphyry as a rock name (especially the obsolete *porphyrite*) and replace it with PORPHYRITIC as a textural prefix to any micro-plutonic rock type.

Portland cement An artificially produced hydraulic CEMENT comprising a compound of clay (alumina/silica) and lime that has been pulverized and burnt. *See also* PUZZOLAN CEMENT.

Portlandian The stratigraphic name for one of the uppermost stages of the Upper JURASSIC, named from Portland in Dorset, England. Portland stone (a white OOLITE) is an important FREESTONE.

positive feedback FEEDBACK.

Post-Glacial That part of geological time which commenced at the termination of the PLEISTOCENE ice age. It is synonymous with HOLOCENE but the date of its commencement differs from region to region according to the time at which the ice-sheets disappeared. In Britain it is usually taken as the end of the LATE-GLACIAL in 10,300 BP but in higher latitudes the date is later. Some scientists believe that post-glacial time merely represents an INTERGLACIAL within the Pleistocene. *See also* FLANDRIAN, MEDIOCRATIC, MESOLITHIC, NEOLITHIC, PROTOCRATIC, TERMINOCRATIC. [*81*]

potamology The scientific study of rivers.

potamoplankton PLANKTON.

potash The carbonate of potassium (K_2CO_3), occurring frequently in association with salt both in modern and in former SALT LAKES.

potassium–argon (K/Ar) dating RADIOMETRIC DATING.

potential energy The energy stored by a body in relation to a fixed datum, such as sea-level or the surface of the Earth. It can be measured by the amount of work required to move that body from its original position to a new position. Thus the potential energy of a river is expressed as $Ep = Wz$, where W is the weight of the water and z is the head of water, i.e. height above BASE-LEVEL. A body of stationary air, high in the atmosphere, possesses a high potential energy because of its height above the Earth's surface. *See also* GEOPOTENTIAL, KINETIC ENERGY.

potential evapotranspiration (PE) EVAPOTRANSPIRATION, EVAPOTRANSPIROMETER.

potential instability of an air mass The original state of an AIR MASS before it moves into a state of CONDITIONAL INSTABILITY. This is normally achieved by raising the air mass at a FRONT or at a topographic barrier.

potential model A MODEL which measures the force exerted by a given object on a spatial POINT, by reference to the same object located at all other points on the spatial field under investigation.

potential temperature The temperature readily obtained from an aerological diagram (AEROLOGY) relating a body of air in the process of being uplifted or subsiding at the DRY ADIABATIC LAPSE-RATE to a standard pressure, generally 1,000 mb.

pot-hole **1** A more or less circular hole in the rocky bed of a stream, carved by the scouring and grinding effect of pebbles rotated in an eddy in a stretch of rapids. *See also* GLACIER MILL, P-FORMS. **2** A steep-sided shaft in limestone terrain (KARST) down which a surface stream disappears. It appears to have originated in Yorkshire, England, as a local name for a SWALLOW-HOLE, but is now used throughout the British Isles and has given its name to the colloquialism *pot-holing* (SPELEOLOGY). *See also* AVEN, PONOR.

pound A unit of mass used in the British system for 1,000 years prior to metrication. Two different pounds were recognized: **1** The *avoirdupois* (7,000 grains); 453.592 grams or 27.692 in^3 of water at 4° C. There was a slight difference between the *Imperial Standard Pound* (0.453592338 kg) and the pound used in the USA (0.4535924277 kg). **2** The *troy* (5,760 grains); 373.2418 grams. [*145*]

powder snow A meteorological term, particularly common in skiing terminology, to denote the state at which snow crystals are dry and loose. This condition occurs at very low temperatures and is most common in polar latitudes and in high mountain areas, where it is particularly welcomed by skiers.

power function A value multiplied by itself a stated number of times, e.g. $\chi \times \chi = \chi^2$ (χ to the power of two); $\chi \times \chi \times \chi = \chi^3$ (χ to the power of three).

power of test The ability of a statistical test to discriminate correctly between a true and false hypothesis.

pradoliny A Polish equivalent to URSTROMTAL.

prairie **1** The extensive, treeless grassy plains of N America to the east of the Great Plains. It is a vegetation CLIMAX developed under 250–500 mm rainfall, most of which occurs in summer, and characterized by high summer temperatures and low winter temperatures. It is sometimes subdivided into the short-grass prairies of the W and the long-grass prairies of the E. The Canadian and US prairies are one of the world's great cereal-producing regions, particularly wheat, so that little of this mid-latitude natural grassland now survives. The counterpart of the prairie is the PAMPA of S America, the STEPPE of Eur-Asia and the VELD of S Africa. **2** A French term referring to a meadow.

prairie soil A dark-coloured soil occurring in the mid-latitude subhumid climatic zone beneath the temperate grasslands. It is similar to a CHERNOZEM in which some leaching has occurred but the sesquioxides have not been translocated and the carbonate has not been deposited in the B-horizon. It is classified in the SEVENTH APPROXIMATION as a MOLLISOL, but it is also known as a *brunizem*, *brown steppe soil* and *degraded chernozem*.

Pratt's hypothesis ISOSTASY.

Pre-Boreal The first of the POST-GLACIAL climatic periods in N Europe, immediately succeeding the LATE-GLACIAL. In the British Isles it is a relatively brief 850-year spell between 10,300 and 9,450 BP, when increasing warmth after the melting of the ice-caps caused the growth of birch–pine forest in England and Wales and solely birch forests in Scotland. It is referred to as zone IV in the FLANDRIAN pollen-zone chronology. The Pre-Boreal witnessed the emergence of the MESOLITHIC culture in Britain. [*81*] *See also* BLYTT–SERNANDER MODEL.

Pre-Cambrian The entire span of geological time prior to the CAMBRIAN and all the rocks which were formed before 570 million years BP. There is no clear agreement on the length of the Pre-Cambrian but it is thought to exceed 4,000 million years during which time there were several orogenies (OROGENY). Most Pre-Cambrian rocks, therefore, have been altered on numerous occasions and comprise, in the main, complex series of METAMORPHIC ROCKS which are exposed in SHIELD areas. Some fossils have been discovered in a few of the unmetamorphosed rocks but the rock series are difficult to correlate from region to region because of their considerable alteration, although RADIOMETRIC DATING has helped. In the British Isles rocks of this age are exposed mainly in NW and W Ireland, NW Scotland, W Wales and the Welsh borderlands. *See also* ARCHAEAN, ARCHAEOZOIC, CHARNIAN, DALRADIAN, EOZOIC, HURONIAN, KEWEENANWAN, LAURENTIAN, LAXFORDIAN, LEWISIAN, LONGMYNDIAN, MALVERNIAN, MOINIAN, ONTARIAN, PROTEROZOIC, SCOURIAN, TORRIDONIAN, URICONIAN.

precession of the equinoxes The slow change of the relative positions of the Earth's EQUATOR and the ECLIPTIC, in which the celestial N Pole (CELESTIAL SPHERE) appears to describe a complete circle in space every 26,000 years. At the same time the EQUINOXES complete one movement around the ecliptic. This form of motion results from the gravitational attraction of the Sun and the Moon on the Earth's EQUATORIAL BULGE, which causes the Earth's axis to trace out a conical figure in the heavens.

precipitable water An expression used by meteorologists as a theoretical index of the moisture content of air above a specified point. It assumes that all the vapour in a standard column of air would be condensed on to a given horizontal surface, but in fact precipitation processes are not efficient enough to produce this theoretical state. The index is expressed as:

$$Mw = \frac{1}{g} \frac{p^1}{p^2} rdp$$

where: Mw is the precipitable water; g is the acceleration of free fall; p^1 and p^2 represent the pressures in millibars at the top and bottom, respectively, of the columns of air; r is the MIXING RATIO; and dp is the depth of the individual layer of mixing ratio.

precipitation The deposition of water in a solid or a liquid form on the Earth's surface from atmospheric sources. It includes DEW, DRIZZLE, HAIL, RAIN, SLEET and SNOW. Its formation depends on the processes of coalescence (*see also* COLLISION THEORY OF RAINFALL, LANGMUIR THEORY OF RAINDROP GROWTH), or on the incremental growth of ice crystals within a cloud (BERGERON–FINDEISEN THEORY OF PRECIPITATION), or on a combination of both processes. *See also* CONVECTIONAL RAIN, CYCLONE, FREEZING NUCLEI, FRONTAL RAINFALL, NUCLEUS, OROGRAPHIC RAINFALL.

precipitation-day A period of 24 hours during which at least 0.25 mm of PRECIPITATION are measured. Since this includes all types of precipitation it is of wider use than the RAIN DAY.

precipitation efficiency (effectiveness) A technique whereby the efficiency or usefulness of rainfall for crop growth, water supplies or industrial potential can be measured. A *precipitation-efficiency index* was devised by C. W. Thornthwaite in 1931, as part of his first attempt to draw up a classification of CLIMATIC REGIONS. His formula is $P/E = 11.5(p/[T - 10])10/9$, where p is the monthly mean precipitation in inches and T is the monthly mean temperature. *See also* MOISTURE INDEX, THERMAL-EFFICIENCY INDEX, THORNTHWAITE'S METHOD OF CLIMATIC CLASSIFICATION.

predation The killing of an animal for food by another animal.

predictive A term regularly used by geographers, especially in relation to MODEL BUILDING, stating what will happen. *See also* PRESCRIPTIVE, PROSCRIPTIVE.

Pre-Ludhamian The first recognized stage in the PLEISTOCENE deposits of E Anglia, England, but found only in boreholes at Stradbroke and Sizewell, Suffolk. Its pollen spectrum indicates a cool temperate climate dominated by pine forest and the FORAMINIFERA of the lowest part of the stage closely resemble those of the RED CRAG (Waltonian), although its lower limit (i.e. the Plio-Pleistocene boundary) is uncertain. [*197*]

Pre-Pastonian A stage name of the Lower Pleistocene in Britain. It is represented by a deposit of gravels, sands, silts and muds exposed on the foreshore at Beeston, Norfolk, England and interpreted as a colder climatic phase between the milder BRAMERTONIAN and the PASTONIAN. It has been suggested, however, that the *Pre-Pastonian* of Norfolk may be correlated with the BAVENTIAN stage as recognized in the borehole at Ludham, Norfolk. [*197*]

preprocessing A REMOTE SENSING term which refers to the processing of data received from the remote SENSOR into a form acceptable to the subsequent processing functions. It usually includes geometric and radiometric calibration, IMAGE ENHANCEMENT and other transformations. *See also* IMAGE CLASSIFICATION, PATTERN RECOGNITION.

prescriptive A term used by geographers, especially in relation to MODEL BUILDING, stating what ought to happen. *See also* PREDICTIVE, PROSCRIPTIVE.

pressure ATMOSPHERIC PRESSURE.

pressure gradient (barometric gradient) The amount of change in ATMOSPHERIC PRESSURE between two points. It is depicted on a synoptic chart by the differential spacing of the ISOBARS – a wide spacing indicates a weak or gentle gradient (and therefore light winds) while a narrow spacing represents a steep gradient (and strong winds). The pressure variations at a given horizontal level create a force (the *pressure-gradient force*) which is the major motivating force of air movement in the atmosphere. *See also* GEOSTROPHIC FLOW. [*90*]

pressure head HEAD.

pressure melting-point The temperature at which ice and minerals may be induced to melt by the introduction of pressure. The temperature is normally the FREEZING-POINT (0° C) at the surface of glacier ice, but at depth within glaciers this temperature fractionally declines because of the increased pressure (weight) of the ice. Ice-sheets in middle and low latitudes generally exhibit uniform pressure melting-point throughout their thickness, but polar glaciers (COLD GLACIER) have lower pressure melting-points because of the very low air temperatures which prevail in these latitudes. *See also* REGELATION.

pressure pan A subsurface soil horizon with a higher density and a lower POROSITY than the horizons directly above or below it. It results from the pressure of agricultural machinery during tillage operations and has to be artificially broken to avoid GLEYING. It is sometimes referred to as INDUCED PAN.

pressure-plate anemometer ANEMOMETER.

pressure release in rocks DILATATION, PSEUDO-BEDDING.

pressure ridge A small crumple of the surface of floating ice at the junction of two ICE-FLOES. It should not be confused with a RIDGE OF HIGH PRESSURE. *See also* SASTRUGI.

pressure system A pattern of ISOBARS denoting an atmospheric circulation system of either low pressure (CYCLONE, DEPRESSION) or high pressure (ANTICYCLONE). *See also* RIDGE OF HIGH PRESSURE, SECONDARY DEPRESSION, TROUGH.

pressure tendency The rate of surface-pressure change at a selected observation station. It is a useful indicator of the probable future behaviour of a PRESSURE SYSTEM. *See also* ISALLOBAR.

pressure-tube anemometer ANEMOMETER.

pressure wave COMPRESSIONAL WAVE.

prevailing wind The wind which blows most frequently at any location. This normally depends on the particular situation of the site in the Earth's ATMOSPHERIC CIRCULATION, but the pattern may be affected by seasonal changes (MONSOON) and locally by topographic factors (KATABATIC WIND, SEA BREEZE). The British Isles are affected mainly by the WESTERLIES, in which the prevailing wind is normally from the SW, but this may be temporarily replaced by winds from other directions, as seasonal pressure patterns change. A distinction should be made between the prevailing wind and the DOMINANT WIND.

primärrumpf A German term, introduced by the geomorphologist W. Penck, to describe the initial progressively expanding dome in which the tectonic uplift is so slow that denudation is able to keep pace with it. Thus he postulated that there would be no actual net rise of the surface nor increase in its relief. Accelerated uplift would elevate the primärrumpf above sea-level, even though it undergoes degradation. The end stage of degradation is marked by declining uplift and leads to the production of Penck's ENDRUMPF. The only difference between W. M. Davis's concept of a PENEPLAIN and the *endrumpf* of Penck is that the latter considers the endrumpf to have been achieved directly from the primärrumpf without any of the intermediate sequential stages of peneplain formation. *See also* PIEDMONTTREPPEN. [*193*]

primary A term used in early geological literature but with different senses: **1** Initially, to describe rocks of PRE-CAMBRIAN age. **2** Secondly, to describe rocks of Lower PALAEOZOIC age. **3** Finally, to describe rocks of Palaeozoic age, as a forerunner of the succeeding secondary (MESOZOIC), TERTIARY and QUATERNARY. The term is now obsolete.

primary energy Energy contained in fossil fuels such as COAL and PETROLEUM and energy derived from renewable sources such as the Sun, wind and ocean waves. *See also* USEFUL ENERGY.

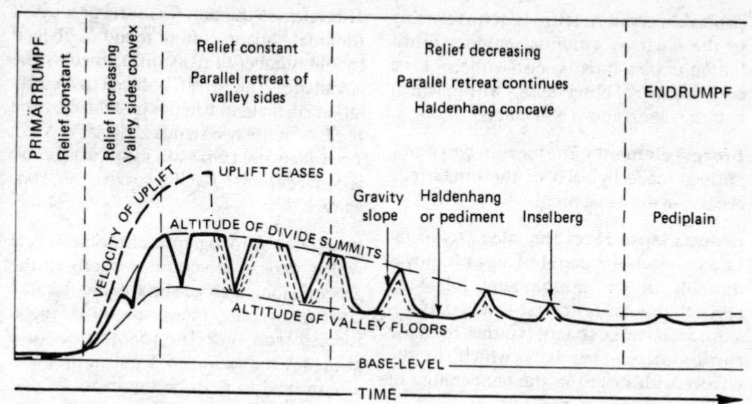

Figure 193 *Penck's cycle of erosion*

primary succession The ecological succession which begins as a 'pioneer' on a surface that has not been previously occupied by a community of organisms, i.e. on a lava-flow or a newly exposed glaciated rock surface. *See also* SECONDARY SUCCESSION.

primary (p) wave SEISMIC WAVES.

primates The order of MAMMALIA that includes man, apes, monkeys and lemurs. Monkeys appeared in the OLIGOCENE and man-like forms in the PLIOCENE.

prime meridian The meridian on the Earth's surface from which LONGITUDE is measured. Since 1884 the prime meridian has been the GREENWICH MERIDIAN, by universal agreement. [*176*]

primitive equations Physical equations used by meteorologists when building MODELS to simulate ATMOSPHERIC CIRCULATION.

principal directions Two ORTHOGONAL directions at any point on a sphere or spheroid which remain as orthogonal directions when transferred to a plane map of that point.

principal meridian The central MERIDIAN on which a rectangular GRID system is based.

principal stresses The maximum, minimum and intermediate intensities of STRESS along each of three mutually perpendicular axes in a given material.

principle of superposition A stratigraphic principle introduced by William Smith (1769–1839) and upon which the whole of geological chronology is based. It states that underlying rock strata must be older than overlying strata where there has been neither inversion nor overthrust. Sometimes referred to as the *law of superposition.*

prisere SERE.

prismatic compass COMPASS.

prismatic structure of soil A type of SOIL STRUCTURE in which the aggregates and particles have vertical axes much longer than the horizontal axes. *See also* PLATY STRUCTURE OF SOIL. [*236*]

probability A statistical term referring to the absolute value of the chance that an event will actually occur. Several statistical tests have been designed to predict the probability of many types of relationships occurring or not occurring. Probability is assessed by dividing the number of occurrences by the total number of cases.

probability distribution A distribution of values of a variable indicating the odds (probability) of encountering each of the values.

probable error A statistical term referring to the range in a normal random distribution of data (NORMAL DISTRIBUTION), for a large number of observations within which half the observations will fall.

process elements The measurable contributions made by each of the forces in a PROCESS–RESPONSE MODEL.

process lapse-rate The rate of temperature decrease of a parcel of air as it moves upwards in the atmosphere, following either the DRY ADIABATIC LAPSE-RATE or the SATURATED ADIABATIC LAPSE-RATE. It differs from the ENVIRONMENTAL LAPSE-RATE, which is the change with height in the temperature of the atmosphere. *See also* LAPSE-RATE.

process–response model A representation of the PROCESS–RESPONSE SYSTEM to illustrate the ways in which forms are related to processes. *See also* INPUTS AND OUTPUTS, MODEL.

process–response system A combination or intersection of a CASCADING SYSTEM and a MORPHOLOGICAL SYSTEM, with the linkages between the two being provided by morphological components which are the same as, or closely correlated with, storages (STORE) or REGULATORS, which are fundamental parts of the cascading system. When a process–response system is modified by human intervention, as in a river authority controlling a drainage basin, it becomes a CONTROL SYSTEM, because its INPUTS AND OUTPUTS have been regulated. The process–response system shows the manner in which form is related to process. [*249*] *See also* SYSTEMS ANALYSIS.

proclimax A stable plant community the original establishment of which is thought to have taken place under climatic conditions different from those of today. *See also* CLIMAX.

product-moment coefficient of correlation The most powerful test of correlation of variables in which the *correlation coefficient* (*r*) is calculated from the formula

$$r = \frac{1/n\Sigma(a - \bar{a})\ (b - \bar{b})}{\sigma a \cdot \sigma b}.$$

This refers to the sum (Σ) of the product of the total variations from $\bar{a}$ and $\bar{b}$, divided by the number of pairs (*n*), known as the covariance. The latter is reduced to *r* (correlation coefficient) when it is divided by the product of the two standard deviations (σ). The widely used SPEARMAN RANK CORRELATION is derived from it. *See also* CORRELATION, STATISTICAL.

profile **1** A topographical profile is one which depicts a vertical section of the ground along a given surface line (*see also* COMPOSITE PROFILE, PROJECTED PROFILE, SUPERIMPOSED PROFILE). **2** The LONG PROFILE OF A RIVER. **3** A SOIL PROFILE. **4** A pollen profile is a term used to describe the increases and decreases of all types of POLLEN in diagrammatic form. [*195*]

profile of equilibrium A section of a slope, a depositional shoreline or a LONG PROFILE OF A RIVER, depicting the stage at which they are in a state just capable of maintaining a balance between erosion, transport and deposition. *See also* EQUILIBRIUM, GRADE.

pro-glacial channel A type of MELTWATER CHANNEL which is formed across the slope that declines ahead of an ice front. It usually forms at right angles to the ice margin but can also occur parallel to the ice front. The channel may carry meltwater for only a limited period before the wasting ice-sheet or glacier shifts its position thereby causing its outlet streams to carve fresh pro-glacial channels.

pro-glacial deposit **1** Any sediment which is laid down in a water body having been carried by meltwater streams beyond the limits of a glacier or ice-sheet. **2** Deposits of GLACIOFLUVIAL origin that are laid down in front of an ice margin, e.g. a SANDUR.

pro-glacial lake A body of water that accumulates in a basin as a result of damming by ice-sheets as they advance into an ice-free area. The ice front will form one of the lake margins but elsewhere the waters will be impounded by drainage divides (ridges, scarps, etc.). In Britain there are many examples of former pro-glacial lakes

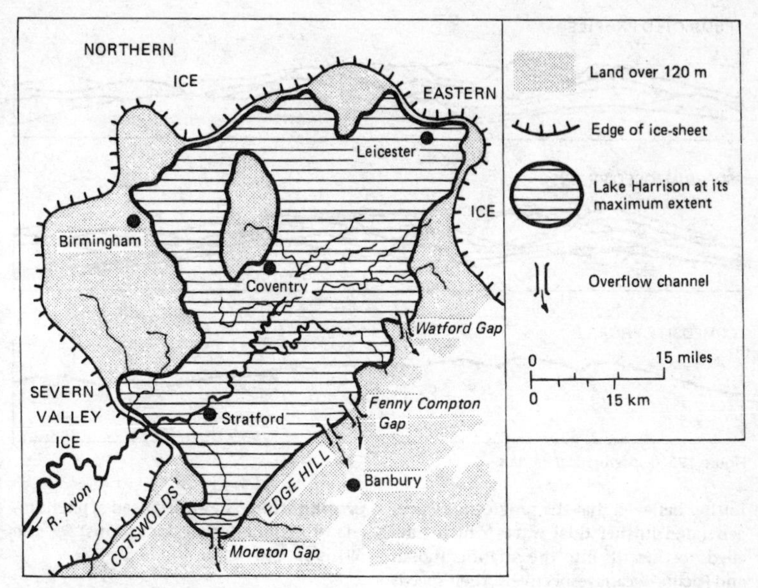

Figure 194 *Pro-glacial L. Harrison*

impounded by the Pleistocene ice-sheets, e.g. L. Harrison in the English Midlands and L. Pickering in E Yorkshire. In Scotland the PARALLEL ROADS were formed at the margins of pro-glacial lakes. [*194*]

progradation The process in which a shoreline advances seawards owing to the inshore deposition of sediments accumulating at a faster rate than that of sediment removal. The sediment accumulates from both estuarine and longshore sources (LONGSHORE DRIFT). *See also* RETROGRADATION.

programming The procedure by which a sequence of events or operations is systematically organized or planned, often by using a computer (*computer programming*). *See also* LINEAR PROGRAMMING, NON-LINEAR PROGRAMMING.

progressive failure The process of slow loss of strength of a material over time, owing to the gradual increase of its water content, which leads to its becoming progressively softer. This usually occurs in overconsolidated clays, e.g. London Clay (EOCENE). The softening process is aided by fissures which not only aid percolation of water and weathering processes but also help to concentrate the stress at certain points, ultimately leading to FAILURE of the slope.

progressive wave A wave propagated along a water-filled channel or across an open ocean of theoretically infinite length. Its speed of progression across the surface of the water body exceeds the speed at which the water particles themselves advance.

progressive wave theory of tides An early concept which attempted to explain tidal behaviour but which has now been replaced by the OSCILLATION THEORY OF TIDES. The earlier theory postulated that in the Southern Ocean two large water waves (TIDAL WAVES) of low amplitude and great breadth travelled westwards around the globe every 24 hours and 50 minutes. One was thought to follow the orbit of the Moon and the other was thought to be following the first but in the opposite hemisphere. It was

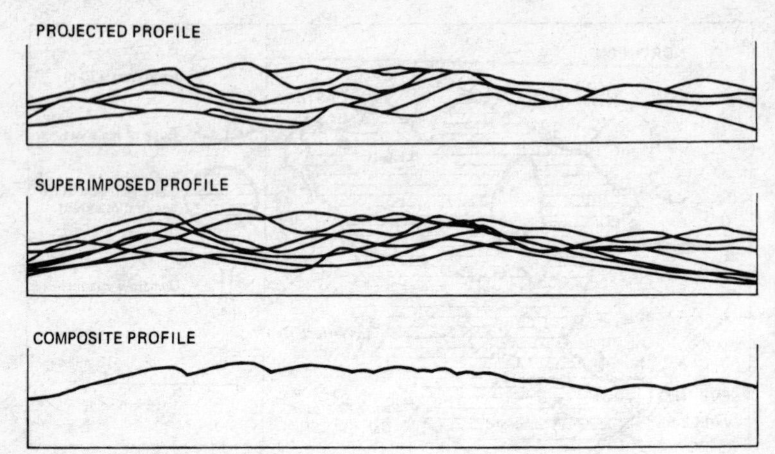

Figure 195 *Topographical profiles*

further believed that the progressive waves generated further tidal waves which travelled northwards into the Atlantic, Indian and Pacific oceans respectively, their speeds being governed by the shape of the oceans and the depths of water rather than by the lunar period, thereby reaching particular coastlines at progressively later times. *See also* AMPHIDROMIC SYSTEM, TIDE.

projected profile One of several topographical profiles that, in one method, are placed at regularly spaced intervals one behind the other to form a three-dimensional model of a land surface. Alternatively the profiles may be accurately drawn on a single diagram which, however, includes only those parts of the profiles not obscured by the higher land depicted on the intervening profiles. The latter method produces a type of landscape drawing with a panoramic effect. [*195*] *See also* COMPOSITE PROFILE, SUPERIMPOSED PROFILE.

projection A diagram of three-dimensional space relations constructed by extending lines from various points to their intersection with a plane surface. *See also* MAP PROJECTION.

promontory A rocky coastal headland projecting significantly into the sea. *See also* NESS, POINT.

propagation The areal spread of phenomena of material or abstract forms. *See also* DIFFUSION, DISPERSION.

properties of fluids The characteristics, or physical properties, of moving fluids in relation to hydrodynamic flow in an open channel (OPEN-CHANNEL FLOW).

proportional dividers An instrument used to obtain detailed lengths from a map in order to produce a smaller or larger version of a particular feature depicted on the map.

proscriptive A term used by geographers, especially in relation to MODEL-BUILDING, stating what ought not to happen. *See also* PREDICTIVE, PRESCRIPTIVE.

protalus rampart A linear or curvilinear ridge formed from the accumulation of frost-shattered debris at the lower margin of a snow patch. The broken rock fragments from the backwall of a NIVATION CIRQUE fall on to the top of the snow bank whence they slide down to its lower edge. Its appearance and composition is similar to that of a MORAINE, although it has a tendency to form a convex-in-plan ridge facing towards the backwall of the hollow, in contrast to that of a glacier-formed cirque moraine which produces a feature curving in the opposite direction when seen in plan. [*196*]

protectionist hypothesis A concept introduced by E. J. Garwood in 1910 in which he argued that glaciers are not as effective erosive agents as streams and that a glacier lying within a valley actually protects it from the more aggressive sculpturing by stream erosion. *See also* GLACIAL STAIRWAY, HANGING VALLEY.

Proterozoic **1** A synonym for the PRE-CAMBRIAN in the usage of the US Geological Survey. **2** Most geologists regard it as the youngest of the three aeons of the Pre-Cambrian, preceded by the EOZOIC and the ARCHAEOZOIC.

protocratic A term introduced by von Post (1946) and adopted by Iversen (1958) to describe the period of increasing warmth in the post-glacial (FLANDRIAN) climate of NW Europe prior to the *Climatic Optimum*. It is thought to be synonymous with the PRE-BOREAL period and the BOREAL period, i.e. from 10,300 to 7,450 BP, and with pollen zones IV–VI inclusive. *See also* MEDIOCRATIC, MESOLITHIC, TERMINOCRATIC. [*81*]

protore A MINERAL deposit which, by the action of further natural processes, may be upgraded to an ORE.

Protozoa Microscopic unicellular organisms including FORAMINIFERA and RADIOLARIA.

provenance The source area from which have been derived the different materials which make up a sedimentary rock.

province A geographic area of some considerable extent, smaller than a continent but larger than a region, which is unified by some or all of its characteristics and which can therefore be studied as a whole. A faunal province, for example, has a particular assemblage of animal species, which differs from assemblages in different contemporaneous environments elsewhere.

proximal trough A topographical depression surrounding a steep rocky outcrop caused by the increase in velocity occurring when moving ice, water or wind flows around an obstruction.

psammitic A geological term referring to an ARENACEOUS rock that has been metamorphosed, although it was once used synonymously with arenaceous. *See also* PELITIC, PSEPHITIC.

psammophyte A plant which is adapted to live in a PSAMMOSERE, e.g. marram grass.

psammosere The succession of vegetation (SERE) that develops in an environment of moving sand-dunes.

psephitic A geological term once used synonymously with RUDACEOUS ROCKS but now used only when referring to their metamorphosed equivalents. *See also* PELITIC, PSAMMITIC.

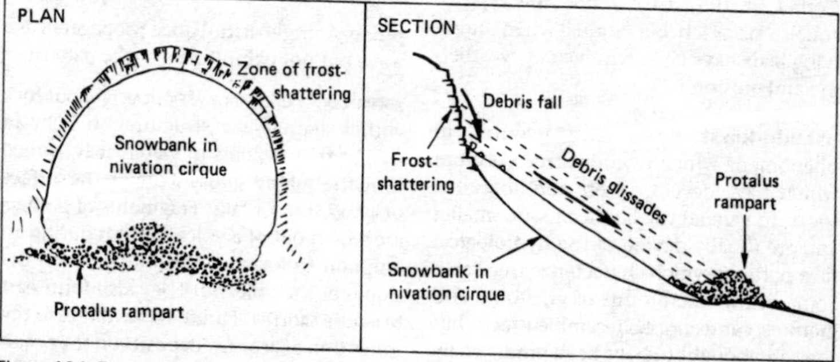

Figure 196 *Formation of a protalus rampart*

pseudo-azimuthal projections A class of MAP PROJECTIONS based on variants of the AZIMUTHAL PROJECTION. They consist of concentric circular parallels and curved meridians which converge at the pole at their true angular value.

pseudo-bedding Structures in igneous rocks which resemble the bedding of sedimentary rocks. The divisions are not strata separated by bedding planes but sheets separated by fractures, hence the alternative expression of SHEETING to describe this phenomenon. Such fractures are due to expansion of the rock on pressure release (*unloading*), resulting from removal of overlying materials by denudation. The sheets are usually concentric and follow the shape of the ground surface. *See also* DILATATION.

pseudo-conical projections A major class of MAP PROJECTION in which the principal scale is preserved along one standard parallel and along the central meridian. In the normal aspect, the parallels are represented by concentric circular arcs and the meridians by concurrent curves. *See also* CONICAL PROJECTION.

pseudo-cylindrical projections A major class of MAP PROJECTION in which the principal scale is preserved along the equator and along the central meridian. In the normal aspect, the parallels are represented by a system of parallel straight lines and the meridians by concurrent curves.

pseudo-erratic A term introduced by J. Corbel in 1957 to describe the type of PERCHED BLOCK left as a residual when underlying beds have been removed by weathering and erosion.

pseudo-karst The term referring to phenomena which resemble true KARST but which have developed on non-limestone rocks. In general the landforms are smaller and are usually devoid of the hydrological flow patterns which characterize true karst. Some of the microforms of gritstones and granites, for example, resemble KARREN but they are thought to have been produced by mechanical erosion rather than by solution.

pseudomorph **1** A mineral that has replaced another mineral and assumed its external form, e.g. GYPSUM replacing anhydrite. **2** A fossil which has gradually been replaced by secondary material to form a 'cast' that has preserved its former shape.

Psychozoic A little used term, derived from the Greek *psyche* (= the soul), referring to the span of geological time since *Homo sapiens* appeared on the scene.

psychrometer An instrument used to measure the RELATIVE HUMIDITY of the air. A type of portable HYGROMETER. All models involve the use of WET- and DRY-BULB THERMOMETERS, but in the aspirated *Assman psychrometer* air is driven past the wet bulb by means of a motorized fan, while in the *whirling* variety maximum air circulation past the wet bulb is achieved manually by rapid rotation of the instrument.

Pteridophyta A phylum of vascular plants that reproduces from spores instead of seeds. Among the pteridophytes are ferns, club mosses and horsetails.

pteropod ooze OOZE.

puddingstone A colloquial name for a CONGLOMERATE in which the structure resembles a plum pudding. In SE England the *Hertfordshire puddingstone* is thought to represent the coarse facies of the material known as SARSEN.

pulsation theory A concept suggesting that EUSTATIC movements of sea-level cause a world-wide TRANSGRESSION followed by a REGRESSION of all epicontinental seas.

pulse A single disturbance propagated as a wave but not exhibiting a cyclic pattern.

pumice A cellular or VESICULAR igneous rock with a sponge-like structure, so light in weight that it floats in water. It is formed from the bubbly, glassy scum on the surface of a LAVA flow or lake. Fragments of pumice are blown out of a volcanic vent during an eruption to form an accumulation on the slopes of the volcano. They also form part of a NUÉE ARDENTE. Pumice contributes to the formation of RED CLAY deposits on the ocean floor.

puna The name given to the higher intermontane plateaux of the Andes, lying between 3,600 m and 4,500 m and characterized by extremely large seasonal and diurnal ranges of temperature. There is only a brief rainy season, resulting in a sparse grass cover and a montane vegetation of xerophytic character (XEROPHYTE). Above the puna, extending up to the snow-line, is the so-called *puna brava*, with virtually no vegetation other than a few Arctic–Alpine species (PARAMOS).

punctuated equilibrium An expression referring to the *theory of allopatric speciation* which sees the evolution of species taking place not gradually but by a series of sudden leaps after lengthy periods of little change, i.e. equilibrium punctuated by saltatory leaps (SALTATION). The theory of allopatric speciation states that: **1** New species arise by the splitting of lineages. **2** New species develop rapidly. **3** A small subpopulation of the ancestral forms gives rise to the new species. **4** The new species originates in a very small part of the ancestral species' geographic extent – in an isolated area at the periphery of the range. Punctuated equilibrium is in complete contrast to GRADUALISM. *See also* CATASTROPHISM, EVOLUTION.

Purbeckian A stratigraphic stage name for the uppermost strata of the Upper JURASSIC, named from Purbeck in Dorset, England.

push moraines The landforms produced by the bulldozing effect of an ice-sheet advancing across the glacial DRIFT from an earlier glaciation. Such features have been described in Canada, USA, the Netherlands and Germany, where the fabrics (FABRIC, TILL) of the disturbed material bear little relationships to the direction of the ice-sheet responsible for the tectonics of the newly formed push moraines. Many of the latter exhibit thrust-faults, suggesting that the drift was in a frozen state when it was affected by the onset of the new ice-sheet.

push wave COMPRESSIONAL WAVE.

puszta The Hungarian name for the grasslands (STEPPE) of central Hungary.

puy A French term for a small volcanic PLUG, especially in the Auvergne region of south-central France, e.g. the Puy de Dôme. The plugs are composed of breccia and/or lava but are generally in the form of steep-sided, rounded, craterless hills rising dramatically from the surrounding plateaux.

Puzzolan cement A type of CEMENT formed by mixing powdered slaked lime with blast-furnace slag or volcanic ash. *See also* PORTLAND CEMENT.

p-wave SEISMIC WAVES.

pyramidal peak A steep-sided isolated mountain summit produced by the convergence of the BACKWALLS of adjoining CIRQUES. The precipitous cliffs of this horned peak, or *horn*, are separated by steep *arêtes* on the intervening ridges. The Matterhorn of the Swiss Alps is the most striking example.

pyranometer The term given to any instrument used for measuring scattered RADIATION on a horizontal surface. *See also* SOLARIMETER.

pyrheliometer An instrument for measuring direct solar RADIATION, by means of metal plates or *thermopiles* exposed at right angles to the Sun's rays. These convert the absorbed energy into an electric current (voltage) which can be measured. *See also* INSOLATION.

pyrites IRON PYRITES.

pyroclastic material The fragmental rock products ejected by a volcanic explosion having been 'broken by fire' – the literal translation of *pyroclastic*. The material may be thrown out as molten lava, which rapidly chills into larger volcanic bombs (BOMB, VOLCANIC) and smaller fragments (2–64 mm diameter) known as lapilli. The latter comprise tiny glass droplets (PELÉ'S TEARS) and glass filaments (PELÉ'S HAIR) together with vesicular clinker-like material termed SCORIA or volcanic cinders. Other material, which is ejected in solidified form, includes ASH, IGNIMBRITE, and PUMICE, which together form bedded layers known as TUFF. An alternative name for pyroclastic material is *tephra*.

pyrometer 1 An instrument used for measuring the temperature of molten lava. One model is based on light intensity emitted by the white-hot magma (*optical pyrometer*) while another depends on the changes of electrical resistance produced by the changes of heat and expansion of gases. The temperatures are, of course, too high to be measured by a normal mercury THERMOMETER. **2** An instrument for measuring temperatures of furnaces.

pyrophytic Plants which have become adapted to withstand fire. Giant redwood trees, for example, have developed very thick bark.

pyroxenes A group of silicate rock-forming minerals, found most abundantly in BASIC ROCKS and ULTRABASIC ROCKS.

Q

qaid A desert dune massif in which the summits rise high above the general level of the DUNES to form irregularly shaped conical hills, their steep sides being dimpled with hollows and terraces.

quadrangle The name given to a single map sheet in the USA. The map is bounded on the right- and left-hand margins by meridians, and on the top and bottom by parallels which are a specified number of minutes or degrees apart.

quadrant **1** A quarter of a circle, i.e. an arc of 90°. **2** An early type of navigation instrument used for measuring the angle of the Sun above the horizon. It has been superseded by the SEXTANT.

quadrat A measured area of any shape and size used as a sample in a vegetation survey.

quadratic surface The second degree of surface fitted in TREND-SURFACE ANALYSIS. The surface has one change of curvature in each direction. [*259*]

quadrature The times at which the Moon and Sun are so situated that rays drawn from each to the Earth make an angle of 90°. Such a situation occurs twice every SYNODIC month and coincides with NEAP TIDES. *See also* CONJUNCTION, OPPOSITION, SYZYGY.

quagmire A colloquial term for any BOG or MORASS, but strictly for any soft wet ground that yields under the weight of a person.

quantification The objective processes which deal with quantities of those data that can be handled by different SCALES OF MEASUREMENT as opposed to the data that cannot be put in a numerical scale. It also includes statistical techniques for explaining, solving and testing problems as well as methods of classification. It contrasts with subjective, descriptive and qualitative work.

quarry An open pit or void from which building stone, sand, gravel or mineral wealth is obtained by excavation.

quartile One of four equal values into which a data distribution may be divided around the MEDIAN value.

quartz The crystalline form of the mineral silicon dioxide (SiO_2). It is lustrous and sufficiently hard to scratch glass (HARDNESS SCALE). In its commonest form it is transparent and uncoloured but there are several varieties, including amethyst, yellow quartz (citrine), rose quartz, rock crystal (watery quartz), and smoky quartz (cairngorm).

quartzite A hard, impermeable whitish or greyish quartzitic rock cemented by silica. It can be either a METAQUARTZITE or an ORTHOQUARTZITE.

quartzitic schist SCHIST.

quartz porphyry A rock containing PHENOCRYSTS of quartz and alkali feldspar. *See also* PORPHYRY.

quasi-climax A phase in a plant SUCCESSION very close to a CLIMAX but which never quite achieves the state of dynamic equilibrium with its environment.

quasi-equilibrium A state of near EQUILIBRIUM, reached when a system moves towards a STEADY-STATE EQUILIBRIUM, but where absolute equilibrium is never actually achieved in the face of a constantly changing gross energy environment.

Table 197 *Quaternary (British sequence)*

Stage	*Type locality*	*Notes on boundaries, lithostratigraphy, etc.*
Flandrian	(Post-glacial, Holocene)	Begins at base of pollen zone IV (Fl. I)
Devensian	Four Ashes, Staffordshire (pit) SJ 916082	*Late* 25,000 BP to end of Loch Lomond Stadial; Dimlington Stadial *Middle* 50,000 BP to 25,000 BP; includes Upton Warren Interstadial complex *Early* That part of the Weichselian preceding 50,000 BP. Includes Chelford Interstadial
Ipswichian	Bobbitshole, Ipswich (excavation) TH 148414	Ipswichian lake muds
Wolstonian	Wolston, Warwickshire (pit) SP 411748	Includes Baginton–Lillington gravels, Baginton sand, Wolston series, Dunsmore gravels. Lower limit at base of Baginton–Lillington gravels
Hoxnian	Hoxne, Suffolk (pit) TH 175767	Hoxnian lake muds
Anglian	Corton, Suffolk (cliff) TH 543977	Lowestoft Stadial: Lowestoft Till Corton Interstadial: Corton Sands Gunton Stadial: Norwich Brickearth; Cromer Till Base at bottom of lower till of the cliff section
Cromerian	West Runton, Norfolk (cliff) TC 188432	Estuarine sands, silts, fresh-water peat (Upper Fresh-water Bed). Base at beginning of zone C.1
Beestonian	Beeston, Norfolk (cliff) TG 169433	Gravels, sands, silts, (Arctic Fresh-water Bed). Base at beginning of zone Be. I.
Pastonian	Paston, Norfolk (foreshore) TG 341352	Estuarine silts, fresh-water peat. Base at beginning of zone P.I
Pre-Pastonian	Beeston, Norfolk (foreshore)	Gravels, sands, silts and muds. Base at beginning of zone Pre-Pa.a
Bramertonian	Blake's Pit, Bramerton, Norfolk TG 298060	Marine shelly sand. Base at beginning of *Alnus–Quercus–Carpinus* biozone
Baventian	Easton Bavents, Suffolk (cliff) TR 518787	Marine silt. Base at beginning of zone L.4 (pollen)
Antian	Ludham, Norfolk (borehole) TG 385199	Marine shelly sand. Base at beginning of zone L.3 (pollen), LV (Foraminifera)
Thurnian	Ludham, Norfolk (borehole) TG 385199	Marine silt. Base at beginning of zone L.2 (pollen), L.III (Foraminifera)
Ludhamian	Ludham, Norfolk (borehole) TG 385199	Shelly sand. Base at beginning of zone L.I (pollen), L.I (Foraminifera)
Pre-Ludhamian	Stradbroke, Suffolk (borehole) TM 122738	Red Crag. Lower limit uncertain. Basal Crag (Waltonian) may be Pliocene

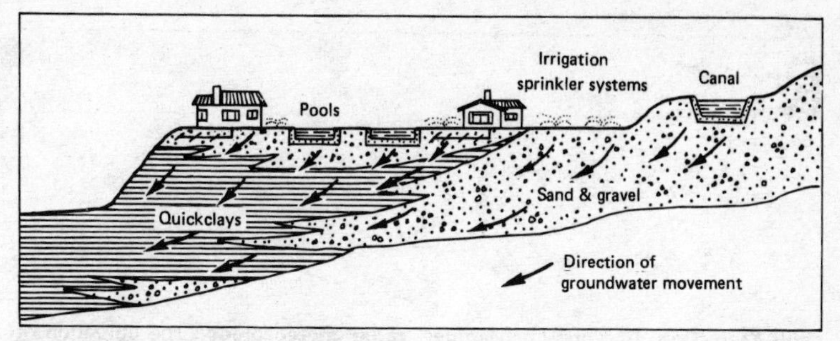

Figure 198 *Quickclays: potential hazard caused by groundwater seepage*

Quaternary **1** The younger of the two periods of the CAINOZOIC era. It comprises two epochs, the PLEISTOCENE and the HOLOCENE, and has so far lasted some 1.8 to 2 million years (LONG TIME-SCALE). **2** Some geologists regard it as a division of the TERTIARY. It was the era which saw the appearance of mankind. The divisions of the British Quaternary are shown in [*197*].

queuing theory A mathematical technique used to deal with problems in the relationship between INPUTS AND OUTPUTS caused by bottlenecks in a system. It helps to decide the chances of a given input (e.g. rainfall) producing a given output (e.g. discharge) in a given system (e.g. a drainage basin) over a given time. The disposition of soil moisture, for example, can be treated as a queuing problem in which assumed inputs of rainfall of given magnitude, distribution and time duration are *queued* into soil storages and 'serviced' by evapotranspiration and other moisture losses.

quickclays Clays analogous to a QUICKSAND because of their lack of COHESION under certain conditions of stress. They are initially the products of the grinding action of ice and have subsequently been deposited in a body of standing water to form stratified or varved sediments. They have a propensity to liquefy when disturbed by a sudden vibration, such as an earthquake shock, thereby giving rise to severe slope instability. Considerable loss of life was incurred, for example, by landslipping in the quickclays beneath parts of Anchorage, Alaska, during an earthquake in 1964. [*198; see also 146*]

quickflow OVERLAND FLOW.

quicksand An unstable layer of sand through which water moves upwards so fast that the sand is held in suspension by the water, thereby giving it no load-bearing capacity. Sudden vibrations can cause the sand to liquefy, resulting in flow and slope collapse.

Q-wave (Love wave) SEISMIC WAVES.

R

r and K selection An expression referring to a theory that NATURAL SELECTION may favour either individuals with high reproductive rates and rapid development (*r* selection) or individuals with low reproductive rates and better competitive ability (*K* selection). The terms are used definitively when r = maximum intrinsic rate of natural increase of a population; and K = the CARRYING CAPACITY of the environment in question.

race **1** A group of persons, animals or plants, connected by common descent or origin. **2** A rapid flow of water through a restricted channel, either through a watermill (*mill-race*) or in the sea (*tidal-race*).

radar An ACTIVE REMOTE-SENSING SYSTEM operating at MICROWAVE wavelengths. Radar is an acronym short for 'radio detection and ranging'. The two better-known systems are SLAR (sideways-looking airborne radar) and SAR (synthetic aperture radar). SLAR uses a short antenna to pulse microwaves down towards the Earth's surface and to receive them at the aircraft. In the majority of SLAR systems the returned pulse is written on to film to produce an image of the ground at a spatial RESOLUTION of around 60 m. SAR produces a synthetically long antenna by storing and comparing the doppler signals received while the aircraft travels along its flight path. In the majority of SAR systems, digital data on the terrain are recorded at a resolution of around 5 m. Radar has proved to be of value in geological, terrain and land-use studies. It has been used in several humid tropical countries owing to its ability to penetrate and 'see through' cloud. The US SEASAT satellite carries radar.

radar meteorology The utilization of radar technology by meteorologists to assist in weather forecasting. Radar is used to detect precipitation patterns and cloud distributions and to measure wind characteristics in the upper atmosphere. Scanning in the horizontal plane, by means of a *plan position indicator*, assists in plotting areal distributions of atmospheric phenomena (e.g. thunderstorm extent); scanning in the vertical plane, by means of a *range-height indicator*, allows the vertical extent of the phenomena to be measured (e.g. height of cumulonimbus cloud).

radar winds Air streams of the upper atmosphere detectable only by radar transported on a free balloon.

radial drainage A pattern of outflowing rivers away from a central point, analogous with the spokes of a wheel. It tends to develop on the flanks of a dome or a volcanic cone. The Lake District of N England is the best British example of an almost radial drainage pattern, thought to have been formed by superimposition (SUPERIMPOSED DRAINAGE) from a cover of sedimentary rocks that have subsequently been removed.

radial dyke A DYKE of igneous rock which radiates from a volcanic VENT, following fracture lines due to the stress imposed by crustal swelling prior to the eruption. *See also* DYKE-SWARM.

radian A unit of angular measurement equal to the angle subtended at the centre of a circle by an arc the length of which is equal to the radius. One radian = 57.29578°. There are 2π radians in a circle.

radiance A REMOTE SENSING term referring to the spatial distribution of radiant-power density. It is measured in watts/m^2 per STERADIAN and carries the symbol *L* in modern terminology (symbol *N* in older publications). Spectral radiance refers to the radiance at a particular wavelength of ELECTROMAGNETIC RADIATION. *See also* IRRADIANCE, RADIANT EXITANCE, RADIANT FLUX.

radiant energy The ENERGY originally transferred from the Sun as the *solar radiant*. It may subsequently be reradiated. *See also* RADIANT EXITANCE, RADIANT FLUX.

radiant exitance A REMOTE SENSING term referring to the measure of RADIANT ENERGY per unit area that leaves the object or surface which is being monitored or 'sensed'. It is measured in watts/m^2 and carries the symbol *M* in modern terminology (symbol *W* in older publications). Spectral radiant exitance refers to the radiant exitance at a particular wavelength of ELECTROMAGNETIC RADIATION. *See also* IRRADIANCE, RADIANCE, RADIANT FLUX.

radiant flux A REMOTE SENSING term referring to the time-rate of the flow of RADIANT ENERGY. Also referred to as *radiant power*. It is measured in watts and carries the symbol Φ in modern terminology (symbol *P* in older publications). *See also* IRRADIANCE, RADIANCE, RADIANT EXITANCE.

radiation 1 The transfer of energy emitted by the Sun and the Earth through a medium by means of electromagnetic waves of different wavelengths (from 70 µm to 0.1 µm) [*70*]. The transfer can take place through a vacuum. The term is used in meteorology with reference to *solar radiation*, a process by which the Sun's energy reaches the Earth (INSOLATION), and also with reference to *terrestrial radiation*, by which energy received from the Sun is reradiated at a longer wavelength (*thermal* INFRARED RADIATION) from the Earth's surface [*98*]. The atmospheric gases absorb much of the terrestrial radiation (ABSORPTION) except any radiation with wavelengths between 8.5 µm and 14 µm, which passes through the atmosphere and is lost into space (ATMOSPHERIC WINDOW). Owing to absorption the atmosphere is warmed from below and in turn reradiates energy back to the Earth's surface (GREENHOUSE EFFECT), thereby helping to maintain higher temperatures than would otherwise be expected. *See also* ELECTROMAGNETIC RADIATION, PLANCK'S LAW, RADIATION BALANCE, SOLAR CONSTANT, STEFAN'S LAW. **2** In surveying, the term is used in PLANE-TABLING, whereby points are fixed by use of an ALIDADE to determine direction and a measuring-tape to determine distance.

radiation balance A term used in meteorology to denote the net effect of the difference between incoming and outgoing RADIATION at a given point on the Earth's surface. In general there is a positive radiation balance at the Earth's surface during the day and a negative radiation balance at night owing to cooling (especially on cloudless nights). The atmosphere maintains a negative radiation balance at all times but there is an annual positive balance over the surface of the Earth as a whole. If the Earth and the atmosphere are taken as a whole there is an overall equilibrium, but this is made up of a *radiation deficit* polewards of latitude 38° and a *radiation surplus* equatorwards of this latitude. In addition, there is a radiation surplus over the oceans and a radiation deficit over the continents. *See also* NET RADIATION.

radiation fog A shallow layer of FOG formed near to the ground surface owing to nocturnal cooling and loss of heat due to terrestrial radiation (RADIATION). Loss of heat by the surface cools the air in contact with it, especially during weather with little turbulent mixing and clear skies, and the DEW-POINT is soon reached whereupon CONDENSATION occurs. Since the cold air flows downhill gravitationally there is a tendency for radiation fog to develop in hollows and valleys. In summer the fog will disperse quickly after sunrise but in autumn and winter it may be trapped beneath an INVERSION OF TEMPERATURE and may persist for much of the day; in some urban and industrial areas the radiation fog could develop into a SMOG under these conditions. *See also* ADVECTION FOG, ICE-FOG.

radiative forcing An alteration in the functioning of a climatic system which alters the RADIATION BALANCE. Among the many reasons for such a change are perturbations in INSOLATION, the GREENHOUSE EFFECT and dust veils (DUST VEIL INDEX).

radioactive decay The gradual breakdown of an element by the emission of charged particles from the nuclei of its atoms, leading to the formation of stable ISOTOPES. By measuring the rates of radioactive decay in certain rocks it is possible to determine their age (RADIOMETRIC DATING). *See also* HALF-LIFE, LEAD-RATIO.

radiocarbon dating A method of determining the age of an organic material by measuring the proportion of the C-14 ISOTOPE contained within its carbon content. This radioactive carbon isotope enters the Earth's living organisms from the atmosphere, continuing to be assimilated until the organism dies or is buried beneath sediments, at which point RADIOACTIVE DECAY begins. The HALF-LIFE of C-14 is 5,570 years, a relatively short half-life, which makes it particularly useful for dating objects or material up to *c.* 70,000 years, but accuracy diminishes beyond 30,000–40,000 years. The ABSOLUTE AGE of the material is correct to ±5%, having been determined by a knowledge of the initial radiocarbon concentration, its constant rate of decay, and its present proportion. Radiocarbon dating is one method of RADIOMETRIC DATING.

Radiolaria An order of PROTOZOA, characterized by internal siliceous skeletons, which are important constituents of marine PLANKTON. They are of no value as zone fossils since they are easily destroyed under alkaline conditions. *See also* FORAMINIFERA.

radiolarian ooze A type of deep-sea siliceous OOZE, containing the minute skeletal remains of RADIOLARIA. It is not very commonplace when judged at a global scale, occurring largely in the central Pacific and in the SE sector of the Indian Ocean.

radiolarite A type of CHERT containing a high proportion of RADIOLARIA.

radiometer A REMOTE SENSING term referring to a device for measuring the RADIANCE of a surface or the IRRADIANCE on to a surface. A SPECTRORADIOMETER records ELECTROMAGNETIC RADIATION in narrow wavelength intervals. A *bandpass radiometer* records electromagnetic radiation in broad wavelength intervals. A radiometer that scans in order to build up an image of a scene in a number of wavebands is termed a MULTISPECTRAL SCANNER. A handheld radiometer is often used to provide GROUND INFORMATION to aid in the interpretation of remotely sensed aerial or SATELLITE IMAGES. *See also* NET RADIOMETER.

radiometric dating A method of dating rocks by measuring the rates of RADIOACTIVE DECAY of their radioactive elements. It is based on the principle that a radioactive 'parent' element breaks down into a 'daughter' element at a constant rate (λ), with the time period necessary for one half of the 'parent' element atoms to decay being known as the HALF-LIFE (T). The relationship between T and λ is $T = 0.693/\lambda$. Among the most commonly used types of radiometric dating are: **1** RADIOCARBON C-14 DATING. **2** Uranium–lead, with uranium-238 yielding lead-206 + 8 helium-4 (half-life 4,498 million years) and uranium-235 yielding lead-207 + 7 helium-4 (half-life 713 million years). **3** Lead–lead, in which lead-207 accumulates six times as fast as lead-206. **4** Potassium–argon (K/Ar), in which 11% of potassium-40 yields argon-40, the remaining 89% yielding calcium-40, the half-lives of these two processes being 11,850 million years and 1,470 million years respectively. **5** Rubidium–strontium, in which rubidium-87 yields strontium-87, with a half-life of 50,000 million years. This very long half-life makes this method particularly useful for Lower Palaeozoic and Pre-Cambrian rock dating. The most reliable radiometric dates are derived from igneous rocks, but the methods can also be applied to metamorphic rocks. This type of dating is of very limited use in sedimentary rocks. *See also* ELECTRON SPIN RESONANCE.

radiosonde An instrument used by meteorologists to record humidity, pressure, temperature and winds in the upper atmosphere. It comprises an automatic recording apparatus linked to a small radio-transmitter, both of which are carried to heights of 20,000 m (50 mb) by a hydrogen-filled balloon which is tracked by radar. *See also* RAWINSONDE, ROCKETSONDE.

radon A colourless, odourless gas (Ra), the only radioactive gaseous element, and some eight times denser than air. It is produced by the decay of URANIUM, with a HALF-LIFE of 3.8 days and is measured with a RADON EMANOMETER. It is especially common in granitic rocks and was once thought to create a natural hazard to inhabitants of granite-built houses and people who dwelt in such granitic terrains as Cornwall and Devon in Britain. Recent research, however, has shown that this threat is not as serious as was once believed.

rafting The transportation of material (seeds, fauna, soil, ERRATICS) by means of floating ice or by means of attachment to floating vegetable matter.

rag, ragstone The name given in Britain to rocks of a hard, rubbly nature, e.g. Coral Rag of JURASSIC age.

rain The condensed water vapour of the atmosphere occurring in drops large enough to fall under the influence of gravity. 'The Rains' is a term for the rainy season in MONSOON regions. *See also* PRECIPITATION.

rainbow An optical phenomenon in the form of a luminous arc across the sky comprising the spectral colours with (in a *primary* bow) red on the outer side and violet on the inner (when a *secondary* bow occurs the colour sequence is reversed). It is caused by refraction and internal reflection of sunlight through falling RAINDROPS (or waterfall spray), when the observer stands with his back to the Sun. The greater the magnitude of the drops the more spectacular are the colours. *See also* FOGBOW.

rain day There are three uses of this expression: **1** In the UK, a period of 24 hours from 09.00 h GMT, within which 0.2 mm or more of rainfall is recorded. **2** On the continent of Europe, a period of 24 hours within which a minimum of 1 mm of rainfall is recorded. **3** In the USA, any day with measurable (>0.01 inches) precipitation. *See also* WET DAY.

raindrop A small globule of RAIN, the size of which varies according to rainfall intensity. In the least-intense rainfall (DRIZZLE) the drops are less than 0.5 mm in diameter, have a terminal velocity of 0.7–2.0 m/sec, and are just heavy enough to fall out of a cloud. In an intermediate-intensity rainfall the majority of drops will be less than 0.75 mm in diameter and have a terminal velocity of 4 m/sec. In heavy-rainfall situations drops with diameters over 5.0 mm in diameter will occur, achieving terminal velocities higher than 10 m/sec. Unless aided by a strong DOWNDRAUGHT, however, it is unlikely that very large drops will reach the ground surface because they will break up under aerodynamic forces. Falling raindrops undergo evaporation in the unsaturated air beneath the cloud base. A droplet of 0.2 mm diameter falling into air of 90% relative humidity at a temperature of 5° C will fall only 150 m before it evaporates; a droplet of 2 mm diameter would descend 42 km before evaporating. *See also* COLLISION THEORY OF RAINFALL, LANGMUIR THEORY OF RAINDROP GROWTH.

raindrop erosion A type of SOIL EROSION created by the impact of large raindrops falling at high velocities. It is most intensive under tropical rainstorms in areas which have been recently cleared of vegetation. Soil particles are mechanically detached and transported by surface RUNOFF, especially as the fine material chokes the voids in the soil. *See also* RAINSPLASH, RAIN-WASH.

rain factor A climatological index devised by R. Lang in an attempt to classify climatic regions by calculating their degree of aridity. The rain factor is expressed by the equation:

$$rf = \frac{\textit{annual precipitation}\ \text{(mm)}}{\textit{mean annual temperature}\ (^\circ\text{C})}.$$

rainfall The water equivalent of all forms of atmospheric PRECIPITATION received in a RAIN-GAUGE, assuming that there is no loss by EVAPORATION, PERCOLATION or RUNOFF. The term includes RAIN, DEW, HOAR FROST and RIME;

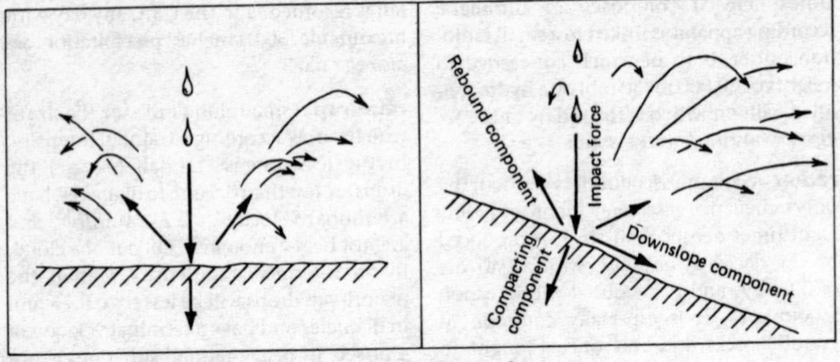

Figure 199 *Rainsplash trajectories on level and sloping ground*

parts of the two latter will, on melting, find their way into the rain-gauge. Rainfall is classified into four types: **1** CONVECTIONAL RAINFALL; **2** *cyclonic rainfall* (CYCLONE); **3** FRONTAL RAINFALL; and **4** OROGRAPHIC RAINFALL. *See also* INTENSITY OF RAINFALL, RUNOFF.

rain-forest EQUATORIAL RAIN-FOREST.

rain-gauge An instrument used to measure RAINFALL at a given point on the ground. It comprises a collecting funnel, the rim of which is a standard 30 cm from the ground surface, and an interchangeable aperture leading into a collecting container. The accumulated water is periodically emptied and measured in a calibrated cylindrical flask. *See also* HYETOGRAPH, NATURAL-SIPHON RAINFALL RECORDER, RAINFALL, TIPPING-BUCKET RAIN-GAUGE.

rain-making A process which attempts to stimulate *rainfall* by artificial means. Experiments, mainly in N America, have shown that only certain types of clouds at specific temperatures can be used successfully for rainfall generation (CLOUD-SEEDING) and that in some instances the results increase the likelihood of floods, crop damage and even loss of life. DRY ICE and silver iodide are injected as FREEZING NUCLEI to promote the BERGERON–FINDEISEN process.

rain pits, rain prints Tiny craters preserved in sedimentary rocks, having been formed by the impact of raindrops on a soft sediment exposed at the surface.

rain shadow The area on the leeward side of a topographic barrier, where precipitation amounts are smaller than those on the windward side. It results from the fact that a moist air stream, causing heavy *rainfall* (OROGRAPHIC RAINFALL) on the windward slopes, is dried out as it descends the lee slope. *See also* ADIABATIC, CHINOOK, FÖHN.

rain-spell A period of fifteen consecutive days on each of which at least 0.25 mm of RAINFALL are recorded.

rainsplash The process of raindrop impact which contributes to RAINDROP EROSION. On level ground it merely rearranges the soil particles but on a slope there is a net transport of material owing to a longer downslope 'flight path' for rebounded droplets and to the downslope component of the impact force. The latter is related to raindrop size and rainfall intensity. Once OVERLAND FLOW is initiated rainsplash becomes unimportant, since raindrop impact is cushioned by the layer of water. Thus, rainsplash is probably only significant in short, sharp storms where little runoff is generated, or close to an interfluve where there is insufficient catchment to generate flow. *See also* RAIN-WASH. [*199*]

rain-wash A type of SURFACE WASH characterized by the movement of soil particles and loose materials down a hillside as a result of RAINSPLASH, RUNOFF and OVERLAND FLOW. It can ultimately lead to SOIL EROSION. *See also* SHEET EROSION.

raised beach A shoreline and its littoral deposits elevated above present sea-level either by positive ISOSTATIC MOVEMENTS or by a fall in sea-level (EUSTASY). It may be backed by a cliffline and associated with an ABRASION PLATFORM. If several movements of sea-level have occurred a succession of raised beaches of different ages may be found. Some examples include abandoned sea caves, stacks and arches, while others have wide shingle deposits. Many raised beaches, mostly of Pleistocene age, have been identified around the coasts of the British Isles, ranging in elevation from a few metres to 45 m above present sea-level. In certain parts of the world coastlines are elevated rapidly by earthquakes, e.g. in the 1964 earthquake in Alaska.

raised bog A variety of BOG, corresponding approximately to the German *hochmoor* (MOORLAND), and contrasting in character with a BLANKET BOG, which is much thinner and occurs in wetter, cloudier climatic zones. The raised bog starts its growth in an OLIGOTROPHIC environment of shallow lakes in which a FEN or CARR vegetation may eventually create a basin of fen peat. As the surface of this peat bog rises above the neutral or alkaline groundwater the acidity of the decaying vegetation is no longer neutralized, resulting in an upper accumulation of acid peat. It then passes into an OMBROTROPHIC state in which the surface of the bog takes on a convex form sloping gently towards the perimeter of the basin where a watercourse receives the drainage of the dome-shaped raised bog, which may approach 7 m thickness. It generally occurs in areas receiving lower rainfall totals than those occupied by blanket bogs, e.g. raised bogs occur in the Central Plain of Ireland while the blanket bogs occur in the wetter western coastlands. [*108*] REGENERATION COMPLEX.

RAM An acronym for Random Access Memory. A type of computer memory used for temporary storage of data which is lost if the power is turned off. It contrasts with ROM.

randkluft The term given to the headwall gap between the glacier and the rock face (BACKWALL) at the back of a CIRQUE. It is similar to a BERGSCHRUND but differs from it in so far as the latter has both of its sides composed of ice while the randkluft has only the lower (distal) side of glacier ice. It is thought to be formed by melting caused by radiation of heat from the backwall.

randomness A state in which there is an equal chance of any number of events occurring or of any variable being chosen. Several statistical tests have been devised to test whether or not a sequence of events may be considered random. These are known as *randomization* tests. *See also* NEAREST-NEIGHBOUR INDEX.

random numbers A table of numbers devised to occur in random form, used in the technique of *random sampling* (RANDOM SAMPLE).

random-point distribution A pattern of spatial RANDOMNESS in the *distribution of*

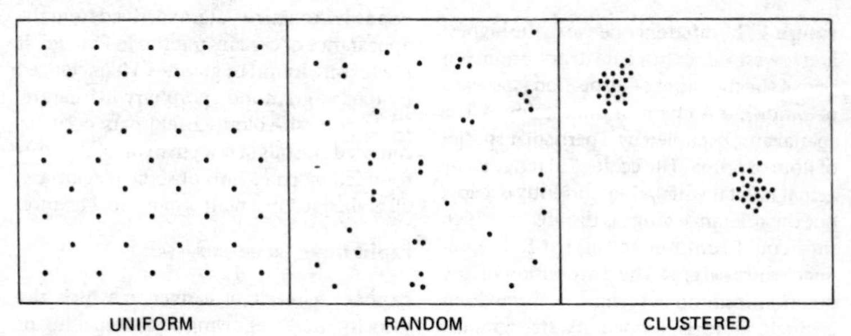

Figure 200 *Distribution patterns*

points, lying part way between a uniform distribution and a clustered distribution. If the pattern is random a situation occurs in which a number of points are fairly widely distributed but there is also a tendency to cluster [*200*]. A test for a *non-random distribution* is termed the NEAREST-NEIGHBOUR INDEX.

random sample A data SAMPLE in which any one individual measurement or count in the data population is as likely to be included as any other, i.e. it is selected without bias. If, however, the data exhibit a marked clustering there may be a tendency for a biased sample to be taken. This can be avoided by adopting the technique known as *stratified random sampling* in which the data are divided into classes (strata) before taking a random sample within each *stratum*. *See also* SAMPLING FRAME. [*218*]

random-walk model A type of SIMULATION MODEL that represents the RANDOMNESS in the spatial progression of a physical process. It is used especially in geomorphology to denote the degrees of randomness exhibited in the growth of a drainage pattern, whereby no geological or other controls are allowed to dominate the development of the network. The simulation is based on squared paper with each square being referred in sequence to a table of RANDOM NUMBERS, and each compass direction also being allotted a number. Thus, the direction of stream flow from square to square can be generated entirely at random, or it can be biased by allotting more random numbers to one direction than to the others.

range 1 The difference between the highest and lowest values in a data distribution. It is one of the measures of dispersion (STANDARD DEVIATION). **2** A chain of mountains. **3** The spatial area occupied by a particular species of flora or fauna. The ecological range is the actual habitat which they currently occupy, but the tolerance range is the area in which they could continue to survive and reproduce (TOLERANCE). **4** The distribution of any taxonomic group of organisms throughout geologic time is termed its stratigraphic range. **5** A term used in the USA to denote the unfenced grasslands of the Great Plains. *See also* TIDAL RANGE.

ranging rod A straight wooden staff about 2 m in length used by surveyors when setting out a straight line (BASELINE) during a topographical survey.

rank 1 In statistics, one of the orders in a RANKING procedure. **2** A particular category in the stage reached in the formation of coal, from lignite through bituminous coal to anthracite. The criteria used to differentiate coals are both chemical and physical, but usually the carbon content is taken as a measure of its rank. The differences are due to varying degrees of metamorphism.

ranker The name given to a soil when it reaches a stage in its formation just beyond the original primary soil stage. It is marked by the initial development of an upper horizon (A-HORIZON) in which a thin layer of organic material begins to accumulate and mix with the underlying mineral matter of the unaltered parent material. It is found, therefore, in recently glaciated areas, on recently deposited alluvium and on sand-dunes, i.e. environments where the surface is relatively young in terms of its geomorphic history. In Britain it might be classified as a LITHOSOL and in the US SEVENTH APPROXIMATION as an ENTISOL or an INCEPTISOL.

ranking The method by which any data are arranged in ORDER according to a given criterion or set of criteria, e.g. size, hardness, etc., to produce an ordinal scale (SCALES OF MEASUREMENT).

rapakivi texture A term derived from the appearance of certain granites in Finland. It is a texture found in granites, characterized by a coarse grain and a porphyry-like nature. Its large flesh-coloured feldspars occur as rounded crystals (a few cm in diameter) surrounded by paler rims of sodic plagioclase, all embedded in a matrix of normal texture.

rapid flow FROUDE NUMBER.

rapids A reach of a river in which the velocity increases owing to a steepening of the gradient. It is marked by broken water, chutes, small waterfalls and exposures of bedrock. *See also* CASCADE, CATARACT.

rare earths An expression given to those metallic oxides with atomic numbers between 57 and 71. They are chemically similar and therefore difficult to separate. Although they are widely distributed they are relatively scarce minerals.

raster A GIS term referring to a data structure based on the squares (cells) of a GRID.

rated section A hydrological term referring to a method that obtains a continuous record of river DISCHARGE at a given section. The measurements required to rate the section are obtained by using the VELOCITY AREA METHOD.

rating curve A graph used in hydrology to assist in the estimation of stream-flow. It is constructed by plotting the height of the water surface (stage) against DISCHARGE, measured in cusecs or cumecs. A rating curve allows an estimation to be computed from stage records and used in the prediction of FLOODS, in the planning of flood-protection structures and in river-management schemes. [*201*]

ratio The relationship between two similar magnitudes in respect of quantity, determined by the number of times one contains the other.

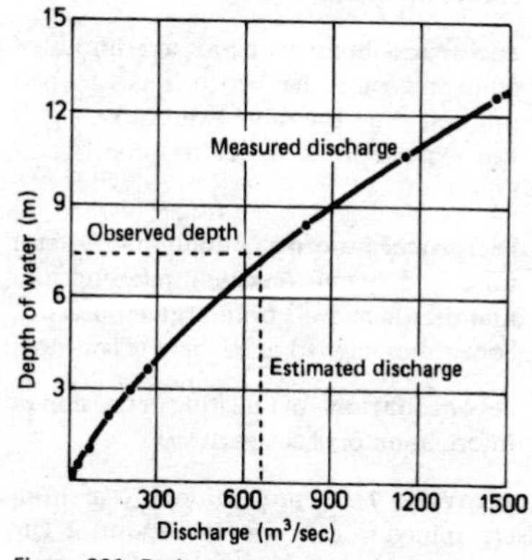

Figure 201 *Rating curve*

rational formula A hydrological formula to estimate RUNOFF (Q) in terms of a coefficient (*C*) which is dependent upon a measure of precipitation (*I*) and an index of the basin area (*A*) and is expressed as

$$Q = CIA$$

ratio scale SCALES OF MEASUREMENT.

rattan A species of climbing palm (*Calamus*) characterized by a long, thin pliable stem. It grows chiefly in Indonesia but to a small extent in Africa and Australia, where it is used for basket-weaving, rope- and net-making.

ravine A deep, narrow river valley but without the precipitous sides of a GORGE, which it resembles in stature. It is bigger than a GULLY. *See also* ARROYO.

raw humus MOR.

rawinsonde A French term (an abbreviation for radar wind-sounding) used in meteorology. It refers to a self-recording and transmitting instrument attached to a radar target and which is carried aloft by a hydrogen balloon, whence it measures air flow patterns and speeds. *See also* RADIOSONDE, ROCKETSONDE.

raw soil INCEPTISOL.

Rayleigh number (Ra) A dimensionless number given by the expression $(l^3 g\gamma\Delta\theta p)/nk$, where: l = length; g = acceleration due to gravity; γ = the cubic expansion coefficient; p = density, n = the dynamic viscosity coefficient; k = thermal conductivity; and $\Delta\theta$ = the temperature difference.

Rayleigh scattering DIFFUSE RADIATION, SCATTERING.

Rayleigh test of preferred orientation A statistical test to determine the significance of the distribution of a series of angular measurements when it is necessary to calculate an average value for the ORIENTATION, i.e. the preferred orientation of the sample. For example, the average orientation of the long axes of pebbles in a till (FABRIC, TILL) can be tested by $L = (R/n) \times 100$, where: L = the value required (%); n = the number of observations; and $R = \sqrt{[(\Sigma f \sin 2\theta)^2 + (\Sigma f \cos 2\theta)^2]}$.

Rayleigh wave SEISMIC WAVES.

RBV A term used in REMOTE SENSING as an abbreviation of *return beam vidicon* camera. The LANDSAT series of SATELLITES carry a number of RBV cameras operating in the green, red and near INFRARED spectral regions. The IMAGE is formed on the photo-sensitive surface of a vacuum tube that is itself scanned with an electron beam for transmission to receiving stations on the Earth's surface. Modern RBV cameras produce data with high spatial but low spectral RESOLUTION. These images are of increasing value in geological, soil, terrain and land-use mapping.

reach 1 A portion of a river, used especially to describe a section between bends or locks. **2** A narrow sea channel extending inland.

reaction The response to an action, or the result of an input into a system (INPUTS AND OUTPUTS). The time period which elapses between the input and the beginning of the change in the operation of the system is known as the *reaction time*.

reaction, soil pH INDEX.

Réaumur scale An arbitrary scale for the measurement of temperature, invented by a French physicist, R. A. F. de Réaumur, in 1730. It was divided into 80 units rather than 100 because he noted that the alcohol in his thermometer increased its volume 8 per cent between measurements taken in melting ice and boiling water. It is now obsolete. °R = 9/4 F° (–32°) = 5/4 C°; 80 R° = 180 F° (212° – 32°). *See also* FAHRENHEIT SCALE, CENTIGRADE SCALE.

Recent 1 In Britain a synonym for the HOLOCENE. **2** Considered by some US geologists to be the last subdivision of the PLEISTOCENE, equivalent to an interglacial following the WISCONSIN glacial.

recentred map projection Any MAP PROJECTION in which the origin or central meridian is repeated in order to reduce the peripheral distortion. This expression is preferable to that of *interrupted map projection*.

recession limb of a hydrograph A hydrological expression referring to that portion of a stream discharge HYDROGRAPH, which refers to the time after precipitation has ceased when DISCHARGE gradually falls and is unaffected by rainfall. It is termed a RECESSION CURVE, itself expressed as:

$$Q_t = Q_{o}e^{-at}$$

where Q_t is discharge at time t; Q_o is the initial discharge, a is a constant, t is the time interval and e is the base of natural logarithms. *See also* BASE-FLOW RECESSION CURVE.

recessional moraine STADIAL MORAINE.

recession col An expression sometimes used to describe a WIND-GAP on the crest of a CUESTA, created when scarp recession breaks into the valley of a dip-slope stream.

recession curve BASE-FLOW RECESSION CURVE, RECESSION LIMB OF A HYDROGRAPH.

recharge The action by which water is added to an AQUIFER. This may occur naturally by percolation from the ground surface of precipitation or snowmelt, or it may be carried out artificially by pumping water upwards to the flanks of the *recharge area*, i.e. the upper slopes of an ARTESIAN BASIN. Artificial recharge of the chalkland aquifers is commonplace in S England.

reciprocal In mathematics, a reciprocal of an expression is that which equals 1 when multiplied by the same expression, e.g. if the expression is 4, its reciprocal is ¼ (¼ × 4 = 1).

reciprocal bearing An alternative term for a *back bearing* (BEARING), referring to a line drawn at 180° from a given bearing. Sometimes referred to as a *reverse bearing*.

recompilation An updating or revision of information on a COMPILED MAP.

recovery 1 The proportion of coal, iron, etc. mined from a mineral deposit. **2** The return of a SYSTEM towards its original state after a short-term input (INPUTS AND OUTPUTS) has ceased.

rectangular drainage pattern A drainage network characterized by right-angled bends and right-angled junctions between tributaries and the trunk stream [*202*]. It is thought to result from the structural control imposed by the jointing or fault pattern of the underlying rocks. It differs from a pattern of TRELLIS DRAINAGE in so far as it is more irregular and its tributary streams are not as long or as parallel as in trellis drainage.

rectangular polyconic projection A type of MAP PROJECTION, similar to a simple POLYCONIC PROJECTION but in which the principal scale is not preserved along all the parallels. Instead, the equator is taken as the standard parallel and the meridians are drawn to intersect all parallels at right angles.

rectilinear slope A slope, or part of a slope, which has a constant angle, i.e. it is straight and has uniform conditions on all its parts. It is claimed, however, that over time a rectilinear slope declines in gradient for the following reason: more soil has to cross the lower part of the slope than the upper part, leading to a thickening of the regolith on the lower part, thereby giving the latter a greater protection from weathering. Thus, the rate of weathering declines progressively downslope, so long as the regolith loss is in whole or in part by surface transport, the rate of which does not vary with distance from the crest. Rectilinear slopes occur most commonly below an upper convex-slope element and above a lower concave-slope element. *See also* SLOPE DECLINE.

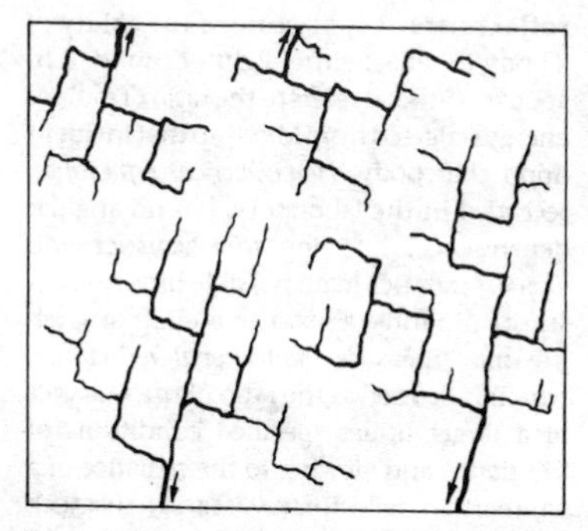

Figure 202 *Rectangular drainage pattern*

recumbent fold An overturned *fold* in which the axial plane is virtually horizontal. *See also* FOLD. [*84*]

recurrence interval The average period of time within which a specified amount or intensity of rainfall can be expected to occur once. These data are of value in FLOOD forecasting where the RETURN PERIOD of a particularly hazardous flood level can be expected at intervals of, say, once in twenty-five years, or once in fifty years. This is not to say that the rainfall will appear at regular intervals, so that exact predictions of timing are impossible.

recurved spit A coastal SPIT the distal end of which is curved inward as a result of wave action.

red beds A term referring to sedimentary deposits or soils whose red colour results from the CLASTS being coated with ferric oxide, usually HAEMATITE. Only deposits having hues redder than 5YR on the MUNSELL COLOUR SYSTEM should be described as red beds.

red clay An ocean-floor deposit of the ABYSSAL ZONE, comprising a fine-grained hydrated silicate of alumina, coloured brownish-red by ferric iron and manganese compounds, and occasionally containing quantities of manganese oxide NODULES, of some economic value. It also contains sporadic occurrences of whales' ear bones and sharks' teeth. It is derived from a combination of volcanic ash, meteoric dust and material dumped from melting icebergs. It is the most widespread of the deep-sea deposits, covering about 28% of the ocean floor. *See also* OOZE, PELAGIC, RED MUD.

Red Crag Series A term given to the lowest formation of the pre-glacial PLEISTOCENE sediments in East Anglia, England. It comprises: **1** a lower Red Crag member (PRE-LUDHAMIAN), including the formerly termed Butley Crag, Newbourn Crag (NEWBOURNIAN) and Walton

Crag (*Waltonian*); **2** an upper Ludham Crag member (LUDHAMIAN). The formation is made up essentially of shelly marine sands (*crags*) with minor amounts of clay and gravel, laid down in a period of climatic cooling.

red earths A term referring to tropical soils coloured by iron compounds. The more specific term LATOSOLS is to be preferred.

red mud A clay/silt deposit occurring on the CONTINENTAL SLOPE. It should be distinguished from the RED CLAY of the abyssal zone in that the red mud is derived directly from the continents.

redox potential A measure of the gain of an electron (REDUCTION) or the loss of an electron (OXIDATION) in a chemical reaction, i.e. electron transfer, which is often dependent on the PH INDEX. *See also* CHEMICAL WEATHERING.

red rain Rainfall coloured by red sand or dust transported on high-level winds from arid to more humid zones. The red rain and red snow of S and W Europe (e.g. SE England, 1 July 1968) is usually connected with an incursion of dust from North Africa.

reduction **1** A chemical process in which oxygen is removed from a compound. This occurs in soils owing to the continuous presence of groundwater, which tends to convert ferric iron (Fe_2O_3) into ferrous iron (FeO), in which state it becomes more mobile because it is more soluble. It manifests itself in a soil profile by replacing the red and yellow colours of OXIDATION with the greys and blues of GLEYING. *See also* ANAEROBIC, CHEMICAL WEATHERING, MOTTLING. **2** A term used in cartography, in which the scale of a map is reduced by means of a PANTOGRAPH or by photography.

reef **1** A line of rocks in the tidal zone of a coast, submerged at high water but partly uncovered at low water. It can be composed of any type of rock. **2** A CORAL REEF, which may be in the form of an ATOLL [*19*], a BARRIER REEF or a FRINGING REEF. **3** A quartz vein containing gold or other precious metal. **4** A gold-bearing conglomerate of Pre-Cambrian age in S Africa. **5** A line of hills in the SW of the USA.

reef-flat **1** A wide expanse of eroded coral reef littered with rock pools and coral debris, uncovered at low tide and thought to be a wave-cut platform abraded across dead reef rock. **2** A dune-covered accumulation zone on which vegetation has become established. It occurs on the inner side of a FRINGING REEF and comprises a platform covered with coral sand and coral fragments.

reef-knoll A dome-like mass of limestone that has grown upwards from a REEF in order to keep pace with the deposition of surrounding sediments. When the strata with which it is associated become part of a land mass the reef-knoll may be exposed by denudation. Because of its poorly developed joint system and its shape it tends to resist erosion and to form a cone-shaped hill.

re-entrant The name given to any marked recess or indentation into an escarpment, ridge or coastline. It contrasts with a SALIENT.

reference net A network of squares printed on a map to assist in the location of a particular point. It is numbered along one side and lettered along the other, so that a combination of the two will indicate a particular square, although with no precision within the square itself. It is frequently used in books of street plans, in gazetteers and in indexes of atlases. *See also* GRATICULE.

reflectance **1** A measure of the ability of a body to reflect either light or sound. **2** In REMOTE SENSING it refers to the ratio of radiant energy reflected from a body to that incident upon that body. *Hemispherical reflectance*, recorded in the laboratory, has no angular dependence, referring to IRRADIANCE and RADIANT EXITANCE in all possible directions. A SENSOR mounted aboard an aircraft or satellite may measure *bidirectional reflectance*, which is defined as the ratio of the RADIANCE of a target under specified conditions of irradiance and viewing to the radiance of a Lambertian reflector (LAMBERTIAN SURFACE). *See also* RADIOMETER.

reflection profile A term given to the seismic profile produced during a SEISMIC SURVEY.

reflective beach (swell beach) A BEACH characterized by low frequency SWELL waves, low beach WATER-TABLE, low BACKWASH and no SURF zone circulation along the shore. It results from the action of CONSTRUCTIVE WAVES which create a steep-fronted profile. It is opposite to a DISSIPATIVE BEACH.

refolded folds Complex FOLDS which have undergone more than one period of folding, giving rise to cross-folding (polyphase folding) in which the axes will intersect each other at an angle. Such *refolding* may produce such complex structures as a CULMINATION and a DEPRESSION.

reforestation Replanting of trees in an area which has been denuded of forest by natural causes or by artificial felling. It is also known as *reafforestation*, which should not be confused with the term *afforestation* which in turn refers to the act of tree-planting in an area that has not formerly been forested.

refraction A term referring to the deflection of wave propagation when waves pass obliquely from one region of velocity to another. *See also* WAVE REFRACTION.

refraction diagram A drawing which shows the changing positions of the waves and/or the ORTHOGONALS over a given area for a given WAVE PERIOD and direction. It is also known as a *wave-refraction diagram*. *See also* WAVE REFRACTION. [*278*]

refractory The ability to withstand the action of heat. Some clays possess this property (FIRE CLAY).

refugium A HABITAT in which formerly widespread organisms survive in a relatively small, isolated location that possesses distinctive ecological, geological, geomorphological or micro-climatic characteristics. A notable British example is the survival of a rare Arctic-Alpine flora on a particular type of metamorphosed limestone in Upper Teesdale in north-east England.

reg An Arabic term referring to a bare, stony, horizontal desert surface, comprising a layer of tightly packed gravel overlying sand and clayey sand. It is formed by wind removal of the finer surface materials, leaving behind a concentration of coarse and fine pebbles, which frequently become cemented into a DESERT PAVEMENT by mineral solutions drawn to the surface by CAPILLARITY. *See also* HAMADA.

regelation The process of refreezing of ice subsequent to melting caused by pressure within an ice mass. It has been suggested that pressure converts ice crystals into molecules of water which then flow (often downslope) to points of lower pressure, where they recrystallize by freezing. It is thought that regelation is a major factor in the mechanism of downslope movement of a GLACIER. *See also* PRESSURE MELTING-POINT.

regelation layer A thin ice band at the base of a glacier, thought to have been formed by alternating pressure melting and refreezing as the glacier moves forward over an obstacle on the rock floor.

regeneration complex An expression referring to a small-scale mosaic pattern of vegetation on the surface of a RAISED BOG, in which the bog vegetation (*Sphagnum*, *Calluna*, *Eriophorum*) colonized the hollows and hummocks, respectively, and regenerated itself in an AUTOGENIC cyclic fashion. This concept of self-perpetuating regeneration, introduced by Scandinavian botanists in the 1920s, has recently been rejected in favour of close climatic control of the bog hydrology.

regime, regimen **1** In general, the total economy of a natural SYSTEM. **2** The characteristic movements of a stream as it attempts to remain adjusted to its channel. A stream is said to be in regime (or regimen) if its channel has achieved a stable form as a result of the stream's flow characteristics. **3** The set of processes by which glaciers are nourished, expanded and ultimately reduced by melting. **4** The alternating seasonal pattern of climatic changes in a given region is the climatic regime of that region.

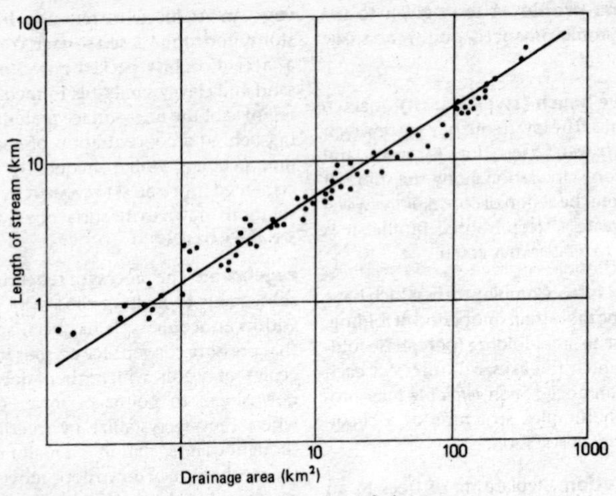

Figure 203 *Linear regression: relation of stream lengths to drainage areas*

regional metamorphism The large-scale action of both pressure and heat (*see also* DYNAMIC METAMORPHISM, THERMAL METAMORPHISM) over a very large area of the Earth, in association with an OROGENY. A wide range of new rocks and minerals is thereby formed.

registration A REMOTE SENSING term referring to the process of geometrically aligning two or more sets of IMAGE data, such that objects on a single area of ground can be superimposed. Data being registered may be: **1** of the same type but in different wavebands – for example, LANDSAT satellite COLOUR COMPOSITE images are produced by registering green, red and INFRARED wavebands together; **2** from different types of sensor – for example, Landsat satellite, and RADAR imagery have been superimposed for geological investigations; **3** from different dates, assisting in the study of transient phenomena, e.g. SOIL EROSION.

regolith The layer of loose, broken, rocky material mantling the surface of the undecomposed bedrock. It comprises all types of rock waste together with the superficial deposits of LOESS, ALLUVIUM, PEAT, volcanic ASH, wind-blown sand and glacial DRIFT, in addition to the soil layers which themselves represent the uppermost biochemically weathered horizons of the regolith. *See also* COLLUVIUM, SCREE.

regosol One of the three divisions of the ENTISOL soil order (SEVENTH APPROXIMATION). It is one of the two sub-groups of AZONAL SOILS of the old classification (the other being the LITHOSOL). It is a thin, immature soil which develops on newly deposited unconsolidated deposits of wind-blown sand, LOESS, or TILL. It is characterized by the lack of a B-horizon.

regression **1** The retreat of a shallow sea from a land mass owing to a negative movement of sea-level or a positive isostatic movement of the land. A marine regression leads to the formation of RAISED BEACHES and contrasts with a marine TRANSGRESSION. **2** The degeneration of organisms, by the loss of advanced characteristics. **3** A statistical technique which expresses the relationship between two or more variables in a graphic form. It comprises the fitting of a *regression line* through a scatter of points in such a way that the sum of the squares of the distance between the points and the line is reduced to a minimum, i.e. the 'best fit' is achieved. This method is termed *linear regression* (*see also* LEAST-SQUARES METHOD). Other methods include: *non-linear regression* (when the regression line is replaced by a mathematically constructed curve); *multiple regression* (where more than one independent variable is involved). The object of *regression analysis* is not only to discover the degree of relationship between VARIABLES: it can also be used for quantitative prediction or forecasting by extrapolation. [*203*]

regulator A term referring to any component which tends to stabilize a SYSTEM from within. When the negative-feedback (FEEDBACK) process operates within a system it is known as *self-regulation*.

regur A Hindustani word referring to a dark-coloured, fertile, loamy soil in India, developed on the basaltic lavas of the Deccan. It is synonymous with BLACK COTTON SOIL. *See also* BASISOL.

reinforcement A term referring to the state of a CORRELATION SYSTEM when the combined effects of the causative factors are greater than the sum of the individual effects, owing to the increase of effectiveness of one of the variables as a result of its operation in conjunction with another, i.e. the correlation system is non-additive (ADDITIVITY). Reinforcement, therefore, differs from REPLICATION in which the combined effect of the causative factors is usually less than the sum of the individual effects.

rejuvenation The state of a river when its energy has been revitalized, thereby leading to an increase in its erosional capability. Such a change of activity may result from **1** a fall in BASE-LEVEL (DYNAMIC REJUVENATION) or **2** an increase in stream discharge (STATIC REJUVENATION). A change in base-level may be caused by a fall in sea-level (EUSTASY) or an uplift of the land (ISOSTASY), either of which will initiate a new CYCLE OF EROSION. In general this will cause the river to incise itself into its former valley, leading to the formation of INCISED MEANDERS, KNICKPOINTS and RIVER TERRACES.

rejuvenation head KNICKPOINT.

relative age Refers to the dating of rocks merely in comparative rather than in absolute terms (ABSOLUTE AGE). It may be determined by: **1** the PRINCIPLE OF SUPERPOSITION; **2** a comparison of FOSSILS or fossil assemblages; **3** the LAW OF INCLUDED FRAGMENTS. **4** the LAW OF CROSS-CUTTING RELATIONSHIPS.

relative co-ordinates A GIS term referring to a pair of CO-ORDINATES that are measured from another point in the co-ordinate system, rather than the origin. *See also* ABSOLUTE CO-ORDINATES.

relative humidity An index of the amount of water vapour present in the atmosphere. It is the actual VAPOUR PRESSURE expressed as a percentage of the saturation vapour pressure which would be possible at the same air temperature. Relative humidity is an attempt to measure the readiness with which vapour will condense from the air, and is concerned with two variables: the actual water vapour in a given mass of atmosphere and the temperature of that mass of air, since this determines the capacity of the air to hold the water vapour. The value of relative humidity varies inversely with temperature and therefore usually rises during the night, because temperature falls, even though the amount of water vapour may remain constant. It is measured by a HYGROMETER. *See also* ABSOLUTE HUMIDITY, CONDENSATION, DEW-POINT.

relative relief AVAILABLE RELIEF.

relative roughness The ratio between the depth of flow in a river channel and the magnitude of the PARTICLE SIZE of the BEDLOAD (termed the *roughness elements*). *See also* ROUGHNESS.

relative vorticity The three-dimensional rotation of air associated with the clockwise, negative spinning of an ANTICYCLONE and with the anticlockwise, positive spinning of a DEPRESSION or CYCLONE in the N hemisphere, according to BUYS BALLOT'S LAW. It is balanced by PLANETARY VORTICITY in order to maintain ABSOLUTE VORTICITY [*1*].

relaxation The change of the SYSTEM from one state of EQUILIBRIUM to another following a change of input (INPUTS AND OUTPUTS). If the adjustment involves passing through a sequential number of states the system is said to have followed a *relaxation path*. Once the original system has been disturbed by changes in its controlling factors, the time that elapses before the re-establishment of a state of equilibrium is referred to as the *relaxation time*, which may vary considerably according to the scale of the change.

release joints An expression referring to those JOINTS which lie parallel to the axial plane of a fold, and which are thought to have been formed by release of pressure at right angles to the axis of greatest stress.

reliability diagram An indication in the margin of a map of the dates and the quality of the source material from which the map has been compiled.

relic sediments A term used in oceanography, referring to those sea-floor deposits which are incompatible with contemporary marine environments, in which *modern sediments* are being formed. The best examples of relic sediments are the glacigenic deposits (GLACIOFLUVIAL material, TILL) deposited by Pleistocene glaciers upon the continental shelf of NW Europe and along the NE coasts of N America. They are still undergoing sorting by waves and tides.

relict landform A landform that was created by processes which are no longer operative or which play only a minor role in its present fashioning. In Britain's upland areas, for example, the features created by glacial erosion (CIRQUE, U-SHAPED VALLEY) are relict landforms because glaciers are no longer extant in Britain. It has been suggested by some writers that owing to the considerable CLIMATIC CHANGE which has occurred since Tertiary times most world landforms exhibit relict features, e.g. inselbergs in deserts, dry valleys of Pleistocene age in chalklands. An INSELBERG and a MONADNOCK were once referred to as *relict mountains*, but this term should be discouraged and replaced by the term RESIDUAL.

relief The character of the land surface of the EARTH. It comprises a wide variety of LANDFORMS which can be grouped into different types of TERRAIN. Its genesis and MORPHOLOGY are studied by geomorphologists (GEOMORPHOLOGY) and measured by MORPHOMETRY and TERRAIN ANALYSIS. The term relief is not synonymous with TOPOGRAPHY. *See also* AVAILABLE RELIEF.

relief map A type of map, showing the inequalities of surface and the variations in elevation of the landforms. The different shapes, heights and degrees of slope are depicted by using CONTOURS, FORM-LINES, HACHURES, HILL-SHADING, LAYER-TINTING, SPOT HEIGHTS.

relief model A three-dimensional model constructed from wood, plaster, plastic, etc. to depict the RELIEF features of a region. *See also* PLASTIC RELIEF MAP.

relief rainfall OROGRAPHIC RAINFALL.

remanié A French term meaning 're-handled' and used to describe any material that has been caught up, transported and redeposited, thereby becoming incorporated in crustal rocks or superficial deposits, e.g. the boulders incorporated in a TILL, the clastic fragments of a BRECCIA. It is synonymous with *reworking*.

remote sensing The scientific collection of mass data (information) relating to objects without being in physical contact with them. The term is usually restricted to methods that record ELECTROMAGNETIC RADIATION which is reflected or radiated from the objects, but excludes magnetic and gravity surveys in which force fields are recorded. The data are gathered by electronic scanning devices carried in aircraft or SATELLITES, with the SENSORS operating at considerable altitudes above the Earth's surface (MULTISPECTRAL SCANNERS, RADAR, RADIOMETERS, THERMAL LINESCANNERS). *See also* AERIAL PHOTOGRAPH.

rendzina The name given to a group of INTRAZONAL SOILS which is characterized by the darkness of the surface horizons. The rendzina develops on calcareous parent material (especially in CHALK lands) and this has weathered to provide a layer of limestone or chalky fragments. The overlying A-horizon is black to dark brown, owing to the high percentage of humus and carbonates, and typically it supports a grassland cover or a grass/tree vegetation, e.g. the English Downlands. Some pedologists have suggested that rendzinas develop only on softer calcareous rocks, being replaced by TERRA ROSSA on harder limestones. It differs from the terra rossa by its lack of free sesquioxides. In the SEVENTH APPROXIMATION soil classification it would be included as a type of MOLLISOL.

renewable energy sources Those energy sources which do not rely on finite reserves of fossil or nuclear fuels. They comprise: solar energy, wind energy, wave energy, tidal energy, hydroelectric power and geothermal energy.

renewal of exposure The process by which bare rock is continually re-exposed, generally as a FREE FACE, by DENUDATION.

replacement **1** In general, the process by which one constituent in a system is progressively substituted by another. **2** In geology: the replacement of one crystal by another (PSEUDOMORPH); the substitution of another material for the skeletal material of a FOSSIL; the replacement of a limestone by chert; the HYDROTHERMAL replacement of rock bodies by deposits of sulphide minerals; the replacement of COUNTRY-ROCK by the emplacement of granite (GRANITIZATION); the replacement of country-rock by a non-dilation vein rather than the mechanical invasion of the rock by a dilation vein. **3** In palaeontology, the replacement of one fauna by another, horizontally in space or vertically in time.

replication **1** An act of repetition. **2** In correlation analysis (CORRELATION, STATISTICAL) the term refers to the inclusion of the same mathematical information in two or more of the variables. *See also* REINFORCEMENT.

repose ANGLE OF REPOSE.

repose slope One of the terms given to a slope the gradient of which is controlled by the ANGLE OF REPOSE of the debris layer, e.g. SCREE, TALUS SLOPE or BOULDER-CONTROLLED SLOPE.

representative fraction (RF) The ratio between distance on the ground and distance on a map, expressed as a fraction. The former Ordnance Survey map series of one mile to one inch (1 : 1 mile) has an RF of 1 : 63,360. It has been replaced by a metric series of 0.5 km to 1 cm (2 cm : 1 km) which has an RF of 1 : 50,000.

reptile A member of the vertebrate class of *Reptilia*, consisting of egg-laying animals with great diversity of form and size. Reptiles dominated the Earth's fauna until the end of the TRIAS. Modern forms include snakes, lizards, crocodiles, turtles and terrapins. *See also* AMPHIBIAN.

resection The determination of the location of the observing station during a PLANE-TABLE survey. It involves the drawing of rays from easily identifiable points after the plane-table has been orientated by use of a compass. A minimum of three rays will be necessary, for where these intersect a TRIANGLE OF ERROR will be formed within which the observing station can be accurately fixed.

resequent drainage A stream pattern which flows in the same direction as that of the initial CONSEQUENT STREAM but which develops in response to a new BASE-LEVEL. The recognition of a resequent-drainage pattern involves interpretations of former stream patterns that are not always valid or demonstrable, so that many geomorphologists avoid the expression.

resequent fault-line scarp A FAULT-LINE SCARP which develops when denudation along an *obsequent fault-line scarp* creates an escarpment facing in the original direction of the initial fault-line scarp, i.e. across the downthrow block [*77*]. Since it is difficult to demonstrate that such a chain of events has occurred the expression is rarely used.

reservoir **1** A rock stratum at depth characterized by good permeability and high porosity, thereby serving as a natural underground container of oil, water and natural gas. **2** An artificially constructed surface receptacle for water.

residence time The mean length of time that a water molecule remains in a particular section of the HYDROLOGICAL CYCLE, e.g. in the river, in the ocean, in the atmosphere or in the groundwater. It is calculated by dividing the volume of water in a particular section by the rate at which the water passes through it. Residence time will be lengthened by increasing the volume or capacity of any particular part of the above system, or by decreasing the rate at which water moves through it. The concept has also been applied to sediment transport on the surface of the Earth.

residual **1** In general, any residuum remaining after the removal of the majority of the original matter. **2** In geomorphology, a remnant of a formerly extensive mass of rock or land surface that has been partly destroyed by denudation. *See also* INSELBERG, MESA, MONADNOCK. **3** In geophysics, the resultant of a measurement upon a quantity, minus the most probable value of that quantity. **4** In statistics, the difference between an observed and a computed value. In plotting a REGRESSION, the residuals (anomalies) will be shown by the different degrees of deviation from the regression line exhibited by some of the points. Residuals will not occur where there is a perfect correlation.

residual debris The material remaining above the BASAL SURFACE OF WEATHERING after a lengthy period of CHEMICAL WEATHERING. It forms part of the REGOLITH and is the decomposed material which surrounds a CORESTONE during the formation of a TOR or a NUBBIN. *See also* BAUXITE, LATERITE.

residual map A map of non-systematic, relatively small-scale variations in any areal distribution of data. Such distributions are residual when the trend component is subtracted from the observed pattern.

residual soil A soil formed *in situ* by the decomposition of the underlying solid rock by weathering. It contrasts with a soil that develops on parent material that has been transported (e.g. till). The term has replaced that of *sedentary soil*.

residual strength A term commonly used in SOIL MECHANICS, and sometimes termed the RESIDUAL SHEAR STRENGTH, refers to the strength of a soil (usually clay soil) which is below the peak SHEAR STRENGTH. It is determined by means of a SHEARBOX.

resistance **1** In general, the power or capacity of an object or a material to withstand the force imposed upon it. **2** In physics, that which limits the steady electric current in a conductor. Electrical resistivity methods are used in geophysical surveying (GEOPHYSICS).

resistant rock Any rock that withstands the forces of denudation more successfully than a *less-resistant rock*, because of differences in jointing, hardness, compactness, cementation and mineral composition. *See also* DIFFERENTIAL WEATHERING.

resistivity method A type of geophysical exploration (GEOPHYSICS), which investigates the underground distribution of electrical resistivity by means of electrodes. It is widely used in ARCHAEOLOGICAL DATING, civil engineering, and HYDROGEOLOGY.

resolution **1** The separation of a VECTOR into its components. **2** A REMOTE SENSING term which has three separate applications: (a) *Spatial resolution*, which refers to the ability of a SENSOR to distinguish between objects that are spatially close to each other. It is a measure of the smallest angular or linear separation between two objects. A smaller-resolution parameter denotes greater resolving power, e.g. LANDSAT 1, 2 and 3 MULTISPECTRAL SCANNER IMAGES have a nominal resolution of 79 m. (b) *Spectral resolution* (SPECTRAL), which refers to the ability of a sensor to distinguish between objects which are spectrally similar. It is a measure of both the discreteness of wavebands and the sensitivity of the sensor to distinguish between ELECTROMAGNETIC RADIATION intensity levels. For example, early Landsat satellites had three sensors sensitive over 0.1 μm wavelength ranges with an ability to distinguish a maximum of 128 intensity levels of electromagnetic radiation, while the fourth sensor was sensitive over a 0.3 μm range and could distinguish a maximum of 64 intensity levels. (c) *Thermal resolution*, which refers to the ability of a sensor to distinguish between objects with a similar temperature. Many thermal sensors (INFRARED THERMOGRAPHY) have a resolution of ±0.5° C.

resources, natural Those parts of the environment that are capable of satisfying the needs of mankind. Natural resources are dynamic, and become available through a combination of increased human knowledge and expanding technology in addition to changing individual and societal objectives. A division is recognizable between *renewable* and *non-renewable* resources: *renewable resources* are parts of functioning natural SYSTEMS that maintain a sustained yield, in which they are created at approximately comparable rates to those of their consumption; *non-renewable resources* are those that have a finite distribution or are created at rates considerably slower than those of their use, e.g. oil is being created in modern marine sediments at least 10^{-6} times slower than the current rate of extraction. The total flow of a resource from its state in nature through its period of human contact to its disposal (perhaps in a form available for re-use) is termed a *resource process*.

response An action or feeling which answers to stimulus (PROCESS–RESPONSE MODEL). The features produced by the operation of the stimulus are termed *response elements*.

response time The delayed response of a SYSTEM to a change of input, e.g. the delayed reaction of glaciers and ice-sheets to an alteration in their *regime* as a result of CLIMATIC CHANGE. The response time will vary according to the size of the input.

resultant The distance and direction between the first and last points of a VECTOR analysis. The single force represented by the resultant produces the same result as two or more forces.

resultant wind The mathematically determined overall velocity of a number of wind velocities. It is achieved by summing the two individual components of the actual wind (i.e. its zonal (east) and meridional (north) components). The components are then squared, added together and the square root taken:

$$\sqrt{[(\Sigma Vn)^2 + (\Sigma Ve)^2]}$$

where: V = velocity; n = meridional (north) component; e = zonal (east) component.

resurgence The reappearance at the surface of a stream which originated on impermeable rock but which disappears underground on reaching calcareous strata.

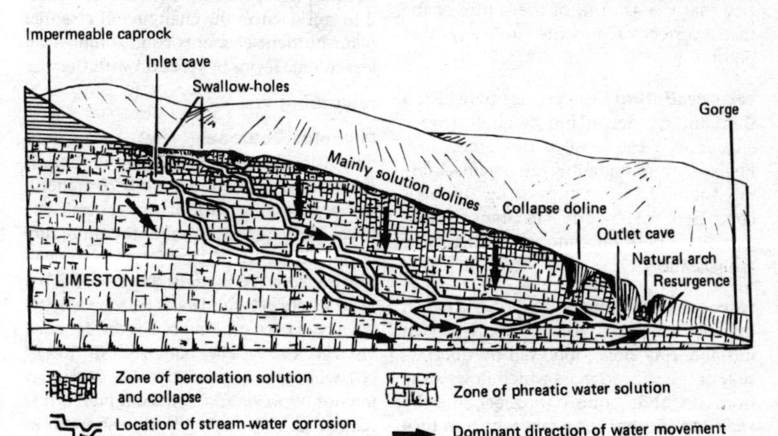

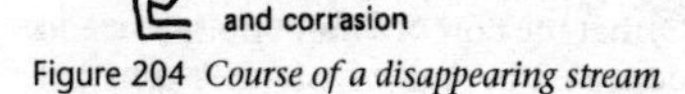

Figure 204 *Course of a disappearing stream*

A resurgence marks the emergence of a substantial flow of KARST WATER that had previously travelled some distance underground in limestone rocks, having commenced its journey as a surface stream. Although it is larger than a SPRING it is sometimes referred to as a Vauclusian spring. *See also* RISING. [*204*]

retarding basin An artificially maintained flood-control area into which floodwaters are deflected by means of a dam placed across a river channel. It consists of a FLOODPLAIN, OXBOW LAKE or WASHLAND, where there is no risk of damage to settlement. *See also* WATER MEADOW.

retention curve RETENTIVITY PROFILE.

retentivity profile A graph depicting the retaining capacity of a soil as a function of depth. The retaining capacity may relate to water (retained water) held in interstices by surface tension against gravity, or to any other substances held by soils. The soil moisture *retention curve* is found by plotting soil moisture content against soil moisture SUCTION. *See also* CAPILLARITY.

retreating coast RETROGRADATION.

retro-azimuthal projection A MAP PROJECTION possessing the fundamental property that the AZIMUTH of the centre of the map is correctly represented from any other point.

retrogradation The process by which a shoreline recedes inland because of wave erosion, e.g. the coast of Holderness in E England is retrograding at a remarkably rapid rate (2.75 m per year) because it is composed of poorly resistant glacial deposits. The term contrasts with that of PROGRADATION.

retrogressive failure A process which involves an initial ROTATIONAL SLIP on an unstable clay slope, followed by LIQUEFACTION of the slipped mass which flows away from the slide zone (LANDSLIDE), thereby exposing another steep face which in turn slips and liquefies. *See also* FAILURE, QUICKCLAYS.

return period The average length of time separating events of a similar magnitude (RECURRENCE INTERVAL).

reversed drainage The term given to a stream pattern whose direction of flow is altered by river capture (CAPTURE, RIVER).

reversed polarity MAGNETIC REVERSALS.

reverse-fault A fault along which the HANGING WALL has been raised relative to the FOOT WALL. *See also* NORMAL-FAULT, THRUST-FAULT. [*76*]

reversing dune A DUNE that grows upwards but stays in virtually the same position due to seasonal changes in wind direction causing it to move backwards and forwards over time.

reversing thermometer A thermometer specially designed for measuring temperatures at depth in water bodies, especially the oceans. It operates on the principle that a constriction in the glass tube containing the column of mercury will cause a break in the column once the thermometer is reversed (inverted), thereby indicating the precise reading of the water layer at depth.

reversion **1** The action of returning to a former condition or ancestral type, e.g. natural vegetation tends to regain its former position when human interference ceases. **2** In soil science, the changing of essential plant-nutrient elements from soluble into less-soluble forms by reaction with the soil.

reworking REMANIÉ.

Reynolds number (Re) A non-dimensional parameter of fluid motion. It refers to the extent by which VISCOSITY modifies a fluid-flow pattern. In general, the lower the Reynolds number the more predominant are the viscous forces in controlling the flow. It is expressed as $Re = pfvL/\mu = vL/\nu$, where pf is fluid density, v is velocity, L is a characteristic length, μ is the dynamic viscosity and ν is kinematic viscosity ($\nu = \mu/pf$), which is a measure of molecular interference between adjacent fluid layers. It is named after O. Reynolds who showed, in 1883, that the flow of a fluid in a tube could be described by a dimensionless constant

(*Re*). When *Re* exceeds 750 turbulent flow (TURBULENCE) will prevail for smooth boundary conditions in river channels; when *Re* is less than 500 LAMINAR FLOW will prevail. *See also* FROUDE NUMBER.

Rhaetic A transitional rock series spanning the boundary between the TRIAS and the Lower JURASSIC. European geologists favour placing the dark shales, sandstones, marls and thin limestones, which comprise the Rhaetic series, in the Trias, but in Britain the sediments are usually placed in the LIAS. The name is derived from the Rhaetic Alps in central Europe.

rheid A body of rock which exhibits a flow structure, e.g. RHYOLITE.

rheidity The process by which a material deforms by plastic flow (PLASTIC DEFORMATION). It is an important mechanism in both GLACIER FLOW and rock folding. *See also* RHEOLOGY.

rheological model A type of model representing the different degrees of deformation of a material as it is subjected to STRAIN, in passing from one state of equilibrium to another following the application of a new STRESS.

rheology The scientific study of the flowage of materials, especially that of plastic solids (GLACIER FLOW). *See also* PLASTIC DEFORMATION.

rhexistasis A breakdown in the state of biological equilibrium between soil, vegetation and climate at a particular site (BIOSTASIS), perhaps owing to erosion or climatic change. *See also* HOMOEOSTASIS.

rhizosphere The soil close to plant roots (*rhizomes*) where there is usually an abundant and specific microbiological population (MICRO-ORGANISMS IN THE SOIL).

rhodolith A type of NODULE of calcareous red ALGAE occurring both in shallow REEF-FLATS and in the deeper water of fore-reef terraces. *See also* CORALLITH.

rhombochasm A term referring to a major parallel-sided fracture in the Earth's continental crust (SIAL), probably created by SEA-FLOOR SPREADING (PLATE TECTONICS). It allows ocean crust (SIMA) to infill the gap caused by the fracture.

rhourd A large pyramidal DUNE formed when smaller dunes intersect each other.

rhumb line A line of constant bearing which crosses all MERIDIANS at the same angle. It is sometimes referred to as a *loxodrome* or *line of constant bearing*. *See also* GREAT CIRCLE [97].

rhyolite A pink or yellow fine-grained, acid VOLCANIC ROCK, mineralogically similar to GRANITE but possessing more quartz and having its ferromagnesian minerals less obvious than in the corresponding *plutonic* rocks. It is divided into a sodic form and a potassic form according to the type of feldspar present. The volcanic equivalent of granodiorite is a very fine-grained rhyolitic rock known as DACITE which, with decreasing quartz, becomes an ANDESITE. Both rhyolites and andesites are characteristic of *island arcs* and orogenic regions, while strongly alkaline rhyolites (rich in sodium, poor in aluminium) are commonly found in RIFT VALLEYS of continental areas. Some rhyolitic lavas have a glassy structure (OBSIDIAN, PITCHSTONE) and many have a distinct FLOW STRUCTURE, termed flow banding, formed when the viscous lava was extruded. Nevertheless, owing to their high viscosity, modern rhyolitic lavas never form extensive flows around the volcanic vents.

rhythmic sedimentation A sedimentary sequence which changes character progressively from one extreme type to another, followed immediately by an abrupt return to the original type. A typical sedimentary *rhythmic unit* would be:

Marine/Swamp/Deltaic/Lagoonal/Marine/Swamp

Rhythmic unit

Some *rhythmic units* are as large as cyclic units (CYCLE OF SEDIMENTATION), but the majority are of very small-scale rhythms. *See also* CYCLOTHEM.

rhythmite A finally laminated sediment in which two or three different lithologies

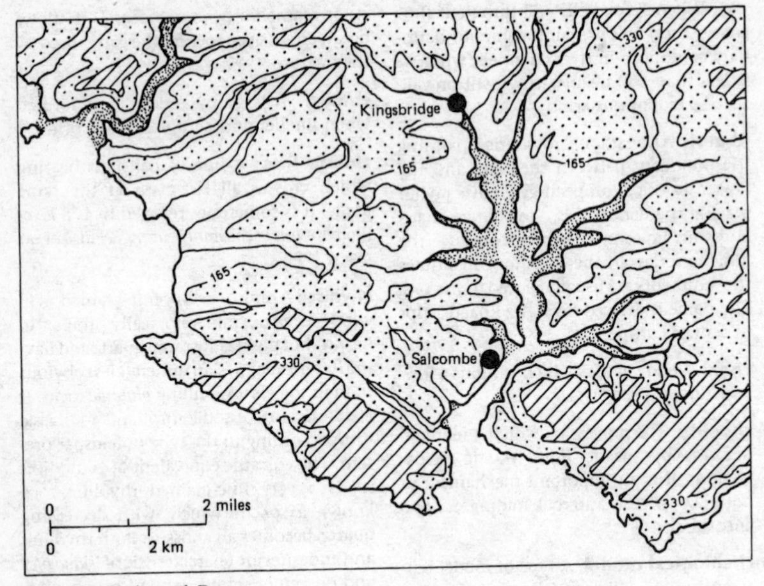

Figure 205 *Ria in south Devon*

are regularly repeated. It is commonly found in the sedimentation process occurring in a GLACIAL LAKE.

ria A Spanish term widely adopted to describe a submerged coastal valley or estuary resulting from a rise of sea-level (EUSTASY). Most rias were formed by the post-glacial (FLANDRIAN) drowning of DISCORDANT COASTS, e.g. in SW Ireland, and are also found in SW England and NW Spain. They are typically V-shaped in cross-profile and deepen progressively as they are traced seawards, in contrast to the profiles of FIORDS, which have been glacially modified. [*205*]

ribbed moraines ROGEN MORAINES.

ribbon diagram A single continuous vertical geological section along a curved or sinuous line illustrating the relations of rock strata in a sequential series. It is especially useful when trying to illustrate the relationships of stratigraphic data revealed from a number of disparate bore-holes. [*206*]

ribbon lake A temporary narrow body of water impounded between the front of an ice-sheet and its recessional moraines. Meltwater streams debouching into such a lake will produce a KAME TERRACE. As ribbon lakes increase in size they will coalesce to form a PRO-GLACIAL LAKE.

Richter denudation slope The name given to the smooth rock slope of uniform

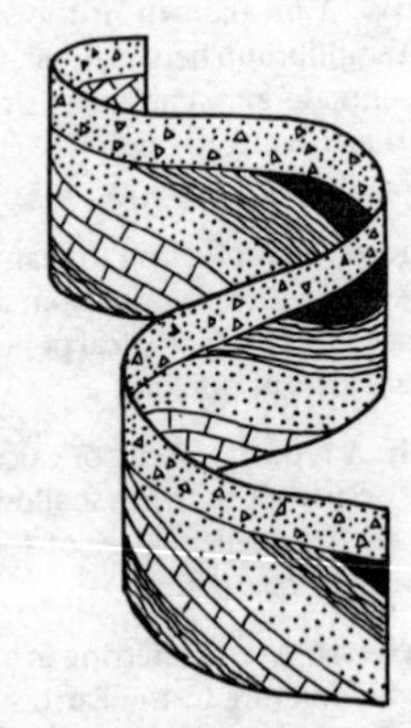

Figure 206 *A ribbon diagram*

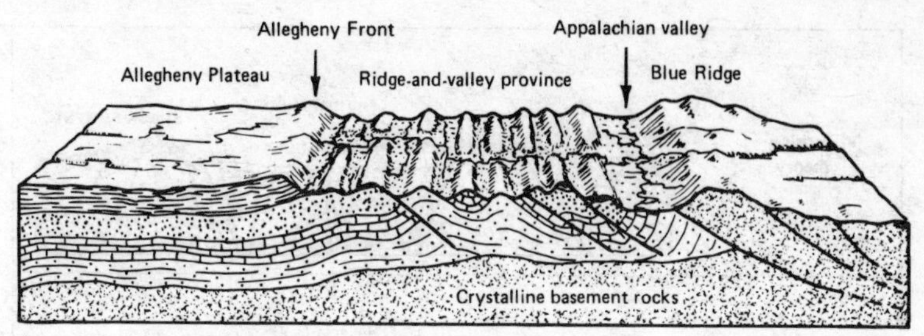

Figure 207 *Ridge-and-valley terrain in the Appalachians*

inclination (32–36°) underlying the talus slope. It is formed by the same processes which create the so-called CONSTANT SLOPE. It is synonymous with the term REPOSE SLOPE. *See also* SLOPE.

Richter scale A logarithmic scale, devised in 1935 by an American, C. F. Richter, to identify the MAGNITUDE OF AN EARTHQUAKE. Designed to compare the magnitudes of Californian earthquakes, it is now universally adopted and extended. It is more precise than the MERCALLI SCALE, which refers to the INTENSITY OF AN EARTHQUAKE. The Richter scale reflects the total amount of elastic energy released when overstrained rocks suddenly rebound to give a seismic shock. The relationship is given by: $\log E = a + bM$, where: E is energy expressed in ergs, M is magnitude, a is 5.8, and b is 2.4. The scale is open-ended, ranging from zero up to 8.9, which is the largest recorded magnitude (Chile, 1960), and which produced energy of $10^{27.2}$ ergs. It has been calculated that the average annual energy release from all earthquakes ranges from 10^{25} to 10^{27} ergs. Moderate-to-strong intensities on the Mercalli scale (IV and V) are thought to be equivalent to a Richter magnitude of 4.3 to 4.8. Destructive-to-disastrous intensities (VIII, IX and X) are possibly equivalent to magnitudes of 6.2 to 7.3.

ridge **1** A linear, steep-sided upland. **2** A narrow spur of a mountain which can be transformed into an ARÊTE by glacial modification. **3** A median rise on the ocean floor (MID-OCEANIC RIDGE).

ridge-and-ravine terrain A type of relief characteristic of the EQUATORIAL RAIN-FOREST zone, where both CHEMICAL WEATHERING and stream erosion are rapid. A high drainage density creates valley slopes with narrow, smoothly curved convex crests and lengthy slopes of 30°–40° gradients. The pattern of INSEQUENT DRAINAGE creates a uniform drainage density and a similarity of slope angle in different valleys, suggesting that landform evolution is by uniform ground loss.

ridge-and-vale terrain A type of relief which develops in an area of gently dipping sedimentary rocks in which there is a recurrent pattern of parallel ridges formed by CUESTAS on the harder rocks, separated by linear vales carved by rivers along the outcrops of less resistant strata. One of the best examples of this terrain is that of SE England with its scarplands of limestone and chalk and its intervening clay vales. *See also* RIDGE-AND-VALLEY TERRAIN, SCARP-AND-VALE TERRAIN. [*222*]

ridge-and-valley terrain A type of relief that develops in an area of parallel folding, where parallel ridges and river valleys develop on the contrasting lithologies. Its type-site is the ridge-and-valley province of the Appalachians, USA, which varies in width between 40 and 120 km. The valleys form in the easily eroded rock formations, whether they form anticlines, synclines or the flanks of folds. [*207*]

ridge of high pressure An elongated area of high pressure extending out from an ANTICYCLONE and bounded by areas of low pressure on either flank. It is broader than a WEDGE OF HIGH PRESSURE. Its weather is similar

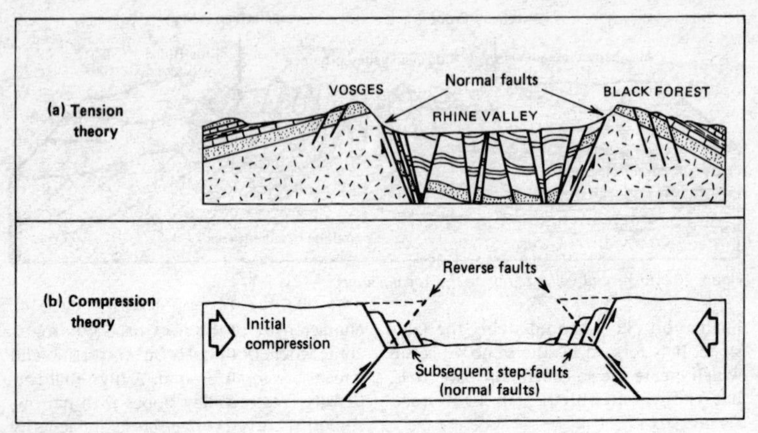

Figure 208 *Rift valleys*

to but less persistent than that of an anticyclone. *See also* TROUGH.

ried A German term for a marshy floodplain tract in the Rhine valley near Strasbourg. It has been extensively drained but is still prone to seasonal flooding.

riegel A German term for a transverse rock bar or rock step which causes an irregularity in the long profile (THALWEG) of a glacially eroded valley. It usually marks the outcrop of a highly resistant rock layer and is often associated with a waterfall or zones of rapids in the modern stream. The riegel bar is the point where previous glacier flow is thought to have been extending (EXTENDING FLOW). Where there are a series of bars a GLACIAL STAIRWAY may develop, possibly associated with a pattern of PATERNOSTER LAKES, after the glacier has disappeared. Most riegel bars and steps have been glacially smoothed on the up-valley side and plucked on the down-valley side, in similar fashion to that of a ROCHE MOUTONNÉE. [*178*]

riffle A depositional bar on the channel floor of a river. As with many dynamic depositional forms (e.g. a POINT BAR) a riffle comprises a collection of sedimentary particles formed into a characteristically rippled surface, although individual particles move intermittently downstream from one riffle to another. *See also* POOL AND RIFFLE. [*192*]

rift valley A linear depression or trough created by the sinking of the intermediate crustal rocks between two or more parallel strike-slip FAULTS. The structure is known as a GRABEN and the accompanying morphological feature as a rift valley. Many examples of the latter are known, ranging in magnitude from those of the MID-OCEANIC RIDGES, and the Red Sea graben, to the East African Rift and the Rhine graben [*208a*]. Almost all are associated with vulcanicity and some are bounded by HORSTS. Although rift valleys resulting from crustal doming have been simulated by models there is no complete agreement over their mode of formation. The most commonly held belief is that of the *tensional rift*, produced when two large crustal masses move apart, a genesis which is particularly apposite to the ideas of sea-floor spreading (PLATE TECTONICS). At a smaller scale it has been suggested that local doming of a CRATON may produce a graben of limited length, e.g. the Rhine graben (280 km). An alternative theory is that of the *compressional rift* bounded by reverse faults [*199b*], in which the central block is forced down by the overthrust marginal masses.

rigidity modulus A term used in soil mechanics and rock mechanics referring to the ability of a material to withstand SHEARING. It is the ratio of STRESS to STRAIN when the former is a shear, and it is calculated by dividing the tangential force per unit area by the angular deformation.

rill A narrow, steep-sided watercourse of small (centimetric) scale. It is an ephemeral feature, considerably smaller than a GULLY. It carries water intermittently, generally during storms, and is thought to be the intermediate stage between OVERLAND FLOW and the development of a permanent gully network which itself forms the headwaters of a stream system. *See also* RILL EROSION, RILL MARKS.

rillenkarren KARREN.

rill erosion The removal of surface material, usually soil, by the action of running water. The process creates numerous tiny channels (rills), a few centimetres in depth, most of which carry water only during storms. The head of the rill system may not extend all the way up to the watershed divide, thereby leaving a zone of no rills across which the depth of OVERLAND FLOW is insufficient to develop an erosive force equal to the forces of cohesion which hold the soil particles in place. It has been suggested by R. E. Horton that the initial parallel rills are enlarged by a process of coalescence known as CROSS-GRADING. Further enlargement leads on to the formation of a GULLY and ultimately to severe SOIL EROSION.

rilling The US term for RILL EROSION.

rill marks Small furrows or channels running down the slope of a beach. It is thought that they are formed partly by the seepage of fresh-water springs near high-water level but mainly by the seaward flow of sea water that has percolated into the upper part of the beach zone during a high tide after a storm. During the period of low tide or after a storm the water will return to the sea by cutting the RILLS across the face of the beach.

rill wash SURFACE WASH.

rimaye A French term for BERGSCHRUND.

rime A deposit of opaque, white, rough-textured ice crystals, formed when a fog of SUPERCOOLED WATER droplets is brought by a light wind into contact with surfaces the temperatures of which are below FREEZING-POINT. Because the ice particles accumulate on the windward side of the surface (e.g. trees, fences, rocks, telegraph poles, etc.) the rime builds out to form *frost-feathers*. It is uncommon at low altitudes where super-cooled fogs are rare. It should not be confused with GLAZED FROST. *See also* ACCUMULATION.

rimstone A FLOWSTONE deposit of calcium carbonate that accumulates around the rim of a pool of standing karstic water (*rimstone pool*) which forms on the floor of a cave in KARST terrain.

ring complex An expression referring to a composite circular structure of igneous intrusive rocks made up essentially of CONE-SHEETS and RING-DYKES. *See also* CAULDRON SUBSIDENCE.

ring-dyke A circular injection of igneous rock which is injected as a fluid MAGMA into the country-rock along vertical fractures produced by the upward pressure of the magma in the chamber below [*36*]. There are many examples of Tertiary ring-dykes in the Inner Hebrides, Scotland, e.g. Mull. *See also* CONE-SHEET.

rinnen karren KARREN.

rinnental A German term for an ice-walled channel (plural *rinnentäler*). *See also* TUNNEL VALLEYS, URSTROMTAL.

rip A turbulent body of sea water created by the meeting of tidal currents, or where wind-generated waves meet a tidal current head on. It is usually associated with a RIP CURRENT. The term *rip tide* should not be used for it is a misnomer for *rip current*.

rip current A strong seaward-moving current which develops as a compensation mechanism, whereby sea water that is piled up along coastlines by wave action can be returned seawards. In plan the current is seen to run initially parallel to the coast before flowing out through the BREAKER zone

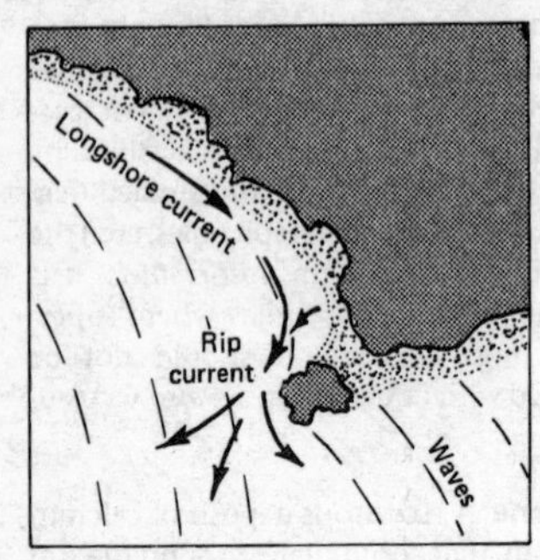

Figure 209 *Rip current*

[*209*]. Rip currents are particularly dangerous to swimmers.

ripple bedding A type of small-scale CROSS-BEDDING formed by the rapid deposition of sediment as sinuous-crested or straight-crested *ripples* which slowly migrate. It is sometimes referred to as *ripple-drift bedding*. [*52*]

ripple marks Patterns produced in unconsolidated sediments by the fluid motion of waves, currents or wind. *Symmetric ripple marks* are produced by an oscillatory current, but where the current direction is well-defined *asymmetric ripple marks* will be formed. [*210*(c)]. In either of these cases their crests may be sinuous or straight, but a particular type of *lobate ripple mark* can be either *lunate* (pointing up-current) or *linguoid* (pointing down-current) [*210*(b)]. Ripple marks are in more or less perpendicular alignment with the flow direction and are found in the intertidal zone, on the sea-bed below low-water mark, or on a stream-bed (all formed by moving water). Alternatively, ripple marks may be caused by wind blowing across dune sand. Ripple marks tend to be larger in deeper offshore water and in exposed surface areas. [*14*]

riprap A term referring to large fragments of broken rock tipped at high-water mark along a shoreline to protect the coast against wave erosion.

rise **1** In oceanography, a broad, gently sloping, elevated portion of the sea-floor, similar to a MID-OCEANIC RIDGE but without the median rift valley. **2** A name applied occasionally to a RESURGENCE.

rise pit A term referring to a spring which rises by artesian forces (ARTESIAN WELL) through the alluvium of a valley floor. It is usually fringed on all but the stream-outlet

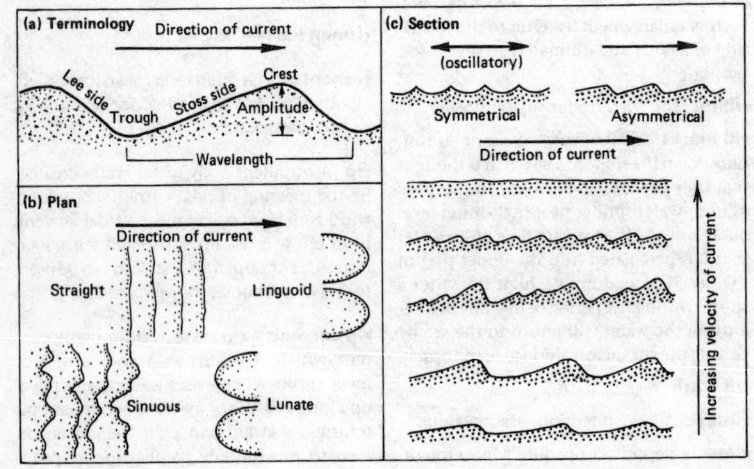

Figure 210 *Ripple marks*

side by a small perimeter ridge similar to a levee.

rising The RESURGENCE of an underground watercourse at the base of a KARST limestone massif. The water may issue freely or under pressure as a *Vauclusian spring*. [204]

Riss The penultimate glacial in the Pleistocene chronology of central Europe. It was recognized in 1909 by A. Penck and E. Brückner in the valley of the R. Riss to the north of the Alps, where glaciofluvial outwash features could be linked to the glacial moraines of the former ice-sheets. It is older than the WÜRM but younger than the MINDEL, and is usually equated with the SAALE in N Europe, the WOLSTONIAN of Britain and the ILLINOIAN of N America.

river A stream of water which flows in a channel from high ground to low ground and ultimately to a lake or the sea, except in a desert area where it may dwindle away to nothing. It is the main transporting agent in the FLUVIAL CYCLE. A river and all its tributaries within a single basin is termed a *drainage system* and in combination several drainage systems will form a variety of DRAINAGE PATTERNS.

river capture CAPTURE, RIVER.

river channel CHANNEL, CHANNEL GEOMETRY.

river flow FLOW REGIMES.

river load LOAD.

river profile CROSS-PROFILE, LONG PROFILE OF A RIVER.

river terrace A portion of the former FLOODPLAIN of a RIVER, now abandoned and left at a higher level as the stream downcuts, usually from REJUVENATION. The term is used to describe both the bench-like form and the alluvial deposits of the former floodplain, and is applied to the scarp and to the flat tread above and behind it. Some authorities regard the term river terrace as synonymous with ALLUVIAL TERRACE, but such a limited definition would exclude terraces cut on bedrock (*strath terraces*). If periods of incision and aggradation occur repeatedly a number of terraces may be formed (PAIRED TERRACE) but it is dangerous to assume that there is a simple relationship between the height of a terrace and its age. Higher usually means older but not invariably since, during a period of AGGRADATION, the alluvial fill may be either *inset* or *overlapping* [211]. The later of the two fills (shaded in figure *211b*) may

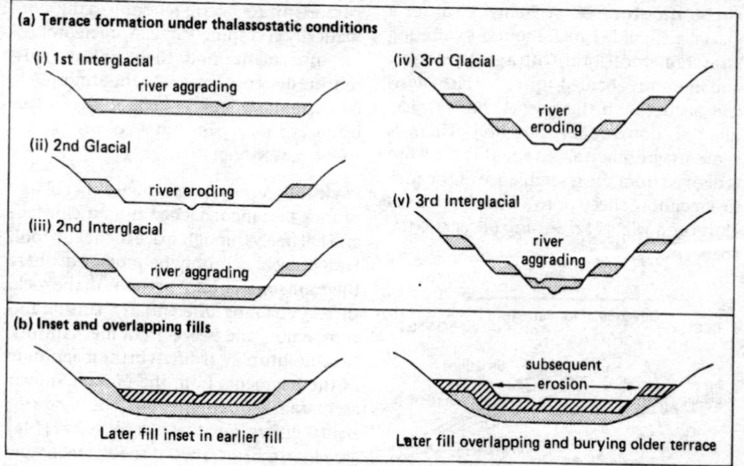

Figure 211 *Formation of river terraces*

simply be inset at a lower level in a trench cut into the older fill, or it may have been of sufficient volume to overlap the trench cut into the earlier fill, thereby burying the older terrace. [*See also 174*] *See also* THALASSOSTATIC.

roadstead, roads An offshore anchorage in a bay or estuary where some degree of protection from storms is provided for any shipping unable to get into a harbour.

Roaring Forties The region of the S hemisphere between latitudes 40°S and 50°S where few land masses occur as barriers to an uninterrupted flow of the WESTERLIES, which therefore blow with great regularity and strength across the open oceans. The frequency of depressions brings storminess, cloudiness and rough seas – a notoriously hazardous shipping zone.

Robin effect A sub-glacial mechanism in which the removal of rock at the edge of steps beneath a glacier and just above a glacial cavity is reinforced by a heat pump process. Such a process results from the variable STRESS distribution in the glacier ice (GLACIER FLOW) as it moves over an uneven surface. *See also* EXARATION, ICE-PLUCKING, RIEGEL.

roche moutonnée A French term for a glacially moulded rock mound exhibiting an asymmetrical form, with a gently sloping and smoothly abraded, up-valley face (*stoss*) contrasting with the steeper, broken, ice-plucked, down-valley face (*lee*). There is some disagreement about whether the word is derived from the resemblance of the rock to a reclining sheep or to a mutton-dressed 18th-cent. wig. [*212*] *See also* ONSET AND LEE, ROBIN EFFECT.

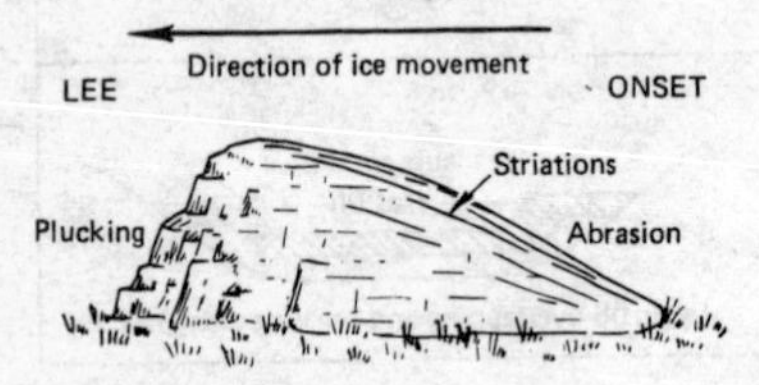

Figure 212 *Roche moutonnée*

rock 1 A coherent, consolidated and compact mass of mineral matter. It may be classified by its age, by its hardness or by its mode of formation (*see also* IGNEOUS ROCKS, METAMORPHIC ROCKS, SEDIMENTARY ROCKS). **2** A place-name for a prominent cliff, peak or sea stack, e.g. Ayers Rock, Australia, the Rock of Gibraltar, Tusker Rock, S Wales.

rock bar RIEGEL.

rock creep The movement of rock fragments across an inclined surface by means of intermittent slip along a plane between the fragment and the ground surface. Such movement may be initiated by heating and cooling or by the growth of ice crystals beneath the rock fragment. It is also referred to as *surficial rock creep*. The definition by C. F. S. Sharpe of the movement of boulders away from an outcrop on the surface of, or embedded within, the soil, is, strictly, soil creep. *See also* CREEP, TALUS SHIFT. [*146*]

rock crystal QUARTZ.

rock dome A type of INSELBERG, characterized by its vertical sides and 'sugar-loaf' shape. *See also* RUWARE.

rock drumlin A whale-backed hillock resembling a true DRUMLIN but comprising a bedrock core veneered with TILL.

rocketsonde A term referring to the automatically recording cameras, meteorological instruments and their linked transmitting devices carried into the atmosphere by *rocket*. They later descend to the surface by means of a parachute. *See also* RADIOSONDE, RAWINSONDE.

rock-fall A process by which small blocks of rock become detached from a cliff face and fall freely through space to the cliff foot. Loose blocks can only be produced where there are suitable joint patterns in the rock, thereby enabling differential expansion and contraction, and freeze–thaw mechanisms, to come into play. It differs in the magnitude of the fragments from the process known as GRANULAR DISINTEGRATION. The mode of FAILURE differs from that of a ROCKSLIDE. [*146*] *See also* MECHANICAL WEATHERING, BERGHLAUP, BERGSTURZ, STURZSTROM.

rock flour **1** Fine-grained rock material with a PARTICLE SIZE akin to that of silt or clay produced by the grinding action of boulders held in the basal ice of a glacier. It is formed mainly when boulders grind against each other. Unless the rocks producing the rock flour are themselves argillaceous the flour will not behave as a normal clay. **2** Fine-grained rock material produced by crushing along a fault (GOUGE) in association with fault breccia.

rock glacier A glacier-like tongue of angular TALUS extending out from a cirque but in which no ice may be seen at the surface. Boreholes prove that the interstices of the rubble are filled by ice, suggesting that the mobility of the rock glacier is due to the interstitial ice. Nevertheless, the movement of most rock glaciers is extremely slow and many of them appear to be stagnant. It has been suggested that they have formed in cirques abandoned by Pleistocene glaciers and that they exhibit a differentiation into an upper layer of larger angular blocks overlying a lower layer of smaller angular blocks, sand and silt, all with interstitial ice. In a few cases a basal layer of buried glacier ice occurs, thought to be a remnant of a true glacier, and some examples actually grade into true glaciers at their upper ends. It is also called a *talus glacier*. *See also* GLACIER.

rock mass strength A quantitative measure, devised by M. J. Selby (1980), to illustrate the resistance of a rock mass to EROSION. [*213*]

rock mechanics The scientific study of the mechanical properties of rock. It includes a knowledge of its COHESION, ELASTIC LIMIT, FRICTION, COEFFICIENT OF, PERMEABILITY, POROSITY, SHEAR STRENGTH, in order to determine its load-bearing capability and its likelihood of FAILURE. *See also* STABILITY ANALYSIS.

rock milk A white paste or crumbly white powder occurring on floors and walls of limestone caves. It is thought to be formed either by rock disintegration by microorganisms or by freezing of the limestone by ice in which CO_2 is expelled and a *milky rock fluid* formed. It usually comprises a mixture of magnesium and calcite minerals. It is also referred to as *moon milk*.

rock pavement A term introduced by L. C. King to describe a low, whaleback rock exposure in the flat alluvial plains of tropical areas. It has been suggested that it represents the initial exposure of a BORNHARDT, during an early stage of stripping away of the regolith to reveal the BASAL SURFACE OF WEATHERING [*282*]. Some geomorphologists have suggested an alternative hypothesis in which the rock pavement is the final stage in the destruction of a bornhardt (shield inselberg) rather than the first stage of its exposure from beneath the mantle of weathered rock waste. This low, gently sloping, rocky hill is also termed a *ruware*.

rock pendant A protuberance of limestone on the roof of a cave. Although resembling a STALACTITE, it is formed from the differential solution of the surrounding bedrock. *See also* KARST, ROOF PENDANT.

rock salt HALITE.

rockslide A type of rock FAILURE in which part of the plane of failure passes through intact rock and where material collapses *en masse*, not in individual blocks. It is thought that pore-water pressures in the joint systems play a critical part, but slides can also occur in homogeneous, unjointed rock. [*146*] *See also* ROCK-FALL.

rock step RIEGEL.

rock stream An expression used in the USA as a synonym for a ROCK GLACIER.

roddon The name given to a river LEVEE in the Fenland of E England. It began as a natural feature but has been artificially altered over the centuries and, owing to the gradual lowering of the surrounding countryside by drainage and the shrinkage of the peaty soils, the roddon now stands high above the field surfaces.

Rogen moraines A Swedish term given to a field of MORAINES, 10–30 m in height and spaced at 100–300 m intervals, which have formed sub-glacially transverse to the direction of ice advance. They are sometimes called *ribbed moraines*.

Table 213 *Rock mass strength (r)*

Variable	Weighting (%)	Very strong	Strong	Moderate	Weak	Very weak
Intact rock strength (Schmidt hammer rebound value)	20	100–60 *r* = 20	60–50 *r* = 18	50–40 *r* = 14	40–35 *r* = 10	35–10 *r* = 5
Weathering	10	Unweathered *r* = 10	Slightly weathered *r* = 9	Moderately weathered *r* = 7	Highly weathered *r* = 5	Completely weathered *r* = 3
Joint spacing	30	>3 m *r* = 30	3–1 m *r* = 28	1–0.3 m *r* = 21	300–50 mm *r* = 15	<50 mm *r* = 8
Joint orientations	20	Very favourable. Steep dips into slope, cross joints interlock *r* = 20	Favourable. Moderate dips into slope *r* = 18	Fair. Horizontal or nearly vertical dips (Hard rocks only) *r* = 14	Unfavourable. Moderate dips out of slope *r* = 9	Very unfavourable. Steep dips out of slope *r* = 5
Joint width	7	<0.1 mm *r* = 7	0.1–1 mm *r* = 6	1–5 mm *r* = 5	5–20 mm *r* = 4	>20 mm *r* = 2
Joint continuity and infill	7	None, continuous *r* = 7	Few, continuous *r* = 6	Continuous, no infill *r* = 5	Continuous, thin infill *r* = 4	Continuous, thick infill *r* = 1
Groundwater outflow	6	None *r* = 6	Trace *r* = 5	Slight <40 ml s^{-1} m^{-2} *r* = 4	Moderate 40–200 ml s^{-1} m^{-2} *r* = 3	Great >200 ml s^{-1} m^{-2} *r* = 1
Total rating		100–91	90–71	70–51	50–26	<26

Source: Modified from Selby, 1980 Table 6, pp. 44–5

roller A colloquial term for a large ocean wave that breaks on exposed ocean coastlines where FETCH is considerable. Rollers occur regularly on the coasts of oceanic islands even during calm weather, and are highly regarded in Australia and California because of their importance to the sport of surfing.

ROM An acronym for Read Only Memory. A type of semiconductor computer memory the data of which have been fixed into the ROM during manufacture and cannot be altered. It does not lose its power when the computer is switched off in contrast to a RAM.

roof pendant A protuberance of country-rock projecting downwards from the roof of a BATHOLITH, and therefore almost surrounded by the igneous rocks of the latter. *See also* CUPOLA, ROCK PENDANT.

root 1 The steepest and lowest part of a recumbent fold or NAPPE. [*159*] **2** The downward projection of SIAL into the underlying SIMA in the Earth's crust [*121*]. **3** The permanent underground stock of a plant, shrub or tree which serves both to anchor it and to derive nourishment from the soil. **4** In mathematics, a number, quantity or dimension which, when multiplied by itself a requisite number of times, produces a given expression (e.g. square root, cube root).

root action 1 The processes by which the roots of vegetation penetrate the soil and help to break up the underlying bedrock by pressure exerted in the fissures. **2** The mechanisms by which the root growth of vegetation is able to hinder soil movement because of its fixing and binding properties. In many instances this may stabilize sand-dunes and may make scree slopes somewhat steeper. It is an important aid in any attempt to combat soil erosion. In mangrove swamps the root system plays a significant role in stabilizing unconsolidated fine-grained sediments.

ropy lava PAHOEHOE.

rose diagram A circular graph used for plotting the dip or strike of planar features, e.g. dykes, joints or the orientation of particles within a sediment (FABRIC, TILL). [*214*]

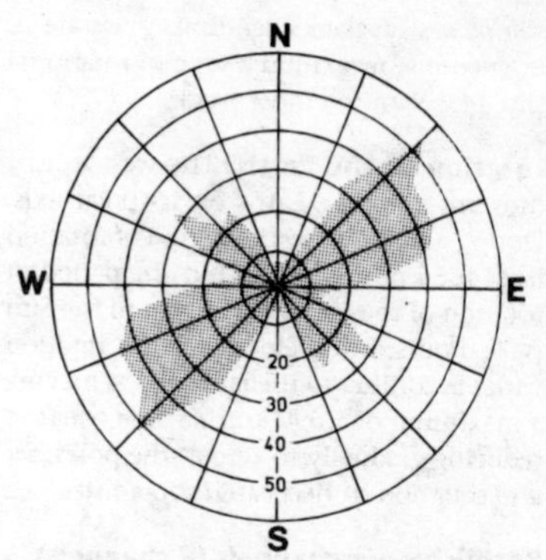

Figure 214 *Rose diagram: orientation of stones in till as evidence of direction of ice movement*

Rossby wave (long wave) A smooth, wave-shaped undulation in the airflow of the middle and upper TROPOSPHERE, with a wavelength of about 2,000 km. A varying combination of four, five or six waves makes up the complete westerly circulation of the atmosphere at middle and high latitudes of the hemispheres [*215*]. It represents a response to a combination of the ROTATION OF THE EARTH (PLANETARY VORTICITY) and the latitudinal variation of the CORIOLIS FORCE, balanced by RELATIVE VORTICITY. It has an important effect on surface CONVERGENCE and DIVERGENCE. *See also* ABSOLUTE VORTICITY, INDEX CYCLE.

Rossi–Forel scale A ten-point scale for rating the intensity of EARTHQUAKE shocks, devised in 1878 by M. S. de Rossi and F. A. Forel. It was replaced by the modified MERCALLI SCALE in 1930. *See also* RICHTER SCALE.

rotational slip **1** A type of LANDSLIDE, characterized by the downhill movement of a mass of surface material along a curvilinear plane of FAILURE or *slip plane*. The slipped blocks rotate about a point termed the slip-plane centre [*131*] and this type of failure usually occurs on a rectilinear slope of finite height and homogeneous material. Slip failure occurs when the shearing STRESS exceeds the SHEAR STRENGTH of the material, and if the latter has a uniform character it is likely to take place along a curved surface, concave to the slope and passing through its slope or base (COULOMB'S FAILURE LAW). **2** The curved

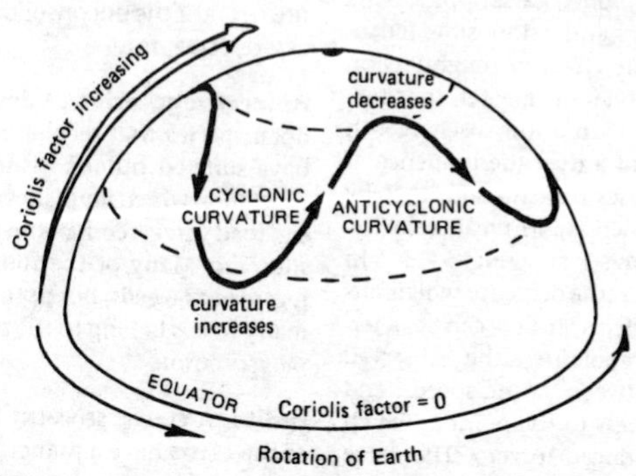

Figure 215 *Rossby waves*

slip planes of glacier ice during a state of EXTENDING FLOW exhibit a type of rotational slip. [45] *See also* TOREVA BLOCKS.

rotation of the Earth The west to east movement of the Earth on its polar axis. During a SIDEREAL DAY the period of rotation is 23 h 56 min 4.09 sec, but the period of rotation of the Earth in relation to the Sun is 24 hours. The velocity of the rotation varies according to the latitudes, achieving a maximum of 1,690 km/h at the equator, reducing gradually to zero at the poles. *See also* EQUATION OF TIME, MEAN SOLAR TIME.

Röthlisberger channels (R channels) A series of conduits at the base of a glacier that are incised upwards into the ice, probably by the heat generated by the TURBULENCE of the glacial MELTWATER. *See also* NYE CHANNELS.

rotor streaming A form of turbulent air motion in a vertical plane, characteristic of a LEE WAVE. It sometimes produces a LENTICULAR CLOUD.

rotting CHEMICAL WEATHERING.

roughness **1** In general, the degree of irregularity of a surface. **2** In hydrology, it refers to the amount by which the flow of a natural river is impeded owing to resistance at the boundary surface of the river channel (MANNING EQUATION). The resistance is generated by roughness elements, which include the magnitude of BEDLOAD in addition to a variety of bed and bank features, such as ANTIDUNES, bars, ripples, bank undulations and channel sinuosities (CHANNEL GEOMETRY). The effect of roughness on flow is proportionately greater in small streams or those with a low discharge. In the long profile of a river the influence of roughness decreases downstream, thereby resulting in an increase in mean velocity. *See also* RELATIVE ROUGHNESS. **3** In MICROMETEOROLOGY, the degree to which surface roughness affects air-flow can be determined by extrapolating the observed relationship between wind-speed and height of roughness to the point at which wind-speed is reduced to zero. The value known as *roughness length* is then taken as one-tenth of the true height of the *roughness elements*. *See also* CHEZY EQUATION.

roundness index A measure of the degree of surface curvature exhibited by an individual fragment of a CLASTIC SEDIMENT, notwithstanding the shape of the fragment. In general, increasing degrees of roundness are usually well correlated with duration and length of transport from source. The index can be applied to any type of PARTICLE SIZE but is most extensively adopted to describe the roundness of pebbles, although it is not synonymous with SPHERICITY. The method of measurement was introduced by A. Cailleux in 1945 and is expressed by $(2R)/a \times 1{,}000$, where: R is the minimum radius of curvature in the principal plane; and a is the length of the long axis of the pebble. *See also* FLATNESS INDEX.

***r*-selection** *R* AND *K* SELECTION.

rubble Any accumulation of loose, angular fragments that have not been waterworn. Once it becomes compacted it may result in a BRECCIA. The term is usually reserved to describe accumulations on a horizontal surface (REGOLITH) rather than those of SCREE or TALUS SLOPES.

rubidium–strontium dating RADIOMETRIC DATING.

rudaceous rocks A group of detrital sedimentary rocks in which the PARTICLE SIZE is >2 mm. The two main divisions are: **1** the compacted rocks; BRECCIA, CONGLOMERATE, TILLITE; and **2** the unconsolidated sediments; GRAVEL, SCREE, TILL.

ruderal vegetation A description of any floral species that recolonize habitats that have suffered human disturbance, or any vegetation which flourishes on waste places, e.g. road verges, embankments, demolition sites, etc. Many of the ruderal species are regarded as weeds, but in the strictest sense many weeds belong to the category of SEGETAL VEGETATION.

rudite A CLASTIC SEDIMENT in which >30% of the CLASTS have a PARTICLE SIZE of >2 mm. It is one of the RUDACEOUS ROCKS.

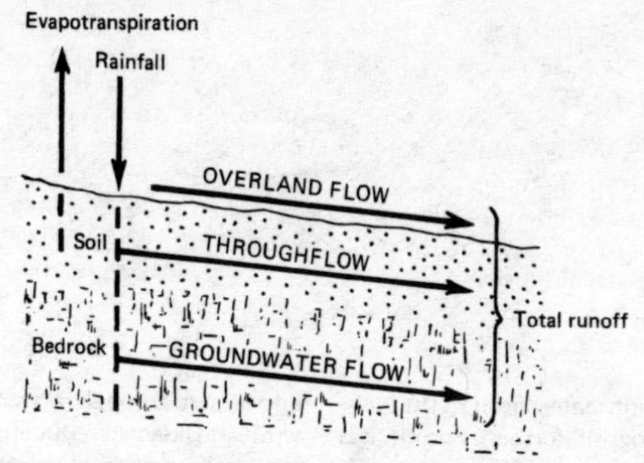

Figure 216 *Runoff*

run A term used in the USA to describe a small, swift-flowing stream or CREEK.

rundkarren KARREN.

runnel An alternative term for the linear depression (swale) which lies between parallel ridges (bars) on a shoreline beach or on a river floodplain. *See also* BAR-AND-SWALE.

running mean A statistical procedure designed to smooth out irregularities in a time series of numbers or values, especially when adjoining values exhibit great contrasts. The procedure is to take each value in turn and convert it to the *mean* of the sum of itself plus its two nearest neighbours (for example, in the series 15, 12, 26, 7, 25 the running mean for 26 would be 12 + 26 + 7 ÷ 3 = 15). If four nearest neighbours were required, to cover a longer time-interval, the running mean for 26 would be 15 + 12 + 26 + 7 + 25 ÷ 5 = 17. The running mean can then be used to produce a *moving average*, which may be plotted for each time-interval on a graph, by omitting the first value and summing the remainder. Thus, a series of running means can be obtained for a specified time-period, helping to illustrate the overall trend by smoothing out the extreme values.

runoff 1 The water leaving a drainage area. It is normally regarded as the rainfall minus the loss by evaporation. *See also* WATER BALANCE. **2** The water running across the land surface, as opposed to that running through the soil (THROUGHFLOW), but for this usage it is preferable to use the term OVERLAND FLOW. [*216*] *See also* HYDROLOGICAL CYCLE, INFILTRATION-RATE, INFILTRATION VELOCITY, OVERLAND FLOW, PERMEABILITY, RATIONAL FORMULA, THROUGHFLOW.

runoff coefficient IMPERMEABILITY FACTOR.

ruware ROCK PAVEMENT.

S

Saale The penultimate glacial in the Pleistocene chronology of northern Europe. It is older than the WEICHSEL, from which it is separated by the EEMIAN interglacial, but younger than the ELSTER glaciation. It has been equated with the RISS glacial of the Alps, the WOLSTONIAN of Britain and the ILLINOIAN of N America. It appears to have been the most far-reaching ice advance in the western reaches of the N European Plain and it left a series of outwash deposits stretching from Utrecht in the Netherlands to Krefeld in W Germany. [*109*]

sabkha SEBKHA.

saddle **1** A low point or COL on a ridge connecting two summits. **2** A structural feature associated with a sag in the crest of an anticline (termed a *saddle fold* or DEPRESSION).

saddle reef A bedded deposit of gold-bearing quartz (REEF) either in the form of a saddle-shaped structure along the sagging crest of an anticline, or as an inverted saddle associated with a synclinal structure. *See also* PHACOLITH.

sagebrush A greyish shrub vegetation in the arid and semi-arid regions of SW USA and Mexico.

Sahel The name given to a narrow east–west zone of semi-desert in N Africa along the southern fringes of the Sahara, stretching from the Red Sea to the Atlantic coast in Senegal. It receives between 150 and 500 mm annual rainfall but droughts are frequent occurrences, bringing regular crop failures and famines (DESERTIFICATION). In its modern usage it is a vegetation zone in which the SAVANNA tropical grasslands phase into desert through a transition zone of scrubland (XEROPHYTE), but its original meaning is 'seashore' (from the Arabic), although some writers suggest that it simply means 'north'.

St Elmo's Fire An electrical discharge (known as a brush discharge) which occurs when there is a strong electrical field in the atmosphere, and named after a 4th-cent. Italian saint. This blue-green coloured discharge manifests itself during thundery weather on protuberances such as aircraft wing tips, weather vanes, and ships' masts, and is accompanied by a crackling noise. It has been suggested that St Elmo's Fire may have caused the destruction of the German airship *Hindenberg* in 1937 as it was preparing to land in the USA.

salar The name given to a PLAYA in the SW USA. *See also* SABKHA, SHOTT.

salcrete A surface crust, mainly of sodium chloride, which cements a sand surface on a BEACH by the evaporation of seaspray. *See also* CALCRETE, SILCRETE.

salient **1** A projecting spur or headland which protrudes from a line of hills or a coastline. It is the opposite of a RE-ENTRANT. **2** That part of an orogen (MOBILE BELT) which curves towards the FORELAND, e.g. the European Alps comprise a number of arcuate salients with their convexity towards the central European foreland.

salina A Spanish term used extensively in S America, Central America, Mexico and the SW states of the USA to describe saline deposits in salt-marshes, PLAYAS and SALARS.

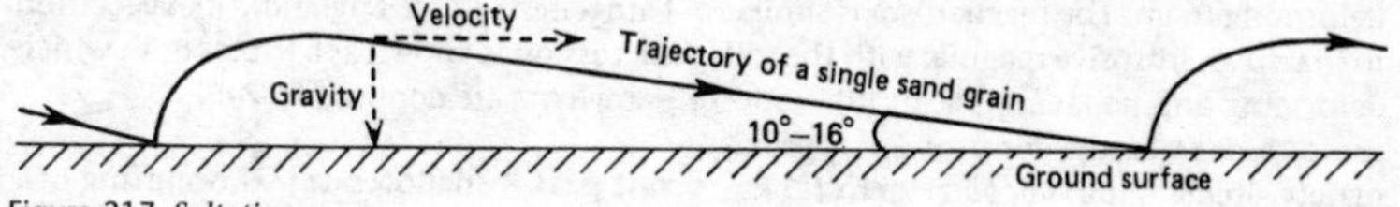

Figure 217 *Saltation*

saline-alkali soil A soil of high alkalinity and one containing appreciable quantities of soluble salts, with sufficient exchangeable sodium to interfere with the growth of all but the most specially adapted plants. Its pH is usually around 8.5. It is sometimes called *saline-sodic soil* (SOLONETZ). *See also* HALOMORPHIC SOILS.

saline soils A group of INTRAZONAL SOILS containing high proportions of soluble salts which are drawn to the surface in solution by CAPILLARITY and where they create a greyish crust. Thus, they are commonly found in regions of considerable seasonal evaporation, in arid and semi-arid environments and around the fringes of PLAYAS. Irrigation is needed to flush the salts down through the soil to prevent them from accumulating. Saline soils are non-alkaline, are synonymous with the SOLONCHAKS and form one of the three major groups of the HALOMORPHIC SOILS. *See also* SOLOD, SOLONETZ.

salinity The degree to which water contains dissolved salts. A fundamental property (together with oxygen content and temperature) of sea water. It is usually expressed in parts per thousand or in grams per thousand grams, and normal sea water has a salinity of 34.33‰. This rises to 40‰ in the enclosed seas of the Persian Gulf and the Red Sea owing to high evaporation, high summer temperatures and few discharging rivers. In high latitudes, however, with less evaporation and dilution from melting icesheets, salinity falls to 34‰ or less in the polar oceans. The soluble salts in sea water comprise 23 parts sodium chloride, 5 parts magnesium chloride, 4 parts sodium sulphate, 1 part calcium chloride, 0.7 parts potassium chloride and 0.63 parts minor trace elements.

salinization The process of salt accumulation in soil or in water. *See also* DESALINIZATION.

salpausselka The name given to the long morainic ridges, of late Pleistocene age, that run east to west in S Finland, some 63 km inland from the coast. They were formed during the FINIGLACIAL STAGE of ice retreat.

salt As a workable economic resource, common salt, or sodium chloride (NaCl), occurs: **1** in certain sedimentary rocks, especially those of PERMIAN and TRIASSIC age; **2** in SALT DOMES; **3** in SALT LAKES; **4** in SALT FLATS; **5** in solution as BRINE. It is synonymous with *halite*. *See also* EVAPORITE.

saltation **1** A mechanism by which sediment is transported by bouncing or hopping along the surface. It is the most common form of wind transport in arid environments (AEOLIAN) but is also significant in the movement of solid particles along the bed of a stream. The term is derived from the Latin *saltare* (= to jump) and is used to describe the movement of any particle which is too heavy to remain in suspension. The size of the particles which can be transported by saltation is correlated with the current velocity and the density of the current, with water being able to lift larger particles than air. Nevertheless, because of the viscosity of water a particle lifts more slowly in water (which also cushions its fall) than in air, and the height of the saltatory jumps is therefore considerably greater in air than in water. The use of the term saltation in freeze–thaw processes and in the context of long-shore drift is to be discouraged. **2** The term has been used in the context of PUNCTUATED EQUILIBRIUM, the concept that evolution progresses by jumps rather than by steady progression (*saltatory evolution*). [*217*]

salt dome A mass of salt which is injected as a diapir (DIAPIRISM) into overlying sedimentary rocks, thereby piercing and deforming them. The mechanism is similar to that of an intrusive magma, with the salt deforming and behaving plastically under pressure. It is in the form of an approximately circular plug (*salt plug*) up to 1.5 km in diameter but of considerable depth (down to 10 km). It is of great economic importance because it assists in the formation of a 'trap' structure for oil accumulation, in addition to its associated deposits of ANHYDRITE, GYPSUM and SULPHUR. In the few instances where salt domes reach the surface a mass of mobile salt, termed a *salt glacier*, flows outwards from the centre of the exposure, e.g. around the Persian Gulf. *See also* HALOKINESIS.

salt-earth podzol SOLOD.

salt flat The dried-up bed of a former SALT LAKE, sometimes referred to as a *salt prairie* in the USA. *See also* ALKALI FLAT.

salt glacier SALT DOME.

salt lake A lake containing water of high salinity, which occurs in arid or semi-arid environments where evaporation is high. It is usually found in a basin of inland drainage (ENDORHEIC), especially within a PLAYA, and is surrounded by *salt flats*. The SALINITY of a salt lake is considerably higher than that of sea water, rising to 330‰ in L. Van, Asia Minor, 238‰ in the Dead Sea, and 220‰ in the Great Salt L., Utah, USA. *See also* SALAR, SALINA, SALINE SOILS, SALT PAN, SEBKHA, SHOTT.

salt-marsh A coastal marsh formed in a LAGOON, behind a coastal bar, BARRIER-BEACH or SPIT, or in an estuary. It may be tidal and inundated daily by sea water or it may be reclaimed (as in Romney Marsh behind Dungeness, Kent, England). Its vegetation succession is known as a HALOSERE, in which HALOPHYTES are dominant. [*240*]

salt pan A shallow SALT LAKE occurring in a small enclosed basin, smaller in its dimensions than a true salt lake.

salt pillow The protuberance forming on a salt layer when HALOKINESIS has commenced.

saltwater intrusion GHYBEN–HERZBERG PRINCIPLE

salt-weathering HALOCLASTY.

samoon, samun An Iranian term for a hot, dry wind of FÖHN-like character, descending from the highlands of Kurdistan. [*83*]

sample A part of some total quantity or statistical population, in general selected deliberately in order to investigate the properties of the parent population. The selection of the items to be sampled is normally carried out according to particular statistical methods designed to ensure that the items are representative of the population as a whole. *See also* RANDOM SAMPLE.

sampling The technique of obtaining a SAMPLE to provide adequate information on a particular population. In order that every member of the population will have an equal chance of being selected, bias must be excluded from the procedure. This may be done by *systematic sampling*, e.g. by taking every tenth member of the population, by *random sampling* (RANDOM SAMPLE) or by *nested sampling* (NESTING). Examples are illustrated in [*218*].

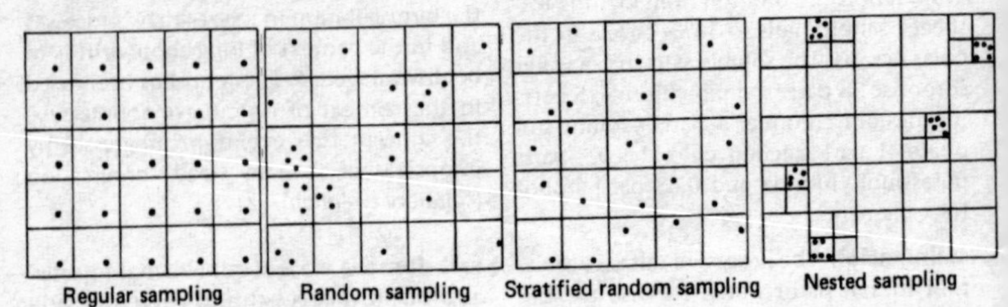

Figure 218 *Point sampling designs*

sampling frame The term given to the entire statistical population from which a SAMPLE is taken.

sand One of the grades of detrital material, composed largely of quartz, and having a PARTICLE SIZE in the British Standard classification between silt and gravel (i.e. 0.06 to 2.0 mm in diameter). Sand is also subdivided into: *coarse sand* (0.6–2 mm); *medium sand* (0.2–0.6 mm); and *fine sand* (0.06–0.2 mm). There are slight size differences between the above classification, the WENTWORTH SCALE and the International scale of SOIL TEXTURE, in which a sandy soil is subdivided according to the percentages of sand of various grades within the sample. In general, wind-blown desert sand is more rounded than marine or fluvial sand. Windblown sand falls largely in the categories of *fine sand* and *medium sand*, with a particle size of 0.15–0.3 mm.

sand bank **1** An accumulation of sand in an ESTUARY or in a coastal environment, with its linear or parabolic form created by the tidal flow working on an abundant supply of land-derived sediments. It is usually uncovered at low tide, to reveal a complex pattern of ebb and flood channels (EBB CHANNEL). **2** A bank or ridge of sand built by currents in a river channel, in which context it is preferable to use the term bar (POINT BAR).

sand-dune BARCHAN, DUNE.

sand lens A discontinuous sand layer in a sedimentary sequence. It represents the remnant of a former CHANNEL-FILL DEPOSIT or a channel-margin sediment.

sand ribbon A very thin linear accumulation of sandy deposits on the sea-floor in environments where there are abundant supplies of detrital sediment and strong tidal streams. The ribbon runs parallel with the direction of the strongest tidal stream and its boundaries are more clearly defined the greater the velocity of the tidal stream. Some ribbons near the S coasts of Britain have been found to be over 2 km long, 100 m wide and a few cm thick.

sand-rose A circular HISTOGRAM, illustrating the amount of sand moved at a given point by winds blowing from various directions. *See also* WIND-ROSE.

sandstone A type of SEDIMENTARY rock of the ARENACEOUS variety. It has rounded sand particles which differ from the angular and sub-angular particles of GRITSTONE, which it otherwise resembles. Its particles are mainly quartz, although mica, feldspar and other minerals also occur; and it is often classified according to the agent which cements the particles: *calcareous sandstones* have a dolomitic type of cement; *siliceous sandstones* (ORTHOQUARTZITE) are bonded by silica; *ferruginous sandstones* are cemented by LIMONITE. It is also characterized by a wide variety of colours owing to different degrees of HYDRATION and OXIDATION and to varying amounts of iron or other minerals, e.g. high proportions of glauconite produce a GREENSAND. Colours range from dark brown through red, yellow and grey to white. Sandstones can also be subdivided, according to their PARTICLE SIZE, into coarse, medium and fine varieties, some of which make excellent FREESTONES. *See also* CARSTONE, CROSS-BEDDING, NEW RED SANDSTONE, OLD RED SANDSTONE, TORRIDONIAN.

sandstone dyke A near-vertical, sheet-like body of sandstone, less than 10 cm in thickness, that cuts through bedrock in the manner of an igneous DYKE. It is thought to be formed by the infilling of a fissure from above or by injection from below (INJECTION STRUCTURES).

sandstorm A movement of relatively coarse sand driven by a strong wind in an arid environment. It differs from a DUSTSTORM in that the sand particles do not rise more than a few m above the ground and are not transported such great distances. *See also* AEOLIAN.

sand stream A broad tract (tens of kilometres wide) of sand moving parallel to the coast below low-water mark, driven by strong tidal streams. *See also* SAND RIBBON.

sandur (sandr, sandar) An Icelandic term for an outwash plain formed from

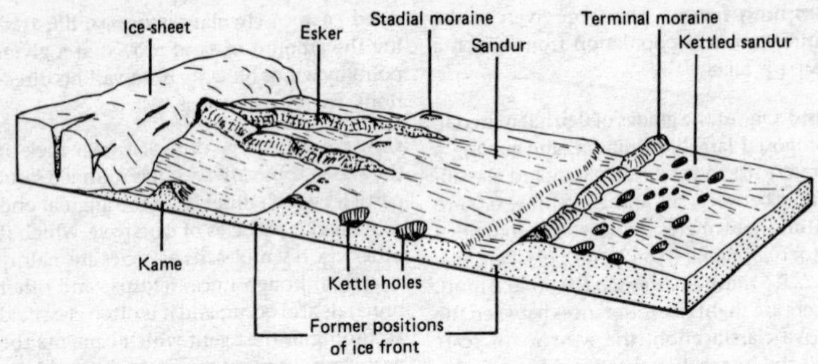

Figure 219 *Glaciofluvial landforms*

GLACIOFLUVIAL material carried out from the front of an ice-sheet by meltstreams. Sandurs appear as extensive accumulations of gravel, sand and silt, crossed by BRAIDED STREAMS, in which variations in discharge and debris-supply cause alternating periods of net erosion or net accumulation, in part related to seasonal melting and in part to JÖKULHLAUPS. Changes in the relative position of the ice front may create ridges or steps across the sandur surface parallel to the ice front and these may represent moraines or ice-contact slopes. *See also* KETTLED SANDUR. [*219; see also 267*]

sand volcano A small (<50 m) mound which simulates a true volcano but from which sand is ejected rather than lava. It is a type of INJECTION STRUCTURE formed in the same way as a MUD VOLCANO.

sand-wave ANTIDUNE, DUNE.

sand wedge The infilling by sand of a fissure in the ground surface formerly occupied by an ICE WEDGE. The sand takes on the shape of the ice-wedge cast and is regarded as evidence of formerly frozen ground in a PERIGLACIAL environment.

sandy loam A soil which contains between 50 and 70% sand and between 10 and 20% clay. This mixture produces a free-draining, friable soil. *See also* CLAY LOAM, LOAM, SILTY LOAM, SOIL TEXTURE.

Sangamon The last INTERGLACIAL of the Pleistocene ice age in N America. It is equivalent in age to the IPSWICHIAN of Britain and the EEMIAN of N Europe. It is succeeded by the WISCONSIN glacial stage.

Sanson–Flamsteed projection A MAP PROJECTION of the PSEUDO-CYLINDRICAL PROJECTION type and with equal-area properties. It is a particular case of BONNE'S PROJECTION and it is also known as a sinusoidal projection. The equator is the standard parallel and is drawn as a straight line truly divided, with the central meridian also truly divided and also drawn as a straight line: parallels are drawn as straight lines at equal distances apart, and the meridians drawn as sine curves through corresponding points on each of the parallels (hence the term *sinusoidal*, derived from *sine*). It was designed by Sanson in 1650 and adopted by Flamsteed in 1729. Despite its equal-area attributes its shape is distorted towards the margins, especially in high latitudes where the meridians are very oblique to the parallels. It is used for world maps and especially for maps of Africa. [*220*] It forms the basis for GOODE'S INTERRUPTED HOMOLOSINE PROJECTION.

Santa Ana (Anna) A Spanish name for the FÖHN-type wind which blows southwestwards as a hot, desiccating blast of air from the American Sierra Nevada down into the arid coastlands of S California. It brings clouds of dust, lowers the humidity and raises the temperature. If it occurs in spring it causes severe damage to fruit-trees when the young buds are being formed. *See also* SIROCCO.

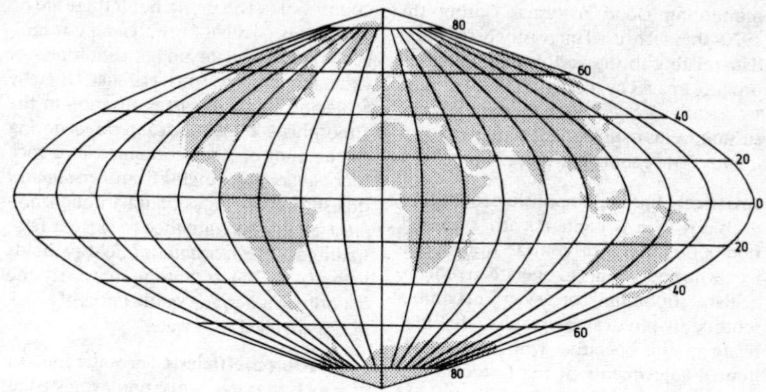

Figure 220 *Sanson–Flamsteed projection*

sapping The undermining of the base of a cliff, thereby leading to the collapse of the cliff face. It can be the result of a stream's LATERAL EROSION, by the action of a water WAVE, or by SPRING SAPPING.

saprolite The residual part of the weathering profile of deeply rotted bedrock. Since the weathered material has remained *in situ* the original structures (joining, veining, etc.) of the parent rock may have survived. It is a type of REGOLITH.

sapropel An aquatic ooze, highly charged with carbonaceous matter, that forms in SWAMPS and shallow marine basins, as a result of rotting. It has a higher percentage of HYDROCARBON than peat and is thought to be a source of natural gas and petroleum.

saprophyte A plant that flourishes in an environment of putrefaction and draws its nourishment from rotting vegetable matter, e.g. fungus.

sarn A causeway of rocks and boulders jutting out from a coastline and exposed at low tide. The sarns of Cardigan Bay in central Wales are thought to be remnants of old glacial MORAINES.

sarsen The local name in S England for a block of extremely hard siliceous sandstone, thought to have been derived from the break-up of a former cover of sedimentary rocks of early Tertiary age (PALAEOCENE). The sarsen blocks are found either high on the chalk downlands where they are termed GREYWETHERS, or along the floors of DRY VALLEYS where they have been carried by mass-movement during one or more PERIGLACIAL phases of the Pleistocene, or they are found incorporated in river terraces or buried in SOLIFLUCTION deposits. Some writers believe that they are remnants of a Tertiary DURICRUST formed by the INDURATION and silicification of sands and gravels, with the finer fraction creating the sarsen and the pebbly fraction the so-called Hertfordshire PUDDINGSTONE which is a type of conglomerate. Sarsens have been widely used in S England and N France in megalithic monuments, e.g. Stonehenge. Termed MEULIÈRE in the Paris basin. *See also* SILCRETE.

sastrugi A Russian term for the surface irregularities of snowfields caused by the scouring action of the windborne ice particles whipped up by the prevailing Arctic and Antarctic winds. The furrows and sharp ridges impede the movement of sledges across the 'grain' of the sastrugi patterns. A plural noun (singular: *sastruga*).

satellite, artificial A space vehicle, housing various scientific instruments, that is placed into orbit around the Earth or other planets in order to collect and transmit data. Since their inception in 1959 satellites have greatly assisted weather forecasting by monitoring cloud patterns. During the 1970s they were used increasingly to collect data relating to the Earth's surface and to produce images by means of REMOTE SENSING TECHNIQUES. *See also* HEAT-CAPACITY MAPPING MISSION, LANDSAT, NOAA/TIROS SATELLITES, SEASAT I, SKYLAB, SPACE SHUTTLE, SPOT.

satisficer A person responsible for making decisions who is content with a limited, non-optimum goal, in contrast to an optimizer, who endeavours to operate in order to achieve optimum or even maximum returns. In physical geography, this may relate to the operative who makes only limited adjustments in the face of a perceived natural hazard, a response referred to as risk minimization.

saturated adiabatic lapse-rate (s.a.l.r.) A measure used to describe the changing temperature of a moving body of air with height. When a parcel of saturated air rises through the atmosphere it cools by expansion, but the amount of LATENT HEAT released with condensation depends on the amount of water vapour and therefore on the temperature of the air. At very low temperatures little heat is released and the rate of cooling is similar to that of the DRY ADIABATIC LAPSE-RATE (about 1° C for 100 m) but always below it. At high temperatures more latent heat is released because so much water vapour is present and the saturated adiabatic lapse-rate could be as low as 0.4° C for 100 m. Thus, the s.a.l.r. varies between 0.4° C and 0.9° C for 100 m of ascent. The term *wet* or *moist* is sometimes used in place of *saturated*. *See also* ADIABATIC, CONDITIONAL INSTABILITY, ENVIRONMENTAL LAPSE-RATE, LAPSE-RATE, PROCESS LAPSE-RATE. [*132*]

saturation 1 The state of the atmosphere in which a parcel of moist air is in equilibrium with an open water surface at the same pressure and temperature, i.e. there is a balance between the number of molecules of water entering the air and those returning to the water surface. The ability of air to retain water vapour depends on its pressure and temperature, so that saturation can be achieved either by cooling or by adding more water vapour. If nuclei are present CONDENSATION will occur, but if there are no nuclei it is possible for water-vapour molecules to coagulate under conditions of extremely high VAPOUR PRESSURE, thereby achieving a state of SUPERSATURATION in the atmosphere. **2** A geological term, denoting the amount of silica contained in a rock (*silica saturation*): ranging from *oversaturated* (free silica occurring as quartz) through *saturated* (all silica is combined to exclude feldspathoids) to *undersaturated* (feldspathoids present). **3** The condition in which the majority of pores, cracks and joints of a rock or soil are filled with water.

saturation coefficient The water content of a rock after free SATURATION expressed as the percentage of the maximum water content. It is important in the testing of the MECHANICAL WEATHERING of rocks.

saturation deficit The amount of water vapour that is required to convert a parcel of non-saturated air at a given pressure and temperature to the state of SATURATION.

saturation overland flow The type of OVERLAND FLOW that occurs when all the soil pores are filled with water after a long period of relatively low-intensity rainfall. It is the most important cause of overland flow in humid temperate regions. *See also* INFILTRATION-EXCESS OVERLAND FLOW.

savanna (savannah, savana) A Spanish term now universally adopted to describe the world's tropical grasslands. The vegetation zone, midway between the deserts and the tropical rain-forests, comprises a number of combinations of scattered trees, shrubs and various heights of grass according to the length of the dry season (SAVANNA CLIMATE). The trees are not forest trees but are gnarled, thorny XEROPHYTES with reduced leaves (e.g. acacia) or bulbous trunks (e.g. baobab) that reduce moisture loss, especially towards the desert borders. Most of the trees are deciduous, losing their leaves in the dry season, at which time the tall grasses die back. It has been claimed that large tracts of tropical grassland owe their character to human interference, and are kept at a PLAGIOCLIMAX by burning and grazing. The husbandry of the savanna is largely pastoral,

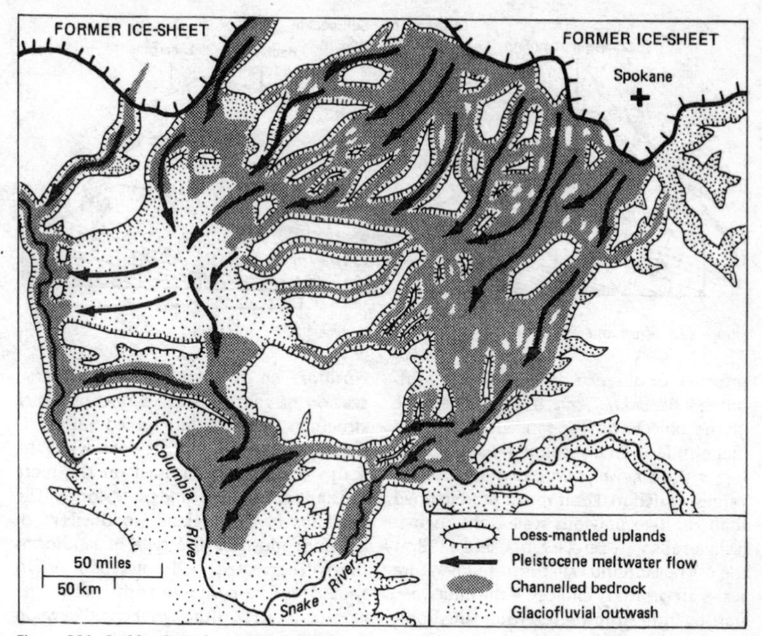

Figure 221 *Scablands in the north-west USA*

although good crops of millet, maize, tobacco, cotton, etc. can be grown with efficient management and adequate water supply. *See also* CAMPO, LLANO.

savanna climate The climate of that part of the tropical zone which is characterized by a marked dry season and a rainy season of monsoonal attributes as the INTERTROPICAL CONVERGENCE ZONE passes overhead. It takes its name from the vegetation type (SAVANNA).

scablands The term given to the terrain of bare rock surfaces with thin soil cover and scanty vegetation which characterizes parts of the basaltic plateau country of the NW USA. Much of the regolith and soil mantle has been carried away by glacial floodwaters, which also deeply channelled the mantle of LOESS. [*221*]

scale 1 In cartography, the ratio of the distance on a map, globe, mode, or vertical section to the actual distance on the ground. Most scales are linear, i.e. they comprise a subdivided line showing linear distances at a given scale, but they can be expressed as a REPRESENTATIVE FRACTION (RF), or expressed in words. In Britain the ORDNANCE SURVEY distinguishes between *large-scale maps* (>1 : 10,560) and *small-scale maps* (<1 : 25,000). The Directorate of Overseas Survey, however, classifies its maps into *large-scale* (>1 : 25,000), *small-scale* (<1 : 125,000) and *medium-scale* (1 : 25,000 to 1 : 125,000). **2** Any graduated means of measuring the magnitude of an object, mechanism or process, e.g. a temperature scale or earthquake-magnitude (RICHTER) scale. *See also* SCALES OF MEASUREMENT.

scales of measurement A means of expressing the character of a wide variety of phenomena in an objective and standardized way. Four such scales, in increasing order of exactitude, are: **1** the *nominal scale*, which differentiates between objects by placing them in classes according to the

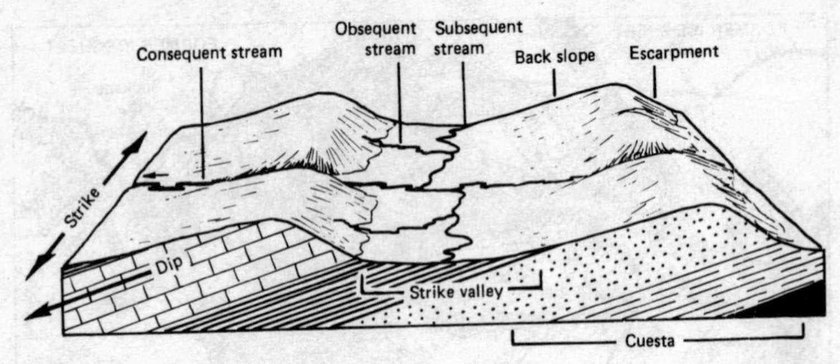

Figure 222 *Scarp-and-vale terrain*

presence or absence of recognizable attributes; **2** the *ordinal scale*, in which the value of the objects can be ranked in order of magnitude, even if absolute values are lacking; **3** the *interval scale*, based on absolute values, which makes it much more refined than the two previous scales, although it lacks an absolute zero, e.g. 0° C and 32° F on the CENTIGRADE and FAHRENHEIT temperature scales are arbitrary choices; **4** the *ratio scale*, starting from an absolute zero and incorporating equivalence of ratios, which is the most precise of all, e.g. measurements on the ABSOLUTE TEMPERATURE scale (KELVIN SCALE) mean that 20° K is twice as hot as 10° K. Nominal and ordinal scales can be used only in NON-PARAMETRIC STATISTICAL TESTS while interval and ratio scales can be used for PARAMETRIC STATISTICAL TESTS.

scanner An instrument for DATA CAPTURE, which converts an ANALOGUE image or map into digital RASTER form by systematic line-by-line sampling. It can be a FLAT-BED SCANNER or a DRUM (cylindrical) SCANNER. *See also* PLOTTER.

scar A local term for a rocky cliff in N England, e.g. Gordale Scar, Nab Scar.

scar limestone A type of massive limestone of Carboniferous age which, in N England, forms conspicuous cliffs where it is exposed at the surface. It is termed the Great Scar Limestone in the Pennines.

scarp ESCARPMENT.

scarp-and-vale terrain The alternating pattern of CUESTAS and intervening lowland corridors which is created by DIFFERENTIAL EROSION of gently tilted sedimentary rock structures. The vales are carved out in the less-resistant rocks, such as clays, while the scarps (cuestas) are formed on the more resistant limestones and sandstones. The lowlands of SE England (to the NW of London) illustrate the type of landforms that will be created [*222*]. This type of terrain should not be confused with that of the RIDGE-AND-VALLEY, although it could develop from a BELTED-OUTCROP PLAIN.

scarp-foot spring A spring located at the base of an ESCARPMENT, usually where permeable strata overlie less permeable rocks. The line of springs is an important factor in the location of settlements along the foot of the escarpment.

scarp retreat PARALLEL RETREAT OF SLOPES.

scattergram A graphic representation of the degree of correlation between two sets of statistical data. One set is plotted along the *x* axis (ABSCISSA), the other along the *y* axis (ORDINATE), and a 'best fit' line is drawn (REGRESSION). If there are no residuals or anomalies the correlation is perfect, but if the scatter of points does not lie on a straight line there is imperfect correlation. The deviations of the residuals from the regression line can be shown by drawing residual lines. It is also termed a *scatter diagram.* [*203*]

scattering, Rayleigh scattering The break-up of solar RADIATION by dust and molecules of water vapour in the atmosphere. Scattering is more effective in the shorter (blue) wavelengths than in the longer (red) wavelengths; the reduction of the blue light makes the Sun appear as a red or yellow object to an observer on Earth. *See also* DIFFUSE RADIATION, DUST, INSOLATION.

schattenseite The German term for the shaded slope of a valley (UBAC). It contrasts with SONNENSEITE.

scherm The name given to the small, blunt-ended bays along the coast of the Red Sea. Some writers believe that the *scherm* (plural *schurum*) represents the partly drowned seaward end of a WADI.

schist A well-foliated, medium- to coarse-grained rock created by the effects of REGIONAL METAMORPHISM, which causes a marked segregation of minerals into thin layers, together with a reorientation of the platy minerals, such as MICA, into a characteristic texture known as *schistosity*. All traces of former bedding are destroyed and the schist splits easily along the *schistosity* planes, which are often wavy. Greenschists, like SLATES and PHYLLITES, are formed from clay sediments altered by low-grade metamorphism (GRADE). Crystalline schists are formed at higher metamorphic grades, producing quartzitic schists from sandstones, while *mica schists* sometimes include garnets. Basalt and gabbro are changed into *hornblende schists* and *biotite schists*.

schlieren A German term (meaning *streaks*) applied to elongated patches in igneous rocks that differ in composition and colour from the host rock. They are thought to be mineral segregations or modified inclusions (XENOLITH) created during an earlier phase of crystallization.

schott SHOTT.

schuppenstruktur The German term for IMBRICATE STRUCTURE.

sciophilous The response shown by organisms that have adapted to live entirely in the shade. A plant that thrives in a well-shaded environment is termed a *sciophyte*.

sciophyte SCIOPHILOUS.

sclerophyllous A term referring to those species of evergreen shrubs and trees that have adapted to lengthy seasonal drought by producing tough, leathery leaves to cut down moisture loss by TRANSPIRATION. They are commonly found in regions with a MEDITERRANEAN CLIMATE. Olive and cork oak are typical examples.

scoria A type of PYROCLASTIC MATERIAL ejected from a volcano. It is characterized by a dark, vesicular structure of partly glassy and partly cindery material, known as volcanic slag, and is formed by the rapid cooling of the blisters and bubbles of lava that has been fragmented by a volcanic eruption.

scotch mist A fine DRIZZLE (sometimes referred to as '*mizzle*') falling from thick cloud, the base of which is near to the ground-level, characteristically on British hills and mountains. It is formed mainly when saturated but stable air masses are forced to rise over topographic barriers (OROGRAPHIC RAINFALL) thereby causing considerable amounts of CONDENSATION. *See also* MIST.

scour 1 The erosive power of moving water, either in a river channel, where the current excavates a depression near the outside of a bend, or by tidal currents in estuaries and narrow straits, due largely to the VENTURI EFFECT. The material scoured from depressions is redeposited when the current slackens, a process termed *scour and fill*. **2** The term is sometimes used when referring to the abrasive effects of boulders, incorporated in the base of an ice-sheet, upon the underlying rocks (ABRASION).

Scourian A term given to the oldest known episode of folding and metamorphism in the Pre-Cambrian rocks of the LEWISIAN complex of NW Scotland between 2,200 and 2,900 million years ago. It is named from Scourie in Sutherland. *See also* LAXFORDIAN.

scree An accumulation of fragmented rock waste below a cliff or rock face, formed as a result of disintegration, largely by MECHANICAL WEATHERING of a rock exposure. It is usually of coarse PARTICLE SIZE, has varying degrees of soil or vegetation cover and exhibits a concave slope profile. The steeper upper part of the *scree slope* is normally between 30° and 38° while the gentler lower slope lies between 25° and 30°. The ANGLE OF REPOSE of screen differs according to the character of its constituent material, the general rule being that for imperfectly sorted material, the larger the average particle size the steeper the slope gradient. Spherical or round-edged fragments rest more steeply than angular or platy fragments. In most screes larger blocks travel farther down the slope because volume, and therefore mass and momentum, increase as the cube of the diameter, whereas surface area, and therefore friction, increase only as the square of the diameter. Slow CREEP, which has been measured on the surface layers of some scree slopes, is due partly to FROST-HEAVING or to lateral expansion by heating (INSOLATION). Some writers differentiate scree from *talus*, with the former being regarded as all loose material lying on a hill slope and the latter being loose material accumulated only beneath a FREE FACE. In the USA, however, the two terms are regarded as being synonymous. *See also* CONSTANT SLOPE. [*49*]

screen, meteorological STEVENSON SCREEN.

scroll A low, curving ridge on a FLOODPLAIN running parallel with the loop of a MEANDER. It is a type of POINT BAR formed during a stage of high-water flow which overtops the river banks (OVERBANK FLOW STAGE OF A RIVER). It is sometimes referred to as a *scroll bar*, and is part of the floodplain micro-relief known as BAR-AND-SWALE, marking the former channel margin during an earlier phase of meander growth.

scrub A vegetation type associated with poor soils, exposed locations or semi-arid environments, in which the species are stunted, gnarled or specially adapted to seasonal drought (XEROPHYTE). *See also* ASSOCIATION, BRIGALOW SCRUB, CHAÑARAL, CHAPARRAL, GARRIGUE, MAQUIS, MESQUITE, MULGA, SAGEBRUSH, SCLEROPHYLLOUS, SPINIFEX.

scud An alternative name for *fracto-stratus* cloud (FRACTUS), which describes the ragged, torn masses of cloud which are driven rapidly by the wind beneath the main cloud layer (usually NIMBOSTRATUS) during stormy conditions.

S-curve A term representing the surface runoff HYDROGRAPH caused by an effective rainfall of intensity I/T mm^{h-1} where T is the duration of the effective rainfall.

sea **1** In general, the marine section of the globe as opposed to that of the land. **2** The name given to a body of salt water smaller than an ocean and generally in proximity to a continent. **3** In a few instances, the term is used to describe large enclosed saline water bodies (e.g. Caspian Sea, Dead Sea) that are cut off from the open ocean. **4** In occasional usage as a colloquialism to describe an extensive stretch of desert sand-dunes – a sand sea (ERG).

sea arch ARCH.

sea breeze A local air movement, reaching its greatest intensity during the afternoon, blowing from the relatively cooler sea to the warmer land during any period of time when low pressure develops over the land owing to solar heating and CONVECTION. It is most pronounced during calm weather but is a commonplace occurrence in equatorial latitudes, where it brings some amelioration to the hot, humid coastlands. Elsewhere it is generally a summer phenomenon only. *See also* LAKE BREEZE, LAND BREEZE [*130*].

sea cave CAVE.

sea cliff CLIFF.

sea-floor spreading The movement in PLATE TECTONICS whereby some ocean floors are extended laterally as plates move apart from a zone of separation (SPREADING ZONE) which corresponds to a MID-OCEANIC RIDGE. These ocean floors are underlain by OCEANIC CRUST. [*223*] *See also* SUBDUCTION ZONES.

sea-level MEAN SEA-LEVEL.

sea-level change EUSTASY.

sea loch LOCH.

seamount An isolated mountain rising abruptly some 1,000 m from the ocean floor but without extending above sea-level. It is

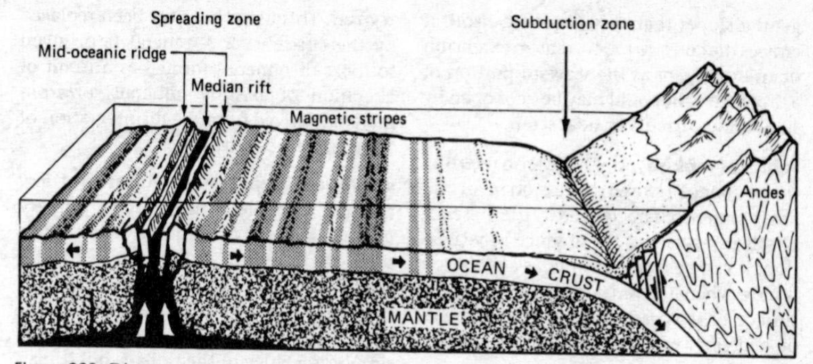

Figure 223 *Diagrammatic representation of sea-floor spreading in the East Pacific*

probably of volcanic origin and its conical summit contrasts with that of a GUYOT, which it otherwise resembles. Sea-mounts can be isolated, clustered into *seamount groups*, or in a linear pattern as a *seamount range*. They occur most frequently on the floor of the Pacific Ocean.

Seasat I The first of a series of SATELLITES launched by NASA (National Aeronautics and Space Administration) of the USA as an aid to oceanographic research. It was launched in 1978 and during its ninety-nine operational days it obtained RADAR imagery of the Earth's surface using an SAR system with a ground RESOLUTION of around 50 m. *See also* HEAT-CAPACITY MAPPING MISSION, LANDSAT, NOAA/TIROS SATELLITES, SKYLAB, SPACE SHUTTLE, SPOT.

seascarp A submarine escarpment resulting from a tectonic movement of a FRACTURE ZONE.

season A time period generally based on differences in duration and intensity of solar radiation. The resulting contrasts of daylight hours and temperature differences in middle and high latitudes are used to divide the year into seasons, often linked with the farming calendar, i.e. the life cycle of cultivated crops. Because of the small seasonal temperature changes and the fairly constant daylight hours of low latitudes, however, the equatorial seasons are usually defined not by temperature but by rainfall regimes (MONSOON). In the N hemisphere four seasons are recognized, of different durations according to the latitude: spring – the sowing season; summer – the growing season; autumn (fall) – the harvesting season; winter – the dormant season. An astronomical definition of the seasons (in the N hemisphere) gives a precision which is not found in nature: SPRING – 21 March to 21 June; SUMMER – 22 June to 23 September; AUTUMN (fall) – 23 September to 22 December; WINTER – 22 December to 21 March. In very high latitudes (the polar regions) the spring and autumn seasons are barely noticeable, so that a two-season year (winter and summer) prevails. [*73*]

sea stack STACK.

seat-earth A name given to the bed of FIRE CLAY beneath a coal seam. It is a lithified PALAEOSOL in which the coal forests formerly grew. [*224*]

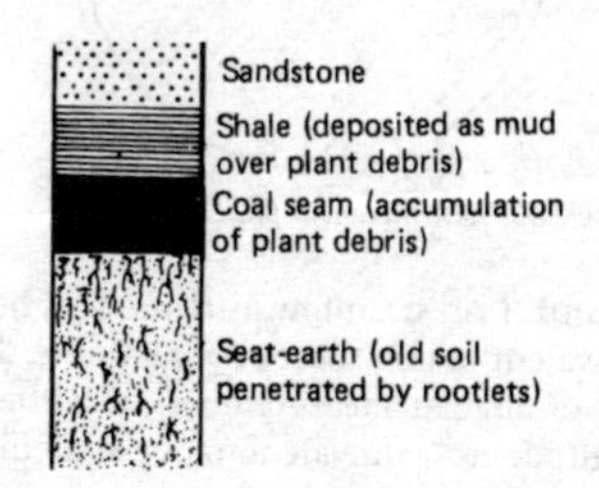

Figure 224 *Seat-earth*

sea valley A linear depression on the sea-floor with a broader cross-section and gentler slopes than a SUBMARINE CANYON. It crosses the CONTINENTAL SHELF as an extension of an estuary, or as the seaward portion of a drowned valley, and may be kept open by submarine currents or tidal scour.

sebkha (sabkha) An Arabic term referring to the floor of a closed depression in an arid environment, characterized by the presence of salt deposits and the absence of vegetation. It can be formed either as a DEFLATION hollow where the WATER-TABLE is close to the surface, or as a coastal lagoon on the fringes of a desert, where EVAPORITES are usually found.

secant conical projections **1** In general, CONICAL PROJECTIONS with two standard parallels which have the geometric interpretation of the conical surface intersecting the surface of the GENERATING GLOBE at those two parallels. **2** In the strictest sense the true secant conical projection must have its two standard parallels separated by the secant distance, not the arcuate distance. [*225*]

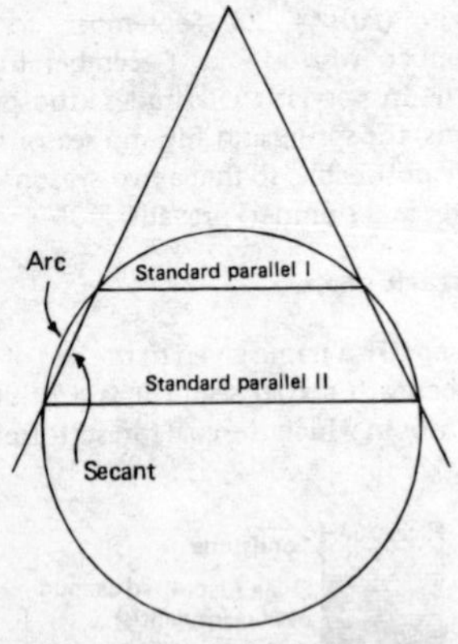

Figure 225 *Secant conical projection*

second **1** An SI unit measurement of time equivalent to one-sixtieth of a minute. **2** A unit of angular measurement, of latitude, longitude etc., equivalent to one-sixtieth of a minute of arc. *See also* MINUTE.

secondary **1** The term originally given to the second of the two eras of geological time, i.e. following the PRIMARY and preceding the TERTIARY. This usage has now been replaced by the MESOZOIC. **2** A general term given to rocks or minerals formed as a result of alteration of existing minerals, or from material derived from the disintegration of rocks (CLASTIC SEDIMENTS).

secondary depression An area of low pressure, smaller than the primary DEPRESSION on the margin of which it forms prior to travelling around its perimeter (anticlockwise in the N hemisphere; clockwise in the S hemisphere). A secondary depression may start life, often at a FRONT, simply as a bulge in the isobars, but it may develop into a distinct pressure system in its own right, with closed isobars and with a central pressure sometimes lower than that of the associated primary. In the latter case the secondary depression may be associated with stronger winds and heavier rainfall than the primary and ultimately it may incorporate the original depression.

secondary response A change in a SYSTEM which is not immediately evident owing either to a pronounced operational time lag or to the very small magnitude of the change.

secondary succession The ecological succession that begins on an area that has previously been occupied by a community of organisms but one which has been destroyed; e.g. by burning.

secondary (s) wave SEISMIC WAVES.

section **1** A vertical cut through rock, soil or landform to illustrate the disposition and character of the subsurface material, e.g. a *geological section* shows the details of the rock strata. A section may be artificial or natural, as in a sea cliff. **2** The graphic representation of the vertical cut in the form of a generalized drawing (BLOCK DIAGRAM) or more accurately with little or no vertical exaggeration. **3** A thin slice of a material prepared for examination under a microscope (THIN SECTION). **4** A faunal or floral division of a GENUS.

secular trend **1** In general, a consistent tendency to change in a particular direction as a result of long-term forces. **2** In climatology, a slow change of climatic character over a long period of time, generally associated with climatic variation (CLIMATIC CHANGE). The trend may indicate a persistent tendency for the mean value of any climatological ELEMENT to increase or decrease.

secular variation Changes over a lengthy time period of the properties of the Earth's MAGNETIC FIELD.

sedentary soil RESIDUAL SOIL.

sediment Solid particles and grains of rock material that have been transported and deposited. In the strictest sense a sediment is the solid material which settles after being suspended in a liquid, but the term has been extended to include all detrital material deposited by fluvial, marine, glacial and aeolian agencies in the process of *sedimentation*. It is usually reserved for unconsolidated material. *See also* SEDIMENTARY ROCKS, TILL.

sedimentary iron ores The main source of the world's iron. They are formed from iron-rich fluids in either a fresh-water environment (BLACKBAND IRONSTONE, BOG ORE, CLAYBAND), in a marine environment (OOLITIC LIMESTONE), or as a residual deposit from the weathering of rocks (LATERITE).

sedimentary rocks Rocks formed by the accumulation of material derived from pre-existing rocks or from organic sources. They are deposited in a layered sequence and may be consolidated by DIAGENESIS or unconsolidated, although most geologists prefer to use the term SEDIMENT for material which has not been lithified. They are usually classified, according to their mode of formation, into: **1** CLASTIC SEDIMENTS, which are mechanically formed (*see also* ARENACEOUS, ARGILLACEOUS, RUDACEOUS); **2** ORGANIC DEPOSITS (*see also* CALCAREOUS ROCKS, CARBONACEOUS, DIATOMACEOUS EARTH, IRONSTONE, PHOSPHATE); **3** rocks formed by chemical processes (*see also* DOLOMITE, EVAPORITE, FLINT, GYPSUM, HAEMATITE, LIMONITE, ROCK SALT). Alternatively, sedimentary rocks may be classified on a geochemical basis into: *resistates* (the arenaceous and rudaceous rocks); *hydrolysates* (the argillaceous rocks); *oxidates* (the sedimentary iron and manganese ores); *reduzates* (the sedimentary sulphides, coal and petroleum); *precipitates* (the chemically formed limestones); *evaporates* (the evaporites).

sediment discharge ratio An expression used in hydrology, referring to the ratio between the total discharge of a stream and the discharge of the transported sediment. It is expressed by the SEDIMENT-RATING CURVE. [*226*]

sedimentology The scientific study of SEDIMENTS, SEDIMENTARY ROCKS and of the process by which they were formed.

sediment-rating curve The empirical expression of the relationship between stream discharge and stream load at a given point. It is shown as: $T = Qc = KQ^n$, where: T is the rate of transport, Q is the discharge, c is the mean sediment concentration, and K and n are constants. The sediment load (a combination of transport rate and mean sediment concentration) is shown to increase in relation to the increasing discharge [*226*], which would take place after a thunderstorm, for example.

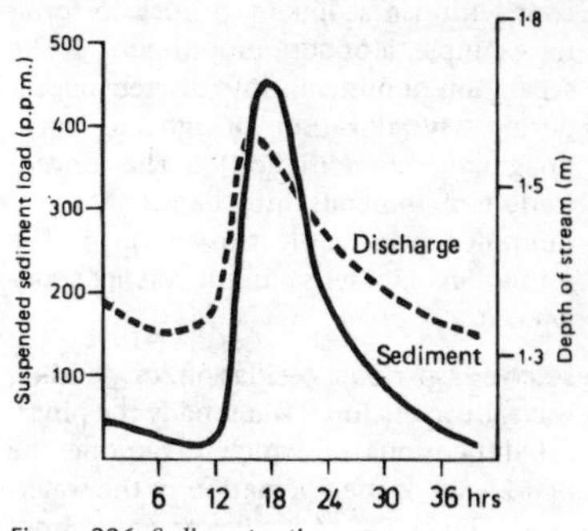

Figure 226 *Sediment rating curve*

sediment yield The mean sediment load carried by a stream, giving some measure of the rate of erosion in a drainage basin in addition to the CAPACITY of the stream itself. The sediment yield comprises material transported both as BED LOAD and SUSPENDED LOAD and is expressed as weight per unit area.

seeding of clouds CLOUD-SEEDING.

seepage **1** The gradual soaking away of surface water into the soil. **2** The slow emission of groundwater at the surface when the rock structures (joints, bedding planes) permit, but with insufficient volume for it to constitute a SPRING. The area on a hill slope where seepage is concentrated is sometimes referred to as the *seepage zone*. When seepage is concentrated into a subsurface channel (PIPE) it is known as *tunnelling* or *piping*. The underground course is marked by waterlogged soils and perhaps a line of HYDROPHYTES (PERCOLINE). **3** The oozing out of mineral oil at the surface.

seepage step A micro-SCARP, up to 1 m in height, formed in the regolith of the SEEPAGE zone. It may run parallel with the contours, be concave downslope or crescent-shaped, but the step moves progressively upslope as surface regolith is slowly removed.

segetal vegetation A type of semi-natural vegetation growing where there has been human interference with the natural growth. *See also* CULTURAL VEGETATION.

segregation **1** The concentration of minerals within a sedimentary rock to form, for example, a DOGGER or a NODULE. **2** The separation of minerals into discrete masses during crystallization of igneous rocks (magmatic differentiation). **3** The concentration of minerals into bands in metamorphic rocks (DIFFERENTIATION). **4** The formation of ice lenses in the soil (ICE SEGREGATION).

seiche A periodic oscillation, or standing wave, in an enclosed water body the physical dimensions of which determine the periodicity of the fluctuation of the water-level. The period of oscillation varies from a few minutes to several hours and is set up by rapid changes in wind direction, in atmospheric pressure, or occasionally by an EARTHQUAKE.

seif-dune An Arabic term meaning *sword-dune* and adopted to describe a knife-edged ridge of sand or *longitudinal-dune*. Its axis lies parallel with the direction of the prevailing wind and may extend for many kilometres. Groups of seif-dunes form long ridges with intervening wind-scoured troughs and dune chains often extend over 100 km in length and rise to heights of 200 m. There is no agreement over their formation, some writers believing that they grow simply as a downward extension of a sand drift behind an obstacle, others that they represent the coalescence of the 'horns' of partly destroyed BARCHANS which have suffered a blow-out. Their spacing may be due to VORTEX FLOW [*273*].

seismic focus, seismic origin FOCUS.

seismicity The study of the frequency and INTENSITY OF AN EARTHQUAKE in an area. *See also* SEISMOLOGY.

seismic survey The scientific investigation of subterranean rock structures by measuring ground movements with such devices as a SEISMOGRAPH. The techniques are derived from the study of EARTHQUAKES, and include those of *seismic reflection* and *seismic refraction*. The former determines the rock structure by measuring the time interval which elapses between the generation of a surface pulse and its return to the surface after it has been reflected from a seismic discontinuity within the crustal rocks. The latter technique determines the structure by measuring the shortest-time travel path from a seismic source to a set of receivers (*seismometers* or *geophones*) distributed around the rock structure being investigated. Data from seismic surveys are used in conjunction with those from GRAVIMETRIC SURVEYS and MAGNETIC SURVEYS to construct a complete picture of underground rock structures that may yield petroleum, minerals and ores of economic value.

seismic waves **1** Earthquake shock waves generated from the FOCUS within the Earth's crust. Three major divisions have been recognized: (a) *Primary* (*p*) *waves*, of short wavelength and high frequency, that are longitudinal waves which travel not only through the solid crust and mantle but also through the liquid part of the core of the Earth. (b) *Secondary* (*s*) *waves*, of short wavelength and high frequency, that are transverse waves which travel through all the

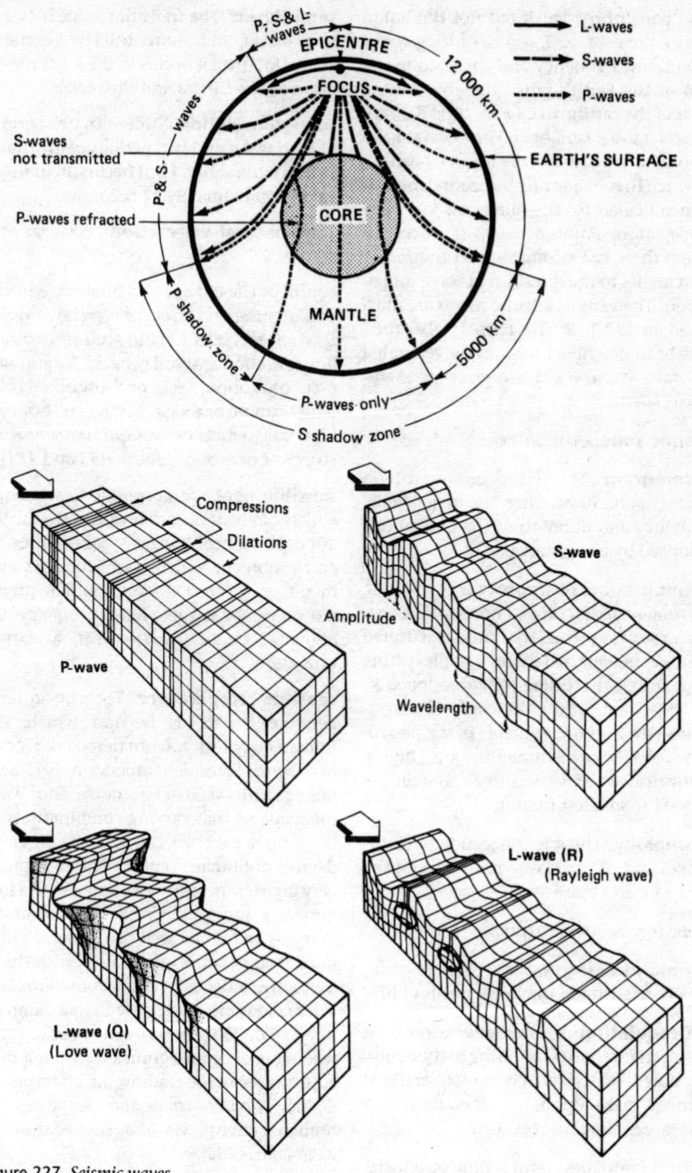

Figure 227 *Seismic waves*

solid parts of the Earth but not the liquid part of the core. (c) *L-waves*, of long wavelength, low frequency and confined to the skin of the Earth's crust, thereby causing most of the earthquake's structural damage (MERCALLI SCALE, MODIFIED). The *L-waves* have been subdivided into: *Love waves* (*Q-waves*), in which there is a strong horizontal motion perpendicular to the direction of wave propagation; *Rayleigh waves* (*R-waves*) in which there is a strong vertical motion at right angles to the direction of wave propagation. The various seismic waves are illustrated in [*227*]. **2** The term is also used loosely to describe a seismically generated sea-wave (TSUNAMI). *See also* CORE, EARTHQUAKE, MANTLE.

seismic zone EARTHQUAKE.

seismogram The record produced by a seismograph, illustrating the magnitude, frequency and duration of the oscillations produced by an earthquake.

seismography, seismometer A scientific instrument designed to record SEISMIC WAVES. The prototypes were very unsophisticated Chinese models based on simple tilting mechanisms, but these were developed in the 19th cent. into more complex instruments based on the principle of the pendulum. Modern instruments are highly sophisticated and sensitive enough to record the smallest tremors.

seismology The scientific study of earthquakes and of the elastic properties of the Earth. *See also* SEISMICITY.

selection NATURAL SELECTION.

self-maintaining system A living system, usually referring to the lower forms of life.

self-regulation An alternative expression for negative FEEDBACK, referring to the damping down of the effects which external changes make upon the operation of a system. *See also* HOMOEOSTASIS.

selva A Portuguese term for the EQUATORIAL RAIN-FOREST of the Amazon basin, now also applied to similar types of vegetation elsewhere in the world.

semi-desert The transition zone between true DESERT and more thickly vegetated zones. In Africa it occurs in the SAHEL region between the Sahara and the SAVANNA.

semi-diurnal tide A tide with two periods of high water and two periods of low water within a LUNAR DAY. Each occurs at an interval of approximately 12 h 25 min.

semi-natural vegetation CULTURAL VEGETATION.

senility The old age and final stage in the cyclic concept (GEOGRAPHICAL CYCLE) of W. M. Davis. It can refer to land sculpture, shoreline evolution and soil genesis. *See also* ARID CYCLE OF EROSION, CYCLE OF EROSION, CYCLE OF UNDERGROUND DRAINAGE, GLACIAL CYCLE OF EROSION, MARINE CYCLE OF EROSION, MATURITY, NORMAL CYCLE OF EROSION, YOUTH. [*53* and *181*]

sensible heat Atmospheric heat energy which can be transferred within the Earth/atmosphere heat-budget system either by CONVECTION, in which warm air rises and mixes, or by direct CONDUCTION. The physicist's term for sensible heat is *enthalpy*. *See also* HEAT BALANCE, LATENT HEAT, RADIATION BALANCE.

sensible temperature The critical temperature thresholds beyond which the human body feels too hot or too cold (COMFORT ZONE). Sensible temperature is closely linked with RELATIVE HUMIDITY and WIND intensity, so that varying combinations of these three variables can produce different degrees of human comfort. For example, a very high temperature associated with a low humidity and a stiff breeze will not prove as trying to human well-being as a slightly lower temperature associated with high humidity under calm conditions. Similarly a dry cold is more tolerable than a damp or 'raw' cold. Thus, a WET-BULB THERMOMETER reading is of more significance than a DRY-BULB THERMOMETER reading, in attempts to define *effective temperature* relating to human activity. *See also* BIOCLIMATOLOGY, WIND-CHILL. [*280*]

sensor A REMOTE SENSING term referring to any device that is sensitive to ELECTROMAGNETIC RADIATION and can convert its measurements into a form suitable for input into an information-gathering system. Some examples of sensors are: aerial photographic cameras, MULTISPECTRAL SCANNERS, RADAR, THERMAL LINE-SCANNERS and RADIOMETERS.

sequence A continuity of events, following one after another, i.e. in succession. Some geomorphologists prefer this term to that of CYCLE.

sequential landforms The landforms created by modifications of the INITIAL LANDFORM, no matter what the degree of change that has been wrought by DENUDATION. Some geomorphologists prefer to use the expression in place of the more controversial terms of YOUTH, MATURITY and SENILITY of W. M. Davis's GEOGRAPHICAL CYCLE.

sérac A French term for a pinnacle of glacier ice formed where crevasses intersect at an ice-fall.

seral community One of the temporary development stages in the sequence of plant colonization leading up to a vegetation CLIMAX.

serclimax The terminal phase in a SERE in which all further progress towards a true CLIMAX is inhibited by the constant repetition of a natural intervention, e.g. regular flooding or wetting by salt spray.

sere A sequential development of a plant community or group of plant communities on the same site over a period of time. It develops from a PRISERE into one of a HYDROSERE, PSAMMOSERE, XEROSERE. *See also* LITHOSERE, SUBSERE. [*108*]

series **1** A stratigraphic unit (CHRONOSTRATIGRAPHY) which is part of a SYSTEM and is equivalent to an EPOCH in the timescale. It comprises a number of STAGES. **2** A SOIL SERIES.

serir An Arabic term referring to the stony and gravel-covered deserts of NE Africa. It is equivalent to the REG of NW Africa.

serozem SIEROZEM.

serpentine A rock-forming mineral with a complex chemical composition. It forms from the breakdown of OLIVINE and PYROXENES when basic and ultrabasic rocks are altered by metamorphism. It is mainly of greenish or whitish colour but some serpentine rocks (termed *serpentinites*) also have streaks of red. The variegated colouring, resembling a serpent's skin, produces an attractive stone, often quarried for ornamental purposes and sometimes referred to, wrongly, as marble.

servomechanism A device which automatically monitors the progressive operation of a SYSTEM, correcting any deviation by means of FEEDBACK and thereby maintaining a satisfactory performance within the system, e.g. a thermostat.

set A group of similar objects, elements or members that can be defined according to a definite parameter.

settling velocity The uniform speed attained by a particle as it descends through a fluid, e.g. a dust particle through the atmosphere or a particle of clay through water. *See also* STOKES'S LAW.

Seventh Approximation A soil classification compiled by the US Department of Agriculture in 1960, based on the presence or absence of a specific diagnostic horizon, together with certain other criteria such as soil chemistry, physical properties and morphological features. It is not easily applicable to Britain, where a different classification has been adopted (SOIL CLASSIFICATION OF ENGLAND AND WALES). The US classification includes the following ten groups: ALFISOL; ARIDOSOL; ENTISOL; HISTOSOL; INCEPTISOL; MOLLISOL; OXISOL; SPODOSOL; ULTISOL; and VERTISOL.

sextant An instrument used for measuring the angular distance between two points or objects. It is widely used in marine navigation.

shade temperature The temperature normally taken in meteorological observations, i.e. that obtained in a STEVENSON SCREEN, away from direct insolation or radiation from nearby objects.

shake-hole A local term in England for a limestone depression or DOLINE, formed as the result of surface collapse.

shake-wave A transverse wave or s-wave (SEISMIC WAVES).

shale A well-laminated ARGILLACEOUS rock that splits easily along its bedding planes because of its FISSILITY. It is more fissile than a MUDSTONE, which it otherwise resembles. Some shales are oil-bearing (OIL-SHALE).

shallow focus A description of an EARTHQUAKE the FOCUS of which is less than 65 km from the surface. *See also* DEEP FOCUS, EPICENTRE.

sharp-crested weir V-NOTCH WEIR.

shatter belt A narrow tract of broken crustal rocks (BRECCIA) where one or more faults, often with the development of secondary faulting, have fractured the rocks to such a degree that it has become a zone of relative weakness, less resistant to denudation than the neighbouring rocks. Rivers sometimes pick out the shatter belts, leading to a marked linearity of their valleys in this particular tract.

shearbox An apparatus designed to determine the SHEAR STRENGTH of a soil or unconsolidated rock. It consists of a split metal box, into which the sample is placed, with a load-measuring device (proving ring) on the upper part and a pushing mechanism driven by a motorized ram in the lower part. Thus, while movement of the top half of the sample is resisted by the proving ring, the lower half is pushed forward until shearing occurs [*228*]. The amount of STRESS and the shear strength at FAILURE can be measured. The shearing takes place along a SHEAR PLANE and is expressed by COULOMB'S FAILURE LAW. *See also* TRIAXIAL TEST.

shear fault A fault produced by tangential STRESS which shears one part of a rock past its adjoining part, along a SHEAR PLANE.

shearing The action caused by the application of tangential STRESS to a solid material. It results in the formation of a SHEAR FAULT and a SHEAR PLANE and can also produce slip CLEAVAGE and IMBRICATE STRUCTURES. Although the volume of the material remains the same the two adjacent parts slide past each other, often causing crushing along the *shear plane* (GOUGE).

shear moraine An accumulation of debris on a glacier surface, marking the lines where SHEAR PLANES within the ice (created by COMPRESSING FLOW in the glacier) carry morainic material from the glacier bed diagonally upwards through the ice to the surface. The debris is disgorged as a linear transverse ridge which usually becomes ice-cored (ICE-CORED MORAINE) as the surrounding unprotected glacier surface is lowered by ablation. *See also* THULE–BAFFIN MORAINES.

shear plane The surface along which SHEARING occurs during the application of tangential stress to a material. A shear plane may not be a truly planar surface, particularly in a soil or rock aggregate where the shear surface may follow outlines of the individual particles. It is impossible for a perfectly planar shear plane to pass through granular materials unless grain splitting occurs, owing to the interlocking of particles. Shear planes can also be found in glaciers, where they are the result of COMPRESSING FLOW.

shear strength The maximum RESISTANCE of a material to applied STRESS. The two major

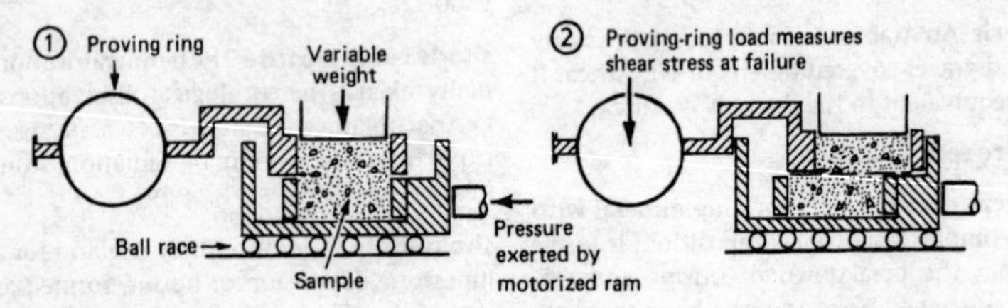

Figure 228 *Shearbox*

sources of resistance are COHESION and FRICTION, but when the shearing stress exceeds the shear strength of a material FAILURE will occur. In a freely drained material the shear strength will be high and is expressed as $s = c + \sigma \tan \theta$, where: s is shear strength, c is cohesion, and σ is normal stress on the shear plane. When the water-table rises, however, the normal stress is altered by changes in PORE-WATER PRESSURE which also changes the values of cohesion and of the friction angles, thereby modifying the shear strength equation, which is now expressed by the MOHR–COULOMB EQUATION. When the water-table is at its highest, shear strength is at its maximum, so that LANDSLIDES often occur after periods of heavy rainfall. One of the ways in which the shear strength of a material may be measured is by use of a RESIDUAL STRENGTH SHEARBOX. [*228*]

shear stresses The two perpendicular stresses (STRESS) applied to the surface of a body and which produce angular deformation (shear strain) in that body. They are perpendicular to the direction of NORMAL STRESS. *See also* STRAIN-RATE. [*229*]

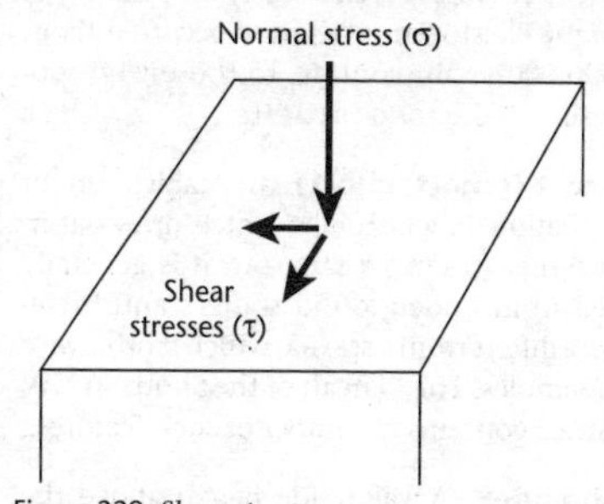

Figure 229 *Shear stresses*

sheet erosion, sheet wash The widespread removal of surface debris (SOIL EROSION) by OVERLAND FLOW on low-gradient slopes, generally at relatively slow speeds and over long periods. It comprises the two processes of raindrop impact and OVERLAND FLOW, with sheet wash carrying away the particles dislodged by the falling rain. When continuously heavy rainfall occurs sheet wash becomes sheet flood, and soil erosion becomes extremely severe as sheet erosion is accelerated.

sheet flood SHEET EROSION.

sheet flow OVERLAND FLOW.

sheeting, sheet jointing The terms given to the process of rock expansion due to DILATATION. This results in the splitting-off of sheets of rock along the upper surface of the outcrop once it has been exposed by denudation and the overlying material removed, thereby leading to pressure release (UNLOADING). These terms are generally reserved for pressure release in massive igneous rocks and should not be confused with the weathering process of EXFOLIATION. *See also* PSEUDO-BEDDING.

sheet lightning The reflection of forked LIGHTNING on surrounding clouds, where the actual flashes are obscured by the clouds themselves, but their illumination is diffused. *See also* THUNDERSTORM.

sheet wash SHEET EROSION.

shelf **1** In general, a ledge or projecting layer. **2** More specifically, the CONTINENTAL SHELF.

shelf edge, shelf break The change of slope occurring between the CONTINENTAL SHELF and the CONTINENTAL SLOPE. [*193*]

shelf ice ICE-SHELF.

shell sand A beach deposit containing a high percentage of shell fragments and therefore highly calcareous rather than siliceous. It is known as MACHAIR in W Scotland, where it is artificially mixed with seaweed and/or peat to form plots of relatively fertile soil in a region of highly acid and leached soils. It is most commonly found on tropical islands.

shelter belt WIND-BREAK.

shield A large stable block of the Earth's crust which has been unaffected by mountain-building for a lengthy period of geological time (CRATON). Shields, which are generally made up of rocks of Pre-Cambrian age, form the central nucleus of each of the continents, e.g. the Laurentian shield of N America, the Baltic shield of N Europe.

shield inselberg BORNHARDT.

shield volcano A broad, dome-shaped volcanic cone with very gentle slopes and extensive circumference owing to the fluid nature of the basic lava from which it is formed. The world's largest volcano, Mauna Loa, Hawaii, is a shield volcano, rising no less than 9,750 m (32,000 ft) from the seafloor with slope gradients between 2° and 10°. Other examples occur in Iceland. [*271*]

shift The maximum relative displacement of rocks on opposing sides of a FAULT away from the fault itself. The net shift is often greater than the amount of net SLIP along the fault plane because of drag and bending of the strata at the actual faultline. The shift may be in the direction of the dip or the strike.

shillow A layer of loosened chippings or flakes of Carboniferous limestone that have been peeled off by weathering from the top surface of a CLINT. Each of these thin flakes has a smooth convex upper surface and a rough, faintly concave lower surface. The flakes often fall into a GRIKE but if the weathering is rapid they may accumulate on the surface of the clint. Shillow-covered LIMESTONE PAVEMENTS can be seen at Helsington Burrows near Kendal, N England. In the Alps the flakes are termed *karrenstein*.

shingle beach A shoreline deposit of waterworn pebbles and cobbles, coarser than GRAVEL.

shoal **1** A bank of coastal sediment that rises almost to the surface of the sea, thereby creating a navigation hazard. **2** The term is also used as a verb to indicate a gradual shallowing of the sea.

shoaling coefficient The ratio between the height of a wave in any depth of water and its height in deep water with the effect of wave REFRACTION eliminated.

shore BACKSHORE, FORESHORE.

shore-face terrace The name given to the accumulation of sedimentary material immediately beyond the outer edge of a marine ABRASION PLATFORM. [*144*]

shoreline The line of contact between a land surface and a lake or ocean surface. It differs from the term COASTLINE, which is not normally used to describe a lake shore. In the marine charts of the US Coast and Geodetic Survey the shoreline is equivalent to the average line of high tides.

shoreline classification A division suggested in 1919 by D. W. Johnson, who recognized the four main categories: **1** EMERGENT SHORELINE; **2** SUBMERGENT SHORELINE; **3** NEUTRAL SHORELINE; and **4** *compound shoreline*, which exhibits characteristics of at least two of the other categories largely as a result of differential earth movements along the same stretch of coastline (COMPOSITE COAST).

shore platform ABRASION PLATFORM.

short time-scale A time-scale relating to the length of the PLEISTOCENE glaciation. It is based not on radiometric dating but on alleged climatic fluctuations caused by periodic perturbations of the Earth's orbit. Some theorists suggest, therefore, that the onset of the Pleistocene glaciation occurred about 600,000 BP, in contrast to those who subscribe to the LONG TIMESCALE.

shott (schott, chott) An Arabic term for a shallow brackish lake which dries out in summer to form a SALT FLAT. It is generally fed from underground sources and therefore differs from a SEBKHA, which it otherwise resembles. The Plateau of the Shotts in NW Africa contains a number of such features.

shoulder A valleyside bench above the level of glacial overdeepening (ALP).

shower A short-lived outbreak of rain or hail, varying in duration and intensity. It is often associated with a CUMULONIMBUS cloud form, which is sometimes referred to as a *shower cloud*. Showers can occur at any time of year but are most frequent in the British Isles during spring and summer when CONVECTION leads to maximum vertical growth of cumulonimbus clouds in unstable air. In some equatorial regions convection clouds produce late-afternoon showers on most days of the year.

Shurin season A Japanese term for the secondary period of precipitation maximum during September and early October in Japan. It occurs when low-pressure systems and TYPHOONS from the Pacific Ocean are able to swing northwards over Japan owing to the eastward contraction of the Pacific subtropical anticyclone. *See also* BAI-U SEASON.

sial That part of the Earth's CRUST which is composed of granitic material rich in *si*lica (Si) and *al*umina (Al). It has a lower density (2.65 to 2.7) than the simatic rocks (SIMA). The sial forms the CONTINENTAL CRUST. [*121*] *See also* ISOSTASY.

siallitic soil A soil containing a medium to high ratio of silica to aluminium. Since chemical weathering produces a greater proportion of aluminium, *siallitic soils* are uncommon in zones of tropical weathering but are typical of the temperate zone.

Siberian (cold) anticyclone A persistent area of high pressure during winter in N Asia, the intensity of which is amplified by the extreme measure of *continentality* (CONTINENTAL CLIMATE) and by the extensive cover of winter snow. It is one of the sources of polar continental AIR MASSES and is characterized by extremely low winter temperatures. When occasional cyclonic storms invade Siberia a strong north-easterly or northerly wind in the rear of the depression brings blizzards (BURAN) before the high pressure re-establishes itself. *See also* ANTICYCLONE.

Sicilian **1** One of the highest of the raised shorelines created by a former sea-level of the Mediterranean during the PLEISTOCENE. It stands at elevations of 90–100 m above present sea-level and the date of its formation is of early Pleistocene age. The name was introduced by C. Depéret in 1918. **2** A name given by M. Gignoux in 1913 to an assemblage of fossil fauna (the *Sicilian fauna*) associated with both the Sicilian and the MILAZZIAN raised shorelines of the Mediterranean. In addition to its Mediterranean species it contains other species now restricted to more northerly (cooler) seas (e.g. *Cyprina islandica*) and a number of PLIOCENE survivals. *See also* MONASTIRIAN, TYRRHENIAN.

sidereal day The time interval equivalent to one complete rotation of the Earth in relation to the stars. Its length of 23 h 56 min 4.09 sec is almost 4 min shorter than that of a mean SOLAR DAY (MEAN SOLAR TIME).

sidereal month MOON.

sidereal year The time interval during which the Earth makes one complete revolution around the Sun with reference to the stars, i.e. 365 days 6 hr 9 min 9.54 sec (365.2564 mean SOLAR DAYS).

siderite **1** An iron ore composed of ferrous carbonate ($FeCO_3$). **2** A METEORITE composed entirely of metal (nickel/iron).

side slip The sliding action of a glacier margin past a stationary moraine or its valley side.

sierozem (serozem) A Russian term for a grey earth or grey desert soil, occurring in semi-arid zones. It is characterized by a lack of leaching, a low organic content, and an accumulation of lime at the level of the interface between the B- and C-horizons. The lime may sometimes extend upwards to the surface owing to CAPILLARITY. The sierozem is included in the ARIDISOL order in the SEVENTH APPROXIMATION soil classification.

sierra A Spanish term (literally = saw) applied to a line of mountains with jagged crests. It is found as a place-name in Latin America, Spain, and the SW USA.

sieve map One of a series of transparent map sheets containing a particular type of information. If the maps are superimposed certain data will be '*sieved* out' to clarify distributions.

sight-rule ALIDADE.

signal **1** In general, an event or phenomenon that conveys information from one place to another. **2** A REMOTE SENSING term referring to the effect of a pulse of ELECTROMAGNETIC RADIATION when conveyed over a communication path or system. In practice a signal is generated by a SENSOR in proportion to the amount of electromagnetic radiation received. The signal is then converted to a form suitable for recording or for transmission to a PREPROCESSING system. For example, in the LANDSAT satellite the MULTISPECTRAL SCANNER records a pulse of electromagnetic radiation for each PIXEL area on the Earth's surface, prior to its conversion to an intensity-level value for transmission back to Earth.

significance, statistical A statistical calculation to demonstrate to what degree a hypothesis is acceptable. A hypothesis will be acceptable if the calculated PROBABILITY exceeds a given value $\propto$, termed the *level of significance*. If $\propto$ is <0.05 (95% limit of confidence) the result is said to be significant; if $\propto$ is <0.01 (99% limit of confidence) the result is highly significant. *See also* CONFIDENCE LIMITS.

silcrete A very tough sandstone that has been silicified, perhaps when it formed part of a DURICRUST. *See also* SARSEN.

silica Chemically, silica is silicon dioxide (SiO_2), but the properties of the silica group of minerals are closely allied to the SILICATES. Among the several varieties comprising the silica group are QUARTZ, chalcedony and opal. *See also* CHERT, FLINT, JASPER.

silicates The most important constituent compounds in the CRUST of the Earth, of which they comprise (including the SILICA group of minerals) some 95%. *See also* AMPHIBOLE, CHLORITE GROUP, CLAY MINERALS, FELDSPAR, OLIVINE, PYROXENES, QUARTZ.

siliceous ooze A fine-grained OOZE containing a high percentage of siliceous skeletal material.

siliceous rocks Rocks which contain an abundance of SILICA. These include siliceous shales, which are normal SHALES altered by SILICIFICATION, SILICEOUS SINTER (GEYSERITE) and silicified wood.

silicification The process by which SILICA replaces existing structures or minerals within an organic or inorganic body. Most commonly quartz, chalcedony (CHERT, FLINT) or opal fill up pores in non-siliceous sediments or replace the structure of woody material ('fossil wood' or silicified wood). The silica is introduced either by groundwater solutions or from igneous sources.

sill 1 An intrusive body of solidified magma, usually DOLERITE, that has been injected as a near-horizontal sheet between the bedding-planes of the crustal rocks. [*230*] It can therefore be said to be concordant with the structure, in contrast to a DYKE, which is discordant. Well-known British examples include the Great Whin Sill of N England, Salisbury Crags in Edinburgh, Scotland and the Fair Head sill in Co. Antrim, N Ireland. In N America the Palisades which flank the Hudson R. represent a tilted sill exposure. **2** A submarine bar or ridge across the mouth of a FIORD.

silt One of the finest of the CLASTIC SEDIMENTS, with a PARTICLE SIZE between 0.002 mm and 0.06 mm, i.e. coarser than CLAY but finer than SAND. When waterborne sediments are deposited in stream channels, lakes and reservoirs, sometimes threatening navigation or the functioning of machinery, the process is often referred to as *silting*, but this is a loose usage since the sediments may be of various sizes other than true silt. *See also* ARGILLACEOUS.

siltation The accumulation of fine-grained sediments in a body of water, which generally leads to it becoming choked up.

siltstone A fine-grained, consolidated ARGILLACEOUS rock, composed largely of silt.

silty loam A soil which contains between 20 and 50% sand, 72 and 88% silt and less than 30% clay. This mixture produces a friable soil with some permeability and with particles of different sizes. *See also* CLAY LOAM, LOAM, SOIL TEXTURE, MEDIUM-TEXTURED SOIL, SANDY LOAM.

Silurian The third of the periods of the PALAEOZOIC, succeeding the ORDOVICIAN and preceding the DEVONIAN, and lasting from 439 million years ago to about 408 million years ago. It was named after an ancient

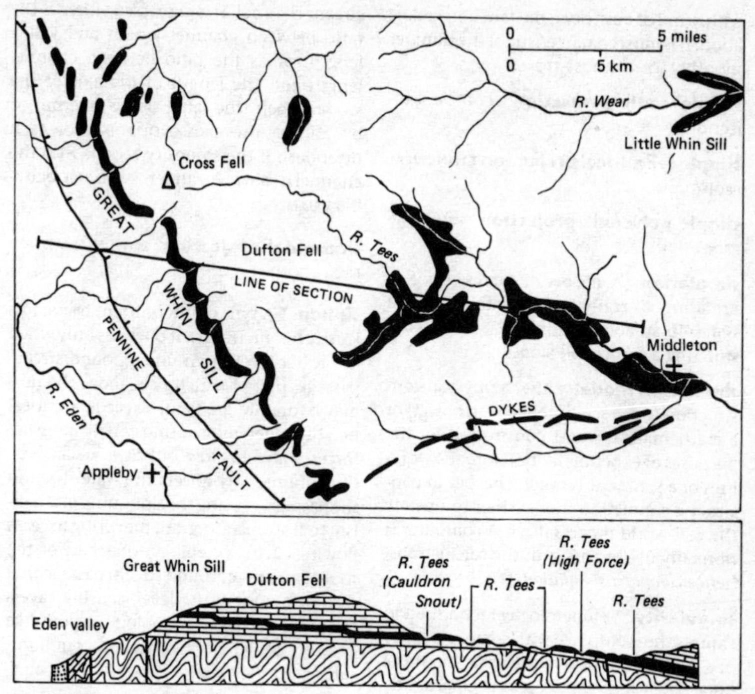

Figure 230 *Great Whin Sill*

tribe which inhabited the Welsh borderland. There is some dispute concerning the upper limit of the Silurian and whether or not it includes the *Downtonian*, which contains fossils of Silurian aspect but with some new forms, akin to the Devonian. Its rocks comprise thick layers of sandstones, shales, mudstones and limestones and are found in Shropshire, central, S and N Wales, Cumbria, the Southern Uplands and eastern Ireland. The Silurian marks the final phase of infilling of the Lower Palaeozoic geosyncline and this is the period in which land plants first appeared. It is subdivided into three series: the Llandovery (Valentian) (the oldest); the Wenlock; and the Ludlow (plus the Downtonian according to some geologists). In continental Europe the Silurian still includes the Ordovician (Lower Silurian) with the term *Gotlandian* being used as an equivalent of the British Silurian.

silver (Ag) A precious metal often found in association with lead and zinc ores, and in veins produced during HYDROTHERMAL ACTIVITY.

sima That part of the Earth's CRUST which is composed of material rich in silica (Si) and magnesium (Mg), with the term being derived from a combination of the two terms (*si*lica and *ma*gnesium). It is denser (2.9 to 3.3) than the SIAL which covers it in some places, but where the sial is missing sima comprises most of the ocean floor, hence its alternative title OCEANIC CRUST. [*121*] *See also* ISOSTASY, PLATE TECTONICS.

simoom An Arabic term for an excessively hot, dust-laden wind (49°–57° C/120°–135° F) in the N Sahara, accompanying an occasional cyclonic storm which invades the desert. It appears largely as a swirling phenomenon with some affinities to a whirlwind or DUST-DEVIL, but is of longer duration. It is most characteristic of the summer months. *See also* DUST-STORM.

simple conic projection CONICAL PROJECTION.

simple cylindrical projection CYLINDRICAL PROJECTION.

simple polyconic projection POLYCONIC PROJECTION.

simulation A means of representing or imitating a real-world situation or real SYSTEM in an abstract form, usually by constructing a SIMULATION MODEL.

simulation model Either a physical construction (e.g. a scale model of an object) or a mathematical MODEL constructed for the purposes of experimentation into the workings of a SYSTEM. It is much cheaper to construct a simulation model than to operate the real-world process itself. A computer is normally used to assist in the calculations, hence *computer simulation*.

singularity A climatological term used to express the recurrence with some regularity of a particular type of weather at about the same date each year. *See also* BUCHAN SPELL.

sinistral-fault FAULT.

sink-hole **1** A funnel-shaped depression, or sink, in the ground surface of a limestone or chalk terrain. It is equivalent to a PONOR of KARST country. It is usually dry or exhibits only minor seepage of surface water and should be distinguished from a SWALLET, which marks the disappearance of a surface stream. A sink-hole is formed by subterranean collapse of a cave system or by surface solution. **2** A deep, broad depression in a SHIELD VOLCANO, caused by downsagging after the withdrawal of subterranean lava.

sinter A deposit of siliceous material around the orifice of a geyser (GEYSERITE) or a hot spring. The expression *calc-sinter* for calcareous deposits around a hot spring should be discouraged and the term TRAVERTINE used instead.

sinuosity ratio A term referring to the amount of meandering exhibited by a stream channel. It is expressed either as the ratio between channel length and valley length, or as the ratio between channel length and the length of the axis of the MEANDER belt (the latter being determined by joining the CROSSOVER POINTS of each meander). If the sinuosity ratio is >1.5 the channel is said to meander. *See also* MEANDERING VALLEY.

sinusoidal projection SANSON–FLAMSTEED PROJECTION.

siphon **1** A type of intermittent SPRING in a karstic terrain (KARST). It operates only when the water-level in an underground stream passage, part of which rises above the normal hydraulic gradient, exceeds a critical height, whereupon atmospheric pressure forces water to flow out at a RESURGENCE. Continuing flow lowers the water-level in the passage, causing the siphoning mechanism to stop operating and the spring to cease flowing. **2** In speleology, that part of the roof of an underground stream passage that dips below the water-level, causing cavers to dive beneath it. **2** In zoology, a spout-like aperture by which a gastropod communicates with the exterior.

sirocco (scirocco) A very hot, dust-laden, south or south-easterly wind blowing from the Sahara in the forefront of advancing depressions moving eastwards in the Mediterranean basin. Although it starts as a dry wind it picks up moisture as it crosses the N African coast and by the time it affects Malta, Sicily and Italy it has become a humid, oppressive wind which causes discomfort. *See also* KHAMSIN.

site MORPHOLOGICAL REGION.

SI units (Système International d'Unités) A *system of units* agreed internationally in 1960 as the rationalized system within which the magnitude of any physical quantity may be expressed. It is a refinement of the MKS (metre–kilogram–second) system and replaces the cgs (centimetre–gram–second) system and the fps (foot–pound–second) system (the Imperial system). There are seven *base units* and two

Figure 231 *SI units*

a. Base and supplementary SI units

Physical quantity	*Name of unit*	*Symbol for unit*
length	metre	m
mass	kilogram(me)	kg
time	second	s
electric current	ampere	A
thermodynamic temperature	kelvin	K
luminous intensity	candela	cd
amount of substance	mole	mol
plane angle*	radian	rad
solid angle*	steradian	sr

*supplementary units.

b. Derived SI units with special names

Physical quantity	*Name of unit*	*Symbol for unit*
frequency	hertz	Hz
energy	joule	J
force	newton	N
power	watt	W
pressure	pascal	Pa
electric charge	coulomb	C
electric potential difference	volt	V
electric resistance	ohm	Ω
electric conductance	siemens	S
electric capacitance	farad	F
magnetic flux	weber	Wb
inductance	henry	H
magnetic flux density (magnetic induction)	tesla	T
luminous flux	lumen	lm
illuminance (illumination)	lux	lx

c. Decimal multiples and submultiples to be used with SI units

Submultiple	*Prefix*	*Symbol*	*Multiple*	*Prefix*	*Symbol*
10^{-1}	deci-	d	10^{1}	deca-	da
10^{-2}	centi-	c	10^{2}	hecto-	h
10^{-3}	milli-	m	10^{3}	kilo-	k
10^{-6}	micro-	μ	10^{6}	mega-	M
10^{-9}	nano-	n	10^{9}	giga-	G
10^{-12}	pico-	p	10^{12}	tera-	T
10^{-15}	femto-	f			
10^{-18}	atto-	a			

supplementary units: *metre*, *kilogram(me)*, *second*, *ampere*, *kelvin*, *candela*, *mole*, together with *radian* and *steradian* [231*a*]. Measurements of all other physical quantities are expressed in *derived units*, which are produced by a combination of two or more *base units*. The fifteen derived units are: *hertz*, *joule*, *newton*, *watt*, *pascal*, *coulomb*, *volt*, *ohm*, *siemens*, *farad*, *weber*, *henry*, *tesla*, *lumen* and *lux* [231*b*]. All SI units are expressed in decimal multiples using a set of standard symbols [231*c*].

sixth-power law COMPETENCE.

skarn A limestone or dolomite which has been altered by thermal METAMORPHISM and in which METASOMATISM has occurred leading to the formation of manganese–silicate minerals alongside the calc–silicate minerals.

skeletal soil A term once used to describe a recently developed soil in which there has been insufficient time for a B-horizon to form (AZONAL SOIL). The term ENTISOL is preferred.

skerries A term derived from the Swedish *skär* and used to describe lines of low, rocky islands just off shore. It is commonly found as a place-name in Scandinavia, Ireland and the Hebrides of Scotland (Gaelic, *sgeir* = *skerry*).

skewness A term which defines the asymmetry of a DISTRIBUTION curve. It is the third of the MOMENT MEASURES which describe the character of a curve.

sky cover The amount of sky obscured by cloud, measured in tenths according to international conventions or in eighths in the UK (OKTA).

Skylab The first US space workshop. This SATELLITE was launched in 1973, carrying a crew of three and an Earth Resources Experimental Package (EREP). The EREP included a six-camera MULTISPECTRAL array, a thirteen-channel MULTISPECTRAL SCANNER and two MICROWAVE systems. The EREP REMOTE SENSING experiments were the first to demonstrate the complementary nature of photography and electronic imaging from space for geographic and geological studies. *See also* HEAT-CAPACITY MAPPING MISSION,, LANDSAT, NOAA/TIROS, SATELLITES, SPACE SHUTTLE, SPOT.

sky light (sky radiation) DIFFUSE RADIATION.

slab failure A type of FAILURE that occurs in hard rocks when weathering opens up the tension JOINTS, thereby altering the rock's EFFECTIVE STRESS. Once a critical PORE-WATER PRESSURE has been exceeded, failure will occur and a slab falls. *See also* TOPPLING FAILURE.

slack **1** A depression among the area of coastal sand-dunes (*dune slack*). **2** An area of greatly reduced velocity in a stream channel, often on the inside of a bend or beyond the stream channel during a stage of flood (OVERBANK FLOW OF A RIVER). It is sometimes referred to as slack water. **3** The state of a tidal current when it is almost motionless (*slack tide*), usually reached in the intermediate period between ebb and flood currents. **4** A local name for a boggy depression in English moorlands.

slaking **1** In general, the crumbling and disintegration of soil or earthy material during a period of drying-out from a state of wetness. **2** More precisely, the term refers to the breakup of dry clay when it is immersed in water.

SLAR (Sideways-looking airborne radar) A remote sensing RADAR system.

slate **1** A fine-grained rock produced from ARGILLACEOUS rocks by METAMORPHISM of a low GRADE. It is characterized by a well-developed CLEAVAGE (*slaty cleavage*), along which it splits quite easily. If it is affected by thermal metamorphism the cleavage may be destroyed, the slate often becomes spotted (*spotted slate*), and HORNFELS may be produced. Slate is found in many of the older rock formations in the British Isles, especially those of Cambrian, Ordovician and Devonian age (CLAY-SLATE, PHYLLITE). It is an excellent roofing material because of its durability. **2** A colloquial term for a thin slab of limestone used for roofing purposes in parts of England, e.g. the Stonesfield 'slates' of the Jurassic rocks in the Cotswolds. **3** A term for any shaley material found in association with coal during a mining operation. NB Only the first of the three usages is strictly correct.

sleet **1** A mixture of RAIN and SNOW or partially melted falling snow (British definition). **2** *Precipitation* consisting of raindrops which have frozen into ice pellets and have partially melted as they continued to fall (US definition).

slickenside A polished and scratched planar surface at a fault plane, produced by

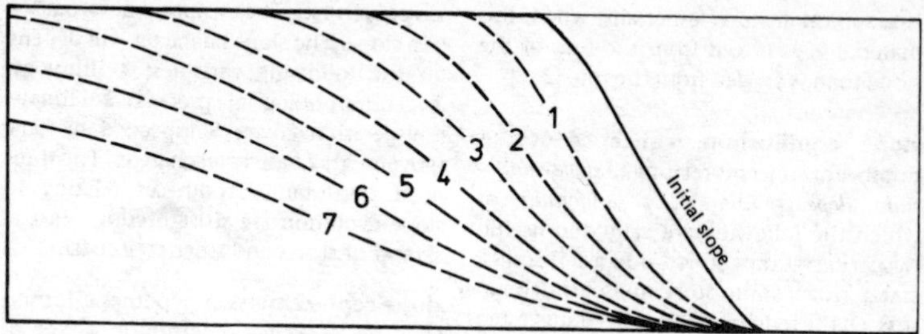

Figure 232 *Seven stages of slope decline*

friction between the opposing sides of a FAULT. Tiny parallel grooves may be formed in some places, but these often stop abruptly in any transverse steps facing away from the direction of movement. The slickenside surface will feel smooth only if one's fingers run over it in the same direction as the former movement; in the opposite direction the surface will feel rough.

slide The downslope movement of rock or superficial material *en masse* in response to gravity (ROCKSLIDE). The material moves as one mass, not as discrete units as in a ROCKFALL. *See also* TRANSLATIONAL SLIDE.

sling psychrometer An alternative name for a WHIRLING PSYCHROMETER.

slip 1 The actual movement of rock bodies in relation to each other on opposite sides of a FAULT. It may be all or only part of the total displacement (SHIFT) and may be measured in the direction of DIP (*dip-slip*) or in the direction of STRIKE (*strike-slip*). **2** A type of LANDSLIDE in which a mass of rock, soil or superficial material moves downhill *en masse* along one or more slip planes. *See also* ROTATIONAL SLIP. **3** The sliding movement of a glacier over its rock floor (BASAL SLIP).

slip-face The leeward slope of a sand DUNE down which sand falls after having been blown up the gentler windward slope. The slip-face is slightly concave in profile, in part owing to the action of a lee EDDY. [*22*]

slip-off slope The gentle slope on the inside bend of a MEANDER, on the opposite side of the channel to a river-cut cliff (BLUFF). The slip-off slope is often marked by a POINT BAR. [*147*]

slip plane ROTATIONAL SLIP, SLIP.

slope An inclined surface, the gradient of which is determined by the amount of its inclination from the horizontal, and the length of which is determined by the inclined distance between its crest and its foot. A slope may be concave (CONCAVITY), straight (RECTILINEAR SLOPE) or convex (CONVEXITY) when seen in profile. *See also* ACCUMULATION SLOPE, BOULDER-CONTROLLED SLOPE, CONSTANT SLOPE, DENUDATION SLOPE, FOOT SLOPE, FREE FACE, NON-RELICT SLOPES, PARALLEL RETREAT OF SLOPES, REPOSE SLOPE, RICHTER DENUDATION SLOPE, STABLE SLOPE, STATIC SLOPE, THRESHOLD SLOPE, TRANSPORTATION SLOPE, WANING SLOPE, WASH SLOPE, WAXING SLOPE, WEATHERING-LIMITED SLOPE.

slope decline A hypothesis referring to a type of slope evolution in which the steepest part of the slope progressively decreases in angle, accompanied by the development of an upper convexity and a lower concavity. The original concept of slope decline was formulated by W. M. Davis as part of his CYCLE OF EROSION, and can be contrasted with the later hypothesis of PARALLEL RETREAT OF SLOPES. Slope decline was once thought to be characteristic only of the NORMAL CYCLE OF EROSION in a humid temperate climate, but it has subsequently been shown that slope decline is also dependent on lithology and on processes operative under different climates; in clays, for example, slope decline is a typical feature, especially where less material is removed from the foot of the slope than is eroded from the top. [*232*]

slope equilibrium, concepts of A number of concepts relating to *slope equilibrium*, *slope grade* and *slope uniformity*, of which the following are valid and useful: PARALLEL RETREAT OF SLOPES, NON-RELICT SLOPES, STABLE SLOPE, STATIC SLOPE, UNIFORM GROUND LOSS. Of the remainder it has been suggested by A. Young that their utility is questionable and that the concepts are not capable of field testing. These include the so-called *profile of equilibrium* (H. Baulig); *external equilibrium* (F. Ahnert); *endogenetic equilibrium* (W. Penck); the concept of a *slope as an open system in a* STEADY STATE (L. von Bertalanffy). Two remaining concepts are valid but are virtually untestable: **1** the concept of *graded slope* (W. M. Davis) refers to one possessing a continuous REGOLITH cover, without rock outcrops; **2** the concept of *equilibrium balance of denudation* (A. Jahn) refers to a condition on a slope, or part of one, in which the regolith cover remains unchanged in thickness with time.

slope genesis The evolution or formation of a slope. The slope character will depend on the following variables: **1** lithology; **2** geomorphological process; **3** climate; **4** vegetation cover; **5** aspect; **6** tectonic movement; **7** base-level changes. The three most important hypotheses relating to slope evolution are SLOPE DECLINE, PARALLEL RETREAT OF SLOPES and SLOPE REPLACEMENT.

slope replacement A hypothesis, formulated by W. Penck, referring to a type of slope evolution in which the maximum angle decreases through replacement from below by gentler slopes, causing the greater part of the profile to become occupied by the concavity which itself may be either segmented or smoothly curved. *See also* PARALLEL RETREAT OF SLOPES, SLOPE DECLINE.

slope retreat PARALLEL RETREAT OF SLOPES, SLOPE DECLINE, SLOPE REPLACEMENT.

slope unit The individual element of a curvilinear (concave or convex) slope or the segment of a RECTILINEAR SLOPE. When examined in detail most slopes are found to be composed of a number of slope units with

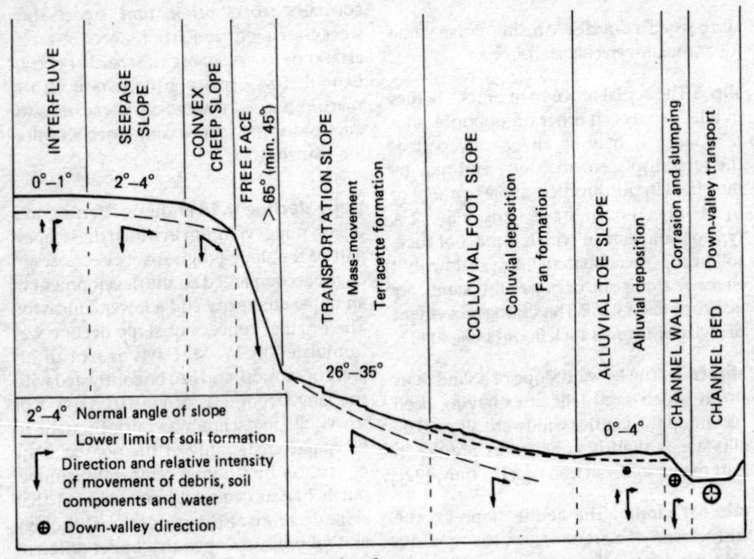

Figure 233 *Slope units on a hypothetical land surface*

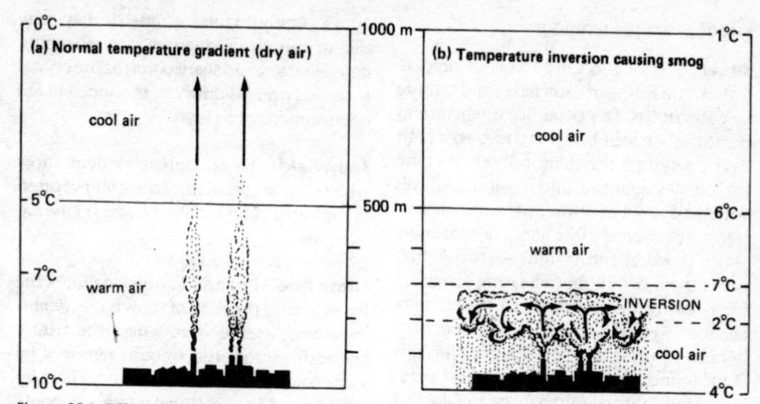

Figure 234 *Effect of vertical temperature gradients on smoke dispersion or smog formation*

different angles, rather than of a single unit of uniform gradient. [*233*]

slope wash SURFACE WASH.

slope wind ANABATIC WIND, KATABATIC WIND.

sludging SOLIFLUCTION.

sluice **1** A channel constructed to take a rapidly flowing stream. **2** A vertically sliding gate used to control the flow of water in a stream channel (*sluice gate*). **3** To wash away earth or silt by directing a high velocity jet of water, especially in mining operations.

slump **1** A mass of surface rocks or superficial material that becomes detached from a hillside along a slip plane (SLIP) and moves downslope, often as a ROTATIONAL SLIP. **2** The term is also used as a verb to indicate the movement itself. *See also* EARTH-FLOW, LANDSLIDE, MASS-MOVEMENT.

slurry A fluid mixture of water and fine solid material which turns into a highly mobile flow of mud.

slush **1** FRAZIL, ICE RIND. **2** A colloquial term to describe a wet, watery mass of snow which clogs roads, pavements, etc. during a period of thaw following a snowfall.

slush-flow A flow of water-saturated snow (*slush*) along a stream-course, caused by a rapid spring thaw in a PERIGLACIAL environment. It is capable of transporting surface soil and often takes on the appearance of a MUD-FLOW.

slush zone The zone in which slushy snow, saturated by summer meltwater, avalanches across the upper basin of a glacier above the FIRN LINE.

small circle A circle on the Earth's surface whose plane does not pass through the centre of the Earth. Of the lines of latitude all but the equator are small circles.

smog Literally, a contraction of '*smoke-fog*' in its original usage but in the USA and elsewhere the term has become synonymous with PHOTOCHEMICAL FOG. The original use of the term (from 1905) was intended to describe the thick RADIATION FOG which blanketed urban and industrial areas under certain weather conditions. Thus, smoke, soot and other AEROSOLS, together with noxious gases such as sulphur dioxide and carbon monoxide were unable to escape from beneath the 'lid' of the INVERSION OF TEMPERATURE, thereby giving a sickly yellow colour and an acrid odour to the smog. Smogs were commonplace in most industrial cities until the smoke-abatement laws of the mid-20th cent. led to cleaner air, but not before there had been several disasters, the greatest of which was the London smog of 1952 when some 4,000 died from bronchial disorders exacerbated by the smog. [*234*]

snout A GLACIER terminus.

snow One of the solid forms of precipitation, consisting of snowflakes and minute spicules of ice. The latter occur when temperatures are well below FREEZING-POINT but as temperatures rise to near 0° C (32° F) the ice spicules aggregate into larger snowflakes which display a multitude of geometric patterns. At extremely low temperatures snow will be powdery and dry but near to freezing-point there will be an increasing tendency for thawing to take place so that it may become wet, structureless and compact. Because of these different densities it is difficult to measure snowfall with great accuracy. Since normal rain-gauges are blocked by snowfall, measurements are generally made by inserting a graduated ruler vertically into a flat layer of undrifted snow. On average 300 mm of newly fallen snow will produce 25 mm of water (precipitation or rainfall equivalent); 100–150 mm of wet, compact snow will produce 25 mm of water; 500–750 mm of dry, powdery snow are needed to produce 25 mm of water. However, SNOW SAMPLERS provide more accurate water-equivalent values for the snow pack. *See also* ACCUMULATION, HAIL, SLUSH, SNOW CRYSTAL.

snow barchan A crescentic dune-form created by wind-driven snow in polar regions, where wind direction is fairly constant. Its mode of formation is identical to that of a sand BARCHAN.

snow crystal The fine, delicate ice spicule of which snowflakes are made. First described by Claus Magnus, a Swede, in about 1555, the exact forms of the snow crystal were not clearly identified until the invention of photography. By 1931 some 6,000 photomicrographs of differently shaped snow crystals had been collected by an American, W. A. Bentley. Crystal shapes vary according to the temperature at which they fall. They range from thin hexagonal needles at temperatures near to freezing (0° to –5° C), through dendritic crystals (–12° to –50° C). The needle crystals grow only in moist air; the dendritic forms in moderately moist air; the columnar prisms only in dry air. The finely textured geometric shapes are due to complex sequences of EVAPORATION, CONDENSATION and SUBLIMATION as the crystal is carried up and down in the supercooled environment of a cloud.

snowfield An area of permanent snow which has accumulated in a hollow, often in a mountainous terrain. *See also* ACCUMULATION, NÉVÉ.

snow-line The altitude which delimits the lower level of permanent snow is the *climatic snow-line*, i.e. the zone where the winter snowfall exceeds the amount removed by summer melting and evaporation. The term is also used in a seasonal sense to denote the lower limit of snow-covered highland during the winter, but where snow cover disappears completely in the summer. Because of slope steepness snow does not accumulate everywhere above the snow-line. The snow-line varies in altitude according to latitude and degree of *continentality* (CONTINENTAL CLIMATE). It is higher away from maritime influences and also in the tropics. On tropical mountains the snow-line may be as high as 5,000 m, but when traced polewards it descends to 2,700 m in the European Alps, to 600 m in S Greenland, and to sea-level near the poles. Locally, variations in snow-line altitude will depend on factors such as differences in precipitation, aspect and shelter. *See also* FIRN-LINE, URAL GLACIER.

snow-patch erosion NIVATION.

snow sampler A set of lightweight jointed tubes for taking *snow samples*, in association with a spring-balance, which reads directly the water equivalent corresponding to a given weight of snow.

soapstone (steatite) A massive impure variety of TALC that lends itself to carving. It is used by Eskimos to produce figurines for the commercial market.

soda lake A lake possessing a high sodium content, characteristic of the rift valleys of East Africa, for example.

soft hail GRAUPEL.

soft water HARDNESS SCALE.

soil The naturally occurring thin layer of unconsolidated material on the Earth's surface that has been influenced by PARENT MATERIAL, climate and relief, in addition to physical, chemical and biological agents (MICRO-ORGANISMS IN THE SOIL) to produce a medium suitable for the growth of land plants. Because these soil-forming factors act over time they will produce soils that may differ from the materials from which they were derived, e.g. they may become severely leached (LEACHING). Soils will exhibit differences in their physical and chemical characteristics (SOIL STRUCTURE, SOIL TEMPERATURE, SOIL TEXTURE, SOIL WATER) as well as in their capability for growing crops (FERTILITY, SOIL). Many of the world's soils have been considerably affected by human activity, especially by cultivation (MAN-MADE SOILS). Classification of soils has changed from one based on the SOIL PROFILE (which depended on SOIL FORMATION factors) to one based on the PEDON (which depends more on SOIL HORIZON arrangement, SOIL STRUCTURE and SOIL TEXTURE) as in the SEVENTH APPROXIMATION. *See also* AZONAL SOIL, CATENA, CHELUVIATION, DECALCIFICATION, ELUVIATION, GLEYING, HUMIFICATION, ILLUVIATION, INTRAZONAL SOILS, LATERITE, LATERIZATION, LESSIVAGE, PEDOLOGY, PODZOL, PODZOLIZATION, SOIL SCIENCE, SOLUM, ZONAL SOILS.

soil acidity ACID SOIL, PH INDEX.

soil association A grouping of soils based on their spatial proximity, usually for the purpose of their depiction on a map, even though they may exhibit different profile characteristics.

soil classification of England and Wales A 1973 soil classification based on SOIL PROFILES which are related to such environmental attributes as terrain, geology, climate and vegetation and are grouped according to properties that affect land-use capability. It is divided into six major groups of mineral soils and one major group of organic (peaty) soils. Each group is further subdivided into 108 subgroups. The major groups are: LITHOMORPHIC SOILS; BROWN SOILS; PODZOLIC SOILS; PELOSOLS; GLEY SOILS; MAN-MADE SOILS; and PEAT SOILS. *See also* SEVENTH APPROXIMATION.

soil colour MUNSELL COLOUR SYSTEM.

soil conservation CONSERVATION.

soil consistency (consistence) CONSISTENCE.

soil creep CREEP.

soil erosion The removal of soil at a greater rate than its replacement by natural agencies. Some soil erosion occurs without the intervention of human activities but the latter often accelerate the natural processes or set them in train, e.g. vegetation clearance, over-grazing, some land-drainage schemes. Removal may be in the form of SHEET WASH, RILL EROSION, GULLYING or DEFLATION. It may be checked by such management practices as contour ploughing and crop rotation. Human-induced erosion should be termed *accelerated erosion*.

soil family SOIL TAXONOMY.

soil fertility FERTILITY, SOIL.

soil formation The combination of natural processes by which soils are formed. It is also known as *pedogenesis*. The most important soil-forming factors are PARENT MATERIAL, TERRAIN (including elevation), CLIMATE, ASPECT, vegetation cover, MICRO-ORGANISMS IN THE SOIL and the age of the land surface. Some pedologists would add to this list the influence of human activities. All the factors exhibit varying degrees of interrelationship and some are more important than others, with climate often being singled out as the most important (ZONAL SOILS). *See also* CALCIFICATION, CHELUVIATION, DECALCIFICATION, ELUVIATION, GLEYING, HUMIFICATION, ILLUVIATION, LATERITE, LATERIZATION, LEACHING, LESSIVAGE, MAN-MADE SOILS, PODZOL, PODZOLIZATION, RUBIFACTION, SALINIZATION.

soil group (major soil group) A collection of soils that occur over wide areas and have similar soil climates (SOIL WATER and SOIL TEMPERATURE), thereby leading to the development of similar SOIL HORIZONS. It is the third-ranked category in the hierarchy of SOIL TAXONOMY.

soil horizon A layer of soil more or less parallel to the ground surface, but differing from adjacent genetically related layers in any of the following properties: SOIL COLOUR, CONSISTENCE, SOIL STRUCTURE, and SOIL TEXTURE – in addition to biological, chemical and mineralogical differences. Horizons result in part from the differing magnitudes at which a variety of processes operate within the soil. Although four horizons are recognized (A-HORIZON, B-HORIZON, C-HORIZON, D-HORIZON) only horizons A and B represent the true soil (SOLUM), while C is the subsoil of weathered parent material, and D the relatively unweathered parent material or bedrock. The *A-horizon* (zone of ELUVIATION) may be subdivided, as may the *B-horizon* (zone of ILLUVIATION) as illustrated in [*235*].

soil mechanics The scientific study of the physical properties of *soils*. *See also* COHESION, FAILURE, FRICTION, PORE-WATER PRESSURE, SHEAR STRENGTH, STABILITY ANALYSIS, STRESS.

soil moisture SOIL WATER.

soil-moisture deficit The amount by which soil water falls below FIELD CAPACITY. [*79*]

soil order The highest level of generalization in soil classification, in which all the included soils have one or more characteristics in common. For example, in Canada eight orders are recognized: Brunisolic, Chernozemic, Gleysolic, Luvisolic, Organic, Podzolic, Regosolic, and Solonetzic. In the SEVENTH APPROXIMATION ten orders are recognized (ALFISOL, ARIDISOL, ENTISOL, HISTOSOL, INCEPTISOL, MOLLISOL, OXISOL, SPODOSOL, ULTISOL, VERTISOL).

soil phase An expression used in soil classifications to denote the lowest of the hierarchical subdivisions. It is based on such properties as soil depth, stoniness, amount of erosion, etc.

soil profile A vertical section of soil in which all the SOIL HORIZONS are shown. If it is necessary to study the entire profile in a laboratory a section may be removed in a metal box *en masse* without disturbing the

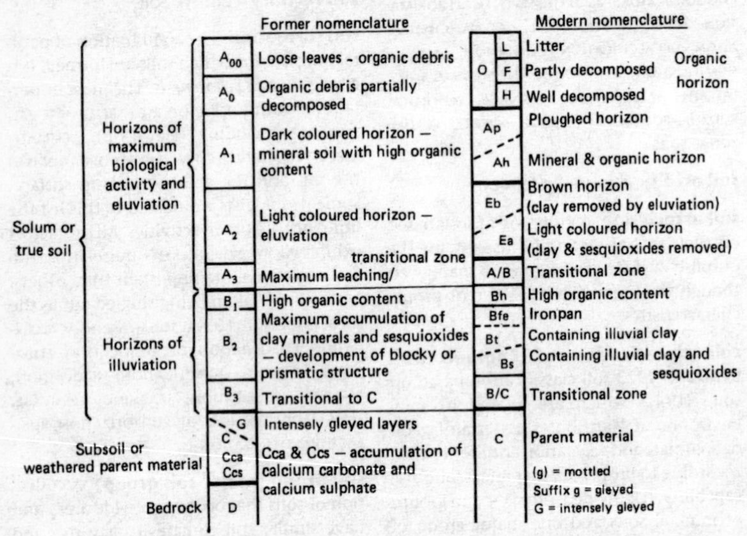

Figure 235 *Soil horizons*

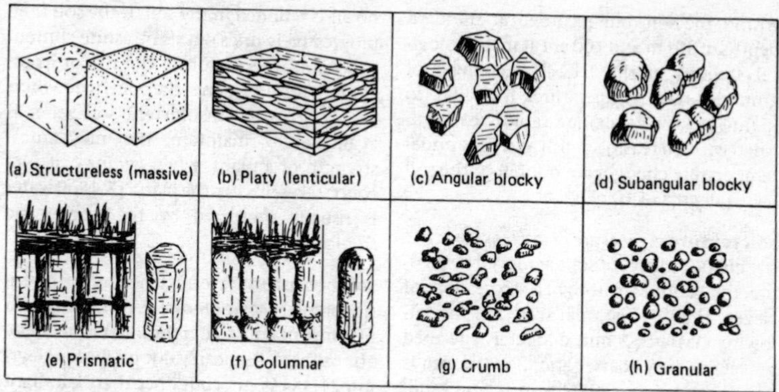

Figure 236 *Soil structure*

horizonation or structure. The latter sample is referred to as a *soil monolith*.

soil science The scientific study of soils in their entirety. It includes PEDOLOGY, soil chemistry and SOIL MECHANICS, in addition to studies of soil FERTILITY and the degree to which soils may be affected by degradation and SOIL EROSION.

soil series A grouping of soils based on the similarities of all the properties except those relating to its erosional state and the texture (SOIL TEXTURE) of the surface horizon (SOIL HORIZON). It is the lowest ranked category in the hierarchy of SOIL TAXONOMY and is the basic unit in soil mapping. *See also* SOIL TYPE.

soil structure The character of a soil expressed in terms of its mode of aggregation or the binding properties or ability of its individual particles to form a secondary unit (AGGREGATE, PED). The degree of aggregation varies from strong through moderate and weak to structureless, whereby structures are broken down by cultivation inasmuch as this tends to remove the organic content which is one of the chief binding agents. Since structure affects other soil properties, e.g. POROSITY, it has an important effect on SOIL FERTILITY. The following types of soil structure have been recognized: **1** structureless, in which there is no observable aggregation either in the massive amorphous state of a coherent mass or in the loose incoherent mass of individual particles (e.g. dune sand); **2** platy, in which the particles are flat, thin and flaky (PLATY STRUCTURE OF SOIL); **3** angular blocky, in which the faces are rectangular and the edges sharply angular; **4** subangular blocky, in which the faces are subrectangular and the edges oblique or rounded; **5** prismatic, in which the vertical faces and sharp edges are well defined (PRISMATIC STRUCTURE OF SOIL); **6** columnar, in which the prismatic units have flat or rounded tops (COLUMNAR STRUCTURE); **7** CRUMB STRUCTURE OF SOIL; **8** granular, in which the faces are subrounded and the edges rounded. *See also* SOIL TEXTURE. [*236*]

soil subgroup SOIL TAXONOMY.

soil suborder SOIL TAXONOMY.

soil taxonomy A hierarchical classification of soils ranging from the most general grouping (SOIL ORDER) to the most detailed (SOIL SERIES): orders (10); suborders (47); great groups (SOIL GROUP) (180); subgroups (960); families (4,700); series (11,000). The figures in brackets refer to the number of classes recognized in the USA; while the first three classes are expected to remain relatively unchanged, the remaining categories will be greatly increased as more research and mapping is carried out in other continents.

soil temperature The temperature obtained by a soil thermometer at any level within the soil, but recorded at standard depths of 30 cm and 100 cm at meteorological stations in the UK. The changes of ground-surface temperatures from day to night (DIURNAL RANGE) are rapidly extinguished with increasing depth and the annual temperature changes are not felt in the soil below depths of 10 m.

soil texture A measure of different PARTICLE SIZE characteristics of soils achieved by passing the samples through a succession of sieves. The International scale has the following classes: >2 mm diameter is termed gravel and very coarse sand; 2 to 0.2 mm is coarse sand, 0.2 to 0.02 mm is fine sand; 0.02 to 0.002 mm is silt; <0.002 mm is clay. Once the various percentages are worked out the soil can be named by reference to the TEXTURAL TRIANGLE [*251*]. The sizes differ slightly between the WENTWORTH SCALE, the British Standard classification (PARTICLE SIZE) and the International scale noted above. *See also* SOIL STRUCTURE.

soil type A subdivision of a SOIL SERIES, based on textural variations of the A-HORIZON (SOIL TEXTURE). Such variations are usually related to the microrelief or to small changes in slope gradient.

soil water One of the most important elements involved in pedological processes and plant growth. There are three basic forms: **1** Water adhering in thin films by molecular attraction to the surface of soil particles and not available for plants is termed HYGROSCOPIC WATER. **2** Water forming thicker films and occupying the smaller pore spaces is termed CAPILLARY WATER. Since it is held against the force of gravity it is permanently available for plant growth and it is this type of soil water which contains plant nutrients in solution. **3** Water in excess of HYGROSCOPIC and CAPILLARY WATER is termed *gravitational water*, which is of a transitory nature because it flows away under the influence of gravity. When the excess has drained away the amount of water retained in the soil is termed its FIELD CAPACITY, when some of its pore spaces are still free of water. A state of waterlogging is achieved when all pore spaces are filled and no air is retained in the soil. If the soil then progressively dries out its moisture content will fall in proportion to the rise of its air content until it reaches the point (the *wilting coefficient*) when insufficient soil water is available to maintain the mechanical strength of a plant, when *wilting* (WILTING-POINT) takes place in the plant. [*79*] Soil water is usually measured by the GRAVIMETRIC METHOD.

solar constant A term denoting the intensity of solar RADIATION received on a unit area of horizontal surface at the outer limit of the atmosphere. Its mean value is 1.94 langleys/min (1,353 W m^{-2}) but since there is a slight variation there is debate about whether the term *constant* is a misnomer. The value is virtually constant except in the shortest part of the spectrum, where the amount is lower, but it should also be noted that values up to 2.04 langleys/min (1,423 W m^{-2}) have been recorded. *See also* INSOLATION.

solar day The time interval between two successive transits of the Sun across a given meridian. The mean solar day is 24 hours: the time varies slightly, with latitude, since the Earth's orbit is ellipsoid and inclines towards the equator. *See also* LUNAR DAY, MEAN SOLAR TIME, SIDEREAL DAY.

solar energy Any form of energy which has its origin in the Sun (INSOLATION), including that used in PHOTOSYNTHESIS. It can be exploited directly by conversion to thermal, electrical or chemical energy, or by development of solar heating. Indirectly, the energy derived from winds, waves and the thermal gradient of the oceans is also derived from the Sun by the effect of insolation on the atmosphere and the oceans. Direct conversion systems include: **1** *solar cells*, which convert directly to electricity by thermoelectric and photovoltaic means; **2** *flat-plate collectors*, in which a fluid, such as water, is heated; **3** *focusing collectors*, which concentrate direct radiation on to a small area; **4** *chemical and biochemical conversion systems*, which are largely at the experimental stages. It is noteworthy that although solar radiation travels through space without energy loss the intensity of radiation within a beam of given cross-section decreases inversely as the square of the distance from the Sun. Thus the Earth intercepts only about two-billionths of the solar-energy output.

solarimeter An instrument designed to measure the intensity of solar RADIATION received at the Earth's surface: direct (*Q*) and DIFFUSE (Q) RADIATION. A glass dome protects the sensor from wind and rain, and only transmits shortwave radiation between 0.35 and 2.8 μm (i.e. from the longest ultraviolet, through the visible, to the infrared parts of the spectrum). The receiving surface has two sensing elements, one blackened to absorb incident solar radiation, the other white to reflect the radiation. The temperature difference between the two elements is proportional to the absorbed solar radiation, and the voltage produced is continuously recorded on volt-time integrators and printers. Upward and downward facing solarimeters represent an *albedometer* and measure the ALBEDO of the surface. *See also* PYRANOMETER.

solar radiation INSOLATION.

solar system The collection of the nine PLANETS plus the ASTEROIDS which revolve around the Sun following almost circular orbits.

solar year The time period during which the Earth completes one orbit around the Sun. It comprises 365.2422 mean SOLAR DAYS or 365 days 5 h 48 min 45.51 sec (diminishing by *c*. 5 sec/millennium). *See also* MEAN SOLAR TIME.

solclime A term used occasionally to describe the climate of a soil (SOIL TEMPERATURE, SOIL WATER).

sole **1** The lowest THRUST-PLANE in a region of thrust faulting (THRUST-FAULT) in which strong compressional movements of the crust have caused rock masses to ride forward one above the other in the direction of pressure, often with the development of IMBRICATE STRUCTURES. **2** The base of a glacier.

solfatara An Italian term for a late phase of volcanic activity prior to the extinction of the volcanic cone (VOLCANO). The phase is characterized by a gentle emission of sulphurous gases and steam but without violent volcanic explosions. It is named from a small cone in the Phlegrean Fields, near to Naples, Italy.

solid geology The study of the solid rocks in contrast to the superficial deposits (DRIFT). The Institute of Geological Sciences in Great Britain produces maps of both the solid geology and the *drift geology*. The former category portrays the disposition of the solid rocks as if the drift had been stripped away, while the latter gives a truer picture of the actual disposition, with the solid being hidden to a greater or lesser degree by the drift cover.

solifluction (solifluxion) A term, literally meaning 'soil flow', which can be defined as the slow downslope flowage of masses of surface waste saturated with water. Some writers have used the term to include soil creep (CREEP), MUD-FLOWS and DEBRIS FLOWS, but most confine its use to the movement of fine-grained material on relatively gentle slopes. Although in more general usage solifluction can occur under a variety of climatic conditions it is most effective in PERIGLACIAL regions, where it is said to be the chief agent of denudation. Periglacial processes provide considerable instability in the slope debris and seasonal thawing in the active layer of the soil (MOLLISOL) produces sufficient water to assist the downslope movement. It does so not by lubricating the soil particles but by causing them to lose their SHEAR STRENGTH, owing to loss of friction and cohesion, thereby leading to slow flowage, although vegetation is an important restraining factor. Solifluction in a PERMAFROST environment has been termed CONGELIFLUCTION by J. Dylik, while K. Bryan has included solifluction within his term CONGELITURBATION, which comprises all types of MASS-MOVEMENT under periglacial conditions. Creep due to frost-heave is usually distinguished from solifluction, although it undoubtedly renders the soil more susceptible to flow by increasing the void ratio between the soil particles. Rates of downslope movement by creep are considerably less than those due to solifluction. *See also* ALTIPLANATION, COOMBE ROCK, CRYOPLANATION, GELIFLUCTION, GELIFLUCTION LANDFORMS, GOLETZ TERRACES, HEAD, INVOLUTION, PLOUGHING BLOCK, STONE-BANKED TERRACE, TURF-BANKED TERRACE. [*146*]

sol lessivé A soil in which clay is moved downwards (translocated) in suspension by the process of LESSIVAGE, to form a horizon in which the aggregates are conspicuously coated in oriented clay films. It is sometimes referred to as a leached brown soil and is classified as an ALFISOL in the SEVENTH APPROXIMATION soil classification of the USA.

solod (soloth) A *salt-earth podzol* produced by LEACHING of a SOLONETZ soil, either by natural processes or by improved irrigation owing to human intervention. The bleached A-horizon is typical of true PODZOLS and it overlies a gleyed and mottled, dark brown B-horizon which is characterized by a columnar and prismatic structure (SOIL STRUCTURE). It is one of the HALOMORPHIC SOILS. *See also* INTRAZONAL SOILS.

solonchak A *white alkali soil*, taking its name from a Russian term applied to a soil in which sodium chloride and sodium sulphate make up more than 0.3% of the soil and saline groundwater is within 2 m of the surface. The latter not only causes GLEYING in the lower C-horizon but provides the supply of soluble salts which move up to the surface by CAPILLARITY, in response to the high degree of surface evaporation. It is characterized by lack of a B-horizon. A pH of >8 makes these soils of little value for agriculture although they will support some salt-tolerant plants (HALOPHYTE). Solonchaks are characteristic of interior deserts of Eur-Asia and N America, where they may have been artificially induced by excessive irrigation. They are classified in the ARIDISOL order of the SEVENTH APPROXIMATION soil classification and are a category of the HALOMORPHIC SOILS. *See also* INTRAZONAL SOILS, TAKYR.

solonetz A *black alkali soil*, taking its name from a Russian term for a soil dominated by sodium carbonate in which some LEACHING of the soluble salts has taken place owing to increased rainfall or irrigation. Thus the solonetz is a partly leached SOLONCHAK, while excessive leaching produces a SOLOD. In contrast to a solonchak, the solonetz has a brown or grey B-horizon enriched with colloids. The alkaline solutions which accumulate at the surface mix with humic material during spells of wet weather before drying out to form black surface crusts. In the SEVENTH APPROXIMATION soil classification the solonetz falls into both the ARIDISOLS and the MOLLISOLS. There is some disagreement about whether it should be regarded as part of the ZONAL SOILS or the INTRAZONAL SOILS, with most writers preferring the latter. They are one of the categories of HALOMORPHIC SOILS.

soloth SOLOD.

solstice The time during winter or summer when the overhead midday Sun reaches its minimum or maximum declination (angular distance) from the equator. The summer solstice for the N hemisphere is about 21 June, when the midday sun is vertically overhead at the Tropic of Cancer (23°27'N). The winter solstice for the N hemisphere is about 22 December, when the midday sun is vertically overhead at the Tropic of Capricorn (23°27'S). The summer solstice has the year's longest day, while the winter solstice has the shortest. [*73*]

solum A term for the true soil, which consists only of the A- and B-horizons. The underlying C-horizon (subsoil) and D-horizon (unweathered bedrock) are not part of the solum. *See also* SOIL HORIZON. [*235*]

solutes Organic or inorganic substances which are easily dissolved in a liquid (*solvent*). Although the vast majority of solutes occur in the oceans, which make up 97% of the HYDROSPHERE, those solutes which are contained in the remaining water associated with the terrestrial phase of the HYDROLOGICAL CYCLE exhibit considerable variety. The solutes contained in this 'terrestrial' water are comprised very largely of the following IONS: calcium (Ca^{2+}), magnesium (Mg^{2+}), sodium (Na^{+}), potassium (K^{+}), chlorine (Cl^{-}), hydrogencarbonate (HCO_3^{-}), sulphate (SO^{2-}), nitrate (NO_3^{-}) and silica (SiO_2). There are marked contrasts in the

Figure 237 *Average solute content of precipitation*

	Concentration (mg l^{-1})					
	Ca^{2+}	Mg^{2+}	Na	K	Cl^-	SO_4^{2-}
Coastal	0.29	0.45	3.45	0.17	6.0	1.45
Inland	0.43	0.19	0.37	0.15	0.75	1.73

Source: Meybeck, 1983

solute content of precipitation in coastal and inland (>100 km) environments. [*237*] *See also* AEROSOLS, SOLUTION.

solution 1 The process by which matter is changed from a solid or gaseous state into a liquid state by combination with a solvent. It is one of the less important forms of CHEMICAL WEATHERING, in which solid rocks are dissolved by water. It is of some significance in the case of calcareous rocks acted upon by rainwater, which has become slightly acidic owing to its accumulation of carbon dioxide from the atmosphere. Thus it is one of the several agents responsible for the production of KARST landforms, although weathering of limestone rocks is partly due to HYDROLYSIS and CARBONATION. Of the minerals most vulnerable to solution, calcium, magnesium, potassium and sodium are the most important. **2** A fluid containing IONS.

solution basin A small depression, a few metres in diameter and several centimetres deep, occurring in a level surface of calcareous rocks. It is developed initially by standing water, which deepens the hollow by solution, following which the sides, in contrast to the smooth floor, become increasingly fluted or fretted by free-flowing water. Solution basins, or *kamenitzas*, are frequently aligned along joints and often become filled with organic sediments. The latter are a type of LAPIÉ.

solution collapse An expression referring to the subsidence of the surface into a crater-like DOLINE in a non-calcareous rock owing to the removal by solution of an underlying layer of limestone. *See also* KARST WINDOW.

solution pipe PIPE.

sonnenseite A German term for a sunny slope, synonymous with ADRET.

sonograph, sonoprobe The graphic record produced by the use of an echo-sounding device, or sonoprobe, which is a continuously recording acoustic reflection instrument. It is used by oceanographers to determine the shape of the ocean floor in addition to the thickness, character and disposition of marine sediments. It is important in the siting of oil rigs, laying of submarine cables and pipelines and in offshore mineral exploration.

sorting 1 The process by which materials are graded according to one of their particu-

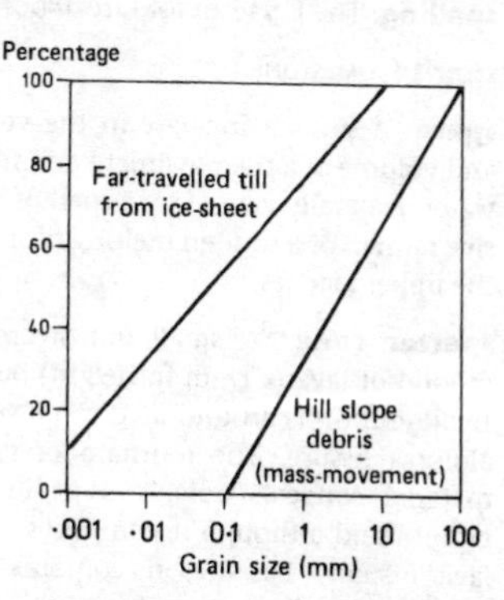

(a) Sorting of two sediments moved by non-fluid transport

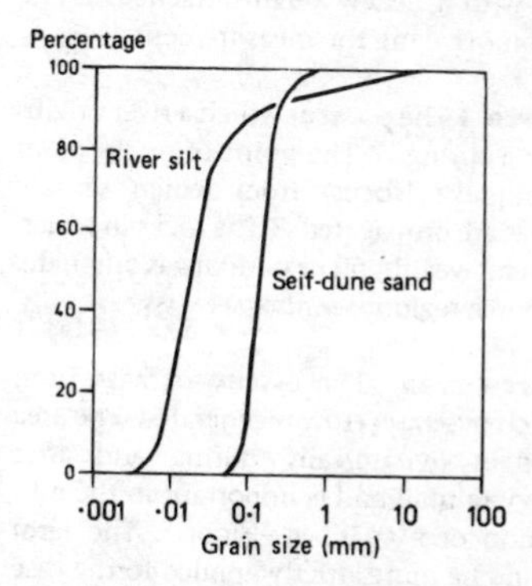

(b) Sorting of two fluid-transported sediments

Figure 238 *Sorting*

lar attributes, i.e. shape, size, density, etc. In sedimentology, the natural sorting of sediments by PARTICLE SIZE is taken as a basis for classification into *well sorted* or *poorly sorted* deposits. Sediments which have been repeatedly reworked by marine waves or have travelled a long distance downstream are usually *well sorted* into different-sized particles, because different particle sizes have different settling velocities. In general, sediments in fluid transfer tend to be *well sorted*. Sediments transported by glaciers and mass-movement are *poorly sorted* [*238*]. In zones subjected to FREEZE–THAW ACTION a considerable degree of sorting occurs during the formation of PATTERNED GROUND. **2** In statistics, a measure of the spread of a data distribution on either side of an average.

sound 1 A waterway connecting two larger areas of water, rather wider than a STRAIT, e.g. the Inner Sound which separates Raasay Is. from the mainland of W Scotland. **2** An inlet of the sea, e.g. Plymouth Sound in S England. **3** A lagoon on the eastern coast of the USA, e.g. Pamlico Sound. **4** A pressure wave with a frequency detectable by the human ear (*c*. 20 Hz to 20 kHz).

sounding 1 A measure of sea- or lake-water depth, expressed in FATHOMS. An echo-sounding device has replaced the use of the relatively inaccurate *sounding-lead*, a hemp line with a heavy weight attached. **2** The action of taking the measurement.

source 1 The point at which a river originates; a spring. **2** The point of origin of an earthquake (FOCUS) from which seismic waves are propagated. **3** The rocks in which mineral wealth, oil or natural gas originate. **4** Source region. *See also* AIR MASS.

source area The CATCHMENT AREA from which OVERLAND FLOW is generated. The area changes dynamically during and after heavy rainfall and is important in the estimation of PARTIAL AREA MODELS. The term should be more strictly applied to the case when overland flow is produced by subsurface saturation rather than the INFILTRATION-RATE having been exceeded.

South-East Trades TRADE WINDS.

Southerly Burster An outbreak of cold, polar air behind a trough of low pressure crossing SE Australia, manifesting itself as a strong, dry southerly wind which causes rapid falls of temperature with magnitudes of about 10° C. It is experienced generally in spring and summer, when the cold wave 'bursts' on to the coastlands of New South Wales which are experiencing mild weather conditions. *See also* AIR MASS, BRICKFIELDER.

southern oscillation (Walker circulation) A climatological term referring to a PERTURBATION in the intertropical general circulation, especially in the Indian Ocean and the southern Pacific. The term was introduced by G. Walker who noted that when pressure over the Pacific is high, it tends to be low over the Indian Ocean, with important consequences on precipitation amount.

Space Shuttle A NASA (National Aeronautics and Space Administration) SATELLITE possessing the capacity for repeated round-trip flights into space. It was first launched in 1981 as a maintenance vehicle for Earth resource satellites and as a base for a variety of REMOTE SENSING instruments. It has a design life of 500 space missions. *See also* HEAT-CAPACITY MAPPING MISSION, LANDSAT, NOAA/TIROS SATELLITES, SEASAT I, SKYLAB, SPOT.

spalling The US term for EXFOLIATION.

sparite LIMESTONE.

spate A sudden increase in the velocity and volume of a river owing to an influx of water, generally caused by a spell of intensive rainfall or a sudden melting of snow in the upper reaches.

spatter cone A small but steep-sided mound of lava (< 15 m in height) built by small eruptions around a volcanic vent or along a fissure. The fountain of molten material congeals quickly as it hits the ground and although it comprises mainly lava (usually basaltic) it contains some SCORIA and PELÉ'S HAIR.

Spearman rank correlation CORRELATION, STATISTICAL.

speciation The arrival of new SPECIES when the genetic difference between populations that have been subjected to different patterns of NATURAL SELECTION is sufficiently different to prevent interbreeding. *See also* DARWINISM, EVOLUTION.

species A collection of similar organisms that are capable of interbreeding to produce fertile offspring. It is the basis of any taxonomic classification of organisms and depends on such attributes as morphological similarity, physiological compatibility, ecological association, geographical distribution and continuity in time, all of which will relate to their ability to interbreed. The name of a species is always italicized, preceded by the name of the GENUS to which it belongs and often succeeded by the abbreviated name of the scientist who first described it. *See also* TAXONOMY.

species-area curve A graph used by biogeographers to denote the relationships between animal and plant numbers and the area of sample plots. It is a useful means of determining which particular QUADRAT size would be most effective in a SAMPLING exercise.

species-energy theory A theory formulated by D. H. Wright (1983), which proposes that the modern species richness of flora and fauna in large regions can be explained in terms of available solar energy, given that there are sufficient water supplies.

specific gravity The ratio of the mass of a body to the mass of an equal volume of water at a given temperature. *See also* DENSITY.

specific heat The amount of heat necessary to raise the temperature of 1 g of a given substance through 1° C: e.g. to raise 1 g of water from 0° to 1° C requires 1 calorie.

specific humidity The ratio of the weight of water vapour in a parcel of the atmosphere to the total weight of moist air (i.e. including water vapour). It is stated as g of water vapour per kg of *moist* air (expressed as g kg^{-1}), in contrast to the MIXING RATIO. Very humid warm air in the equatorial zone can have specific humidities between 16.0 and 18.0, whereas cold, dry air in a polar latitude may exhibit a specific humidity of less than 1.0.

specific retention A hydrological term referring to the volume of water that is retained in a soil or rock (against the influence of gravity) following drainage after SATURATION. The difference between POROSITY and specific retention is known as the SPECIFIC YIELD, which itself represents the volume of water released under gravitational influences.

specific yield SPECIFIC RETENTION.

spectral A REMOTE SENSING term referring to a small part of the total range of ELECTROMAGNETIC RADIATION and usually given in wavelength intervals. For example, a MULTISPECTRAL SCANNER records the spectral RADIANCE of the Earth's surface in a number of discrete *spectral wavebands*. *See also* MULTISPECTRAL.

spectral signature A REMOTE SENSING term referring to the SPECTRAL characteristics of an object or class of objects on the Earth's surface. Although it is a widely used term its use is to be discouraged since the spectral properties of a natural object (recorded by a SENSOR) are continually changing. Thus, such objects have no absolute signature.

spectrum 1 An image created by dispersing a beam of radiant energy so that its rays are arranged in the order of their wavelengths (ELECTROMAGNETIC RADIATION). **2** In biogeography, the percentage distribution of the life-forms in the flora of a given area. **3** In palynology, the percentage distribution of the pollens found within a specific horizon. **4** In oceanography, a wave spectrum consists of all the individual waves that are present at any one place in the ocean, although they may be of different heights and periods. It can be analysed by FOURIER ANALYSIS.

specular iron ore HAEMATITE.

specular reflector A REMOTE SENSING term referring to a surface from which ELECTROMAGNETIC RADIATION is reflected without scat-

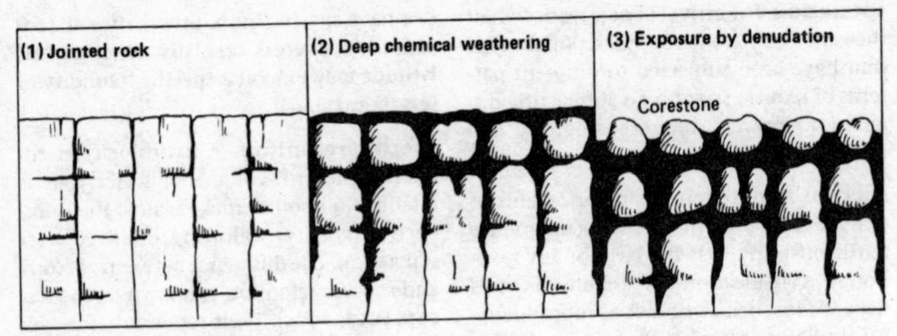

Figure 239 *Spheroidal weathering*

tering or diffusion. A water body is a good example of a specular reflector. *See also* DIFFUSE REFLECTOR.

speleology The scientific study of the origin of caves and cave life.

speleothem The generic term for chemically precipitated deposits in cave environments. The most common are formed from CALCITE, including cave coral, cave drapery, cave onyx, cave pearl, curtains, FLOWSTONES, HELICITES, RIMSTONES, STALACTITES and STALAGMITES. Speleothems are less commonly made of SILICA, GYPSUM and HALITE.

sphagnum A genus of mosses growing in wet places, some species of which form spongy cushions and which contribute to BOG formation.

sphere A solid figure any point on the surface of which is equidistant from its centre.

sphericity The amount by which the shape of a fragment approaches the form of a SPHERE. A measure of sphericity suggested by W. C. Krumbein is $3\sqrt{(bc/a^2)}$, where a = the long axis, b = the intermediate axis, and c = the short axis of the fragment. *See also* FLATNESS INDEX, ROUGHNESS, ROUNDNESS INDEX.

spheroid Any figure that is almost a SPHERE, e.g. the Earth.

spheroidal weathering A type of subterranean CHEMICAL WEATHERING in which jointed rock masses are slowly rounded by the gradual removal of their concentric outer shells to leave an internal spherical boulder. The rock shells are loosened by the process of HYDROLYSIS, with the disintegration products being finally removed by groundwater. It is termed *concentric weathering* in the USA. Spheroidal weathering at depth will produce *boulders of decomposition*, which may be found at the ground surface following exhumation, where they will resemble spherical boulders produced by the similar hydrolysis process in EXFOLIATION. [*239*]

spilites Basaltic lavas in which the primary mafic minerals have been altered to chlorite and epidote. They are commonly found as PILLOW LAVAS interbedded with marine sediments in a GEOSYNCLINE, suggesting that they were formed from submarine volcanoes.

spillway, glacial OVERFLOW CHANNEL.

spinifex A type of herbaceous vegetation found in the arid lands of Australia. It has sharp spiny leaves and grows in isolated grass-like clumps separated by bare ground.

spit A narrow and elongated accumulation of sand and shingle projecting into a large body of water, usually the sea [*240*]. It grows out from a coastline as a result of LONGSHORE DRIFT, often at a location where the line of the coast changes direction, as at the mouth of an estuary where spits are common. Many spits are characterized by a curved termination at the distal end, thought to be a product of wave REFRACTION.

spitskop An Afrikaans term for a hill with a sharply pointed top, in contrast to a TAFELKOP. *See also* KOPJE.

spitzkarren KARREN.

splash erosion RAINDROP EROSION, RAINSPLASH.

spodosol One of the ten orders in the SEVENTH APPROXIMATION soil classification. It is characterized by a B-horizon containing an accumulation of organic material (HUMUS) and compounds of aluminium and iron carried down from the overlying bleached A-horizon by the process of ELUVIATION. The ash-grey to white A-horizon is strongly acid and low in plant nutrients and is typical of true PODZOLS and PODZOLIC SOILS which have developed in humid cool temperate climatic zones, often under coniferous forest cover.

spongework A pattern of shallow depressions and intervening rim-like partitions occurring on the walls and ceilings of caves in a KARST terrain. They are occasionally interrupted by deeper depressions known as pockets. The arrangement is thought to be caused by the differential solubility of limestone when submerged in the groundwater flow (phreatic flow).

SPOT An abbreviation for Satellite Probatoire pour l'Observation de la Terre. It is France's first EARTH RESOURCES satellite (SATELLITE), the first mission of which carried a number of SENSORS, similar to those carried by LANDSAT 4, in order to obtain remotely sensed data for use in a variety of geological, geographical and agricultural projects. *See also* HEAT-CAPACITY MAPPING MISSION, NOAA/TIROS SATELLITES, REMOTE SENSING, SEASAT I, SKYLAB, SPACE SHUTTLE.

spot height An exact point on a map with an accompanying elevation printed alongside. The dot represents height above a given DATUM, in Britain the ORDNANCE DATUM (OD). There is no indication of a spot height on the ground, in contrast to a BENCH-MARK.

spreading zone The zone along which new OCEANIC CRUST is formed on either flank of the MID-OCEANIC RIDGES. It is always found at constructive plate margins (PLATE TECTONICS) and is the zone from which new ocean floor moves laterally towards the destructive plate margins (SUBDUCTION ZONES) by the action of SEA-FLOOR SPREADING. [*223*]

spring 1 The season which occurs between winter and summer, usually regarded as the three months of March, April and May by the inhabitants of the N hemisphere. In astronomical terms, however, it spans the

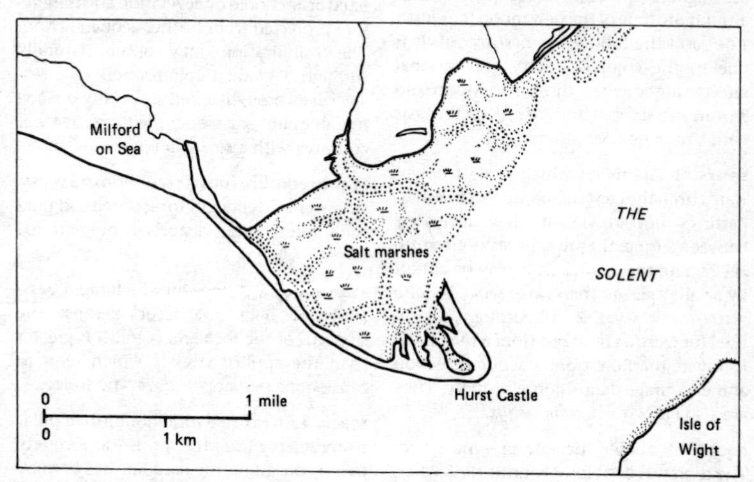

Figure 240 *Hurst Castle Spit, Hampshire*

period from the vernal (spring) equinox (*c.* 21 March) to the summer solstice (*c.* 21 June). **2** A natural flow of water from the ground at the point where the WATER-TABLE intersects the surface. *See also* FAULT-SPRING, INTERMITTENT SPRING, RISING, SCARP-FOOT SPRING.

spring-head alcove The arcuate cliff surrounding a RISING, formed by progressive headward retreat due to SPRING-SAPPING and to cavern collapse. It is typical of the margins of KARST limestone terrain and can eventually form a sizeable amphitheatre.

spring-line A linear distribution of SPRINGS along the zone where the water-table intersects the surface, either at a fault (FAULT-SPRING) or at the foot of an escarpment (SCARP-FOOT SPRING). It was an important locating factor for early settlements.

spring-sapping An erosional process around the point where a spring issues from the ground. The removal of material causes a headward extension of the stream channel, dissection of the slope and the gradual retreat of the SPRING-HEAD ALCOVE.

spring tide A tide with a range considerably increased from that of the mean tidal levels (i.e. low tides are lower and high tides are higher). It occurs twice each month around the time of the new moon (CONJUNCTION) and the full moon (OPPOSITION). It is due to the complementary gravitational effects caused when the Earth, Moon and Sun are in a straight line (SYZYGY). It contrasts with a NEAP TIDE. *See also* TIDE.

spurs **1** The ridges which project downwards from the crests of mountains as waterpartings (INTERFLUVE). If they intervene between cirques they may be fashioned into ARÊTES, and their lower ends may be cut off by a valley glacier (TRUNCATED SPUR). *See also* INTERLOCKING SPURS. **2** In hydrology, a term used for GROYNES built out from a river bank to divert the flow from a scoured section and encourage deposition elsewhere. They are also known as *training walls*.

squall A sudden increase of wind-speed. The term was used loosely until 1962 when a precise definition stated that for a duration of at least one minute the wind-speed must rise by at least 8 m/sec and reach 11 m/sec (16 knots to 22 knots) before quickly dying away. *See also* LINE SQUALL.

stability **1** In general, the capacity to resist displacement and an ability to recover the original state after displacement. **2** In meteorology, the state of equilibrium reached in the atmosphere when vertical air movement cannot be sustained. This is achieved where the ENVIRONMENTAL LAPSE-RATE of a particular air mass is less than the SATURATED ADIABATIC LAPSE-RATE, so that a rising parcel of air, produced by thermal activity, will become cooler than the surrounding air and will subside to its former level. This is known as a state of STABLE EQUILIBRIUM, typical of conditions prevailing within a summer ANTICYCLONE. *See also* ABSOLUTE STABILITY, HUMILIS, SUBSIDENCE. [*132*]

stability analysis A term used in soil mechanics and rock mechanics to denote the procedure for examining potential FAILURE of a slope. It is important to know such variables as COHESION, FRICTION and PORE-WATER PRESSURE before this type of analysis can be carried out. *See also* RESIDUAL STRENGTH, SHEAR STRENGTH.

stabilized dune A dune that has become fixed or anchored by vegetation and is therefore protected from further aeolian action. The stabilization may occur naturally through a gradual colonization (PSAMMOSERE) or be brought about artificially by sowing marram grass and planting trees. It contrasts with a MOBILE DUNE.

stable equilibrium A condition in a SYSTEM when it has a tendency to recover its original state after being disturbed by external forces. [*72*]

stable slope A slope on which rapid MASS-MOVEMENT does not occur, because the strength of the rock and regolith is greater than the applied stresses which tend to cause slope FAILURE. *See also* STATIC SLOPE.

stack **1** An isolated rock monolith or pillar rising steeply from the sea. It was formerly part of the adjoining land but has become isolated from it by wave erosion, probably after having formed part of a marine ARCH. The Old Man of Hoy (137 m) in the Orkney Isles of Scotland is Britain's most spectacular example. **2** The Gaelic spelling *stac* usually refers to steep offshore islands, e.g. Stac an Armin (191 m), St Kilda, but is also used in certain mountain names in W Scotland, e.g. Stac Poly. **3** A seismic record produced by adding together a number of different SEISMOGRAMS.

stade A period of time of no fixed length referring to a glacial substage when glaciers temporarily waxed or stopped retreating, usually owing to climatic change. It is marked by glacigenic deposits and often by a STADIAL MORAINE (recessional moraine). A stade differs from a stratigraphic substage (CHRONOSTRATIGRAPHY) because it does not represent a true rock unit and has a variable time from place to place.

stadial moraine A moraine marking a recessional phase or STILL-STAND in the overall decline of an ice-sheet or glacier. [*219*]

staff (Sopwith staff) LEVELLING-STAFF.

staff gauge A hydrological instrument for measuring the depth of water at a given station on a river. It consists of a tall post fixed in the river bed (or on the river bank) and is painted with elevation marks at precisely measured intervals, thereby enabling the gauge to be read by eye at all stages of river flow. It is commonly seen at locks on the river.

stage 1 In CHRONOSTRATIGRAPHY, a division of rock in the Standard Stratomeric scheme of stratigraphic classification. It is a subdivision of a SERIES and comprises a number of CHRONOZONES. It represents the body of rock formed during an AGE. *See also* BIOSTRATIGRAPHY, LITHOSTRATIGRAPHY. **2** A division of the Pleistocene; the time equivalent of a stratigraphic unit, such as a FORMATION, or a grouping of units which may be of glacial or interglacial age. **3** The point to which a landform has evolved during a CYCLE OF EROSION (youth, maturity, senility). This concept has been questioned because it has been shown that landforms do not adjust steadily through equally long stages over time. **4** In the USA, the level of water in a river channel above a given datum (RATING CURVE).

stage-discharge curve RATING CURVE.

stagnant-ice topography DEAD-ICE FEATURES.

stagnogley soils One of the subdivisions of the GLEY SOILS, which are themselves part of the SOIL CLASSIFICATION OF ENGLAND AND WALES. They are typically non-alluvial, non-calcareous loamy or clayey soils with a relatively impermeable subsurface horizon (causing IMPEDED DRAINAGE) but without a humose or peaty topsoil.

stagnohumic gley soils One of the subdivisions of the GLEY SOILS, which are themselves part of the SOIL CLASSIFICATION OF ENGLAND AND WALES. They are typically non-calcareous, with a clayey, impermeable subsurface horizon that causes considerable impedance of drainage (IMPEDED DRAINAGE). Because of their wetness throughout the year and their ANAEROBIC conditions they are characterized by accumulations of peat in the topsoil. In this respect they differ from STAGNOGLEY SOILS, which are devoid of the peat horizon, although they are similar in most other respects.

stagnopodzols One of the subdivisions of the PODZOLIC SOILS, which are themselves part of the SOIL CLASSIFICATION OF ENGLAND AND WALES. They are characterized by a peaty topsoil, a periodically wet, gleyed bleached horizon which in turn overlies a thin IRONPAN and/or a brown or ochreous, relatively friable subsurface horizon.

stalactite A tapering pendant of concretionary material descending from a cave ceiling, created by the re-precipitation of carbonate in CALCITE form from percolating groundwater in a KARST environment. A type of SPELEOTHEM. *See also* ROCK PENDANT, STALAGMITE.

stalagmite A columnar concretion ascending from the floor of a cave. It is formed from the re-precipitation of carbonate in CALCITE form perpendicularly beneath a constant source of groundwater that drips off the lower tip of a STALACTITE or percolates through the roof of a cave in a KARST environment. It may eventually combine with a stalactite to form a pillar. A type of SPELEOTHEM.

stand An area occupied by a collection of plants or trees which are structurally and floristically homogeneous. It is usually applied to forests, where stands of similar trees facilitate logging operations.

standard atmosphere The idealized average condition of the atmosphere in terms of its temperature and pressure at successive altitudes. It is used for the calibration of an ALTIMETER and is particularly important in the measurement of aircraft performance, despite the differences in standards accepted in various countries. The most commonly accepted standard atmosphere is that based on a surface temperature of 15° C and a surface pressure of 1013.25 mb, with a lapse-rate of 6.5° C/km up to the TROPOPAUSE. One standard atmosphere corresponds to a BAROMETRIC PRESSURE of 760 mm (29.9213 in) of mercury of density 13.595×10^3 kg m^{-3}, where the acceleration due to gravity is 9.80655 m/sec^{-2}. It is also equal to 14.691 lb/in^2.

standard deviation In statistics, the second of the MOMENT MEASURES. It indicates the spread of all the values on either side of the arithmetic MEAN in a frequency distribution, i.e. it is a measure of dispersion. It is the square root of the VARIANCE, and is calculated by obtaining the arithmetic mean and then by measuring how much each value differs from it, either positively or negatively. Each difference is squared, the squares are summed and the total divided by the number (n) and values (x). The formula for the standard deviation (σ) is: $\sigma = \sqrt{\Sigma(x - \bar{x})^2/n}$. [165]

standard error **1** The standard error of the mean (S.E. $\bar{x}$) is a measure which estimates the average dispersion of the means (MEAN) of statistical samples of a given size about the mean of the population. It is expressed as: S.E. $\bar{x} = \sigma/\sqrt{n}$, where n = sample size. It is comparable with the STANDARD DEVIATION (σ). **2** The *standard error of estimates* is a measure of the average departure of the observed values of the dependent variable from the calculated values.

standard hill slope A SLOPE comprising four elements as defined by A. Wood (1942) and redefined by L. C. King (1962). The elements are: **1** *Crest*, a convex profile (the *waxing slope*) where weathering and CREEP are dominant processes. **2** *Scarp*, the steepest part of the slope (FREE FACE), where bedrock outcrops and where ROCKFALLS, LANDSLIDES and RILL EROSION are the dominant processes. **3** *Debris slope*, formed by detritus fallen from the scarp above. It is also known as the CONSTANT SLOPE, the angle of which is determined by the ANGLE OF REPOSE. One of the main processes of this element is SUBSURFACE WASH. **4** PEDIMENT, a broadly concave profile, which is termed the *waning slope*, and which extends at a relatively low angle down to the stream or alluvial plain. It is a rock-cut feature, cut mainly by SURFACE WASH, although it is often veneered with detritus. It has been claimed by King that the standard hill slope is the natural product of slope evolution produced by fluvial activity and/or mass-movement, and is independent of climatic control. But a strong bedrock is required to produce element **2** and if the rocks are incoherent or easily eroded this element is usually missing [*49*].

standard parallel The line of LATITUDE, selected during the construction of a MAP PROJECTION, in which the principal scale is preserved.

standard time The mean time at a MERIDIAN that is located more or less centrally over a country, and which is used for timekeeping throughout the country instead of local time (*apparent time*) based on the Sun's position (*local solar time*) as shown on a sun-dial. In general, the world is divided into a number of time zones, differing from the time at the GREENWICH MERIDIAN by multiples of 7½° or 15°, i.e. by an exact number of half-hours or hours. The *time zones* extend 7½° on either side of the standard meridian, and all places within that zone are governed by the same standard time. In countries with a great E–W extent there are several *standard*

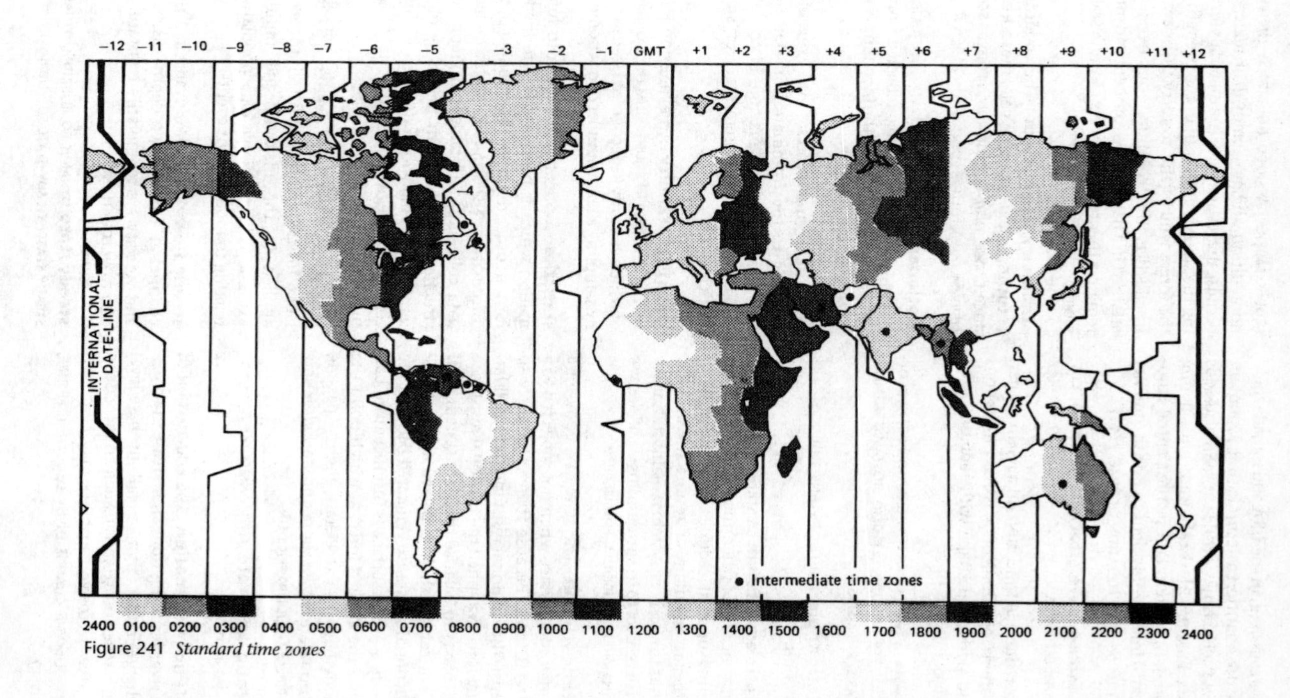

Figure 241 *Standard time zones*

time zones, e.g. in the USA and Canada there are the Atlantic, Eastern, Central, Mountain and Pacific zones, while the USA also adds the Yukon, Alaska–Hawaii and Bering zones. In western Europe the Central European Time is equivalent to British Summer Time. *See also* EQUATION OF TIME. [*241*]

standing wave (stationary wave) CLAPOTIS, LEE WAVE.

star-dune A sand DUNE, in a relatively permanent position, of massive proportions and pyramidal shape with radiating sand ridges.

stasis A term denoting a period of no evolutionary change. *See also* BIOSTASIS, HOMEOSTASIS, RHEXISTASIS.

static angle ANGLE OF PLANE-SLIDING FRICTION.

static equilibrium A state achieved by a body or SYSTEM when the balance of forces acting upon it requires no movement in order to maintain or restore the balance. Thus the body or system remains stationary with respect to its surroundings. [*72*] *See also* DYNAMIC EQUILIBRIUM, EQUILIBRIUM.

static friction The state of a body on a horizontal surface when not subjected to a lateral force. Because there is no force to resist, the frictional resistance is zero. If lateral force is applied the frictional resistance increases to balance the force system until a critical threshold is passed at which the frictional resistance generated between the body and the surface is insufficient to balance the applied force, at which point the body begins to accelerate, thereby passing into a state of DYNAMIC FRICTION. *See also* ANGLE OF PLANE-SLIDING FRICTION.

static lapse-rate ENVIRONMENTAL LAPSE-RATE.

static rejuvenation The REJUVENATION of a river resulting from an increase in its discharge rather than a fall in base-level (DYNAMIC REJUVENATION). Such an increase in a river's erosive capability may be caused by CLIMATIC CHANGE or by capture (CAPTURE, RIVER).

static slope A slope on which there is no addition and no removal of material, because the resistance of the materials forming the slope is greater than the strength of the forces which tend to carry them away. In practice, this condition can never be attained in a natural environment and must remain a theoretical concept. *See also* STABLE SLOPE.

stationary equilibrium The condition occurring when opposing velocities are balanced. *See also* EQUILIBRIUM, POISED EQUILIBRIUM.

stationary wave **1** A *standing wave* formed in the ocean when a wave reflected from a sea cliff exactly matches the incoming wave (CLAPOTIS). **2** A LEE WAVE.

statistical mechanics An expression referring to the formulation of statistical laws relating to the behaviour of complex SYSTEMS comprising assemblages of small particles.

statistical stability An expression referring to the state of EQUILIBRIUM the existence of which can only be recognized statistically by sampling a large number of phenomena.

statistics A scientific approach to information which is presented in a numerical form in order to clarify the meaning of the information. **1** *Descriptive statistics* includes data collection and table and graph construction, together with frequency distributions. **2** *Analytical statistics* includes correlation, regression and tests of significance.

statistics of extreme events The statistics utilized when statistical theory is applied to the maximal and minimal extreme values of large data populations relating to independent events.

steady flow The flow of water in a river channel at a time when depth, discharge and velocity are temporarily constant. *See also* FLOW EQUATIONS.

steady state DYNAMIC EQUILIBRIUM, DYNAMIC STEADY STATE, STEADY-STATE EQUILIBRIUM.

steady-state equilibrium The condition of an OPEN SYSTEM in which properties are invariant when considered with reference to a given time-scale, but wherein its instantaneous condition may oscillate owing to the presence of interacting variables. [*72*]

steady time One of the categories of the 'geomorphological time-scale' proposed by S. A. Schumm and L. W. Lichty in 1965. It is the time when minor landform features are created, i.e. a period of short duration. *See also* CYCLIC TIME, GRADED TIME.

steam fog A type of ADVECTION FOG formed when cold air passes over a body of fresh water, the temperature of which is considerably higher. Because the moisture condenses to form visible water droplets in the air the surface of the fresh-water body appears to steam. *See also* ARCTIC SEA SMOKE.

steering A meteorological term referring to the way in which another atmospheric factor controls the speed and the direction of movement of atmospheric pressure systems.

Stefan's law (Stefan–Boltzmann law) A law relating to the intensity of solar radiation (INSOLATION) received at the Earth's surface and to the amount of terrestrial radiation (RADIATION). It is based on the assumption that both the Sun and the Earth behave individually as black bodies (BLACK BODY). Stefan's law states that the radiant energy of a black body is emitted at a rate *F*, where $F = \sigma\tau_4$, (i.e. at a rate proportional to the fourth power of its ABSOLUTE TEMPERATURE measured on the KELVIN SCALE). Consequently, the solar flux (6,000° K) peaks at 2.0 langleys min^{-1} (SOLAR CONSTANT), compared with the terrestrial flux (300° K), which peaks at 0.05 langleys min^{-1}. *See also* PLANCK'S LAW, WIEN'S LAW.

Stefan's method A means of measuring the depth of PERMAFROST by using STEFAN'S LAW. A knowledge of ice density, the LATENT HEAT of fusion of ice, the thermal conductivity (CONDUCTION) of frozen soil and its thermal gradient are all required before the depth of penetration of the COLD WAVE into the ground can be determined.

steinpolygonen POLYGONS.

steinstreifen STONE STRIPES.

stem flow A biogeographical term describing the drainage of intercepted precipitation down the stems of plants to reach the ground surface. *See also* INTERCEPTED MOISTURE, THROUGHFALL.

step-fault A NORMAL-FAULT which, when repeated by a series of parallel faults each with an increased throw in the same direction, will produce a *stepped slope*. Step-faults are common in a FAULT-BLOCK TOPOGRAPHY and may be present on the flanks of a RIFT VALLEY. [*208*]

step functions An expression referring to rapid changes which occur between one variable and another, with one of these variables normally being time. Before and after the rapid change uniform conditions exist. Step functions occur when a SYSTEM passes across a threshold, thereby initiating a positive-FEEDBACK mechanism and leading to a state of disequilibrium. At the end of the step a new equilibrium state will occur only with the establishment of a stabilizing negative-feedback mechanism. *See also* CATASTROPHISM.

stepped topography The terrain produced by step-faulting (STEP-FAULT). *See also* RIEGEL, STEPS.

steppes A Russian term for the mid-latitude grasslands which extend from central Europe eastwards to Siberia. They are rolling or flat treeless plains, equivalent to the N American PRAIRIE and the S American PAMPA.

step-pool system A staircase type of system in a mountain stream in which each pool is separated from the next one below by a 'step' consisting of boulders, cobbles and occasionally fallen tree trunks or branches.

steps A type of micro-relief created by frost-action on slopes (PATTERNED GROUND). They consist of either STONE-BANKED TERRACES or TURF-BANKED TERRACES, which have been derived from CIRCLES, NETS or POLYGONS rather than developing independently. Their

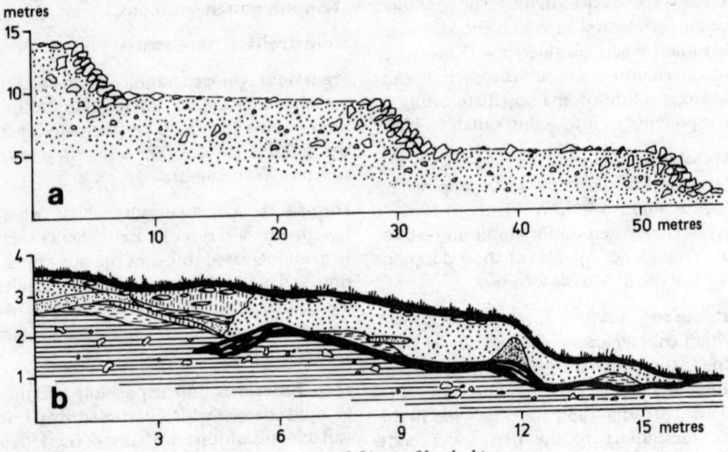

Figure 242 *Steps: (a) a stone-banked terrace and (b) a turf-banked terrace*

treads are bare and have gradients gentler than that of the general slope. Some sorted steps are seen to be an intermediate stage between polygons and STRIPES. [*242*]

steptoe A term referring to an isolated hill or tract of land that has been completely surrounded by a lava-flow.

steradian A geometrical term which can be defined as the unit solid angle which cuts a unit area from the surface of a sphere of unit radius centred at the vertex of the solid angle. There are 4π steradians in a SPHERE. The term is used in REMOTE SENSING to define the angle over which REFLECTANCE from the Earth's surface is measured.

stereographic projection The CONFORMAL member of the AZIMUTHAL PROJECTION class of MAP PROJECTIONS. Both meridians and parallels are circles and are projected on to a standard plane at a point on the opposite extremity of the diameter. Thus the map has the special property that all circles on the globe appear as circles on the projection. It is a very ancient projection, and is attributed to Hipparchus (160–125 BC). It is of most use for any map on which the world needs to be depicted in two hemispheres, but area becomes increasingly exaggerated away from the central point. [*243*]

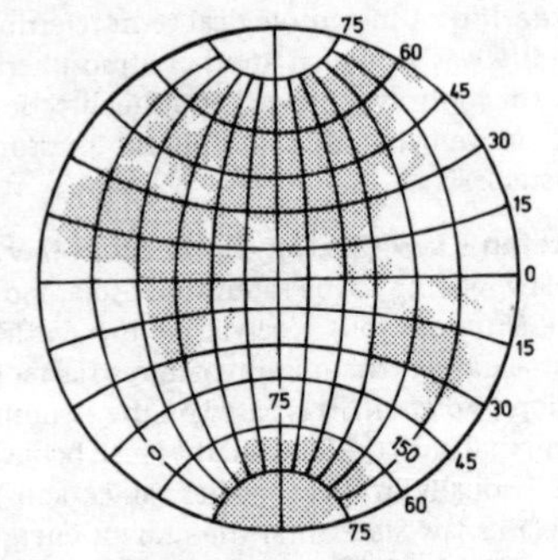

Figure 243 *Stereographic projection*

stereoscope A binocular optical instrument for viewing pairs of overlapping photographs (*stereograms*) of the same landscape taken from slightly different angles. A three-dimensional impression of the topography is obtained, enabling the viewer to make an interpretation of AERIAL PHOTOGRAPHS. The construction of accurate maps from the latter is termed *stereo-plotting*.

Stevenson screen A specially designed housing for meteorological instruments designed by the father of Robert Louis Stevenson. It comprises a white wooden box, raised on legs so that its base is 1 m from the ground, and louvred on all sides to enable air to circulate freely around the instruments without them being exposed to direct sunlight. Access is provided by the hinged, north-facing side of the screen. The ordinary version houses the DRY-BULB THERMOMETER, the WET-BULB THERMOMETER and the MAXIMUM AND MINIMUM THERMOMETER. The larger pattern also accommodates the HYGROGRAPH and THERMOGRAPH.

stick-slip **1** The behaviour of a geological FAULT characterized by alternations of slow STRAIN RATE and rapid SLIP when the STRESS overcomes the friction (FRICTION, COEFFICIENT OF) on the fault-plane. **2** The changing motion of a glacier when it moves over bedrock. Long-term phases when no movement occurs alternate with sudden slip movements of a few centimetres. *See also* BASAL SLIP, GLACIER FLOW.

stilling well A term given to the mechanism that measures the still-water surface level in a river channel. It comprises a large-diameter vertical tube on the bank of the river, connected to the water in the channel by intake pipes and enclosing a FLOAT RECORDER. [*244*]

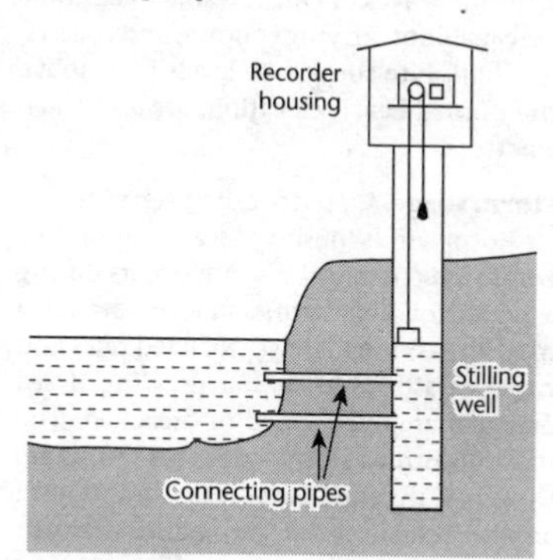

Figure 244 *Stilling well*

still-stand A time period during which the relative level of land and sea remains constant, i.e. there is no ISOSTATIC or EUSTATIC change.

stochastic A term used for a method of obtaining a solution to a problem in which a random variable is present and in which, because of the presence of this element of randomness, various degrees of probability (CONFIDENCE LIMITS) need to be stated. A stochastic process contrasts with a DETERMINISTIC PROCESS.

stochastic-process model MODEL.

stock A small body of intrusive igneous rock similar to a BATHOLITH, of which it may be the upstanding CUPOLA, but defined as having a surface exposure of less than 100 km^2 (*c.* 40 sq miles). *See also* BOSS.

Stokes's law An equation, stated in 1850 by G. G. Stokes, to express the rate of settling of spherical particles in a viscous fluid of known density and viscosity. It is applied in PARTICLE SIZE analysis of soils and sediments, by the use of pipette, hydrometer or centrifuge methods. Its formula is $V = (2gr^2)d_1 - d_2)/9\eta$, where: V is velocity of fall (cm sec^{-1}); g is acceleration of gravity (cm sec^{-2}); r is radius of particle; d_1 is density of particle (g cm^{-3}); d_2 is density of fluid (g cm^{-3}); η = viscosity of fluid (dyne sec cm^{-2}). It is used particularly to determine the diameters of the *fines* which cannot conveniently be sieved.

stone **1** Hard mineral matter. **2** A small piece of rock. **3** A gem or precious stone. In general, the term should only be used as a suffix (e.g. sandstone, limestone, ironstone) when describing different rock types or when making reference to a building stone, e.g. freestone. *See also* STONE AGE, STONINESS.

Stone Age The expression given by archaeologists to the period of time when early Man used stone implements (*artefacts*) during the later part of the Pleistocene and that part of the Post-Glacial prior to the BRONZE AGE. The earliest part of the Stone Age is referred to as the *Eolithic*, which is followed by the *Old Stone Age* (PALAEOLITHIC), *Middle Stone Age* (MESOLITHIC) and the *New Stone Age* (NEOLITHIC).

stone-banked terrace A hill-slope terrace consisting of fine soliflucted (SOLIFLUCTION) material flanked by coarser material. It is part of a stepped hill-slope profile, with lengthy 'treads' (*c.* 30 m from front to rear) interrupted by steep 'risers' (*c.* 3–5 m high), occurring on slopes of moderate gradient (10°–25°). It has been suggested that the stones are heaved to the surface by frost (FROST-HEAVING) where they then tend to travel downslope faster than the slow flowage of the finer material. At certain places their movement will be arrested by gradient changes, vegetation or rock exposure, causing the coarser material to accumulate as a nucleus for an increasing pile of stones against which the finer material builds up. There may be little further movement beyond a critical point, except for CREEP. They are sometimes referred to as *stone garlands* or *stone-banked lobes*. *See also* GOLETZ TERRACES, STEPS, TURF-BANKED TERRACE. [*242*]

stone-line A layer of angular or subangular stones at the base of the A-horizon in a tropical-soil profile. It is thought to be a type of lag-deposit (LAG) formed by SURFACE WASH and subsequently buried by soil CREEP.

stone pavement An accumulation of flat-lying cobbles and boulders fitted together in a mosaic form at the ground surface in a PERIGLACIAL environment. Although superficially similar to a DESERT PAVEMENT the latter is composed of smaller stones. There is no clear agreement on its mode of formation, although in general it is thought to result from the combined action of: UPFREEZING of stones; ground saturation and the removal of FINES by meltwater; the rotation and shifting of stones in saturated ground under their own weight and the weight of overlying snow; *See also* GELIFLUCTION; AUFEIS.

stone polygon POLYGONS.

stone stripe STRIPES.

stoniness The relative proportion of stones in or on a soil. A property used in soil classification.

stoping **1** A mechanism that has been suggested by some geologists to account for the emplacement of large masses of igneous rocks. It is suggested that MAGMA is forced up into fissures of the COUNTRY-ROCK, causing blocks to break off and sink into the magma by a process similar to ENGULFMENT. Other geologists raise objections to such a mechanism on the grounds that the detached blocks would not be assimilated into the magma owing to differences in specific gravity. **2** A method of ore extraction in a mine which proceeds by working upwards or downwards rather than laterally along a vein.

storage, store **1** In general usage, a device capable of retaining mass, energy or information. In the USA it is termed STORAGE, in the UK it is often used synonymously with STORE. **2** A hydrological term referring to the reservoirs of water contained within the HYDROLOGICAL CYCLE. These include surface-water storage (RESERVOIR), SOIL WATER storage and GROUNDWATER storage.

storm A severe meteorological disturbance, officially designated as force 10 (storm) and force 11 (violent storm) on the BEAUFORT SCALE, in which wind-speeds range between 55 and 72 miles/h (25 and 32.6 m/sec). *See also* DUST-STORM, STORM SURGE, THUNDERSTORM.

storm beach A linear mass of coarse, rounded material that has accumulated at a high elevation above the foreshore during a storm. Waves of considerable magnitude are capable of carrying BOULDERS and COBBLES, in addition to shingle, to levels high above the normal beach elevation. *See also* BEACH RIDGE.

storm surge A rapid rise of sea-level caused by storm winds 'pushing' sea water on to a windward coast. If the storm occurs during a period of high SPRING TIDE the sea-level may surpass the highest predicted tides and cause catastrophic coastal flooding if sea defences are overtopped or breached. The most disastrous storm surges on record are those of Galveston, Texas, USA, in 1900 and around the shores of the southern North Sea in 1953.

stoss **1** A German term for the slope facing the direction from which a glacier moves. **2** The less steep side of a RIPPLE. The opposite slope is termed the *lee*. *See also* ONSET AND LEE. [*210*]

stoss-and-lee terrain An arrangement of small, persistently asymmetric hills in a glaciated region, in which all exhibit a relatively gentle abraded slope on the side (stoss) facing the direction from which the ice-sheet advanced, while the lee slope is steeper, rougher and more plucked. The hills have the same general form as, but are somewhat larger than, a ROCHE MOUTONNÉE. *See also* KNOCK-AND-LOCHAN.

stow MORPHOLOGICAL REGION.

strain The deformation of a body as a result of an applied force (STRESS). The change may be in shape and/or volume. Within elastic limits strain is proportional to stress. When a body deforms equally in all directions, thereby resulting in planar structures, it is termed *homogeneous strain*. If it does not deform equally in all directions, thereby causing linear structures to become curved, it is termed *heterogeneous strain*.

strain-rate The speed at which a material deforms when subjected to STRESS. It is usually expressed as the percentage change of the original length per second.

strain-slip cleavage A type of secondary CLEAVAGE which has deformed the original cleavage planes, leading to flexing and FOLIATION.

strait A relatively narrow waterway linking two larger bodies of water, e.g. the Strait of Dover links the North Sea with the Channel.

strandflat A Norwegian term for their western coastal wave-cut platform (ABRASION PLATFORM) now raised above sea-level (RAISED BEACH) by ISOSTATIC MOVEMENTS following the removal of the weight of the Pleistocene ice-sheets. It extends more than 50 km from the former cliff lines in some areas. [*80*]

strath A broad flat-floored river valley deeply filled with alluvial deposits and often with glacigenic deposits. It is commonly used as a prefix in Scotland, where it is regarded as being wider than a glen.

strath terrace RIVER TERRACE.

stratification 1 In geology, the structure produced when sediments or volcanics are laid down in horizontal layers (STRATUM). **2** In biogeography, the arrangement of vegetation in layers. **3** In pedology, the development of layers within a soil (if soil horizons are being described the terms *horizonation* is preferred). **4** In hydrology and oceanography, the formation of distinct temperature layers in a body of water (THERMOCLINE).

stratified drift UNSTRATIFIED DRIFT.

stratified random sample RANDOM SAMPLE.

stratified scree GRÈZES LITÉES.

stratiform cloud A cloud form characterized by its horizontal layers (STRATUS, of which it is the adjectival form).

stratigraphy The branch of GEOLOGY which deals with the composition, sequence, spatial distribution, classification and correlation of the stratified rocks (SEDIMENTARY ROCKS, VOLCANIC ROCKS). *See also* BIOSTRATIGRAPHY, CHRONOSTRATIGRAPHY, LITHOSTRATIGRAPHY.

stratocumulus A type of cloud (symbol: Sc), grey or whitish in colour, in which the horizontal layers of the cloud sheet have developed into rolls and rounded masses but without any breaks. It extends from about 500 m to 2,000 m. *See also* CLOUD. [*43*]

stratomere Any segment of a sequence of rock (LITHOSTRATIGRAPHY) of no standard thickness, deposited during no standard interval of geological time. *See also* CHRONOMERE.

stratomeric standard terms CHRONOZONE, SERIES, STAGE, SYSTEM.

stratopause The boundary between the STRATOSPHERE (below) and the MESOSPHERE (above), at an elevation in the atmosphere of about 50 km. *See also* MESOPAUSE, TROPOPAUSE. [*17*]

stratosphere One of the concentric layers of the ATMOSPHERE, occurring between the TROPOSPHERE (below) and the MESOSPHERE (above). It is bounded by the TROPOPAUSE (below) and the STRATOPAUSE (above). It extends from 50 km above the Earth's surface down to the tropopause, which itself varies in elevation from 8 to 16 km. That part of the JET STREAM known as the *polar-night jet stream* occurs in the lowest layer of the

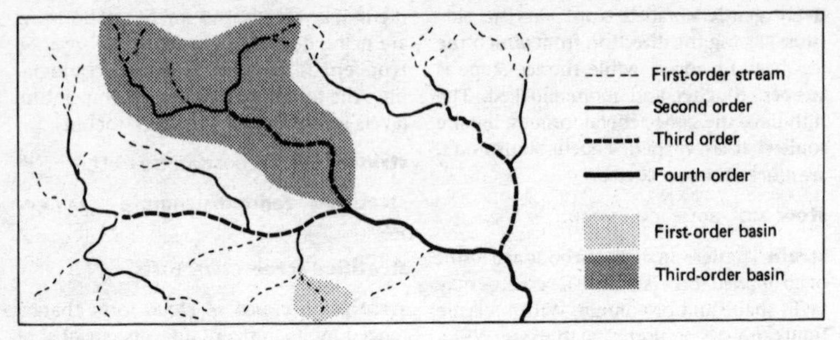

Figure 245 *Stream orders*

stratosphere. The lower part of the stratosphere is virtually isothermal, but in the upper layers temperature increases with height (between 20 and 50 km) to reach its secondary maximum at 0° C at the *stratopause*. A great part of the atmospheric OZONE is found in the stratosphere, but very little water vapour or AEROSOLS, except from volcanic explosions. [*17*]

strato-volcano COMPOSITE VOLCANO.

stratum A bed of sedimentary or pyroclastic material, regardless of thickness, consisting of a number of layers. Its plural form is *strata*.

stratus A type of CLOUD (symbol: St), grey in colour and with a uniform base. It occurs in thin sheets at any height from the surface up to 500 m, above which it changes to other STRATIFORM CLOUD forms. Often associated with drizzle. *See also* ALTOSTRATUS, CIRROSTRATUS, NIMBOSTRATUS, STRATOCUMULUS. [*43*]

stream abstraction ABSTRACTION.

stream capacity CAPACITY.

stream derangement DERANGED DRAINAGE.

stream-flood A rarely used term, introduced by the US geomorphologist, W. M. Davis, to indicate a FLASH-FLOOD confined to a normally dry channel in arid terrain. *See also* FLOOD.

stream-flow FLOW, FLOW EQUATIONS, FLOW REGIMES, FLUVIAL, GAUGING STATION, HYDRAULICS, HYDRODYNAMICS.

streamline **1** A line parallel to the direction of *all* parcels of air in the wind field, measured at a given moment at *all* points along the line, i.e. it gives an instantaneous indication of air movement at any specific time. In tropical areas the mechanism of normal GEOSTROPHIC FLOW is not generally applicable, so that synoptic charts which depict isobars cannot be usefully employed, with streamlines often being utilized in their place. Note that the term TRAJECTORY is not synonymous with streamline, for it has a different meaning. **2** EQUIPOTENTIAL.

stream order A morphometric classification (MORPHOMETRY) of a drainage system according to a hierarchy of orders of magnitude of the channel segments. [*245*]. The scheme of *stream ordering* was devised by R. E. Horton (HORTONIAN ANALYSIS) and later modified by A. N. Strahler. Within a single drainage basin the unbranched channel segments which terminate at the stream head are designated as *first order*. Where two *first-order* segments meet, a channel of the *second order* is produced, extending down to meet another *second-order* channel to create a *third-order* channel and so forth. It is noteworthy that where a *first-order* channel meets a *second-* or *third-order* channel there will be no increase of order at the junction. *See also* BASIN ORDER, BIFURCATION RATIO, LAW OF STREAM LENGTHS, LAW OF STREAM NUMBERS, LAW OF STREAM SLOPES, ORDER.

strength of sediments The maximum resistance of any sedimentary material to applied STRESS. The two main sources of resistance are COHESION and FRICTION.

stress The system of internal forces per unit area of a body produced as a reaction to externally applied forces which tend to deform the body. If two forces are applied towards each other *compressive stress* will result; if two forces act away from each other they will create *tensional stress*; if the forces act tangentially to each other *tangential stress*, or SHEARING, will occur. When materials are subjected to stress they will suffer STRAIN. Stress may be found by dividing the total force by the area to which it is applied.

stress ecology A term introduced by G. W. Barrett (1981) to describe the impact of natural or artificial perturbations on the structure and function of an ECOSYSTEM, e.g. fire, flood, pesticides and radiation.

striated soil NEEDLE ICE.

striation, stria A tiny groove or scratch on the surface of an ice-abraded rock, produced by the scoring action of rocks frozen into the base of a glacier. Its plural form is *striae*, hence *glacial striae*, which demonstrate the direction of former ice movement. [*212*]

strike 1 The direction taken by a horizontal line on the plane of an inclined rock stratum, fault or cleavage. It is at right angles to the TRUE DIP of the rock structure. **2** It is used adjectivally to describe the direction of a structure (STRIKE-FAULT) or the direction of a morphological feature developed along a structure (STRIKE VALLEY). [*62*]

strike-fault A fault whose STRIKE is parallel to the strike of the rock strata through which it passes. It differs from a *strike-slip fault* (TEAR-FAULT) which trends transversely to the strike of the folded strata.

strike line A line joining points of equal elevation on a rock stratum above or below a selected stratigraphic datum. It is essentially a structure CONTOUR.

strike-slip fault TEAR-FAULT.

strike valley A valley which has been eroded along structural lines that lie parallel with the regional strike of an area. [*222*]

string bogs (strangmoore) Areas of MUSKEG that consist of ridges (<2 m high) of peat and vegetation interspersed with depressions often filled with shallow ponds, all occurring on terrain with gradients of <2°. All the peat ridges contain ice lenses and are therefore typical of a PERIGLACIAL environment, especially the sub-Arctic TAIGA zone, but occur slightly further south than the PALSAS of the N hemisphere. There is no clear agreement on their mode of origin and the following hypotheses have been proposed: SOLIFLUCTION and wrinkling of the bog surface; hydrostatic pressure rupturing the bog surface and frost-thrusting raising ridges; growth of ice lenses during a cold period and subsequent solifluction during climatic warming; differential thawing of PERMAFROST leading to local bog collapse; a combination of biological (vegetation) factors and hydrological factors.

stripes A type of PATTERNED GROUND oriented down the steepest slope gradient of a terrain influenced by marked thermal activity in the soil. They can be divided into sorted and non-sorted stripes according to their degree of SORTING. *Sorted stripes* (*stone stripes*) consist of parallel lines of stones and intervening strips of FINES, with the latter tending to be several times wider than the coarser stripes. Any tabular stones, especially in the coarse fraction, are usually on edge and stone size generally decreases with depth. *Non-sorted stripes* consist of parallel lines of vegetation-covered ground and intervening strips of relatively bare ground, with the latter being sometimes wider than the former. Both categories occur on slopes of more than 3° gradient and have been recorded on gradients up to 30°. It has been suggested that they are downslope developments, respectively, of sorted and non-sorted CIRCLES, NETS and POLYGONS, and that their origins may be found in a single process or a combination of processes including DESICCATION-CRACKING, DILATION-CRACKING, FROST-CRACKING and FROST-HEAVING (*see also* CONGELIFLUCTION, CONGELIFRACTION, GELIFLUCTION); differential mass displacement; differential thawing and ELUVIATION; RILL EROSION. They are not confined to the PERMAFROST zone or the true PERIGLACIAL zone, but range from polar to warm arid environments. [*246; see also 179*]

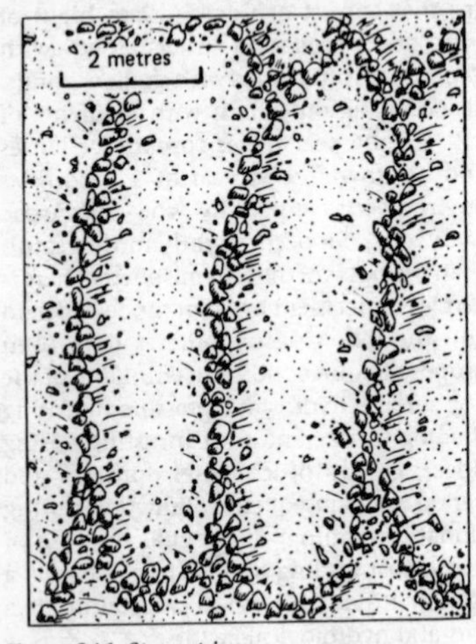

Figure 246 *Stone stripes*

stripped plain An area of low relief developed upon gently tilted or horizontally bedded sedimentary rocks from which less-resistant sediments have been removed thereby revealing a more resistant stratum that appears to have controlled the depth of erosion. The plane surface thus exposed by *stripping* is also known as a *stripped surface* or *structural surface*, but this may be uplifted to form an upland plain. *See also* EXHUMATION.

stripped surface STRIPPED PLAIN.

Stromatolites, stromatoliths Laminated, calcareous structures in some sedimentary rocks. They were first described by E. Kalkowsky (1908) as microbially formed structures in the BUNTER of North Germany. They are thought to have been developed in marine, swamp, or lacustrine environments when sediment was trapped and calcium precipitated as a result of the metabolic activity of such micro-organisms as ALGAE. Stromatolites vary greatly in form (branching, columnar, laminar, bulbous, etc.) and in size (MESOSCOPIC to MACROSCOPIC).

Strombolian eruption A type of volcanic eruption in which molten lava and gases escape at frequent intervals but without violent explosions, i.e. without a great build-up of pressure within the cone. It is named after the volcano on the island of Stromboli off N Sicily. [*271*]

stroph KATABATIC WIND.

structural geology The scientific study of rock structures, including their form, genesis and spatial distribution, but not their composition.

structural surface STRIPPED PLAIN.

structure 1 In general, an organized body or combination of connected elements. **2** In geology, the overall relationships between rock or TILL masses together with their large-scale arrangements and dispositions. *See also* DIP, FAULT, FOLD, STRIKE, UNCONFORMITY. **3** In PETROLOGY, the more detailed internal relationships between the different parts of rocks or tills. *See also* BANDED STRUCTURE, BED, CLEAVAGE, FLOW STRUCTURE, JOINT, LAMINA, PETROFABRIC ANALYSIS, TILL FABRIC. **4** In biogeography, the spatial distribution of the vegetation in an ASSOCIATION. **5** A SOIL SURFACE. **6** In civil engineering, anything of human construction.

sturzstrom A term introduced by K. J. Hsü (1975) to describe very large ROCK-FALLS from high cliff faces and which travel long distances (5–30 km) at high velocities (90–350 km h^{-1}) due to their low coefficient of friction (FRICTION, COEFFICIENT OF). *See also* BERGHLAUP, BERGSTURZ.

subaerial Any feature, substance or process which occurs or operates on the Earth's surface, as distinct from subterranean or submarine phenomena.

sub-alluvial bench A rock surface of marked convexity formed beneath alluvial deposits (ALLUVIUM) at the base of a retreating scarp slope. It was postulated by A. C. Lawson in 1915 that the process of ALLUVIATION, commencing on the lower part of slopes, would progressively extend towards the uplands thereby creating a sub-alluvial bench of solid rock [*247*] beneath the surface of the alluvial fan as the scarp progressively retreated.

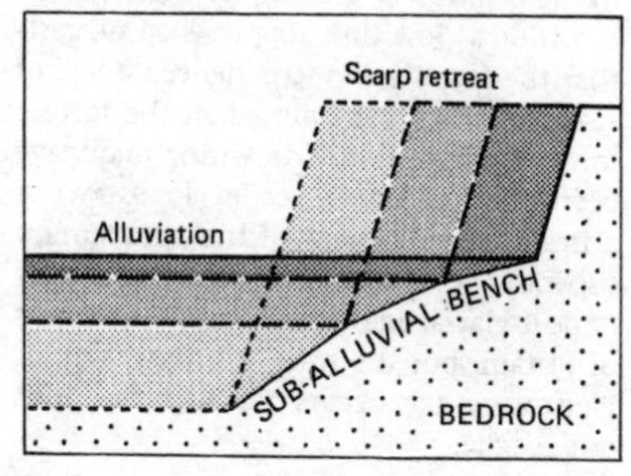

Figure 247 *Sub-alluvial bench*

sub-Arctic 1 A term relating to the latitudinal zones which occur adjacent to but outside the Arctic Circle. **2** A term used by W. Köppen (KÖPPEN'S CLIMATIC CLASSIFICATION) to group together some of the climatic types of his cold zone (D) in which there are some months with temperatures above 10° C (50° F) and both precipitation and evaporation are low. **3** A term occasionally used as a synonym for the PRE-BOREAL climatic phase.

Sub-Atlantic period An expression used to describe the later phase of decreasing warmth which commenced in NW Europe about 2,450 BP, in post-glacial (FLANDRIAN) times. It is referred to as part of the Late-Temperate (Fl.III) or pollen zone VIII in Britain. It was marked by a deterioration of summer temperatures from those of the preceding SUB-BOREAL PERIOD, while the increasing rainfall helped in the regeneration of peat bogs (termed the Upper Turbarian phase in Scotland). Although lime forests declined, ash and birch continued to increase while beech and hornbeam became dominant in southern Britain. The period saw the Middle and Late BRONZE AGE cultures replaced by those of the IRON AGE. *See also* TERMINOCRATIC. [*81*] *See also* BLYTT–SERNANDER MODEL.

Sub-Boreal period The early phase of decreasing warmth which succeeded the so-called Climatic Optimum (ATLANTIC PERIOD) in NW Europe about 4,450 BP in post-glacial (FLANDRIAN) times. It is the earlier part of the Late-Temperate (Fl.III) or pollen zone VIIb in Britain, characterized by cooler but drier conditions which saw the drying out of the Atlantic Period peat bogs and a spread of ash, birch and pine forests. It has been suggested that wind-blown deposits were laid down more frequently during this drier period of some 2,000 years before the increasing wetness of the SUB-ATLANTIC PERIOD replaced it in about 2,450 BP. Some of the wind-blow may have been caused by the introduction of agriculture and by the increased forest clearance by the peoples of NEOLITHIC culture who settled in the British Isles during this period. After 4,000 BP the latter culture was gradually replaced by that of the BRONZE AGE. It was a time of *isostatic* uplift in N Britain, which witnessed the growth of forests (the Upper Forestian phase) on uplifted coastal marine clays (the carse clays of Scotland) prior to their demise and renewed marine flooding in the succeeding Sub-Atlantic period. [*81*] *See also* BLYTT–SERNANDER MODEL.

sub-climax The stage reached by any seral vegetation community (SERE) that has been permanently checked in its progression towards a CLIMAX by natural factors, e.g. exposure, poor drainage. It may be indistinguishable from a PLAGIOCLIMAX, which is due to human interference.

subduction zones The linear areas which represent the zones where crustal plates (PLATE TECTONICS) are overridden by other plates and are forced down into the underlying MANTLE along an oblique plane of thrusting [*223*]. They mark collision zones or destructive plate margins characterized by a zone of EARTHQUAKE activity, an orogenic belt, an ISLAND ARC [*81*] and volcanic activity. As the crustal sediments of the descending plate margin reach greater depths they are melted, thereby becoming MAGMA which in turn rises in the form of BATHOLITHS in the core of the orogenic belt. *See also* DEEP, OCEAN. [*188*]

subglacial Pertaining to the environment beneath a glacier. It refers to the processes by which a glacier moves across its floor (GLACIER FLOW) in addition to the movements of meltwater at the base of the ice. *See also* BASAL MELTING, ENGLACIAL, GROUND MORAINE, SUPRAGLACIAL.

subglacial channel A type of MELTWATER CHANNEL cut by glacial meltwaters beneath a glacier or ice-sheet. It may take the form of a *subglacial* CHUTE which descends down the slope of an ice-buried valley, or it may be a normal basal channel or one with an UP–DOWN PROFILE. Most subglacial channels are now streamless or occupied by tiny watercourses which clearly were not responsible for their formation.

subhumid A climatological term referring to any area lying just outside the humid zones (where precipitation is adequately distributed seasonally) and the desert zones where ARIDITY prevails. It is a zone too dry for tree growth but humid enough for grass growth. *See also* SAVANNA CLIMATE, STEPPE.

sublimation The process whereby a solid is converted into a vapour (or vice versa) with no liquid state intervening. Chemically speaking, the term can only refer to the conversion of ice to water vapour, but in its wider use it has been adopted to describe the formation of ICE-CRYSTALS directly from water vapour during a phase of condensation occurring below freezing-point. *See also* ABLATION, FREEZING NUCLEI, BERGERON–FINDEISEN THEORY OF PRECIPITATION.

sublimation till TILL.

submarginal drainage channel MARGINAL CHANNEL.

submarine canyon A steep-sided V-shaped valley incised into the CONTINENTAL SHELF and the CONTINENTAL SLOPE before opening out on to the deep-ocean floor. This exit is often marked by a large fan-shaped mass of sediment (*submarine fan*) deposited by TURBIDITY CURRENTS and possibly submarine slumps due to earthquakes. The canyon may have been formed partly by turbidity currents, or by the effects of subaqueous SPRING-SAPPING, or simply by fluvial erosion by rivers flowing across the continental shelf during periods of lower sea-level.

submarine channel SEA VALLEY, SUBMARINE CANYON.

submerged forest The expression given to a peat layer, with eroded tree stumps in the position of growth, exposed along a coastline at low tide. Its presence suggests that the locality has experienced a rise of sea-level since the time when the former forest flourished. The drowning may have been due to a rise in sea-level (EUSTASY) or a negative movement of the land surface (ISOSTASY). Several submerged forests, mostly of post-glacial age, occur around the coasts of Britain, but the so-called FOREST BED, of Pleistocene age, occurs near high-tide level in E Anglia.

submergent shoreline A coastline that has resulted from a rise of sea-level (EUSTASY) or a negative movement of the land surface (ISOSTASY). It will be characterized by drowned estuaries, submerged river valleys (DROWNED VALLEY), RIAS, SUBMERGED FORESTS and, if the area has been glacially overdeepened, FIORDS. *See also* SHORELINE CLASSIFICATION.

subpolar glacier One of the three types of glacier identified by H. W. Ahlmann (*see also* COLD GLACIER, TEMPERATE GLACIER). It differs from a *cold glacier* (*polar glacier*) mainly because some melting occurs in the subpolar glacier's accumulation zone during summer, despite the fact that its ice remains well below the PRESSURE MELTING-POINT at depths of 10 to 20 m.

subsequent stream A stream that has developed its valley along an outcrop of less-resistant rocks or a regional fault or joint pattern, at right angles to the drainage which is consequent upon the slope of the land (CONSEQUENT STREAM). Because they are developing along relatively weak structures subsequent streams develop mainly by HEADWARD EROSION. [*167* and *222*]

subsere, secondary sere Any point in the seral development of a vegetation CLIMAX in which the succession towards a SERE has been halted temporarily by non-climatic arresting factors. These controls may be topographic, edaphic or biotic, but once these have been removed the normal pattern of seral development can again be resumed.

subsidence **1** In a climatological sense the term refers to the widespread downward movement of air associated with DIVERGENCE at the surface, as in an ANTICYCLONE. It is invariably associated with dry weather. *See also* INVERSION OF TEMPERATURE. **2** In a topographical sense the term refers: (a) to a small-scale lowering of the surface following cave-roof collapse in a KARST area; (b) to the gradual settling of the ground resulting from the extraction of mineral deposits, e.g. salt or coal subsidence; or (c) to a large-scale structural readjustment of the Earth's surface, as in basin downwarping or RIFT VALLEY formation. *See also* SOLUTION COLLAPSE.

subsoil The weathered parent material which lies above the bedrock and below the true soil (SOLUM). It corresponds to the C-HORIZON. *See also* SOIL HORIZON.

subsurface wash An expression covering a group of processes relating to sediment transport within the REGOLITH, with the actual flow of water being referred to as THROUGHFLOW. The ratio of throughflow to *surface flow* (OVERLAND FLOW) is greatest where the regolith is thick and has a high INFILTRATION RATE. The ratio of throughflow to PERCOLATION depends on the PERMEABILITY of the regolith and the underlying bedrock. Some throughflow may emerge lower down the slope to continue as SURFACE WASH. When throughflow is concentrated into SEEPAGE lines it is termed TUNNELLING, which takes place through a PIPE. If fine soil particles are transported downslope through the regolith by the action of throughflow the pedological process is termed *lateral eluviation* (ELUVIATION).

subtropical A term used to describe the latitudinal zones adjacent to but just outside the Tropics of Cancer and Capricorn, more specifically extending as far polewards as latitudes 40°N and 40°S. It is used in various climatic classifications to describe those areas exhibiting no month with a mean temperature below 6° C (43° F). A *subtropical high* is one of the ANTICYCLONES in the zone of atmospheric high pressure which is a semi-permanent feature of the *subtropics*.

Subtropical Jet Stream (STJ) The high-velocity air flow of the JET STREAM at about latitude 32°.

succession **1** The vertical sequence of rock strata at a particular locality. **2** The sequential development of changes within a plant community as it progresses towards a CLIMAX. *See also* SERE.

suction The state of SOIL WATER under unsaturated conditions. Soil moisture suction is a negative pressure potential created by the capillary (CAPILLARY WATER) and adsorptive (ADSORBED WATER) forces. This results from the soil matrix holding its moisture at a pressure less than atmospheric pressure. *See also* CAPILLARITY.

Sudd A floating mass of vegetation, comprising plants that have broken away from neighbouring swamps, which regularly blocks the main channel of the White Nile in S Sudan. It has long been a serious navigational hazard and has caused the river to bifurcate and form ANASTOMOSING channels, marshes and lakes.

suffosion A process of erosion occurring on the upper surface of limestone buried by ALLUVIUM or glacial deposits, whereby percolating water creates voids in the limestone surface into which fine material from the overlying drift is washed.

sugar limestone A term given to exposures of easily weathered, crumbly limestone after it has been affected by thermal METAMORPHISM.

sugar-loaf A fanciful term for an INSELBERG, e.g. that which rises steeply from the city of Rio de Janeiro, Brazil.

sulphation A term referring to a type of CHEMICAL WEATHERING when sulphur dioxide (SO_2) reacts with calcium carbonate ($CaCO_3$) in humid and often polluted environments. It is an especially significant process in the decay of calcareous building stones (FREESTONE) where a layer of GYPSUM can form on surfaces protected from rain-wash. If AEROSOLS settle on such layers an unsightly BLACK CRUST may form, ultimately leading to a break-up of the stonework.

sulphur A non-metallic yellow-coloured element (S) occurring in the vicinity of volcanic vents, hot springs and fumaroles and also in association with IRON PYRITES.

summer The three months of June, July and August in the N hemisphere and the warmest of the seasons (SEASON); the period between the summer SOLSTICE and the autumn EQUINOX. (About 21 June to about 22 September in the N hemisphere.) In the S hemisphere the dates are 21 December to 20 March and there the three summer months are December, January and February. Paradoxically, in the N hemisphere, the first day of summer (21 June) is also called Midsummer Day.

summit plane ACCORDANT SUMMITS.

Sun The central body of the SOLAR SYSTEM. It has a diameter of *c.* 1,392,000 km (865,000 miles) and is believed to consist of a liquid inner section surrounded by an incandescent gaseous outer covering. Its surface temperature is thought to be of the order of 6,000° C and it is the source of solar radiation (INSOLATION) and SOLAR ENERGY. It is on average 150 million km (*c.* 93 million miles) from the Earth and this distance is traversed by the Sun's ELECTROMAGNETIC RADIATION in about 9 min 20 sec. The Sun completes a single rotation in 24.5 days at its equator.

sun cracks DESICCATION CRACK.

sunshine An element comprising the direct radiation (INSOLATION) received from the SUN, the amount varying according to latitude and to degrees of cloudiness. Some of the world's hot deserts receive 80–90 per cent of possible sunshine hours, amounting to more than 3,600 hours annually. In Britain the amount ranges from about 1,850 hours along the south coast to some 1,200 hours in N Scotland. *See also* DIFFUSE RADIATION, ISOHEL, SUNSHINE RECORDER.

sunshine recorder An instrument to measure the amount of bright sunshine received at a point on the Earth's surface. The most widely used device is the *Campbell–Stokes recorder*, comprising a glass sphere which focuses the Sun's rays and leaves a trace on a linear card, by means of burning, whenever bright sunshine occurs. Thus hazy sunshine may not be recorded.

sunspots Dark spots on the surface of the SUN, representing disturbances of the Sun's MAGNETIC FIELD. Their numbers and recurrence fluctuate according to the solar cycle, but have an average periodicity of some 11 years. Recently, a 100-year cycle has been recognized, thought by some to account for the LITTLE ICE AGE of the 16th to 19th century. It is believed by some climatologists that there is a strong correlation between sunspot activity and fluctuations of climate.

Supan's temperature zones A classification of the globe into CLIMATIC REGIONS according to mean annual temperature differences. The zones were suggested by A. Supan in 1896, who suggested that the mean annual isotherm of 20° C should delimit the hot climates and that the polar climates should be separated from the temperate climates by the isotherm of 10° C for the warmest month. Although the boundaries appear to have some biological significance, e.g. the polar limit of tree growth, they are probably more coincidental than causal. KÖPPEN'S CLIMATIC CLASSIFICATION.

supercooled water Water which has been cooled to a temperature below its normal FREEZING-POINT (0° C), but which remains in liquid form if undisturbed. It is a common element in clouds, where water droplets below 0° C cannot form ice because of the absence of CONDENSATION NUCLEI. If supercooled water comes into contact with a body such as an aircraft, freezing will take place, leading to ICING. *See also* BERGERON–FINDEISEN THEORY OF PRECIPITATION, RIME.

superficial deposits Materials of Pleistocene or Holocene age that lie in an unconsolidated form on the land surface of the Earth. They are formed independently of the underlying BEDROCK and have usually been moved to their present position by natural agencies, with the exception of PEAT,

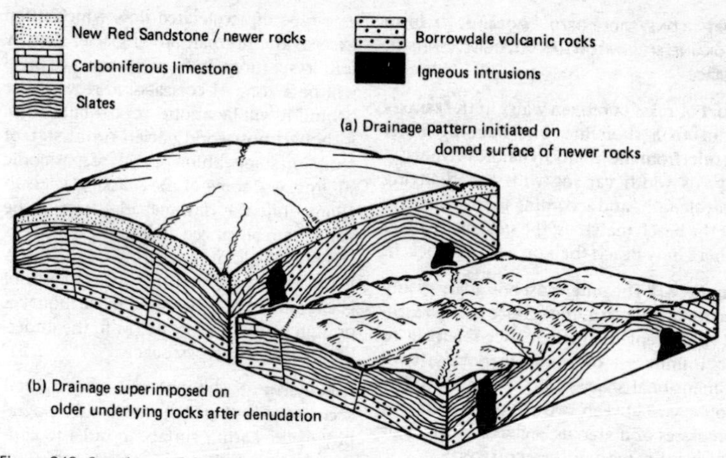

Figure 248 *Superimposed drainage in the Lake District*

which has accumulated *in situ*, and of CLAY-WITH-FLINTS, which is often classed as a superficial deposit but which, since it has developed from the solid rocks, is strictly speaking a weathering product. Among the deposits that have been previously transported by various means are: ALLUVIUM; COLLUVIUM; DRIFT, GLACIAL; blown sand (DUNE); GRAVEL TRAINS; HEAD; PLATEAU GRAVEL; LOESS; RAISED BEACH deposits; SOLIFLUCTION deposits.

superglacial SUPRAGLACIAL.

superimposed drainage A drainage system which exhibits a discordance (DISCORDANT DRAINAGE) with the underlying rock structures because it originally developed on a cover of rocks that has now disappeared owing to denudation. Consequently, river directions relate to the former COVER ROCKS and, as the latter were being eroded, the rivers have been able to retain their courses notwithstanding the newly exposed structures. Although there are several instances where *superimposition* is thought to explain the discordance of certain drainage patterns in Britain, e.g. the Lake District rivers, it has been suggested that in some instances, such as the discordant rivers of the Weald, the concept of ANTECEDENT DRAINAGE may be a more satisfactory explanation. [*248*]

superimposed ice A term given to the ice formed when water from melting snow comes into contact with the cold surface of a GLACIER.

superimposed profile One of several topographical profiles that are drawn on a single diagram one on top of another to indicate where there is a marked coincidence of EROSION SURFACES, etc. [*195*]. *See also* COMPOSITE PROFILE, PROJECTED PROFILE.

superposition, law of PRINCIPLE OF SUPERPOSITION.

supersaturation 1 The state of a parcel of air when its RELATIVE HUMIDITY exceeds 100%, i.e. when it contains more water vapour than is required to saturate it. This state would normally lead to CONDENSATION but this will only occur if CONDENSATION NUCLEI are present. **2** Supersaturation of oxygen in a water body can occur by PHOTOSYNTHESIS during daylight hours.

supplementary contour An extra CONTOUR included on certain relief maps where the terrain is very irregular and where the normal CONTOUR INTERVAL would be insufficient to illustrate the character of the relief.

supraglacial Pertaining to the environment at the surface of a glacier. ABLATION,

DIRT CONE, ICE-CORED MORAINE, MEDIAN MORAINE, MELTWATER CHANNEL, OGIVE, PENITENT, SÉRAC.

surf A mass of broken water in the BREAKER zone of a shoreline. The foaming water results from the highly dynamic wave activity, its width varying with the changing wavelengths and according to the character of the beach materials; the *surf zone* is wider on a sandy beach than on a shingle beach.

surface 1 The outermost boundary of any material body, e.g. the surface of the Earth. **2** A concept in mathematics referring to an infinite set of points marking a two-dimensional space of no thickness. **3** The BOUNDARY between two contrasting media, processes or materials, e.g. BASAL SURFACE OF WEATHERING, FRONTAL SURFACE (FRONT).

surface detention The process by which an amount of water is retained in temporary STORAGE both during and immediately after a rainstorm before moving downslope as OVERLAND FLOW. *See also* DEPRESSION STORAGE, SURFACE STORAGE.

surface flow OVERLAND FLOW.

surface storage Water stored on the ground surface within a drainage basin. This will include water in lakes, marshes, reservoirs, glaciers and snow patches.

surface tension The force required per unit length to pull the surface of a liquid apart, i.e., to increase the surface by unit area. The tension results from the attractive forces between the molecules of the liquid.

surface wash The transport of surface soil and regolith downslope across the ground surface through the agency of running water. It consists of two processes, RAIN-SPLASH and OVERLAND FLOW, and results in GULLYING, RILL EROSION, SHEET EROSION and SOIL EROSION.

surfzone DISSIPATIVE BEACH.

surge 1 A down-glacier movement of a KINEMATIC WAVE, representing the glacier's response to a change in its MASS BALANCE, and manifesting itself as a sudden advance of the snout. The wave moves down the glacier as a type of accelerated flow which often exceeds the normal rate of glacier flow by four to six times. In front of the wave there will be a zone of COMPRESSING FLOW, while behind it will be a zone of EXTENDING FLOW. Although most world glaciers are in a state of recession some exhibit periods of spasmodic surging, e.g. some of the Alaska glaciers in 1966, while the Hassanabad glacier of the Karakoram advanced 10 km in 2½ months at a rate of 130 m/day during a 1953 surge. It has been suggested that other causes of glacier surges may be found in earthquakes or high geothermal heat flow in the underlying rocks. **2** A STORM SURGE.

surveying 1 The precise measuring and accurate recording of the relief and topography of the Earth's surface in order to construct maps and plans. *See also* LEVELLING, PLANE-TABLE, TACHEOMETRY, TRAVERSE, TRIANGULATION. **2** Geodetic surveying (GEODESY). **3** The term is used more loosely to denote the collection of information for other types of map construction, e.g. geological-survey maps, land-use survey maps.

suspended load That part of a river's LOAD which is carried in SUSPENSION. It consists of the finest solid particles that can be supported by the water without recourse to SALTATION. *See also* BEDLOAD, DISSOLVED LOAD.

suspension The process by which light-weight materials (FINES) are transported by moving water in the zone of turbulent flow (TURBULENCE). A great part of a river's LOAD is carried by suspension; comparisons based on the Mississippi statistics suggest that the suspended load is 8½ times greater than that carried by the process of SALTATION and 2½ times greater than that part of the load dissolved in the water and carried by SOLUTION. Much fine material is also held in suspension shoreward of the breaking-point of marine waves, i.e. in the SURF ZONE.

sustained yield A management policy designed to regulate the recycling system of certain resources that are renewable in order to maintain their yield at a desired level for the foreseeable future. Such maintenance requires a thorough understanding of the ECOSYSTEM in question.

swale BAR-AND-SWALE, RUNNEL.

swallet SWALLOW-HOLE.

swallow-hole, swallet The point at which a surface stream disappears underground in a limestone terrain (KARST) prior to commencing its underground journey. It is synonymous with a 'water sink' in the USA, with the French *embut*, and with a PONOR in Yugoslavia. Some swallow-holes have no topographical expression, merely a point where a stream-flow gradually dries up as it percolates through its gravel; others are marked by spectacular vertical shafts with waterfalls (sometimes referred to as POT-HOLES, in Britain), e.g. Gaping Gill, Yorkshire, England, while some streams simply disappear into a cave before tumbling down a steeply sloping shaft. [*204*]

swamp A permanently waterlogged area in which there is often associated tree growth, e.g. MANGROVE and *swamp-cypress* varieties. Some writers have attempted to differentiate swamp from the periodically inundated MARSH and the partially decayed waterlogged vegetation of BOG. *See also* CARR, FEN, WETLAND.

swash The movement of a turbulent layer of water up the slope of a beach as a result of the breaking of a wave. It is capable of moving beach material of substantial size, i.e. pebbles and cobbles, and is an important element in LONGSHORE DRIFT. Waves with a relatively flat profile create the greatest swash (CONSTRUCTIVE WAVE). Most of the water returns to the sea as BACKWASH, but some infiltrates into the beach material. [*140*]

S-wave SEISMIC WAVES.

swell **1** In geology, a domed area of considerable areal extent. **2** A regular movement of marine waves created by wind stress in the open ocean which travels considerable distances away from the generating field and into another wind field. The waves are characterized by relatively smooth, generally unbroken, crests and a fairly regular wavelength, but swell increases in wavelength and decreases in wave height as it moves away from the generating area. Local wind waves may be superimposed upon swell waves as they approach a coastline, thereby creating sharper crests and a choppy sea (CHOP).

swell beach REFLECTIVE BEACH.

sword-dune SEIF-DUNE.

syenite A group of coarse-grained INTERMEDIATE ROCKS of igneous origin characterized by alkali FELDSPARS and occurring as the plutonic equivalent of the volcanic PHONOLITE.

symbiosis **1** In general terms, referring to any two different organisms which live together. **2** More specifically, two distinct organisms which live together to their mutual advantage, e.g. LICHENS.

symbolic model MODEL.

sympatric A term referring to vegetation or animal taxa (TAXON) that overlap spatially, either in part or over most of their distribution area.

syncline A downfold or basin-shaped FOLD of crustal rocks in which the strata DIP inwards towards a central axis, in contrast to those of an ANTICLINE. *See also* SYNFORM.

synclinorium A complex SYNCLINE of great lateral extent in which many minor folds occur. *See also* ANTICLINORIUM.

synecology A term referring to the ECOLOGY of whole plant and animal communities as opposed to single species or individual organisms (AUTOECOLOGY). The level of study is less intensive than that of the ECOSYSTEM or the BIOCOENOSIS.

synform A general term denoting a downfold of crustal strata in which the precise stratigraphic relationships of the rocks are not known. In an area of NAPPE FORMATION denudation may have worn down the structures to reveal a downfold, but since the stratigraphical succession may be inverted owing to overfolding, it is not possible to say whether the fold expressed at the ground surface is a true SYNCLINE or merely a synform.

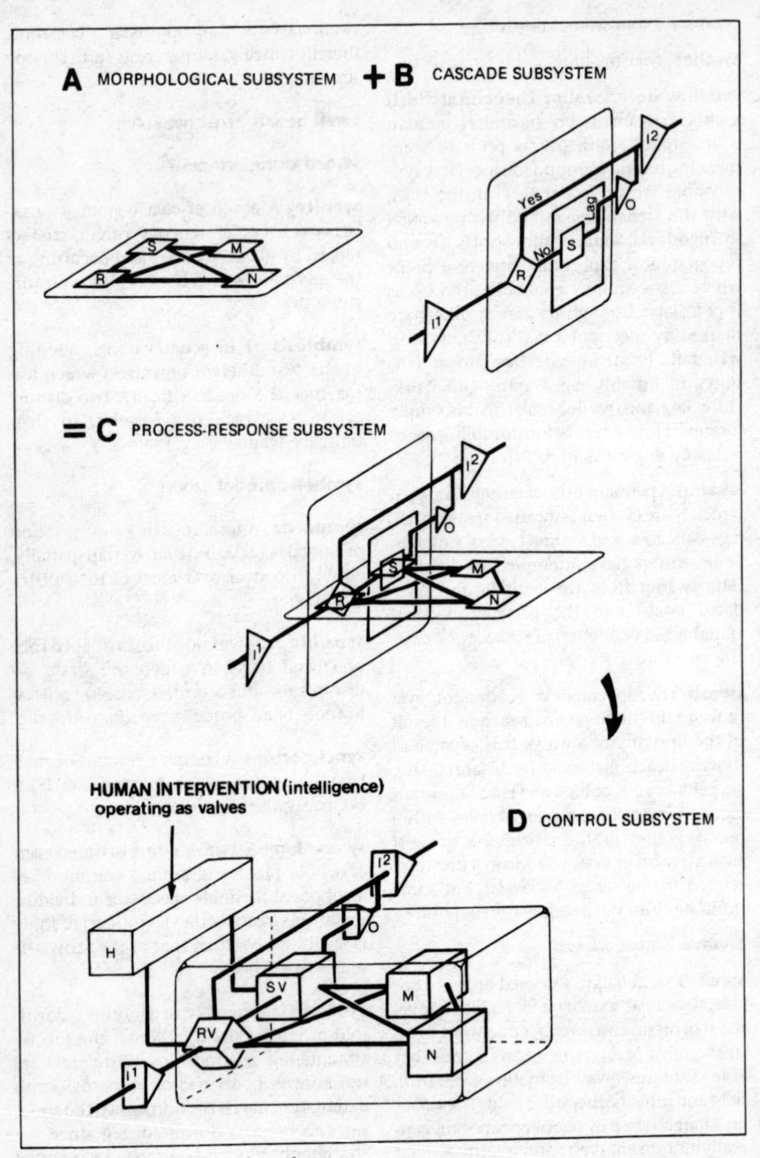

Figure 249a *Systems analysis: four subsystems*

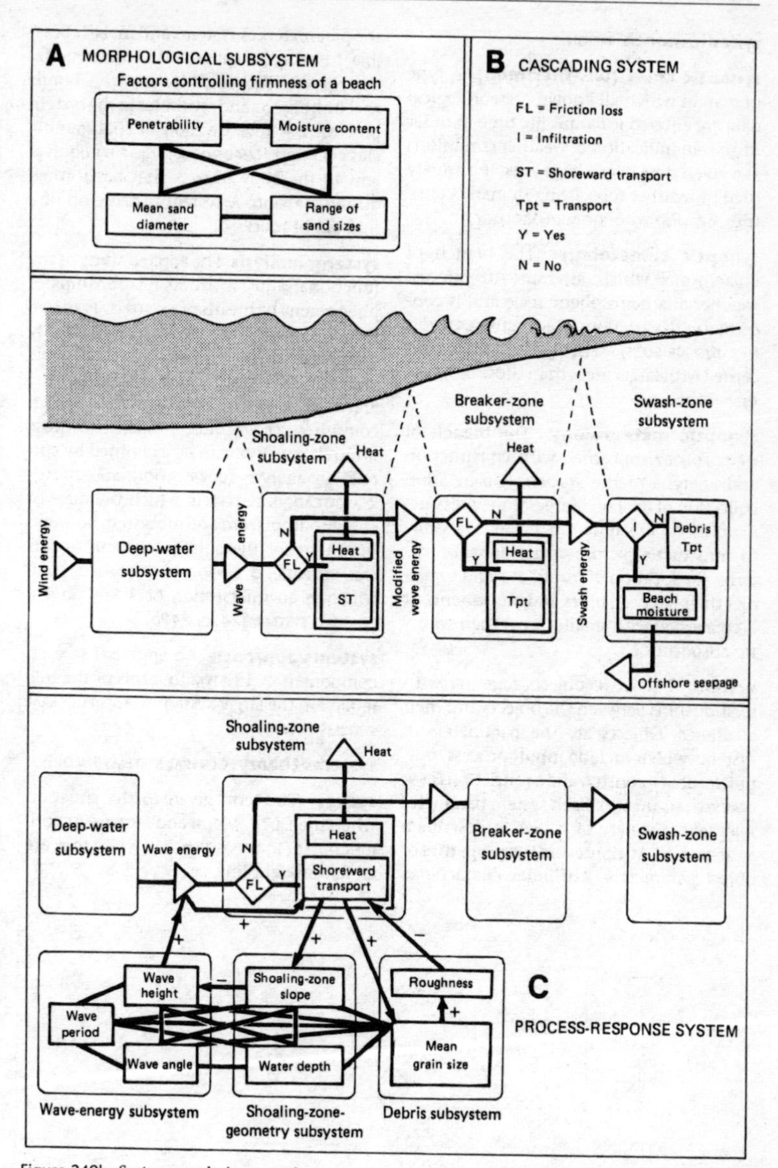

Figure 249b *Systems analysis: examples*

synodic month MOON.

synoptic chart (weather map) A type of map in which all known meteorological data are entered for a specific time in order to give an indication of weather conditions at a given time over a large area. It is widely used in weather forecasting in many countries. *See also* SYNOPTIC METEOROLOGY.

synoptic climatology The branch of CLIMATOLOGY which attempts to forecast weather at a hemispheric scale and is concerned with variations in advective controls (*see also* AIR MASS, PRESSURE SYSTEM). It is concerned with larger areas than those of SYNOPTIC METEOROLOGY.

synoptic meteorology The branch of METEOROLOGY concerned with construction and analysis of the SYNOPTIC CHART. From collection of data on temperature, pressure, wind speed, precipitation, etc. it is possible to construct a picture of the pressure patterns, the location of fronts, etc. and to predict their likely changes and movements. It is concerned with smaller areas than SYNOPTIC CLIMATOLOGY.

system **1** A set of objects together with relationships between the objects and their attributes. Objects are the parameters of systems which include input, process, output, feedback control, and a restriction (GENERAL SYSTEMS THEORY). Each system parameter may take a variety of values to describe a system state. Attributes are the properties of object parameters. Attributes characterize the parameters of systems and make it possible to assign a value and a dimensional description. Relationships are the bonds that link objects and attributes in the system process. **2** A rock division of stratigraphic classification (CHRONOSTRATIGRAPHY) equivalent to the body of rock that has formed during a PERIOD. A system is made up of a number of SERIES.

systems analysis The applied study of the functional and structural relationships of phenomena by means of a SYSTEMS APPROACH. Physical geographers are concerned mainly with four types of system, which become increasingly sophisticated and more integrated: **1** a MORPHOLOGICAL SYSTEM, which comprises recognizable properties, whose interrelationships can be examined by correlation analysis (CORRELATION, STATISTICAL); **2** a CASCADING SYSTEM, in which the mass or energy output from one subsystem becomes the input for the adjacent system (INPUTS AND OUTPUTS); **3** a PROCESS–RESPONSE SYSTEM, which is an intersection of **1** and **2**; **4** a CONTROL SYSTEM. [*249a, 249b*]

systems approach An appraisal of each component of a SYSTEM in terms of the role it plays in the larger system. *See also* GENERAL SYSTEMS THEORY.

systems theory GENERAL SYSTEMS THEORY.

syzygy The term given to the situation when the Earth, Moon and Sun are in CONJUNCTION or in OPPOSITION, i.e. when they are all in a straight line.

T

tableland A flat-topped or undulating area of high relief, usually bounded by steeply descending slopes. *See also* MESA.

tablemount GUYOT.

tabular iceberg An immense horizontal floating mass of ice in the Southern Ocean, often exceeding 20 km in length (icebergs up to 100 km long have been recorded). It is part of the Antarctic ICE-SHELF which has broken away and floated north.

tacheometry (tachymetry) A surveying technique developed to produce rapid determinations of direction, elevation and distance by the use of a specially adapted THEODOLITE, known as a *tacheometer* (*tachymeter*).

tafelkop An Afrikaans term for a flat-topped or table-topped hill (BUTTE), in contrast to a pointed SPITSKOP. *See also* KOPJE.

tafoni Hollows occurring on rock surfaces, especially GRANITE, due to cavernous weathering. They are thought to be formed by HALOCLASTY and wind scour on exposed rock or where a case-hardened surface (CASE HARDENING) has been breached by mechanical weathering, thereby revealing an underlying less resistant layer.

taiga A Russian term for the belt of coniferous forest which circles the land masses of the N hemisphere between the temperate grasslands and the TUNDRA. Most writers regard it as synonymous with the BOREAL forests, although in parts of N Canada the taiga zone is also occupied by large expanses of MUSKEG. STRING BOGS are also characteristic of the taiga zone. The conifers have to withstand extremely low temperatures and lengthy periods of snowfall.

tail-dune An accumulation of wind-blown sand leeward of an obstacle. *See also* DUNE. [*65*]

tailings The waste portion remaining after a metallic ore has been washed and divided into the materials valuable enough to be smelted, known as concentrates, and the debris regarded as uneconomic to work (tailings).

takyr A Russian term for a clay-floored depression in which seasonal lakes accumulate. In the dry seasons the clay surface becomes desiccated and cracked as strong evaporation creates deposits of EVAPORITE in the soil (SOLONCHAK).

talc A whitish mineral, comprising a hydrous magnesium silicate ($Mg_3(Si_4O_{10})(OH)_2$). It has a greasy feel and can easily be scratched with the finger-nail, i.e. it has a hardness of only 1 on the Mohs scale (HARDNESS SCALE). It occurs among some ultrabasic igneous rocks, e.g. PERIDOTITE, and is also formed by low-grade thermal metamorphism of certain dolomites. *See also* SOAPSTONE.

talik A layer of unfrozen ground occurring between the PERMAFROST and the seasonally frozen layer (active layer). Russian scientists distinguish three types: **1** open taliks (*skvoznoy*); **2** closed taliks (*zamknuty*); **3** interpermafrost taliks (*mezhmerzlotny*), possibly due to local heat sources, including groundwater circulation. A fourth type has also been described: the unfrozen ground beneath the permafrost. Where there is a marked change of thermal regime within the frozen ground the permafrost becomes degraded and THERMOKARST may be present. [*182* and *187*]

talsands A German term for 'valley sands', referring to large-scale cappings of wind-blown sands overlying fluvially deposited sediments in former ice-marginal valleys. *See also* RINNENTÄLER, URSTROMTÄLER.

talus Generally regarded as a synonym for SCREE, but a distinction has sometimes been made, with the latter term being regarded merely as the debris constituent of the landform (talus).

talus shift The slow downslope movement of a superficial mass of rock fragments, the interstices of which are predominantly true voids rather than infilled by fine soil particles. Since only a proportion of the individual rock fragments move by the mechanism of CREEP, the expression is preferred to that of *talus creep*. It is also distinguished from the movement of a ROCK GLACIER, where interstitial ice is present. Nevertheless, it has been claimed that talus shift may be initiated by ice-wedging (CONGELIFRACTION) but also by heating and cooling mechanisms, by the impact of rockfalls and by slippage on moistened surfaces. [*146*]

talus slope CONSTANT SLOPE.

talwind An up-valley wind that usually occurs at the same time as an ANABATIC WIND (an upslope wind). It is generally very light and tends to develop during a warm afternoon when vertically expanding air is laterally constricted and tends to blow up the axis of the valley. At night the reverse process occurs when cold air drains down the valleys in the form of a mountain wind blowing down the valley axis at the same time as the KATABATIC WIND (downslope wind) develops. *See also* BERG WIND.

tape A surveying device for measuring distance. It consists of a graduated ribbon of steel, linen or an alloy of iron, nickel etc. (invar), hand wound or spring-coiled into a steel or leather case when not in use.

taphrogenesis Vertical movements of the Earth's crust leading to the formation of major faults and RIFT VALLEYS (TAPHROGEOSYNCLINES).

taphrogeosyncline A synonym for a RIFT VALLEY.

tarn A local name for a small lake, especially in the Lake District of N England. It is often, but not necessarily, a CIRQUE lake [*41*].

tar sands Sedimentary formations of loosely grained rock material bonded by heavy bituminous tar (BITUMEN). Processing of tar sands produces crude oil, but often more expensively. It is synonymous with OIL-SAND.

taxon A named group or systematic unit of organisms of unspecified rank; i.e. a FAMILY, GENUS or SPECIES, etc. Its plural form is *taxa*.

taxon cycle A name given by E. O. Wilson (1961) to a concept which illustrates the stage by stage colonization of an island or ARCHIPELAGO by an invading SPECIES. It demonstrates how populations adapt to new HABITATS and undergo evolutionary diversification (EVOLUTION).

taxonomy The scientific classification of plants and animals into hierarchical groups according to their similarities. The ascending hierarchy is: SPECIES, GENUS, FAMILY, ORDER, CLASS, PHYLUM (fauna), DIVISION (flora), KINGDOM.

Taylor–Görtler flow VORTEX FLOW.

TDS DISSOLVED SOLIDS.

tear-fault A strike-slip fault that trends transverse to the STRIKE of the folded rocks. [*76*]

tectogene **1** A down-buckled belt of crustal sediments within a EUGEOSYNCLINE [*91*]. **2** The deep sialic 'roots' (SIAL) below a fold mountain range (MOBILE BELT).

tectogenesis The processes by which the Earth's crustal rocks are deformed and their structures created: FAULT, FOLD, CLEAVAGE, JOINT, THRUST-PLANE. When such processes act regionally they contribute to mountain-building (*orogenesis*: OROGENY). *See also* CYMATOGENY, DIASTROPHISM, EPEIROGENESIS.

tectonic Pertaining to the internal forces which deform the Earth's crust, thereby affecting the patterns of sedimentation and the landforms which are created by warping or fracturing. Such landforms are referred to as *tectonic relief*, e.g. FAULT-BLOCK TOPOGRAPHY, RIFT VALLEY.

tectonic earthquake EARTHQUAKE.

tectonic relief TECTONIC.

tektite A black, glassy, rounded object, thought to be of extraterrestrial origin. *See also* METEORITE.

teleconnections A meteorological term referring to atmospheric events that occur simultaneously but in areas geographically distant from each other, e.g. the SOUTHERN OSCILLATION.

telluric Pertaining to the Earth, e.g. *telluric currents* are natural electric currents that flow as sheets near to the crustal surface.

tellurometer A surveying instrument of considerable accuracy for precise measurement of distances. It is based on the transmission of radio microwaves the velocities of which can be measured and corrected for slope, altitude and especially for meteorological conditions.

temperate **1** A term used loosely as a synonym for moderate. **2** An adjectival qualification for a mid-latitude climate or vegetation zone. **3** One of the three standard temperature zones of the Earth (FRIGID ZONE, TEMPERATE ZONE, TORRID ZONE). *See also* KÖPPEN'S CLIMATIC CLASSIFICATION.

temperate glacier One of the three types of glacier identified by H. W. Ahlmann (*see also* COLD GLACIER, SUBPOLAR GLACIER). It is characterized by having all but the surface 10–20 m of its ice at the PRESSURE MELTING-POINT. The surface layer can have temperatures below 0° C for at least part of the year but this cold layer must disappear before the end of the summer if the glacier is to be classed as temperate. Because of the considerable amounts of meltwater at all levels of the ice and especially at the base of the ice there is considerable erosion from ICE PLUCKING and ABRASION as the basal ice slides relatively easily over the underlying rock. Thus the rate of movement is usually faster than in cold and subpolar glaciers.

temperate zone One of three climatic zones recognized by early geographers, based on differences in temperature. It occurs between the TORRID ZONE and the FRIGID ZONE. In modern usage the term temperate has come to be used to denote moderate climates with no great extremes of temperature. But since the early defined temperate zone includes areas with large temperature extremes, especially in continental interiors, it is preferable to use the term *mid-latitude* when describing a zonal area. The term *temperate* has survived in its adjectival form to describe vegetation types (e.g. *temperate grassland*) and to distinguish glacier types (TEMPERATE GLACIER). The term is also used in KÖPPEN'S CLIMATIC CLASSIFICATION.

temperature The index of the heat contained in any substance, albeit temperature and heat cannot be directly equated because the specific heat capacity varies considerably between different substances. In climatology, temperature is one of the most important elements for it represents the amount of sensible heat or cold within the atmosphere. It can be measured on a variety of scales, the most common being CELSIUS (*centigrade*), although FAHRENHEIT is still used in the USA. *See also* ABSOLUTE TEMPERATURE, KELVIN, RÉAUMUR.

temperature gradient LAPSE-RATE.

temperature–humidity index An index devised to indicate the effects of different weather types on human comfort. *See also* COMFORT ZONE, SENSIBLE TEMPERATURE.

temperature inversion INVERSION OF TEMPERATURE.

temporal change An alteration that occurs over a period of time.

tensiometer A device for measuring the TENSION (negative pressure) of water in an *in situ* soil. It consists of a porous ceramic cup connected through a tube to a vacuum-gauge.

tension A stress caused by forces in opposition, i.e. pulling apart. It contrasts, therefore, with COMPRESSIONAL MOVEMENTS. Tension in materials creates: JOINTS in crustal upfolds (ANTICLINE); gashes or fissures (CREVASSE, GULL); *fractures* (NORMAL-FAULT).

tensional fracture NORMAL-FAULT, TENSION.

tepee An overthrust sheet of LIMESTONE with the form of an inverted V when seen in section. It is named from the hide-covered dwellings of native Americans. Such phenomena are found in tidal environments in CALCRETE and around PLAYAS and are caused by DESICCATION and deformation resulting from fluctuating water levels or changes in chemical precipitation.

tephigram A diagram used by meteorologists on which the changing properties of moving air parcels are plotted at various isobaric levels in the atmosphere. It comprises a display of ISOTHERMS, ISOBARS, dry adiabats (DRY ADIABATIC LAPSE-RATE), saturated adiabats (SATURATED ADIABATIC LAPSE-RATE) and saturation MIXING RATIO lines. [*250*]

tephra An alternative name for PYROCLASTIC MATERIAL.

tephrochronology The dating of past events by means of their association with layers of volcanic ash from former volcanic eruptions.

terminal fan A type of ALLUVIAL FAN in which the volume of surface RUNOFF decreases downslope due to EVAPORATION, resulting in the virtual disappearance of surface water on the fan itself.

terminal grades A term given to the finest particles (FINES) of a disintegrating glacial TILL when it has broken down to mineral-sized fragments after which it is thought that very little further size reduction will occur.

terminal moraine A linear ridge of glacial debris (MORAINE) marking the maximum limit of an ice-sheet or glacier. Its proximal (inner) slope is usually steeper than its distal (outer) slope since the former represents the ICE-CONTACT SLOPE. [*219*]

terminal velocity The maximum velocity which can be attained by a body falling freely in a fluid (usually air or water). That relating to spherical particles is measured by STOKES'S LAW, and is an important factor in *soil mechanics* and in fluvial studies measuring the transportation of solid material in a stream current where it is being carried in SUSPENSION.

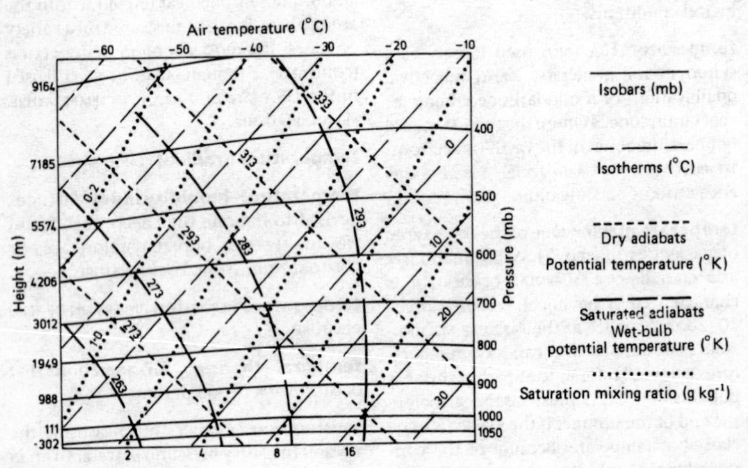

Figure 250 *Tephigram*

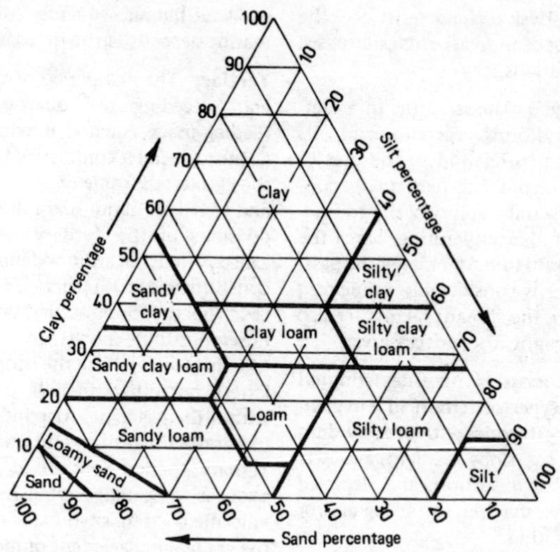

Figure 251 *Textural triangle: a ternary diagram for determining soil-texture classes*

terminocratic A term introduced by von Post (1946) and adopted by J. Iversen (1958) to describe the period of decreasing warmth in the post-glacial (FLANDRIAN) climate of NW Europe after the *Climatic Optimum*. It is thought to be synonymous with the SUB-BOREAL PERIOD and the SUB-ATLANTIC PERIOD (4,450 BP to the present), and with pollen zones VIIb and VIII. *See also* MEDIOCRATIC, PROTOCRATIC. [*81*]

termite activity The effect of termites on the landforms of an area, especially the creation of termite mounds in tropical regions.

ternary diagram A triangular graph used for plotting data simultaneously on three scales in order to illustrate the relative dominance of each. It may be used when the values are percentages with each of the three sides of the equilateral triangle being graduated from 0 to 100. Provided that the three values total 100% they may be represented by a single point on the diagram. A ternary diagram is frequently used in the study of soil characteristics, e.g. soil texture (TEXTURAL TRIANGLE [*251*]).

terrace A flat or gently inclined land surface bounded by a steeper ascending slope on its inner margin and a steeper descending slope on its outer margin. *See also* ALLUVIAL TERRACE, KAME TERRACE, MARINE-BUILT TERRACE, MEANDER TERRACE, PLUVIAL TERRACE, RIVER TERRACE.

terracette A type of MICRO-RELIEF on a hillside, sometimes referred to as 'sheep tracks'. It comprises a narrow, horizontal step with almost level tread (usually tens of cm wide) bounded by risers that are steeper than the overall slope gradient. It is usually one of a subparallel series that characterizes steep slopes, mainly those under grassland. It is thought to be formed by preferential trampling by cattle and sheep along the sides of a hill and has not been found in areas inaccessible to animals. Nevertheless, some writers still regard terracettes as natural CREEP features in the soil, and others as micro-landslips.

terra fusca A brown clay-loam soil developed under a warm, seasonally dry climate, on limestone. It exhibits a neutral or slightly alkaline pH and is commonly found beneath a MAQUIS type of vegetation in some parts of the Mediterranean. It has been suggested that it may represent a degraded TERRA ROSSA.

terrain The physical characteristics of the natural features of an area, i.e. its LANDFORMS, vegetation and soils.

terrain analysis The scientific interpretation of the landforms, vegetation and soils of a given area in relation to the uses to which it may be put. It is particularly concerned with the trafficability of the ground, especially in military planning. Since the ability of a terrain to assist or deter the passage of vehicles is considerably influenced by the weather, the climatic factors are also taken into account where necessary.

terrain-type map A map depicting all the different types of terrain in a region, classified into categories with broadly similar attributes, e.g. slope gradient, AVAILABLE RELIEF, degree of DISSECTION, and depicted with distinctive shading, colouring and/or numerical notation.

terra rossa A reddish clay-loam soil developed under a warm seasonally dry climate, on limestone, especially in the KARST terrain of Yugoslavia. It is rich in iron sesquioxides and has a low BASE status. Although it is thought to be a RESIDUAL SOIL there is no agreement on its exact origin, but since it appears typically under GARRIGUE vegetation in the Mediterranean region it may possibly result from deforestation. The general lack of humus in the terra rossa would support such a possibility. It is related to the red PODZOLIC SOILS. *See also* RENDZINA.

terra roxa A term referring to a soil developed on basic igneous rocks in Brazil. It is a type of LATOSOL and must not be confused with a TERRA ROSSA.

terrestrial environment The continental as distinct from the marine and atmospheric environments. It is the environment in which terrestrial organisms live and in which TERRIGENOUS SEDIMENTS are formed.

terrestrial magnetism GEOMAGNETISM.

terrestrial radiation RADIATION.

terrigenous sediments Sedimentary inorganic deposits that have either been formed and laid down on a land surface (e.g. blown sand) or that have been derived from the land but subsequently mixed in with marine deposits of the LITTORAL ZONE.

Tertiary The first period of the CAINOZOIC era, preceding the QUATERNARY period. (Earlier usage equated it with the entire Cainozoic era.) It commenced at the end of the CRETACEOUS, some 65 million years ago, and lasted for about 63 million years. The divisions of the Tertiary were formerly called periods but are now defined as EPOCHS, consisting of the PALAEOCENE, EOCENE, OLIGOCENE, MIOCENE and PLIOCENE. Many geologists now group these epochs into the PALAEOGENE (the three former) and the NEOGENE (the two latter), but others include the *Pleistocene* within the Neogene, thereby ignoring the division of the Tertiary and Quaternary. The Tertiary period witnessed the ALPINE OROGENY, a spatially extensive episode of vulcanicity (TERTIARY VOLCANIC PROVINCE) and the gradual replacement of more primitive organisms by modern invertebrates and mammals. ANGIOSPERMS became the dominant plants.

Tertiary volcanic province The name sometimes given to those parts of NW Scotland and NE Ireland where widespread vulcanicity occurred during the PALAEOGENE. In addition to the emplacement of granite batholiths in Arran, Skye and the Mourne Mts, there were several RING COMPLEXES formed in places such as Skye, Mull, Ardnamurchan and N Ireland. Extensive spreads of BASALT were poured out in Skye, Mull and county Antrim and these now form LAVA PLATEAUX. *See also* CAULDRON SUBSIDENCE, CONE-SHEET, DYKE-SWARM, GABBRO, SILL.

tessellation A GIS term referring to the subdivision of a two-dimensional plane surface (or three-dimensional volume) into disjoint congruent polygonal TILES.

Tethys A broad GEOSYNCLINE of Mesozoic age which divided GONDWANALAND from LAURASIA. Within it enormous thicknesses of sediment accumulated and these were later to be compressed and uplifted to form the mountain ranges of S Europe and S central Asia during the ALPINE OROGENY. The Mediterranean Sea is now a vestige of this former ocean. *See also* PLATE TECTONICS.

tetrahedral theory A hypothesis formulated in 1875 by L. Green, in which it is envisaged that the shrinking globe of planet Earth would tend to assume the form of a *tetrahedron* (a four-faced figure, with its base at the N Pole and its apex at the S Pole, and each face having equal intercepts on all three axes). There have been many geophysical objections and the theory is now considered to be untenable.

textural triangle A triangular graph for determining soil-texture classes (TEXTURE). It is constructed using data obtained from a mechanical analysis of the 'fine-earths' fraction (FINES) of a soil sample [*251*].

texture **1** In geology, a term referring to the geometrical aspects of the component particles of a rock, and including their size, shape and arrangement. **2** In geomorphology, a term referring to the relative spacing of rivers in an area of FLUVIAL erosion. **3** In REMOTE SENSING, a term describing the frequency of change and arrangement of tones in a remotely sensed IMAGE, relating either to an AERIAL PHOTOGRAPH or to a SATELLITE image. Adjectival descriptions of texture include 'coarse', 'medium', 'fine', 'stippled' and 'mottled'. **4** In soil science, the geometrical relationships of the constituent particles of a soil. A fine-grained soil (CLAY) has particles of <0.002 mm in diameter and feels sticky; an intermediate-grained soil (SILT) has particles ranging from 0.002 mm to 0.02 mm in diameter and feels silky; a coarse-grained soil (SAND) has particles between 0.2 mm and 2 mm in diameter and feels gritty. NB A LOAM comprises a mixture of particles of many sizes. *See also* TEXTURAL TRIANGLE.

thalassostatic A term introduced by F. E. Zeuner in 1959 to describe river terraces caused by fluctuations of sea-level (EUSTASY). A fall in sea-level causes a KNICKPOINT and a terrace which diverges from the newly created river profile (THALWEG) in the downstream direction and ends abruptly at the coast. A rise in sea-level creates a funnel-shaped estuary or fiord with a tendency for its gradual infilling by river-borne detritus. Such an aggradational process will be induced during a Pleistocene interglacial and will form a thalassostatic terrace when the river is rejuvenated during the next phase of sea-level lowering. [*211a*]

thalweg (talweg) A German term (literally, valley way) for the long profile of a river valley. [*128*]

thawing The action which changes ice or snow into water as a result of temperatures rising above freezing-point. The period during which the process operates is referred to as the *thaw*. When thawing occurs in regions of frozen ground it will have important effects, including the formation of PATTERNED GROUND, ASYMMETRICAL VALLEYS, STRING BOGS, THAW LAKES and THERMOKARST. *See also* SLUSH-FLOW.

thaw lakes Water bodies enclosed in basins which have been formed or enlarged by thawing of frozen ground. Thus they are a type of THERMOKARST feature. They also occur in collapsed PINGOS and on the flat floors of ALASES. [*253*] When they have a parallel alignment they may be described as ORIENTED LAKES [*170*].

thematic map A map designed to demonstrate particular features or concepts, as opposed to a TOPOGRAPHIC MAP. Examples include climatic maps, economic maps and population maps.

theodolite An accurate optical instrument for measuring horizontal and vertical angles and used extensively in surveying. It consists of a telescope, mounted so that it can be rotated in either the horizontal or vertical planes, a spirit-level and a compass, all of which are attached to a tripod. In the USA it is the refined development of an instrument termed a 'transit'.

thermal **1** a buoyant pocket of air that rises vertically in the atmosphere owing to a steep ENVIRONMENTAL LAPSE-RATE or intense solar heating of the Earth's surface. Because of differences in texture, colour, etc. some parts of the surface will heat more rapidly than others, thereby giving rise to locally rising air currents or updraughts. If the vertical motion is great enough the thermal may reach the CONDENSATION LEVEL and cumulus-cloud formation will occur, possibly leading to convectional rainfall and even a THUNDERSTORM. In other cases the rising air current may lose its buoyancy before reaching the condensation level. **2** The term thermal is also used in an adjectival sense when applied to TEMPERATURE descriptions of air or water.

thermal depression, thermal low An area of low pressure resulting from very strong solar heating of the Earth's surface, which in turn leads to convectional rising of air currents (THERMAL) and a fall in atmospheric pressure at the surface. It can range in size from local vortices (DUST-DEVIL, SIMOOM) through the summer low of central Spain, to the larger monsoonal lows of N India (e.g. Thar Desert low). Some may be associated with local rainstorms but the monsoonal low of NW India is overlain by a strong anticyclone which prevents the thermals reaching the CONDENSATION LEVEL.

thermal-efficiency index An empirical index devised by a climatologist, C. W. Thornthwaite, as part of his climatic classification (THORNTHWAITE'S METHOD OF CLIMATIC CLASSIFICATION) in conjunction with his index of PRECIPITATION EFFICIENCY. It is a measure of the amount of heat received by a particular area in terms of the *potential evapotranspiration* that would be produced. It is expressed as: $(T - 32)/4$, where T is the mean monthly temperature in degrees F. *See also* EVAPOTRANSPIRATION.

thermal equator (heat equator) A line encircling the low latitudes of the globe joining up the point on each meridian which exhibits the highest mean surface air temperature. Its average position lies to the north of the true EQUATOR because there are larger land masses in the N hemisphere and land absorbs more heat than do the oceans. If plotted on a monthly basis, however, it will be seen that the thermal equator moves north and south following the apparent movement of the Sun between the tropics. In July it reaches 20°N over the land masses but in December it moves south again to approximately the position of the true equator, except in S America where it lies in the S hemisphere. *See also* INTERTROPICAL CONVERGENCE ZONE.

thermal fracture The action by which rocks crack when subjected to rapid changes of temperature. These cause stress fractures resulting from the varying coefficients of expansion of the rock constituents.

thermal linescanner A SENSOR used in REMOTE SENSING to enable a thermal IMAGE of the Earth's surface to be produced. It is used extensively in INFRARED THERMOGRAPHY.

thermal metamorphism CONTACT METAMORPHISM.

thermal pollution The POLLUTION of freshwater due to artificial increases in temperature. It is due mainly to the discharge of condenser cooling water into rivers and lakes after it has been utilized in electricity-generating power stations. For example, cooling water released into a lake from Trawsfynydd nuclear power station in North Wales produced changes in the lake fauna due to increased temperatures.

thermal stratification of water The division of water temperatures in the ocean or a deep lake. Three layers can usually be recognized: the uppermost layer, affected by wind and convection currents (EPILIMNION); the middle layer, exhibiting a rapid decrease of temperature with depth (THERMOCLINE); and the lowest layer, with the lowest temperatures, the most stagnant water and the least oxygen (HYPOLIMNION). At low latitudes all year round and in middle latitudes in summer the surface-layer temperatures may rise to 20° C, while temperatures in the lowest layer average 4° C. In Arctic and Antarctic seas, however, the thermal stratification breaks down and the three-layer system is replaced by a single layer of cold water. [*252*]

thermal wind, vector wind In general, the shear or change in direction of the GEOSTROPHIC FLOW with increasing elevation, determined by the distribution of mean temperature in a specified atmospheric layer. The thermal wind blows parallel to the mean ISOTHERMS with the effect that low mean temperatures remain on the left in

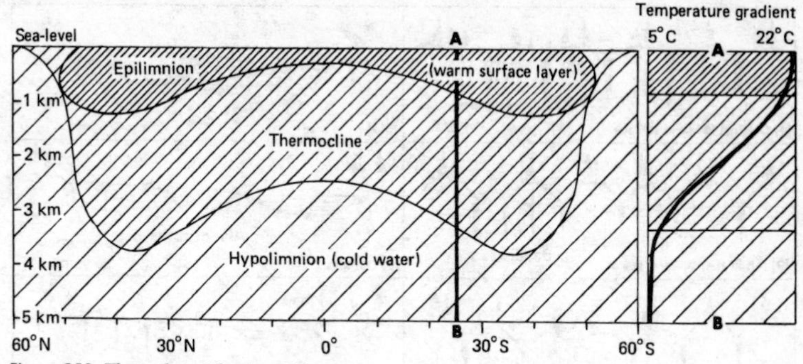

Figure 252 *Thermal stratification of the ocean*

the N hemisphere and on the right in the S hemisphere. Consequently, the poleward decrease of temperature in the TROPOSPHERE causes a large westerly component in the upper winds (JET STREAM).

thermistor A small instrument for measuring temperature. Since the latter varies with changing electrical resistance in a substance the thermistor monitors changes in resistance.

thermoclasty (thermoclastis) An alternative term for insolation weathering in which rocks are allegedly broken down by alternating expansion and contraction due to diurnal temperature changes. Despite 19th-cent. reports of rock disintegration under thermal heating, many modern geomorphologists are unconvinced that this type of WEATHERING process exists, since none of their experiments to simulate thermoclasty has been successful. *See also* HALOCLASTY.

thermocline An intermediate layer of water in an ocean or a deep lake, lying between the disturbed upper layer (EPILIMNION) and the relatively stagnant lower layer (HYPOLIMNION). It represents the zone, at depths generally between 1 and 3 km, where the maximum vertical temperature gradient occurs. It is fairly sharply defined from the epilimnion but grades into the hypolimnion. This threefold division breaks down in the Arctic and Antarctic oceans where only a single layer of cold water is found [*252*]. *See also* THERMAL STRATIFICATION OF WATER, WATER MASSES.

thermocouple A device for measuring TEMPERATURE by means of thermo-electric effects. It measures the temperature differences between two junctions, each composed of two different metals. One junction is held at 0° C while the other acts as the sensor.

thermodynamic equation A productive equation used in meteorology. It expresses the speed of change of TEMPERATURE of a parcel of air moving through the ATMOSPHERE. It shows that the temperature change is related to both DIABATIC heating and to either ADIABATIC expansion or compression.

thermodynamic equilibrium The tendency within an isolated SYSTEM to move towards a condition of maximum ENTROPY. Thus there will be an equal probability of encountering given states, events or energy levels throughout the system. Additionally, there will be a slow decay of free energy leading to an extinction of energy differences capable of performing work within the system. [*72*] *See also* EQUILIBRIUM.

thermodynamics The scientific study of the movement induced by changes in heat. The *first law of thermodynamics* is the law of the conservation of energy. The *second law* states that heat cannot be transferred from a colder to a warmer body without an outside source of energy.

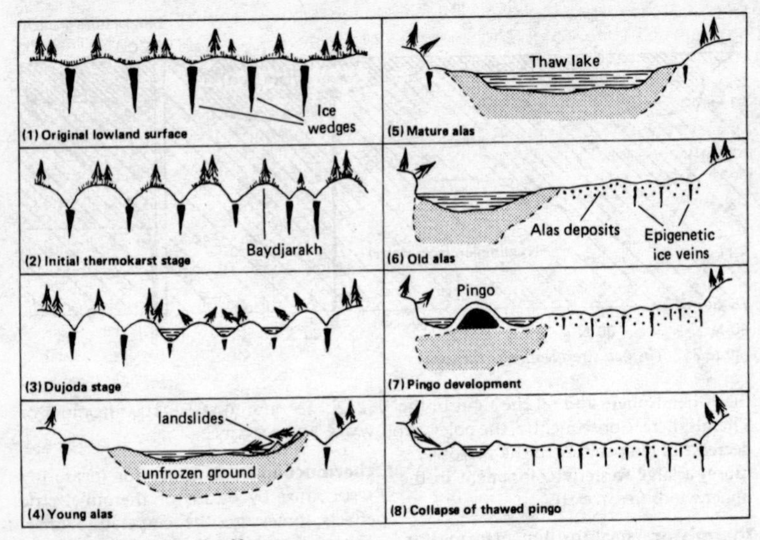

Figure 253 *Thermokarst landforms*

thermogenic soils Soils with properties that have been influenced primarily by high soil temperatures as the dominant soil-forming factor, e.g. some desert soils.

thermograph BIMETALLIC THERMOGRAPH.

thermohaline circulation Ocean-water circulation resulting from differences of temperature and salinity. Thermohaline convection is one of the chief mechanisms behind the operation of OCEAN CURRENTS.

thermokarst A term referring to the ground-surface depressions which are created by the thawing of ground ice in the PERIGLACIAL ZONE. The subsidence creates some features which may resemble true KARST, but only thermal processes are involved, there is no chemical weathering and it does not depend on the presence of limestone. The features include ALAS, BEADED DRAINAGE, THAW LAKES and collapsed PINGOS. [*253*]

thermoluminescence The light emitted by many minerals when heated below the temperature of incandescence. It results from the release of energy stored as electron displacements in the mineral's crystal lattice, and the amount of thermoluminescence is proportional to the number of electrons. By monitoring the amount it is possible to estimate the value of a constant rock temperature imposed by a former climatic environment and therefore to calculate the length of time since the change of temperature occurred in the rock's environment. It is suggested that the technique may be refined to assist in the calculation of rates of erosion of the land surface.

thermomeion A term applied to an area that has a high negative ANOMALY of temperature. *See also* THERMOPLEION.

thermometer An instrument for measuring temperature. The most common model consists of a column of mercury enclosed in a sealed glass capillary tube rising from a small bulb which acts as a reservoir for the mercury. The glass tube is graduated so that changes in the length of the column can be read off as the mercury expands and contracts with changing temperature. The calibrated scale can be in degrees CELSIUS (*centigrade*), FAHRENHEIT or RÉAUMUR. When low temperatures have to be measured the mercury is replaced by alcohol. Thermometers are used to determine RELATIVE HUMIDITY by means of the DRY- AND WET-BULB THERMOMETERS. A BLACK-BULB THERMOMETER is used to measure solar radiation temperatures. Other types of thermometer include the *resistance thermometer*, to measure electrical resistance (THERMISTOR) and a *gas thermometer* to measure the pressure variations in a gas kept at constant volume. *See also* PYROMETER.

thermoperiodism The response of plants to rhythmic fluctuations in temperature. Most plants tend to thrive in an environment of varying temperatures and are kept in check where conditions are constant.

thermophile An organism that grows best at temperatures of 50° C or higher.

thermopile An instrument for detecting heat RADIATION by means of a series of THERMOCOUPLES.

thermopleion A term applied to an area that has a high positive ANOMALY of temperature. *See also* THERMOMEION.

thermosphere The layer of the atmosphere which occurs above the MESOPAUSE, some 80 km above the Earth's surface, in which temperature increases with height. Some scientists regard the thermosphere and the IONOSPHERE as being synonymous but others would regard the ionosphere only as the layer between 100 and 300 km, where electron density is very high. [*17*]

therophyte One of the six major floral lifeform classes recognized by the Danish botanist, Raunkiaer, based on the position of the regenerating parts in relation to the exposure of the growing bud to climatic extremes, and without respect to TAXONOMY. The therophytes are annual plants: after producing seeds the parent plant dies and the seeds lie dormant until favourable weather conditions return.

thickness **1** In general usage, a dimension of depth between opposite surfaces. **2** In synoptic meteorology, a term referring to the height difference above mean sea-level of two PRESSURE surfaces above a given reference point. Such differences are obtained by means of RADIOSONDE observations and are depicted on a CONTOUR CHART. They are also proportional to the mean TEMPERATURE of the layer under examination, so that small values indicate cooler air and larger values relatively warm air.

Thiessen polygon A term used by meteorologists to define the horizontal area which is nearer to one RAIN-GAUGE than to any other in a rain-gauge network. It is used to allow for the uneven distribution of rain-gauges.

thin section A thin slice of rock or TILL ground to a standard thickness of 0.3 mm after having been mounted on a glass slide. Finally, it is covered with another glass slide in order for its minerals to be examined under a microscope.

thixotropic A term applied to certain solid/liquid systems which are solid when stationary but which become mobile when affected by shearing stresses, e.g. QUICK-CLAYS.

tholoid A name given to a new volcanic dome created by the slow upheaval of viscous acid lava within the crater or caldera of a volcano.

Thornthwaite's index of potential evapotranspiration A climatological method of defining potential evapotranspiration (PE) (EVAPOTRANSPIRATION) devised by C. W. Thornthwaite in 1948. It gives a reasonable result in temperate latitudes but is not as universally applicable as PENMAN'S FORMULA. Thornthwaite obtained both an index of aridity (Ia) and an index of humidity (Ih) by determining the annual moisture surplus and deficit (MOISTURE INDEX). He used the latter as a basis for a climatic classification (THORNTHWAITE'S METHOD OF CLIMATIC CLASSIFICATION).

Thornthwaite's method of climatic classification A system of classification of climates devised by C. W. Thornthwaite between 1931 and 1948, based on the ability of a climate to support certain plant communities. He used the variables of temperature, rainfall, PRECIPITATION EFFICIENCY, THERMAL EFFICIENCY and MOISTURE INDEX. The classification recognized the following zones: Perhumid (A) with moisture index (MI) >100; Humid (B with four subdivisions) with MI 20 to 100; Moist subhumid (C_2) with MI 0 to 20; Dry subhumid (C_1) with MI –20 to 0; Semi-arid (D) with MI –40 to –20; Arid (E) with MI <–40.

threshold **1** A term used to describe the submerged rock bar at the mouth of a FIORD (SILL). **2** A low neck of sand dividing a lake into two basins. **3** A condition marking the transition from one state of operation of a SYSTEM to another, thereby complicating the simple self-regulation mechanism of negative FEEDBACK. The passage of a system across a threshold is often irreversible and even the change of one variable may force the entire system to adjust to a radically different DYNAMIC EQUILIBRIUM.

threshold angle The limiting angle of STATIC FRICTION, used in the context of loose debris on a hill slope. It can also be termed the *critical angle of stability*.

threshold slope A term introduced by M. A. Carson in 1976 for straight slope segments between upper convexities and lower concavities, and which represent the maximum GRADIENT, for temporary stability, permitted by the strength and PORE-WATER PRESSURE of the soil cover. They are common in areas of rapid uplift and stream incision, and therefore prone to LANDSLIDES, but will be ultimately eliminated by soil CREEP and SURFACE WASH. *See also* COEFFICIENT OF FRICTION.

throughfall A term referring to all that PRECIPITATION which reaches the ground through a vegetation cover, either directly or by the drip from INTERCEPTED MOISTURE, but it does not include STEM FLOW.

throughflow The downslope movement of water through the REGOLITH in contrast to that of surface flow (OVERLAND FLOW). When the PERMEABILITY of the regolith decreases downwards, most commonly at the base of the soil's A-HORIZON, some of the downward-percolating water cannot penetrate to the lower horizons (SOIL HORIZON) fast enough and is therefore deflected laterally as *throughflow* [*216*]. When the latter process carries fine soil particles through the regolith of a slope it is a recognized pedological process known as lateral ELUVIATION. *See also* LATERAL FLOW, RUNOFF, SUBSURFACE WASH.

throughput A term referring to the mass, energy or information which passes through a SYSTEM.

throw FAULT, DOWNTHROW, UPTHROW.

thrust-fault A type of REVERSE-FAULT that is characterized by a low angle of inclination with reference to the horizontal plane. *See also* NORMAL-FAULT. [*254*]

thrust-plane The planar surface across which the overturned upper limb of a RECUMBENT FOLD travels at a THRUST-FAULT, or

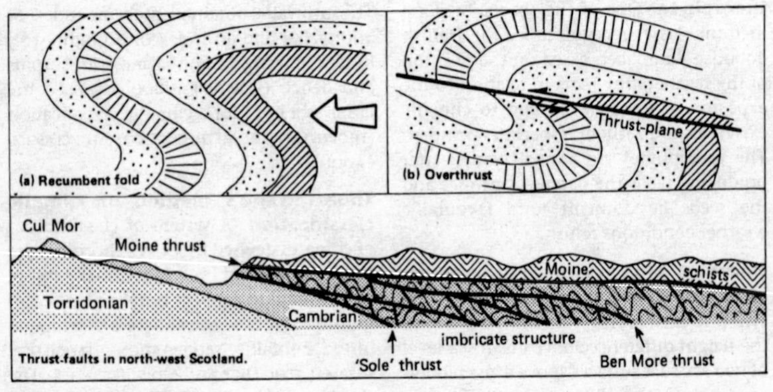

Figure 254 *Thrust-faults*

over which the upper strata are driven at a REVERSE-FAULT, which is characterized by a low angle of inclination with reference to the horizontal plane. [*254*]

thufur An Icelandic term for an earth HUMMOCK.

Thule–Baffin moraines An expression referring to certain ice-cored TERMINAL MORAINES at the margins of 'cold' ice-sheets (COLD GLACIERS) near to Thule (Greenland) and in Baffin Island, Canada. Morainic debris brought to the ice surface along shear planes produces a thick TILL cover along the glacier snout. This protects the latter from ablation by direct insolation, thereby creating a type of ICE-CORED MORAINE.

thunder The noise which accompanies a LIGHTNING FLASH. It results from the sudden heating and expansion of the air by the flash, causing sound-waves. As the speed of sound is considerably less than the speed of light the noise of thunder always follows the flash. A rough calculation of distance from the flash is that of 1 mile (1.6 km) for every 5 sec interval between lightning and thunder. *See also* CONVECTIONAL RAIN, THUNDERSTORM.

thunderstorm A storm associated with THUNDER, LIGHTNING and very heavy precipitation of rain or hail. It represents extreme conditions of INSTABILITY, often associated with the passage of a COLD FRONT or due to intense heating of the ground surface and the resulting convectional uplift of air. If this rapidly rising air current rises above the condensation level CUMULONIMBUS (*thunderhead*) clouds will probably form, since these are the principal thunderstorm clouds. Very high electrical differences occur within the clouds and the separation of the electrical charge is the precursor to a lightning flash. It has been shown that a positive charge collects at the top of a cumulonimbus cloud and a negative charge near its base, with a second, smaller positive charge low down in the cloud associated with the falling rain, although the exact mechanism of the charge separation is not fully understood. Because there is little convection or surface heating in polar regions cumulonimbus clouds rarely form, so that thunderstorms are almost unknown. The greatest thunderstorm activity occurs in low latitudes where there is both intense heating and moisture availability. Java has a high frequency as also does Kampala, Uganda. *See also* CONVECTIONAL RAIN.

Thurnian One of the earliest stages of the PLEISTOCENE in Britain, known only from East Anglia, where a borehole was drilled at Ludham, Norfolk. Its deposits consist of marine silts, the Foraminifera of which indicate cold sub-Arctic conditions. They overlie deposits of LUDHAMIAN age and are succeeded by deposits of ANTIAN age.

tidal current A horizontal movement of sea water in response to the rise and fall of the TIDE. Two types of current are recognizable: **1** The movement in and out of an estuary or bay (referred to as a *tidal stream*), produced by the vertical oscillation to the tide (FLOOD TIDE, EBB TIDE). This type of tidal current carries large amounts of sediment across the continental shelf. **2** The movement of water between two points affected by different tidal regimes, thus giving rise to height differences in the level of the water at each of the points. This type of hydraulic tidal current, which operates merely to equalize water levels, is especially common in straits, e.g. the Pentland Firth in N Scotland and the Menai Strait, N Wales.

tidal datum A reference plane (DATUM LEVEL) used as a base for topographical survey (ORDNANCE DATUM) or as a basis for navigational charts (CHART DATUM). It is determined by calculating a mean tidal figure from a lengthy series of records.

tidal glacier The term given to a valley glacier which debouches into an ocean, thereby calving into ICE-FLOES at its snout.

tidal inlet A submarine channel between segments of a BARRIER-BEACH, through which TIDAL CURRENTS flow. Its shape is controlled by the relative magnitude of the energy of the WAVES and that of the tidal SCOUR.

tidal palaeomorph A term introduced by W. F. Geyl (1985) to describe an oversized, frequently meandering valley believed to

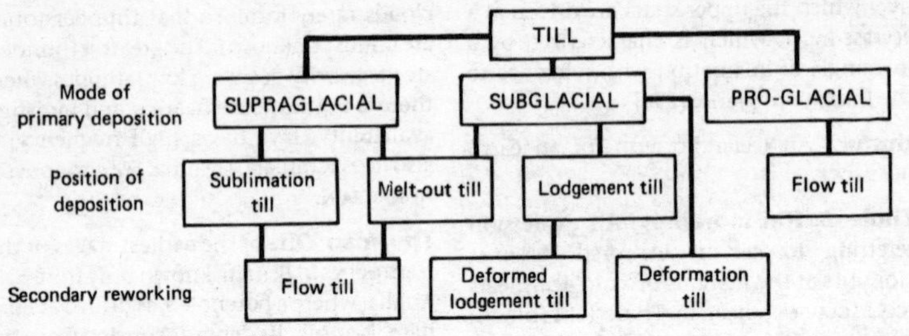

Figure 255 *Classification of till*

have been formed as a tidal channel when sea-level was higher than at present. The course of the R. Thames through London has been given as an example.

tidal prism The amount of sea water (measured in m^3) that a FLOOD TIDE carries into, or an EBB TIDE carries out of, an estuary or bay. The tidal prism is an important calculation in the determination of the tidal equilibrium by a coastal authority for it helps it to decide on the degree of dredging which may be necessary in a particular harbour or tidal inlet. The flood tide, reinforced by LONGSHORE CURRENTS, tends to cause deposition, whereas the ebb tide, reinforced by fresh-water discharge, tends to assist tidal scour. NB The tidal prism excludes the volume of fresh-water flow.

tidal range The difference between the level of sea water at high tide and low tide. The range cannot be fixed, because of the differences between NEAP TIDES and SPRING TIDES.

tidal scour SCOUR.

tidal stream The TIDAL CURRENT which responds to the flood and ebb tides and which helps to develop the FLOOD CHANNELS and EBB CHANNELS of a littoral environment (SCOUR). [*67*]

tidal wave TSUNAMI.

tide The regular rise and fall of water level in the world's oceans, resulting from the gravitational attraction that is exerted upon the Earth by the Sun and the Moon. The tidal-producing force of the Moon is slightly more than twice that of the Sun. *See also* AMPHIDROMIC SYSTEM, APOGEAN TIDES, NEAP TIDE, OSCILLATION THEORY OF TIDES, PERIGEAN TIDE, PROGRESSIVE WAVE THEORY OF TIDES, SPRING TIDE.

tierra A Spanish term that has been adopted by climatologists to define specific climatic regions based on altitude in S America, Central America and Mexico: **1** *Tierra helada*, the mountain zone of snow-covered peaks. **2** *Tierra fria*, the upland zone above 1,800 m (average monthly temperatures 12°–18° C). Some coniferous forest but mainly pastures and temperate crop-growing; most favourable to European settlement. **3** *Tierra templada*, the intermediate zone between 900 and 1,800 m (average monthly temperatures 18°–24° C); since rainfall is not very high only a few tropical crops are grown. **4** *Tierra caliente*, the lowest, coastal zone, below 900 m (average monthly temperatures >24° C). In the wetter areas dense tropical forest occurs; crops include sugar, coffee, cacao, banana.

Tiglian A term for the Lower Pleistocene in Britain.

tile A GIS term for a logical rectangular set of DATA covering a chosen area of a large digital map DATA SET, which divides it into more manageable units for interpretation.

till **1** A type of sediment in which the components have been brought into contact by the direct agency of glacier ice. Although the till may suffer subsequent glacially induced flow (FLOW TILL) it does not become disaggregated. The term has now replaced that of *boulder clay* because of the latter's implication of too precise a composition. In 1980 British scientists drew up a classification of tills based on a combination of sediment-forming processes and the position of the depositional environment, i.e. SUBGLACIAL, SUPRAGLACIAL or pro-glacial [*255*]. It has been suggested that the term *ablation till* should be abandoned; that the term *sublimation till* be used to describe material laid down in a supraglacial environment; that the term *melt-out till* should refer to one produced by the top melting of a debris-rich buried ice mass in which the structures inherited from the parent ice are preserved after all the ice has melted. Till which has been deposited by the sole of the glacier in a subglacial environment is referred to as *lodgement till*, but if this is glacially reworked following initial deposition it should be termed *deformed lodgement till* or *deformation till*. Till is one type of glacial *drift*, the other is termed a GLACIOFLUVIAL deposit. **2** To plough or cultivate the soil (*tillage*).

till fabric FABRIC, TILL.

tillite A former TILL that has become compacted and lithified to form a tough sedimentary rock.

tilloid A chaotic often partly indurated sedimentary mixture of large sub-angular rocks in a clay-rich matrix resulting from the action of LANDSLIDES, MUDSLIDES and GLACIAL OUTBURSTS.

till-plain A wide area of low relief created by a *till-sheet*, which masks all irregularities in the bedrock relief. Thus the glacial deposits vary greatly in thickness, being greatest over the bedrock hollows and thinnest over ridges and hills.

till-sheet TILL-PLAIN.

tilted blocks Crustal blocks that have been tilted in a region of BLOCK-FAULTING. They are one of the characteristic features of a BASIN-AND-RANGE TERRAIN. [*26*]

tilth The physical state of the soil that determines its capability for plant growth. Although there are no regular rules or strict measures the tilth of a soil is based on a subjective estimate of the relationships between its texture, structure, consistence and pore space.

tiltmeter A device for measuring the amount of tilt of the ground surface as a result of TUMESCENCE, due either to the updoming of a volcano by magnetic pressure or to the crustal doming created by the 'locking' of major tear faults, e.g. the Palmdale swelling on the San Andreas fault, California, USA.

timber-line TREE-LINE.

time **1** The period through which an action, condition or state continues. **2** The time indicated by the apparent motion of the Sun, i.e. that indicated on a sun-dial is known as the *apparent time, local time* or *local solar time*. *See also* EQUATION OF TIME, GREENWICH MEAN TIME, INTERNATIONAL ATOMIC TIME, MEAN SOLAR TIME, STANDARD TIME, UNIVERSAL TIME.

time-dependent form A landform in which the form of a particular slope depends on the time that has elapsed since the inception of its evolution. The CYCLE OF EROSION, outlined by W. M. Davis, implies that slope form is time-dependent.

time-independent form A landform, particularly a slope, in which the relative form remains unchanged with time even though the altitude of the entire slope may be lowered, i.e. the position of each point on the profile relative to the other points remains unaltered over time. This is in contrast to the concept of TIME-DEPENDENT FORM.

time-lag The length of time separating two correlated physical phenomena. The length will vary, e.g. the separation between a period of heavy rainfall and the rise of a river to flood conditions may be a matter of hours, where in rock weathering the span between the initial loosening of material and its subsequent removal may be several millennia.

time-scale, geological CHRONOSTRATIGRAPHY, GEOCHRONOLOGY.

time series Successive observations of the same phenomenon over a period of time. The data series is arranged in order of occurrence but with equal time intervals being placed between each value. Much of the analysis of time series involves the use of graphs, in which the horizontal scale measures time and the vertical scale the magnitude of the variable being examined. A more sophisticated extension of time-series analysis is FOURIER ANALYSIS.

time-transgressive An expression usually referring to a sedimentary formation the sediments of which were deposited in an environment that shifted geographically with time, such that its age varies from one place to another.

time zone A time division devised for convenience and usually comprising 15° of longitude (STANDARD TIME). [*241*]

tin The metal derived from cassiterite or tin oxide (SnO_2). It is associated with lodes in igneous rocks, in veins created by HYDROTHERMAL ACTIVITY or as a PLACER DEPOSIT.

tipping-bucket rain-gauge A self-recording RAIN-GAUGE, which empties itself when full. Two buckets are pivoted on a knife edge and are calibrated to tip/empty after a fixed amount of rain has accumulated; an electrical contact is made when tipping so that the specific rain totals are monitored on an adjacent recorder. The first model was made by Sir Christopher Wren, the architect, in 1662, but the design was later perfected by W. H. Dines, a meteorologist, at the end of the 19th cent. *See also* NATURAL-SIPHON RAINFALL RECORDER.

TIROS NOAA/TIROS SATELLITES.

titration A technique used in quantitative measurements of the concentration of a fluid SOLUTION.

tjaele (tjäle, taele) A Swedish term meaning 'frost table' and describing any frozen surface in the active layer as it thaws downwards to the PERMAFROST TABLE. It is also applicable to a frozen surface in seasonally frozen ground in a non-permafrost environment. It should not be confused with PERMAFROST or the permafrost table.

toadstones A term used in N England, but thought to be derived from the German *todstein* (literally a dead stone), to describe the various types of igneous rocks, especially extrusive BASALT sheets and intrusive SILLS, which occur among the Carboniferous limestone rocks in the Pennines. Although the rocks themselves are devoid of metallic ores they have caused mineralization in the surrounding rocks and mineral veins occur in their vicinity. In some instances they have created a PERCHED WATER-TABLE within the limestone.

tolerance The resistance of an organism to the excess or the deficiency of an element or a condition in its environment. All organisms exhibit limits of tolerance beyond which they will not survive. Botanists refer to tolerance as a hardiness factor, enabling an organism to withstand harmful conditions inside its cell tissues, distinguishing it from *avoidance* and *evasion*. A XEROPHYTE, for example, may avoid drought by the development of a thick cuticle on its leaves, it may evade drought through dormancy, or it may tolerate drought by developing a wide range of physiological controls.

tombolo A type of sandy or shingly coastal SPIT which links an island to the mainland. It may be produced by LONGSHORE DRIFT of beach material along a coastline, creating a connecting bar with a neighbouring island. There are examples of triple tombolos linking an island to a coast (Monte Argentario, W coast of Italy), of double tombolos (Castlegregory, Co. Kerry, Ireland) and of several islands being interconnected by a number of tombolos (Nantasket Beach, Massachusetts, USA). [*256*]

ton, tonne A unit of mass in both the foot–pound system and the metric system. In the UK the *long ton* is equal to 2,240 avoirdupois pounds. In the USA the *short ton* is equal to 2,000 pounds. In the metric system the *tonne* (1,000 kilograms) is equal to 2,205 pounds. [*145*]

tonalite A quartz-diorite igneous rock intermediate in its quartz content between a DIORITE and a granodiorite.

toppling failure A type of slope FAILURE in which large masses or columns of rock suddenly become detached from a cliff face and overbalance as they tilt outwards. It can

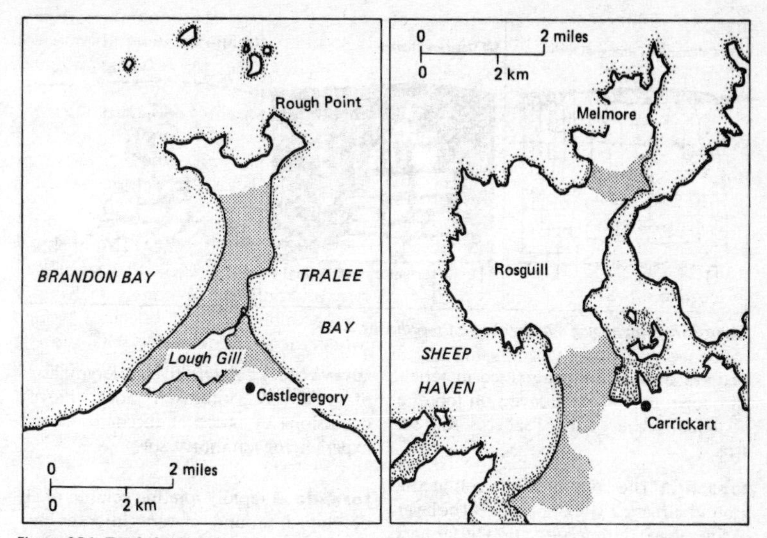

Figure 256 *Tombolas in Ireland*

be caused by pressures generated by FREEZING or by water after heavy rainfall. It is common where BEDDING PLANES in the rock dip down the slope to a maximum of 35°, beyond which SLAB FAILURE is more likely to occur. *See also* ROCK-FALL.

topoclimatology The study of the interrelationship between TOPOGRAPHY and CLIMATOLOGY at the local scale. This is of particular importance in urban areas and in areas of broken relief with very little flat land and with slopes of various gradients and aspects. Examination of the variations in solar-radiation receipt, water balance, wind flow, etc. may indicate that they are very different from those recorded in an urban environment in a less hilly terrain. Thus, regions with their own topoclimates may be distinguished from each other. *See also* ADRET, INVERSION OF TEMPERATURE, KATABATIC WIND, UBAC.

topographic map A map whose principal purpose is to portray and identify the features of the Earth's surface as faithfully as possible within the limitations imposed by the scale. The features include both those of the cultural landscape as well as those of the TERRAIN and the RELIEF. *See also* RELIEF MAP, TOPOGRAPHY.

topography The surface features of the Earth's surface, including the RELIEF, the TERRAIN, the vegetation, the soils and all the features created in the landscape by human endeavour. It is not synonymous merely with relief.

topology **1** The scientific study of a particular locality. **2** A branch of geometry referring to certain spatial relationships between phenomena, e.g. relative position and contiguity, rather than those of actual distance or direction. It differs from the rigidity of Euclidean geometry because it is more elastic. It is useful in drawing sketch maps of geographical distributions, for example, where interfering symbols (NOISE) can be eliminated in order to clarify the map's main intentions, but it cannot be used in navigation charts, where correct scale and orientation are essential.

toposequence A sequence of related soils that differ from each other primarily because RELIEF has been the most influential soil-forming factor. *See also* CATENA.

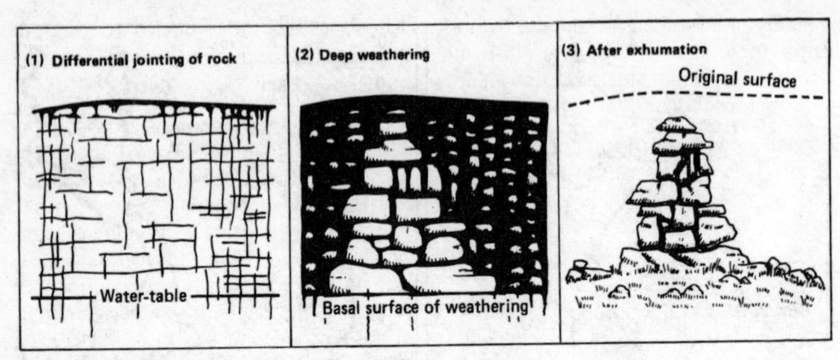

Figure 257 *Formation of a tor by chemical weathering*

top-set beds The fine-grained materials laid down in horizontal layers on top of a DELTA [*85*]. *See also* BOTTOM-SET BEDS, FORE-SET BEDS.

topsoil **1** The layer of fertile soil at the ground surface. **2** The A-HORIZON. **3** The layer of soil moved during cultivation by farmers or gardeners.

tor A small castellated hill or exposure of well-jointed rock rising abruptly from a relatively smooth hilltop or slope. It is composed of a stack of well-jointed blocks projecting from a platform of solid rock, often surrounded by a mass of collapsed blocks. Tors can form in any massively jointed rock (JOINT), although most of the tors in SW England are made of granite. In the tropics similar landforms are known as KOPJES. There is no agreement on their origin, with some authors favouring a PERIGLACIAL origin, others supporting a hypothesis based on deep CHEMICAL WEATHERING extending down to the BASAL SURFACE OF WEATHERING prior to a period of EXHUMATION. The latter involves the stripping of the rotted, weathered material, which is deepest in zones of narrowly spaced jointing, and the subsequent exposure of the rectangular or sub-rounded CORESTONES, in the zones where jointing is more widely spaced, and it is these that contribute to the character of the newly uncovered tor [*257*]. Most tors exhibit horizontal divisions between the blocks, due either to BEDDING PLANES (sedimentary rocks) or PSEUDO-BEDDING (igneous rocks).

toreva blocks A term given to large masses of relatively unbroken rock moved downslope by LANDSLIDE mechanisms and experiencing ROTATIONAL SLIP.

tornado A rapidly rotating column of air developed around a very intense low-pressure centre. It is associated with a dark funnel-shaped cloud and with extremely violent winds (>300 km/h) blowing in a counterclockwise spiral, but accompanied by violent downdraughts. The precise mechanisms are not fully understood but the following atmospheric conditions appear to be necessary for tornado development: a layer of warm moist air at low altitude; a layer of dry air at higher altitude with an INVERSION OF TEMPERATURE at about 1,000 m; a triggering mechanism, usually in the form of an active, intense cold front or solar heating of the ground which will create a VORTEX (THERMAL). Such conditions are found in many low- and mid-latitude countries (even Britain) but are very common in the mid-west of the USA, where great damage is caused by tornadoes in most years. The greatest material damage appears to be caused when the extremely low pressure at the centre of the vortex causes buildings to 'explode' outwards. The swathe of destruction is very localized, no more than 150 m across, but the tornado is often unpredictable in its behaviour, rapidly changing direction, 'leap-frogging' certain tracts as the funnel loses contact with the ground, and occasionally back-tracking. It is generally regarded as having the greatest wind intensity of all the meteorological hazards hence there is a well-organized tornado warning system in the US mid-west. *See also* DUST-DEVIL, WATERSPOUT.

torrent A high-velocity flow of water down a stream channel (FRESHET). The term has also been used to describe a fast-moving flow of lava.

Torridonian A division of the PRE-CAMBRIAN in Scotland, younger than the LEWISIAN but the unmetamorphosed equivalent of the MOINIAN. It comprises thick stratified deposits of shale, grit and sandstone (the Torridon sandstone), all resting unconformably on the underlying Lewisian gneiss. Its tough sandstones form rugged mountain peaks in NW Scotland.

torrid zone Synonymous with the TROPICS; that area lying between 23.5°N and 23.5°S of the equator. One of the three temperature zones recognized by early geographers. *See also* TEMPERATE ZONE.

torsion The action of twisting or turning a body spirally by the operation of contrary forces acting at right angles to its axis.

tower karst TURMKARST.

trace elements Those elements in the Earth's CRUST which occur in very small quantities, e.g. boron, cadmium, cobalt, in contrast to the eight major rock-forming elements which occur in abundance (i.e. aluminium, calcium, iron, magnesium, oxygen, potassium, silicon and sodium). Some trace elements in the soil may have a deleterious effect on livestock and on human health, e.g. molybdenum in Carboniferous black shales in SW England produces bone deformities in cattle. Toxic levels of trace elements can also be found in sewage sludge and metal mine TAILINGS.

trace fossil A sedimentary structure caused by biological activity in an earlier period of geological time, i.e. the fossilized remnant of the effects of an organism rather than the FOSSIL organism itself, e.g. tracks, footprints, burrows.

tracer A substance, often a coloured dye, introduced into DISAPPEARING STREAMS to determine their subterranean course and their point of RESURGENCE.

trachyte A fine-grained, extrusive igneous rock of intermediate character (INTERMEDIATE ROCKS), belonging to the alkaline basalt volcanic suite. It is the extrusive equivalent of SYENITE and is characterized by a FLOW STRUCTURE (from its former molten state) which is referred to as a *trachytic texture*. It is commonly found in areas of continental rifting, e.g. the East African Rift Valley, and in the Puy-de-Dôme region of France. *See also* KERATOPHYRE, PHONOLITE, RHYOLITE.

tract **1** The continuance of some action or state. **2** A stretch or expanse of land with no areal definition. **3** One of the divisions in D. L. Linton's MORPHOLOGICAL REGION classification.

traction load Synonymous with BEDLOAD. *See also* LOAD.

tractive force CRITICAL TRACTIVE FORCE.

Trade Winds The predominantly easterly winds that blow from the subtropical high-pressure cells towards the EQUATORIAL TROUGH, winds that are more pronounced over the oceans than over the continents. The claim that their title is derived from the part they once played in assisting mercantile marine commerce is, however, a fallacious one, since their name is derived from the Latin *trado*, meaning constant 'direction', and has given rise to the phrase 'to blow *trade*', i.e. consistently in a constant direction. The NE Trades and SE Trades are more constant on the eastern side of the oceans than over the western sides, where there is some interference from pressure disturbances on the windward shores. As a result, the eastern coastlines and offshore islands in the Trade Wind belt exhibit very arid conditions (e.g. Canary Is.) owing to the dryness of the air mass and the strong low-level INVERSION OF TEMPERATURE, which prevents vertical cloud development. Conversely, on windward coasts, the moisture-laden air mass and the much weaker temperature inversion sees stronger convection developed with accompanying rainstorms. The Trade Winds play a crucial part in the Earth's ATMOSPHERIC CIRCULATION where they are part of the PLANETARY WIND system. Their main function is to remove surplus heat from the subtropical high-pressure belt by evaporating great quantities of heat from the tropical oceans, thereby helping to maintain the global HEAT BALANCE as they return surplus air to the equatorial zone. [*18*]

trafficability The capacity of a soil or a type of terrain to support moving vehicles. It is an important component of TERRAIN ANALYSIS.

trajectory **1** A term used in meteorology to describe a line indicating the path followed by a particle of air over a given time period. It contrasts, therefore, with the term STREAMLINE. **2** A term used to describe a series of SYSTEM states.

tranquil flow FROUDE NUMBER.

transcurrent-fault A TEAR-FAULT of massive proportions in which the net slip is in the direction of the fault strike. It is the type of fault which frequently offsets MID-OCEANIC RIDGES. *See also* MEGASHEAR.

transect **1** A transverse section across a region. It is regularly employed by physical geographers to show the spatial relationships existing between vegetation, soils and relief. **2** In morphological mapping, a transect is a line across the ground surface that does not follow the true slope. Angle measurements along a transect, therefore, show only the apparent slope.

transection glacier A glacier that totally occupies a valley system to such an extent that it overspills divides to create breaching by GLACIAL DIFFLUENCE and GLACIAL TRANSFLUENCE. [*92*]

transfluence GLACIAL TRANSFLUENCE.

transform-fault A type of massive TEAR-FAULT on a continental scale in which the fault terminates sharply at a place where the movement is transformed into a structure of another type. The term was introduced in 1965 by J. T. Wilson to explain the transformation of strike-faults into MID-OCEANIC RIDGES, ISLAND ARCS or FOLD-MOUNTAIN BELTS. *See also* TRANSCURRENT-FAULT.

transgression An advance of the sea across a former land area, caused by a positive eustatic movement (EUSTASY) or a negative isostatic movement (ISOSTASY). It will lead to the formation of FIORDS, RIAS, FIARDS and SUBMERGED FORESTS. It contrasts with a marine REGRESSION. *See also* FLANDRIAN TRANSGRESSION.

transit **1** The passage of a heavenly body across a particular meridian of the globe. **2** A US term for a THEODOLITE.

translational slide The collapse and movement of rock or soil along JOINTS and BEDDING PLANES that lie more or less parallel to the slope of the ground. It is a type of MASS-MOVEMENT.

translocation The migration of material in solution or suspension from one soil HORIZON to another.

transmission capacity The capacity of soil to allow water to pass through it. When the transmission capacity is less than the INFILTRATION-RATE of the upper soil horizons, the latter will become saturated and OVERLAND FLOW will occur. *See also* PERCOLATION.

transpiration The process by which plants lose water vapour through the stomata (pores) on their leaves, thereby extracting soil moisture and returning it to the atmosphere. *See also* EVAPOTRANSPIRATION.

transport, transportation **1** The movement of material by a natural agent within a geomorphological system between a point of EROSION and a site of DEPOSITION. The means of transport may be running water (SALTATION, SOLUTION, SUSPENSION, TRACTION), wind (DEFLATION), glacier ice, marine waves (LONGSHORE DRIFT), tides and currents. Movement induced simply by gravity (LANDSLIDE, MASS-MOVEMENT, ROCKFALL) is not accepted as transport by some geomorphologists, although geologists speak of *gravity transport* as including CREEP, SOLIFLUCTION, EARTH-FLOWS, MUD-FLOWS and slumping (SLUMP). The transported material is termed the LOAD, which causes ABRASION and itself suffers

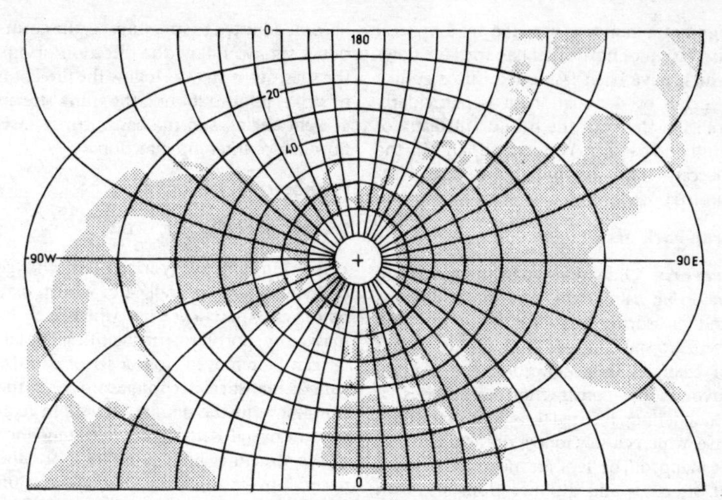

Figure 258 *Transverse Mercator projection: northern hemisphere*

ATTRITION. The maximum size of particles transported by ice is greater than that for water, which in turn is greater than that for wind. Considerable particle SORTING occurs during transport, causing the material to become graded (GRADE). **2** The displacement of one part of a rock in relation to another by thrusting (THRUST-FAULT) is known as *tectonic transport*, the scale of which can vary between a few centimetres and tens of kilometres.

transportation slope A type of SLOPE on which there is neither a net loss nor a net gain of ground, because at each point the amount of material brought in from upslope is balanced by the amount of material carried away downslope. *See also* ACCUMULATION SLOPE, CONSTANT SLOPE, DENUDATION SLOPE, FOOT SLOPE, SLOPE EQUILIBRIUM, CONCEPTS OF, TRANSPORT. [*233*]

transverse coast DISCORDANT COAST.

transverse dune An asymmetrical sand-dune which stands at right angles to the direction of wind blow. A pattern of regular spacing may be due to WAVE FLOW. [*277*]

transverse Mercator projection The transverse aspect of the MERCATOR PROJECTION in which the line of zero distortion is coincident with a meridian. The cylinder is tangential to the globe along a meridian, not, as in the normal case, along the equator, which means that the projection has been turned transversely through 90°. The scale error increases away from the CENTRAL MERIDIAN, which is divided truly. The LOXODROMES are curved lines, not, as in the normal case, straight lines. The projection is used chiefly to depict small areas which extend in a predominantly north–south direction. It is the basis of the US UTM GRID and the British NATIONAL GRID and is the one from which the maps of the ORDNANCE SURVEY are constructed. It is synonymous with the Gauss conformal projection. [*258*]

transverse valley Any valley which cuts at right angles across a ridge (WATER-GAP).

trap **1** A crustal structure in which oil or natural gas may collect. **2** A name once used for a fine-grained igneous rock (*trap-rock*), especially with reference to basalt. **3** The point at which an underground cave roof dips below water-level in a subterranean river system. This is also called a *sump*.

trap landscape A terrain characterized by a stepped appearance (*trappa* in Swedish signifies a staircase), created by a vertical succession of horizontal basaltic lava-flows which have been fashioned into step-like benches by denudation. It is particularly well illustrated in the basaltic plateaux of Mull and Skye in W Scotland, and in the Deccan traps of India. *See also* PIEDMONTTREPPEN.

trap-rock TRAP.

traverse A line surveyed by the method of *traversing*, using either a prismatic COMPASS and measuring TAPE or a THEODOLITE and LEVELLING-STAFF. Instead of obtaining a series of triangles (TRIANGULATION), the method involves the construction of a series of straight lines, referred to as 'legs' of the traverse, which closely follow the actual line to be mapped. The 'legs' are measured in terms of direction and distance from a known starting-point closing on a known finishing-point to produce a *closed traverse*. If only one of the stations, at the beginning or end of the traverse, is of accurately known position, the traverse is referred to as an *open traverse*. The method suffers from measurement error, which is cumulative during the operation. The cumulative error is usually distributed proportionately along the traverse. *See also* TACHEOMETRY.

travertine A light-coloured concretionary deposit of calcium carbonate deposited by precipitation from highly impregnated groundwater around a hot spring. It is sometimes referred to as calcareous TUFA or calcareous SINTER (calc sinter). Banded varieties are sought after as so-called onyx marble.

tree-line The upper limit of tree growth. Its altitude varies with latitude, being lower in the temperate zone than in the tropics. Its elevation is also affected by local factors of soil and exposure in addition to ASPECT. The tree-line is lower on the shaded side of a mountain than on the sunny side. It is known as the *timber-line* in the USA.

trellis drainage, trellised drainage A rectilinear drainage pattern formed where two sets of structural controls occur at right angles. It is most commonly found in a SCARP-AND-VALE TERRAIN where drainage has become adjusted to structure, i.e. the CONSEQUENT STREAMS follow the direction of dip, the SUBSEQUENT STREAMS follow the direction of strike, prior to joining the trunk stream at right angles, and the OBSEQUENT STREAMS flow in a counter-dip direction.

tremor EARTH TREMOR.

trench, oceanic DEEP, OCEAN.

trend 1 In climatology, a long-term change of temperature, rainfall, etc. which may reflect CLIMATIC CHANGE but which may be masked by short-term irregularities. The latter can be removed from a set of data by various statistical techniques, one of the simplest being the RUNNING MEAN. **2** In geology, the overall pattern or general tendency of the structural lines in an area. *See also* GRAIN. **3** In statistics, the general direction in which a series of observations is increasing or decreasing. *See also* TIME SERIES, TREND-SURFACE ANALYSIS.

trend surface A mathematical expression fitted to spatially distributed data, by means of a least-sum-of-squares REGRESSION, to produce a three-dimensional surface. *Trend-surface analysis* is a method by which height or depth data at points on a surface can be plotted by constructing surfaces of increasing complexity, i.e. a development from a *linear surface* through a *quadratic surface* and *cubic surface* to a higher-degree *polynomial surface*. [259]

trend-surface analysis TREND SURFACE.

trend-surface map A map of the systematic, relatively large-scale, regional pattern of an areal distribution. It can be constructed either by: **1** *filter mapping*, when all the data are available and a *response surface* is achieved by gridding the area, determining the values for each grid square and then filtering (FILTER) the surface to discover local anomalies; or by **2** *nested sampling* (NESTING), when the data are incomplete and several areas of equal size are taken at random (RANDOM SAMPLES) and randomly split into smaller units. By using methods of VARIANCE ANALYSIS it is possible to calculate the values for each sampling level.

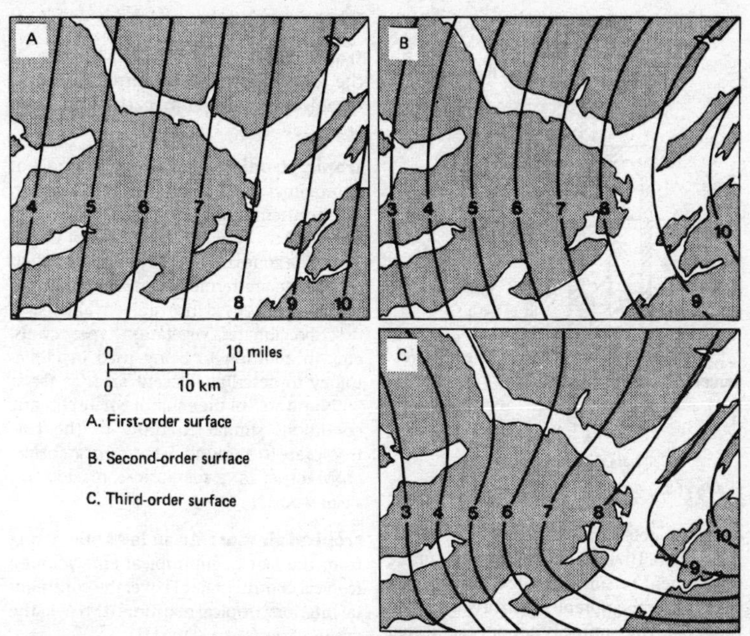

Figure 259 *Trend surfaces for the Main Rock Platform fragment averages, Mull*

treppen concept PIEDMONTTREPPEN.

triangle of error The triangle produced in a PLANE-TABLE survey when rays are drawn back from three known objects in order to determine the observer's position. If the measurement is not totally accurate the rays will not intersect exactly. *See also* RESECTION.

triangular diagram TERNARY DIAGRAM.

triangulation A surveying term referring to the measurement of a land area by constructing a geodetic framework based on a system of triangles. Once the BASELINE has been accurately measured it is taken as one side of the primary triangle and the angular values are subsequently measured by a THEODOLITE. The lengths of the sides of the other triangles are ascertained by trigonometry as the surveying instrument is set up in turn at all the points that are to be fixed. Triangulation is the basis of all topographical surveying.

Triassic (Trias) The first geological period of the MESOZOIC era, extending from about 245 million years ago to about 208 million years ago. It succeeded the PERMIAN and preceded the JURASSIC. Because of the difficulty of defining its lower limit it is often linked with the Permian to form the NEW RED SANDSTONE in Britain. Its type locality is in Germany, where marine facies occur, but in Britain only continental deposits occur, consisting mainly of shales, bright-red desert sandstones, marls and pebble beds. The full stratigraphical sequence in W Europe comprises: **1** the BUNTER (oldest); **2** the MUSCHELKALK; and **3** the KEUPER (youngest). Although many of the British sediments of Triassic age are unfossiliferous the period witnessed the evolution of the reptiles and contains the earliest known dinosaur remains. The evaporites of this age are of considerable economic importance.

triaxial test A test used in SOIL MECHANICS to measure the strength of sediments in

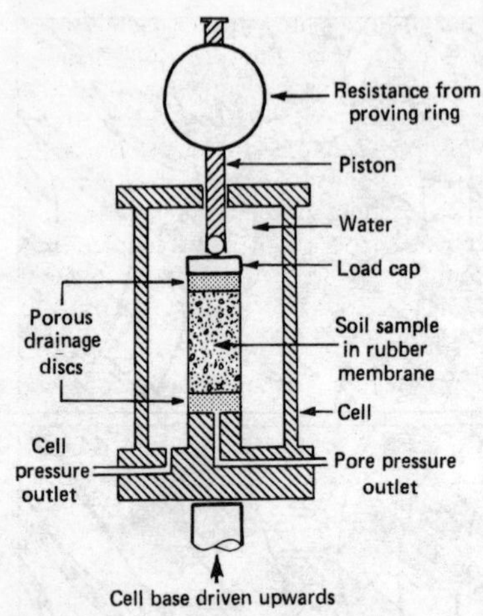

Figure 260 *Triaxial apparatus*

order to determine the point at which materials fail (FAILURE) under STRESS. A sample of sediment or soil is enclosed in a watertight rubber membrane within a water-filled cell [*260*]. The sample is not only subjected to a constant confining pressure of the cell fluid but also to a vertical load which is increased at a fixed rate by driving the sample upwards on to the proving ring. The pressure is applied so that only normal stresses are operative and no shear stresses exist on the sample surface. As the major principal stress becomes steadily greater than that of the constant confining pressure of the cell the sample becomes strained (STRAIN) until it shears along a plane inclined to the directions of principal stress, i.e. when shear stress has exceeded SHEAR STRENGTH. *See also* MOHR–COULOMB EQUATION.

tributary A stream which drains into a larger one, thereby contributing water to it. *See also* DISTRIBUTARY.

trim line A line separating vegetated lateral moraine on a valley side from the bare moraine lower down the slope. It marks the former extent of the glacier margin and gives some indication of the former ice thickness.

trittkarren KARREN.

trophic level The level at which energy in the form of food is transferred from one organism to another as part of the FOOD CHAIN. [*30*]

tropic, tropics A line of latitude at approximately 23.5°N and S of the equator. The northern tropic is the Tropic of Cancer, the southern tropic is the Tropic of Capricorn. The zones between these two parallels of latitude are termed the tropics and the term *tropical* is used in an adjectival sense to describe climates, vegetation types, crops, etc., in addition to being used in climatology to describe adjacent areas as far as 30°N and 30°S of the equator where climatic conditions similar to those of the true tropics are to be found. *See also* INTERTROPICAL CONVERGENCE ZONE, SUBTROPICAL, TORRID ZONE, TRADE WINDS.

tropical air mass An air mass originating from the belt of subtropical anticyclones: tropical continental (cT) over the continental interiors; tropical maritime (tM) over the oceans. *See also* AIR MASS. [*5*]

tropical black soil A dark coloured, heavy-textured soil, characterized by the presence of MONTMORILLONITE, which causes swelling during the rainy seasons but drying out and cracking during dry periods. It is synonymous with a GRUMUSOL, which itself is included among the VERTISOLS in the SEVENTH APPROXIMATION soil classification. The tropical black soil is also similar to a REGUR but has less organic material and a different A-horizon than the CHERNOZEM, because of the higher temperatures and greater evaporation rates of the tropics. It has a neutral or alkaline pH. *See also* BLACK COTTON SOIL.

tropical climate An imprecise expression referring to the climate experienced within the tropics, but one which can be subdivided into a number of distinct climatic regimes, largely on the seasonal distribution of rainfall. It is distinguished from other climates by the lack of a cool season, with no month experiencing a mean temperature below 20° C (68° F) in KÖPPEN'S CLIMATIC CLASSIFICATION. The definition of hot climates by Supan (1896) was based on a mean annual isotherm of 20° C (68° F), although A. A. Miller chose 21° (70° F) to define hot climates, which are generally accepted as being synonymous with tropical climates. The TORRID ZONE is also regarded as coinciding with the zone of a tropical climate. *See also* MONSOON, SAVANNA, TROPICAL RAIN-FOREST.

tropical cyclone A system of low pressure occurring in tropical latitudes, characterized by its very strong winds (CYCLONE). It is a general term for any type of cyclonic storm at low latitudes, for which many other terms are used, depending on the intensity of the winds or on local names (*see also* HURRICANE, TYPHOON, WILLY-WILLY). Tropical cyclones follow well-defined tracks, moving at 15–25 km/h along clockwise courses in the N hemisphere, and counter-clockwise courses in the S hemisphere. They occur only in oceanic areas where the sea temperatures exceed 27° C and where plentiful supplies of water vapour can be acquired by evaporation from the ocean surface to provide the necessary energy. In order to achieve the rotational component the tropical cyclone must be spawned at least 5°N or S of the equator, for in the equatorial zone the CORIOLIS FORCE is zero. Although the precise triggering mechanism is debatable it is thought that in an area where differential heating of land and water occurs there will be an increasing tendency for pressure to fall rapidly as convectional heating accelerates. Anticyclonic divergence in the upper air appears to be an essential prerequisite for CYCLOGENESIS at the surface. The developed cyclone has a diameter of 150 to 300 km and is accompanied by towering cumulonimbus clouds, torrential rainfall, violent winds and often thunder and lightning.

tropical grassland SAVANNA.

tropical rain-forest The forest which resembles the EQUATORIAL RAIN-FOREST in its vegetation structure and soil characteristics. But because its global distribution is more extensive than the latter (between 7° and 23½° latitude on windward coasts) there is a marked seasonality. A long wet season alternates with a season of lower rainfall (but not drought) in which temperatures also are lower, resulting in a season of much slower tree growth. Thus, there are fewer tree species and fewer LIANAS than in the equatorial rain-forest, although EPIPHYTES are equally abundant because of the constantly high humidity of the coastlands. Tropical rain-forest is typical of the coastlands of Burma, W India, Vietnam, the Philippines, E Malagasy, E Brazil, Central America and the West Indies.

tropical year The time interval occurring between successive apparent passages of the SUN through the Earth's equatorial plane at the vernal EQUINOX. *See also* SOLAR YEAR, YEAR.

tropopause The boundary between the TROPOSPHERE (below) and the STRATOSPHERE (above), at an elevation in the atmosphere of about 18 km above the equator, 9 km at latitude 50°N and S, and 6 km at the poles. Its altitude, which is not constant either in space or in time, is correlated with sea-level pressure and temperature, which vary according to season, latitude, etc. It is marked by an abrupt change in the LAPSE-RATE which, while falling rapidly with elevation throughout the troposphere, remains constant at the tropopause for several km of height (ISOTHERMAL LAYER) before rising with elevation in the lower stratosphere. [*17* and *18*] *See also* MESOPAUSE, STRATOPAUSE.

tropophyte A plant which adopts alternating growth patterns in unison with changing seasonal weather, i.e. it acts as a XEROPHYTE during the dry season but becomes a HYGROPHYTE during the rainy season.

troposphere The lowest of the concentric layers of the ATMOSPHERE, occurring between the Earth's surface and the TROPOPAUSE. It is the zone where atmospheric TURBULENCE is at its greatest and where the bulk of the Earth's weather is generated. It contains almost all the water vapour and AEROSOLS and three-quarters of the total gaseous mass of the atmosphere. Throughout the troposphere temperature decreases with height at a mean rate of 6.5° C/km (3.6° F/1,000 ft) and the whole zone is capped by either an INVERSION OF TEMPERATURE or an ISOTHERMAL

LAYER at the tropopause. *See also* MESOSPHERE, STRATOSPHERE. [*17*]

trough **1** A system of low atmospheric pressure characterized by a much greater length than width. Not all troughs are frontal but all FRONTS occupy troughs. **2** A valley that has been overdeepened by GLACIAL EROSION and which is termed a U-SHAPED VALLEY. It is characterized by HANGING VALLEYS and TRUNCATED SPURS [*101*]. Troughs have been classified by D. L. Linton as: (a) *Alpine types*, in which large areas of high ground overlook their accumulation areas; (b) *Icelandic types*, in which troughs are cut by glaciers descending from ice accumulations on a plateau surface; (c) the five *Composite types* which form when the pre-glacial valley system is unable to cope with the discharge of a complex valley glacier system, thereby creating watershed breaching by GLACIAL DIFFLUENCE and GLACIAL TRANSFLUENCE; (d) the *Intrusive or Inverse type*, in which an ice-sheet pushes up-valley against the direction of pre-glacial drainage. **3** The lowest part of a wave form between two crests. [*169* and *276*] **4** A term referring to CROSS-BEDDING (trough-bedding) in sedimentary structures.

trough end The precipitous rock wall which closes off the upper end of an over-deepened glacial valley and around which a group of CIRQUES frequently occurs. [*261*] It has been suggested that it is the coalescence of glacier ice from these surrounding cirques that has increased the eroding power of the valley glacier by considerably increasing its volume and its morainic material, thereby leading to more effective downwearing at the valley head. [*80*]

trough lake FINGER LAKE.

trowal A trough of warm air in the upper atmosphere but which is not in contact with the ground; characteristic of an *occluded front* (OCCLUSION).

true bearing The horizontal angle between any survey line and the TRUE NORTH. *See also* BEARING.

true dip The maximum DIP of a rock stratum in contrast to that of the APPARENT DIP. [*62*]

true north The direction of geographic north from the observer, i.e. the direction along a meridian towards the North Pole, in contrast to the GRID NORTH and to the *magnetic north* (MAGNETIC POLES).

truncated conical projections Those *conical projections* in which the geographic pole is represented by the arc of a circle of finite length.

truncated spur A valley-side SPUR which has been abruptly cut off at its lower end by

Figure 261 *Trough end*

the erosive action of a valley glacier, thus creating a sudden steepening. Many truncated spurs are precipitous cliffs which provide sport for rock climbers, e.g. Dinas Mot in the Llanberis pass, N Wales, and El Capitan, Yosemite, USA. [*101*]

trunk valley The main valley occupied by the dominant stream of a river system (the *trunk stream*) or the largest glacier of a valley glacier complex (*trunk glacier*).

tschernosem CHERNOZEM.

tsunami A Japanese term which has been universally adopted to describe a large seismically generated sea wave which is capable of considerable destruction in certain coastal areas, especially where submarine earthquakes occur. Although in the open ocean the wave height may be less than 1 m it steepens to heights of 15 m or more on entering shallow coastal water. The wavelength in the open ocean is of the order of 100 to 150 km and the rate of travel of a tsunami is between 640 and 960 km/h (400–600 miles/h). It may travel considerable distances, with tsunamis generated around the seismically active coasts of the Pacific ocean reaching Hawaii on several occasions. Tsunamis can also be generated by violent volcanic explosions at or below sea-level, e.g. Krakatoa in 1883. They have been incorrectly referred to as *tidal waves*.

t-test One of the best-known PARAMETRIC STATISTICAL TESTS, used to establish the SIGNIFICANCE of the difference in the means of two samples of data measured on an interval scale. It can be used to compare differences in both independent samples and paired samples and to decide whether to reject or retain the NULL HYPOTHESIS.

tufa, calc tufa A sedimentary deposit formed around a spring of calcareous groundwater and comprising calcium carbonate ($CaCO_3$) derived by solution of calcium bicarbonate. It is found mainly in limestone regions where it infills cavities, builds STALACTITES and STALAGMITES and cements superficial gravel to produce CALCRETE. In the vicinity of hot springs a type of tufa is known as TRAVERTINE. Tufa is deposited when water saturated with $CaCO_3$ and CO_2 is subjected to an increase in temperature or a decrease in pressure. Loss of water by evaporation will also cause it to be deposited.

tuff A volcanic rock composed of compacted, medium- to fine-grained PYROCLASTIC MATERIAL (grain size <4 mm in diameter). Waterlain tuffs often show excellent bedding, while welded tuffs are termed IGNIMBRITES.

tumescence The swelling or updoming of the ground surface in the vicinity of a VOLCANO prior to an eruption, owing to the upwelling magma.

tundra A Lapland term for the extensive treeless plains of N Siberia, N Scandinavia and N Canada and Alaska. The tundra lies between the polar region of perpetual snow and ice and the northern limit of tree-growth (TAIGA). Its surface is stony or marshy and is characterized by mosses, lichens and a low vegetation of dwarf shrubs (e.g. cranberry, cloudberry) and occasional stunted Arctic willows. It experiences a short summer with no mean monthly temperature exceeding 10° C (50° F), but with sufficient warmth to melt the surface snows and thaw out the surface layer of the PERMAFROST. Thus, the badly drained soils and marshy hollows become a breeding-ground for countless insects, especially mosquitoes, midges and blackflies. Of the large mammals in the tundra the reindeer and the caribou are the most notable. The *tundra zone* was classified climatically as the ET type by W. Köppen (KÖPPEN'S CLIMATIC CLASSIFICATION). NB The term cannot be used to describe climate or vegetation at high altitudes in low latitudes. *See also* MUSKEG, SUB-ARCTIC.

tungsten (W) A heavy, grey, ductile and infusible metal that is used to harden a metal tool or cutting-edge because of its toughness. It occurs as veins in metamorphosed sedimentary rocks.

tunnelling PERCOLINE, PIPE, SEEPAGE, SUBSURFACE WASH.

tunnel valleys Valleys in N Germany and Denmark formed by subglacial streams during the later stages of the Pleistocene glaciation. They have narrow flat floors and steep sides which rise up to 100 m from the valley bottom. In long profile they have an irregular form with deeper basins (often lake-filled) being interspersed with higher tracts, created by the subglacial streams, which were under considerable pressure and were therefore able to erode upslope. Some tiny underfit streams may follow parts of these *tunneldales* but other stretches may be streamless. *See also* MELTWATER CHANNEL, RINNENTAL.

Turbarian The term given to periods of peat-bog growth in Scotland during post-glacial times. It is derived from the Latin *turbaria* (= turf). The Lower Turbarian succeeded the Lower Forestian (mixed oak/hazel, pine forest) after 7,450 BP and itself was terminated in about 4,450 BP by the SUB-BOREAL, being roughly equivalent in age to the ATLANTIC PERIOD. The Upper Turbarian succeeded the Upper Forestian (mixed ash/birch forest) in about 2,450 BP and was roughly equivalent in age to the SUB-ATLANTIC PERIOD. *See also* BOG.

turbidite A sediment deposited in water by a TURBIDITY CURRENT. Its PARTICLE SIZE decreases upwards to create a type of GRADED BEDDING, and it is most commonly formed in a GEOSYNCLINE. A GREYWACKE is now known as a turbidite.

turbidity **1** The stirring up of sediment by the movement of water. **2** A measure of the reduced transparency of the atmosphere caused by the addition of volcanic DUST and pollution from human activities, especially the combustion of fuels. *See also* AEROSOLS, DIFFUSE RADIATION.

turbidity current A submarine flow of a dense mixture of water and sediment (SLURRY), capable of moving at very high speeds (50–80 km/h) for distances of tens of kilometres, as a type of density current. It can be triggered off by earthquakes or submarine slides on the CONTINENTAL SLOPE and has been strong enough to break undersea cables. It transfers large volumes of sedimentary material into the deeper zones of the ocean and is capable of eroding channels on the ocean floor. The sediment which finally settles out on the sea-bed is termed a TURBIDITE.

turbulence, turbulent flow **1** In meteorology, the term describes the net forward movement of air in an irregular, eddying flow, in contrast to smooth LAMINAR FLOW. Thus air molecules can be carried from one layer to another within the turbulent flow by means of complex patterns of eddies (EDDY) which themselves create hydrodynamic lift. It is an important atmospheric mechanism for *mixing* and *dispersal*, but there is no general agreement on the mathematical definition or the physical explanation of turbulence. It is probably linked with unequal heating of the Earth's surface giving rise to INSTABILITY and the creation of 'air-pockets' which give bumpiness to aircraft flights (THERMAL). *See also* CLEAR-AIR TURBULENCE. **2** In hydrology, turbulence is an integral part of the mechanism of STREAM-FLOW and is especially important in the TRANSPORT of sediment by SUSPENSION or SALTATION, assisted by hydrodynamic lift. Turbulent flow replaces laminar flow when the viscosity of the water falls as temperature increases, but turbulent flow can also be generated by an increase in stream-bed roughness or stream velocity or by a decrease in stream depth. These latter factors are especially significant in a zone of rapids and waterfalls, combining to produce a type of turbulence known as *shooting flow*, which accelerates erosion. *See also* FLOW REGIMES. **3** In oceanography, it is a feature of an ocean DRIFT.

turbulent flow TURBULENCE.

turf **1** A layer of grass and herbaceous plants which, together with their root system, create a mat of vegetable material on the soil surface. **2** A block of peat used as a fuel.

turf-banked terrace A hill-slope terrace consisting of fine, soliflucted material (SOLIFLUCTION), resembling a STONE-BANKED TERRACE in shape and size but differing in its cover or part-cover of TURF. [*242*] Its genesis is thought to be similar to that of the stone-banked terrace, but with the turf taking the place of the stones as a means of restraining the downslope movement. It has been

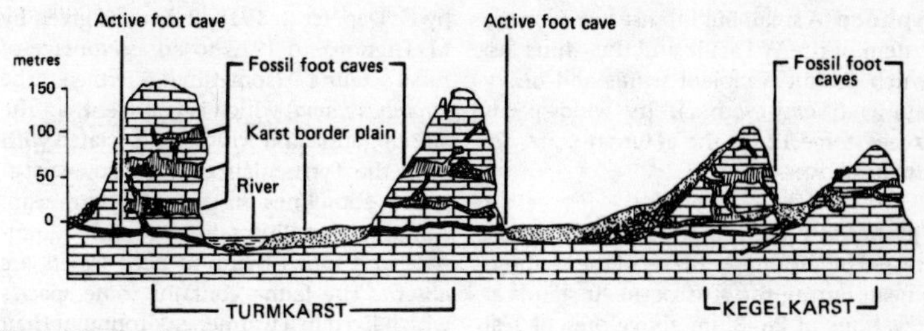

Figure 262 *Turmkarst*

claimed that turf-banked terraces tend to occur on somewhat gentler slopes (5°–20°) than their stonier counterparts. Where the turf cover ruptures, a flow of fine soil material is released which spreads down as a small fan on to the next lower terrace. There is evidence to show that some turf-banked terraces have overridden vegetation and old humus layers, possibly at a rate of about 2 mm per annum, but this is somewhat slower than the downslope movement of stone-banked terraces. It is sometimes referred to as a *turf-banked lobe*. *See also* GOLETZ TERRACES.

turlough, terlough A small, enclosed surface depression in a limestone terrain in which intermittent water bodies occur after heavy rainfall has raised the water-table and increased surface runoff. It is named from the Irish *tuar loch* (= dry lake), and is a common feature in the lowlands of Clare and Galway, Ireland. In the drier summer season most turloughs dry up as the water-table falls. Similar features are found in the Everglades area of Florida, USA.

turmkarst One of the two major forms of tropical KARST terrain (*see also* KEGELKARST) and often referred to as *tower karst* or *pinnacle karst*. It consists of steep-sided hills or pinnacles, termed MOGOTES, rising abruptly from an alluvial plain to heights of 200–300 m. It differs from kegelkarst largely because its genesis depends mainly on lateral solution caused by temporary or permanent water-levels either in lakes or where ALLOGENIC rivers cross the limestone terrain. The zone where the precipitous limestone walls abut on to the flat alluvial plain is an area of cave formation caused partly by SPRING-SAPPING at the water-table. Tower karst is well developed in Puerto Rico, central China, Vietnam, Sarawak and Indonesia. [*262*]

twilight The faint glow which illuminates the sky before sunrise and after sunset, owing to the reflection of sunlight on to the Earth from the upper layers of the atmosphere where the Sun is below the horizon. The length of twilight differs according to latitude and to season. It is shorter in the tropics than in polar latitudes where, since the Sun is never more than 18° below the horizon in midsummer at high latitudes, the twilight lasts from sunset to sunrise.

type The general form, structure or character which distinguishes a group or class of objects or organisms. Among the terms commonly employed are: **1** *holotype*, meaning a single specimen chosen in order to illustrate the main character of a SPECIES; **2** *paratype*, referring to additional specimens chosen along with the holotype to show other species' characteristics not exhibited by the holotype; **3** *syntypes*, which are a series of type specimens of equal status, chosen to illustrate the range of variation within a species.

type locality **1** The actual site of a rock outcrop that has been chosen to define a stratigraphical division. **2** The location from which a TYPE SPECIES fossil has been taken.

type species A SPECIES chosen as being typical of the GENUS to which it belongs. This is often referred to as a *holotype* (TYPE).

typhoon A small but intense low-pressure system of the W Pacific and the China Sea which produces violent winds and heavy rain (TROPICAL CYCLONE). Its wind-speeds exceed force 12 on the BEAUFORT SCALE. *See also* HURRICANE.

Tyrrhenian 1 One of the raised shorelines created by a former sea-level of the Mediterranean during the PLEISTOCENE. It stands at elevations of 28–32 m above present sea-level and the date of its formation has been tentatively correlated with the Great Interglacial (HOXNIAN). The name was introduced by C. Depéret in 1918. **2** A name given by M. Gignoux in 1913 to an assemblage of fossil fauna (sometimes termed the *Strombus fauna*) which is younger than the *Silician* fauna and which is associated with both the Tyrrhenian and the MONASTIRIAN raised shorelines in the Mediterranean. Among the molluscs, littoral species dominate and forms living in deep waters are absent. The fauna contains some species which lived in a warmer environment than that enjoyed by the present-day littoral fauna in the same localities. *See also* MILAZZIAN.

U

ubac A French term used to describe the shaded slope in a mountainous terrain (specifically, north-facing slopes in the Alps). In general it carries minimal settlement and improved land in contrast to the ADRET.

ultisol One of the orders of the SEVENTH APPROXIMATION soil classification. It refers to the ultimate stage achieved by tropical soils with reference to their degree of weathering. Ultisols form most characteristically under forests in the humid tropics in areas that have been subjected to prolonged denudation, leaching and considerable CHEMICAL WEATHERING. Thus the A-horizon contains residual iron oxides, the B-horizon has a marked accumulation of clay and there is a deficiency of base minerals. The soil is relatively infertile and has suffered from the system of shifting cultivation in many underdeveloped countries. Ultisols are similar to LATOSOLS and include the red-yellow PODZOLS. They are sometimes referred to as FERRISOLS, but are more weathered and leached than FER(RI)SIALLITIC SOILS.

ultrabasic rocks Igneous rocks, sometimes referred to as *ultramafic rocks* or *ultramafites*, consisting primarily of ferromagnesian minerals and having a low percentage of silica and feldspar. Most ultrabasic rocks are plutonic, e.g. PERIDOTITE, and there are relatively few fine-grained volcanic varieties.

ultraviolet radiation That part of the solar RADIATION spectrum (*70*] which lies adjacent to the blue end but which is invisible. It has a wavelength between 10^{-2} μm and 10^{-1} μm. Only a small percentage of *ultraviolet* rays reach the Earth's surface; the majority are intercepted by OZONE molecules in the atmosphere. *See also* ELECTROMAGNETIC RADIATION.

unaka An alternative name for a MONADNOCK, but one which has not been universally adopted. It refers to a RESIDUAL of very large size rising from a PENEPLAIN.

unavailable water Water that is present in the soil but which cannot be taken up by plant roots because it is strongly adsorbed (ADSORPTION) on to the surface of soil particles.

unconfined aquifer A term given to an AQUICLUDE and in which the upper limit of saturation is, therefore, the WATER-TABLE.

unconformity A geological term referring to a hiatus or break in a sequence, marked by a major break in sedimentation or by a structural planar surface separating younger rocks above from older rocks below. The surface of unconformity may be a surface of denudation or a surface of non-deposition; in either case it represents an unspecified period of unrecorded time. The following types of unconformity have been recognized [*263*]: **1** an *angular unconformity*, in which the beds above and below the 'break' exhibit different dips and strikes from each other, most commonly when older folded strata are overlain by younger unfolded strata; **2** a DISCONFORMITY (*parallel unconformity*), where the upper and lower beds dip in the same direction and by the same amount; **3** a *non-depositional unconformity*, where it is almost impossible to distinguish a break in a sedimentary sequence, unless there is a change of fossil zones or a

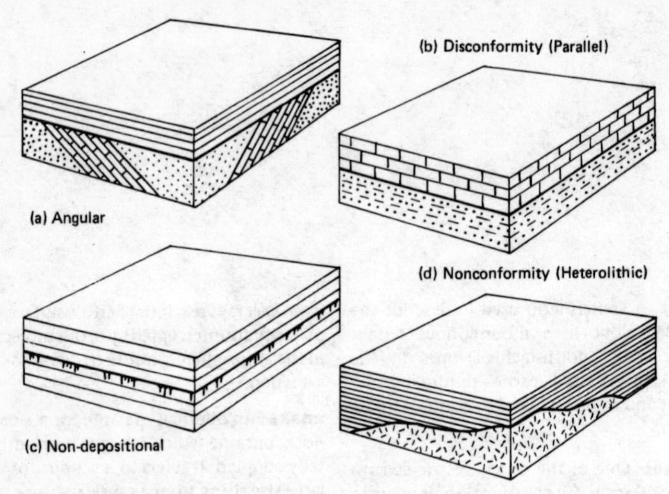

Figure 263 *Types of unconformity*

line of borings by marine organisms along a particular horizon; **4** a NONCONFORMITY (*heterolithic unconformity*), in which sediments have been deposited on a denuded surface of igneous and/or metamorphic rocks. Somewhat confusingly, the term *nonconformity* has also been used as a synonym for *angular unconformity*, whence the alternative *heterolithic unconformity* is preferred. *See also* OVERLAP, OVERSTEP.

underclay A layer of FIRE CLAY beneath a coal seam, often referred to as a SEAT-EARTH. [*224*]

undercutting The erosive action of a river or an ocean wave at the base of a cliff. The term has also been used to describe the mechanical wearing effect (ABRASION) of wind-carried sand on the base of a rock pillar (ZEUGEN).

underfit stream MISFIT STREAM.

underflow A hydrological term referring to the flow of GROUNDWATER parallel to and below a river channel. *See also* THROUGHFLOW.

underground drainage CYCLE OF UNDERGROUND DRAINAGE, KARST WATER.

underplating A term given to the process during which MAGMA generated above a PLUME in the Earth's MANTLE, thickens the CRUST by adhering to its underside. Such a process is usually related to a HOT SPOT. *See also* PLATE TECTONICS, ISOSTASY.

undertow, marine A flow of water in a seaward direction occurring beneath the zone of breaking waves on a beach. It is thought to result from the return flow of water along the sea-bed after the SWASH has pushed sea water up a beach and the BACKWASH has returned it down the beach. It is thought by some to be the initial stage of a RIP CURRENT.

unequal slopes, law of A principle defined by G. K. Gilbert in 1877 referring to an asymmetric ridge which has unequal stream lengths and slope gradients on opposite sides of its divide. He recognized that on an asymmetrical ridge (CUESTA) the stream flowing down the steeper slope erodes its valley at a faster rate than one flowing down the gentler slope, thereby causing the divide to migrate away from the more rapidly eroding stream. *See also* PARALLEL RETREAT OF SLOPES.

uniclinal The British equivalent of the term *homoclinal* (HOMOCLINAL TERRAIN).

uniclinal shifting The act of a river, which occupies a STRIKE VALLEY in a region of gently

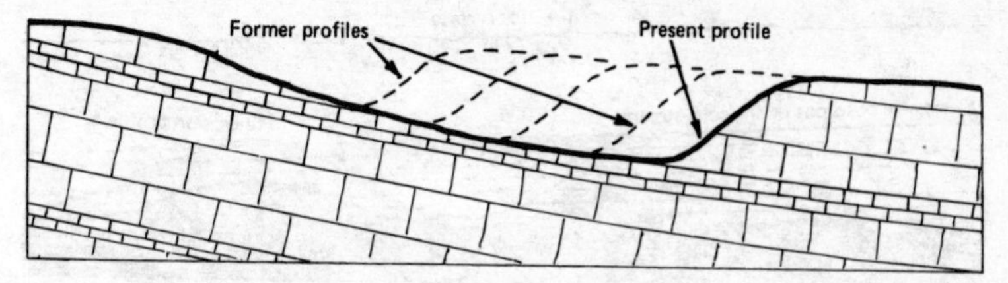

Figure 264 *Cross-section of a river valley showing uniclinal shifting*

dipping rocks, when continued downcutting leads to a gradual migration of the river in the direction of DIP. [*264*] It has been suggested that the reach of the R. Thames between Pangbourne and Henley, in SE England, has moved down the dip slope of the chalk CUESTA of the Chilterns by means of uniclinal shifting.

uniform ground loss A method of slope formation in which ground loss measured perpendicularly to the surface takes place at an equal rate at all points on the slope, or on the particular unit involved. It may be compared with the method of slope evolution in which ground loss is measured in a horizontal direction, i.e. PARALLEL RETREAT OF SLOPES, although also occurring at the same rate over the whole of the slope, or of the part involved. *See also* SLOPE EQUILIBRIUM, CONCEPTS OF.

uniformitarianism A school of thought based on the view that processes currently acting upon the Earth are virtually the same as those which have been acting throughout geological time. The idea was first outlined by J. Hutton in the late 18th cent. and developed by C. Lyell in the early 19th cent. It contrasts with the ideas of CATASTROPHISM but is supported by the scientific concept of GRADUALISM. In the context of GENERAL SYSTEMS THEORY the uniformitarianist would support an OPEN SYSTEM rather than a CLOSED SYSTEM.

uniform steady flow A term referring to that state of stream-flow when DISCHARGE flow path, width, cross-section area and velocity are constant at every section in a channel reach. The water surface, bed profile and ENERGY GRADE LINE are all parallel. *See also* CHANNEL GEOMETRY, UNSTEADY FLOW.

unimodal distribution A statistical expression denoting that the frequency curve representing the data DISTRIBUTION has a single maximum. *See also* BIMODAL DISTRIBUTION.

unit Any specified quantity or dimension adopted as a standard of measurement. Most countries have agreed to base scientific measurement on the *Système International d'Unités* (SI UNITS), although certain sciences continue to use older systems, such as the so-called Imperial units. *See also* ATMOSPHERIC PRESSURE, KNOT, LENGTH, TEMPERATURE, VOLUME.

unit hydrograph An expression referring to the conceptual process in a drainage basin by which precipitation is converted into stream-flow. It is based on certain assumptions, namely: **1** that the RUNOFF from effective rainfall events of the same duration (produced by isolated storms in the same drainage basin) causes *hydrographs* of equal lengths of time; **2** that ordinates of the unit hydrograph are proportional to the total volume of direct runoff from rainfalls of uniform intensity and equal duration, irrespective of the total volume of rain. These assumptions have been shown not to be completely justified in natural drainage basins, but the theoretical concept remains a useful one for predicting the time and volume of peak discharge, which in turn is of value in flood forecasting for basins with no records. *See also* HYDROGRAPH.

Universal Polar Stereographic Grid System (UPS) UPS GRID.

Universal soil loss equation (USLE) A widely used term in soil science, which

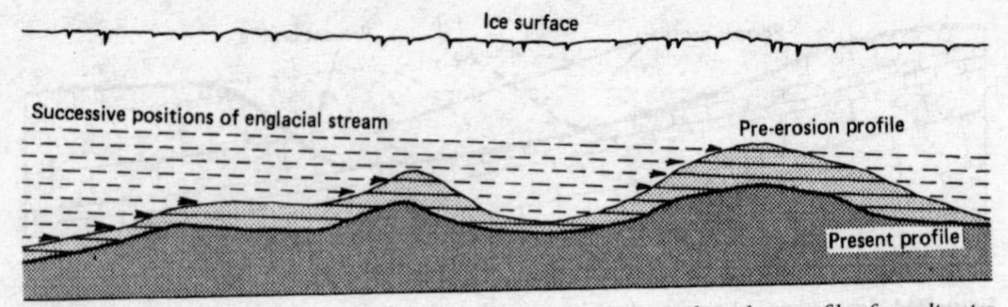

Figure 265 *Superimposition of an englacial stream to produce an up–down long profile of a meltwater channel*

refers to the theoretical prediction of soil loss (*A*):

$$A = RKSCP$$

where *R* is the rainfall erosivity factor, *K* is the soil erodibility factor, *S* is the slope gradient factor, *C* is the crop management factor and *P* is the erosion control practice factor. *See also* SOIL EROSION.

Universal Time (UT) The scale used, regardless of location, to define TIME on Earth. It is essentially the same as GREENWICH MEAN TIME. With the advent of atomic clocks, however, a new measurement of time was introduced (INTERNATIONAL ATOMIC TIME, TAI) which was more accurate than that based on the motion of the Earth. Broadcast time signals, therefore, are now based on TAI plus or minus a whole number of seconds. Such a scale never varies by more than 1 second from Universal Time.

Universal Transverse Mercator Grid System (UTM) UTM GRID.

unloading A term referring to pressure release in rocks (DILATATION) resulting from the removal of overlying materials by erosion. It should not be confused with EXFOLIATION, which is a weathering process. The structures formed by unloading are known as PSEUDO-BEDDING or sheets (SHEETING). NB In the USA, *unloading, sheeting* and *dilatation* are synonymous with *exfoliation*.

unpaired terrace (non-paired terrace) PAIRED TERRACE.

unstable channel A stream or tidal CHANNEL which constantly shifts its position due to sudden changes in erosional and/or depositional processes.

unstable equilibrium A condition in a SYSTEM when a small displacement leads to an even greater displacement. This condition is usually terminated by the achievement of a new stage of STABLE EQUILIBRIUM [*72*]. It can be illustrated by the behaviour of an air parcel under certain weather conditions when the ENVIRONMENTAL LAPSE-RATE becomes greater than the DRY ADIABATIC LAPSE-RATE. It is characteristically found, for example, in CUMULONIMBUS cloud formation. *See also* INSTABILITY.

unsteady flow A term referring to that state of flow in a river or canal when it is affected by a flood wave or surge. Thus the DISCHARGE, flow depth and velocity change over time at various sections of a channel reach. Because of such changes FLOW EQUATIONS become redundant and are replaced by FLOOD ROUTING methods of flow analysis. *See also* UNIFORM STEADY FLOW.

unstratified drift A term used to distinguish the angular/subangular, unsorted MORAINE and TILL deposits (unstratified drift) laid down by ice from the washed, generally rounded and sorted sand and gravel (GLACIOFLUVIAL) deposits laid down by meltwater (*stratified drift*).

up-doming UPWARPING.

up–down profiles An expression used to describe the long profiles of certain MELTWATER CHANNELS which cross spurs or ridges and which consist of an uphill segment and a downhill segment (i.e. they are humped midway along their profile). They are thought to have been produced englacially when meltwaters, subjected to considerable hydrostatic pressure within a thick body of ice, had sufficient power to force a stream to erode a channel when flowing uphill across a rock feature aligned at right angles to the ENGLACIAL meltwater stream. This would be possible if the englacial stream had been gradually superimposed upon the underlying ridge. [*265*]

updraught A powerful, ascending current of air (THERMAL), often associated with CUMULONIMBUS clouds and THUNDERSTORM conditions. Strong updraughts, in the form of whirling air within the vortex, are the most characteristic features of all whirlwinds (DUST-DEVIL, TORNADO, WATERSPOUT). Updraughts are of considerable use to glider pilots seeking to gain height rapidly and are characteristically found in unstable air masses (INSTABILITY) with strong CONVECTION. *See also* DOWNDRAUGHT.

upfreezing The upward movement of blocks of rock or stones from within the finer fraction (FINES) of the soil, brought about by FROST-HEAVING. There is no agreement on the mechanisms involved but three suggested ones are: **1** ground expansion during *freezing*, in which all sized particles are carried upwards but only the finer particles, which adhere to each other, fall back on contraction following thawing; **2** the raising of all materials equally on *freezing* but with finer material falling into the voids left by the stones, thereby preventing the latter returning to their former positions; **3** the raising of all materials equally but with the stones kept from sinking back by the preservation of ice pedestals at their base.

uplift EPEIROGENESIS, ISOSTATIC MOVEMENTS, OROGRAPHIC CLOUD, OROGRAPHIC RAINFALL.

Upper Greensand A lithostratographic unit of the CRETACEOUS in Britain and NW Europe. *See also* GREENSAND.

UPS grid The GRID system used by the USA for military map projections polewards of the 80°N and 80°S lines of latitude, i.e. polewards of the UTM GRID. The term refers to the *Universal Polar Stereographic Grid System*, which is superimposed upon a polar stereographic map projection within the circle formed by the 80° line of latitude. Its horizontal grid lines are parallel with meridians of 90°E and 90°W longitude, while its vertical grid lines are parallel with meridians of 0° and 180° longitude, thereby giving a grid divided into 500-km squares [*266*]. The pole is the ORIGIN of the grid, situated at the centre of the projection, but the FALSE ORIGIN is transferred 2,000 km W and 2,000 km S in the N hemisphere and 2,000 km W and 2,000 km N in the S hemisphere. *See also* STEREOGRAPHIC PROJECTION.

upslope fog A type of ADVECTION FOG common in upland areas, where moist air is forced upwards by the terrain and its tem-

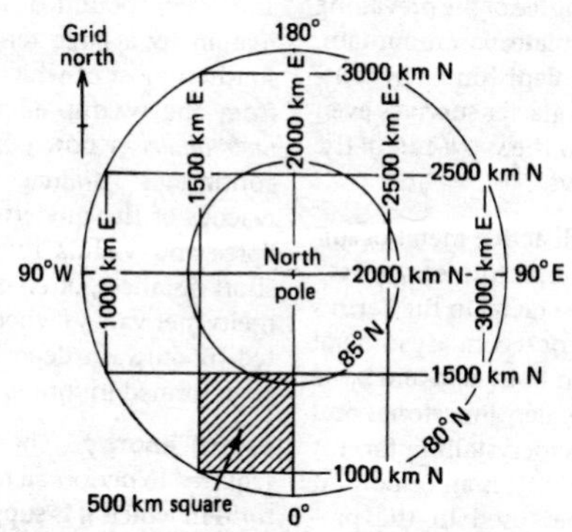

Figure 266 *UPS grid: north zone*

perature is cooled below the saturation point. The fog appears to be a type of STRATUS cloud which drifts slowly upslope only to be replaced by new fog layers at the CONDENSATION level.

upthrow The side of a FAULT that has moved upwards or has been left unaffected by the relative displacement of the downthrow block (DOWNTHROW) on the other side of the fault-plane.

upwarping **1** The slow upheaval of the Earth's crust by large-scale epeirogenetic forces (EPEIROGENESIS) or as part of the updoming mechanism involved in CYMATOGENY. **2** At a more local scale tracts of land may be uplifted by the intrusion of an igneous body (BATHOLITH); by volcanic TUMESCENCE; by the injection of a salt diapir (SALT DOME); by cryostatic processes (PINGO); by seismic uplift (EARTHQUAKE); or by GLACIO-ISOSTASY (RAISED BEACH). *See also* DOWNWARPING.

upwelling The transfer of colder water from lower layers upwards to the higher layers of a water body by means of a current. Such transfers are common in the zones of DIVERGENCE of ocean currents, thereby replacing the warmer surface layers by upwelling colder water from below (OCEAN CURRENTS) and giving rise to cold currents, rich in nutrients, which support important fisheries, e.g. BENGUELA CURRENT, PERU CURRENT.

ural glacier A term given to a small type of GLACIER that forms in the lee of the prevailing winds blowing over a plateau or mountain. Because of constant replenishment by drifting snow this type of glacier survives even though it stands below the SNOW-LINE of the region. *See also* LEE WAVE.

uranium A grey radioactive metal occurring in a number of isotopic forms (ISOTOPES). It never occurs as a free metal in the Earth's crust but as a primary or secondary mineral associated with ores in both acid and basic igneous rocks or in shales, limestones and sandstones. In its microcrystalline form it is known as PITCHBLENDE. It is an important source of the isotopes used in the production of nuclear energy.

uranium–lead dating RADIOMETRIC DATING.

urban climate The local climate associated with those expanses of the built environment in which the MACROCLIMATE has been considerably modified by the built form. Thus, temperatures are influenced by the masses of concrete and masonry, by the emissions of industrial and domestic heating (HEAT 'ISLAND'); the RUNOFF is affected by the large areas of impermeable ground surface, streets, etc.; the wind velocities are affected, either by frictional retardation owing to the broken surface form, or by being funnelled between tall buildings to increase the gustiness (VENTURI EFFECT); fogs are exacerbated by the considerable volume of AEROSOLS; SMOG may be generated by vehicle exhaust; and rainfall and clouds appear to be more frequent.

Uriconian A stratigraphic name for the volcanic rock series of Pre-Cambrian age which occurs beneath the LONGMYNDIAN rocks of Shropshire, England.

urstromtal, urstromtäler A German name for a broad shallow valley cut largely into sandy outwash plains (SANDUR) by glacial meltwater streams which flowed along the fronts of successive ice-sheets that occupied the North European Plain during the PLEISTOCENE. Five major *urstromtäler* (plural form) can be traced as east–west valley systems between Wrocław (Poland) and the Baltic coast, each marking a former ice front and each bounded along its northern margin by a large terminal moraine, the southernmost of which is thought to date from the WARTHE advance. None of the *urstromtäler* is now occupied by a line of continuous drainage, although some reaches of the modern Elbe, Havel, Oder, Spree and Vistula rivers follow them for short distances, often breaking out of these meltwater valleys where they are interrupted by outwash deposits thought to have been formed in former RINNENTÄler. [*267*]

useful energy The quantity of ENERGY required to perform a task, regardless of the form in which it is supplied. It is an important component in energy-planning strat-

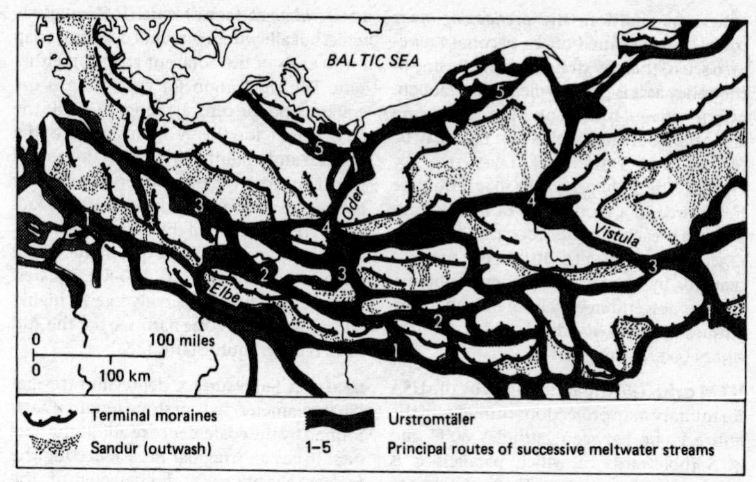

Figure 267 *Glaciofluvial features of the north European plain*

egy for it can be used to measure the effects of changes in the fuel-supply mix on the demand for PRIMARY ENERGY.

U-shaped valley A valley that has been overdeepened and widened by GLACIAL EROSION, thereby severely modifying its preglacial form (V-SHAPED VALLEY). The slopes of the former fluvial valley are considerably oversteepened by the glacier, which occupies the entire valley rather than merely the channel on the floor of the former river

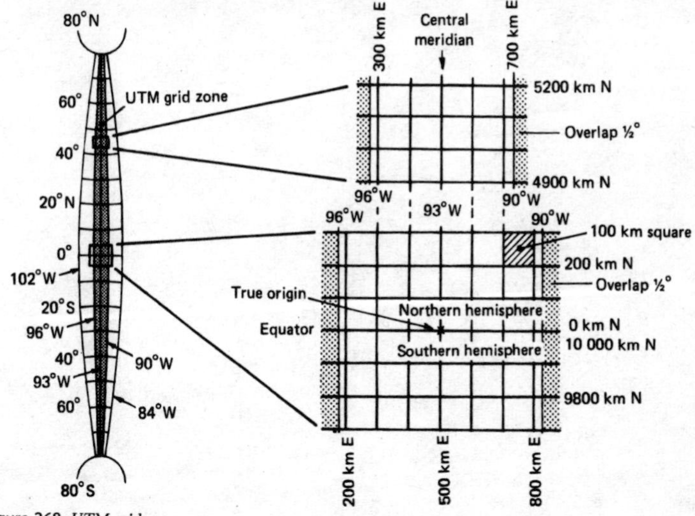

Figure 268 *UTM grid*

valley. In addition, the *overlapping spurs* (OVERLAP) are planed-off to become TRUNCATED SPURS, thereby straightening the line of the valley as it is transformed into a TROUGH. Minor irregularities on the floor are smoothed out, although RIEGEL bars may be formed and PATERNOSTER LAKES may be created. The U-shaped valley may terminate headwards in a TROUGH END or it may form part of a breached divide (GLACIAL DIFFLUENCE, GLACIAL TRANSFLUENCE). Its flanks may be marked by DEBRIS FANS where tributary streams descend steeply from neighbouring cirques or from HANGING VALLEYS. *See also* FINGER LAKE, FIORD. [*101*]

UTM grid The *grid* system used by the USA for military map projections throughout the entire globe between latitudes 80°N and 80°S (polewards of which parallels it is replaced by the UPS GRID). The term refers to the *Universal Transverse Mercator Grid System* (TRANSVERSE MERCATOR PROJECTION). It consists of 60 *grid zones*, each of which is 6° longitude wide, but allowing a one-half degree overlap on to each of the adjacent zones on either side. The true origin for each zone is the equator and the central meridian, while the false origin for the N hemisphere is the equator and a point 500 km W of the central meridian. For the S hemisphere the equator is given an arbitrary northing of 10,000 km. Each grid zone is subdivided latitudinally into 8° sections and these quadrilaterals are further subdivided into 100-km squares. Because the meridians converge in higher latitudes the grid zone narrows, but the grid lines remain equidistant. [*268*]

uvala A large surface depression (several km in diameter) in limestone terrain (KARST) formed by the coalescence of adjoining DOLINES. It has an irregular floor reflecting the former elevations and character of the degraded slopes of the dolines, and its floor is not as smooth as that of a POLJE. [*64*]

V

vadose water Underground water which moves freely under the influence of gravity above the level of saturation (WATER-TABLE) in permeable rocks. It is especially characteristic of underground water in karstic limestone (KARST). It contrasts with PHREATIC WATER.

valley A linear depression sloping down towards a lake, sea or inland depression. It is initially created by fluvial erosion but may have been subsequently modified by glacial erosion (U-SHAPED VALLEY). *See also* HANGING VALLEY, LONGITUDINAL VALLEY, RIFT VALLEY, TRANSVERSE VALLEY, V-SHAPED VALLEY.

valley-and-ridge terrain RIDGE-AND-VALLEY TERRAIN.

valley bulge The name given to the arching of an incompetent rock stratum (INCOMPETENT BED) which is forced up into the floor of a valley by the weight of the overlying rocks forming the hill masses on either flank [*33*].

valley fill The unconsolidated rock waste (ALLUVIUM, COLLUVIUM) derived from the denudation of surrounding uplands and deposited in a valley.

valley-floor steps RIEGEL.

valley glacier A glacier which occupies a pre-glacial valley, either as a result of the coalescence of several cirque glaciers, in which case it is termed an *Alpine type* of valley glacier, or one formed at the edge of an ice-cap or ice-sheet, when it is termed an OUTLET GLACIER [*92*].

valley train A linear accumulation of glaciofluvial outwash following a valley for a considerable distance beyond the ice front. Most Pleistocene valley trains have been dissected by post-glacial fluvial erosion, leaving the remnants in the form of terraces (GRAVEL TRAIN).

valley wind A wind associated with diurnal solar heating (ANABATIC WIND) and nocturnal cooling (KATABATIC WIND) in a mountainous region. The air-flow is funnelled by the topography and becomes most marked in the valleys [*125*]. *See also* TOPOCLIMATOLOGY.

vallon de gélivation A term adopted from the French to describe a small valley carved predominantly by ice and frost action (CONGELIFRACTION) acting on the JOINTS of bedrock rather than by FLUVIAL processes.

valve 1 A device for regulating the flow of a fluid substance through a passage. **2** A component of a CONTROL SYSTEM that can be varied to a certain extent by artificial means. [*249a*]

vanadium (V) A metal widely used for toughening steel and making it less prone to metal fatigue.

Van Allen belts Two bands of the outermost layer of the upper atmosphere, termed the MAGNETOSPHERE, at heights of 3,000 km and 16,000 km above the Earth's surface, in which ionized particles are concentrated, trapped by the Earth's MAGNETIC FIELD.

Van't Hoff's law A law referring to the rate of increase of chemical reaction with increasing temperature when a system is in a state of thermodynamic equilibrium.

vapour concentration ABSOLUTE HUMIDITY.

vapour pressure The pressure exerted by the molecules of WATER VAPOUR as part of the overall pressure exerted by the atmosphere (ATMOSPHERIC PRESSURE). The maximum vapour pressure at any temperature occurs when the air is saturated. It can be measured by using dry- and wet-bulb temperatures in conjunction with a set of statistical tables. It varies from 30 mb at the equator to about 10 mb in the UK. Vapour pressure gradients between ice crystals and supercooled droplets are important in the formation of PRECIPITATION (BERGERON–FINDEISEN THEORY OF PRECIPITATION).

vapour trail CONDENSATION TRAIL.

variable A measurable item or number capable of taking different values. The following have been recognized: **1** a *dependent variable* is one whose value is determined or constrained by the values assumed by other variables; **2** an *independent variable* is one whose value can be freely chosen, within a permitted range, and whose variation constrains or determines the value assumed by the *dependent variable*; **3** an *endogenous variable* is one whose value is to be determined by forces operating within the model under consideration; **4** an *exogenous variable* is one whose value is determined by forces outside the model and is unexplained by the model.

variance 1 A number which measures the DISPERSION within a data population or sample, expressed as the arithmetic mean of the squares of the deviation from the distribution mean. The greater the variance the bigger the dispersion. **2** The number of internal variables that can be altered independently and arbitrarily (within certain limits) to bring a SYSTEM into a new state of equilibrium without causing a phase change is known as *variance of a system* or DEGREES OF FREEDOM.

variance, analysis of An expression referring to a group of statistical techniques used to test whether more than two independent samples are derived from the same population. They enable one to apportion the VARIANCE into **1** that which arises from variation between the means of the samples, and **2** that arising from variation within the sample. All the techniques are based on the *variance ratio (F) test.*

variate One measure or estimate among a set of values for a single VARIABLE.

variation 1 The angle by which a compass needle deviates from TRUE NORTH (MAGNETIC DECLINATION). **2** In statistics, the variety of values assumed by a VARIATE. The *coefficient of variation* refers to the variability within a sample or population and is calculated as 100 times the standard deviation divided by the mean.

Variscan orogeny The mountain-building episode of Carbo-Permian times, and the Central European equivalent of the ARMORICAN OROGENY of Western Europe. It was named after a Germanic tribe (the Varisci) by E. Suess in 1888. It is part of the HERCYNIAN OROGENY of Eur-Asia. The suffix *-oid* (*Variscoid*) indicates that direction only is implied, whereas the termination *-an* or *-ian* should be used to indicate that the folding belonged strictly to the orogeny in question, e.g. CALEDONIAN (Caledonoid). [*104*]

varnish DESERT VARNISH.

varve A thin laminar bed of sediment divided into a thicker lower, lighter-coloured band of sand grading upwards into a thinner upper, darker-coloured band of silt. It is formed in PRO-GLACIAL LAKES by glacial meltwaters with the summer melt season being represented by the lower, coarser sediment while the remainder of the year is represented by the slow accumulation of fine silt as the degree of melting is considerably reduced. Since the two bands, which comprise the varve, are equivalent to one year's sedimentary accumulation, the age of the deposit can be calculated by counting the varves. The technique of using varves to reconstruct Pleistocene chronology was pioneered by a Swede, de Geer, in the early part of the present century.

vasque The name given to a large but shallow solution pan in the intertidal zone of tropical seas. It is believed that the combined action of marine organisms and the effects of brine are responsible for its formation.

Vauclusian spring RESURGENCE, RISING.

vector 1 A physical quantity that has both magnitude and direction. A value which forms the link between the decision function and the structural system of a FEEDBACK CONTROL SYSTEM is termed a *decision vector*. Values which form part of the negative-feedback loop in a feedback control system is known as a *structural state vector*. A vector is usually depicted as an arrowed line drawn stage by stage from an initial point to a final point (e.g. to mark a tidal current's velocity and flow direction over time). Such a technique of monitoring change of magnitude and direction is termed *vector analysis*. **2** The relative displacement direction across a FAULT. *See also* WIND ROSE.

vector wind THERMAL WIND.

veering (of wind) A term used to describe a change of wind direction when it moves progressively in a clockwise direction, e.g. from N to NE to E. BACKING.

vegetation The total plant cover of an area, comprising one or more floral communities (COMMUNITY). It is controlled by several factors acting in unison: climate, soil (edaphic factor), terrain, the effect of organisms (biotic factor) and the results of human interference (CULTURAL VEGETATION). *See also* RUDERAL VEGETATION, SEGETAL VEGETATION, SERE.

vein The scientific term for a mineral-bearing LODE.

veld (veldt) An Afrikaans name for open grassland in S Africa. In addition to a distinction being made by altitude (High Veldt, Middle Veldt and Low Veldt) it is also classified according to the character of the grassland, e.g. Grass Veldt, Bush Veldt and Sand Veldt (i.e. semi-arid).

velocity The rate of motion or rapidity of action of a process or mechanism.

velocity-area method A technique used by hydrologists to measure river DISCHARGE. It employs a series of verticals spaced at intervals of no greater than 1/15th of the channel width, and these are used to subdivide the river's cross-section into a number of segments. The discharge of each segment is calculated as the product of average *velocity* (measured by using a CURRENT METER) and the cross-sectional *area*. *See also* RATED SECTION.

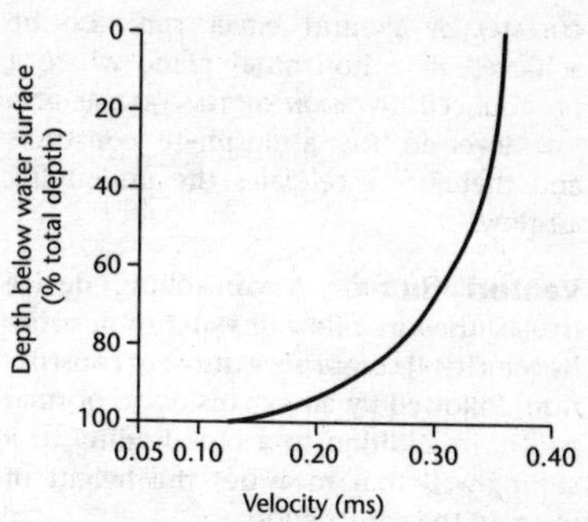

Figure 269 *Velocity profile*

velocity profile The graph which depicts the variation of VELOCITY in relation to the depth of a river, usually measured by means of a variety of CURRENT METERS. Because of varying degrees of FRICTION on the channel floor (BED ROUGHNESS) and on the river banks, maximum velocities will usually be found near the surface at the centre of the channel. [*269*]

vent The orifice of a VOLCANO, consisting of a vertical passage leading down from the surface into the magma chamber. It is the opening through which a volcanic ERUPTION takes place and is usually surrounded by a CONE. It is also termed a CONDUIT. *See also* CENTRAL-VENT VOLCANO.

ventifact A loose stone or pebble that has been polished and faceted by wind-blown sand in an arid environment. *See also* ABRASION, DREIKANTER, EINKANTER, ZWEIKANTER.

Venturi effect The increase in velocity of a current of air or liquid induced by the constriction of the aperture or channel through which the current is passing. The effect is particularly marked in valleys (VALLEY WIND) and in urban areas in narrow streets between high-sided buildings (URBAN

CLIMATE). A Venturi effect can also be achieved in a horizontal plane where a pronounced INVERSION OF TEMPERATURE at a low level in the atmosphere constricts and therefore accelerates the underlying air-flow.

Venturi flume A controlling device (FLUME) used in a flow of water to measure its velocity. It comprises a throat or constriction, followed by an expansion to normal width, in addition to a pipe leading to a stilling-well that measures the height of water in the constriction.

verglas A thin envelope of clear ice which covers exposed rock surfaces after freezing or rainfall or moisture from melting snow. *See also* GLAZED FROST.

vermiculite One of the platey CLAY MINERALS closely related to MONTMORILLONITE and with a considerable degree of expansion when heated. It is widely used to assist thermal insulation.

vernal A term referring to the season of spring.

vernalization The renewal of plant growth due to the increasing temperatures during the spring months, following the dormant winter period.

vernier An auxiliary scale on a measuring device, used to obtain a more detailed reading (i.e. one more significant figure or decimal place). It comprises a graduated scale which slides along the measuring scale of the main instrument, e.g. slide rule, barometer and theodolite.

vertebrates (Vertebrata) The most advanced subphylum of the phylum Chordata (possessing an internal skeleton rod). Vertebrates possess an internal skeleton rod of bone or cartilage and, in addition, an axial skeleton supporting symmetrically arranged appendages and a cranium (skull) for the protection of the brain. Their name comes from their jointed backbone of *vertebrae*. The vertebrates are divided into several classes: **1** Pisces (fish); **2** Amphibia; **3** Reptilia; **4** Aves (birds); and **5** MAMMALIA.

vertex **1** In general, the highest point of a feature. **2** The junction of two lines at a point opposite to the base of a plane or solid figure. It is used especially to describe the topmost angular point of a triangle. **3** A point in the sky directly above a given place.

vertical exaggeration (VE) A term used to express the ratio of the vertical to the horizontal scale on a section or relief model. The deliberate exaggeration of the vertical scale is adopted in order to emphasize its details, which might otherwise be less discernible if they were drawn at the true scale.

vertical interval (VI) The vertical height difference between two points. It is synonymous with the CONTOUR INTERVAL.

vertical motion A general meteorological term referring to the upwards and downwards movements of air in the atmosphere. Vertical motion in the TROPOSPHERE is usually upwards in CYCLONES and in weather FRONTS, but downwards in ANTICYCLONES. During intense CONVECTION the velocity of vertical motion may increase in a normal CUMULUS cloud from about one metre per second to several tens of metres per second in a CUMULO-NIMBUS cloud.

vertical photograph AERIAL PHOTOGRAPH.

vertical stability/instability *See also* STABILITY, ABSOLUTE STABILITY, ABSOLUTE INSTABILITY, CONDITIONAL INSTABILITY, LAPSE RATE.

vertisol One of the orders of the SEVENTH APPROXIMATION soil classification. Vertisols are characterized by a high cation exchange capacity (BASE EXCHANGE) and a weakly developed horizonation (SOIL HORIZON) but most of all by their high clay content, giving vertical swelling, shrinking and cracking according to the amount of water present. In fact, they have been subdivided into four suborders according to different soil-moisture regimes which promote degrees of cracking for varying durations of time. The vertisol forms under grasslands having a marked dry season. The order includes the BLACK COTTON SOIL, REGUR, TROPICAL BLACK SOILS and GILGAI SOILS.

vesicle A small, circular cavity in a PYROCLASTIC rock or solidified lava-flow. It is formed by volcanic gas bubbles coming out of solution during the cooling phase. PUMICE is extremely vesicular (full of vesicles), hence its ability to float.

vesicular Pertaining to a substance which contains many small cavities or VESICLES, e.g. a solidified lava-flow. The A-horizon of certain desert soils may also develop a vesicular structure because the formation of a superficial crust traps air beneath the surface.

vicariance biogeography In general, a term referring to the study of fauna and flora whose individual groups have descended from common ancestors but which are currently geographically isolated from each other (DISJUNCT DISTRIBUTION). A more recent use of the term has seen it restricted to that part of biogeography which looks specifically at ALLOPATRIC speciation and those studies which link CLADISTICS and TAXONOMY with BIOGEOGRAPHY.

Villafranchian A stratigraphic name approximately equivalent to the Calabrian stage of the Lower Pleistocene of S Europe. It was originally thought to be of Upper Pliocene age.

virga The veils or shreds of CLOUD which trail beneath the cloud base as it passes overhead. They are composed of water droplets or ice crystals, which evaporate before reaching the ground surface.

virgation The term given to the spreading out of several mountain ranges from a central mountainous KNOT.

viscosity The properties of a fluid which render it slow-flowing because of internal friction due to molecular cohesion.

viscous Pertaining to a substance which is sticky, adhesive and although it flows it is slowed down, e.g. asphalt, wax and certain types of LAVA. In the case of lavas, those rich in silica (ACID LAVA) are the most viscous and do not flow far from the vent, thereby building steep-sided volcanic cones.

visibility **1** In surveying, the term refers to the distance which can be seen by an observer, depending on the amount of DEAD GROUND and the height above sea-level. **2** In meteorology, it refers to the clarity of the atmosphere, which depends in part on the type of AIR MASS (e.g. *polar maritime* air is characteristically clear) and in part on the amount of human-induced pollution. Visibility in polar regions is usually excellent, because of the lack of pollutants in the atmosphere. It is lowest in conditions of FOG, MIST, HAZE or during a DUST-STORM.

visualization A term increasingly used in CARTOGRAPHY, especially in computer graphics, to assist in an exploration and search for patterns amidst a mass of data in an attempt to answer questions and develop hypotheses. The THEMATIC MAP is regarded as the first of the visualization tools, but high-interaction computing has demanded more than this single-map view. There is now a potential for COMPUTERS to link geographic visualization with other powerful tools for analysis, synthesis and discovery, and to explore spatial and temporal variations of data simultaneously rather than separately. Visual thinking is said to be exploratory; visual communication is explanatory. Visualization allows map users to experiment with different categories, SCALES, symbols, AZIMUTHS, PROJECTIONS, etc., by means of interaction and animation. Visualization is now widely employed in MODEL BUILDING, by tracking the model's performance, by steering it to effect changes and by post-processing the output by high-speed networks to specialized graphics workstations. The visual variables and their effectiveness in signifying the three levels of measurement of graphic data are shown in [*270*].

vitrification A term referring to the formation of glassy rocks during the rapid cooling of MAGMA. Vitrified forts found on the Celtic fringes of the British Isles were subjected to artificial heat (possibly deliberately burned) so that their building stones were fused together.

vlei An Afrikaans term for a South African swamp or shallow lake which has a permanent water supply and is often connected with a river system.

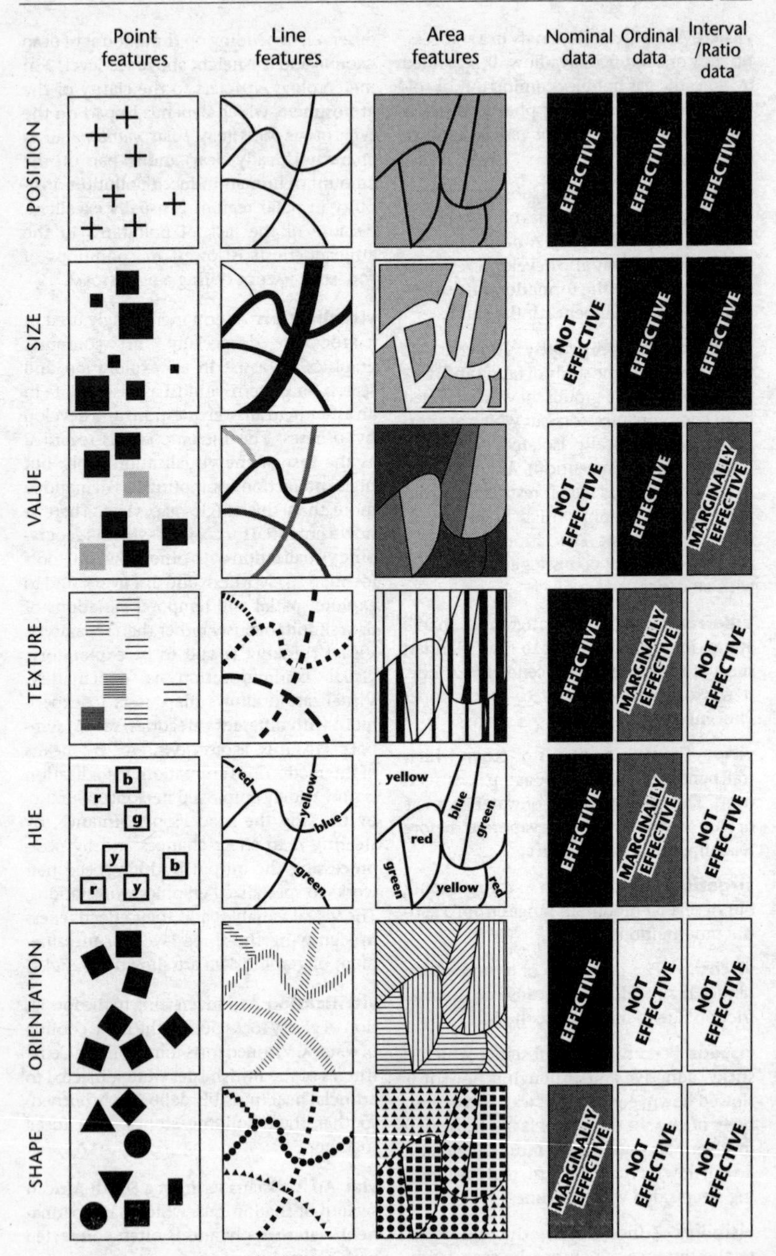

Figure 270 *Visual variables*

V-notch weir A triangular type of WEIR to measure the flow of a stream, the triangular form of which is of particular use in the accurate measurement of small DISCHARGES. The angles of the most common notches are 90° and 120°. Its discharge formula is:

$$Q = K \tan(\theta/2)H^{5/2}$$

where Q is the discharge, θ is the angle of the V-notch, H is the head of water above the V-notch and K is the weir coefficient. An alternative version is the *sharp-crested weir* with a rectangular shape. Its discharge formula is: $Q = KbH$, where Q is the discharge, b is the width of the weir crest, H is the head of water above the crest and K is the weir coefficient, which is dependent on the dimensions of the installation itself.

void ratio The ratio of the volume of void space to the volume of solid particles in a given mass of soil.

volcanic ash ASH.

volcanic bomb BOMB, VOLCANIC.

volcanic earthquake An EARTHQUAKE associated with an area of active vulcanicity, probably caused by crustal stresses related to the rising MAGMA and changes in HYDROSTATIC PRESSURE.

volcanic neck NECK.

volcanic plug PLUG.

volcanic rocks Those IGNEOUS ROCKS that have been ejected or poured out at the surface. They are synonymous with *extrusive rocks* (EXTRUSION) and contrast with PLUTONIC ROCKS. *See also* ANDESITE, BASALT, LAVA, OBSIDIAN, PILLOW LAVA, PUMICE, PYROCLASTIC MATERIAL, RHYOLITE, SCORIA.

volcanic tuff TUFF.

volcano An opening in the crust of the Earth, connected by a conduit to an underlying magma chamber, from which molten lava, volcanic gases, steam and PYROCLASTIC MATERIAL are ejected. It is usually in the form of a peak which may be cone-shaped (CONE) or dome-shaped (DOME) according to the character of the materials ejected. Volcanoes may be: **1** Extinct, as in the TERTIARY VOLCANIC PROVINCE of Britain, in which case their original relief form may have been almost obliterated or considerably diminished by denudation (PLANÈZE). **2** Dormant, as in the cases of Tristan da Cunha (prior

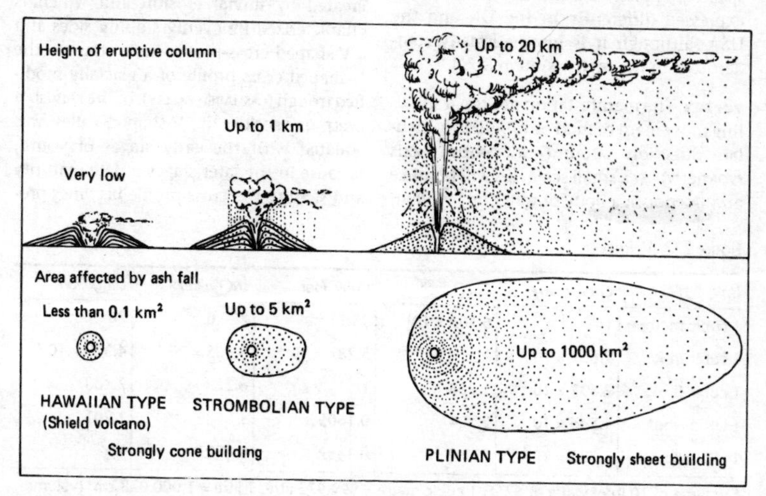

Figure 271 *Types of volcanic eruption*

to 1961) and Kilimanjaro. Many volcanic cones which are thought to be extinct may only be dormant. **3** Active, which applies to several hundreds of the world's volcanoes, including Mt St Helens, USA, which had been dormant for a century. The form of the volcano depends on the type of ERUPTION and on its magnitude [*271*], with the least explosive (HAWAIIAN ERUPTION) producing a SHIELD VOLCANO, and the most explosive (PLINIAN ERUPTION) creating a high and extensive cone. The collapse of a volcano into its magma chamber creates a CALDERA. *See also* AGGLOMERATE, ASH, BOMB, CENTRAL-VENT VOLCANO, CINDER CONE, CRATER, DUST, FISSURE ERUPTION, LAVA, LAVA CONE, LAVA DOME, NUÉE ARDENTE, PAROXYSMAL ERUPTION, PELÉAN ERUPTION, PIPE, PLUG, PUMICE, PUY, SCORIA, STROMBOLIAN ERUPTION, THOLOID, TUFF, TUMESCENCE, VOLCANIC EARTHQUAKE, VULCANIAN ERUPTION.

volt The unit of electrical potential. It is equal to the difference in electrical potential between two points on a conductor carrying a constant current of one ampere when the power dissipated between these points is one watt. It is named after A. Volta, an Italian physicist.

volume A measure of bulk, capacity or space. It applies to fluids and solids and is expressed differently in the UK and the USA, although it is internationally standardized in the metric system [*272*].

vortex The manifestation of a rapid spiralling movement of fluid or vapour in a circular direction around a central axis (VORTICITY). At a small scale this will produce a WHIRLPOOL in a body of water or a violent, revolving WHIRLWIND in the atmosphere (DUST-DEVIL, TORNADO). But larger vortices include all spiralling low-pressure systems such as CYCLONES and HURRICANES.

vortex flow A pattern of turbulent wind-flow that follows a corkscrew motion parallel to the ground and is thought to be responsible for the equal spacing between SEIF-DUNES [*273*]. It is also known as *Taylor-Görtler flow*. *See also* WAVE-FLOW.

vorticity The rotational action of a fluid about an axis, either in the form of WHIRLPOOLS in a turbulent river or as *vortices* in the atmosphere. Vorticity in the atmosphere plays an essential part in the Earth's HEAT BALANCE or ENERGY BALANCE mechanism, by moving air horizontally or vertically from one place to another. It varies in magnitude from the small-scale WHIRLWINDS to the larger spiralling cloud-systems of HURRICANES, CYCLONES and DEPRESSIONS which are the major distributors of energy through the Earth's atmosphere. Vorticity is defined as positive if it is cyclonic (rotating in the same direction as the Earth) and negative if it is anti-cyclonic. *See also* ABSOLUTE VORTICITY, PLANETARY VORTICITY, RELATIVE VORTICITY.

V-shaped valley A valley whose form is created by fluvial erosion and which is characterized by evenly sloping sides and a V-shaped cross-profile in contrast to the U-shaped cross-profile of a glacially modified trough (U-SHAPED VALLEY). In the Davisian CYCLE OF EROSION the V-shaped valley was equated with the early stages of youth, because in the later stages of late maturity and senility the cross-profile becomes pro-

Figure 272 *Volume*

Unit	*Cubic metre*	*Cubic inch*	*Cubic foot*	*UK gallon*	*US gallon*
1 cubic metre =	1	6.102×10^4	35.31	220.0	264.2
1 cubic in =	1.639×10^{-5}	1	5.787×10^{-4}	3.605×10^{-3}	4.329×10^{-3}
1 cubic ft =	2.832×10^{-2}	1,728	1	6.229	7.480
1 UK gallon* =	4.546×10^{-3}	277.4	0.1605	1	1.201
1 US gallon =	3.785×10^{-3}	231.0	0.1337	0.8327	1

* Volume of 10 lb of water at 62° F. 1 cubic metre = 999.972 litre. 1 litre = 1,000.028 cm^3 (volume of 1 kg of water at maximum density).

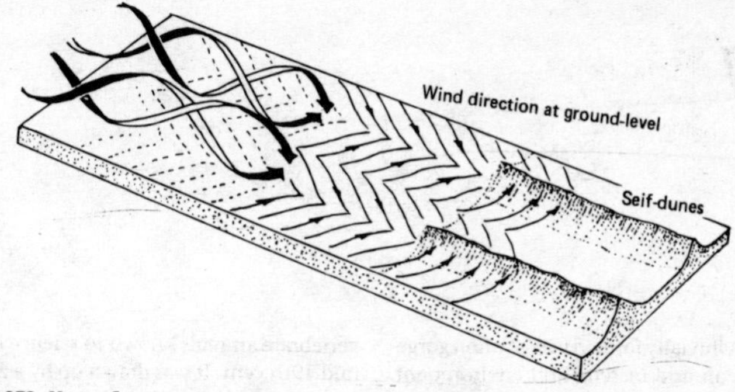

Figure 273 *Vortex flow*

gressively less V-shaped. It is now recognized, however, that most unglaciated river valleys exhibit a V-form in their upper reaches. In practice the angle of the V will be influenced by a number of variables: **1** climate, for rapid mass-movement is more common in humid climates, tending to widen the V-shape: in arid climates the walls of the valley may be steeper owing to the slower rate of weathering; **2** rock structure and the resistance of the rocks to weathering and erosion, e.g. valleys cut in cohesive silts or clays (LOESS) have a tendency to create deeply cut ravines with steep slopes (BADLANDS); **3** aspect, with shaded and sunny slopes experiencing different microclimates and different magnitudes in the operation of slope processes; **4** the rapidity at which the river cuts down, e.g. a rapid period of uplift will ensure that a narrow V-shape will be maintained; **5** the location of the cross-profile on the LONG PROFILE of the river, for in its lower reaches the valley will be extremely broad and its bluffs a great distance apart, thus creating a cross-profile which cannot be described as V-shaped.

vug, vugh A small irregular cavity in an intrusive rock or carbonate sedimentary rock, which may be lined with precipitated minerals. *See also* DRUSE.

Vulcanian eruption A volcanic eruption in which the lava surface rapidly solidifies because of its high VISCOSITY. This results in a build-up of pressure beneath the lava crust and a continuous series of violent explosions during which large quantities of PYROCLASTIC MATERIAL are ejected violently from the vent. *See also* VOLCANO.

vulcanicity, volcanicity Pertaining to volcanic activity; the state of being volcanic.

vulcanology, volcanology The scientific study of volcanoes and volcanic phenomena.

W

wadi A fluvially formed valley, often gorge-like, in an arid or semi-arid environment. Because of the rare fluvial activity the valley walls are steep and are mantled with thick layers of weathered material. It may be permanently dry, having been cut during a former climatic period of greater precipitation, or it may carry an INTERMITTENT STREAM, especially after an occasional storm, when the torrent is often capable of considerable erosion. Many of the wadis in Arabia and N Africa are partly sand-filled.

wake 1 A line of trailing bubbles and eddies (EDDY) behind an object moving through water. **2** A trail of eddies or vortices (VORTEX) in the flow of air over and around an obstacle, developed on the leeward side.

wake-dune An isolated sand-DUNE which forms to the leeward end of a LATERAL DUNE. [*65*]

Wallace's line A line drawn by A. R. Wallace, a 19th-cent. zoologist, to demarcate the floral and faunal distinctions between Australasia and SE Asia. The line originally separated Sumatra from Java, Borneo from Celebes and the Philippines from New Guinea, but was later modified by Huxley so as to include the Philippines within the Australasian realm (WALLACE'S REALMS). The sharp division is thought to have followed a narrow zone of crustal disturbance of Mesozoic age.

Wallace's realms A broad zoogeographical classification of the world's fauna into six regions on a basis of the spatial distribution of all the vertebrate and non-vertebrate animals known to science in the mid 19th cent. It was drawn up by a zoologist, A. R. Wallace, who recognized the following realms: (a) Palaearctic; (b) Ethiopian; (c) Oriental; (d) Australasian; (e) Nearctic; (f) Neotropical [*274*]. The detailed boundaries of the realms have been gradually modified as further discoveries have been made. This is particularly true in the SW Pacific where the Australasian realm has been limited merely to Australia and New Zealand and the Oriental realm extended to include all Indonesia, Polynesia and Melanesia. *See also* FLORISTIC REALMS.

wall-sided glacier A valley glacier that has extended out on to the adjoining lowlands as a narrow tongue but without losing its steep-sided form.

Walker circulation SOUTHERN OSCILLATION.

Waltonian RED CRAG SERIES.

wanderblöcke PLOUGHING BLOCK.

waning slope The low-angle FOOT SLOPE of gentle concavity which lies beneath the CONSTANT SLOPE as described by A. Wood following the ideas of W. Penck. It is produced by SURFACE WASH and is sometimes referred to as the *wash slope* or *wash-controlled slope*. Although it is usually covered with a veneer of debris it is nevertheless a feature cut in solid rock and is therefore synonymous with a PEDIMENT, which extends from the base of the other slope elements across to the alluvial plain and stream channel. It has been claimed that the waning slope becomes more extensive as the CYCLE OF EROSION proceeds, with the coalescence of pediments becoming commonplace as the

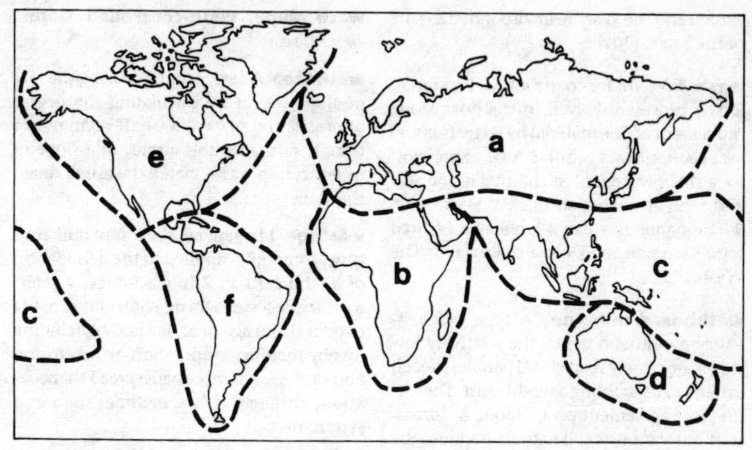

Figure 274 *Wallace's realms*

process of backwearing (PARALLEL RETREAT OF SLOPES) becomes dominant. [*49*] *See also* WAXING SLOPE.

warm front The boundary in a DEPRESSION between an advancing mass of warm air where it is overriding and rising above a mass of colder air which it is slowly overtaking. Thus it represents a marked thermal gradient in the atmosphere. The frontal surface is at a very low angle (< 1°) so that the weather which characterizes the approaching warm front may stretch over a broad zone of tens of kilometres [*44*]. The earliest indication of the front's arrival is the spreading of CIRRUS clouds across a clear sky, followed in succession by CIRROSTRATUS, ALTOSTRATUS, and finally NIMBOSTRATUS from which the intermittent DRIZZLE turns into heavy, persistent rainfall. Meanwhile, the cloud base has become progressively lower and the wind has gradually backed (BACKING) before the warm front arrives, finally VEERING as the front passes through. The *warm sector* of the depression is heralded by the warm front and it is characterized by rising temperature and higher relative humidity. Just as a warm front is the leading boundary of the *warm sector* so a COLD FRONT marks its terminal boundary. [*60*] *See also* ANAFRONT, FRONTOGENESIS, KATAFRONT, OCCLUSION.

warm glacier An alternative term for a TEMPERATE GLACIER.

warm sector WARM FRONT.

warp A term used to describe fine-grained sediments deposited in tidal estuaries. The expression *warp soils* is used by the Soil Survey of England and Wales to describe alluvial deposits (ALLUVIUM).

warping 1 The gentle deformation of the Earth's crust over a widespread area, but without major faulting or folding. It may be due to ISOSTASY or GLACIO-ISOSTASY and can result in tilted peneplains, deformed raised beaches, etc. **2** The process of deposition of fine-grained sediments in a tidal estuary to form warpland (WARP).

Warthe The name given to a glacial stage in the Upper Pleistocene chronology of N Europe. It is named after a river which passes through Poznań (Poland) *en route* to the R. Oder, and its moraines run eastwards from E Germany into Poland. There is no agreement over its age, with some authorities regarding it as a shortlived readvance or still-stand in the recession of the SAALE ice-sheets, while others regard it as the maximum limits of the WEICHSEL advance. This latter view has now been generally discarded and the Warthe incorporated as part of the Saale. [*109*]

wash **1** A term for coarse alluvial material. **2** The process involved in the downslope movement of fine material by water (SUBSURFACE WASH, SURFACE WASH). **3** A US expression for a shallow streamless channel in the arid and semi-arid lands of the SW USA. [*189*] **4** The name given to a large area of tidal sand banks on the E coast of England: The Wash.

washboard moraine A type of cross-valley moraine in which the relatively low ridges are closely spaced and parallel to each other, i.e. resembling a washboard. There is no close agreement on its mode of formation but various explanations include: the seasonal pushing effect of a glacier front; the concentration of ground moraine along the zones where thrust-planes reach the base of the ice mass; the squeezing out of subglacial TILL into marginal troughs along the ice front where it terminates in a proglacial lake. *See also* DE GEER MORAINES.

washland A river FLOODPLAIN surrounded by artificial embankments into which the river is diverted in times of flood in order to alleviate further flooding downstream in more vulnerable settlement zones. It is the main type of RETARDING BASIN.

wash-out **1** In general, the destruction caused by a torrential river in spate following a storm. It refers to bank erosion, levee breaching, dam bursting and bridge destruction. **2** In geology, a channel cut through a seam of coal during or after its formation and subsequently infilled by a non-carbonaceous sediment, usually sandstone. The term has been extended to include a channel cut into any pre-existing sediment, and the feature is commonplace in deltaic beds.

washplain A virtually flat plain of ALLUVIUM overlying layers of deeply weathered BEDROCK where seasonal floods lack large volumes of sediment and abrasive BEDLOAD. Thus, the floodwaters are incapable of DOWNCUTTING.

wash slope, wash-controlled slope WANING SLOPE.

wash trap A device for measuring the rate of slope retreat by calculating the degree of erosion by SURFACE WASH. It comprises a trough sunk into the surface of a slope in order to trap waste material washed down the slope.

wastage **1** In general, the loss of bulk over time, sometimes applied to the denudation of a land surface. **2** In glaciology, it refers to the processes which contribute to the overall shrinkage of an ice body, including loss by melting, evaporation, wind erosion and calving. It is more widespread than ABLATION, although it is sometimes used as a synonym.

waste fan An alternative term for an ALLUVIAL FAN.

waste mantle REGOLITH.

water balance A term used in meteorology to denote the cyclical movement of water between the atmosphere and the ground surface. The water balance takes into account the complex interrelationships of all atmospheric processes which relate to moisture, e.g. PRECIPITATION, EVAPORATION and HUMIDITY, in addition to land-surface processes such as EVAPOTRANSPIRATION, OVERLAND FLOW and THROUGHFLOW. At a global scale it also includes data on ocean currents and sea temperatures but since information from ocean areas is relatively scanty the reliability of the water balance is greatest over the continents. *See also* HYDROLOGICAL-BALANCE BUDGET, HYDROLOGICAL CYCLE.

water cycle HYDROLOGICAL CYCLE.

waterfall A point in the long profile of a river where the water descends vertically. It marks the position of a KNICKPOINT in the river-course, but the position of the waterfall will slowly move upstream owing to erosion at the lip of the fall and undercutting in the zone of the plunge pool at the foot of the fall, where CAVITATION is a significant process. Waterfalls may be produced by: **1** resistant rock bands, especially those created by a SILL; **2** descent from the edge of a plateau; **3** descent from a HANGING VALLEY into a glacial trough; **4** crossing of a FAULT-SCARP, created by an earthquake; crossing of a FAULT-LINE SCARP; descent from a marine-cut cliff into the sea. The world's highest falls are the Angel Falls, Venezuela (979 m/3,212 ft). The highest waterfall in Britain is the Eas a'Chual Aluinn in NW Scotland (200 m/658 ft).

water-gap A pass or valley through a mountain or ridge that is followed by a river. It contrasts with a WIND-GAP.

waterlogging A term referring to the character of the soil where a state of IMPEDED DRAINAGE exists. It leads to GLEYING, the formation of GLEY SOILS and the growth of PEAT.

water masses Large bodies of water exhibiting virtually uniform characteristics, especially those of temperature (T) and salinity (S). Other characteristics include nutrients and dissolved oxygen amounts. Their major properties are achieved primarily from contact with the atmosphere in source regions (AIR MASS). Their patterns are largely controlled by OCEAN CURRENTS. Upper water masses are separated from deeper water masses by a THERMOCLINE above and below which different characteristics occur. [*275*] *See also* THERMAL STRATIFICATION OF WATER.

water meadow A zone of grassland along the floodplain of a river where the growth of herbage is stimulated by periodic flooding, by either natural or artificial means. It is termed a WASHLAND in the USA. *See also* RETARDING BASIN.

watershed **1** In Britain watershed is used synonymously with DIVIDE, i.e. to mean a water-parting from which headstreams flow to separate river systems. **2** In the USA, the term refers to the entire catchment area of a single DRAINAGE BASIN. **3** PHREATIC DIVIDE.

watershed breaching GLACIAL DIFFLUENCE.

watershed model A physical or mathematical MODEL of the physical processes operating within a DRAINAGE BASIN.

water sink SWALLOW-HOLE.

waterspout The marine equivalent of a TORNADO, although of shorter duration and lower intensity. Its water content is derived in part from cooling and condensation within the VORTEX and in part from the water sucked up by the UPDRAUGHT from the sea or lake surface, which is released in an immense deluge if the waterspout's track crosses a shoreline. It is due to intense local heating creating small-scale low-pressure systems and their associated cumulonimbus clouds. *See also* CONVECTION, VORTICITY.

water-table The upper level of the zone of groundwater saturation in permeable rocks (PERMEABILITY). It represents the surface at which the pressure in the groundwater is equal to the atmospheric pressure. The water-table's elevation varies seasonally according to the amount of precipitation but it also depends on other factors, such as EVAPOTRANSPIRATION and the amount of PERCOLATION through the soil. The slope (HYDRAULIC GRADIENT) of the water-table is inversely proportional to the permeability of the AQUIFER. The water-table marks the change in the GROUNDWATER zone between the zone of aeration, where some pores are open, and the underlying zone of saturation, in which water fills all the spaces in the rocks. Where the water-table reaches the ground surface it is marked by a SPRING or by SEEPAGE. *See also* ARTESIAN WELL, PERCHED WATER-TABLE, PIEZOMETRIC SURFACE.

water vapour The non-visible state of moisture in the atmosphere. It exists as an independent gas in variable quantities and exerts variable pressures between 5 and 30 mb (VAPOUR PRESSURE). The presence of water vapour is a prerequisite for any form of PRECIPITATION, but it also fulfils an important role in the Earth's HEAT BALANCE mechanism, for, together with carbon dioxide, it forms a type of global 'blanket' which prevents excessive heat loss from terrestrial radiation – were it not present the Earth's temperature would be considerably lower (GREENHOUSE EFFECT).

watt A unit of power, expressed as the power dissipated when one joule is

Figure 275 *Major water masses of the world oceans and their T-S characteristics*

Atlantic Ocean

	Central waters		Intermediate waters			Deep and bottom waters		
Name	North Atlantic	South Atlantic	Atlantic subarctic	Mediterranean intermediate	Antarctic intermediate	North Atlantic deep and bottom	Antarctic deep	Antarctic bottom
T(°C)	20.0	18.0	2.0	11.9	2.2	2.5	4.0	–0.4
S(%)	36.5	35.9	34.9	36.5	33.8	34.9	35.0	34.66

Indian Ocean

	Central waters			Intermediate waters			Deep and bottom waters
Name	Bay of Bengal	Equatorial	South Indian	Red Sea intermediate	Timor Sea intermediate	Antarctic intermediate	Antarctic deep and bottom
T(°C)	25.0	25.0	16.0	23.0	12.0	5.2	0.6
S(%)	33.8	35.3	35.6	40.0	34.6	34.7	34.7

Pacific Ocean

	Central waters			Intermediate waters				Deep and bottom waters
Name	Western North Atlantic	Eastern North Pacific	Equatorial Western South Pacific	Pacific Subarctic	North Pacific intermediate	South Pacific intermediate	Antarctic intermediate	Antarctic deep and bottom
T(°C)	20.0	20.0	20.0	5–9	4–10	9–12	5.0	1.3
S(%)	34.8	35.2	35.7	33.5–33.8	34.0–34.5	33.9	34.1	34.7

Source: Tolmazin, 1985. Table 7.1. After Mamayev, 1975; Sverdup *et al.*, 1942

expended in one second. It is named after the British engineer James Watt (1736–1819).

watten (wadden) A German term for the tidal marshes lying between the North Sea coast of Germany and the Frisian Is. In the Netherlands the neighbouring coastal marshes and mud-banks are known as *wadden*.

wave **1** In general, a disturbance of the equilibrium of a body or a medium, whereby the disturbance is propagated from point to point with a continuous recurring motion (SEISMIC WAVES). **2** In hydrology and oceanography, a deformation of a water surface in the form of an oscillatory movement which manifests itself by an alternating rise and fall of that surface. The oscillation is generated by the wind's pressure on the water surface and the wave magnitude is in proportion to the speed of the wind, its duration and the length of FETCH. Some waves, however, are caused by submarine earthquakes (TSUNAMI) or landslides into water bodies. A wave can be described by: its height, between crest and trough (AMPLITUDE); its length, between successive crests; its velocity; its steepness; its period, defined as the time taken for it to travel one WAVELENGTH; and its energy. [*276*]. Waves and their incorporated sediments are the principal cause of MARINE ABRASION, but they are also capable of erosion by hydraulic action and can play a part in rock solution. They also play an important part in deposition of sediments, especially by the process of LONGSHORE DRIFT. *See also* BACKWASH, BOUNDARY WAVES, BREAKER, CONSTRUCTIVE WAVE, DESTRUCTIVE WAVE, DOMINANT WAVE, EDGE WAVE, ORBITAL MOTION, OSCILLATORY WAVE, PROGRESSIVE WAVE, STATIONARY WAVE, SURF, SWASH, SWELL. **3** In meteorology, *atmospheric waves* include EASTERLY WAVES and ROSSBY WAVES. *See also* WAVE DEPRESSION.

wave base The lowest limit of orbital motion beneath waves and the depth at which wave action ceases to disturb the sea-floor sediments. It is therefore an important concept in problems of sediment transport.

wave-built terrace MARINE-BUILT TERRACE.

wave-cut platform ABRASION PLATFORM.

wave depression DEPRESSION.

wave diffraction The term given to the spreading out of energy as water waves pass through an opening in an obstacle such as a breakwater, usually leading to a considerable reduction in wave height. Some of the energy may affect the water in the lee of the obstacle, i.e. inside a harbour. The term should not be confused with WAVE REFRACTION.

wave erosion MARINE ABRASION, WAVE.

wave-flow The passage of air over a topographic obstacle to produce a series of regularly spaced zones of faster and slower flow together with reversing flow, in the form of rotors, near the ground surface in the lee of the obstacle. The appearance of this kind of

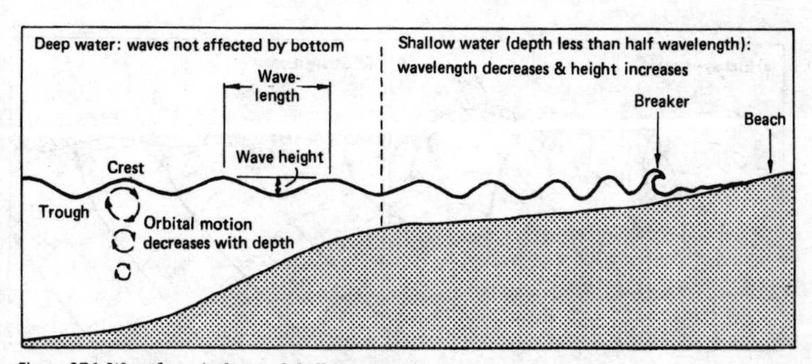

Figure 276 *Wave forms in deep and shallow water*

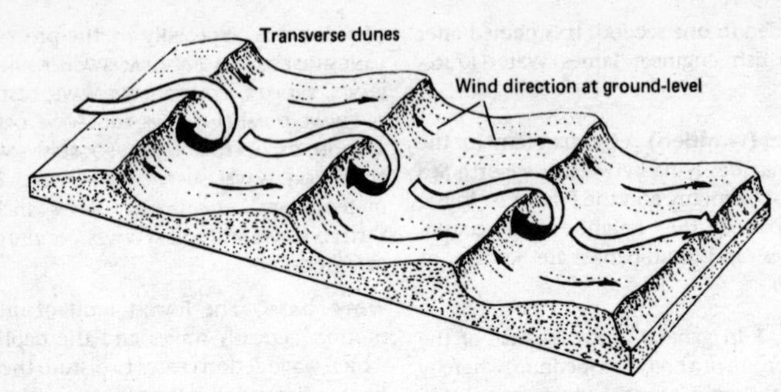

Figure 277 *Wave-flow*

flow depends on the height and the shape of the obstacle in relation to the vertical temperature and wind profile. It has been suggested as a mechanism which could account for the spacing of TRANSVERSE DUNES. [*277*]. *See also* DUNE, EDDY.

wavelength The distance between successive crests or other equivalent points in a harmonic wave. The symbol is λ; λ equals velocity/frequency.

wave period **1** The time that elapses during which a wave form travels the distance of one WAVELENGTH. **2** The time during which two successive wave crests pass a fixed point.

wave refraction The process by which wave crests change direction as they approach a shoreline, owing to the shallowing of water. In deep water the wave crests are parallel but as the depth of water decreases so the velocity of the wave is diminished. On a coastline of bays and headlands the shallowing, and therefore slowing, will take place initially in front of the headlands, while in the deeper water in front of the bays retardation occurs much later in the wave travel. Thus, the wave front becomes curved (refracted) and approaches the shoreline from a different direction than its initial line of advance, thereby concentrating wave energy on the headlands [*278a*]. Wave refraction also operates when waves approach a relatively straight shoreline at an oblique angle, for they are refracted so as to break almost parallel with the shoreline [*278b*]. Wave refraction can be plotted on REFRACTION DIAGRAMS.

wave-refraction diagram REFRACTION DIAGRAM.

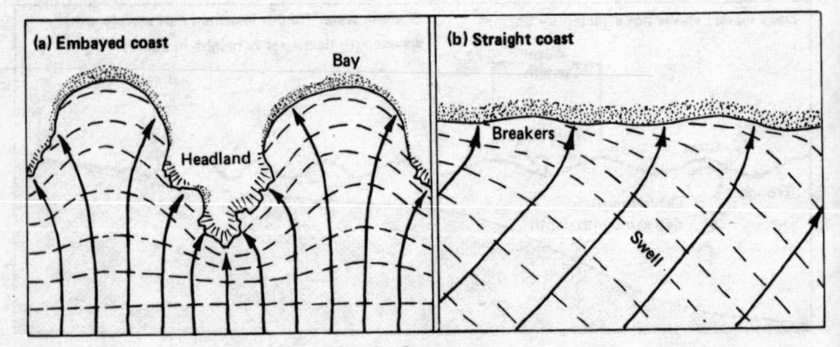

Figure 278 *Refraction of waves approaching the shore*

wave spectrum **1** The complete array of water waves at a given place at a particular time. **2** A classification of waves according to their WAVE PERIODS. They range from tiny capillary waves (a few cm long) to very lengthy waves over 100 km in length.

way-up structures An expression for any geological criteria which assist in establishing the original orientation of a body of rock, particularly in determining sequences in sedimentary rocks and interbedded lavas, where the PRINCIPLE OF SUPERPOSITION may subsequently have been affected by tectonic movements. In sediments the most important criteria are: **1** CROSS-BEDDING; **2** GRADED BEDDING; **3** RAIN PITS; **4** DESICCATION CRACKS; **5** RIPPLE MARKS; **6** TRACE FOSSILS; **7** WASHOUTS. In addition, any fossils in the position of growth (e.g. tree roots) help to identify the top and bottom of the stratum. In lavas, vesicles are usually concentrated at the top of the flow.

waxing slope The convex element in the upper part of a SLOPE PROFILE, situated above the CONSTANT SLOPE, as described by A. Wood following the ideas of W. Penck [*49*]. This convex summit slope is called the *crest* by L. C. King, who believes that it dominates the hill-slope profile during the early stages of the CYCLE OF EROSION when valley incision is greater than backwearing of slopes (PARALLEL RETREAT OF SLOPES). The major processes at work on the waxing slope are weathering and CREEP (in humid climates) and RAINSPLASH and SURFACE WASH (in semi-arid climates). *See also* WANING SLOPE.

Wealden A stratigraphic name for the deltaic and lacustrine sedimentary rocks of Lower CRETACEOUS age which occur in the Weald region of SE England. They are thought to be equivalent to Neocomian facies elsewhere in Britain. The Wealden consists of over 600 m of clays, silts and sands, grouped into CYCLOTHEMS, each of which represents the southward advance of a delta into a large fresh-water lake, and its later submergence by lake muds. The succession is as follows:

Weald Clay	Upper Weald Clay
	Horsham Stone
	Lower Weald Clay
Hastings Beds	Upper Tunbridge Wells Sand
	Grinstead Clay
	Lower Tunbridge Wells Sand
	Wadhurst Clay
	Ashdown Beds

weather The state of the ATMOSPHERE at a specific time and at any one place. It takes into account all the atmospheric elements (ELEMENT), the total synthesis of which, judged over a long period of time (>30 years), constitutes the CLIMATE of that place. The scientific study of weather is termed METEOROLOGY. Attempts to predict future changes are known as WEATHER FORECASTING. There have been numerous efforts to modify weather by artificial means (e.g. CLOUD-SEEDING) and many studies to show man's unintentional modification of weather (*see also* HEAT 'ISLAND', PHOTOCHEMICAL FOG, URBAN CLIMATE).

weather forecasting ANALOGUE WEATHER FORECASTING, FORECAST, NUMERICAL WEATHER FORECASTING, SYNOPTIC CLIMATOLOGY, SYNOPTIC METEOROLOGY.

weathering The breakdown and/or decay of rock on the surface or very near to the surface of the Earth, whereby a mantle of waste (REGOLITH) is created which will remain *in situ* until agents of EROSION cause it to be moved (TRANSPORT). Thus, processes of weathering do not involve transport, but they can be regarded as making up the essential early stages of the overall DENUDATION process. Weathering can be subdivided into CHEMICAL WEATHERING, MECHANICAL WEATHERING and ORGANIC WEATHERING. *See also* DEEP WEATHERING, DIFFERENTIAL WEATHERING, INSOLATION WEATHERING, WEATHERING INDEX, WEATHERING RIND.

weathering front BASAL SURFACE OF WEATHERING.

weathering index A quantitative measure of the intensity of CHEMICAL WEATHERING and of MECHANICAL WEATHERING. Chemical weathering is expressed as the ratio of the chemically more mobile to the chemically less mobile materials:

$$W_1 = \frac{\text{Proportion of chemical reactants}}{\text{Proportion of residual products}}$$

or, more simply, as the ratio of unweathered to weathered minerals in a given volume. Mechanical weathering is variously expressed as: the softening effect of water on materials; the swelling due to absorption of water; changes in hardness and strength of materials; the ability of unweathered material to absorb water in relation to that of weathered material.

weathering-limited slope A hillslope whose form is controlled by its bedrock's resistance to WEATHERING. It is governed by the environmental conditions which determine that the weathered material's potential rate of removal by erosion is greater than the rate at which weathered material is produced.

weathering rind A layer of partly weathered rock, extending a few centimetres below the surface (and therefore contrasting with the much thinner DESERT VARNISH). It occurs on exposed surfaces or where there is prolonged contact with groundwater just below the surface. Its orange, red or yellow colouration is due to OXIDATION of iron minerals in the rock.

weather map SYNOPTIC CHART.

wedge failure The process of rock collapse generated by the intersection of two or three high-angle joints on a FREE FACE and where the fractures dip down the face of the cliff. *See also* SLAB FAILURE, TOPPLING FAILURE.

wedge of high pressure A system of *high atmospheric pressure* dividing two DEPRESSIONS, and similar to a RIDGE OF HIGH PRESSURE but somewhat narrower, thereby bringing only a brief spell of fine weather.

wedges ICE WEDGE, SAND WEDGE.

Weichsel The final glacial stage of the Pleistocene on the N European Plain, succeeding the EEMIAN interglacial. It is equivalent in age to the WÜRM of the Alps, the DEVENSIAN in Britain and the WISCONSIN of N America. It has been subdivided into the Brandenburg moraines, the Frankfurt moraines and the Pomeranian moraines, each of which marks a recessional stage in the N European ice-sheet. [*109*]

weighted mean percentage silt/clay A method devised by S. A. Schumm (1960) to define the PARTICLE SIZE of sediment in the cross-section perimeter of a river channel. Channels with low values are generally wide and shallow, carrying most of their sediment as BEDLOAD. Conversely, those with high values are usually narrow and deep, carrying most of their load in SUSPENSION. The formula is:

$$M = [(Sb \times w) + 2(Sc \times d)]/(w + 2d)$$

where M is the weighted mean percentage, Sb is the percentage silt and clay in the river banks, w is the channel width, and d is the channel depth.

weir A dam across a stream over which the water is allowed to flow, although it raises the water-level. It is used to measure the flow, control the depth and to provide a head of water for a water wheel. *See also* V-NOTCH WEIR.

welded tuff IGNIMBRITE.

welt A narrow but steep line of upfolding or uplift.

Wentworth scale A scale used to classify the PARTICLE SIZE of sediments, outlined by C. K. Wentworth in 1922, and widely used by geologists. The list of grades is as follows: boulders (>256 mm diameter), cobbles (64–256 mm), pebbles (2–64 mm), sand (0.0625–2 mm), silt (0.004–0.0625 mm), clay (<0.004 mm). *See also* SOIL TEXTURE.

Westerlies The wind system which dominates the zones between latitudes 40° and 70°N and S of the equator. The flow of air from the subtropical high-pressure cells to the temperate zone of low pressure (FERREL CELL, PLANETARY WINDS), thereby maintaining the fundamental meridional heat-exchange mechanism (ATMOSPHERIC CIRCULATION). The prevailing wind direction in the N hemisphere is from the SW and in the S hemisphere from the NW. DEPRESSIONS are most common in this wind system, which is one of the strongest and most undisturbed airflows among the planetary winds. The strength of the Westerlies increases with altitude (JET STREAM), reaching average high-altitude speeds of 45–67 m/sec (100–150 miles/h), which double in velocity for short spells in the winter. The S hemisphere Westerlies are stronger than their northern counterparts because the broader expanses of ocean and dearth of land masses preclude the development of stationary pressure systems; their constancy and strength leads to them being termed ROARING FORTIES in the southern oceans. The term Westerlies has also been applied, in a more limited sense, to the zone of generally westerly winds that occur in equatorial latitudes between the two belts of TRADE WINDS, especially over the continents, for the INTERTROPICAL CONVERGENCE ZONE does not move sufficiently far from the equator to permit westerly development over the Pacific and Atlantic.

West-Wind Drift A slow eastward movement of ocean water in the zone of the WESTERLIES. In the N hemisphere it occurs between 35°N and 45°N in the Pacific, where it is referred to as the North Pacific current. In the North Atlantic it is deflected far to the north as the NORTH ATLANTIC DRIFT. In the Southern Ocean it becomes a continuous circumpolar current owing to the dearth of land masses. In the South Pacific it is a relatively warm current flowing in a zone between latitudes 45°S and 60°S, while in the South Atlantic it is a cooler current between 40°S and 65°S. [*168*]

wet adiabatic lapse-rate SATURATED ADIABATIC LAPSE-RATE.

wet analysis The MECHANICAL ANALYSIS of those soils with PARTICLE SIZES less than 0.06 mm diameter (i.e. the coarse and fine sands, silt and clay). It is carried out by mixing the sample in a measured volume of water and determining the density of the mixture at regular intervals with a HYDROMETER. The analysis is governed by STOKES'S LAW. *See also* SETTLING VELOCITY.

wet-bulb thermometer An instrument for measuring temperature in which the bulb is enclosed in a moist muslin bag, thereby causing a loss of LATENT HEAT owing to EVAPORATION. Thus the temperature recorded by the wet-bulb thermometer is lower than that of the DRY-BULB THERMOMETER, the difference being referred to as the *wet-bulb depression*. By using the instrumental readings in conjunction with a set of statistical tables or the humidity slide rule, the RELATIVE HUMIDITY, DEW-POINT and VAPOUR PRESSURE of the air can be calculated. *See also* HYGROMETER, PSYCHROMETER, THERMOMETER.

wet day A period of 24 hours (commencing at 09.00 h GMT) during which at least 1 mm (0.04 in) of rainfall is recorded. It is noteworthy that this terminology applies only to the UK, for a rainfall of this amount and duration on the continent of Europe would be termed a RAIN DAY, for which Britain has a different definition.

wetland A term referring to an area that has developed a specially adapted vegetation (ECOSYSTEM) because it has been long dominated by water and whose processes are largely controlled by water, static or flowing, brackish or saline. *See also* BOG, FEN, MANGROVE, MARSH, PEAT, SWAMP.

wet snow A type of SNOW which occurs when temperatures are high enough to bond crystals into large flakes by REGELATION. It has a density of about 0.3 g/cm^3 (in contrast to that of 0.1 g/cm^3 of DRY SNOW) and is characteristic of maritime coastal areas in latitudes 40°–60°. It does not form a very good skiing material and is more difficult to remove by snow-ploughs and blowing machines since it solidifies by regelation inside the machinery. Wet snow accelerates the FIRNIFICATION process.

wet spell A period of at least fifteen successive WET DAYS. It is a term which is only of common usage in the UK, and has not been internationally adopted.

wetted perimeter The total length of the cross-section at the interface between a channel bed and the stream water which occupies it. *See also* HYDRAULIC GEOMETRY.

wetting-and-drying action A type of mechanical SORTING process in the soil when periodic wetting and drying leads to differential expansion and contraction, particularly where MONTMORILLONITE is dominant among the clay minerals. In semi-arid areas this can lead to the formation of GILGAI.

wetting front The lowest limit to which a soil becomes saturated by percolating precipitation, especially in arid regions.

Weybourne Crag ICENIAN.

whaleback **1** A term used to describe a glacially smoothed, elongated rock exposure in a MAMMILLATED TOPOGRAPHY. **2** An elongated RUWARE. **3** A very large longitudinal dune characterized by a flat top on which smaller BARCHANS or SEIF-DUNES may occur. It has been suggested that it may be a coalescence of several seif-dunes.

whinstone A local term from N England for any fine-grained, dark-coloured igneous rock, usually BASALT or DOLERITE. It is derived from the Great Whin Sill, which occurs over an area of 3,900 km^2 in NE England.

whirling psychrometer (hygrometer) An instrument for measuring the RELATIVE HUMIDITY of the air. It consists of a DRY-BULB THERMOMETER and a WET-BULB THERMOMETER clipped side by side to a wooden frame which itself is pivoted on a handle. It is operated manually by whirling the instrument (in the manner of a football rattle) in order to obtain a regular maximum flow of air across the two thermometer bulbs. *See also* HYGROMETER, PSYCHROMETER.

whirlpool A violent, circular, swirling movement in a water body, either at the base of a waterfall or in the open sea where two tidal currents meet, e.g. the Corrievreckan whirlpool between the islands of Scarba and Jura in W Scotland.

whirlwind A rapidly rotating column of air, or VORTEX, revolving around a local low-pressure centre. It is due to surface heating leading to instability and convectional air currents. The term is generally confined to small-scale vortices, such as DUST-DEVILS or TORNADOES. *See also* WATERSPOUT.

white-box system A system in which most of the internal structures (storage, flows, etc.) are known. Thus is it possible to have a detailed knowledge of the linkages between a given input and given output (INPUTS AND OUTPUTS). *See also* BLACK-BOX SYSTEM, GREY-BOX SYSTEM.

white-out A state of the atmosphere, occurring during or after a BLIZZARD, when normal visual perception of distance and shape is lost or blurred because shadows disappear, owing to multiple reflection from cloud and snow crystals, and the snow cover merges into the sky without a perceptible horizon. It is a common occurrence in polar regions and leads to loss of balance and sense of direction.

white smoker A plume of HYDROTHERMAL fluid containing finely dispersed particles of white sulphates issuing from a submarine vent on a MID-OCEANIC RIDGE. *See also* BLACK SMOKER.

width-depth ratio An equation defining part of the CHANNEL GEOMETRY of a river. It is expressed as:

$$d = \frac{a}{w}$$

where d is derived as an average value of depth by dividing the cross-sectional area (a) by the width (w).

Wien's law A law relating to the amount of solar RADIATION received at the surface of the Earth together with the amount of terrestrial radiation reradiated into the atmosphere. The wavelength of maximum emission (λ max) varies and Wien's law states that it varies inversely with the absolute temperature (T) of the radiating body:

$$\lambda\ \text{max} = \frac{2897}{T} \times 10^{-6}\text{m (where: } 10^{-6}\text{m} = 1\ \mu\text{m)}$$

Solar radiation is very intense and consists mainly of short waves (0.2–4.0 μm), with a peak intensity of 0.48 μm, while the much weaker terrestrial radiation ranges between 4.0 μm and 100 μm, with a peak intensity of 10 μm (infrared). *See also* STEFAN'S LAW.

Wilcoxon test A NON-PARAMETRIC STATISTICAL TEST for comparing paired-data samples on an interval scale, in which the differences between pairs are ranked. Thus, the test is

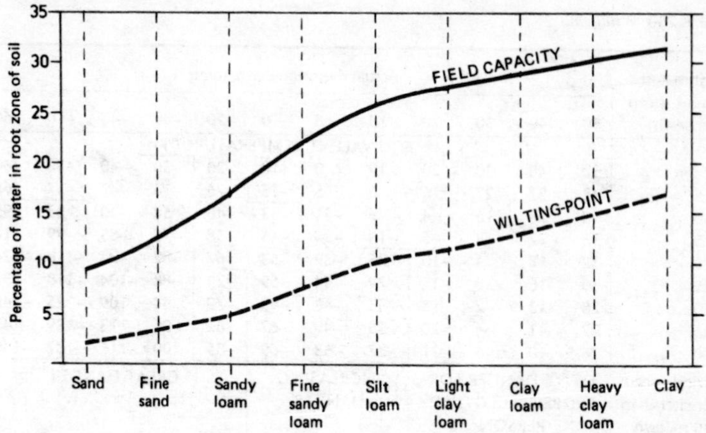

Figure 279 *Field capacity and wilting-point of different soil textures*

made distribution-free, i.e. its validity does not depend upon the interval-scale values in the two samples being distributed normally (NORMAL DISTRIBUTION) or in any other special way, as in the case of the T-TEST. *See also* MANN–WHITNEY U-TEST.

wilderness Strict usage of the term is confined to a wild, uncultivated region untouched by the impact of mankind. Since few parts of the world remain as such, with the exception of the polar regions, the term is more loosely used to describe uninhabited and uncultivated areas. In Britain, there are no true wilderness areas but the remoter parts of Scotland are described as 'wilderness'.

wildfires, bushfires, brushfires Conflagrations occurring in desiccated vegetation encouraged by hot, dry weather conditions and spread by strong winds. They can be triggered by lightning strikes or by careless human activity. They occur at regular intervals in southern France, California and south-east Australia, especially in GARRIGUE and MAQUIS scrub-types of vegetation.

willy-willy A type of TROPICAL CYCLONE, characteristic of the coastline of NW Australia, and deriving its name from the Aboriginal term *willi-willi*.

wilting-point The point at which the rate of moisture loss from the leaf surface is greater than the uptake by the plant roots. The wilting-point differs in various soil types according to their dominant PARTICLE SIZE. *See also* FIELD CAPACITY. [*279*]

wind A horizontal movement of air in relation to the surface of the Earth. Such a movement is caused by variations of pressure resulting from differential heating of the global surface (PLANETARY WINDS) or by dynamic factors in the atmosphere itself (CONVERGENCE, DIVERGENCE). Although during periods of TURBULENCE winds can exhibit a vertical component, this is uncommon at the surface, so the vertical movement of air is generally referred to as a CURRENT (THERMAL). Wind is one of the basic elements of weather and its direction (shown by a WIND VANE) and its velocity (measured by an ANEMOMETER) have been carefully monitored for centuries. Because of its importance for early ocean transport, for energy (through windmills) and for its varying influences on local weather, there are numerous local names for winds throughout the world (e.g. *föhn*, *khamsin*, *Trade Winds*). Its intensity can vary from nearly absolute calm to very high speeds (0 to >74 miles/h; >119 km/h; >64 knots), the various forces of which are defined in the BEAUFORT

Table 280 *Wind-chill*

Estimated wind-speed (miles/h)	Actual thermometer reading °F											
	50	40	30	20	10	0	−10	−20	−30	−40	−50	−60
	EQUIVALENT TEMPERATURE (°F)											
Calm	50	40	30	20	10	0	−10	−20	−30	−40	−50	−60
5	48	37	27	16	6	−5	−15	−26	−36	−47	−57	−68
10	40	28	16	4	−9	−21	−33	−46	−58	−70	−83	−95
15	36	22	9	−5	−18	−36	−45	−58	−72	−85	−99	−112
20	32	18	4	−10	−25	−39	−53	−67	−82	−96	−110	−124
25	30	16	0	−15	−29	−44	−59	−74	−88	−104	−118	−133
30	28	13	−2	−18	−33	−48	−63	−79	−94	−109	−125	−140
35	27	11	−4	−20	−35	−49	−67	−82	−98	−113	−129	−145
40	26	10	−6	−21	−37	−53	−69	−85	−100	−116	−132	−148
Wind-speeds greater than 40 miles/h have little additional effect	LITTLE DANGER FOR PROPERLY CLOTHED PERSON				INCREASING DANGER			GREAT DANGER				
					DANGER FROM FREEZING OF EXPOSED FLESH							

To use the chart, find the estimated or actual wind-speed in the left-hand column and the actual temperature in degrees F in the top row. The equivalent temperature is found where these two intersect. (*After the National Science Foundation, Washington DC*)

SCALE. *See also* ANGULAR MOMENTUM, ABSOLUTE VORTICITY.

wind action AEOLIAN, WIND EROSION.

wind-blown deposits DUNE, LOESS.

wind-break Any artificially generated barrier that reduces surface wind-speeds, the degree of shelter depending upon the height and thickness of the barrier, and, if it is a SHELTER BELT, the degree to which the wind is reduced by friction according to the tree species, thereby affecting the dimensions of the WIND SHADOW.

wind-chill The effect of cold winds on living creatures which can lead to a lowering of body temperature and the onset of hypothermia. In general, the stronger the wind the greater the heat loss from exposed skin, so that absolute air temperature is not a true guide to potential exposure. A creature may survive low temperatures in calm air more successfully than somewhat higher temperatures in strong winds (SENSIBLE TEMPERATURE). A wind-chill *index*, using formulae derived from wind-speeds and air temperatures, was devised by the US Army to assist in the design of cold-weather clothing. It is measured in kilo-calories per m^2 of exposed flesh. *See also* BIOCLIMATOLOGY, COMFORT ZONE. [*280*]

wind-drift current An ocean current produced by wind stress on the surface of the water. That part of the wind energy which is not used in producing waves is responsible for the surface currents. *See also* OCEAN CURRENTS.

Windermere Interstadial The name given to the interstadial stage of the LATE-GLACIAL phase of the DEVENSIAN in Britain. It is based on sedimentary stratigraphy in L. Windermere, Cumbria, in which the pollen sequence shows a marked rise and subsequent fall in birch and juniper, marking respectively the beginning and the end of the interstadial. It preceded the LOCH LOMOND STADIAL and is thought to have lasted from 13,000 to 11,360 BP. It is the British equivalent of the ALLERØD INTERSTADIAL (pollen zone II), the OLDER DRYAS (zone IC) and the *Bølling Interstadial* (zone IA–IB) in combination.

wind erosion The action of sediment removal by AEOLIAN forces, depending on the character of the air motion and the properties of the surface materials. Wind passing over a stable surface is retarded by friction at its base but above this layer the wind velocity increases with height above the surface. Loose particles on the surface begin to move once a critical wind velocity, termed the *fluid threshold* by R. A. Bagnold, has been exceeded, but wind-tunnel experiments have shown that sediments composed only of smaller materials become more difficult to move as GRAIN SIZE decreases. This is due to lower values of surface roughness, greater moisture retention and increased interparticle cohesion by chemical bonds. Once particle movement has commenced the surface is bombarded with grains, thereby initiating further movement. Three types of wind pressure are exerted on the surface particle: impact pressure (positive) on the windward side; viscosity pressure (negative) on the leeward side; and static pressure, a negative pressure on the top of the particle, creating both drag and lift. The properties of surface materials which affect their resistance to wind erosion include: **1** the binding agents of the silt, clay and organic matter; **2** the amount of soil moisture, with the rate of soil movement varying approximately inversely as the square of effective surface soil moisture; **3** surface roughness, with the greater the roughness the less its susceptibility to erosion; **4** vegetation cover, which increases surface roughness and inhibits erosion. *See also* ABRASION, DEFLATION.

wind-gap A valley or col in a ridge through which no stream passes, i.e. it contrasts with a WATER-GAP. Although it may appear to resemble a dry MELTWATER CHANNEL, the term wind-gap is usually confined to a dry valley produced by river CAPTURE, in which the stream that formerly occupied the valley has been incorporated into another drainage system by piracy.

window FENSTER.

wind-rose A diagram to illustrate the frequency with which wind blows from the various points of the compass (usually eight or twelve points are included). The length of the projecting arm is related to the frequency over a specified time period. The number of calms is generally recorded in a circle drawn around the focal point of the diagram. In addition each projecting arm of the wind-rose can be used to illustrate the frequency of different wind speeds by appropriate subdivisions of its length. [*281*]

wind-row dunes A name given to elongated sand-dunes in the arid south-western

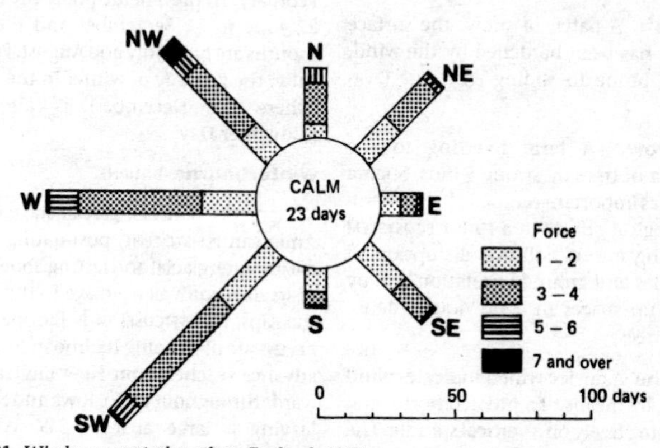

Figure 281 *Wind-rose: central southern England, 1978*

states of USA because of their resemblance to the wind-rows of a hayfield (rows of mown grass).

wind set-down The lowering effect of wind on the still-water level of a body of water. This occurs on the upwind flank as a result of the stresses caused by wind on the water surface. On large lakes the wind may carry surface water to the farthest (downwind) end and cause differences in water-level of several cm (WIND SET-UP).

wind set-up The raising effect of wind on the still-water level of a water body. It occurs on the downwind flank as a result of the stresses caused by wind on the water surface, leaving the level at the upwind end a few cm lower (WIND SET-DOWN).

wind shadow A zone of sheltered air in the LEE of an obstacle. The area, although sheltered from the direct force of the wind, will probably be affected by an EDDY, the effects of which can be seen in the building of a BARCHAN or a snow-drift.

wind shear The action and rate of change of wind velocity, either in the horizontal or in the vertical flow of air. The term is most frequently used to describe a right-angled change in a horizontal air-flow, but the gradient of wind velocity with height is referred to as the *vertical wind shear*, the amount of shear being dependent upon the temperature gradient of the air.

wind slab A patch of snow, the surface of which has been hardened by the wind, which is prone to sliding *en masse* (AVALANCHE).

windthrow A term referring to the uprooting of trees by strong winds. Such a process has important ecological and microclimatological effects in a forest ECOSYSTEM by creating minor hollows, disruption of soil profiles and ground vegetation and by opening up spaces in a previously dense stand of trees.

wind vane A device which indicates wind direction by means of a pivoted horizontal arm rotating freely on a vertical spindle. The 'tail' of the arm is broadened to form a broad surface or fin upon which the wind pressure acts, swinging its opposite end to point in the direction from which the wind is blowing.

windward The upwind side or the side exposed to the direction from which the wind blows. The opposite to *leeward* (LEE).

Winkel's projections A number of MAP PROJECTIONS introduced in 1921. They include: **1** an equidistant combined PSEUDO-CYLINDRICAL PROJECTION comprising the arithmetic mean of the PLATE CARRÉE PROJECTION and the SANSON–FLAMSTEED PROJECTION; **2** a *pseudo-cylindrical projection* comprising the arithmetic mean of the Plate Carrée projection and the MOLLWEIDE PROJECTION; **3** a POLYCONIC PROJECTION comprising the arithmetic mean of the Plate Carrée projection and the Hammer–Aitoff projection (HAMMER'S PROJECTIONS). It is also known as the *Tripel projection*, used regularly to show global distributions of resources, population, natural phenomena, etc.

winnowing The action of removing soil or other sedimentary PARTICLES by the wind or by currents of water.

winter The coldest SEASON of the year; the period from the winter SOLSTICE to the spring EQUINOX (21 December to about 20 March in the N hemisphere), although the three winter months in the N hemisphere are generally regarded as December, January and February. In the S hemisphere the dates are 22 June to 21 September and the winter months are June, July and August. Paradoxically, the first day of winter in the N hemisphere (21 December) is also called Midwinter Day.

winterbourne BOURNE.

Wisconsin The last glacial stage of the N American PLEISTOCENE, post-dating the SANGAMON interglacial and lasting about 60,000 years. It is equivalent in age to the WÜRM of the Alps, the WEICHSEL of N Europe and the DEVENSIAN of Britain. Its line of maximum advance reaches from New England westwards throughout Ohio, Iowa and S Dakota, leaving a large area of SW Wisconsin unglaciated, but in places pushing farther S than the much earlier KANSAN glacial limits. Its area was in fact greater than that of all the other N American glacial stages combined. It has been subdivided into a number of separate phases as follows (oldest first): **1** Iowan glaciation; **2** Peorian interstadial; **3** Tazewell glaciation; **4** Cary glaciation; **5** Two Creeks interstadial; **6** Valders glaciation; **7** Mankato glaciation.

woebourne BOURNE.

wold **1** In British usage an unwooded area of rolling chalk downland. It occurs frequently as a place-name. **2** In the USA, a CUESTA.

Wolstonian The penultimate glacial stage of the British Pleistocene, named from its type site at Wolston, Warwickshire. It succeeded the HOXNIAN interglacial and lasted from 347,000 BP to 128,000 BP, when it was terminated by the IPSWICHIAN interglacial [*197*]. The ice-sheets of this age impounded a large pro-glacial lake in the English Midlands, known as Lake Harrison, which stretched from the site of modern Leicester south-westwards to beyond the modern Stratford-upon-Avon, and overflowed through SPILLWAYS across the scarp of the Cotswolds. The eastern limits of the ice-sheet crossed East Anglia, where it deposited the GIPPING TILL, while its western limits impinged on the N coast of Cornwall. In the severe climate of S England the Palaeolithic stone industries continued, with the ACHEULIAN hand-axe culture surviving from the Hoxnian into the Wolstonian. It is the British equivalent of the SAALE of N Europe and the RISS of the Alps.

Woodgrange interstadial A warm phase of the Irish late PLEISTOCENE (14,000–10,500 BP), recognized as a warm phase by the analysis of lacustrine deposits at Woodgrange, Co. Down.

woolsack A colloquial term for a CORESTONE of unweathered rock amidst a mass of *in situ* unweathered rock.

World Weather Watch (WWW) The programme planned by the World Meteorological Organization (WMO) as a co-ordinated attempt to give a global coverage of data relating to atmospheric phenomena.

wrench-fault TEAR-FAULT.

Würm The last glacial stage of the PLEISTOCENE in the Alps, post-dating the RISS glacial, from which it is divided by the last interglacial. It was recognized in 1909 by A. Penck and E. Brückner, who correlated the fresh glacial moraines with glaciofluvial outwash in the valleys of S Germany to the north of the Alps. It is equivalent in age to the WEICHSEL of N Europe, the DEVENSIAN of Britain and the WISCONSIN of N America.

X

xenocrysts Crystals occurring in igneous rocks which, although resembling PHENOCRYSTS, are foreign to the rock body, because they have been introduced into the magma from an extraneous source.

xenolith An inclusion of pre-existing rock within an igneous rock. It may be a block of COUNTRY-ROCK, or a fragment of metamorphic rock from the surrounding aureole. It can also be a piece of the igneous rock itself which had previously solidified and had been only partly assimilated.

xerophyte A plant species which has developed means of combating drought by changes in its anatomical and morphological structures. These include: the reduction of the leaf surface; the substitution of spines or thorns for leaves; the development of hard, glossy leaves; the ability to produce waxy secretions; the growth of thick bark; the formation of thick bulbous stems and other water-storing devices; the growth of deep roots. Cactus, acacia and baobab are examples of *xerophilous* species.

xerosere The sequential development of a plant community (SERE) in a very dry habitat. If it develops on bare, well-drained scree or on a glacially smoothed rock it is termed a LITHOSERE, but if it develops in a dune environment it is known as a PSAMMOSERE.

xerosphere A general term for the arid lands of the world, in which XEROPHYTES and XEROSERES are found.

xerothermic index A theoretical measure of the effects of aridity on plant growth. The index indicates the number of 'biologically' dry days in a dry season and was originally defined to describe the BIOCLIMATOLOGY of MEDITERRANEAN CLIMATES, but is now used more universally.

Y

yard The fundamental unit of LENGTH in the long-established Imperial system of measurement. In Britain it has now been officially replaced by the metric system. The British yard = 0.9144 metres, while in the USA the yard = 0.91440183 metres, i.e. 0.7 × 10^{-4} inches longer than the British yard [*136*].

yardang A sharp-crested linear ridge of rock, aligned in the direction of the prevailing winds in a desert environment. The upwind face of the feature is rounded and the downwind extension takes the form of a lengthy, sharply crested ridge. It varies in size from a few metres to 1 km in length, up to 6 m in height and 35 m in width. It is separated from its neighbouring yardang by a wind-scoured groove. Although in most cases there can be little doubt that the yardangs are wind-eroded features there is no agreement concerning the degree to which other agencies may have contributed to their form. *See also* ZEUGEN.

Yarmouth Interglacial The penultimate interglacial stage of the N American PLEISTOCENE, following the KANSAN glacial and preceding the ILLINOIAN glacial. It is thought to be equivalent in age to the HOLSTEIN of N Europe and the HOXNIAN of Britain.

yazoo A tributary stream of the DEFERRED JUNCTION type, prevented from joining the main river because of the LEVEES which flank the latter. It is named from the R. Yazoo, which is a tributary of the lower Mississippi.

year A measure of the time taken for the Earth to complete one revolution in its orbit around the Sun. The period of time taken to complete the cycle of the SEASONS. Distinctions must be made between the following: **1** The *anomalistic year*, which represents the time between two successive PERIHELIONS, and comprises 365.25964 mean SOLAR DAYS. **2** The *calendar year* (Gregorian), comprising 365 mean solar days each of 24 hours' duration. To compensate for the extra 0.2422 mean solar days an extra day is inserted in the February of each fourth year (*leap year*), which therefore comprises 366 days. **3** The SIDEREAL YEAR, comprising 365.2564 mean solar days. **4** The SOLAR YEAR, comprising 365.2422 mean solar days. The *solar year* is also known as the *astronomical year*, the *equinoctial year*, the *nature year* and the *tropical year*.

yield limit ELASTIC LIMIT.

Yoredale Beds A facies of the Lower CARBONIFEROUS in Britain, with its type area being in Wensleydale in the N Pennines of England. In essence the succession is a cyclic sequence or CYCLOTHEM, in which a marine limestone is succeeded upwards by marine shales, shales and silty shales, current-bedded sandstones and finally a thin coal. In places the lower limestones rest on the Great Scar Limestone (SCAR LIMESTONE).

younger drift NEWER DRIFT.

Younger Dryas The youngest of the five phases of the LATE-GLACIAL episode of the DEVENSIAN stage in Britain and NW Europe. It is named from the Alpine plant mountain avens (*Dryas octopetala*) which flourished at lower elevations as the mountain glaciers waxed and the final Scottish ice-sheet extended its limits during the so-called LOCH

LOMOND STADIAL. Elsewhere during this period of climatic deterioration solifluction increased in intensity until the 1,000-year span of the Younger Dryas (zone III in the pollen zonation) was terminated in about 10,300 BP by the increasing warmth of the post-glacial (zone IV) which marked the end of the Late-Glacial and the beginning of the FLANDRIAN.

youth The earliest stage in the CYCLE OF EROSION and a concept which has been extended to all types of land sculpture, shoreline evolution and soil formation. *See also* ARID CYCLE OF EROSION, CYCLE OF UNDERGROUND DRAINAGE, GLACIAL CYCLE OF EROSION, MARINE CYCLE OF EROSION, MATURITY, NORMAL CYCLE OF EROSION, SENILITY. [*53* and *181*]

Z

zenith The point on the CELESTIAL SPHERE that is vertically above the observer. It is used to calculate the input of solar RADIATION to the Earth.

zenithal equal-area projection AZIMUTHAL EQUAL-AREA PROJECTION.

zenithal equidistant projection AZIMUTHAL EQUIDISTANT PROJECTION.

zenithal projection AZIMUTHAL PROJECTION.

zeolites A large group of tekto SILICATES which contain true water of crystallization. They occur in cavities in basic volcanic rocks and are thought to be a product of HYDROTHERMAL ACTIVITY. They probably indicate a very low-grade metamorphism (GRADE).

zero curtain An expression denoting the zone immediately above the PERMAFROST TABLE, in which, according to S. W Muller's definition (amended by A. L. Washburn), 'zero temperature (0° C) lasts a considerable period of time (as long as 115 days a year) during freezing and thawing of overlying ground'. It is caused by the fact that LATENT HEAT of fusion is released when water freezes, thus maintaining soil temperatures around 0° C for an extended period of time. The higher the moisture content at the permafrost table the more difficult it becomes for the latent heat to escape.

zeugen A German term variously used to describe desert rock pillars, rock mushrooms or those YARDANGS that are considerably undercut because of differential abrasion by wind of their less-resistant beds. It appears that all zeugen (singular = *zeuge*) are characterized by the presence of a resistant cap-rock, but some are elongated in the direction of the prevailing wind.

zibar A type of low sand DUNE often occurring in the corridors between higher dunes. It is suggested that it has a higher percentage of coarser grains than other dunes.

zincblende Zinc sulphide (ZnS), the major ore from which zinc is obtained. It occurs in rocks where METASOMATISM and HYDROTHERMAL ACTIVITY have occurred. It is also termed *sphalerite*.

zonal circulation, zonal flow The movement of air in a dominant west–east or east–west direction (i.e. latitudinal in contrast to longitudinal or MERIDIONAL FLOW). The most significant winds of the zonal circulation are the TRADE WINDS of the tropics and the WESTERLIES of mid latitudes. *See also* ZONAL INDEX.

zonal index A measure of the strength of the ZONAL CIRCULATION over a given length of time in a particular region. A low zonal index refers to large wave amplitudes (i.e. strong MERIDIONAL FLOW) whereas a high zonal index refers to waves of small amplitude (i.e. weak meridional flow). *See also* INDEX CYCLE, ROSSBY WAVE.

zonal inselberg A type of INSELBERG, with the term having been introduced by J. Budel to distinguish a feature produced by PARALLEL RETREAT OF SLOPES and PEDIPLANATION (i.e. a zonal inselberg) from one formed when the BASAL SURFACE OF WEATHERING is exhumed by denudation to produce a BORNHARDT (i.e. a *shield inselberg*). [*282*]

zonal soils Soils whose characteristics are dominated by the influence of climate and

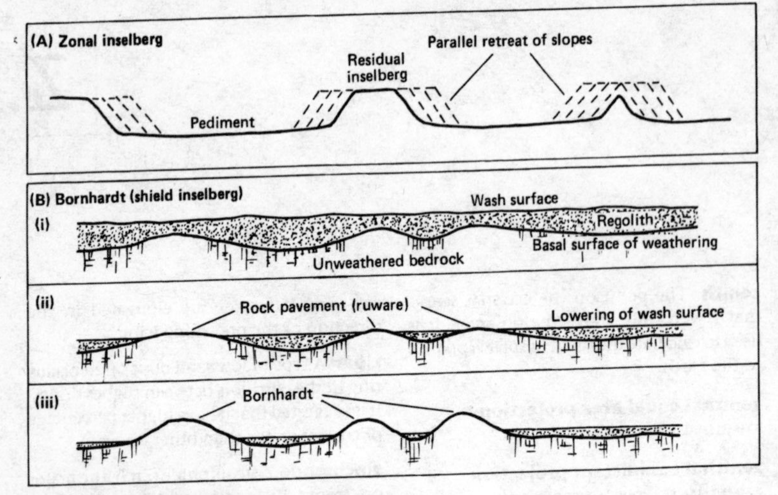

Figure 282 *Types of inselberg*

vegetation. *See also* AZONAL SOILS, INTRAZONAL SOILS, PEDALFER, PEDOCAL.

zonda **1** A hot, sultry northerly wind in Uruguay and Argentina, bringing enervating conditions. It is similar to the BRICKFIELDER of Australia. **2** A warm, dry FÖHN wind blowing down the eastern slopes of the Andes in Argentina.

zone, zonation **1** In general, a region defined by specific limits. **2** A climatic distinction based on the global temperature belts (FRIGID ZONE, TEMPERATE ZONE, TORRID ZONE). **3** In BIOSTRATIGRAPHY, a fundamental division demarcated by fossils (INDEX FOSSIL) used to correlate rock successions. **4** A spatial division of an area of metamorphic rocks, based on index minerals, used to define areas of different metamorphic GRADE and depicted on a map by ISOGRADS. **5** The term *zone* is regularly used with other words to form expressions relating to different SOIL HORIZONS, to different layers and structures in solid rocks (e.g. BENIOFF ZONE, FRACTURE ZONE) and to varying depths of the ocean (e.g. ABYSSAL ZONE, LITTORAL ZONE). *See also* ACCUMULATION ZONE, INTERTROPICAL CONVERGENCE ZONE.

zone fossil INDEX FOSSIL.

zoogeography The study of the geographical distribution of animals. *See also* VICARIANCE BIOGEOGRAPHY.

zooplankton The minute marine animal organisms that spend their life floating in shallow seas, particularly where cold currents upwell at the surface. They include RADIOLARIA and FORAMINIFERA, together with larvae and eggs of the animal BENTHOS and NEKTON. They are the animal equivalent of PHYTOPLANKTON. *See also* PLANKTON.

zweikanter A VENTIFACT with two facets. *See also* EINKANTER, DREIKANTER.